W9-CLS-020

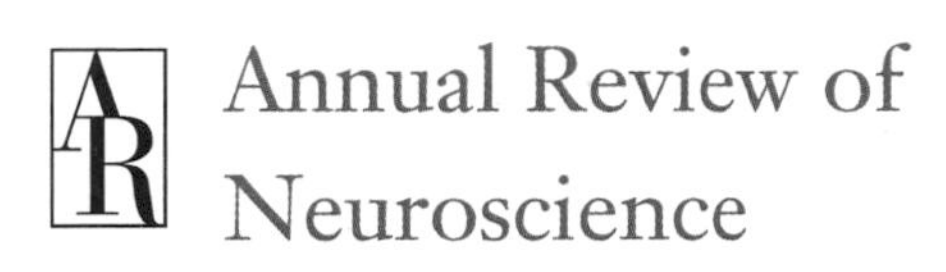
Annual Review of
Neuroscience

Production Editor: Jennifer E. Mann
Managing Editor: Linley E. Hall
Bibliographic Quality Control: Mary A. Glass
Electronic Content Coordinators: Suzanne K. Moses, Erin H. Lee
Illustration Coordinator: Douglas Beckner

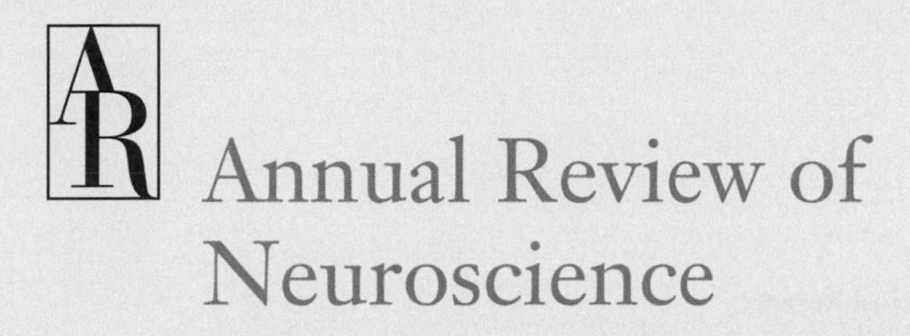

Annual Review of Neuroscience

Volume 35, 2012

Steven E. Hyman, *Editor*
Harvard University

Thomas M. Jessell, *Associate Editor*
Columbia University

Carla J. Shatz, *Associate Editor*
Stanford University

Charles F. Stevens, *Associate Editor*
Salk Institute for Biological Studies

Huda Y. Zoghbi, *Associate Editor*
Baylor College of Medicine

www.annualreviews.org • science@annualreviews.org • 650-493-4400

Annual Reviews
4139 El Camino Way • P.O. Box 10139 • Palo Alto, California 94303-0139

Annual Reviews
Palo Alto, California, USA

International Standard Serial Number: 0147-006X
International Standard Book Number: 978-0-8243-2435-3

TYPESET BY APTARA
PRINTED AND BOUND BY SHERIDAN BOOKS, INC., CHELSEA, MICHIGAN

Preface

When I was invited to write the preface for Volume 35 of the *Annual Review of Neuroscience*, I agreed without ever having read an Annual Reviews preface. And when it was time to start writing, I discovered that the custom of preface writing for the *Annual Review of Neuroscience* series has been more honored in the breech than the observance (a LitCrit friend told me quoting Shakespeare in a preface is always good): in the preceding 34 volumes, only two have had prefaces, a probability of just over 0.05. In Volume 1, Max Cowan, the founding editor, laid out the reasons for having an *Annual Review of Neuroscience*, and after Max died a quarter of a century later, Steve Hyman, the new Editor, wrote a preface that was a celebration of Max's contributions to the field. My job now is to initiate a tradition more often honored in the observance of including invited prefaces for the *ARN*.

In his preface, Max observed that understanding how the brain produces behavior is one of Society's Grand Challenges, and he noted that the new discipline of Neuroscience must breech the disciplinary boundaries that had previously separated the various approaches to studying the nervous system and the behavior it produced. Of course, this sounds like what we would say today, but today's Neuroscience was literally unimaginable at the time the first volume appeared. Since publication of Volume 1 in 1978, we have cloned and sequenced many families of ion channels and receptors and have gotten crystal structures for representative ones; we have learned how to make knock-out and knock-in mice (and other genetic species); we know every gene that is expressed in the brain of many species; confocal and two-photon microscopes are standard laboratory tools; and optical methods for recording the activity of neurons and for producing or preventing activity in particular cell types are available, as are computerized microscopes with super-resolution, commercial patch clamp amplifiers, vastly increased computer power, and the ability to record brain signals related to localized neural activity in awake humans (fMRI). Not only did these things not exist when the first *ARN* volume appeared, but most of them could not even be imagined at the time. This is truly a time of revolutionary technical advances.

At one level, the questions being asked by neuroscientists are the same ones addressed in Volume 1. For example, Volumes 1 and 35 both have reviews on circadian rhythms, patterning in the developing brain, and mechanisms of motor control. Furthermore, Volume 1 offers reviews on the role of dopamine systems, pain, optical reporters of neural activity, and new neuroanatomical methods, all topics that happened not to be covered in Volume 35 but certainly would have fit with our current views of what concerns Neuroscience. But if a modern Rip Van Winkle had read Volume 1 just before he went to sleep and then woke up in time to read reviews in Volume 35, he would have very little idea of what the new reviews were talking about. Of course, Rip would be confused by transcription factor names that are now common in Neuroscience. Beyond confusing nomenclature, though, the way we look at the same problems is so changed, often because questions are framed in the context of multiple new technical approaches, that the modern reviews would be incomprehensible to the 1978 mind.

When Volume 1 appeared, one of the reviews was on the newly evolving methods for cell culture. By Volume 12, cell culture was passing its peak and brain slices were considered more like the real brain so that conclusions from cell culture experiments were starting to be considered suspect. And by Volume 25, slice preparations were giving way to in vivo studies of anesthetized brains, again because these were claimed to be the right model for the real brain. These days, conclusions derived from anesthetized animals are suspect, and the gold standards are recordings and optical interventions in head-fixed, behaving animals. I expect that by Volume 40 or 45 the artifacts and limitations due to head fixation will be matters of concern, and use of now-emerging methods for studying the brains of unrestrained animals will be demanded.

To my mind, one of the dramatic changes since Volume 1 has been the increasing acceptance and sophistication of theoretical approaches to questions in Neuroscience. Volume 35 has a review of a technique from applied mathematics and computer science (called compressed sensing), which was developed only half-a-dozen years ago, and already it is being reviewed in *ARN* for its current and potential applications to the brain. Another dramatic change apparent in Volume 35, for me at least, is the emphasis on concepts—such as empathy, social control of the brain, decision making, awareness, and attention—that were considered outside the reach of neuroscientific investigation when the *ARN* was first published.

Reading the *Annual Review of Neuroscience* is how we keep up with the advances across the vast sweep of studies on the brain and its production of behavior. At the same time, the *ARN* captures the evolution of Neuroscience in a time capsule that preserves the intellectual history of our field.

Charles F. Stevens
Associate Editor

Annual Review of Neuroscience

Volume 35, 2012

Contents

Indexes

Errata

An online log of corrections to *Annual Review of Neuroscience* articles may be found at http://neuro.annualreviews.org/

Related Articles

From the ***Annual Review of Biochemistry***, Volume 81 (2012)

Ubiquitin and Membrane Protein Turnover: From Cradle to Grave
Jason A. MacGurn, Pi-Chiang Hsu, and Scott D. Emr

Ubiquitin and Proteasomes in Transcription
Fuqiang Geng, Sabine Wenzel, and lliam P. Tansey

From the ***Annual Review of Biomedical Engineering***, Volume 13 (2011)

In Vitro Models of Traumatic Brain Injury
Barclay Morrison III, Benjamin S. Elkin, Jean-Pierre Dollé, and Martin L. Yarmush

Technologies for Micromanipulating, Imaging, and Phenotyping Small Invertebrates and Vertebrates
Mehmet Fatih Yanik, Christopher B. Rohde, and Carlos Pardo-Martin

Vision-Based Navigation in Image-Guided Interventions
Daniel J. Mirota, Masaru Ishii, and Gregory D. Hager

Mapping Fetal Brain Development In Utero Using Magnetic Resonance Imaging: The Big Bang of Brain Mapping
Colin Studholme

From the ***Annual Review of Cell and Developmental Biology***, Volume 27 (2011)

Cyclin-Dependent Kinases in Brain Development and Disease
Susan C. Su and Li-Huei Tsai

The Role of MeCP2 in the Brain
Jacky Guy, Hélène Cheval, Jim Selfridge, and Adrian Bird

Optogenetic Control of Cells and Circuits
Gero Miesenböck

From the ***Annual Review of Clinical Psychology***, Volume 7 (2011)

Animal Models of Neuropsychiatric Disorders
A.B.P. Fernando and T.W. Robbins

Diffusion Imaging, White Matter, and Psychopathology
Moriah E. Thomason and Paul M. Thompson

Brain Graphs: Graphical Models of the Human Brain Connectome
Edward T. Bullmore and Danielle S. Bassett

Sleep and Circadian Functioning: Critical Mechanisms in the Mood Disorders?
Allison G. Harvey

Gene-Environment Interaction in Psychological Traits and Disorders
Danielle M. Dick

From the ***Annual Review of Genomics and Human Genetics***, Volume 12 (2011)

Copy-Number Variations, Noncoding Sequences, and Human Phenotypes
Eva Klopocki and Stefan Mundlos

Genetics of Schizophrenia: New Findings and Challenges
Pablo V. Gejman, Alan R. Sanders, and Kenneth S. Kendler

Genetics of Speech and Language Disorders
Changsoo Kang and Dennis Drayna

Recent Advances in the Genetics of Parkinson's Disease
Ian Martin, Valina L. Dawson, and Ted M. Dawson

From the ***Annual Review of Medicine***, Volume 63 (2012)

Huntington's Disease: Advocacy Driving Science
Nancy S. Wexler

The Genetic Architecture of Schizophrenia: New Mutations and Emerging Paradigms
Laura Rodriguez-Murillo, Joseph A. Gogos, and Maria Karayiorgou

Multiple Sclerosis: New Insights in Pathogenesis and Novel Therapeutics
Daniel Ontaneda, Megan Hyland, and Jeffrey A. Cohen

Traumatic Brain Injury and Its Neuropsychiatric Sequelae in War Veterans
Nina A. Sayer

Deep Brain Stimulation for Intractable Psychiatric Disorders
Wayne K. Goodman and Ron L. Alterman

From the ***Annual Review of Pathology: Mechanisms of Disease***, Volume 7 (2012)

Pathology of Demyelinating Diseases
Bogdan F.Gh. Popescu and Claudia F. Lucchinetti

Molecular Mechanisms of Fragile X Syndrome: A Twenty-Year Perspective
Michael R. Santoro, Steven M. Bray, and Stephen T. Warren

RNA Dysregulation in Diseases of Motor Neurons
Fadia Ibrahim, Tadashi Nakaya, and Zissimos Mourelatos

From the ***Annual Review of Pharmacology and Toxicology***, Volume 51 (2011)

Transcriptional and Epigenetic Regulation of Opioid Receptor Genes: Present and Future
Li-Na Wei and Horace H. Loh

Schizophrenia: Treatment Targets Beyond Monoamine Systems
Hisham M. Ibrahim and Carol A. Tamminga

Modulation of Monoamine Receptors by Adaptor Proteins and Lipid Rafts: Role in Some Effects of Centrally Acting Drugs and Therapeutic Agents
Karl Björk and Per Svenningsson

Orexin Receptors: Pharmacology and Therapeutic Opportunities
Thomas E. Scammell and Christopher J. Winrow

From the ***Annual Review of Physiology***, Volume 74 (2012)

Neurotransmitter Corelease: Mechanism and Physiological Role
Thomas S. Hnasko and Robert H. Edwards

Small-Conductance Ca^{2+}-Activated K^+ Channels: Form and Function
John P. Adelman, James Maylie and Pankaj Sah

From the ***Annual Review of Psychology***, Volume 62 (2011)

The Neuroscience of Social Decision-Making
James K. Rilling and Alan G. Sanfey

The Neural Bases of Social Cognition and Story Comprehension
Raymond A. Mar

Neuroscience of Self and Self-Regulation
Todd F. Heatherton

Annual Reviews is a nonprofit scientific publisher established to promote the advancement of the sciences. Beginning in 1932 with the *Annual Review of Biochemistry*, the Company has pursued as its principal function the publication of high-quality, reasonably priced *Annual Review* volumes. The volumes are organized by Editors and Editorial Committees who invite qualified authors to contribute critical articles reviewing significant developments within each major discipline. The Editor-in-Chief invites those interested in serving as future Editorial Committee members to communicate directly with him. Annual Reviews is administered by a Board of Directors, whose members serve without compensation.

Annual Reviews of

Analytical Chemistry
Anthropology
Astronomy and Astrophysics
Biochemistry
Biomedical Engineering
Biophysics
Cell and Developmental Biology
Chemical and Biomolecular Engineering
Clinical Psychology
Condensed Matter Physics
Earth and Planetary Sciences
Ecology, Evolution, and Systematics
Economics
Entomology
Environment and Resources
Financial Economics
Fluid Mechanics
Food Science and Technology
Genetics
Genomics and Human Genetics
Immunology
Law and Social Science
Marine Science
Materials Research
Medicine
Microbiology
Neuroscience
Nuclear and Particle Science
Nutrition
Pathology: Mechanisms of Disease
Pharmacology and Toxicology
Physical Chemistry
Physiology
Phytopathology
Plant Biology
Political Science
Psychology
Public Health
Resource Economics
Sociology

SPECIAL PUBLICATIONS
Excitement and Fascination of Science, Vols. 1, 2, 3, and 4

The Neural Basis of Empathy

Boris C. Bernhardt and Tania Singer

Department of Social Neuroscience, Max-Planck Institute of Human Cognitive and Brain Sciences, Stephanstraße 1a, 04309 Leipzig, Germany; email: bernhardt@cbs.mpg.de, singer@cbs.mpg.de

Annu. Rev. Neurosci. 2012. 35:1–23

The *Annual Review of Neuroscience* is online at neuro.annualreviews.org

This article's doi:
10.1146/annurev-neuro-062111-150536

0147-006X/12/0721-0001$20.00

Keywords

social neuroscience, insula, cingulate cortex, fMRI, emotion

Abstract

Empathy—the ability to share the feelings of others—is fundamental to our emotional and social lives. Previous human imaging studies focusing on empathy for others' pain have consistently shown activations in regions also involved in the direct pain experience, particularly anterior insula and anterior and midcingulate cortex. These findings suggest that empathy is, in part, based on shared representations for firsthand and vicarious experiences of affective states. Empathic responses are not static but can be modulated by person characteristics, such as degree of alexithymia. It has also been shown that contextual appraisal, including perceived fairness or group membership of others, may modulate empathic neuronal activations. Empathy often involves coactivations in further networks associated with social cognition, depending on the specific situation and information available in the environment. Empathy-related insular and cingulate activity may reflect domain-general computations representing and predicting feeling states in self and others, likely guiding adaptive homeostatic responses and goal-directed behavior in dynamic social contexts.

Contents

INTRODUCTION

Empathy is a crucial component of human emotional experience and social interaction. The ability to share the affective states of our closest ones and complete strangers allows us to predict and understand their feelings, motivations, and actions. Extending previous work from philosophy and behavioral psychology (Batson 2009, de Vignemont & Singer 2006, Eisenberg 2000, Hoffman 2000), advances in social neuroscience have provided important new insights into the brain basis of empathy.

In this review, we outline the main results of brain imaging studies that have investigated the neural underpinnings of human empathy. Using mainly functional magnetic resonance imaging (fMRI), the majority of studies suggest that observing affective states in others activates brain networks also involved in the firsthand experience of these states, confirming the notion that empathy is, in part, based on shared networks (de Vignemont & Singer 2006, Keysers & Gazzola 2007, Preston & de Waal 2002). In particular, anterior insula (AI) and dorsal-anterior/anterior-midcingulate cortex (dACC/aMCC) play central roles in vicarious responses in the domain of disgust, pleasant or unpleasant tastes, physical and emotional pain, and other social emotions such as embarrassment or admiration (for recent meta-analyses, see Fan et al. 2011, Kurth et al. 2010, Lamm et al. 2011; for anatomical orientation, see **Figure 1*a***). On the basis of their structural and functional patterns of connectivity, and their involvement in other functional processes at the interface of sensory, affective, and cognitive domains, regions such as AI and dACC/aMCC may generally contribute to the generation of subjective experiences and adaptive responses to actual and predicted states in the self and others. These general processes may then subsume empathy as a special case. We also highlight evidence that additional networks involved in social cognition can be flexibly corecruited during empathic understanding, depending on the particular situation and information available in the environment. Moreover, we summarize studies that have identified multiple factors that modulate or even counteract empathy. For example, initial evidence suggests that empathic responses can be counteracted by opposing motivational systems, such as the desire for revenge or Schadenfreude, closely related to activation in brain areas implicated in reward-processing. Finally, we outline future research avenues.

fMRI: functional magnetic resonance imaging

AI: anterior insula

ACC: anterior cingulate cortex

MCC: midcingulate cortex

DEFINING EMPATHY

Despite a long tradition of philosophy and behavioral psychology research (Batson 2009, Eisenberg 2000, Eisenberg & Fabes 1990, Hoffman 2000, Wispe 1986), empathy has no universally accepted definition, and the different phenomena it subsumes remain debatable (Batson 2009, Blair 2005, de Vignemont & Singer 2006, Preston & de Waal 2002). However, previous conceptual work on empathy has greatly facilitated the design and interpretation of empirical studies that assess empathic traits through self-report measures and empathic states through controlled observational experiments. Findings from these experiments, especially those investigating the brain processes underlying the empathic experience, may deepen our understanding of this phenomenon at the interface of social interactions and internal feeling states and ultimately promise to disentangle the conceptual web surrounding empathy.

A relatively specific notion claims that empathy occurs when the observation or imagination of affective states in another induces shared states in the observer (de Vignemont & Singer 2006, Singer & Lamm 2009). This state is also associated with knowledge that the target is the source of the affective state in the self. This reading of empathy necessarily involves components of affective sharing, self-awareness, and self-other distinction (for other more general notions of empathy see Baron-Cohen & Wheelwright 2004, Blair 2005). Therefore, empathy differs from basic sharing-only phenomena such as emotional contagion and mimicry. Indeed, neither contagion nor mimicry requires a distinction about whether the origin of the affective experience is within the observer or was triggered by another person. Emotional contagion and empathy, for example when watching a friend in distress, can lead to personal distress, a self-centered and aversive response in the observer (Eisenberg & Fabes 1990). In contrast, during empathic concern, sympathy, or compassion, vicarious responses involve a feeling of concern for the other's suffering that induces a motivation to alleviate the suffering, but not necessarily any sharing of feelings. Whereas empathizing with a sad person may result in a feeling of sadness in the self, sympathy and compassion often result in a feeling of loving or caring for that person and a motivation to relieve their suffering (Baumeister & Vohs 2007, Klimecki & Singer 2012, Singer & Steinbeis 2009). This motivation may then be transformed into prosocial behavior (Batson et al. 2007). In our own understanding, emotional contagion underlies affect sharing; this can be followed by other-oriented feelings such as compassion, sympathy, and empathic concern, which may further promote prosocial behavior; conversely, contagion and empathy may also induce aversive distress responses that can lead to withdrawal behavior motivated by the desire to protect oneself from negative emotions (Klimecki & Singer 2012).

Emotional contagion: tendency to automatically adopt the emotional state of another person

Mimicry: tendency to synchronize the affective expressions, vocalizations, postures, and movements of another person

Sympathy: feelings for someone, generally coupled with the wish to see them better off or happier

Compassion: an emotional and motivational state characterized by feelings of loving-kindness and a genuine wish for the well-being of others

Empathic concern: an emotional and motivational state characterized by the desire to help and promote others' welfare

EMPATHY IN THE BRAIN

In their seminal article on empathy in 2002, Preston & de Waal suggested that the observation and imagination of others in a given emotional state automatically activates a corresponding representation in the observer, along with its associated autonomic and somatic responses (Preston & de Waal 2002). This hypothesis was inspired by accounts that suggested a close link between action and perception through common coding schemes (Prinz 1984, 2005). Moreover, the discovery of mirror neurons, a class of neurons in monkey premotor and parietal cortices activated during execution and observation of actions, provided a neural mechanism for shared representations in the domain of action understanding (Gallese et al. 2004, Keysers & Gazzola 2007, Rizzolatti et al. 2001). Subsequent studies, based predominantly on fMRI, have investigated empathic brain responses for a variety of states including pain (Morrison et al. 2004, Singer et al. 2004), disgust (Benuzzi et al. 2008, Jabbi et al. 2007, Wicker et al. 2003), fear (de Gelder et al. 2004), anxiety (Prehn-Kristensen et al. 2009), anger (de Greck et al. 2012), sadness (Harrison et al. 2006), neutral touch (Blakemore et al. 2005, Ebisch et al. 2008, Keysers et al. 2004), pleasant affect (Jabbi et al. 2007), reward (Mobbs et al. 2009), and higher-order emotions such as social exclusion (Masten et al. 2011) and embarrassment (Krach et al. 2011). Based mostly on results from empathy for pain, these studies showed that empathic responses recruit, to some extent, brain areas similar to those engaged during the corresponding first-person state. In the following section, we first highlight findings from studies on empathy for pain and then summarize evidence of empathic responses for other emotions and sensations.

Empathy for Pain

S1: primary somatosensory cortex

S2: secondary somatosensory cortex

Empathy for pain has been studied frequently, owing to the robustness of pain in inducing empathy. The firsthand pain experience is generally aversive; moreover, it motivates behavioral responses to reduce the noxious stimulation (Price 2000) and can induce forms of warning communication to conspecifics (Craig 2004). Furthermore, observing others in pain can motivate helping behavior (Hein et al. 2010) and is often experienced as unpleasant and even painful for the observers themselves. Last, the neural circuits involved in pain are relatively well understood (Apkarian et al. 2005, Bushnell et al. 1999, Craig 2003, Peyron et al. 2000, Rainville 2002; Duerden & Albanese 2012).

Firsthand pain experience consistently activates networks in premotor and prefrontal, primary and secondary somatosensory cortices (S1 and S2), dACC/aMCC, and insula, along with thalamic and brain stem regions such as the periaqueductal gray (PAG) (Apkarian et al. 2005, Bushnell et al. 1999, Derbyshire 2000, Peyron et al. 2000, Rainville 2002, Duerden & Albanese 2012). Activations, albeit less consistently, have also been shown in the amygdala and cerebellum (Apkarian et al. 2005, Duerden & Albanese 2012). Most of these regions receive parallel input from multiple nociceptive pathways (Apkarian et al. 2005). Somatosensory regions and adjacent posterior insula are thought to encode the more sensory-discriminative components of pain. A case study showed that a patient with a large lesion in the postcentral gyrus and parietal operculum, comprising S1 and S2, lost discriminative aspects of pain perception, without overt loss of pain affect (Ploner et al. 1999). S2 responses have been shown to correlate with objective stimulus intensity, but not with affective ratings (Maihofner et al. 2006). Other studies have shown a contralateral bias for pain processing in subregions of S2 and posterior insula, suggesting a representation in these areas of the sensory-discriminative attributes, such as the stimulus location, of painful stimuli relative to body side in these areas (Bingel et al. 2003). Conversely, regions such as AI and dACC/aMCC are thought to encode more affective-motivational dimensions of pain (Price 2000). ACC and insula responses vary not simply as a function of noxious input but rather as a function of subjectively felt pain intensity (Kong et al. 2008). In ACC, activity correlates positively with ratings of pain unpleasantness (Rainville et al. 1997) but does not correlate much with stimulus intensity (Peyron et al. 2000). Moreover, insula and ACC responses to painful stimuli can be influenced by the emotional context, suggesting interaction effects within the affective domain (Phillips et al. 2003).

To investigate brain responses during empathy for pain, Singer and colleagues studied females who were accompanied by their romantic partners (Singer et al. 2004). In one condition, the female, lying in the scanner, received a painful shock via an electrode attached to her hand. In the other condition, the male partner who was seated next to the MRI scanner and whose hand could be seen by the female via a mirror received the shock. In both conditions, abstract visual cues indicated to the female who would receive painful stimulation. The authors observed activity in AI, dACC, brain stem, and cerebellum when females received the shock directly and, most importantly, when they vicariously felt their partners' pain. The presentation of facial expressions of others in pain (Botvinick et al. 2005, Lamm et al. 2007a, Saarela et al. 2007), or of body parts receiving painful stimulation (Jackson et al. 2005, Lamm et al. 2007b), has elicited similar findings. The consistency of activations in parts of the pain networks elicited by firsthand experience as well as during vicariously felt pain has thus been taken to support the hypothesis that empathy involves shared representations.

More specifically, employing statistical conjunction analysis, several studies quantified the extent of shared activations in first-person pain and empathy. Comparing average activation patterns in these two conditions within a group of subjects, overlapping regions were located in insular and cingulate regions (Jackson et al. 2006, Morrison et al. 2004, Singer et al. 2004). To extend these findings, Morrison & Downing (2007) studied fMRI signals of individual subjects in native anatomical space, minimizing confounds introduced by image preprocessing. They observed activation overlaps in 6 of 11 subjects in aMCC, at the transition between otherwise nonoverlapping regions activated by directly and vicariously felt pain. Although these findings also indicate divergent activations underlying firsthand pain and empathy, they further support a role of shared representations in empathy.

However, voxel-wise conjunctions do not necessarily indicate shared representations on the neuronal level. A typical voxel in an fMRI experiment has a resolution of around 3 mm per side, and its signal relates to the activity of thousands of neurons within possibly different neuronal populations. Future studies employing fMRI adaptation (Grill-Spector & Malach 2001, Henson & Rugg 2003) or multivariate pattern analysis may more selectively probe commonalities in activations of specific neuronal populations (Norman et al. 2006). Indeed, in a recent multivoxel pattern analyis, bilateral AI regions exhibited a similar spatial distribution of cortical fMRI activity when seeing another person's hand in pain compared to firsthand pain, provinding relatively strong evidence for similar neuronal populations involved in both conditions (Corradi-Dell'Acqua et al. 2011).

Empathic responses to others' pain in somatosensory regions have been less consistently reported. Using transcranial magnetic stimulation (TMS), Avenanti and colleagues (2005) demonstrated that watching a video of a needle pricking a specific hand muscle reduces motor excitability of the equivalent muscle in the observer, similar to the freezing response that would occur if pain was directly administered. This reduction in motor excitability correlated with pain-intensity ratings, but not with those of pain unpleasantness. Importantly, no effect was seen when participants watched a cotton bud touching the same muscle or when the needle prick was applied to a different part of the hand, a foot, or a tomato. Thus, although not directly showing activations in somatosensory cortices, this study suggested that attentively watching pain applied to the other's body parts interferes with somatosensory processing.

A recent meta-analysis by Lamm and colleagues (2011) on 32 fMRI studies of empathy for pain confirmed that observing pain in others most robustly activated AI, extending into the inferior frontal gyrus (IFG) and dACC/aMCC (**Figure 1*a***). Moreover, by classifying previous experiments into those employing abstract visual cues to signal pain in others (cue-based paradigms, **Figure 1*b***) and those showing pictures of body parts receiving pain (picture-based paradigms, **Figure 1*c***) (Lamm et al. 2011), the study yielded further quantitative insights on the role of somatosensory regions in empathy. Indeed, during cue-based designs, activations in S1 and S2 contralateral to the stimulated hand were observed only in self-related but not in vicarious experiences of pain. Conversely, picture-based designs induced activity in both S1 and S2 during the other-related condition. However, similar activity was also elicited to a large extent by nonpainful control pictures and did not seem to be lateralized to a specific hemisphere. These results thus suggest that somatosensory activation sometimes observed in picture-based empathy for pain paradigms may rather be due to unspecific activation based on the perception of touch and movement of body parts and not due to empathy for pain itself (for similar arguments, see Keysers et al. 2010).

Directly comparing activation patterns of both design types, the meta-analysis of Lamm and colleagues also revealed an important divergence in terms of distributed network coactivations (Lamm et al. 2011). Indeed, cue-based studies preferentially activated regions such as ventral medial prefrontal cortex (PFC), superior temporal cortex (STC), and posterior regions such as the temporo-parietal junction (TPJ) and precuneus/posterior cingulate cortex (PCU/PCC) (**Figure 1*d***). These areas are generally thought to play a role in processes related to Theory of Mind or mentalizing

IFG: Inferior frontal gyrus

PFC: prefrontal cortex

TPJ: temporo-parietal junction

Theory of mind (mentalizing): ability to infer and represent beliefs and desires

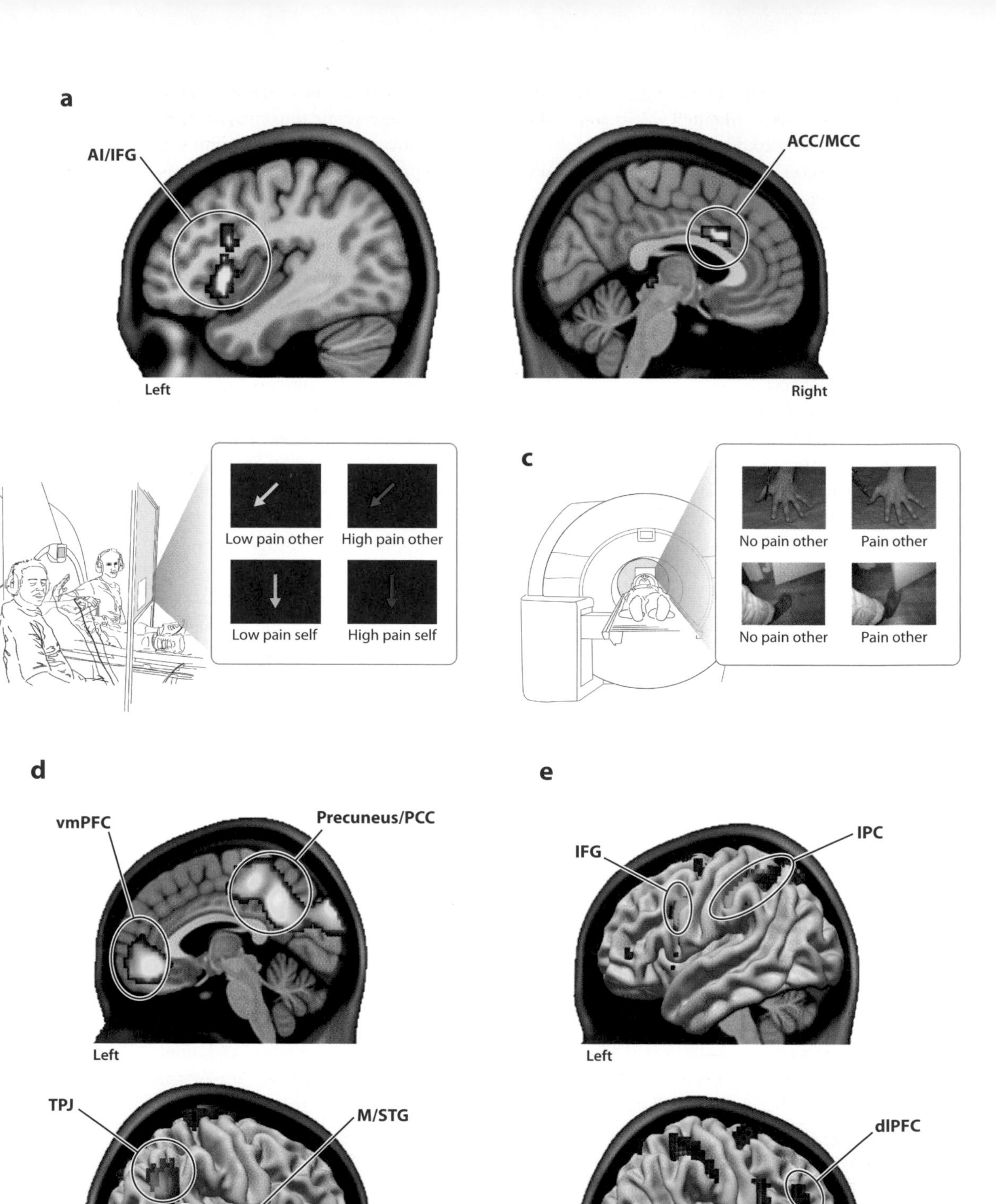
a
AI/IFG
ACC/MCC
Left
Right
b
Low pain other
High pain other
Low pain self
High pain self
c
No pain other
Pain other
No pain other
Pain other
d
vmPFC
Precuneus/PCC
Left
TPJ
M/STG
Right
e
IFG
IPC
Left
dlPFC
AI
Right

(Frith & Frith 2003, Van Overwalle 2009; Saxe & Kanwisher 2003; Mitchell 2009) but have also been associated with self-referential processing and mind-wandering (Buckner & Carroll 2007, Christoff et al. 2009, Mason et al. 2007, Schooler et al. 2012). On the other hand, picture-based studies showed relative increases of activation in dorsolateral and dorsomedial PFC, together with activity in bilateral inferior parietal cortex (IPC) and the IFG (**Figure 1*e***), a network partly overlapping with the human homolog of the monkey's mirror-neuron network (Gallese et al. 2004, Keysers & Gazzola 2007, Rizzolatti et al. 2001). Together with shared networks in empathy, both the so-called mirror-neuron network as well as the mentalizing network represent the most influential accounts currently in social neuroscience to explain how humans succeed at understanding others' actions, intentions, beliefs, or feelings (Frith & Frith 2006, Keysers & Gazzola 2007, Singer 2006). Indeed, computations in mirror-neuron networks in inferior frontal, ventral and dorsal premotor, and inferior parietal regions are thought to generate simulations of movements and goal-directed actions. These perception action-loops may then serve as a basis for depicting the meaning of the presented situation (Gallese et al. 2004, Keysers & Gazzola 2007) and may ultimately be relayed into regions such as AI for predicting the affective consequences of a stimulus. Cue-based paradigms, on the other hand, likely triggered more internally generated processing because the pain of the other is not explicitly shown. Thus the situation and its consequences in turn need to be inferred through mental imagery and prior knowledge, processes associated with activations in medial prefrontal and parietal regions, and temporal and temporo-parietal regions (Frith & Frith 2003, Mitchell 2009, Van Overwalle 2009). Thus, in addition to confirming an important role of insular-cingulate regions for affective sharing, the meta-analytic results of Lamm and colleagues suggest that this process may also involve a flexible activation of either mirror-neuron or mentalizing networks, depending on the particular situation and information available in the environment (Lamm et al. 2011).

Empathy for Other Emotions and Sensations

Studies based on vicarious responses to affective states other than pain, such as social exclusion (Masten et al. 2011), disgust (Jabbi et al. 2008, Wicker et al. 2003), anxiety (Prehn-Kristensen et al. 2009), and taste (Jabbi et al. 2007), have reproduced a central role of AI and ACC/MCC during empathy. Using disgusting odorants Wicker and colleagues observed that subregions of AI and ACC were activated during direct inhalation and when viewing the disgusted faces of people inhaling the probes (Wicker et al. 2003). In a follow-up experiment, the

Figure 1

Meta-analytic findings on empathy for pain. (*a*) A meta-analysis of 32 previous empathy-for-pain studies revealed consistent activations in anterior insula (AI) extending into the inferior frontal gyrus (IFG) and anterior and midcingulate cortex (ACC/MCC) (Lamm et al. 2011). This meta-analysis also classified studies into different experimental paradigms. (*b*) In cue-based paradigms, pain in others is signaled via abstract cues. In the example stimuli, colored arrows indicate whether the other or the self will receive a nonpainful sensation or a painful shock. This paradigm type does not explicitly provide depictions of painful situations, and thus may more likely rely on internally generated processes and exclude effects of emotion contagion. (*c*) In picture-based paradigms, pictures or videos that depict limbs of target persons in painful situations are shown to the observer. In the example stimuli, one image indicates pain in the other, whereas the other image does not (Jackson et al. 2005, Lamm et al. 2007b). In addition to eliciting empathy, this paradigm form may also elicit sensorimotor processes. (*d*) Higher activations during cue-based than during picture-based studies were found in so-called mentalizing or Theory of Mind networks, including temporo-parietal junction (TPJ), ventromedial prefrontal cortex (vmPFC), middle/superior temporal gyrus (M/STG), precuneus and posterior cingulate cortex (PCC) (Lamm et al. 2011). (*e*) Higher activations during picture-based than during cue-based paradigms were found in so-called mirror-neuron networks, such as the inferior-parietal cortex (IPC) and IFG, as well as in AI and dorsomedial and dorsolateral prefrontal cortex (dlPFC) (Lamm et al. 2011). Adapted from Jackson et al. (2005), Lamm et al. (2011), Lamm et al. (2007b), with permission.

authors confirmed common AI activation during the observation and imagination of disgust in others that overlapped with activations when subjects tasted bitter liquids themselves (Jabbi et al. 2008). Similar to findings in the domain of pain, common AI activations were accompanied by differential coactivation across these various conditions. Indeed, while AI activation showed increased functional connectivity only with IFG regions during the observation of disgust, the direct experience and imagination of disgust were related to more extended network coactivations (Jabbi et al. 2008).

Insular and adjacent frontal-opercular regions are also activated when subjects witnessed positive affective states. As in the perception of disgusted facial expression, AI activity was reported when subjects observed pleased facial expressions in others (Jabbi et al. 2007). Moreover, a recent study that induced compassion and admiration reported activation in AI, dACC, and hypothalamus. Interestingly, AI responses had a faster onset when witnessing physical pain compared with social pain or admiration for positive attributes of others (Immordino-Yang et al. 2009). Using meta-analysis, Fan and colleagues (2011) summarized empathic responses across various domains. Although the included studies were mostly based on pain, AI and ACC activation could also be confirmed when subjects observed fear, happiness, disgust, or anxiety in others.

Preliminary evidence also indicates that AI and ACC/MCC may not necessarily be involved in the vicarious sharing of all states. Studies based on the observation of neutral touch reported shared activations in somatosensory cortices, but not in limbic structures (Blakemore et al. 2005, Ebisch et al. 2008, Keysers et al. 2004). Moreover, a recent study that measured subjects who observed socially desirable others being rewarded demonstrated activations in the ventral striatum, a region involved in reward processing (Mobbs et al. 2009). The perceived similarity between the target and observer correlated with increased activity in ventral ACC, possibly mediating an effect of self-relevance in vicarious reward.

INSULA AND ACC: CONNECTIVITY AND FUNCTIONS

Relatively consistent activations of AI and ACC/MCC in empathy suggest an important role of these two regions in vicariously sharing many emotions and sensations. However, joint insular and cingulate activations in vicarious emotions do not imply that these regions are empathy regions per se. Instead, these regions are known to participate in a multitude of sensory, affective, cognitive, and motivational processes (see the 2010 Special Issue on Insula in *Brain Structure & Function*). In Craig's influential model, insular cortex plays a major role in representing and integrating internal and emotional feeling states; ACC, in turn, forms the motivational and action-related counterpart (Craig 2002, 2009). The diverse functional involvement of these regions also suggests that empathy might be a special case of general computational processes related to representing and predicting affective states in the self and others and of guiding adaptive homeostatic and behavioral responses (Singer et al. 2009, Singer & Lamm 2009). In the following section, we describe evidence for the functional implications of these regions based on their patterns of connectivity, their roles across multiple domains, and their frequent coactivation.

Insula

Connectivity and functional data support that the insula plays an important integrative role in sensation, affect, and cognition. Buried within the Sylvian fissure at the interface of frontal, temporal, and parietal lobes (**Figure 2*a*,*b***; Ture et al. 1999), the insula is cytoarchitectonically defined by a rostrocaudal transition from agranular AI to granular PI (Gallay et al. 2011, Mesulam & Mufson 1982a). Tract-tracing experiments in nonhuman primates suggest that AI is densely connected with prefrontal regions, such as orbitofrontal cortex (OFC) and dorsolateral PFC, temporo-limbic regions, such as temporal poles, parahippocampal

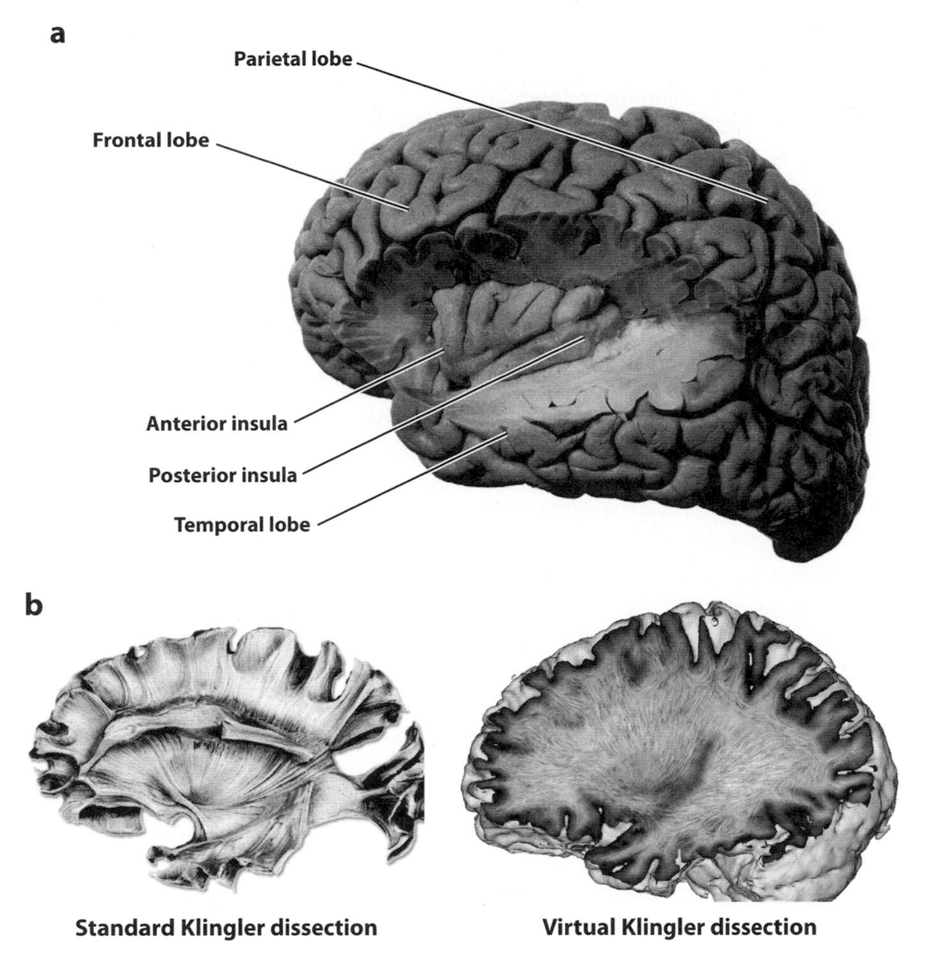

Figure 2

Anatomy of insula cortex. (*a*) Lateral view of the insula cortex in a postmortem human brain, with parts of the lateral cortical surface removed. (*b*) Subinsular fibers shown using Klingler dissection techniques in a postmortem brain *(left)* and using virtual Klingler dissection techniques based on diffusion weighted imaging *(right)*. Adapted from Klingler & Ludwig (1956) with permission, courtesy of Dr. Alfred Anwander.

cortices, amygdala, and cingulate cortex, and subcortical targets in the thalamus, basal ganglia, and brain stem (Amaral & Price 1984, Augustine 1996, Fudge et al. 2005, Mesulam & Mufson 1982b, Mufson & Mesulam 1982). Different segments of the insula are highly interconnected themselves, allowing a bidirectional flow of information between anterior and midposterior segments (Craig 2009). Patterns of connectivity in animals have recently been reproduced in humans using diffusion tractography (Cerliani et al. 2012, Nanetti et al. 2009) and resting-state fMRI signal correlations (Cauda et al. 2011, Deen et al. 2012). These patterns indicate a central role, especially of AI, in integrating interoceptive and affective information (Craig 2009, Critchley et al. 2004, Kurth et al. 2010). According to Craig's model, information of the body's physiological state is mapped to more posterior insular segments and subsequently rerepresented in the AI, where it may become consciously accessible, enabling a subjective affective experience and global feeling state (Craig 2002, 2009).

Cingulate Cortex

In the limbic system, the cingulate cortex represents the motivational-premotor counterpart of the rather sensory-predictive insula (Craig 2009) and has long been acknowledged as a hub region in affective, cognitive, and motor control phenomena (Paus 2001). Encircling the corpus callosum ventrally, dorsally, and posteriorly, it comprises at least four different cytoarchitectonic subregions, namely ACC, MCC, PCC, and retrosplenial cortex (Vogt et al. 2005). These subregions differ in terms of connectivity, as indicated by animal tract-tracing data as well as diffusion tractography and fMRI signal correlations in humans (Beckmann et al. 2009, Margulies et al. 2007, Vogt & Pandya 1987). Within subregions, connection patterns may also vary significantly. Indeed, whereas rostral ACC densely connects to lateral and orbital PFC and temporo-limbic regions (Pandya et al. 1981, Vogt & Pandya 1987), caudal divisions around dACC/aMCC show a relative increase of functional connections to sensorimotor regions (Margulies et al. 2007). This region receives direct projections from ascending pain pathways (Dum et al. 2009) and is interconnected with the insula (Mesulam & Mufson 1982b, Mufson & Mesulam 1982, Vogt & Pandya 1987), amygdala (Morecraft et al. 2007), ventral striatum (Kunishio & Haber 1994), and PAG (Hardy & Leichnetz 1981). In addition to the direct experience of pain and empathy, other negative affective states and anticipation thereof activate dACC/aMCC (Buchel et al. 1998, Ploghaus et al. 1999, Porro et al. 2002). Moreover, dACC/aMCC involvement in various attentionally or cognitively demanding tasks indicates that this region may implement general monitoring and control processes across multiple domains (Paus 2001). A recent meta-analysis of 939 studies showed overlapping activations in dACC/aMCC during negative affect, pain, and cognitive control (Shackman et al. 2011). The authors suggested that this region synthesizes information about unlearned and learned reinforcers. This may then bias adaptive responding in motor centers responsible for expressing affect and executing goal-directed behavior and ultimately guide behavior in uncertain, potentially aversive environments.

Interoceptive Network Interactions

AI and ACC/MCC share a close functional relationship within various flexibly recruited and distributed networks (Craig 2009, Devinsky et al. 1995, Medford & Critchley 2010, Seeley et al. 2007, Sridharan et al. 2008, Taylor et al. 2009). In their seminal article, Devinsky and colleagues acknowledged that AI and ACC/MCC, together with limbic and subcortical regions such as OFC, amygdala, PAG, and ventral striatum, form a coherent network that assesses the motivational content of internal and external stimuli to regulate context-dependent behaviors (Devinsky et al. 1995). Integrating evidence across multiple domains of joint AI and ACC/MCC activation, Medford & Critchley (2010) recently suggested that while the AI forms an input region of a system that is based on self-awareness, these global emotional feeling states are ultimately rerepresented in ACC to control, select, and prepare appropriate responses. Indeed, a close functional relationship between AI and ACC/MCC was recently shown using resting-state fMRI connectivity analysis (Taylor et al. 2009). The authors suggested that this link may enable an integration of interoceptive information with salience. Seeley et al. (2007) also suggested a role in saliency processing in their study observing a correlation between the degree of functional coupling of AI and ACC together with limbic cortical and subcortical regions and anxiety ratings outside the scanner. Using Granger causality analysis of fMRI signals, the study showed that these salience networks may switch between otherwise relatively anticorrelated executive task-activated networks such as dorsolateral PFC and posterior parietal cortex and default-mode networks such as ventromedial PFC and PCU (Sridharan et al. 2008), which are generally more active during stimulus-independent thought, self-projection, and mind-wandering (Buckner & Carroll 2007, Christoff et al. 2009, Mason et al. 2007, Schooler et al. 2012, Smallwood & Schooler 2006).

Recently, Singer, Critchley, and Preuschoff provided a framework that related findings of AI activations during empathy and affective states with neuroeconomic reports of a role of AI in uncertainty processing (Singer et al. 2009). Indeed, these studies have also shown AI activations during the processing of risk, risk prediction error, and uncertainty in decision making (Grinband et al. 2006, Huettel et al. 2006, Kuhnen & Knutson 2005, Paulus et al. 2003, Preuschoff et al. 2008). According to this model, AI integrates modality-specific information from multiple feeling states and uncertainty information with individual risk preferences and contextual information. These computations are thought to contribute to the generation of current and predictive feeling states and may ultimately facilitate error-based learning in the affective domain as a prerequisite for successful decision making under uncertain conditions. These representations in AI enable the formation of affective predictions related to the self but also related to predictions of other people's feeling states. Finally, insula computations can be fed to valuation regions such as the OFC and ventral striatum and also to ACC for response selection and control. Strong interconnections of AI and ACC, and their hub-like position in multiple functional networks, also make them ideally suited to integrate interoceptive information with contextual input into global feeling states, ultimately allowing for modulation of decisions and action responses (Singer et al. 2009, Singer & Lamm 2009). Empathizing with others may thus relate to the involvement of AI and ACC in generating forward models of feeling states for others that, together with certainty computations, may enable one to predict and understand the social and affective behavior of others.

AI and ACC contain a distinctive class of spindle-shaped cells, the Von Economo neurons (Allman et al. 2010, Craig 2009, Von Economo 1926). Their large size and relatively simple dendritic morphology make them suitable for rapid communication between AI and ACC, allowing a fast integration of global affective states, motivation, control, and behavior in dynamic situational contexts (Allman et al. 2010). Comparative histological assessments suggest that these cells are numerous in adult humans; fewer are found in infants, great apes, elephants, and whales (Allman et al. 2010). Also based on the observation that these cells may be selectively destroyed in frontotemporal dementia, a neurodegenerative disorder associated with deficient empathy and socio-emotional functioning (Seeley et al. 2006), some investigators have suggested that Von Economo neurons play a role in empathy, social awareness, and self-control (Allman et al. 2010, Craig 2009).

MODULATION OF EMPATHY

Together with their frequent activation across various situations, patterns of structural connectivity of AI and ACC suggest that these two regions may integrate information from a range of different domains to allow the flexible selection of adaptive responses. Indeed, in the domain of empathy, ample data have shown how vicarious responses in AI and subsequent overt behavior can be modulated by various factors, such as those related to individual traits and situational contexts (**Figure 3**).

Person Characteristics

Individual differences in person characteristics likely affect empathic responses. To measure empathic traits, several relatively easy, reliable, and reproducible self-report questionnaires have been developed, including the Interpersonal Reactivity Index (IRI) (Davis 1983) and the Balanced Emotional Empathy Scale (BEES) (Mehrabian 1997). In empathy for pain, such scales have been correlated with empathic responses. Although results have been mixed (Lamm et al. 2011), some studies have shown a modulation of empathic responses by

Anterior insula (AI)

Nucleus accumbens (NAcc)

a

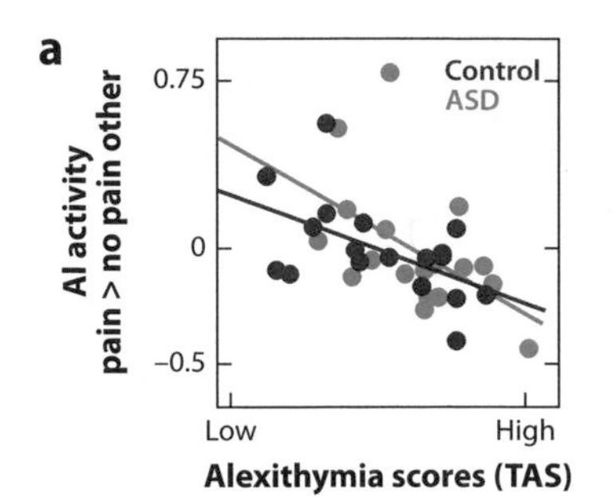

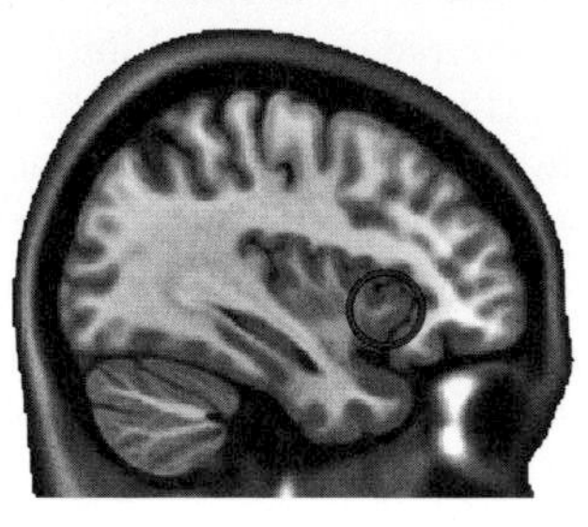

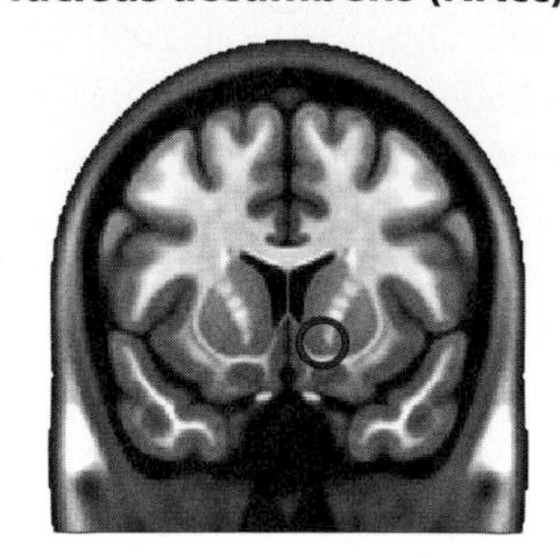

b

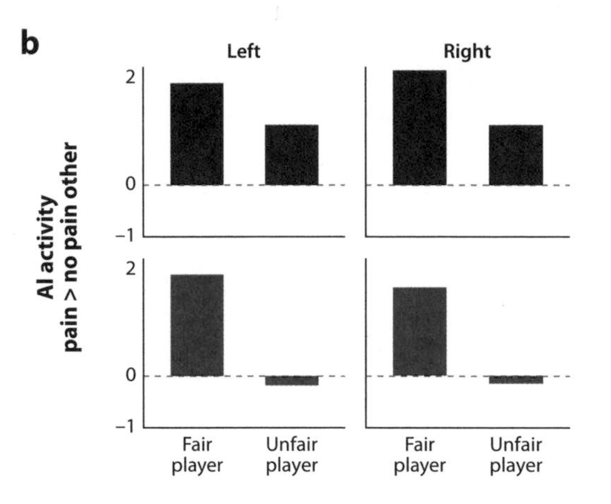

Women

Men

c

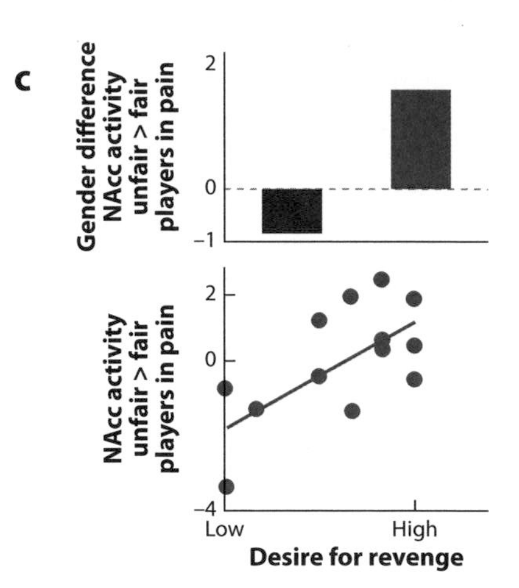

d

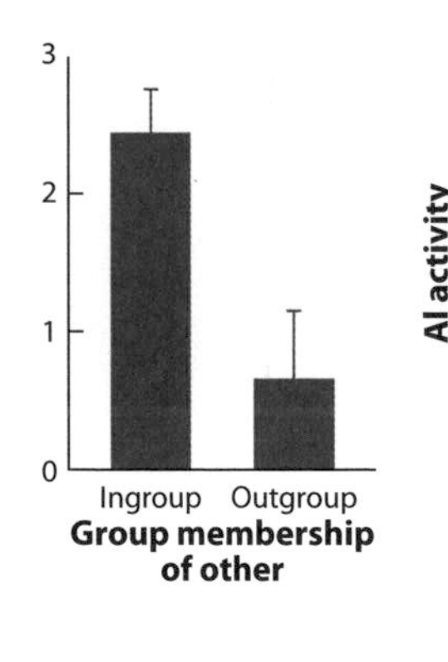

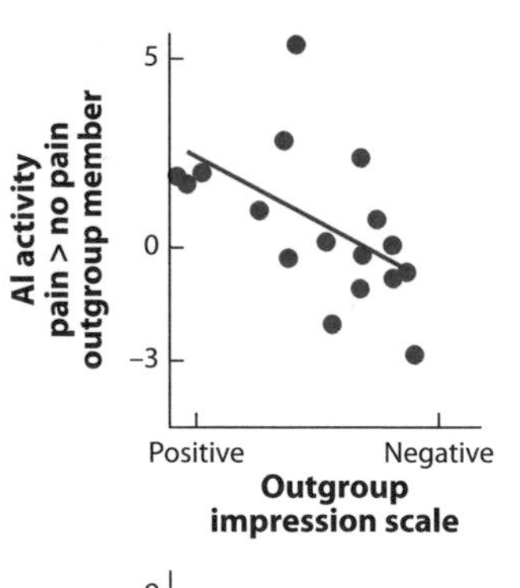

e

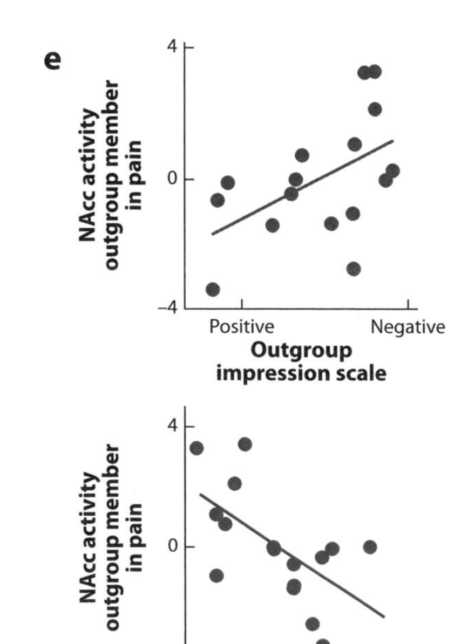

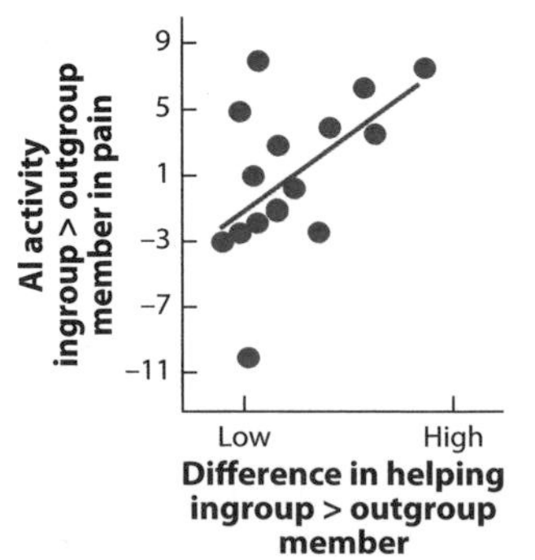

empathic traits (Jabbi et al. 2007; Saarela et al. 2007; Singer et al. 2004, 2006). For example, Singer and colleagues (2004, 2006) reported that scores on the IRI empathic concern subscale and BEES correlated with AI and dACC activity during vicariously felt pain delivered by abstract cues. In another study, researchers displayed faces in pain and found similar correlations in AI and IFG (Saarela et al. 2007). Extending these findings to the domain of taste, Keysers's group observed correlations between fronto-insular activations and several IRI subscales when subjects witnessed disgusted and pleased facial expressions (Jabbi et al. 2007). In conclusion, these experiments suggest that empathy traits may, under some conditions, modulate empathic brain responses. The meta-analysis by Lamm and colleagues (2011) has also summarized data for a modulation of empathic responses by trait measures and observed that, compared with trait measures of empathy, more robust brain-behavior correlations can be depicted when using state measures of felt empathy or unpleasantness in a trial-by-trial fashion or state questionnaires of felt empathic concern in the given situation.

Several studies have shown that empathic responses may be affected by alexithymia. Although this trait is present in ~10% of the general population (Salminen et al. 1999), elevated proportions are found in patients with autism spectrum disorder (ASD), a neurodevelopmental condition associated with communicative and social deficits (Hill et al. 2004). In a study by Silani and colleagues (2008), alexithymia and empathy self-report scores in controls and individuals with high-functioning ASD were found to be correlated with AI activity when subjects had to interocept on their feelings. A similar modulation was also observed during a cue-based empathy for pain design, in which empathy-related brain responses in AI were strongly modulated by the degree of alexithymia in controls and individuals with ASD (Bird et al. 2010). Indeed, the greater the participants' deficits in understanding their own emotions, regardless of whether they were control subjects or patients, the less activation they showed in AI while empathizing with people present in the same room undergoing painful experiences (**Figure 3*a***). These results thus confirm the hypothesis that representations in AI underlie

Alexithymia: subclinical phenomenon related to difficulties in identifying and describing feelings and in distinguishing feelings from bodily sensations. From Nemiah 1977

ASD: autism spectrum disorder

Figure 3

Modulation of empathy. Several cue-based studies have shown a modulation of empathy-related brain responses in fronto-insular regions, such as anterior insula (AI; *left* and *center panels*), associated in some studies with an antagonistic response in the nucleus accumbens of the ventral striatum (NAcc; *right panels*). (*a*) Modulation of empathy by personal traits, such as the degree of alexithymia (Bird et al. 2010). Activity in an independent region of interest in the left AI during empathy for pain in others was shown to correlate with alexithymia scores [Toronto Alexithymia Scale (TAS), a self-report measure of alexithymia] in controls (*gray*) and individuals with autism spectrum disorder (ASD) (*green*). The stronger the participants' deficit was in understanding their own emotions, the lower empathy-related activation in left AI was when witnessing another person suffering pain. This effect was seen in patients and controls alike. (*b*, *c*) Modulation of empathy by perceived fairness (Singer et al. 2006). (*b*) If a target person behaved unfairly in an economic game prior to scanning, men (*blue*) but not women (*purple*) showed reduced bilateral AI activity when the unfair but not fair person is in pain. (*c*) (*Upper panel*) The decrease in activity in AI was paralleled in men but not women by an increase in activation in left NAcc when unfair others receive pain relative to fair others. (*Lower panel*) The degree of activity in left NAcc was correlated to the subjectively expressed desire for revenge in men. (*d*, *e*) Modulation of empathy by perceived ingroup or outgroup membership (Hein et al. 2010). (*d*) (*Upper left panel*) Male soccer fans showed reduced left AI activity when fans favoring an opposing team received pain relative to painful stimulation of fans favoring the same team. (*Upper right panel*) Attitude toward outgroup member furthermore correlates with AI activity, with reduced AI activity for outgroup members that subjects did not view positively. (*Lower panel*) The stronger the participants' AI responses were to ingroup pain relative to outgroup pain, the more they helped the ingroup member relative to the outgroup member in a subsequent session, in which subjects could choose to receive pain in order to reduce the other's pain. (*e*) (*Upper panel*) Increased right NAcc activity in male soccer fans when disliked soccer fans of the opposing team received painful stimulation. (*Lower panel*) Increased NAcc activity when an outgroup member was in pain predicts lack of helping toward an outgroup member in subsequent helping sessions. For the exact region of interest specifications, please see the original articles. Adapted from Bird et al. (2010), Hein et al. (2010), Singer et al. (2006) with permission.

NAcc: nucleus accumbens

representations of our own feeling states, which in turn form the basis for understanding the feelings of others. Thus, understanding your own feeling states may be a prerequisite to engage in vicarious simulation for a better understanding of other people's feeling states. Finally, in addition to providing insight into the modulation of empathic responses in healthy subjects, these studies promote a more differentiated picture of social deficits observed in ASD.

Contextual Appraisal

Attribution of specific traits to the target may influence empathic responses of the observer. In an early study focusing on modulation of empathy-related brain responses in the domain of pain in AI, Singer and colleagues (2006) asked participants to engage in a game with confederates, who followed either fair or unfair strategies, prior to scanning. During subsequent scanning, participants watched cues indicating pain in the confederate. Measuring empathic responses, the authors reproduced well-known patterns of brain activity in fronto-insular and dorsal ACC when fair confederates received painful shocks. However, in males these activations were reduced when unfair players received shocks (**Figure 3*b***); instead, males but not females showed increased activation in the nucleus accumbens of the ventral striatum (NAcc) (**Figure 3*c***, upper panel), which was further correlated with the desire for revenge (**Figure 3*c***, lower panel). These findings thus provided initial evidence for a link between fairness behavior and the degree of affective sharing. Moreover, activation in valence and reward-prediction regions such as NAcc or OFC (O'Doherty et al. 2004, Rolls 2004, Schultz 2000) may provide a neurobiological mechanism that helps reinforce punishment of defectors in social situations (Fehr 2008).

These results were extended in a study focusing on the modulation of empathy-related brain responses in AI as a function of perceived group membership (Hein et al. 2010). As in the case of fairness, participants showed stronger empathy-related brain responses in AI toward ingroup compared with outgroup members (**Figure 3*d***, left panel); this result was also reflected by a correlation between AI activity and their impression of the outgroup members, with stronger AI activation for positively valued outgroup members and reduced AI activations when outgoup members were seen negatively (**Figure 3*d***, upper right panel). Moreover, participants who valued outgroup members more negatively showed increased NAcc activation when observing them in pain, compared to outgroup members who were valued more favorably (**Figure 3*e***, upper panel).

Last, the stronger the participants' AI responses were to ingroup pain relative to outgroup pain, the more they helped the ingroup member relative to the outgroup member in a subsequent session, in which subjects could choose to receive pain in order to reduce the other's pain (**Figure 3*d***, lower right panel). Conversely, increased NAcc activity when observing outgroup members in pain predicted the absence of subsequent helping (**Figure 3*e***, lower panel). Thus, these two studies suggest that activation in reward-related areas such as NAcc in the context of empathy-for-pain paradigms may reflect an antagonistic motivation to empathy, possibly related to feelings of revenge and Schadenfreude, counteracting empathic resonance in regions such as AI. Whenever NAcc activation is high while witnessing the suffering of another person, activation in AI is low. This activation pattern may ultimately be associated with a lack of engagement in prosocial behavior. Therefore, the social evaluation of the suffering person may modulate the balance of the two systems, which in turn motivate either egoistic or altruistic behavior.

Whereas the above-mentioned studies utilized implicit contextual information, several experimenters provided the participants with explicit information that was hypothesized to influence the evaluation of the observed situation and, in turn, empathy (Decety et al. 2009a; Lamm et al. 2007a,b). Lamm and colleagues, for example, showed a series of needle injections and also informed the participants that injections were administered to either normal, pain-sensitive hands or anesthetized hands undergoing biopsy (Lamm et al. 2007b). In another study from the same laboratory, participants watched faces of patients undergoing a special painful sound-based treatment, and they were provided with additional information about the treatment success or failure (Lamm et al. 2007a). As in the case of implicit information in the previous studies (Hein et al. 2010, Singer et al. 2006), these explicit contextual factors modulated hemodynamic responses in regions relevant to empathy, such as AI and ACC.

The above results strongly indicate that individual person characteristics and contextual appraisal modulate empathic brain responses. Moreover, there is also some, albeit less consistent, evidence for a modulation through other factors, including attention (Gu & Han 2007, but also see Avenanti et al. 2006 and Gu et al. 2010), neuropeptides such as oxytocin (Riem et al. 2011, but also see Singer et al. 2008), and perspective taking (Cheng et al. 2010, Lamm et al. 2007a). Studying the effects of perspective taking, Cheng and colleagues (2010) recently observed that adopting the perspective of the self and a loved one in pain, compared with that of a stranger in pain, leads to increased pain ratings (Cheng et al. 2010). Moreover, self- and loved-one perspectives induce AI and ACC activations, whereas adopting the stranger perspective results in superior frontal gyrus and TPJ activity. These findings suggest that imagining oneself, or a loved one, in pain will trigger elevated responses in empathy networks. Conversely, projecting a stranger into the same situation may recruit regions also involved in self-other distinction. These network activations may thus contribute to reevaluating the affective meaning for the self of a given situation and may ultimately be a form of emotion-regulation strategy. Thereby, it can allow the subject to adapt responses to current situational demands, for instance, by increasing empathy for loved ones or to reduce distress when seeing strangers in pain.

CONCLUSIONS AND OPEN QUESTIONS

This review pulls together the growing evidence for the existence of shared representations activated by the firsthand experience and the vicarious experience of sensations or emotions experienced by another person. Recent meta-analyses on empathy fMRI studies revealed a special role of AI and dACC/aMCC in empathy for many, though not all, feelings and sensations. Representation in these interoceptive regions may be involved in integrating current and predictive information of feeling states in the self and others. These representations may ultimately enable adaptive responses to the social and affective behavior of others by promoting forms of other-oriented prosocial helping or self-oriented withdrawal behavior to counteract distress (Klimecki & Singer 2012), possibly via the ACC/MCC.

Questioning early automaticity assumptions of empathic brain responses (Preston & de Waal 2002), several recent studies focusing on the modulation of these empathy-related brain responses revealed that activation in these regions is not fixed but can be dynamically modulated by several factors related to situational context or person characteristics. Furthermore, depending on the particular situation and information available in the environment, empathic responses may furthermore coengage other neural networks relevant for social cognition such as those observed in mentalizing or action observation. Future research should further explore the complex interaction between different routes of social cognition in producing an empathic understanding of other peoples' mental lives.

Additional open questions remain, such as the identification of the computational processes carried out by AI and dACC/aMCC. In this regard, the role of Von Economo neurons in mediating interoceptive signal exchange may be of interest. Moreover, lesion studies and animal recordings, together with methodological advancements in large-scale network modeling, meta-analysis, pattern classification, or

Oxytocin: neuropeptide promoting the formation of social attachments and affiliation. Also attenuates stress by interacting with hypothalamo-pituitary-adrenal axis

repetition suppression study design may help to elucidate further the exact role of different regions involved in empathic networks (Bullmore & Sporns 2009, Caruana et al. 2011, Kurth et al. 2010, Shamay-Tsoory et al. 2009, Yarkoni et al. 2011).

Only initial evidence suggests affective plasticity and long-term alterations of brain systems involved in empathy and related positive affect such as compassion. For example, two recent cross-sectional studies by Lutz and colleagues have shown increased functional activation in insula and mentalizing networks in compassion meditation experts compared with novice meditators (Lutz et al. 2008, 2009). In a recent longitudinal study, Leiberg and colleagues (2011) furthermore observed that even one-day compassion training may enhance prosocial helping behavior, and preliminary evidence suggests that this may relate to altered functional activations in brain areas associated with positive affect and reward such as the mid-insula, medial OFC, putamen, and ventral tegmental area (Klimecki et al. 2011). Training compassionate responses may therefore increase the resiliency to aversive events, possibly by upregulating networks associated with positive affect, reward, and attachment (Klimecki et al. 2011).

Insights into plastic alterations in networks underlying empathy and social cognition in general can also be gained from investigating clinical and subclinical groups associated with deficient empathic ability, such as individuals with conduct disorder, ASD, and alexithymia (Bird et al. 2010, Silani et al. 2008). These studies may provide further insights into abnormal predispositions to unempathic responding, how neural plasticity can go awry, and the influence of deficient empathy on aggressive, impulsive, and selfish behavior (Birbaumer et al. 2005, Boccardi et al. 2010, Decety et al. 2009b, Kiehl et al. 2001, Sterzer et al. 2007, Tiihonen et al. 2008).

Last, despite abundant previous research carried out by developmental behavioral psychologists on the ontogeny of empathy in childhood (Eisenberg 2000, Knafo et al. 2008), the field of social neuroscience has only just started to address important developmental brain changes related to our ability to empathize and the relationship of these changes to moral reasoning and prosocial behavior (Decety & Michalska 2010, Singer 2006).

In the past few years, we have begun to understand better the neural basis of empathy and related states such as compassion. Identifying crucial subcomponents and brain network interactions involved in empathy sheds important light on the generation of this multifaceted experience at the heart of human emotional and social behavior. Ultimately, such insights may guide the development of strategies for circumventing aversive behavior and burnout syndromes in caregivers and physicians (Halifax 2010, Hojat et al. 2009, Klimecki & Singer 2012) and may lead to advances in nourishing socio-affective competences in children and in adults suffering from conduct disorders and ASD.

SUMMARY POINTS

1. Neuroimaging studies on empathy for pain consistently revealed activations in AI and dACC/aMCC when directly experiencing pain as well as when empathizing with the pain of others, suggesting that empathy depends, in part, on shared representations.
2. The vicarious experience of affective states other than pain, such as social exclusion, disgust, anxiety, and taste, also activates AI and dACC/aMCC; nevertheless, initial evidence suggests that these regions may not necessarily be involved in the vicarious sharing of all states.

3. Depending on the situational context and information available in the environment, empathic responses may involve a corecruitment of so-called mirror-neuron networks and regions involved in theory of mind or mentalizing.
4. Empathic brain responses are not fixed but may be modulated by person characteristics such as degree of alexithymia or contextual appraisal such as perceived fairness of another person or group membership.
5. Being crucial hubs in human interoceptive cortex, AI and dACC/aMCC may perform domain-general computations that represent and predict feeling states and guide responses to the emotional experience of the self and others.

DISCLOSURE STATEMENT

The authors are not aware of any affiliations, memberships, funding, or financial holdings that might be perceived as affecting the objectivity of this review.

ACKNOWLEDGMENTS

We thank our colleagues Jonathan Smallwood, Haakon Engen, Cade McCall, and Joshua Grant for their insightful comments on earlier versions of the manuscript. We thank Alfred Anwander, Claus Lamm, and Grit Hein for kindly providing some of the data presented in the figures. We thank Sandra Zurborg for carefully proofreading the final version of this article. We thank Stefan Liebig for providing the illustrations in **Figure 1**. We apologize to all the investigators whose research could not be appropriately cited owing to space limitations. This work was funded by the European Research Council under the European Community's Seventh Framework Program (FP7/2007-2013)/ERC Grant agreement no. 205557 [EMPATHICBRAIN].

LITERATURE CITED

Allman JM, Tetreault NA, Hakeem AY, Manaye KF, Semendeferi K, et al. 2010. The von Economo neurons in the frontoinsular and anterior cingulate cortex. *Ann. N. Y. Acad. Sci.* 1225:59–71

Amaral DG, Price JL. 1984. Amygdalo-cortical projections in the monkey (*Macaca fascicularis*). *J. Comp. Neurol.* 230:465–96

Apkarian AV, Bushnell MC, Treede RD, Zubieta JK. 2005. Human brain mechanisms of pain perception and regulation in health and disease. *Eur. J. Pain* 9:463–84

Augustine JR. 1996. Circuitry and functional aspects of the insular lobe in primates including humans. *Brain Res. Brain Res. Rev.* 22:229–44

Avenanti A, Bueti D, Galati G, Aglioti SM. 2005. Transcranial magnetic stimulation highlights the sensorimotor side of empathy for pain. *Nat. Neurosci.* 8:955–60

Avenanti A, Minio-Paluello I, Bufalari I, Aglioti SM. 2006. Stimulus-driven modulation of motor-evoked potentials during observation of others' pain. *Neuroimage* 32:316–24

Baron-Cohen S, Wheelwright S. 2004. The empathy quotient: an investigation of adults with Asperger syndrome or high functioning autism, and normal sex differences. *J. Autism Dev. Disord.* 34:163–75

Batson CD. 2009. These things called empathy. In *The Social Neuroscience of Empathy*, ed. J Decety, W Ickes, pp. 16–31. Cambridge, MA: MIT Press

Batson CD, Eklund JH, Chermok VL, Hoyt JL, Ortiz BG. 2007. An additional antecedent of empathic concern: valuing the welfare of the person in need. *J. Pers. Soc. Psychol.* 93:65–74

Baumeister RF, Vohs KD. 2007. *Encyclopedia of Social Psychology*. Thousand Oaks, CA: Sage

Beckmann M, Johansen-Berg H, Rushworth MF. 2009. Connectivity-based parcellation of human cingulate cortex and its relation to functional specialization. *J. Neurosci.* 29:1175–90

Benuzzi F, Lui F, Duzzi D, Nichelli PF, Porro CA. 2008. Does it look painful or disgusting? Ask your parietal and cingulate cortex. *J. Neurosci.* 28:923–31

Bingel U, Quante M, Knab R, Bromm B, Weiller C, Buchel C. 2003. Single trial fMRI reveals significant contralateral bias in responses to laser pain within thalamus and somatosensory cortices. *Neuroimage* 18:740–48

Birbaumer N, Veit R, Lotze M, Erb M, Hermann C, et al. 2005. Deficient fear conditioning in psychopathy: a functional magnetic resonance imaging study. *Arch. Gen. Psychiatry* 62:799–805

Bird G, Silani G, Brindley R, White S, Frith U, Singer T. 2010. Empathic brain responses in insula are modulated by levels of alexithymia but not autism. *Brain* 133:1515–25

Blair RJ. 2005. Responding to the emotions of others: dissociating forms of empathy through the study of typical and psychiatric populations. *Conscious. Cogn.* 14:698–718

Blakemore SJ, Bristow D, Bird G, Frith C, Ward J. 2005. Somatosensory activations during the observation of touch and a case of vision-touch synaesthesia. *Brain* 128:1571–83

Boccardi M, Ganzola R, Rossi R, Sabattoli F, Laakso MP, et al. 2010. Abnormal hippocampal shape in offenders with psychopathy. *Hum. Brain Mapp.* 31:438–47

Botvinick M, Jha AP, Bylsma LM, Fabian SA, Solomon PE, Prkachin KM. 2005. Viewing facial expressions of pain engages cortical areas involved in the direct experience of pain. *Neuroimage* 25:312–19

Buchel C, Morris J, Dolan RJ, Friston KJ. 1998. Brain systems mediating aversive conditioning: an event-related fMRI study. *Neuron* 20:947–57

Buckner RL, Carroll DC. 2007. Self-projection and the brain. *Trends Cogn. Sci.* 11:49–57

Bullmore E, Sporns O. 2009. Complex brain networks: graph theoretical analysis of structural and functional systems. *Nat. Rev. Neurosci.* 10:186–98

Bushnell MC, Duncan GH, Hofbauer RK, Ha B, Chen JI, Carrier B. 1999. Pain perception: Is there a role for primary somatosensory cortex? *Proc. Natl. Acad. Sci. USA* 96:7705–9

Caruana F, Jezzini A, Sbriscia-Fioretti B, Rizzolatti G, Gallese V. 2011. Emotional and social behaviors elicited by electrical stimulation of the insula in the macaque monkey. *Curr. Biol.* 21:195–99

Cauda F, D'Agata F, Sacco K, Duca S, Geminiani G, Vercelli A. 2011. Functional connectivity of the insula in the resting brain. *Neuroimage* 55:8–23

Cerliani L, Thomas R, Jbabdi S, Siero J, Nanetti L, et al. 2012. Probabilistic tractorgraphy recovers a rostrocaudal trajectory of connectivity variability in the human insular cortex. *Hum. Brain Mapp.* doi: 10.1002/hbm.21338. In press

Cheng Y, Chen C, Lin CP, Chou KH, Decety J. 2010. Love hurts: an fMRI study. *Neuroimage* 51:923–29

Christoff K, Gordon AM, Smallwood J, Smith R, Schooler JW. 2009. Experience sampling during fMRI reveals default network and executive system contributions to mind wandering. *Proc. Natl. Acad. Sci. USA* 106:8719–24

Corradi-Dell'Acqua C, Hofstetter C, Vuilleumier P. 2011. Felt and seen pain evoke the same local patterns of cortical activity in insula and cingulate cortex. *J. Neurosci.* 31(49):17996–8006

Craig AD. 2002. How do you feel? Interoception: the sense of the physiological condition of the body. *Nat. Rev. Neurosci.* 3:655–66

Craig AD. 2003. Pain mechanisms: labeled lines versus convergence in central processing. *Annu. Rev. Neurosci.* 26:1–30

Craig AD. 2009. How do you feel—now? The anterior insula and human awareness. *Nat. Rev. Neurosci.* 10:59–70

Craig KD. 2004. Social communication of pain enhances protective functions: a comment on Deyo, Prkachin and Mercer 2004. *Pain* 107:5–6

Critchley HD, Wiens S, Rotshtein P, Ohman A, Dolan RJ. 2004. Neural systems supporting interoceptive awareness. *Nat. Neurosci.* 7:189–95

Davis MH. 1983. Measuring individual differences in empathy: evidence for a multidimensional approach. *J. Pers. Soc. Psychol.* 44:113–26

de Gelder B, Snyder J, Greve D, Gerard G, Hadjikhani N. 2004. Fear fosters flight: a mechanism for fear contagion when perceiving emotion expressed by a whole body. *Proc. Natl. Acad. Sci. USA* 101:16701–6

de Greck M, Wang G, Yang X, Wang X, Northoff G, Han S. 2012. Neural substrates underlying intentional empathy. *Soc. Cogn. Affect. Neurosci.* doi: 10.1093/scan/nsq093. In press

de Vignemont F, Singer T. 2006. The empathic brain: how, when and why? *Trends Cogn. Sci.* 10:435–41

Decety J, Echols S, Correll J. 2009a. The blame game: the effect of responsibility and social stigma on empathy for pain. *J. Cogn. Neurosci.* 22:985–97

Decety J, Michalska KJ. 2010. Neurodevelopmental changes in the circuits underlying empathy and sympathy from childhood to adulthood. *Dev. Sci.* 13:886–99

Decety J, Michalska KJ, Akitsuki Y, Lahey BB. 2009b. Atypical empathic responses in adolescents with aggressive conduct disorder: a functional MRI investigation. *Biol. Psychol.* 80:203–11

Deen B, Pitskel NB, Pelphrey KA. 2011. Three systems of insular functional connectivity identified with cluster analysis. *Cereb. Cortex* 21(7):1498–506

Derbyshire SW. 2000. Exploring the pain "neuromatrix". *Curr. Rev. Pain* 4:467–77

Devinsky O, Morrell MJ, Vogt BA. 1995. Contributions of anterior cingulate cortex to behaviour. *Brain* 118(Pt. 1):279–306

Duerden EG, Albanese M-C. 2012. Localization of pain-related brain activation: a meta-analysis of neuroimaging data. *Hum. Brain Mapp.* doi: 10.1002/hbm.21416. In press

Dum RP, Levinthal DJ, Strick PL. 2009. The spinothalamic system targets motor and sensory areas in the cerebral cortex of monkeys. *J. Neurosci.* 29:14223–35

Ebisch SJ, Perrucci MG, Ferretti A, Del Gratta C, Romani GL, Gallese V. 2008. The sense of touch: embodied simulation in a visuotactile mirroring mechanism for observed animate or inanimate touch. *J. Cogn. Neurosci.* 20:1611–23

Eisenberg N. 2000. Emotion, regulation, and moral development. *Annu. Rev. Psychol.* 51:665–97

Eisenberg N, Fabes RA. 1990. Empathy: conceptualization, measurement, and relation to prosocial behavior. *Motiv. Emot.* 14:131–49

Fan Y, Duncan NW, de Greck M, Northoff G. 2011. Is there a core neural network in empathy? An fMRI based quantitative meta-analysis. *Neurosci. Biobehav. Rev.* 35:903–11

Fehr E. 2008. Social preferences and the brain. In *Neuroeconomics: Decision Making and the Brain*, ed. PW Glimcher, E Fehr, C Camerer, A Rangel, RA Poldrack, pp. 215–35. London, UK: Academic/Elsevier

Frith CD, Frith U. 2006. The neural basis of mentalizing. *Neuron* 50:531–34

Frith U, Frith CD. 2003. Development and neurophysiology of mentalizing. *Philos. Trans. R. Soc. Lond. B Biol. Sci.* 358:459–73

Fudge JL, Breitbart MA, Danish M, Pannoni V. 2005. Insular and gustatory inputs to the caudal ventral striatum in primates. *J. Comp. Neurol.* 490:101–18

Gallay DS, Gallay MN, Jeanmonod D, Rouiller EM, Morel A. 2012. The insula of reil revisited: multiarchitectonic organization in macaque monkeys. *Cereb. Cortex* 22(1):175–90

Gallese V, Keysers C, Rizzolatti G. 2004. A unifying view of the basis of social cognition. *Trends Cogn. Sci.* 8:396–403

Grill-Spector K, Malach R. 2001. fMR-adaptation: a tool for studying the functional properties of human cortical neurons. *Acta Psychol. (Amst.)* 107:293–321

Grinband J, Hirsch J, Ferrera VP. 2006. A neural representation of categorization uncertainty in the human brain. *Neuron* 49:757–63

Gu X, Han S. 2007. Attention and reality constraints on the neural processes of empathy for pain. *Neuroimage* 36:256–67

Gu X, Liu X, Guise KG, Naidich TP, Hof PR, Fan J. 2010. Functional dissociation of the frontoinsular and anterior cingulate cortices in empathy for pain. *J. Neurosci.* 30:3739–44

Halifax J. 2010. The precious necessity of compassion. *J. Pain Symptom Manag.* 41(1):146–53

Hardy SG, Leichnetz GR. 1981. Cortical projections to the periaqueductal gray in the monkey: a retrograde and orthograde horseradish peroxidase study. *Neurosci. Lett.* 22:97–101

Harrison NA, Singer T, Rotshtein P, Dolan RJ, Critchley HD. 2006. Pupillary contagion: central mechanisms engaged in sadness processing. *Soc. Cogn. Affect. Neurosci.* 1:5–17

Hein G, Silani G, Preuschoff K, Batson CD, Singer T. 2010. Neural responses to ingroup and outgroup members' suffering predict individual differences in costly helping. *Neuron* 68:149–60

Henson RN, Rugg MD. 2003. Neural response suppression, haemodynamic repetition effects, and behavioural priming. *Neuropsychologia* 41:263–70

Hill E, Berthoz S, Frith U. 2004. Brief report: cognitive processing of own emotions in individuals with autistic spectrum disorder and in their relatives. *J. Autism Dev. Disord.* 34:229–35

Hoffman ML. 2000. *Empathy and Moral Development: Implications for Caring and Justice*. New York: Cambridge Univ. Press

Hojat M, Vergare MJ, Maxwell K, Brainard G, Herrine SK, et al. 2009. The devil is in the third year: a longitudinal study of erosion of empathy in medical school. *Acad. Med.* 84:1182–91

Huettel SA, Stowe CJ, Gordon EM, Warner BT, Platt ML. 2006. Neural signatures of economic preferences for risk and ambiguity. *Neuron* 49:765–75

Immordino-Yang MH, McColl A, Damasio H, Damasio A. 2009. Neural correlates of admiration and compassion. *Proc. Natl. Acad. Sci. USA* 106:8021–26

Jabbi M, Bastiaansen J, Keysers C. 2008. A common anterior insula representation of disgust observation, experience and imagination shows divergent functional connectivity pathways. *PLoS One* 3:e 2939

Jabbi M, Swart M, Keysers C. 2007. Empathy for positive and negative emotions in the gustatory cortex. *Neuroimage* 34:1744–53

Jackson PL, Brunet E, Meltzoff AN, Decety J. 2006. Empathy examined through the neural mechanisms involved in imagining how I feel versus how you feel pain. *Neuropsychologia* 44:752–61

Jackson PL, Meltzoff AN, Decety J. 2005. How do we perceive the pain of others? A window into the neural processes involved in empathy. *Neuroimage* 24:771–79

Keysers C, Gazzola V. 2007. Integrating simulation and theory of mind: from self to social cognition. *Trends Cogn. Sci.* 11:194–96

Keysers C, Kaas JH, Gazzola V. 2010. Somatosensation in social perception. *Nat. Rev. Neurosci.* 11:417–28

Keysers C, Wicker B, Gazzola V, Anton JL, Fogassi L, Gallese V. 2004. A touching sight: SII/PV activation during the observation and experience of touch. *Neuron* 42:335–46

Kiehl KA, Smith AM, Hare RD, Mendrek A, Forster BB, et al. 2001. Limbic abnormalities in affective processing by criminal psychopaths as revealed by functional magnetic resonance imaging. *Biol. Psychiatry* 50:677–84

Klimecki O, Leihberg S, Lamm C, Singer T. 2011. *Neural and behavioral changes related to compassion training*. Poster presented at "The Social Brain" workshop, Cambridge, UK

Klimecki O, Singer T. 2012. Empathic distress fatigue rather than compassion fatigue? Integrating findings from empathy research in psychology and neuroscience. In *Pathological Altruism*, ed. B Oakley, A Knafo, G Madhavan, DS Wilson, pp. 368–84. New York: Springer

Klingler J, Ludwig E. 1956. *Atlas cerebri humani. Der innere Bau des Gehirns, dargestellt auf Grund makroskopischer Präparate/The Inner Structure of the Brain Demonstrated on the Basis of Macroscopical Preparations*. Basel: Karger

Knafo A, Zahn-Waxler C, Van Hulle C, Robinson JL, Rhee SH. 2008. The developmental origins of a disposition toward empathy: genetic and environmental contributions. *Emotion* 8:737–52

Kong J, Gollub RL, Polich G, Kirsch I, Laviolette P, et al. 2008. A functional magnetic resonance imaging study on the neural mechanisms of hyperalgesic nocebo effect. *J. Neurosci.* 28:13354–62

Krach S, Cohrs JC, de Echeverria Loebell NC, Kircher T, Sommer J, et al. 2011. Your flaws are my pain: linking empathy to vicarious embarrassment. *PLoS One* 6:e18675

Kuhnen CM, Knutson B. 2005. The neural basis of financial risk taking. *Neuron* 47:763–70

Kunishio K, Haber SN. 1994. Primate cingulostriatal projection: limbic striatal versus sensorimotor striatal input. *J. Comp. Neurol.* 350:337–56

Kurth F, Zilles K, Fox PT, Laird AR, Eickhoff SB. 2010. A link between the systems: functional differentiation and integration within the human insula revealed by meta-analysis. *Brain Struct. Funct.* 214:519–34

Lamm C, Batson CD, Decety J. 2007a. The neural substrate of human empathy: effects of perspective-taking and cognitive appraisal. *J. Cogn. Neurosci.* 19:42–58

Lamm C, Decety J, Singer T. 2011. Meta-analytic evidence for common and distinct neural networks associated with directly experienced pain and empathy for pain. *Neuroimage* 54:2492–502

Lamm C, Nusbaum HC, Meltzoff AN, Decety J. 2007b. What are you feeling? Using functional magnetic resonance imaging to assess the modulation of sensory and affective responses during empathy for pain. *PLoS One* 2:e1292

Leiberg S, Klimecki O, Singer T. 2011. Short-term compassion training increases prosocial behavior in a newly developed prosocial game. *PLoS One* 6:e17798

Lutz A, Brefczynski-Lewis J, Johnstone T, Davidson RJ. 2008. Regulation of the neural circuitry of emotion by compassion meditation: effects of meditative expertise. *PLoS One* 3:e1897

Lutz A, Greischar LL, Perlman DM, Davidson RJ. 2009. BOLD signal in insula is differentially related to cardiac function during compassion meditation in experts versus novices. *Neuroimage* 47:1038–46

Maihofner C, Herzner B, Otto Handwerker H. 2006. Secondary somatosensory cortex is important for the sensory-discriminative dimension of pain: a functional MRI study. *Eur. J. Neurosci.* 23:1377–83

Margulies DS, Kelly AM, Uddin LQ, Biswal BB, Castellanos FX, Milham MP. 2007. Mapping the functional connectivity of anterior cingulate cortex. *Neuroimage* 37:579–88

Mason MF, Norton MI, Van Horn JD, Wegner DM, Grafton ST, Macrae CN. 2007. Wandering minds: the default network and stimulus-independent thought. *Science* 315:393–95

Masten CL, Morelli SA, Eisenberger NI. 2011. An fMRI investigation of empathy for 'social pain' and subsequent prosocial behavior. *Neuroimage* 55:381–88

Medford N, Critchley HD. 2010. Conjoint activity of anterior insular and anterior cingulate cortex: awareness and response. *Brain Struct. Funct.* 214:535–49

Mehrabian A. 1997. Relations among personality scales of aggression, violence, and empathy: validational evidence bearing on the risk of eruptive violence scale. *Aggress. Behav.* 23:433–45

Mesulam MM, Mufson EJ. 1982a. Insula of the old world monkey. I. Architectonics in the insulo-orbito-temporal component of the paralimbic brain. *J. Comp. Neurol.* 212:1–22

Mesulam MM, Mufson EJ. 1982b. Insula of the old world monkey. III: Efferent cortical output and comments on function. *J. Comp. Neurol.* 212:38–52

Mitchell JP. 2009. Inferences about mental states. *Philos. Trans. R. Soc. Lond. Ser. B Biol. Sci.* 364:1309–16

Mobbs D, Yu R, Meyer M, Passamonti L, Seymour B, et al. 2009. A key role for similarity in vicarious reward. *Science* 324:900

Morecraft RJ, McNeal DW, Stilwell-Morecraft KS, Gedney M, Ge J, et al. 2007. Amygdala interconnections with the cingulate motor cortex in the rhesus monkey. *J. Comp. Neurol.* 500:134–65

Morrison I, Downing PE. 2007. Organization of felt and seen pain responses in anterior cingulate cortex. *Neuroimage* 37:642–51

Morrison I, Lloyd D, di Pellegrino G, Roberts N. 2004. Vicarious responses to pain in anterior cingulate cortex: Is empathy a multisensory issue? *Cogn. Affect. Behav. Neurosci.* 4:270–78

Mufson EJ, Mesulam MM. 1982. Insula of the old world monkey. II: Afferent cortical input and comments on the claustrum. *J. Comp. Neurol.* 212:23–37

Nanetti L, Cerliani L, Gazzola V, Renken R, Keysers C. 2009. Group analyses of connectivity-based cortical parcellation using repeated k-means clustering. *Neuroimage* 47:1666–77

Nemiah JC. 1977. Alexithymia. Theoretical considerations. *Psychother. Psychosom.* 28:199–206

Norman KA, Polyn SM, Detre GJ, Haxby JV. 2006. Beyond mind-reading: multi-voxel pattern analysis of fMRI data. *Trends Cogn. Sci.* 10:424–30

O'Doherty J, Dayan P, Schultz J, Deichmann R, Friston K, Dolan RJ. 2004. Dissociable roles of ventral and dorsal striatum in instrumental conditioning. *Science* 304:452–54

Pandya DN, Van Hoesen GW, Mesulam MM. 1981. Efferent connections of the cingulate gyrus in the rhesus monkey. *Exp. Brain Res.* 42:319–30

Paulus MP, Rogalsky C, Simmons A, Feinstein JS, Stein MB. 2003. Increased activation in the right insula during risk-taking decision making is related to harm avoidance and neuroticism. *Neuroimage* 19:1439–48

Paus T. 2001. Primate anterior cingulate cortex: where motor control, drive and cognition interface. *Nat. Rev. Neurosci.* 2:417–24

Peyron R, Laurent B, Garcia-Larrea L. 2000. Functional imaging of brain responses to pain. A review and meta-analysis. *Neurophysiol. Clin.* 30:263–88

Phillips ML, Gregory LJ, Cullen S, Coen S, Ng V, et al. 2003. The effect of negative emotional context on neural and behavioural responses to oesophageal stimulation. *Brain* 126:669–84

Ploghaus A, Tracey I, Gati JS, Clare S, Menon RS, et al. 1999. Dissociating pain from its anticipation in the human brain. *Science* 284:1979–81

Ploner M, Freund HJ, Schnitzler A. 1999. Pain affect without pain sensation in a patient with a postcentral lesion. *Pain* 81:211–14

Porro CA, Baraldi P, Pagnoni G, Serafini M, Facchin P, et al. 2002. Does anticipation of pain affect cortical nociceptive systems? *J. Neurosci.* 22:3206–14

Prehn-Kristensen A, Wiesner C, Bergmann TO, Wolff S, Jansen O, et al. 2009. Induction of empathy by the smell of anxiety. *PLoS One* 4:e5987

Preston SD, de Waal FB. 2002. Empathy: its ultimate and proximate bases. *Behav. Brain Sci.* 25:1–20; discussion -71

Preuschoff K, Quartz SR, Bossaerts P. 2008. Human insula activation reflects risk prediction errors as well as risk. *J. Neurosci.* 28:2745–52

Price DD. 2000. Psychological and neural mechanisms of the affective dimension of pain. *Science* 288:1769–72

Prinz W. 1984. Modes of linkages between perception and actions. In *Cognition and Motor Processes*, ed. W Prinz, AF Sanders, pp. 185–93. Berlin: Springer

Prinz W. 2005. Experimental approaches to action. In *Agency and Self-Awareness*, ed. J Roessler, N Eilan, pp. 165–87. New York: Oxford Univ. Press

Rainville P. 2002. Brain mechanisms of pain affect and pain modulation. *Curr. Opin. Neurobiol.* 12:195–204

Rainville P, Duncan GH, Price DD, Carrier B, Bushnell MC. 1997. Pain affect encoded in human anterior cingulate but not somatosensory cortex. *Science* 277:968–71

Riem MM, Bakermans-Kranenburg MJ, Pieper S, Tops M, Boksem MA, et al. 2011. Oxytocin modulates amygdala, insula, and inferior frontal gyrus responses to infant crying: a randomized controlled trial. *Biol. Psychiatry* 70(3):291–7

Rizzolatti G, Fogassi L, Gallese V. 2001. Neurophysiological mechanisms underlying the understanding and imitation of action. *Nat. Rev. Neurosci.* 2:661–70

Rolls ET. 2004. The functions of the orbitofrontal cortex. *Brain Cogn.* 55:11–29

Saarela MV, Hlushchuk Y, Williams AC, Schurmann M, Kalso E, Hari R. 2007. The compassionate brain: humans detect intensity of pain from another's face. *Cereb. Cortex* 17:230–37

Salminen JK, Saarijarvi S, Aarela E, Toikka T, Kauhanen J. 1999. Prevalence of alexithymia and its association with sociodemographic variables in the general population of Finland. *J. Psychosom. Res.* 46:75–82

Saxe R, Kanwisher N. 2003. People thinking about people. The role of the temporo-parietal junction in "theory of mind". *Neuroimage* 19(4):1835–42

Schooler JW, Smallwood J, Christoff K, Handy TC, Reichle ED, Sayette MA. 2011. Meta-awareness, perceptual decoupling and the wandering mind. *Trends Cogn. Sci.* 15(7):319–26

Schultz W. 2000. Multiple reward signals in the brain. *Nat. Rev. Neurosci.* 1:199–207

Seeley WW, Carlin DA, Allman JM, Macedo MN, Bush C, et al. 2006. Early frontotemporal dementia targets neurons unique to apes and humans. *Ann. Neurol.* 60:660–67

Seeley WW, Menon V, Schatzberg AF, Keller J, Glover GH, et al. 2007. Dissociable intrinsic connectivity networks for salience processing and executive control. *J. Neurosci.* 27:2349–56

Shackman AJ, Salomons TV, Slagter HA, Fox AS, Winter JJ, Davidson RJ. 2011. The integration of negative affect, pain and cognitive control in the cingulate cortex. *Nat. Rev. Neurosci.* 12:154–67

Shamay-Tsoory SG, Aharon-Peretz J, Perry D. 2009. Two systems for empathy: a double dissociation between emotional and cognitive empathy in inferior frontal gyrus versus ventromedial prefrontal lesions. *Brain* 132:617–27

Silani G, Bird G, Brindley R, Singer T, Frith C, Frith U. 2008. Levels of emotional awareness and autism: an fMRI study. *Soc. Neurosci.* 3:97–112

Singer T. 2006. The neuronal basis and ontogeny of empathy and mind reading: review of literature and implications for future research. *Neurosci. Biobehav. Rev.* 30:855–63

Singer T, Critchley HD, Preuschoff K. 2009. A common role of insula in feelings, empathy and uncertainty. *Trends Cogn. Sci.* 13:334–40

Singer T, Lamm C. 2009. The social neuroscience of empathy. *Ann. N. Y. Acad. Sci.* 1156:81–96

Singer T, Seymour B, O'Doherty J, Kaube H, Dolan RJ, Frith CD. 2004. Empathy for pain involves the affective but not sensory components of pain. *Science* 303:1157–62

Singer T, Seymour B, O'Doherty JP, Stephan KE, Dolan RJ, Frith CD. 2006. Empathic neural responses are modulated by the perceived fairness of others. *Nature* 439:466–69

Singer T, Snozzi R, Bird G, Petrovic P, Silani G, et al. 2008. Effects of oxytocin and prosocial behavior on brain responses to direct and vicariously experienced pain. *Emotion* 8:781–91

Singer T, Steinbeis N. 2009. Differential roles of fairness- and compassion-based motivations for cooperation, defection, and punishment. *Ann. N. Y. Acad. Sci.* 1167:41–50

Smallwood J, Schooler JW. 2006. The restless mind. *Psychol. Bull.* 132:946–58

Sridharan D, Levitin DJ, Menon V. 2008. A critical role for the right fronto-insular cortex in switching between central-executive and default-mode networks. *Proc. Natl. Acad. Sci. USA* 105:12569–74

Sterzer P, Stadler C, Poustka F, Kleinschmidt A. 2007. A structural neural deficit in adolescents with conduct disorder and its association with lack of empathy. *Neuroimage* 37:335–42

Taylor KS, Seminowicz DA, Davis KD. 2009. Two systems of resting state connectivity between the insula and cingulate cortex. *Hum. Brain Mapp.* 30:2731–45

Tiihonen J, Rossi R, Laakso MP, Hodgins S, Testa C, et al. 2008. Brain anatomy of persistent violent offenders: more rather than less. *Psychiatry Res.* 163:201–12

Ture U, Yasargil DC, Al-Mefty O, Yasargil MG. 1999. Topographic anatomy of the insular region. *J. Neurosurg.* 90:720–33

Van Overwalle F. 2009. Social cognition and the brain: a meta-analysis. *Hum. Brain Mapp.* 30:829–58

Vogt BA, Pandya DN. 1987. Cingulate cortex of the rhesus monkey: II. Cortical afferents. *J. Comp. Neurol.* 262:271–89

Vogt BA, Vogt L, Farber NB, Bush G. 2005. Architecture and neurocytology of monkey cingulate gyrus. *J. Comp. Neurol.* 485:218–39

Von Economo C. 1926. Eine neue art spezialzellen des lobus cinguli und lobus insulae. *Zschr. Ges. Neurol. Psychiat.* 100:706–12

Wicker B, Keysers C, Plailly J, Royet JP, Gallese V, Rizzolatti G. 2003. Both of us disgusted in my insula: the common neural basis of seeing and feeling disgust. *Neuron* 40:655–64

Wispe L. 1986. The distinction between sympathy and empathy: to call forth a concept, a word is needed. *J. Pers. Soc. Psychol.* 50:314–21

Yarkoni T, Poldrack RA, Nichols TE, Van Essen DC, Wager TD. 2011. Large-scale automated synthesis of human functional neuroimaging data. *Nat. Methods* 8(8):665–70

Cellular Pathways of Hereditary Spastic Paraplegia*

Craig Blackstone

Neurogenetics Branch, National Institute of Neurological Disorders and Stroke, National Institutes of Health, Bethesda, Maryland, USA; email: blackstc@ninds.nih.gov

Annu. Rev. Neurosci. 2012. 35:25–47

First published online as a Review in Advance on April 20, 2012

The *Annual Review of Neuroscience* is online at neuro.annualreviews.org

This article's doi:
10.1146/annurev-neuro-062111-150400

Keywords

spasticity, lipid droplet, BMP, cytokinesis, endosome, endoplasmic reticulum

Abstract

Human voluntary movement is controlled by the pyramidal motor system, a long CNS pathway comprising corticospinal and lower motor neurons. Hereditary spastic paraplegias (HSPs) are a large, genetically diverse group of inherited neurologic disorders characterized by a length-dependent distal axonopathy of the corticospinal tracts, resulting in lower limb spasticity and weakness. A range of studies are converging on alterations in the shaping of organelles, particularly the endoplasmic reticulum, as well as intracellular membrane trafficking and distribution as primary defects underlying the HSPs, with clear relevance for other long axonopathies affecting peripheral nerves and lower motor neurons.

Contents

INTRODUCTION

Voluntary movement in humans relies on the pyramidal motor system, a tortuous, multisynaptic pathway in the CNS that extends from the cerebral motor cortex to neuromuscular junctions innervating skeletal muscle. This system is arranged in two main stages (**Figure 1**). First, axons of large pyramidal neurons originating in layer V of the cerebral motor cortex course through the medullary pyramids, where most fibers decussate in the caudal medulla before descending as lateral corticospinal tracts within the spinal cord. Although some corticospinal axons establish synapses directly with lower motor neurons in the spinal cord anterior horn, the vast majority synapse with spinal interneurons, which then establish connections with lower motor neurons. In the next stage, lower motor neurons terminate in specialized synapses at neuromuscular junctions throughout the body to regulate skeletal muscle contractility (**Figure 1**) (Carpenter 1991).

HSP: hereditary spastic paraplegia

Distances traversed by corticospinal and lower motor neurons are among the furthest in the body; their axons extend up to 1 m in length and the axoplasm comprises >99% of total cell volume. This length has evolved to permit the very rapid relay of action potentials, enabling timely voluntary movement, but it comes at great expense to the neuron: Complex intracellular machineries are required for sorting and distributing proteins, lipids, mRNAs, organelles, and other molecules over such long distances. These machineries utilize an elaborate neuronal cytoskeletal scaffold along which motor proteins target and deliver components selectively throughout the cell; axonal transport machineries rely on microtubules in particular, which function as polarized tracks with their plus ends oriented toward the axon terminal. A variety of mechanoenzymes within the kinesin, dynein, and myosin protein superfamilies mediate much of the anterograde and retrograde transport specificity through selective cargo interactions. Additional specificity and regulatory control are contributed by various adaptor proteins. The interaction of intracellular cargoes with these complexes permits tightly regulated, selective allocations of organelles, proteins, lipids, and other molecules to growth cones during axonal development and to specialized axon domains such as branch points, internodal segments, and presynaptic terminals in mature neurons (Goldstein et al. 2008, Arnold 2009, Hirokawa et al. 2010).

Not surprisingly, long axons are an Achilles' heel of the nervous system; length-dependent defects in axon development and maintenance give rise to a host of neurological disorders, both acquired and inherited. Acquired disorders are numerous and highly varied, with etiologies encompassing injuries, nutritional deficiencies, endocrine and metabolic disturbances, infections, and environmental toxins, to name a few; these are not discussed here. The focus of this review is on inherited Mendelian disorders, as exemplified by the hereditary spastic paraplegias (HSPs). Although these are among the most genetically diverse of all diseases, with nearly 50 distinct loci and more than 20 gene products identified to date, they are unified by the defining, predominant clinical feature of progressive lower limb spasticity and weakness, with sparing of the upper limbs to a large extent (Fink 2006, Depienne et al. 2007, Salinas et al. 2008, Dion et al. 2009, Blackstone et al. 2011, Lang et al. 2011).

HSPs are uncommon but not rare, with a prevalence of ~3–9/100,000 in most populations, and thus likely afflict several hundred thousand individuals worldwide. Inheritance can be X-linked recessive, autosomal recessive, or autosomal dominant, and age at onset can vary widely, from early childhood to late in life. HSPs have historically been classified as pure or complicated on the basis of the absence (pure) or presence (complicated) of associated clinical features such as distal amyotrophy, cognitive dysfunction, retinopathy, ataxia, thin corpus callosum, and peripheral neuropathy (Harding 1983). Even in pure forms, urinary symptoms and mild dorsal column sensory deficits are frequently encountered. More recently, a numeric labeling scheme has taken hold, and HSPs are increasingly referred to mainly by their genetic classification, SPG1-48 (Depienne et al. 2007, Salinas et al. 2008, Dion et al. 2009, Blackstone et al. 2011).

Because most patients with HSP have a normal life span, a limited number of neuropathologic evaluations of HSPs have been published, particularly for the most instructive pure forms with a genetic diagnosis. Still, these studies have typically shown evidence of axonal degeneration, principally involving the longest ascending sensory fibers and descending corticospinal tract axons in a distal, "dying-back" manner (DeLuca et al. 2004). Because the longest corticospinal axons control the lower motor neurons innervating muscles of the lower limbs, such findings are concordant with the cardinal clinical features of HSP; sensory manifestations tend to be clinically mild. There is usually little neuronal death even late in the disease course, especially in pure forms, so HSPs are a prototype for understanding disorders that impair axons (Soderblom & Blackstone 2006). Importantly, HSPs are fundamentally diseases of massive scale, affecting

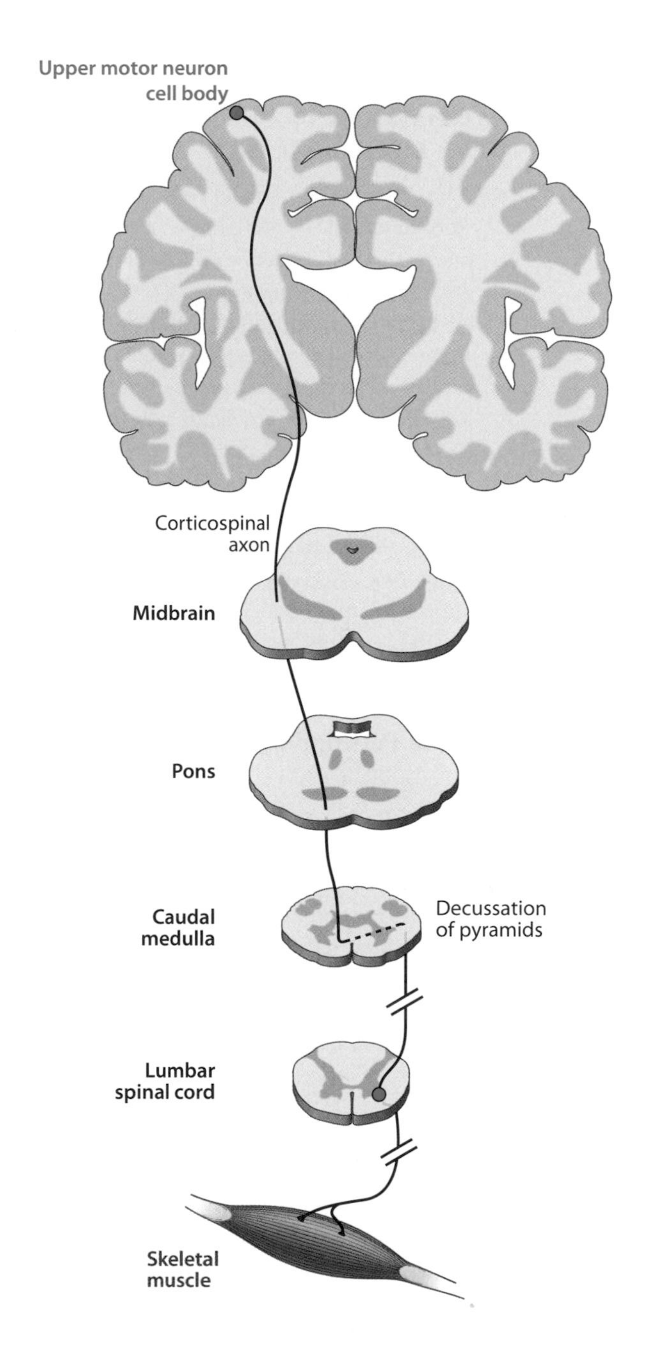

Figure 1

Schematic diagram of the corticospinal tract emphasizing its descent through the CNS. Although most fibers decussate in the caudal medulla, a minority of fibers descend uncrossed as the ventral corticospinal tract (not shown).

Spasticity: increase in muscle tone associated with hyperactive tendon stretch reflexes

Corpus callosum: arched bridge of nerve fibers connecting the left and right cerebral hemispheres

predominantly the longest neurons that are orders of magnitude larger than most other cells.

COMMON CELLULAR PATHOGENIC THEMES

The common clinical and pathological features of different HSPs prefigure a small number of common themes at the cellular level, and the genetic heterogeneity provides a significant advantage in identifying these convergent themes. Indeed, published studies have indicated that HSP disease proteins cluster within a small number of predicted cellular processes (**Table 1** and **Figure 2**) (Soderblom & Blackstone 2006, Depienne et al. 2007, Salinas et al. 2008, Dion et al. 2009, Blackstone et al. 2011). Although we discuss HSP proteins by topic, one must remember that many of them appear to function in a number of different pathways; thus pathogenic groupings may evolve over time.

Axon Pathfinding

Among the first HSP mutations described were in the *L1CAM* gene, encoding a cell surface glycoprotein of the immunoglobulin (Ig) superfamily. Loss-of-function mutations in *L1CAM* are implicated not only in X-linked, early-onset, complicated HSP (SPG1), but also in other X-linked syndromes including MASA (for mental retardation, aphasia, shuffling gait, and adducted thumbs), hydrocephalus, and agenesis of the corpus callosum (Jouet et al. 1994, Weller & Gärtner 2001). Each of these disorders displays clinical and pathological evidence of corticospinal tract impairment; the disorders are considered together along a disease spectrum known as L1 disease or CRASH syndrome (for corpus callosum hypoplasia, retardation, adducted thumbs, spastic paraplegia or shuffling gait, and hydrocephalus) (Soderblom & Blackstone 2006).

The L1CAM protein is more than 1200 amino acid residues in size, with a large extracellular segment harboring 6 Ig-like domains and 5 fibronectin type III domains, a single transmembrane domain, and a short cytoplasmic tail. L1CAM participates in a complex set of extracellular and intracellular interactions, binding not only other L1CAM molecules but also a host of extracellular ligands—including other cell adhesion molecules, integrins, and proteoglycans—as well as intracellular proteins such as ankyrins. Disease mutations are found throughout the protein, and partial or complete loss of L1CAM function seems critical for the L1 disease phenotype. In *L1CAM* null mice, the corticospinal tracts are abnormal, arising from a conspicuous failure to decussate within the medulla (Dahme et al. 1997, Cohen et al. 1998).

How does this pathfinding defect occur? In the developing CNS, L1CAM associates with neuropilin-1 (Nrp1), which itself interacts with Plexin-A proteins to form the Semaphorin3A (Sema3A) receptor complex. Upon Sema3A binding to Nrp1, L1CAM and Nrp1 are cointernalized in a L1CAM-dependent manner. Sema3A is a repulsive guidance cue released from cells in the ventral spinal cord to steer corticospinal neurons away from the midline spinal cord/medullary junction, and L1CAM mutations may affect Sema3A signaling when axons are crossing the midline by interfering with receptor internalization and signaling at growth cones (Castellani et al. 2004). In fact, the association of Nrp1 with L1CAM mediates the activation of a focal adhesion kinase-mitogen-activated protein kinase pathway controlling a critical aspect of the repulsive behavior, the disassembly of adherent zones in growth cones and their subsequent collapse (Bechara et al. 2008). This compelling role of L1CAM in axon pathfinding during development is consistent with the early onset of SPG1.

Myelination

A distinguishing feature of axons in the central and peripheral nervous systems is an insulating myelin sheath, a specialization important for increasing the speed of electrical impulse propagation. Schwann cells supply myelin for peripheral neurons, whereas oligodendrocytes myelinate axons of CNS neurons. Spastic paraplegia as a manifestation of abnormal

Table 1 Identified HSP genes, grouped functionally[a,b]

Disease/gene[b]	Protein name	Inheritance	Cellular functions
Membrane traffic and organelle shaping			
SPG3A/*ATL1*	Atlastin-1	AD	ER morphogenesis BMP signaling
SPG4/*SPAST*	Spastin (M1 and M87 isoforms)	AD	Microtubule severing ER morphogenesis Endosomal traffic BMP signaling Cytokinesis
SPG6/*NIPA1*	NIPA1	AD	Endosomal traffic Mg^{2+} transport BMP signaling
SPG8/*KIAA0196*	Strumpellin	AD	Endosomal traffic Cytoskeletal (actin) regulation
SPG10/*KIF5A*	KIF5A	AD	Microtubule-based motor protein
SPG11	Spatacsin	AR	Endosomal traffic
SPG12/*RTN2*	Reticulon 2	AD	ER morphogenesis
SPG15/*ZFYVE26*	Spastizin/ ZFYVE26/ FYVE-CENT	AR	Endosomal traffic Cytokinesis Autophagy
SPG17/*BSCL2*	Seipin/BSCL2	AD	Lipid droplet biogenesis at ER
SPG18/*ERLIN2*	Erlin2	AR	ER-associated degradation Lipid raft-associated
SPG20	Spartin	AR	Endosomal traffic BMP signaling Cytokinesis Lipid droplet turnover Mitochondrial regulation
SPG21	Maspardin	AR	Endosomal traffic
SPG31/*REEP1*	REEP1	AD	ER morphogenesis ER-microtubule interaction
SPG48/*KIAA0415*	KIAA0415 (AP-5 subunit)	AR	Endocytic adaptor protein complex
AP-4 deficiency/*AP4S1*, *AP4B1*, *AP4E1*	AP-4 S1, B1, and E1 subunits	AR	Endocytic adaptor protein complex
JPLS/*ALS2*	Alsin	AR	Endosomal traffic
Mitochondrial regulation			
SPG7	Paraplegin	AR	Mitochondrial *m*-AAA ATPase
SPG13/*HSPD1*	HSP60	AD	Mitochondrial chaperonin
Myelination and lipid/sterol modification			
SPG2/*PLP1*	Proteolipid protein	X-linked	Major myelin protein
SPG5/*CYP7B*	CYP7B1	AR	Cholesterol metabolism
SPG35/*FA2H*	Fatty acid 2-hydroxylase	AR	Myelin lipid hydroxylation
SPG39/*PNPLA2*	Neuropathy target esterase	AR	Phospholipid homeostasis
SPG42/*SLC33A1*	SLC33A1	AD	Acetyl-CoA transporter
SPG44/*GJC2*	Connexin-47	AR	Intercellular gap junction channel
Axon Pathfinding			
SPG1/*L1CAM*	L1CAM	X-linked	Cell adhesion and signaling

[a]Abbreviations: AD, autosomal dominant; AR, autosomal recessive; BMP, bone morphogenetic protein; ER, endoplasmic reticulum.
[b]When different from disease name.

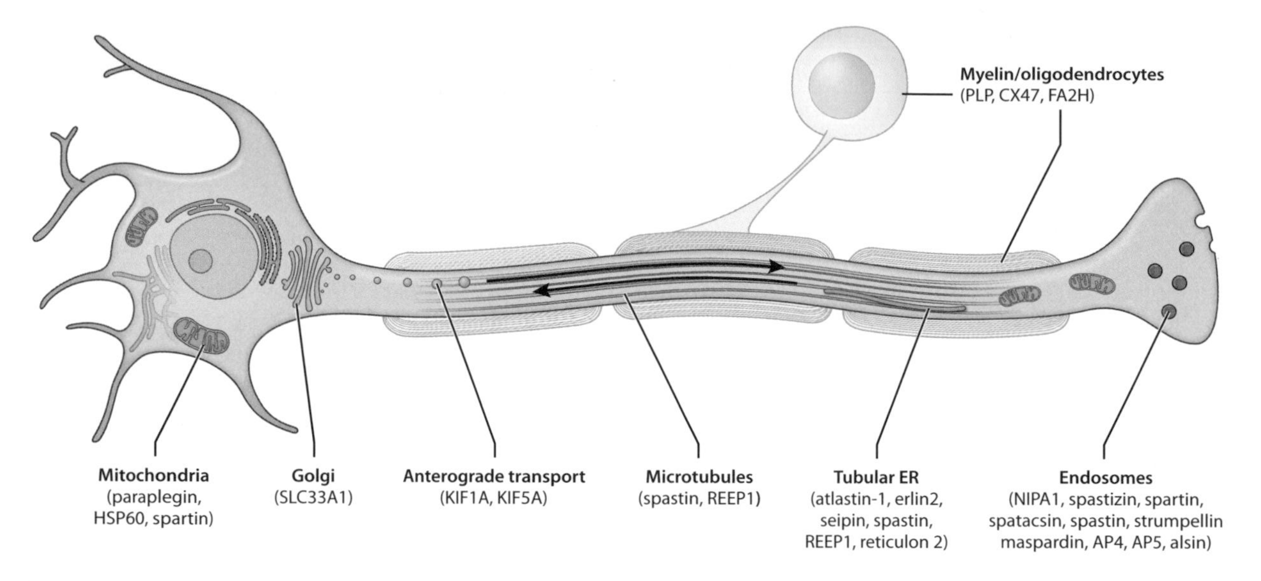

Figure 2

Common pathogenic themes in the HSPs. This schematic representation of a corticospinal motor neuron emphasizes where HSP gene products as listed in **Table 1** are proposed to function. L1CAM is an integral membrane protein localized to the plasma membrane. CYP7B1 and NTE distributions are not shown, pending more detailed studies of their sites of action.

myelination in the CNS is not uncommon; for example, this occurs in multiple sclerosis and a variety of acquired and inherited leukodystrophies. Mutations in the *PLP1* gene encoding the tetraspan integral membrane proteolipid protein (PLP) and its smaller DM20 isoform give rise to two major diseases along a clinical spectrum: a pure or complicated HSP (SPG2) and the generally much more severe Pelizaeus-Merzbacher disease (PMD) (Inoue 2005).

Although PLP and DM20 are the major protein constituents of CNS myelin (~50% of the total protein), *PLP1* duplications paradoxically cause more severe disease than do deletions, whereas complete absence of PLP/DM20 is typically associated with SPG2 or mild presentations of PMD. *Plp1* null mice in particular have been widely studied as a model for SPG2. Unexpectedly, in these mice the myelin sheath maintains its normal thickness, though with subtle anomalies of the intraperiod lines. In the underlying axons, anterograde transport is impaired, and cargoes undergoing retrograde transport become stuck at distal juxtaparanodal regions (Edgar et al. 2004, Gruenenfelder et al. 2011). It seems reasonable to postulate that oligodendrocytes modulate the activity of motor proteins involved in intracellular cargo transport via signaling cascades in the underlying axon and that this modulation is sensitive to PLP/DM20 (Gruenenfelder et al. 2011).

Intraperiod lines: fused outer leaflets of contiguous plasma membranes in the myelin sheath

Mutations in a more recently identified HSP gene similarly define a disease spectrum comprising HSP and PMD-like disease where cell-cell communication is altered. The slowly progressive, complicated SPG44 is caused by homozygous mutations in the *GJC2* gene encoding connexin 47 (CX47). Connexins (typically numbered based on predicted molecular weight) are oligomeric proteins forming gap junction channels, which establish connections between apposed cell membranes to permit the intercellular diffusion of ions and small molecules (typically <1000 Da). CX47 forms connections between astrocytes and oligodendrocytes in concert with CX43. Because CX47/CX43 heterotypic channels appear essential for the maintenance of CNS myelin, alterations in CX47 that result in CX47/CX43 channel dysfunction likely underlie SPG44 (Orthmann-Murphy et al. 2009).

A third HSP with a compelling link to dysmyelination is autosomal recessive SPG35. This disorder sits along a disease spectrum spanning neurodegeneration with brain iron accumulation, leukodystrophy, and HSP and results from loss-of-function mutations in the fatty acid-2 hydroxylase gene *FA2H* (Dick et al. 2010, Schneider & Bhatia 2010). The FA2H protein is a nicotinamide adenine dinucleotide phosphate (NADPH)–dependent monooxygenase that converts free fatty acids to 2-hydroxy fatty acids. These are incorporated into myelin galactolipids containing hydroxy fatty acid as the *N*-acyl chain, which maintains the myelin sheath. *Fa2h* null mice have been developed as a model for SPG35, and these animals exhibit significant demyelination, axon loss or enlargement, cerebellar defects, and spatial learning and memory deficits. Animals lacking *Fa2h* only in oligodendrocytes and Schwann cells do not exhibit memory deficits, indicating that some neurological manifestations may derive from a lack of FA2H in other cell types (Potter et al. 2011).

Taken together, this subgroup of HSPs most exemplifies a noncell-autonomous disease pathogenesis. In this regard, oligodendrocytes from *Plp1* null mice were able to induce a focal axonopathy when transplanted into the dorsal columns of the myelin-deficient *shiverer* mouse (Edgar et al. 2004). Thus, HSP-associated alterations in oligodendrocyte-mediated myelination can directly cause changes in the underlying axon, impairing corticospinal tract function.

Endoplasmic Reticulum Network Morphology

Cellular organelles have diverse but characteristic morphologies that are evolutionarily conserved, indicating that the function of an organelle is fundamentally related to its form. An obvious corollary is that disruption of form can give rise to disease, and in fact, this has been shown for a number of organelles. For instance, mitochondrial morphology is shaped by the opposing processes of fission and fusion, and multiple large GTPases involved in regulating this balance are mutated in autosomal dominant neurological disorders, including optic atrophy type 1 (OPA1) and Charcot-Marie-Tooth type 2A neuropathy (MFN2) (Westermann 2010).

An analogous situation occurs in the endoplasmic reticulum (ER), and the roles of common HSP gene products in shaping the tubular ER network have indicated this may be the most common pathogenic theme (Park et al. 2010, Montenegro et al. 2012). The ER is among the most distinctive organelles because of its large size, morphological heterogeneity, and extension throughout the cell. Although it is a continuous membrane-bound luminal system, it comprises the distinct morphologies of the nuclear envelope (with thousands of specialized pores), peripheral sheet-like structures studded with polyribosomes, and a polygonal network of interconnected smooth tubules distributed widely throughout the cell. Concordant with this structural heterogeneity, the ER is a multifunctional organelle involved in the synthesis, modification, quality control, and trafficking of integral membrane and secreted proteins. It is critical as well for Ca^{2+} sequestration and release, signaling, sterol synthesis, and lipid synthesis and distribution. In neurons, the ER plays crucial roles in the massive polarized membrane expansion that occurs during axon and dendrite genesis and as an intracellular Ca^{2+} store integrated with pre- and postsynaptic signaling pathways (Verkhratsky 2005, Park & Blackstone 2010, Renvoisé & Blackstone 2010, Lynes & Simmen 2011).

The three most common autosomal dominant HSPs—SPG3A, SPG4, and SPG31—as well as the less common SPG12 result from mutations in proteins directly implicated in the formation of the tubular ER network, which is overwhelmingly smooth ER (Park et al. 2010, Montenegro et al. 2012). Mutations in the SPG3A gene *ATL1* are the second most common cause of HSP and are the most common cause of early-onset disease. The SPG3A protein atlastin-1 is a member of a family of large

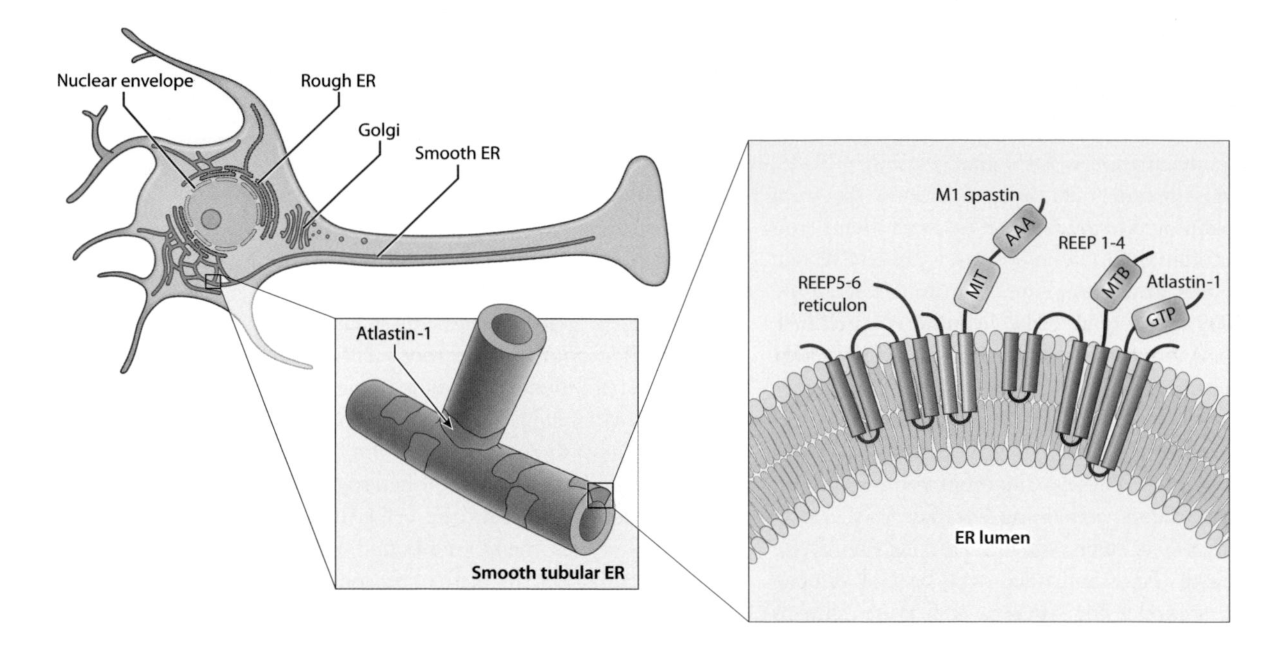

Figure 3

Spastin, atlastin, REEP, and reticulon proteins interact and shape the endoplasmic reticulum (ER) network. (*Left*) Schematic diagram of a neuron showing the distribution of different ER domains. Below this schematic is an enlargement of a three-way tubular ER junction. REEP and reticulon proteins form large oligomers to shape the tubular ER. Atlastin proteins are enriched in puncta along the tubules, including at three-way junctions. (*Right*) Proposed membrane topologies for protein families involved in generating the tubular ER network. GTP, atlastin GTPase domain; MTB, microtubule-binding domain.

oligomeric GTPases related to the dynamin superfamily. Atlastin-1 is one of three homologous proteins in mammals (atlastin-1, -2 and -3) thought to be paralogs, but it is the only form expressed highly in the CNS (Zhu et al. 2003, Rismanchi et al. 2008). Atlastin-related GTPases are found in all eukaryotic cells and include Sey1p in *Saccharomyces cerevisiae* and root hair defective 3 (RHD3) in *Arabidopsis* (Hu et al. 2009). In contrast with mammals, species such as *S. cerevisiae* and *Drosophila melanogaster* have only a single atlastin ortholog. Across species, the atlastins can diverge considerably at the sequence level, but all share a similar domain organization: a large cytoplasmic N-terminal domain containing a tripartite GTP-binding domain, two very closely spaced hydrophobic segments, and a small cytoplasmic C-terminal tail (**Figure 3**). These multimeric, integral membrane GTPases localize predominantly to the tubular ER but are also found in the ER-Golgi intermediate compartment and in the *cis*-Golgi apparatus in some cell types (Zhu et al. 2003, 2006; Rismanchi et al. 2008).

Dynamin superfamily: large, multimeric GTPases involved in membrane fission or fusion

Root hairs: long, thin, tubular outgrowths of plant root epidermal cells that absorb water and minerals from soil

Atlastin GTPase activity is required for the formation of the three-way junctions in ER tubules in a wide range of species by directly mediating homotypic fusion of ER tubules (Rismanchi et al. 2008, Hu et al. 2009, Orso et al. 2009, Bian et al. 2011, Byrnes & Sondermann 2011, Chen et al. 2011, Moss et al. 2011). Consistent with this role, atlastins localize to discrete sites along ER tubules, including at three-way junctions. Depletion of atlastin-1 by shRNA in cultured cortical neurons inhibits axon elongation (Zhu et al. 2006), and there is a link between proper ER morphology and the formation and maintenance of long cellular processes such as axons and plant root hairs (see sidebar, ER Shaping: Plant Roots to Axons).

Mammalian atlastins and yeast Sey1p (synthetic enhancement of *yop1*) interact directly with the reticulon and Yop1p/DP1/REEP families of ER-shaping proteins in the tubular ER (Hu et al. 2009, Park et al. 2010). Members of these families each have two long hydrophobic stretches that form intramembrane hairpin domains predicted to partially span the lipid bilayer, inducing and/or stabilizing high-curvature ER tubules via hydrophobic wedging (Voeltz et al. 2006, Hu et al. 2008, Shibata et al. 2009, West et al. 2011). Mutations in *REEP1* cause an autosomal dominant, pure HSP known as SPG31. REEP1 belongs to a family of related proteins (REEP1–6 in mammals) and was originally identified on the basis of its ability to promote trafficking of olfactory receptors to the plasma membrane surface (Saito et al. 2004). REEP1 localizes to the tubular ER and interacts with atlastin-1 via its predicted hydrophobic hairpin motifs. Some members (REEP1–4) also have an extended C-terminal domain (relative to REEP5–6) that binds microtubules, establishing REEP1 as a member of a subfamily of REEPs (REEP1–4) involved not only in ER shaping but also in interactions of ER tubules with the microtubule cytoskeleton (Park et al. 2010, Blackstone et al. 2011). Very recently, mutations in the *RTN2* gene encoding the ER-shaping protein reticulon 2 have been identified in families with autosomal dominant SPG12 (Montenegro et al. 2012).

ER SHAPING: PLANT ROOTS TO AXONS

In the flowering plant *Arabidopsis*, root hairs are long processes emanating from root epidermal cells. Like axon growth, root hair tip growth is an extreme form of polarized cell expansion regulated by signaling molecules, Ca^{2+} flux, cytoskeletal dynamics, GTPases of the Rab, Arf, and Rho/Rac families, and reactive oxygen species. Loss-of-function mutations in the atlastin/SPG3A ortholog RHD3 have highlighted the importance of proper ER morphology in *Arabidopsis* root hair formation (Wang et al. 1997, Chen et al. 2011). *rhd3* mutant plants have abnormal tubular ER bundles within short, wavy root hairs and an unusually large number of vesicles in subapical (rather than apical) hair regions. This defective polarized expansion may reflect decreased or misplaced deposition of secretory vesicles during root hair elongation. Although ER tubules in plants are oriented mostly along actin fibers, root hair tip growth depends on microtubules also, and ER morphology changes during root hair elongation (Sieberer et al. 2005). Given the speed and adaptability of *Arabidopsis* genetics and conservation of atlastin/RHD3 GTPases as well as ER-shaping reticulon/REEP proteins (Sparkes et al. 2011), continued studies of root hair elongation will likely provide insights into the functions of ER in axon growth and maintenance.

The fundamental link between the tubular ER and microtubules has been appreciated for decades (Terasaki et al. 1986), and although the microtubule cytoskeleton is not absolutely required for ER network formation (Shibata et al. 2009), microtubule-based ER motility is key for the proper organization and distribution of ER tubules. This is achieved through a variety of mechanisms: membrane sliding, in which ER tubules slide along microtubules using motor activity; microtubule movement, in which ER tubules latch on to moving microtubules; and the tip attachment complex, in which ER attaches to growing microtubule plus ends (Waterman-Storer & Salmon 1998).

Impairment of this relationship between ER tubules and the microtubule cytoskeleton as a pathogenic mechanism for HSPs is supported further by the fact that the SPG4 protein spastin, a microtubule-interacting and severing AAA ATPase, binds atlastin-1 and REEP1 as well as the ER-shaping protein reticulon 1 (Evans et al. 2006, Mannan et al. 2006, Sanderson et al. 2006, Connell et al. 2009, Park et al. 2010). Spastin occurs as two main isoforms generated by differential use of AUG start codons: a 60-kDa form and a 67-kDa form (Mancuso & Rugarli 2008). The larger isoform has an additional 86 amino acid stretch at the N-terminus containing a hydrophobic segment predicted to insert in the ER membrane as a partially membrane-spanning hairpin (Park et al. 2010). Interactions of M1 spastin with REEP1 and atlastin-1 appear to be mediated largely through this hairpin, though flanking regions may also participate (Evans et al. 2006, Sanderson et al. 2006). The larger spastin isoform is particularly enriched in the spinal cord, and a dysfunctional M1 spastin polypeptide, but not one representing the shorter M87

Hydrophobic wedging: partitioning the bulk of intramembrane hydrophobic domains within one leaflet of a phospholipid bilayer, generating membrane curvature

Lipid droplet (LD): dynamic cytoplasmic organelle consisting of a phospholipid monolayer surrounding a neutral lipid core; proteins are found within the monolayer and decorating its surface

form (M85 in rodents), was deleterious to axon growth in cultured neurons (Solowska et al. 2008), strengthening the evidence linking M1 spastin to HSP pathogenesis. By comparison, M87 spastin appears involved in cytokinesis, secretion, and possibly endocytosis through its interactions linking microtubule dynamics to membrane modeling in these compartments (Yang et al. 2008, Connell et al. 2009). In sum, there appear to be strong physical and functional links between M1 spastin, atlastin-1, reticulon 2, and REEP1 involved in shaping the tubular ER network in concert with the microtubule cytoskeleton. Even so, the ER has a large number of functions, and it remains unclear which are most pathogenically relevant for HSPs.

Because several other HSP proteins also localize to the ER, understanding these may clarify the role of ER in HSP pathogenesis. Mutations in the Berardinelli-Seip congenital lipodystrophy protein 2 gene *BSCL2*, which encodes an integral membrane of the ER known as BSCL2/seipin, cause two distinct diseases. Heterozygous gain-of-function mutations in an *N*-linked glycosylation site give rise to a disease spectrum encompassing autosomal dominant SPG17 (Silver syndrome), with distal amyotrophy as a significant feature in addition to spastic paraparesis, and distal hereditary motor neuropathy type V, characterized by more prominent distal spinal muscular atrophy (Windpassinger et al. 2004). In contrast, autosomal recessive loss-of-function mutations give rise to Berardinelli-Seip congenital lipodystrophy, without spasticity or amyotrophy. The seipin protein and its yeast ortholog, Fld1p (few lipid droplets 1), regulate the size of lipid droplets (LDs), explaining the loss-of-function lipodystrophy presentation (Cui et al. 2011; Fei et al. 2011a,b; Tian et al. 2011). The missense changes underlying SPG17 have resulted in misfolding of seipin, forming aggregates and triggering ER stress (Ito & Suzuki 2009, Yagi et al. 2011), which may play a role in HSP pathogenesis. LDs accumulate under various cellular stress conditions, and a number of unfolded protein response pathways have been implicated in LD formation (Hapala et al. 2011). However, it will be important to investigate any effects of misfolded seipin on other aspects of ER structure and function.

Recently, the gene for the complicated, autosomal recessive SPG18 was identified in a consanguinous Saudi family as *ERLIN2* (Alazami et al. 2011). A second study independently identified loss-of-function *ERLIN2* mutations in a family with motor dysfunction, joint contractures, and intellectual disability (Yıldırım et al. 2011). The erlin2 protein resides in the ER and contains a SPFH domain—named for its presence in stomatin, prohibitin, flotillin, and HflC/K. SPHF domain–containing proteins share the ability to assemble into large oligomers, and they localize preferentially to cholesterol-rich domains, including lipid rafts (Browman et al. 2006). Erlin2 has been functionally linked to ER-associated degradation (ERAD), a multistep degradative pathway encompassing ubiquitin-proteosome-mediated degradation of ER proteins, thereby regulating levels of proteins such as the inositol trisphosphate receptor (Pearce et al. 2009). Also, erlin2 binds gp78, a membrane-bound ubiquitin ligase similar to HRD1 that mediates sterol-accelerated ERAD of 3-hydroxy-3-methyl-glutaryl (HMG)-CoA reductase, a key enzyme in the biosynthesis of cholesterol (Jo et al. 2011).

Lipid Synthesis and Metabolism

These latter two HSP proteins, erlin2 and seipin, present compelling insights into how defects in ER shaping and distribution might cause HSP. A key function of the ER is the synthesis, metabolism, and distribution of lipids and sterols, employing both vesicular and nonvesicular mechanisms, and other HSP proteins fit into this pathogenic theme. The complicated, autosomal recessive HSP Troyer syndrome (SPG20) is caused by mutations resulting in a loss of spartin protein (Bakowska et al. 2008). Spartin localizes to a variety of cellular structures and has been implicated in a number of functions, including cytokinesis and epidermal growth factor (EGF) receptor trafficking (Robay et al. 2006, Bakowska et al. 2007, Renvoisé et al. 2010, Lind et al. 2011). Spartin also regulates LD biogenesis by promoting atrophin-1 interacting protein 4 (AIP4)-mediated ubiquitination of LD proteins (Eastman et al. 2009, Edwards et al. 2009, Hooper et al. 2010) and by recruiting PKC-ζ via the PKC-ζ-interacting proteins ZIP1 (p62/sequestosome) and ZIP3 to LDs (Urbanczyk & Enz 2011). Little is known about any roles for LDs in axons. However, because the SPG17 protein seipin regulates LD formation, alterations in LD biogenesis or turnover could affect lipid distribution, organelle shaping, or signaling pathways important for axonal health.

Although not directly implicated in LD biogenesis, other HSP proteins are enzymes involved in related lipid and cholesterol biosynthetic pathways. SPG42 is caused by mutations in the *SLC33A1* gene encoding the acetyl-CoA transporter. In animals, acetyl-CoA is essential for maintaining the balance between carbohydrate and fat metabolism. Under normal circumstances, acetyl-CoA from fatty acid metabolism enters the citric acid cycle, contributing to the energy supply of the cell. SLC33A1 transports acetyl-CoA into the Golgi apparatus lumen and has been directly linked to the growth of axons because knock down of *slc33a1* in zebrafish causes defective outgrowth from the spinal cord (Lin et al. 2008).

Last, neuropathy target esterase (NTE) is an integral membrane protein of neuronal ER that is mutated in autosomal recessive SPG39, a complicated HSP with prominent amyotrophy. NTE deacylates the major membrane phospholipid, phosphatidylcholine. Mutation of the NTE gene *PNPLA2* or chemical inhibition of NTE with organophosphates alters membrane composition and causes distal degeneration of long spinal axons in mice and man (Rainier et al. 2008, Read et al. 2009). Another HSP protein, cytochrome P450-7B1 (CYP7B1), is mutated in autosomal recessive SPG5 (Tsaousidou et al. 2008) and functions in cholesterol metabolism; in patients with SPG5 there is a dramatic increase in oxysterol substrates in plasma and cerebrospinal fluid (Schüle et al. 2010). Given the fundamental roles played by lipids and sterols in neuronal functions, it seems very likely that more genes will be identified within this category.

Endosomal Dynamics

Although changes in ER morphology encompass proteins altered in the majority of patients with HSP, more HSP proteins have been implicated in endosomal dynamics; however, an immediate caveat is that the cell seems to co-opt some of these proteins and machineries for other functions (Blackstone et al. 2011). As noted earlier, the SPG20 protein spartin functions in LD turnover, but it is also required for efficient EGF receptor degradation and likely regulates EGF signaling (Bakowska et al. 2007). Spartin may also regulate a variety of signaling pathways via ubiquitin modification through its interactions with E3 ubiquitin ligases such as AIP4 and AIP5. Spartin harbors an MIT domain as well and interacts selectively with IST1, a component of the ESCRT-III complex. ESCRT comprises a series of cytosolic protein complexes, ESCRT-0, ESCRT-I, ESCRT-II, and ESCRT-III, and the sequential activities of these complexes are required to recognize and sort ubiquitin-modified proteins into internal vesicles of multivesicular bodies (Hurley & Hanson 2010). More recently, ESCRT proteins have been implicated in other cellular functions, including viral budding and cytokinesis, and spartin has been shown to participate in cytokinesis (Renvoisé et al. 2010, Lind et al. 2011).

This link to ESCRT-III is shared by the SPG4 protein spastin. Although discussed earlier in the context of its interactions with REEP1 and atlastin-1 to coordinate ER membrane modeling and microtubule interactions (Park et al. 2010), spastin harbors an MIT domain that binds the ESCRT-III subunits CHMP1B and IST1 to couple the severing of microtubules with membrane scission. These ESCRT interactions are crucial for spastin's

MIT domain: conserved domain comprising three α-helices present in microtubule-interacting and trafficking proteins

ESCRT: endosomal sorting complex required for transport

WASH complex: Wiscott-Aldrich syndrome protein and SCAR (suppressor of cAMP receptor) homolog complex

Tubular transport intermediate: small, cigar-shaped organelles that traffic from one membrane compartment to another

role in severing microtubules to complete the abscission phase of cytokinesis (Yang et al. 2008, Connell et al. 2009, Guizetti et al. 2011). ESCRT proteins are also involved in centrosome stability (Morita et al. 2010), raising the possibility that axon genesis and regulation by centrosomes could be a function of these proteins. Other possibilities include roles for the ESCRT-III interactions with spastin and spartin in the delivery and downregulation of cell surface receptors to regulate signaling in axons.

Mutations in the KIAA1096 (SPG8) gene that encodes strumpellin cause a severe, pure HSP (Valdmanis et al. 2007). Strumpellin contains few motifs or domains of identified function, aside from a region of putative spectrin repeats. The main clues to its possible endosomal functions stem from its identification as a subunit of the WASH complex (Derivery & Gautreau 2010). This complex, comprising seven subunits (five core components), connects tubular endosomes of retrograde cargo sorting to the cytoskeleton, and it associates with endosomes via an interaction with vacuolar protein sorting–associated protein 35 (VPS35) (Harbour et al. 2010). Along with VPS26 and VPS29, VPS35 is a component of retromer, an endosomal complex responsible for sorting cargoes from endosomes to the *trans*-Golgi network (Bonifacino & Hurley 2008). Depletion of members of the WASH complex cause increased tubulation at early endosomes, impairing trafficking through early endosomal compartments (Derivery et al. 2009, Gomez & Billadeau 2009, Jia et al. 2010). Several members of the complex—WASH1, FAM21 (family with sequence similarity 21), and actin-capping protein—regulate actin dynamics. The WASH complex helps generate an actin network on early endosomes, for instance by activating the Arp2/3 complex to nucleate new actin filaments branching off extant filaments. The increased tubulation associated with WASH-complex depletion may reflect a lack of actin-mediated forces that are required for fission of tubular transport intermediates from the endosome (Campellone & Welch 2010). Thus, the WASH complex (containing strumpellin) exemplifies another HSP protein functioning in coordinating membrane modeling and cytoskeletal organization. Strumpellin interacts with valosin-containing protein (VCP/p97), an AAA ATPase mutated in frontotemporal dementia with Paget's disease of bone and inclusion body myopathy (Clemen et al. 2010), and *VPS35* is mutated in a number of families with autosomal dominant, late-onset Parkinson disease (Vilariño-Güell et al. 2011, Zimprich et al. 2011). A causative mutation in the WASH subunit SWIP has recently been identified for autosomal recessive intellectual disability (Ropers et al. 2011). Thus, roles of the WASH-retromer axis in neurological disease clearly extend beyond the HSPs and represent a very important area for investigation.

Another emerging HSP-related complex related to endocytic trafficking comprises the SPG15 protein spastizin/FYVE-CENT, the SPG11 protein spatacsin, and the SPG48 protein KIAA0415. Clinically, SPG11 and SPG15 share a number of characteristics; both are frequently associated with thin corpus callosum, and both can present with juvenile parkinsonism. Spastizin and spatacsin colocalize in cytoplasmic structures and were identified as proteins that coprecipitate with the SPG48 protein KIAA0415 (Slabicki et al. 2010, Murmu et al. 2011). KIAA0415 was originally proposed as a DNA helicase on the basis of sequence predictions (Slabicki et al. 2010), but a very recent study provides compelling evidence that it is a subunit of a new adaptor protein complex, AP-5, involved in endosomal dynamics (Hirst et al. 2011). Spastizin/FYVE-CENT contains a FYVE domain and binds the lipid PI(3)P, functioning along with ESCRT proteins in cytokinesis (Sagona et al. 2010). Along these lines, mutations in multiple proteins of the AP-4 complex, which is involved in trafficking of amyloid precursor protein from the *trans*-Golgi to endosomes (Burgos et al. 2010), cause autosomal recessive syndromes with prominent clinical features ranging from intellectual disability to progressive spastic paraplegia (Abou Jamra et al. 2011, Moreno-De-Luca et al.

2011). Thus, these adaptor protein complexes appear highly relevant for pathogenesis of HSPs and other neurological diseases.

Together, the expanding number of HSP genes implicated in endosome dynamics are already revealing new relationships among protein complexes, with implications that extend beyond the primary HSPs. Indeed, a number of patients with familial amyotrophic lateral sclerosis (ALS) resulting from an autosomal recessive mutation in the *ALS2* gene encoding the alsin protein, a guanine nucleotide exchange factor (GEF) for the small GTPases Rab5 and Rac1, have a disease presentation more similar to the HSPs than to ALS. Examination of *Als2* null mice revealed motor impairments and a distal axonopathy of the corticospinal tract. Rab5-dependent endosomal fusion is impaired in neurons from these mice, whereas alsin overexpression in neurons stimulates Rab5-dependent endosomal fusion, resulting in enlarged endosomes (Devon et al. 2006, Deng et al. 2007, Hadano et al. 2007).

Motor-Based Transport

The identification of mutations in the *KIF5A* gene encoding kinesin heavy chain 5A (known also as kinesin-1A) in families with SPG10, a pure or complicated HSP, has provided direct evidence for motor-based transport impairments underlying HSPs (Reid et al. 2002, Goizet et al. 2009). KIF5 proteins are ATP-dependent motors that move cargoes in the anterograde direction along axons, and most mutations are missense changes in the motor domain. *Drosophila* harboring mutations in the *KIF5* ortholog *Khc* have posterior paralysis, with organelle-filled axon swellings jammed with cargoes (Hurd & Saxton 1996). In mammals, the KIF5A motor protein shuttles neurofilament subunits along axons and possibly other anterograde cargoes such as vesicles. KIF5 also regulates transport of cargoes in dendrites and has roles in a number of membrane traffic pathways. The efficiency of cargo transport to the distal axon is thought to be affected either because the mutated KIF5A are slower motors or because they have reduced microtubule binding affinity and act in a dominant-negative manner by competing with wild-type motors for cargo binding (Ebbing et al. 2008).

Amyotrophic lateral sclerosis (ALS): a degenerative disorder affecting corticospinal and lower motor neurons

Mitochondrial Function

Mitochondrial dysfunction has been implicated in a host of developmental and degenerative neurological disorders, manifesting clinically as peripheral neuropathies, movement disorders, visual disturbances, and cognitive disability (DiMauro & Schon 2008). Given this fundamental link to neurological disease, it is surprising that so few HSP genes encode mitochondrial proteins. Two resident mitochondrial proteins mutated in HSPs are paraplegin (autosomal recessive SPG7) and HSP60 (autosomal dominant SPG13). Paraplegin is an *m*-AAA metalloprotease of the inner mitochondrial membrane, where it functions in ribosomal assembly and protein quality control. Muscle tissue from SPG7 patients exhibits defects in oxidative phosphorylation and *Spg7* null mice have axonal swellings with accumulated mitochondria and neurofilaments, indicating that both mitochondrial function and axonal transport are impaired (Ferreirinha et al. 2004). SPG13 is typically a late-onset, pure HSP, and a causative missense mutation (p.V98I) impairs HSP60 chaperonin activity, leading to impaired mitochondrial quality contol (Bross et al. 2008).

RELATED DISORDERS

We have discussed the convergent pathways of many proteins mutated in HSPs, but it has become increasing clear that the HSP presentation is often part of a broader disease spectrum; thus it is important to consider related disorders that may share pathogenic themes. Indeed, the importance of organelle morphology and distribution in maintaining axons is also emphasized by other inherited axonopathies, particularly peripheral nerve disorders such as the Charcot-Marie-Tooth (CMT) neuropathies and hereditary sensory and autonomic neuropathies (HSAN).

Bone morphogenetic protein (BMP): a member of the transforming growth factor-β superfamily

Mutations in the *FAM134B* gene were identified in some patients with HSAN II. The FAM134B protein is a member of the FAM134 protein family, each of which contains a pair of long hydrophobic segments reminiscent of those in ER-shaping reticulon and Yop1p/DP1/REEP proteins. FAM134B is enriched in the *cis*-Golgi apparatus, and its depletion causes prominent changes in Golgi morphology in neurons (Kurth et al. 2009). More recently, homozygous loss-of-function *KIF1A* mutations were identified in an Afghan family with HSAN II (Rivière et al. 2011), and the KIF1A protein is a motor involved in axonal transport of synaptic vesicles. A family has also been recently identified with hereditary spastic paraplegia caused by homozygous mutation in the KIF1A motor domain (Erlich et al. 2011), indicating that HSP and HSANs may, in at least some cases, fall along a phenotypic spectrum. In fact, dominant missense mutations in the SPG3A protein atlastin-1 have been recently reported in hereditary sensory neuropathy (HSN) I (Guelly et al. 2011). More generally, these disorders highlight the implications of morphological defects in the ER and the early secretory pathway as well as distribution defects in the pathogenesis of length-dependent axonopathies.

Pathogenic studies of CMT peripheral neuropathies are also instructive. CMT1 is composed of demyelination disorders, and CMT2 is composed of those that cause axonopathies. Axonal forms of CMT in particular can be caused by mutations in genes that encode proteins that function in organelle morphogenesis and trafficking. CMT2A results from mutations in the gene encoding mitofusin2 (*MFN2*), which regulates mitochondrial morphology by mediating mitochondrial fusion and has also been implicated in mitochondrial connections with the ER (de Brito et al. 2010). The CMT2B protein Rab7, a small GTPase that regulates endosomal vesicle trafficking, interacts with the SPG21 protein maspardin, another HSP-associated protein that localizes to endosomes (Hanna & Blackstone 2009, McCray et al. 2010, Soderblom et al. 2010). Very recently, a large CMT2 pedigree was reported with autosomal dominant mutation in *DYNC1H1*, which codes for the dynein heavy chain 1 involved in retrograde axonal transport (Weedon et al. 2011).

Finally, ER shaping mechanisms may have roles in related neurologic disorders such as familial ALS, in which both corticospinal and lower motor neurons are affected. In the superoxide dismutase 1 (SOD1) G93A transgenic mouse model for ALS, overexpression of the ER-shaping protein reticulon-4A selectively redistributed the ER chaperone protein disulfide isomerase and protected against neurodegeneration. Conversely, loss of reticulon-4A increased disease severity (Yang et al. 2009). Further supporting a role for aberrant ER morphogenesis in neurologic disorders, a mutant variant of vesicle-associated membrane protein–associated protein B (VAP-B) that underlies another familial ALS (ALS8) is associated with the production of a novel form of organized smooth ER (Fasana et al. 2010).

E PLURIBUS UNUM?

These divisions as discussed above are, of necessity, somewhat arbitrary because cellular pathways show a great deal of interdependence, and a number of HSPs can fit into several pathogenic themes. For instance, the shaping phenomenon crosses over a number of categories, from ER to endosomes. A natural question is, then, can these be unified further?

Bone Morphogenetic Protein Signaling

One compelling candidate that crosses HSP categories and is widely implicated in neurodegenerative diseases is bone morphogenetic protein (BMP) signaling (Bayat et al. 2011). HSP-associated mutations are found in at least four proteins—atlastin-1, NIPA1 (nonimprinted in Prader-Willi/Angelman syndrome 1; SPG6), spastin, and spartin—that function as inhibitors of BMP signaling. In *Drosophila* and mammals, BMP signaling functions in regulating axonal growth and synaptic function, and impairment of BMP signaling in *Drosophila* leads to axon transport defects (Wang et al. 2007). In rodents, BMP signaling is upregulated following lesioning of the corticospinal tract, and suppression of this upregulation can promote regrowth of axons (Matsuura et al. 2008).

Of the HSP proteins known to inhibit BMP signaling, the best characterized mechanistically is NIPA1, an integral membrane protein with 9 predicted TMDs that localizes to endosomes and the plasma membrane and functions in Mg^{2+} transport (Goytain et al. 2007). *Drosophila* larvae lacking spichthyin (NIPA1 ortholog) have increased synaptic boutons at neuromuscular junctions and increased phosphorylated MAD (mothers against decapentaplegic), a downstream messenger of BMP signaling. These changes can be suppressed with genetic alterations that inhibit BMP signaling (Wang et al. 2007). NIPA1/spichthyin is thought to inhibit BMP signaling by promoting the internalization of BMP type II receptors and their subsequent lysosomal degradation, and NIPA1 missense changes found in SPG6 patients interfere with this process, upregulating signaling. Similarly, depletion of spartin or spastin, which both localize partially to endosomes, upregulates BMP signaling (Tsang et al. 2009).

Dysregulated BMP signaling linked to axonal abnormalities has also been demonstrated for atlastin. Depletion of atlastin-1 in zebrafish resulted in abnormal spinal motor axon morphology, with increased branching and decreased larval mobility. BMP signaling was upregulated in these larvae, and pharmacological or genetic inhibition of BMP signaling rescued the *atl1* null phenotype (Fassier et al. 2010). In sum, these results suggest that abnormal BMP signaling, probably caused by abnormal BMP receptor trafficking in many cases, could be a unifying mechanism for some classes of HSP, including the two most common, SPG4 and SPG3A, which comprise almost 50% of patients. Investigating relevant HSP animal models will be critical to determine whether inhibition of BMP signaling using small-molecule inhibitors, several of which are available, can rescue disease phenotypes.

Interorganelle Contacts and Communication

Another possibility for linking HSP themes further is via interorganelle contacts. The ER is distributed promiscuously throughout the cell, and to mediate its many functions, it interacts with other organelles at specialized contact sites. Such interactions occur with the plasma membrane, mitochondria, and lysosomes/endosomes. The importance of these connections for functions such as interorganelle exchange of lipids/sterols, signal transduction, and mobilization of Ca^{2+} stores is increasingly appreciated (Carrasco & Meyer 2011, Toulmay & Prinz 2011). In particular, ER-mitochondrial contacts have been intensively studied recently, with MFN2 in mammals and an ER-mitochondrial encounter structure (ERMES) in yeast playing crucial roles (de Brito et al. 2010, Kornmann & Walter 2010). In a model of pulmonary arterial hypertension, there was decreased ER-to-mitochondria phospholipid transfer and intramitochondrial Ca^{2+} (Sutendra et al. 2011). The contacts between mitochondria and ER were disrupted in a manner dependent on increased expression of an ER-shaping protein of the reticulon family, Nogo-B, linking changes in an ER-shaping protein to mitochondrial dysfunction.

CONCLUDING REMARKS

The HSPs have recently been called a "paradigmatic" example of how a disease can foster insights into fundamental cellular processes, particularly with regard to formation of the tubular ER network (De Matteis & Luini 2011). Ongoing studies investigating how the ER network is shaped in neurons using electron microscopy reconstruction or super-resolution confocal microscopy will improve our understanding of the appearance, contacts, and dynamics of ER in axons. With the increasing throughput and falling cost of next-generation sequencing technologies, more genes for HSPs and related disorders will assuredly be uncovered, likely many more.

Important insights into endocytic trafficking pathways in particular seem sure to follow.

With some compelling cellular mechanisms already identified, pharmacologic manipulation of these pathways and evaluations in cellular and animal models will be increasingly important. Pathways such as BMP signaling and microtubule stability (Orso et al. 2005, Yu et al. 2008) currently seem to be particularly attractive targets because they would likely relate to a significant percentage of HSP patients and seem amenable to regulation by small molecules, which could ultimately lead to therapies.

Animal models may be a particular challenge for HSPs because successfully modeling a disease of 1 m axons in small rodents is not a given. The slow, variable rates of progression in HSP patients will be challenges for assessing the efficacy of therapies, but emerging noninvasive stimulation (triple stimulation technique) and imaging modalites such as diffusion tensor imaging (Duning et al. 2010, Unrath et al. 2010) might be useful biomarkers, particularly because they can detect changes in patients with known HSP mutations at a presymptomatic phase, when disease-modifying therapies would be most useful (Duning et al. 2010). The past several years have yielded remarkable advancements in our understanding of the pathogenesis underlying the HSPs; with increasing interest in the fascinating biology of HSP proteins and technological advances in genetics and imaging moving rapidly, the future is hopeful for those afflicted.

DISCLOSURE STATEMENT

The author is not aware of any affiliations, memberships, funding, or financial holdings that might be perceived as affecting the objectivity of this review.

ACKNOWLEDGMENTS

I thank Alan Hoofring for preparing the figures. Work in the author's laboratory is funded by the Intramural Research Program of the National Institute of Neurological Disorders and Stroke, National Institutes of Health.

LITERATURE CITED

Abou Jamra R, Philippe O, Raas-Rothschild A, Eck SH, Graf E, et al. 2011. Adaptor protein complex 4 deficiency causes severe autosomal-recessive intellectual disability, progressive spastic paraplegia, shy character, and short stature. *Am. J. Hum. Genet.* 88(6):788–95

Alazami AM, Adly N, Al Dhalaan H, Alkuraya FS. 2011. A nullimorphic *ERLIN2* mutation defines a complicated hereditary spastic paraplegia locus (SPG18). *Neurogenetics* 12(4):333–36

Arnold DB. 2009. Actin and microtubule-based cytoskeletal cues direct polarized targeting of proteins in neurons. *Sci. Signal.* 2(83):pe49

Bakowska JC, Jupille H, Fatheddin P, Puertollano R, Blackstone C. 2007. Troyer syndrome protein spartin is mono-ubiquitinated and functions in EGF receptor trafficking. *Mol. Biol. Cell* 18(5):1683–92

Bakowska JC, Wang H, Xin B, Sumner CJ, Blackstone C. 2008. Lack of spartin protein in Troyer syndrome: a loss-of-function disease mechanism? *Arch. Neurol.* 65(4):520–24

Bayat V, Jaiswal M, Bellen HJ. 2011. The BMP signaling pathway at the *Drosophila* neuromuscular junction and its links to neurodegenerative diseases. *Curr. Opin. Neurobiol.* 21(1):182–88

Bechara A, Nawabi H, Moret F, Yaron A, Weaver E, et al. 2008. FAK-MAPK-dependent adhesion disassembly downstream of L1 contributes to semaphorin3A-induced collapse. *EMBO J.* 27(11):1549–62

Bian X, Klemm RW, Liu TY, Zhang M, Sun S, et al. 2011. Structures of the atlastin GTPase provide insight into homotypic fusion of endoplasmic reticulum membranes. *Proc. Natl. Acad. Sci. USA* 108(10):3976–81

Blackstone C, O'Kane CJ, Reid E. 2011. Hereditary spastic paraplegias: membrane traffic and the motor pathway. *Nat. Rev. Neurosci.* 12(1):31–42

Bonifacino JS, Hurley JH. 2008. Retromer. *Curr. Opin. Cell Biol.* 20(4):427–36

Bross P, Naundrup S, Hansen J, Nielsen MN, Christensen JH, et al. 2008. The Hsp60-(p.V98I) mutation associated with hereditary spastic paraplegia SPG13 compromises chaperonin function both *in vitro* and *in vivo*. *J. Biol. Chem.* 283(23):15694–700

Browman DT, Resek ME, Zajchowski LD, Robbins SM. 2006. Erlin-1 and erlin-2 are novel members of the prohibitin family of proteins that define lipid-raft-like domains of the ER. *J. Cell Sci.* 119(Pt. 15): 3149–60

Burgos PV, Mardones GA, Rojas AL, daSilva LLP, Prabhu Y, et al. 2010. Sorting of the Alzheimer's disease amyloid precursor protein mediated by the AP-4 complex. *Dev. Cell* 18(3):425–36

Byrnes LJ, Sondermann H. 2011. Structural basis for the nucleotide-dependent dimerization of the large G protein atlastin-1/SPG3A. *Proc. Natl. Acad. Sci. USA* 108(6):2216–21

Campellone KG, Welch MD. 2010. A nucleator arms race: cellular control of actin assembly. *Nat. Rev. Mol. Cell Biol.* 11(4):237–51

Carpenter MB. 1991. *Core Text of Neuroanatomy*. Baltimore, MD: Wilkins & Wilkins. 4th ed.

Carrasco S, Meyer T. 2011. STIM proteins and the endoplasmic reticulum-plasma membrane junctions. *Annu. Rev. Biochem.* 80:973–1000

Castellani V, Falk J, Rougon G. 2004. Semaphorin3A-induced receptor endocytosis during axon guidance responses is mediated by L1 CAM. *Mol. Cell. Neurosci.* 26(1):89–100

Chen J, Stefano G, Brandizzi F, Zheng H. 2011. *Arabidopsis* RHD3 mediates the generation of the tubular ER network and is required for Golgi distribution and motility in plant cells. *J. Cell Sci.* 124(Pt. 13):2241–52

Clemen CS, Tangavelou K, Strucksberg K-H, Just S, Gaertner L, et al. 2010. Strumpellin is a novel valosin-containing protein binding partner linking hereditary spastic paraplegia to protein aggregation diseases. *Brain* 133(10):2920–41

Cohen NR, Taylor JSH, Scott LB, Guillery RW, Soriano P, Furley AJW. 1998. Errors in corticospinal axon guidance in mice lacking the neural cell adhesion molecule L1. *Curr. Biol.* 8(1):26–33

Connell JW, Lindon C, Luzio JP, Reid E. 2009. Spastin couples microtubule severing to membrane traffic in completion of cytokinesis and secretion. *Traffic* 10(1):42–56

Cui X, Wang Y, Tang Y, Liu Y, Zhao L, et al. 2011. Seipin ablation in mice results in severe generalized lipodystrophy. *Hum. Mol. Genet.* 20(15):3022–30

Dahme M, Bartsch U, Martini R, Anliker B, Schachner M, Mantei N. 1997. Disruption of the mouse *L1* gene leads to malformations of the nervous system. *Nat. Genet.* 17(3):346–49

de Brito OM, Scorrano L. 2010. An intimate liaison: spatial organization of the endoplasmic reticulum–mitochondria relationship. *EMBO J.* 29(16):2715–23

De Matteis MA, Luini A. 2011. Mendelian disorders of membrane trafficking. *N. Engl. J. Med.* 365(10):927–38

DeLuca GC, Ebers GC, Esiri MM. 2004. The extent of axonal loss in the long tracts in hereditary spastic paraplegia. *Neuropathol. Appl. Neurobiol.* 30(6):576–84

Deng H-X, Zhai H, Fu R, Shi Y, Gorrie GH, et al. 2007. Distal axonopathy in an alsin-deficient mouse model. *Hum. Mol. Genet.* 16(23):2911–20

Depienne C, Stevanin G, Brice A, Durr A. 2007. Hereditary spastic paraplegias: an update. *Curr. Opin. Neurol.* 20(6):674–80

Derivery E, Gautreau A. 2010. Evolutionary conservation of the WASH complex, an actin polymerization machine involved in endosomal fission. *Commun. Integr. Biol.* 3(3):227–30

Derivery E, Sousa C, Gautier JJ, Lombard B, Loew D, Gautreau A. 2009. The Arp2/3 activator WASH controls the fission of endosomes through a large multiprotein complex. *Dev. Cell* 17(5):712–23

Devon RS, Orban PC, Gerrow K, Barbieri MA, Schwab C, et al. 2006. *Als2*-deficient mice exhibit disturbances in endosome trafficking associated with motor behavioral abnormalities. *Proc. Natl. Acad. Sci. USA* 103(25):9595–600

Dick KJ, Eckhardt M, Paisán-Ruiz C, Alshehhi AA, Proukakis C, et al. 2010. Mutation of FA2H underlies a complicated form of hereditary spastic paraplegia (SPG35). *Hum. Mutat.* 31(4):E1251–60

DiMauro S, Schon EA. 2008. Mitochondrial disorders in the nervous system. *Annu. Rev. Neurosci.* 31:91–123

Dion PA, Daoud H, Rouleau GA. 2009. Genetics of motor neuron disorders: new insights into pathogenic mechanisms. *Nat. Rev. Genet.* 10(11):769–82

Duning T, Warnecke T, Schirmacher A, Schiffbauer H, Lohmann H, et al. 2010. Specific pattern of early white-matter changes in pure hereditary spastic paraplegia. *Mov. Disord.* 25(12):1986–92

Eastman SW, Yassaee M, Bieniasz PD. 2009. A role for ubiquitin ligases and Spartin/SPG20 in lipid droplet turnover. *J. Cell Biol.* 184(6):881–94

Ebbing B, Mann K, Starosta A, Jaud J, Schöls L, et al. 2008. Effect of spastic paraplegia mutations in KIF5A kinesin on transport activity. *Hum. Mol. Genet.* 17(9):1245–52

Edgar JM, McLaughlin M, Yool D, Zhang S-C, Fowler JH, et al. 2004. Oligodendroglial modulation of fast axonal transport in a mouse model of hereditary spastic paraplegia. *J. Cell Biol.* 166(1):121–31

Edwards TL, Clowes VE, Tsang HTH, Connell JW, Sanderson CM, et al. 2009. Endogenous spartin (SPG20) is recruited to endosomes and lipid droplets and interacts with the ubiquitin E3 ligases AIP4 and AIP5. *Biochem. J.* 423(1):31–39

Erlich Y, Edvardson S, Hodges E, Zenvirt S, Thekkat P, et al. 2011. Exome sequencing and disease-network analysis of a single family implicate a mutation in *KIF1A* in hereditary spastic paraparesis. *Genome Res.* 21(5):658–64

Evans K, Keller C, Pavur K, Glasgow K, Conn B, Lauring B. 2006. Interaction of two hereditary spastic paraplegia gene products, spastin and atlastin, suggests a common pathway for axonal maintenance. *Proc. Natl. Acad. Sci. USA* 103(28):10666–71

Fasana E, Fossati M, Ruggiano A, Brambillasca S, Hoogenraad CC, et al. 2010. A VAPB mutant linked to amyotrophic lateral sclerosis generates a novel form of organized smooth endoplasmic reticulum. *FASEB J.* 24(5):1419–30

Fassier C, Hutt JA, Scholpp S, Lumsden A, Giros B, et al. 2010. Zebrafish atlastin controls motility and spinal motor axon architecture via inhibition of the BMP pathway. *Nat. Neurosci.* 13(11):1380–87

Fei W, Du X, Yang H. 2011a. Seipin, adipogenesis and lipid droplets. *Trends Endocrinol. Metab.* 22(6):204–10

Fei W, Shui G, Zhang Y, Krahmer N, Ferguson C, et al. 2011b. A role for phosphatidic acid in the formation of "supersized" lipid droplets. *PLoS Genet.* 7(7):e1002201

Ferreirinha F, Quattrini A, Pirozzi M, Valsecchi V, Dina G, et al. 2004. Axonal degeneration in paraplegin-deficient mice is associated with abnormal mitochondria and impairment of axonal transport. *J. Clin. Invest.* 113(2):231–42

Fink JK. 2006. Hereditary spastic paraplegia. *Curr. Neurol. Neurosci. Rep.* 6(1):65–76

Goizet C, Boukhris A, Mundwiller E, Tallaksen C, Forlani S, et al. 2009. Complicated forms of autosomal dominant hereditary spastic paraplegia are frequent in SPG10. *Hum. Mutat.* 30(2):E376–85

Goldstein AYN, Wang X, Schwarz TL. 2008. Axonal transport and the delivery of pre-synaptic components. *Curr. Opin. Neurobiol.* 18(5):495–503

Gomez TS, Billadeau DD. 2009. A FAM21-containing WASH complex regulates retromer-dependent sorting. *Dev. Cell* 17(5):699–711

Goytain A, Hines RM, El-Husseini A, Quamme GA. 2007. *NIPA1(SPG6)*, the basis for autosomal dominant form of hereditary spastic paraplegia, encodes a functional Mg^{2+} transporter. *J. Biol. Chem.* 282(11):8060–68

Gruenenfelder FI, Thomson G, Penderis J, Edgar JM. 2011. Axon-glial interaction in the CNS: what we have learned from mouse models of Pelizaeus-Merzbacher disease. *J. Anat.* 219(1):33–43

Guelly C, Zhu P-P, Leonardis L, Papić L, Zidar J, et al. 2011. Targeted high-throughput sequencing identifies mutations in *atlastin-1* as a cause of hereditary sensory neuropathy type I. *Am. J. Hum. Genet.* 88(1): 99–105

Guizetti J, Schermelleh L, Mäntler J, Maar S, Poser I, et al. 2011. Cortical constriction during abscission involves helices of ESCRT-III-dependent filaments. *Science* 331(6024):1616–20

Hadano S, Kunita R, Otomo A, Suzuki-Utsunomiya K, Ikeda J-E. 2007. Molecular and cellular function of ALS2/alsin: implication of membrane dynamics in neuronal development and degeneration. *Neurochem. Int.* 51(2–4):74–84

Hanna MC, Blackstone C. 2009. Interaction of the *SPG21* protein ACP33/maspardin with the aldehyde dehydrogenase ALDH16A1. *Neurogenetics* 10(3):217–28

Hapala I, Marza E, Ferreira T. 2011. Is fat so bad? Modulation of endoplasmic reticulum stress by lipid droplet formation. *Biol. Cell* 103(6):271–85

Harbour ME, Breusegem SYA, Antrobus R, Freeman C, Reid E, Seaman MNJ. 2010. The cargo-selective retromer complex is a recruiting hub for protein complexes that regulate endosomal tubule dynamics. *J. Cell Sci.* 123(Pt. 21):3703–17

Harding AE. 1983. Classification of the hereditary ataxias and paraplegias. *Lancet* 1(8334):1151–55

Hirokawa N, Niwa S, Tanaka Y. 2010. Molecular motors in neurons: transport mechanisms and roles in brain function, development, and disease. *Neuron* 68(4):610–38

Hirst J, Barlow LD, Francisco GC, Sahlender DA, Seaman MNJ, et al. 2011. The fifth adaptor protein complex. *PLoS Biol.* 9(10):e1001170

Hooper C, Puttamadappa SS, Loring Z, Shekhtman A, Bakowska JC. 2010. Spartin activates atrophin-1-interacting protein 4 (AIP4) E3 ubiquitin ligase and promotes ubiquitination of adipophilin on lipid droplets. *BMC Biol.* 8:72

Hu J, Shibata Y, Voss C, Shemesh T, Li Z, et al. 2008. Membrane proteins of the endoplasmic reticulum induce high-curvature tubules. *Science* 319(5867):1247–50

Hu J, Shibata Y, Zhu P-P, Voss C, Rismanchi N, et al. 2009. A class of dynamin-like GTPases involved in the generation of the tubular ER network. *Cell* 138(3):549–61

Hurd DD, Saxton WM. 1996. Kinesin mutations cause motor neuron disease phenotypes by disrupting fast axonal transport in Drosophila. *Genetics* 144(3):1075–85

Hurley JH, Hanson PI. 2010. Membrane budding and scission by the ESCRT machinery: It's all in the neck. *Nat. Rev. Mol. Cell Biol.* 11(8):556–66

Inoue K. 2005. *PLP1*-related inherited dysmyelinating disorders: Pelizaeus-Merzbacher disease and spastic paraplegia type 2. *Neurogenetics* 6(1):1–16

Ito D, Suzuki N. 2009. Seipinopathy: a novel endoplasmic reticulum stress-associated disease. *Brain* 132(Pt. 1):8–15

Jia D, Gomez TS, Metlagel Z, Umetani J, Otwinowski Z, et al. 2010. WASH and WAVE actin regulators of the Wiskott-Aldrich syndrome protein (WASP) family are controlled by analogous structurally related complexes. *Proc. Natl. Acad. Sci.USA* 107(23):10442–47

Jo Y, Sguigna PV, DeBose-Boyd RA. 2011. Membrane-associated ubiquitin ligase complex containing gp78 mediates sterol-accelerated degradation of 3-hydroxy-3-methylglutaryl-coenzyme A reductase. *J. Biol. Chem.* 286(17):15022–31

Jouet M, Rosenthal A, Armstrong G, MacFarlane J, Stevenson R, et al. 1994. X-linked spastic paraplegia (SPG1), MASA syndrome and X-linked hydrocephalus result from mutations in the *L1* gene. *Nat. Genet.* 7(3):402–7

Kornmann B, Walter P. 2010. ERMES-mediated ER-mitochondria contacts: molecular hubs for the regulation of mitochondrial biology. *J. Cell Sci.* 123(Pt. 9):1389–93

Kurth I, Pamminger T, Hennings JC, Soehendra D, Huebner AK, et al. 2009. Mutations in *FAM134B*, encoding a newly identified Golgi protein, cause severe sensory and autonomic neuropathy. *Nat. Genet.* 41(11):1179–81

Lang N, Optenhoefel T, Deuschl G, Klebe S. 2011. Axonal integrity of corticospinal projections to the upper limbs in patients with pure hereditary spastic paraplegia. *Clin. Neurophysiol.* 122(7):1417–20

Lin P, Li J, Liu Q, Mao F, Li J, et al. 2008. A missense mutation in *SLC33A1*, which encodes the acetyl-CoA transporter, causes autosomal-dominant spastic paraplegia (*SPG42*). *Am. J. Hum. Genet.* 83(6): 752–59

Lind GE, Raiborg C, Danielsen SA, Rognum TO, Thiis-Evensen E, et al. 2011. *SPG20*, a novel biomarker for early detection of colorectal cancer, encodes a regulator of cytokinesis. *Oncogene* 30(37):3967–78

Lynes EM, Simmen T. 2011. Urban planning of the endoplasmic reticulum (ER): how diverse mechanisms segregate the many functions of the ER. *Biochim. Biophys. Acta* 1813(10):1893–905

Mancuso G, Rugarli EI. 2008. A cryptic promoter in the first exon of the *SPG4* gene directs the synthesis of the 60-kDa spastin isoform. *BMC Biol.* 6:31

Mannan AU, Boehm J, Sauter SM, Raūber A, Byrne PC, et al. 2006. Spastin, the most commonly mutated protein in hereditary spastic paraplegia interacts with Reticulon 1 an endoplasmic reticulum protein. *Neurogenetics* 7(2):93–103

Matsuura I, Taniguchi J, Hata K, Saeki N, Yamashita T. 2008. BMP inhibition enhances axonal growth and functional recovery after spinal cord injury. *J. Neurochem.* 105(4):1471–79

McCray BA, Skordalakes E, Taylor JP. 2010. Disease mutations in Rab7 result in unregulated nucleotide exchange and inappropriate activation. *Hum. Mol. Genet.* 19(6):1033–47

Montenegro G, Rebelo AP, Connell J, Allison R, Babalini C, et al. 2012. Mutations in the ER-shaping protein reticulon 2 cause the axon-degenerative disorder hereditary spastic paraplegia type 12. *J. Clin. Invest.* 122(2):538–44

Moreno-De-Luca A, Helmers SL, Mao H, Burns TG, Melton AMA, et al. 2011. Adaptor protein complex-4 (AP-4) deficiency causes a novel autosomal recessive cerebral palsy syndrome with microcephaly and intellectual disability. *J. Med. Genet.* 48(2):141–44

Morita E, Colf LA, Karren MA, Sandrin V, Rodesch CK, Sundquist WI. 2010. Human ESCRT-III and VPS4 proteins are required for centrosome and spindle maintenance. *Proc. Natl. Acad. Sci. USA* 107(29):12889–94

Moss TJ, Andreazza C, Verma A, Daga A, McNew JA. 2011. Membrane fusion by the GTPase atlastin requires a conserved C-terminal cytoplasmic tail and dimerization through the middle domain. *Proc. Natl. Acad. Sci. USA* 108(27):11133–38

Murmu RP, Martin E, Rastetter A, Esteves T, Muriel M-P, et al. 2011. Cellular distribution and subcellular localization of spatacsin and spastizin, two proteins involved in hereditary spastic paraplegia. *Mol. Cell. Neurosci.* 47(3):191–202

Orso G, Martinuzzi A, Rossetto MG, Sartori E, Feany M, Daga A. 2005. Disease-related phenotypes in a *Drosophila* model of hereditary spastic paraplegia are ameliorated by treatment with vinblastine. *J. Clin. Invest.* 115(11):3026–34

Orso G, Pendin D, Liu S, Tosetto J, Moss TJ, et al. 2009. Homotypic fusion of ER membranes requires the dynamin-like GTPase atlastin. *Nature* 460(7258):978–83

Orthmann-Murphy JL, Salsano E, Abrams CK, Bizzi A, Uziel G, et al. 2009. Hereditary spastic paraplegia is a novel phenotype for *GJA12/GJC2* mutations. *Brain* 132(Pt. 2):426–38

Park SH, Blackstone C. 2010. Further assembly required: construction and dynamics of the endoplasmic reticulum network. *EMBO Rep.* 11(7):515–21

Park SH, Zhu P-P, Parker RL, Blackstone C. 2010. Hereditary spastic paraplegia proteins REEP1, spastin, and atlastin-1 coordinate microtubule interactions with the tubular ER network. *J. Clin. Invest.* 120(4):1097–110

Pearce MMP, Wormer DB, Wilkens S, Wojcikiewicz RJH. 2009. An endoplasmic reticulum (ER) membrane complex composed of SPFH1 and SPFH2 mediates the ER-associated degradation of inositol 1,4,5-trisphosphate receptors. *J. Biol. Chem.* 284(16):10433–45

Potter KA, Kern MJ, Fullbright G, Bielawski J, Scherer SS, et al. 2011. Central nervous system dysfunction in a mouse model of Fa2h deficiency. *Glia* 59(7):1009–21

Rainier S, Bui M, Mark E, Thomas D, Tokarz D, et al. 2008. Neuropathy target esterase gene mutations cause motor neuron disease. *Am. J. Hum. Genet.* 82(3):780–85

Read DJ, Li Y, Chao MV, Cavanagh JB, Glynn P. 2009. Neuropathy target esterase is required for adult vertebrate axon maintenance. *J. Neurosci.* 29(37):11594–600

Reid E, Kloos M, Ashley-Koch A, Hughes L, Bevan S, et al. 2002. A kinesin heavy chain (*KIF5A*) mutation in hereditary spastic paraplegia (SPG10). *Am. J. Hum. Genet.* 71(5):1189–94

Renvoisé B, Blackstone C. 2010. Emerging themes of ER organization in the development and maintenance of axons. *Curr. Opin. Neurobiol.* 20(5):531–37

Renvoisé B, Parker RL, Yang D, Bakowska JC, Hurley JH, Blackstone C. 2010. SPG20 protein spartin is recruited to midbodies by ESCRT-III protein Ist1 and participates in cytokinesis. *Mol. Biol. Cell* 21(19):3293–303

Rismanchi N, Soderblom C, Stadler J, Zhu P-P, Blackstone C. 2008. Atlastin GTPases are required for Golgi apparatus and ER morphogenesis. *Hum. Mol. Genet.* 17(11):1591–604

Rivière J-B, Ramalingam S, Lavastre V, Shekarabi M, Holbert S, et al. 2011. *KIF1A*, an axonal transporter of synaptic vesicles, is mutated in hereditary sensory and autonomic neuropathy Type 2. *Am. J. Hum. Genet.* 89(2):219–30

Robay D, Patel H, Simpson MA, Brown NA, Crosby AH. 2006. Endogenous spartin, mutated in hereditary spastic paraplegia, has a complex subcellular localization suggesting diverse roles in neurons. *Exp. Cell Res.* 312(15):2764–77

Ropers F, Derivery E, Hu H, Garshasbi M, Karbasiyan M, et al. 2011. Identification of a novel candidate gene for non-syndromic autosomal recessive intellectual disability: the WASH complex member SWIP. *Hum. Mol. Genet.* 20(13):2585–90

Sagona AP, Nezis IP, Pedersen NM, Liestøl K, Poulton J, et al. 2010. PtdIns(3)P controls cytokinesis through KIF13A-mediated recruitment of FYVE-CENT to the midbody. *Nat. Cell Biol.* 12(4):362–71

Saito H, Kubota M, Roberts RW, Chi Q, Matsunami H. 2004. RTP family members induce functional expression of mammalian odorant receptors. *Cell* 119(5):679–91

Salinas S, Proukakis C, Crosby A, Warner TT. 2008. Hereditary spastic paraplegia: clinical features and pathogenetic mechanisms. *Lancet Neurol.* 7(12):1127–38

Sanderson CM, Connell JW, Edwards TL, Bright NA, Duley S, et al. 2006. Spastin and atlastin, two proteins mutated in autosomal-dominant hereditary spastic paraplegia, are binding partners. *Hum. Mol. Genet.* 15(2):307–18

Schneider SA, Bhatia KP. 2010. Three faces of the same gene: *FA2H* links neurodegeneration with brain iron accumulation, leukodystrophies, and hereditary spastic paraplegias. *Ann. Neurol.* 68(5):575–77

Schüle R, Siddique T, Deng H-X, Yang Y, Donkervoort S, et al. 2010. Marked accumulation of 27-hydroxycholesterol in SPG5 patients with hereditary spastic paresis. *J. Lipid Res.* 51(4):819–23

Shibata Y, Hu J, Kozlov MM, Rapoport TA. 2009. Mechanisms shaping the membranes of cellular organelles. *Annu. Rev. Cell Dev. Biol.* 25:329–54

Sieberer BJ, Ketelaar T, Esseling JJ, Emons AMC. 2005. Microtubules guide root hair tip growth. *New Phytol.* 167(3):711–19

Słabicki M, Theis M, Krastev DB, Samsonov S, Mundwiller E, et al. 2010. A genome-scale DNA repair RNAi screen identifies SPG48 as a novel gene associated with hereditary spastic paraplegia. *PLoS Biol.* 8(6):e1000408

Soderblom C, Blackstone C. 2006. Traffic accidents: molecular genetic insights into the pathogenesis of the hereditary spastic paraplegias. *Pharmacol. Ther.* 109(1–2):42–56

Soderblom C, Stadler J, Jupille H, Blackstone C, Shupliakov O, Hanna MC. 2010. Targeted disruption of the Mast syndrome gene *SPG21* in mice impairs hind limb function and alters axon branching in cultured cortical neurons. *Neurogenetics* 11(4):369–78

Solowska JM, Morfini G, Falnikar A, Himes BT, Brady ST, et al. 2008. Quantitative and functional analyses of spastin in the nervous system: implications for hereditary spastic paraplegia. *J. Neurosci.* 28(9): 2147–57

Sparkes I, Hawes C, Frigerio L. 2011. FrontiERs: movers and shapers of the higher plant cortical endoplasmic reticulum. *Curr. Opin. Plant Biol.* 14(6):658–65

Sutendra G, Dromparis P, Wright P, Bonnet S, Haromy A, et al. 2011. The role of Nogo and the mitochondria-endoplasmic reticulum unit in pulmonary hypertension. *Sci. Transl. Med.* 3(88):88ra55

Terasaki M, Chen LB, Fujiwara K. 1986. Microtubules and the endoplasmic reticulum are highly interdependent structures. *J. Cell Biol.* 103(4):1557–68

Tian Y, Bi J, Shui G, Liu Z, Xiang Y, et al. 2011. Tissue-autonomous function of *Drosophila Seipin* in preventing ectopic lipid droplet formation. *PLoS Genet.* 7(4):e1001364

Toulmay A, Prinz WA. 2011. Lipid transfer and signaling at organelle contact sites: the tip of the iceberg. *Curr. Opin. Cell Biol.* 23(4):458–63

Tsang HTH, Edwards TL, Wang X, Connell JW, Davies RJ, et al. 2009. The hereditary spastic paraplegia proteins NIPA1, spastin and spartin are inhibitors of mammalian BMP signalling. *Hum. Mol. Genet.* 18(20):3805–21

Tsaousidou MK, Ouahchi K, Warner TT, Yang Y, Simpson MA, et al. 2008. Sequence alterations within CYP7B1 implicate defective cholesterol homeostasis in motor-neuron degeneration. *Am. J. Hum. Genet.* 82(2):510–15

Unrath A, Müller H-P, Riecker A, Ludolph AC, Sperfeld A-D, Kassubek J. 2010. Whole brain-based analysis of regional white matter tract alterations in rare motor neuron diseases by diffusion tensor imaging. *Hum. Brain Mapp.* 31(11):1727–40

Urbanczyk A, Enz R. 2011. Spartin recruits PKC-ζ via the PKC-ζ-interacting proteins ZIP1 and ZIP3 to lipid droplets. *J. Neurochem.* 118(5):737–48

Valdmanis PN, Meijer IA, Reynolds A, Lei A, MacLeod P, et al. 2007. Mutations in the *KIAA0196* gene at the *SPG8* locus cause hereditary spastic paraplegia. *Am. J. Hum. Genet.* 80(1):152–61

Verkhratsky A. 2005. Physiology and pathophysiology of the calcium store in the endoplasmic reticulum of neurons. *Physiol. Rev.* 85(1):201–79

Vilariño-Güell C, Wider C, Ross OA, Dachsel JC, Kachergus JM, et al. 2011. *VPS35* mutations in Parkinson disease. *Am. J. Hum. Genet.* 89(1):162–67

Voeltz GK, Prinz WA, Shibata Y, Rist JM, Rapoport TA. 2006. A class of membrane proteins shaping the tubular endoplasmic reticulum. *Cell* 124(3):573–86

Wang H, Lockwood SK, Hoeltzel MF, Schiefelbein JW. 1997. The *ROOT HAIR DEFECTIVE3* gene encodes an evolutionarily conserved protein with GTP-binding motifs and is required for regulated cell enlargement in *Arabidopsis*. *Genes Dev.* 11(6):799–811

Wang X, Shaw WR, Tsang HTH, Reid E, O'Kane CJ. 2007. *Drosophila* spichthyin inhibits BMP signaling and regulates synaptic growth and axonal microtubules. *Nat. Neurosci.* 10(2):177–85

Waterman-Storer CM, Salmon ED. 1998. Endoplasmic reticulum membrane tubules are distributed by microtubules in living cells using three distinct mechanisms. *Curr. Biol.* 8(14):798–806

Weedon MN, Hastings R, Caswell R, Xie W, Paszkiewicz K, et al. 2011. Exome sequencing identifies a *DYNC1H1* mutation in a large pedigree with dominant axonal Charcot-Marie-Tooth disease. *Am. J. Hum. Genet.* 89(2):308–12

Weller S, Gärtner J. 2001. Genetic and clinical aspects of X-linked hydrocephalus (L1 disease): mutations in the *L1CAM* gene. *Hum. Mutat.* 18(1):1–12

West M, Zurek N, Hoenger A, Voeltz GK. 2011. A 3D analysis of yeast ER structure reveals how ER domains are organized by membrane curvature. *J. Cell Biol.* 193(2):333–46

Westermann B. 2010. Mitochondrial fusion and fission in cell life and death. *Nat. Rev. Mol. Cell Biol.* 11(12):872–84

Windpassinger C, Auer-Grumbach M, Irobi J, Patel H, Petek E, et al. 2004. Heterozygous missense mutations in *BSCL2* are associated with distal hereditary motor neuropathy and Silver syndrome. *Nat. Genet.* 36(3):271–76

Yagi T, Ito D, Nihei Y, Ishihara T, Suzuki N. 2011. N88S seipin mutant transgenic mice develop features of seipinopathy/BSCL2-related motor neuron disease via endoplasmic reticulum stress. *Hum. Mol. Genet.* 20(19):3831–40

Yang D, Rismanchi N, Renvoisé B, Lippincott-Schwartz J, Blackstone C, Hurley JH. 2008. Structural basis for midbody targeting of spastin by the ESCRT-III protein CHMP1B. *Nat. Struct. Mol. Biol.* 15(12):1278–86

Yang YS, Harel NY, Strittmatter SM. 2009. Reticulon-4A (Nogo-A) redistributes protein disulfide isomerase to protect mice from SOD1-dependent amyotrophic lateral sclerosis. *J. Neurosci.* 29(44):13850–59

Yıldırım Y, Orhan EK, Iseri SAU, Serdaroglu-Oflazer P, Kara B, et al. 2011. A frameshift mutation of *ERLIN2* in recessive intellectual disability, motor dysfunction and multiple joint contractures. *Hum. Mol. Genet.* 20(10):1886–92

Yu PB, Hong CC, Sachidanandan C, Babitt JL, Deng DY, et al. 2008. Dorsomorphin inhibits BMP signals required for embryogenesis and iron metabolism. *Nat. Chem. Biol.* 4(1):33–41

Zhu P-P, Patterson A, Lavoie B, Stadler J, Shoeb M, et al. 2003. Cellular localization, oligomerization, and membrane association of the hereditary spastic paraplegia 3A (SPG3A) protein atlastin. *J. Biol. Chem.* 278(49):49063–71

Zhu P-P, Soderblom C, Tao-Cheng J-H, Stadler J, Blackstone C. 2006. SPG3A protein atlastin-1 is enriched in growth cones and promotes axon elongation during neuronal development. *Hum. Mol. Genet.* 15(8):1343–53

Zimprich A, Benet-Pagès A, Struhal W, Graf E, Eck SH, et al. 2011. A mutation in *VPS35*, encoding a subunit of the retromer complex, causes late-onset Parkinson disease. *Am. J. Hum. Genet.* 89(1):168–75

RELATED RESOURCES

Neuromuscul. Dis. Cent. Web page. *Familial spinal cord syndromes*. **http://neuromuscular.wustl.edu/spinal/fsp.html**. A comprehensive, frequently updated resource for information on all types of HSPs and related disorders.

Reid E, Rugarli EI. 2010. Hereditary spastic paraplegias. In *The Online Metabolic and Molecular Bases of Inherited Diseases*, ed. D Valle, AL Beaudet, B Vogelstein, KW Kinzler, SE Antonarakis, et al., Ch. 228.1. New York: McGraw Hill. **http://dx.doi.org/10.1036/ommbid.266**

Spastic Paraplegia Found. Web page. **http://www.sp-foundation.org**. Valuable information on research efforts focusing on the HSPs.

Functional Consequences of Mutations in Postsynaptic Scaffolding Proteins and Relevance to Psychiatric Disorders

Jonathan T. Ting,[1,2,*] João Peça,[1,*] and Guoping Feng[1,3]

[1]McGovern Institute for Brain Research and Department of Brain and Cognitive Sciences, Massachusetts Institute of Technology, Cambridge, Massachusetts 02139; email: jtting@mit.edu, peca@mit.edu, fengg@mit.edu

[2]Department of Neurobiology, Duke University Medical Center, Durham, North Carolina, 27710

[3]Stanley Center for Psychiatric Research, Broad Institute, Cambridge, Massachusetts, 02142

Annu. Rev. Neurosci. 2012. 35:49–71

First published online as a Review in Advance on April 20, 2012

The *Annual Review of Neuroscience* is online at neuro.annualreviews.org

This article's doi:
10.1146/annurev-neuro-062111-150442

0147-006X/12/0721-0049$20.00

*These authors contributed equally

Keywords

PSD95, AKAP, SAPAP, Shank, Homer, psychiatric disorders

Abstract

Functional studies on postsynaptic scaffolding proteins at excitatory synapses have revealed a plethora of important roles for synaptic structure and function. In addition, a convergence of recent in vivo functional evidence together with human genetics data strongly suggest that mutations in a variety of these postsynaptic scaffolding proteins may contribute to the etiology of diverse human psychiatric disorders such as schizophrenia, autism spectrum disorders, and obsessive-compulsive spectrum disorders. Here we review the most recent evidence for several key postsynaptic scaffolding protein families and explore how mouse genetics and human genetics have intersected to advance our knowledge concerning the contributions of these important players to complex brain function and dysfunction.

Contents

PSD: postsynaptic density

INTRODUCTION

The Anatomy of the Postsynaptic Specialization at Excitatory Synapses

The chemical synapse is a microscopic physical structure that conveys electrical signals from presynaptic to postsynaptic neurons within brain circuits by means of chemical neurotransmitter release and action. Both excitatory and inhibitory neurotransmitter release act in concert under constant fine-tuning to orchestrate the flow of information processing in the nervous system. Most excitatory synapses are formed between presynaptic boutons loaded with glutamate-filled synaptic vesicles and tightly apposed protrusions of postsynaptic receptor-laden dendritic spines.

The dendritic spine is a highly specialized structure that is both complex and elegant. Electron microscopy images of excitatory synapses prominently feature a dense proteinacious matrix at the tip of the spine head immediately underlying the postsynaptic membrane face. This protein mesh is called the postsynaptic density (PSD), and decades of research using primarily biochemical and molecular cloning methods have led to the identification of many prominent PSD constituents. The PSD contains many distinct classes of proteins, including neurotransmitter receptors, cell adhesion molecules, ion channels, signaling molecules, and scaffolding proteins (**Figure 1**). The dynamic nature and precise topographical organization of these components give rise to a supramolecular signal-processing machine.

Scaffolding Proteins Constitute the Structural Core of the Postsynaptic Density

Scaffolding proteins are extremely abundant in the PSD, both in terms of absolute protein copy numbers and the distinct types of scaffolding proteins that have been described to date (Kim & Sheng 2004, Sheng & Hoogenraad 2007). The most well-studied postsynaptic scaffolding proteins include members of the

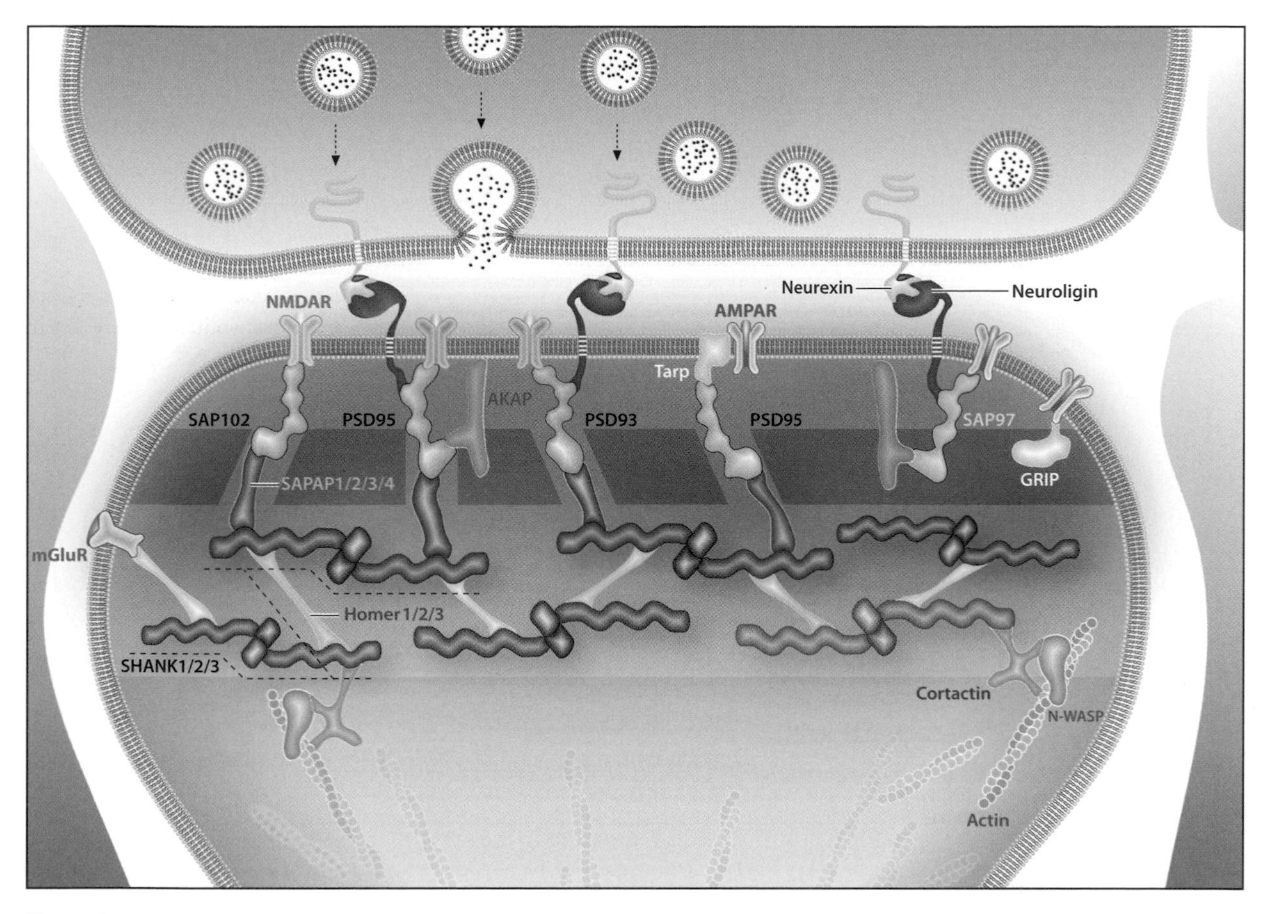

Figure 1

Scaffolding protein networks at the postsynaptic density (PSD). Schematic of the major family members of PSD scaffolding proteins at excitatory synapses. Current information from structural studies suggests that Shank protein can dimerize through C-terminal sterile alpha motif (SAM) domain–SAM domain interaction and form a supramolecular polymeric network with Homer tetramers. This complex may connect to perisynaptic mGluRs and to synaptic NMDA and AMPA-type ionotropic glutamate receptors through the PSD95 and SAPAP (SAP90/PSD95-associated protein) family of proteins. The Shank/Homer platform may also provide key connection points to the spine actin cytoskeleton. A-kinase anchoring protein (AKAP) is another important protein that can anchor kinases and phosphatases (not shown here) in the vicinity of synaptic receptors and ion channels.

PSD95 family, select members of the A-kinase anchoring protein (AKAP) family, the Homer family, the SAP90/PSD95-associated protein (SAPAP) family, and the SH3 and multiple ankyrin repeat domain (Shank) family. Scaffolding-protein families are generally defined by a highly conserved organization of domains for protein-protein interactions, and it is the unique combinations and properties of these domains that impart a specificity of protein-protein interactions exhibited by each of these families (**Figure 2**). Furthermore, postsynaptic scaffolding proteins can interact with multiple binding partners simultaneously to physically link PSD components and, thus, can be viewed as the master organizers within this specialized structure.

Here we aim to highlight evidence from the recent literature concerning the in vivo functional roles served by the major postsynaptic scaffolding protein families. We further examine findings that have emerged from human genetics investigations exploring variations in genes that encode postsynaptic scaffolding

AKAP: A-kinase anchoring protein

SAPAP: SAP90/PSD95-associated protein

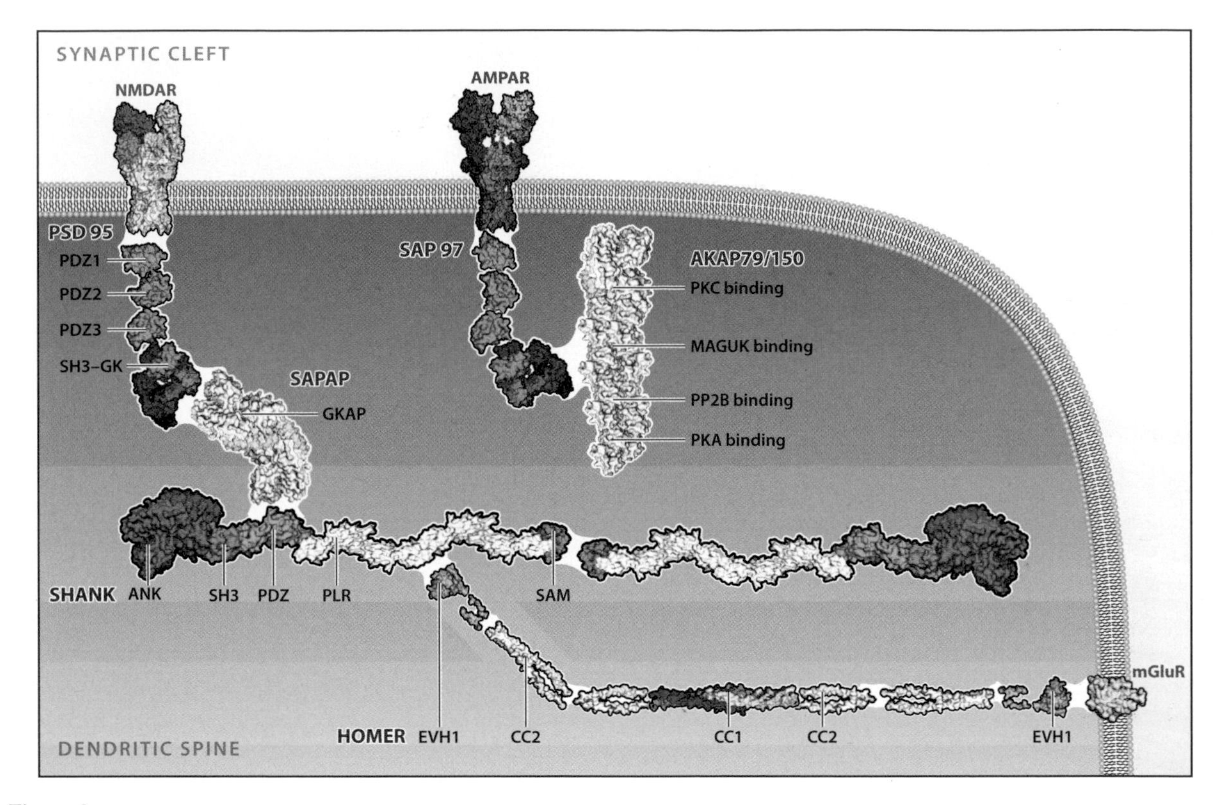

Figure 2

Physical interactions via discrete binding domains in postsynaptic density (PSD) scaffolding proteins. Representation of the known atomic structures from crystallized domains in key PSD scaffolding proteins, highlighting interactions at known binding sites. For simplicity, protein domains that have not been resolved at the atomic level are displayed in white. Represented domains include the PDZ and SH3 domains of PSD95 and Shank family proteins, the GK domain in PSD95 family proteins, the Ank and SAM domains in Shank family proteins, and the EVH1 and coiled-coil (CC) domains in Homer family proteins (information on these structures was obtained through the RCSB Protein Data Bank **http://www.pdb.org**).

MAGUK: membrane-associated guanylate kinase

Ortholog: genes in different species that evolved from a common ancestral gene and have retained the same function

DLG: discs large

PDZ domain: PSD95, Dlg, and ZO-1 domain

proteins in relation to psychiatric disorders (see Diversity of Human Psychiatric Disorders, sidebar below). More comprehensive coverage of other known PSD scaffolding molecules, particularly with respect to structural considerations, can be found elsewhere (Chen et al. 2008, Kim & Sheng 2004, Sheng & Hoogenraad 2007).

PSD95/MAGUK FAMILY

Membrane-associated guanylate kinase (MAGUK) proteins form a superfamily of scaffolding proteins present in several organisms and serving various cellular roles. Here, special consideration is given to the commonly defined PSD95 family of proteins, a subfamily of MAGUKs comprised of synapse-associated protein (SAP)102, SAP97, PSD93, and PSD95. These MAGUKs are orthologs of *Drosophila DLG* (discs large), the first cloned MAGUK (Woods & Bryant 1991).

The PSD95 protein was among the first components to be identified as a part of the PSD (Sampedro et al. 1981). Structurally, members of the PSD95 family share several common protein-protein interaction domains. From the N to C terminus these include an L27 domain, three PDZ domains (PDZ1, PDZ2, and PDZ3, termed after their occurrence in three related MAGUKs, PSD95, Dlg, and ZO-1), an SH3 domain (SRC homology 3 domain), and a

C-terminal catalytically inactive guanylate kinase-like (GK) domain (Kuhlendahl et al. 1998).

PDZ domains are found in a wide variety of eukaryotic proteins and display considerable sequence variation, presumably underlying functional diversity and binding specificities (Sheng & Sala 2001). The majority of known PDZ domains interact with a canonical C-terminal sequence found in the binding partners. Some of the most notable binding partners to the first two PDZ domains of PSD95 include the Shaker-type K^+ channels and NR2A subunits of the N-methyl-D-aspartate type glutamate receptor (NMDAR), both through C-terminal PDZ binding motifs (Kim et al. 1995, Kornau et al. 1995). Neuroligins, a family of cell adhesion molecules located at synapses, also bind to the third PDZ domain in PSD95 through a C-terminal PDZ motif (Irie et al. 1997). The demonstrated interactions were later expanded to include several members of the PSD95 family: PDZ1 and PDZ2 from SAP97 emulate the NR2A/PSD95 interaction, whereas in SAP102 all three PDZ domains can bind to the NR2B subunit of the NMDAR. Finally, PSD93 also interacts and promotes the clustering of NMDAR subunits in heterologous cells (Kim et al. 1996, Niethammer et al. 1996). PDZ domains are also responsible for the regulation of α-amino-3-hydroxy-5-methyl-4-isoxazolepropionic acid type glutamate receptors (AMPARs). SAP97 directly interacts with the GluR1 subunit of the AMPAR and is involved in the trafficking of these channels (Leonard et al. 1998). However, the interaction between GluR subunits and PSD95 is indirectly mediated through transmembrane AMPAR regulatory proteins such as Stargazin (Chen et al. 2000, Schnell et al. 2002, Tomita et al. 2004).

In PSD95, both SH3 and GK domains bind to and promote clustering of the Kainate-type ionotropic glutamate receptors (Garcia et al. 1998). These domains also exhibit intramolecular SH3-GK self-binding (McGee & Bredt 1999), suggesting the possibility that SH3 domains in PSD95 family proteins bind with partners outside of the archetypical SH3 interactions with proline-rich motifs. A prominent binding partner of the GK domain is the SAPAP family of scaffolding proteins (Kim et al. 1997, Satoh et al. 1997, Takeuchi et al. 1997). An additional partner to the GK domain of PSD95 is SPAR (Spine-associated RapGAP). This protein regulates spine morphology and displays actin-reorganization activity (Pak et al. 2001).

DIVERSITY OF HUMAN PSYCHIATRIC DISORDERS

Autism spectrum disorders (ASDs): a group of neurodevelopmental disorders sharing similar core features such as social impairments, language or communication defects, and repetitive behaviors. Examples include autism, Rett syndrome, and Asperger syndrome.

Obsessive-compulsive spectrum disorders: a group of psychiatric disorders sharing core features such as recurrent obsessions, increased anxiety, and compulsive repetitive behaviors. Examples include obsessive-compulsive disorder (OCD), compulsive hair-pulling/Trichotillomania (TTM), Tourette syndrome, and body dysmorphic disorder.

Mood disorders: a group of psychiatric disorders in which the primary symptom is extreme disturbance in mood, such as experiencing either a limited or exaggerated range of feelings. The most prominent examples include major depression and bipolar disorder.

Schizophrenia: a psychiatric disorder characterized by a pervasive disruption in the normal balance of thought and emotion. Symptoms can be divided into three clusters: positive symptoms (hallucinations, delusions, or disorganized speech and/or thoughts), negative symptoms (lack of pleasure or lack of affect), and cognitive symptoms (attention or working memory deficits).

Mutational Analysis of PSD95 Family Function In Vivo

Manipulating the expression levels of PSD95 family proteins has yielded several insights into the role these proteins play at the synapse. Overexpression of PSD95 in dissociated neuron cultures and organotypic slices causes an enhancement of AMPAR-mediated, but

GK domain: guanylate kinase-like domain

NMDAR: N-methyl-D-aspartate receptor

AMPAR: α-amino-3-hydroxy-5-methyl-4-isoxazolepropionic acid receptor

Copy number variations (CNVs): submicroscopic unbalanced structural genomic variations ranging from the kilobase to megabase scale, potentially giving rise to increased or decreased gene copy numbers

not NMDAR-mediated synaptic currents (El-Husseini et al. 2000, Schnell et al. 2002). Conversely, knockdown of PSD95 leads to decreased AMPAR-mediated synaptic currents (Ehrlich et al. 2007). Manipulations of PSD93 and SAP102 protein levels led to similar functional alterations (Elias et al. 2006). Furthermore, the in vivo role of PSD95 family members was probed by the analysis of genetically modified mice harboring mutations in these genes. From these, SAP97 mutant mice are less amenable for study using homozygous germline deletion, given that this perturbation results in perinatal lethality (Caruana & Bernstein 2001). By contrast, SAP102 (Cuthbert et al. 2007), PSD93 (McGee et al. 2001), and PSD95 (Beique et al. 2006, Migaud et al. 1998, Yao et al. 2004) mutant animals manifest only subtle phenotypes. Perhaps the most salient findings have come from a distinct line of PSD95 mutant mice that display augmented sensitivity to the locomotor-stimulating effects of cocaine and enhanced cortical-accumbal long-term potentiation but an absence of cocaine-induced behavioral plasticity (Yao et al. 2004). More recently, further research with PSD95 mutant mice revealed that these animals display several behavioral deficits relevant to autism spectrum disorders (ASDs) (Feyder et al. 2010). Nevertheless, the lack of obvious overt synaptic deficits in PSD95 and PSD93 mice suggests functional redundancy and/or compensation among PSD95 family members. To elucidate this, a *tour de force* study achieved a functional ablation of PSD95, PSD93, and SAP102 by combining PSD95/PSD93 double-knockout animals with SAP102 knockdown (Elias et al. 2006). This work illustrated how synaptic specificity and developmental regulation of AMPARs is influenced by PSD95 and PSD93 in nonoverlapping populations of mature synapses, whereas, SAP102 plays an important role at immature synapses. Moreover, SAP102 is upregulated in response to PSD95/PSD93 deletion, thus contributing to the remarkable functional redundancy within this protein family (Elias et al. 2006). More broadly, this work highlights the difficulties encountered in trying to assess the in vivo functional roles of particular proteins when closely related genes are expressed (or become expressed) in partially overlapping cell populations.

Human Molecular Genetics Data for PSD95 Gene Family

Several groups have reported altered levels of PSD95 family proteins in patients afflicted with mood disorders or schizophrenia (Feyissa et al. 2009, Karolewicz et al. 2009, Kristiansen et al. 2006, Toro & Deakin 2005, Toyooka et al. 2002). Although these data hint at the possibility that altered expression of PSD95 family proteins may play a role in human psychiatric disorders, they in no way address whether such changes are an epiphenomena or if they are in some way causative. Nevertheless, further converging evidence has come from studies examining the involvement of the *DLG1-4* genes (*DLG1*/SAP97, *DLG2*/PSD93, *DLG3*/SAP102, *DLG4*/PSD95) in psychiatric disorders. Of particular note is the association of *DLG1* with the 3q29 microdeletion syndrome—a condition characterized by mild-to-moderate mental retardation, dysmorphic facial features, ataxia, and autism (Willatt et al. 2005). This microdeletion leads to the elimination of PAK2 and DLG1, which purportedly underlie both dysmorphic and neurological symptoms (Willatt et al. 2005). Furthermore, *DLG1* copy number variations (CNVs) have been identified in patients diagnosed with schizophrenia (Magri et al. 2010, Sato et al. 2008). *DLG3* has been strongly implicated in X-linked mental retardation (Tarpey et al. 2004), whereas *DLG4* was recently linked to autistic behaviors and schizophrenia (Cheng et al. 2010, Feyder et al. 2010). Together, these multiple lines of evidence support the hypothesis that the various members of the PSD95 family of proteins collectively contribute toward the healthy functioning of the mammalian brain.

AKAP FAMILY

The A-kinase anchoring protein (AKAP) family is comprised of a broad collection of proteins defined by the ability to anchor protein kinase A (PKA) (Wong & Scott 2004). As such, AKAPs serve critical roles in the spatial and temporal regulation of PKA activity and intracellular signaling cascades. The AKAP family members are classified according to this PKA binding ability rather than on sequence similarity; therefore, AKAPs are structurally very diverse.

The *AKAP5* gene encodes AKAP5, commonly known as AKAP79 in humans and AKAP150 in rodents (jointly called AKAP79/150). In the brain, AKAP79/150 is highly enriched in the PSD of excitatory synapses (Carr et al. 1992) by virtue of an N-terminal polybasic membrane-targeting region (Dell'Acqua et al. 1998). AKAP79/150 also contains distinct sequences that mediate anchoring of the protein kinases PKA and PKC and the protein phosphatase PP2B (also called calcineurin) (Carr et al. 1992, Coghlan et al. 1995, Klauck et al. 1996). In addition to the anchoring of these important signaling molecules, AKAP79/150 interacts directly with the SH3 and GK domains of PSD95 and SAP97 (Colledge et al. 2000). Importantly, PSD95 and SAP97 have specific roles in regulating synaptic localization of NMDARs and AMPARs, respectively. Thus, distinct complexes containing AKAP79/150-PSD95-NMDAR and AKAP79/150-SAP97-AMPAR exist within the PSD region of excitatory synapses (Colledge et al. 2000), providing a molecular basis for differential regulation of the major classes of ionotropic glutamate receptors via scaffolding of unique signaling complexes to different target receptors. AKAP79/150 also interacts directly with and functionally regulates a variety of other ion channels and G protein–coupled receptors (Dart & Leyland 2001, Hall et al. 2007, Hoshi et al. 2003, Lin et al. 2010, Oliveria et al. 2007).

An influential early study in cultured hippocampal neurons showed that cell-wide disruption of PKA binding to AKAPs by an inhibitory peptide (Ht31) led to run-down of evoked AMPAR-mediated currents in a manner identical to infusion of a specific PKA inhibitory peptide (Rosenmund et al. 1994). Ht31 infusion also caused long-term reductions in surface AMPAR subunit GluR1 expression in cultured hippocampal neurons, and it occluded long-term depression evoked by electrical stimulation in acute hippocampal slices (Snyder et al. 2005). However, Ht31 broadly interferes with PKA binding to all AKAPs; therefore, subsequent work was necessary to provide specific evidence for the involvement of AKAP79/150 in the modulation of AMPAR function.

Null mutation: a genetic lesion that ablates gene function completely, most commonly through the functional disruption of mRNA or protein production

Two independent studies used an elegant molecular replacement strategy (depletion of the endogenous AKAP79/150 followed by expression of mutant versions) to show convincingly that expression of PP2B-binding-deficient AKAP79/150 (AKAP79/150ΔPP2B) prevented agonist-induced downregulation of AMPAR currents in cultured hippocampal neurons (Hoshi et al. 2005) and abolished NMDAR-dependent long-term depression in hippocampal slices (Jurado et al. 2010). These results fit well with another report showing that overexpression of AKAP79/150ΔPP2B prevented NMDA-triggered AMPAR endocytosis in cultured hippocampal neurons (Bhattacharyya et al. 2009).

Overall, the anchoring of PKA and PP2B through AKAP79/150 in the PSD region seems to exert influences on AMPAR function and plasticity. These data are largely consistent with the hypothesis that AMPARs are dynamically regulated by phosphorylation and dephosphorylation of GluR1 subunits, mediated by a functional balance of signaling from AKAP79/150-anchored PKA versus PP2B near the receptor substrates. Exactly how anchoring to AKAP79/150 influences PKA and PP2B activities in this context and the relative importance of each to discrete synaptic functions is an open question.

Mutational Analysis of AKAP79/150 Function In Vivo

Two independent laboratories have recently generated *AKAP150* null mice (Hall et al. 2007, Tunquist et al. 2008), providing ample opportunities to investigate the

Knock-in mutation: targeted manipulation in the mouse genome aimed at substituting an endogenous sequence with an altered sequence

physiological functions of this scaffolding protein. Both null mouse lines are viable and fertile, and both groups demonstrated that AKAP150 is the major AKAP in the brain responsible for proper anchoring of PKA within dendritic regions, consistent with the PSD localization of AKAP150. One line also has deficits in motor coordination and strength, consistent with the expression of AKAP150 in the cerebellum (Tunquist et al. 2008).

A third AKAP150 mutant mouse line harboring a knock-in mutation has also been generated by introducing a premature stop codon that results in the deletion of the last 36 amino acids from the C terminus of the AKAP150 protein and fully eliminates PKA anchoring by AKAP150 (i.e., AKAP150ΔPKA), hence the term D36 mice (Lu et al. 2007). D36 mice and *AKAP150* null mice both showed abnormally increased numbers of dendritic spines in vivo and an increased number of functional excitatory synapses in acute hippocampal slices (Lu et al. 2011). These changes are apparent in the early postnatal and juvenile stages but do not persist into adulthood. D36 and *AKAP150* null mice also had larger and more frequent inhibitory synaptic events in acute brain slices from juveniles, which was suggested to be a compensatory change to counteract increased excitatory synaptic function. These findings point to a role of AKAP150-anchored PKA in limiting dendritic spine density in vivo, although these data seem at odds with a portion of earlier results obtained using cultured hippocampal neurons (Robertson et al. 2009). The functions assessed in vivo using mutant mice may have more physiological relevance, although in some cases the potentially confounding influence of compensatory changes may be less of a factor using acute manipulations in vitro.

A surprising finding from multiple studies comparing the *AKAP150* null and D36 mice is that synaptic plasticity and behavioral phenotypes are generally more severe in D36 mice than in the constitutive null mice (Lu et al. 2007, Weisenhaus et al. 2010). For example, long-term potentiation was impaired in young adult D36 mice and long-term depression was impaired in juvenile D36 mice, but no deficits in either form of long-term plasticity were detected in the null mice (but see also Tunquist et al. 2008). Furthermore, reversal learning was impaired in D36 mice but not in null mice. The unique deficits in the D36 mice may partially be explained by the fact that AKAP79/150 normally binds with both PKA and PP2B at synapses; thus, incorporating mutant AKAP150ΔPKA that retains PP2B binding at the PSD may profoundly alter the signaling balance more potently in D36 mice than in the *AKAP150* null mice. The AKAP150ΔPKA deletion also appears to cover the reported binding site for L-type calcium channels in the distal C-terminal portion of AKAP150 (Oliveria et al. 2007), which may further complicate matters in the D36 mice, particularly with respect to the contribution of these channels to postsynaptic calcium entry during synaptic activity and plasticity. Finally, given the recent claim that AKAP150ΔPP2B mutant mice have been established (Sanderson & Dell'Acqua 2011), the detailed characterization of these mutant mice as measured against D36 mice will be of great interest.

Human Molecular Genetics Data on AKAP5

One study reported CNVs in bipolar disorder and schizophrenia cases that mapped to loci containing brain-expressed genes with known roles in neuronal function, including *AKAP5* (Wilson et al. 2006). The copy number increase in *AKAP5* was validated in a single bipolar-disorder sample. A second cohort of 60 samples (15 bipolar disorder, 15 schizophrenia, 15 major depression, and 15 healthy control) was directly tested for CNVs at the identified loci by quantitative PCR. This replication phase revealed three cases with copy number increases in *AKAP5* (one bipolar disorder, one schizophrenia, and one major depression) with no aberrations detected in controls.

A subsequent study called into question the reliability of the high-throughput methodology and provided evidence that the prior study may have generated false-positive CNV results (Sutrala et al. 2007). Two alternative contrasting methodologies were used to test for CNVs in schizophrenia cases for the previously implicated genes. No CNVs in cases or control samples were found for any of the genes examined, including *AKAP5*.

HOMER FAMILY

The Homer family in mammalian species consists of the *Homer1*, *Homer2*, and *Homer3* genes. A wide variety of alternatively spliced transcriptional variants of Homer family members have been described (Shiraishi-Yamaguchi & Furuichi 2007). A short Homer1a form was first identified in the hippocampal brain region as an immediate early gene product that was rapidly and transiently upregulated in neurons in response to seizure (Brakeman et al. 1997). The remaining Homer forms were subsequently identified based on sequence homology with Homer1a and, in particular, by the presence of a conserved N-terminal EVH1 domain found in all family members. Notably, many other family members have a C-terminal coiled-coil domain that is absent in Homer1a; as such, these are referred to as long Homer forms. The predominant long-protein forms isolated from the brain are Homer1b/c, Homer2a/b, and Homer3a/b (Shiraishi-Yamaguchi & Furuichi 2007). The coiled-coil domain mediates multimerization of long Homers into linear tetrameric assemblies in vitro (Hayashi et al. 2006), whereas EVH1 domain mediates interactions with proline-rich motifs. Several important Homer binding proteins have been identified, including group 1 metabotropic glutamate receptors (mGluR1α/mGluR5), IP3 receptors, Ryanodine receptors, TRPC channels, Dynamin3, and Shank proteins (Brakeman et al. 1997; Tu et al. 1998, 1999; Yuan et al. 2003).

The long Homer proteins are found at the PSD of excitatory synapses (Xiao et al. 1998) where they serve as scaffolding proteins linking surface receptors to intracellular signaling pathways, most notably, intracellular calcium signaling (Sala et al. 2005). The multimerization of long Homers into tetramers may be particularly important for linking together a dense matrix of Shanks that form a core structural platform of the PSD specialization (Hayashi et al. 2009). Disruption of tetramerization in neurons using a Homer1b dimeric mutant greatly reduced spine localization of Homer, Shank, and PSD95. Furthermore, these changes correlated with reduced glutamatergic postsynaptic currents, indicating a concerted role of long Homer tetramerization in controlling the structure and function of the postsynaptic compartment. As such, long Homers may be considered the "glue" in the dense Shank network of the PSD, and the tail-to-tail tetrameric arrangement of long Homers with pairs of EVH1 ligand-binding domains at each end can equally well explain an additional role of physically and functionally coupling a range of spatially segregated binding partners in perisynaptic regions.

Synaptic tagging: a hypothetical construct explaining the molecular basis behind conversion of temporary synaptic changes into persistent or "long-term" plasticity at specific synaptic sites

The relationship between the constitutively expressed long Homer forms and activity-inducible Homer1a at the synapse has received much attention. The widely adopted view is that activity-inducible Homer1a may disrupt the assembly of long Homer scaffolding complexes through a competitive EVH1 domain-binding model in response to dynamic neuronal activity. This inferred dominant-negative regulatory mechanism has been demonstrated by direct experimental evidence in a variety of different contexts (Kammermeier 2008; Sala et al. 2001, 2003; Tappe et al. 2006; Tu et al. 1998). Other functional roles for activity-inducible Homer1a at the synapse have also been described (though not mutually exclusive), such as inducing conformation changes in target receptors to influence receptor activity (Ango et al. 2001, Hu et al. 2010), enabling functional crosstalk between metabotropic and ionotropic glutamate receptor classes at the synapse (Bertaso et al. 2010), and synaptic tagging in persistent forms of synaptic plasticity (Okada et al. 2009).

Mutational Analysis of Homer Function In Vivo

Synaptic scaling: a "global" form of homeostatic plasticity where synaptic strength is increased or decreased to counter persistent changes in neuronal activity

Single nucleotide polymorphisms (SNPs): DNA sequence variations at a single nucleotide position within an individual's genome

GKAP: guanylate kinase–associated protein

Homer1 null mice have broad behavioral abnormalities consistent with other animal models of schizophrenia (Szumlinski et al. 2005). Notably, the *Homer1* null allele eliminates both the long and short forms of Homer1 that mediate discrete and, in some cases, opposing functions. This issue largely precludes detailed investigation into the precise roles of Homer1a and Homer1b/c in vivo using these mice. Hu et al. (2010) recently reported a selective Homer1a-deficient mouse and provided convincing evidence that Homer1a is largely indispensable for the induction of homeostatic synaptic scaling. Upregulation of Homer1a facilitated agonist-independent signaling at group 1 mGluRs, which was a requisite step leading to downregulation of synaptic AMPARs.

Both *Homer1* null and *Homer2* null (but not *Homer3* null) mice exhibit behavioral sensitization to the psychostimulant cocaine in the absence of prior cocaine exposure (Szumlinski et al. 2004). Furthermore, the behavioral and neurochemical profiles of *Homer2* null mice closely mirror the numerous changes induced by withdrawal from repeated cocaine administration. Viral expression of Homer2b in the striatum normalized the behaviors of the *Homer2* null mice, thus implicating disruption of striatal Homer2 in enabling cocaine-induced neuroplasticity. How Homer1 and Homer2 are mechanistically coupled to the efficacy of cocaine action in the brain remains unresolved.

Long Homer forms are also expressed at low levels in non-neuronal tissues, and analysis of Homer function in pancreatic acinar cells using *Homer2* and *Homer3* null mice revealed an unexpected role of endogenous Homer2 (but not Homer3) in restricting intracellular calcium oscillations coupled to the activity of G protein–coupled receptors (Shin et al. 2003). The idea of a generalized role for constitutive Homers as buffers of calcium signaling has recently been explored (Worley et al. 2007) and is attractive considering the abundance of binding partners involved in calcium signaling pathways. Such a role may exist in addition to a major scaffolding function, and further work is needed to clarify the relative importance of these functions at excitatory synapses.

Human Molecular Genetics Data on Homers

Evidence on the in vivo roles exerted by Homers at the synapse has led to several hypotheses concerning Homer dysfunction in a wide range of neurological disorders (Szumlinski et al. 2006). In particular, the broad spectrum of generic schizophrenia-like behavioral abnormalities exhibited by *Homer1* null mice have made *Homer1* a good candidate for gene-association studies in schizophrenia. One recent study identified numerous single nucleotide polymorphisms (SNPs) in *Homer* genes, including three variants located in exons (Norton et al. 2003). The evidence for association of a single SNP in *Homer1* with schizophrenia was bordering on statistical significance; however, the authors concluded that *Homers* are most likely not implicated in schizophrenia. Similar nominally significant evidence has suggested linkage of *Homer1* gene variants to major depression (Rietschel et al. 2010), treatment response to antipsychotic drugs in schizophrenia (Spellmann et al. 2011), or *Homer2* gene variants to psychostimulant abuse (Dahl et al. 2005). A large multisite study reported no association of *Homer1* or *Homer2* variants with alcohol dependence (Preuss et al. 2010), which failed to substantiate a hypothesized role of Homer2 in alcohol dependence supported by several prior studies in mice. In all, the available evidence linking Homer variants to psychiatric disorders is tenuous, and the weak evidence for association in small-scale human genetics investigations will require further replication and validation to confirm the suspected links.

SAPAP FAMILY

The SAPAP (also called guanylate kinase–associated protein or GKAP) family is composed of four homologous genes encoding the SAPAP1-4 proteins that are widely yet differentially expressed in the nervous system (Takeuchi et al. 1997, Welch et al. 2004). The SAPAP family was originally identified by a direct interaction with the GK domain of PSD95-family members in yeast two-hybrid screens (Kim et al. 1997, Satoh et al. 1997, Takeuchi et al. 1997). SAPAPs are an abundant component of the PSD (Sheng & Hoogenraad 2007) and interact with a variety of other PSD proteins (Boeckers et al. 1999b, Hirao et al. 2000, Kawabe et al. 1999, Yao et al. 1999), suggesting that SAPAPs are important scaffolding proteins at excitatory synapses.

Mutational Analysis of SAPAP Function In Vivo

SAPAP3 is the only family member strongly expressed in the striatum (Welch et al. 2004), thus offering a unique opportunity to explore the specific function of SAPAP3 at glutamatergic synapses in vivo without potentially confounding effects of functional redundancy arising from other SAPAPs in this brain region. Genetic deletion of *SAPAP3* in mice caused behavioral abnormalities consisting of increased anxiety and compulsive self-grooming to the point of facial hair loss and skin lesions (Welch et al. 2007). These features share similarity with various aspects of core symptoms exhibited by human patients with obsessive-compulsive disorder (OCD), and bare a striking similarity to the phenotypes exhibited by other recently described genetic animal models of OCD-like behaviors (Chen et al. 2010, Shmelkov et al. 2010). Consistent with the localization and predicted function of the SAPAP3 protein, *SAPAP3* null mice also have defects in glutamatergic transmission at cortico-striatal synapses. Remarkably, both synaptic and behavioral defects were rescued by lentivirus-mediated reintroduction of SAPAP3 specifically into the striatum (Welch et al. 2007). This finding establishes the central role of excitatory synaptic dysfunction within cortico-striatal circuitry in the expression of OCD-like behaviors. Additionally, the chronic administration of the selective serotonin reuptake inhibitor (SSRI) fluoxetine successfully alleviated measures of anxiety and compulsive grooming (Welch et al. 2007)—an important distinction given that chronic SSRI treatment is at least partially effective in alleviating symptoms as a first-line treatment in OCD. Thus, the *SAPAP3* null mouse model may serve as a novel tool to identify more effective drugs for the treatment of OCD.

A follow-up study uncovered an altered form of short-term synaptic plasticity expressed at excitatory synapses of striatal medium spiny neurons in acute brain slices from *SAPAP3* null mice (Chen et al. 2011). The mechanism for the anomalous activity-dependent synaptic depression involved a retrograde endocannabinoid signaling pathway through CB1 receptor activation that was engaged under conditions that do not normally activate endocannabinoid signaling in wild-type mice. Further evidence demonstrated the critical involvement of increased group 1 mGluR activity or surface expression as the driving force behind the reduced threshold for engaging endocannabinoid signaling in this experimental paradigm. This study proposes a previously unrecognized role for SAPAP3 in regulating mGluR function in the postsynaptic compartment of excitatory synapses. Further detailed investigation will be required to clarify how this anomalous short-term plasticity at excitatory synapses onto medium spiny neurons in *SAPAP3* null mice may impact synaptic function in vivo and to clarify what implications this has for pinpointing the causal defects underlying compulsive-repetitive behaviors relevant to human OCD. At the synaptic level, the emerging evidence supports the critical involvement of SAPAP3 in controlling both ionotropic (Welch et al. 2007) and metabotropic (Chen et al. 2011) glutamate receptors through a PSD scaffolding role at excitatory synapses.

Human Molecular Genetics Data on SAPAPs

The initial report of OCD-like behaviors in *SAPAP3* null mice has prompted several recent

Paralog: genes that arose by duplication of a gene within the genome; may evolve new functions over the course of evolution

human genetics studies of *SAPAP3* in OCD and obsessive-compulsive spectrum disorders. Zuchner et al. (2009) performed *SAPAP3* gene resequencing analysis in OCD and trichotillomania (TTM), an obsessive-compulsive spectrum disorder, and found an increased frequency of rare nonsynonymous heterozygous *SAPAP3* variants in cases versus controls, thus providing tentative support for a role of *SAPAP3* in OCD and TTM. The majority of the variants represented missense mutations, some of which are predicted to be possibly detrimental to protein function on the basis of bioinformatics analysis. These findings await further validation, including analysis of the functional relevance of these rare *SAPAP3* variants. A second study carried out a relatively large, family-based gene-association study of *SAPAP3* in OCD and grooming disorders (Bienvenu et al. 2009). The preliminary evidence suggests that multiple variations in *SAPAP3* are associated with grooming disorders. No clear association between *SAPAP3* variants and OCD was reported, although grooming disorders without OCD were uncommon in this study, suggesting the possibility that *SAPAP3* variants may be involved in a subtype of OCD involving pathological grooming behaviors. A very recent study of similar design evaluated SAPAP3 as a candidate susceptibility gene in Tourette syndrome, another obsessive-compulsive spectrum disorder, and found a nominally significant association (Crane et al. 2011). A fourth study evaluated SNPs distributed across the SAPAP3 gene to test for association of SAPAP3 variants with TTM and OCD and reported further evidence to link SAPAP3 variants to TTM and early-onset OCD (Boardman et al. 2011). Although the findings of Crane et al. (2011) and Boardman et al. (2011) are represented as supportive of the two earlier studies, these results should be interpreted with caution because statistical correction for multiple testing nullified the nominally significant associations reported in both studies.

Interestingly, in spite of the dearth of evidence on the functional roles of the other SAPAPs, some studies have emerged to suggest involvement of genetic variations in *SAPAP1* and *SAPAP2* in psychiatric disorders. For instance, *SAPAP1* is located in a chromosomal region that was reported to harbor a susceptibility locus for schizophrenia and bipolar disorder (Berrettini et al. 1994, Schwab et al. 1998). This prompted a study to screen for *SAPAP1* mutations in schizophrenia. One SNP was identified in *SAPAP1*, but this SNP was not associated with schizophrenia (Aoyama et al. 2003). In addition, *SAPAP2* was recently identified as one of several novel candidate loci in a large study to search out genome-wide rare CNVs occurring in ASD cases (Pinto et al. 2010). This finding is particularly interesting in light of the demonstrated interaction between SAPAPs and Shank3 (Boeckers et al. 1999b), with strong evidence implicating Shank3 mutations as causative in some ASD cases (Durand et al. 2007, Gauthier et al. 2009, Moessner et al. 2007).

SHANK FAMILY

The SH3 and multiple ankyrin repeat domains (Shank) protein family is coded by three genes (*Shank1-3*) that share a high degree of identity between both paralogs and orthologs. Characterization of this family of genes was initiated by cloning Shank2/CortBP1 (Cortactin binding protein 1) after its identification as a binding partner to Cortactin (Du et al. 1998). Shank1 and Shank3 were subsequently isolated and characterized almost simultaneously by several groups (Boeckers et al. 1999b, Naisbitt et al. 1999, Tu et al. 1999). In the rat brain, the perinatal expression of Shank1-3 is relatively low but rapidly increases during the first weeks of development, peaking at 3–4 weeks (Lim et al. 1999). Expression of Shank1-3 mRNA is prominent in the central nervous system and its protein products are enriched in the PSD (Boeckers et al. 1999a, Lim et al. 1999). Moreover, not only are Shank proteins enriched, they are also some of the earlier elements coalescing at the PSD, predating the arrival of both PSD95 and NMDARs (Boeckers et al. 1999a, Petralia et al. 2005). Finally, the presence of dendritic-targeting elements in the untranslated regions of Shank1 mRNA adds a further level of complexity toward transcript translocation and regulation in neuronal dendrites and spines (Bockers et al. 2004, Falley et al. 2009).

The Shank protein contains several discrete domains including (from N to C terminal) ankyrin repeat domains, one SH3 domain, one PDZ domain, a proline-rich region, and a sterile alpha motif domain (Han et al. 2006, Lim et al. 1999). This abundance of protein-protein interaction domains enables the interaction of Shank with several other synaptic proteins and suggests an important organizational role for these scaffolding proteins. Specifically, Shanks may sit at a convergent point for three independent subcomplexes within the larger PSD. First, Shank proteins interact with the SAPAP family of proteins (Naisbitt et al. 1999); SAPAP then binds to the PSD95 family of proteins, thereby linking ionotropic glutamate receptors to Shank (Naisbitt et al. 1999). Second, the Homer family of proteins is another important Shank binding partner, linking Shanks to metabotropic glutamate receptors and suggesting that Shank proteins may form a molecular bridge between ionotropic and metabotropic glutamate receptors. Third, Shank proteins interact with several partners involved in the regulation of the actin cytoskeleton, including Cortactin (Du et al. 1998, Naisbitt et al. 1999), α-Fodrin (Bockers et al. 2001), and Abp1 (Qualmann et al. 2004). Finally, recent evidence revealed that Shank and Homer may assemble in a macromolecular platform of interleaving Shank3 dimers and Homer tetramers. Owing to the richness of Shank protein-protein interaction domains and binding partners, it is hypothesized that the Shank-Homer matrix plays a pivotal role in the stabilization and organization of the larger PSD (Baron et al. 2006, Hayashi et al. 2009, Tu et al. 1999).

Mutational Analysis of Shank Function In Vivo

Analysis of *Shank1* expression in the rodent brain reveals that Shank1 is highly expressed in cortical regions and the hippocampal formation (Bockers et al. 2004, Peca et al. 2011). *Shank1* null mice exhibit defects in synaptic function and behavioral abnormalities consistent with deficits in hippocampal function and glutamatergic synaptic signaling (Hung et al. 2008). Local abundance of the Shank-interacting proteins Homer and GKAP was reduced at the PSD in mutant animals. Disruption of *Shank1* also led to smaller dendritic spines in hippocampal neurons and a prevalence of thinner PSDs. Furthermore, perturbation of *Shank1* led to a decrease in synaptic strength and a reduction in the frequency of spontaneous postsynaptic excitatory responses, which could be attributed to the presence of spines lacking functional synapses (Hung et al. 2008). At the behavioral level, *Shank1* null mice display an enhanced acquisition of spatial memories but deficiencies in memory retention in the same test. Contextual memory was perturbed in a test of fear conditioning, whereas conditioned response remained intact—again suggesting hippocampal dysfunction (Hung et al. 2008). These defects in spatial and contextual fear memory are consistent with prominent expression of Shank1 in the hippocampus and the proposed role this protein may exert in synaptic and spine maturation (Bockers et al. 2004, Sala et al. 2001). Recent work has attempted to assess if autistic-like phenotypes could be found in Shank1 null mice. These studies showed that, whereas Shank1 mutants display abnormal motor behaviors and communication impairments, reciprocal social interactions in juvenile animals are not impacted (Silverman et al. 2011, Wohr et al. 2011).

Four different groups have independently generated and virtually simultaneously characterized a total of five Shank3 mutant mouse lines (Bangash et al. 2011, Bozdagi et al. 2010, Peca et al. 2011, Wang et al. 2011). Each line was largely aimed at ablating specific exons in the Shank3 gene to induce genetic lesions and perturb expression of Shank3 isoforms. Interestingly, a remarkable amount of converging evidence on the in vivo function of Shank3 was produced. Most notably, all the lines displayed

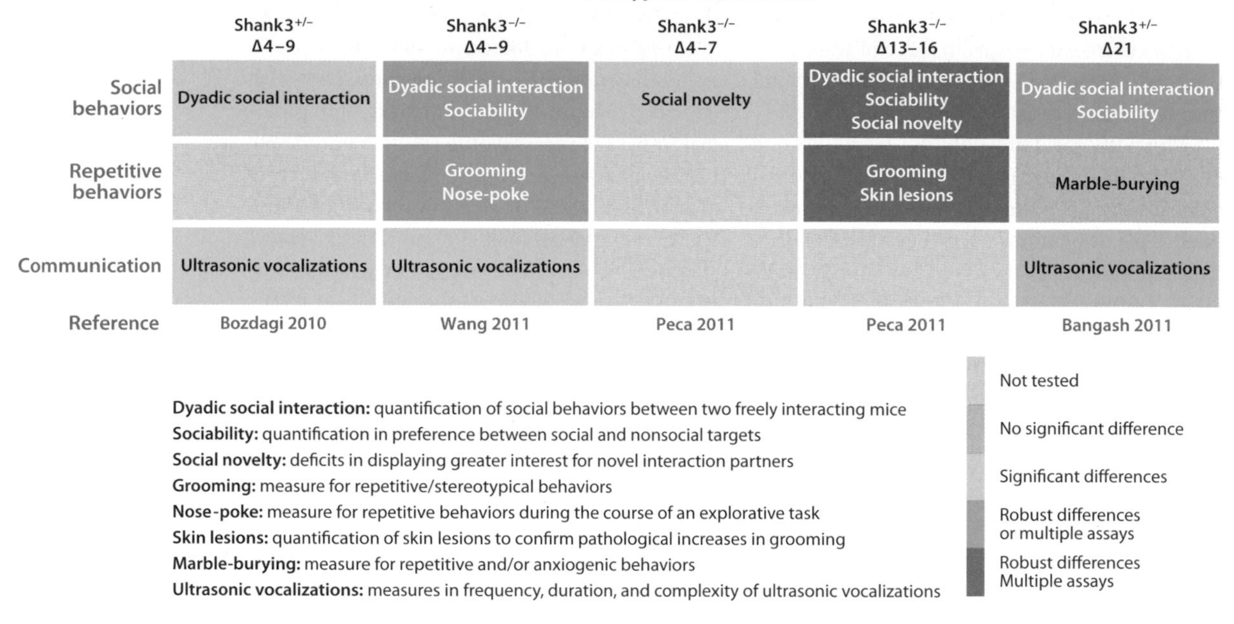

Genotype (deleted exons)
Shank3+/- Δ4–9
Shank3-/- Δ4–9
Shank3-/- Δ4–7
Shank3-/- Δ13–16
Shank3+/- Δ21
Social behaviors
Dyadic social interaction
Dyadic social interaction
Sociability
Social novelty
Dyadic social interaction
Sociability
Social novelty
Dyadic social interaction
Sociability
Repetitive behaviors
Grooming
Nose-poke
Grooming
Skin lesions
Marble-burying
Communication
Ultrasonic vocalizations
Ultrasonic vocalizations
Ultrasonic vocalizations
Reference
Bozdagi 2010
Wang 2011
Peca 2011
Peca 2011
Bangash 2011
Dyadic social interaction: quantification of social behaviors between two freely interacting mice
Sociability: quantification in preference between social and nonsocial targets
Social novelty: deficits in displaying greater interest for novel interaction partners
Grooming: measure for repetitive/stereotypical behaviors
Nose-poke: measure for repetitive behaviors during the course of an explorative task
Skin lesions: quantification of skin lesions to confirm pathological increases in grooming
Marble-burying: measure for repetitive and/or anxiogenic behaviors
Ultrasonic vocalizations: measures in frequency, duration, and complexity of ultrasonic vocalizations
Not tested
No significant difference
Significant differences
Robust differences or multiple assays
Robust differences Multiple assays

Figure 3

Comparison of autistic-like behavioral deficits in five different Shank3 mutant mouse lines.

varying robustness of several forms of behavioral deficiencies relevant to the study of ASDs, such as deficits in social interaction, abnormal vocalization, and compulsive-repetitive behaviors (**Figure 3**). At the cellular level, *Shank3* mutant mice display a pronounced perturbation in synaptic function, more specifically, a decrease in glutamatergic signaling, loss of synaptic strength, or altered synaptic plasticity (Bangash et al. 2011, Bozdagi et al. 2010, Peca et al. 2011, Wang et al. 2011). From these new studies on *Shank3* mutant mice, one study described a *Shank3* genetic lesion that led to a gain-of-function effect through the expression of a form of Shank3 lacking the C-terminal region (Bangash et al. 2011). This mutant protein promotes the recruitment of endogenous full-length Shank3 isoforms and NMDAR subunits for degradation through the proteosomal pathway. This study offered the first insights into a potential mechanistic role played by a discrete set of *Shank3* mutations relevant to *Shank3* mutations in autism (Bangash et al. 2011, Durand et al. 2007). Interestingly, Shank3 and its close interacting partner SAPAP are established targets for ubiquitination at the PSD in response to changing activity levels (Ehlers 2003, Hung et al. 2010). When taking into account that both Shank3 and SAPAP3 mRNA are among the rare transcripts found in dendrites (Peca et al. 2011, Welch et al. 2004), it is tempting to speculate on the importance of rapid bidirectional control of dendritic translation and synaptic localization of both Shank3 and SAPAP3. Moreover, Shank3 and SAPAP3 are both highly expressed in striatal tissue, and the disruption of either gene leads to defects in cortico-striatal synaptic function; thus, these molecular partners may functionally converge on a common pathway in the brain. Dysfunction of this brain circuitry seems to be crucial in the expression and/or gating of compulsive-repetitive behaviors that represent a core feature of both ASDs and obsessive-compulsive spectrum disorder (Peca et al. 2011, Welch et al. 2007).

Human Molecular Genetics Data on Shanks

Phelan-McDermid syndrome (PMS) is a genetic condition characterized in part by delayed or absence of speech and language and a high incidence of autistic behaviors in afflicted children (Phelan et al. 2001). A genetic lesion in the terminal region of human chromosome 22 has been identified in PMS, and in most 22q13 microdeletions a large number of genes, including *Shank3*, are ablated. However, from the multiple genes disrupted in PMS patients, only *Shank3* has been strongly associated with the major neurological complications arising from 22q13 chromosomal aberrations (Bonaglia et al. 2011, Delahaye et al. 2009, Wilson et al. 2003). Also in support of this view, minimal deletions in 22q13.33 that still affect *Shank3* promote the full range of PMS symptomatology, whereas ring chromosome aberrations or 22q13.33 microdeletions that leave *Shank3* intact do not (Jeffries et al. 2005, Misceo et al. 2011). Importantly, mutations in Shank3, including microdeletions, nonsense mutations, and recurrent break points, are found in ASD patients diagnosed outside of PMS, thereby strongly suggesting that a monogenic form of ASDs can be triggered by perturbing this postsynaptic protein (Durand et al. 2007, Gauthier et al. 2009, Moessner et al. 2007). Finally, Shank3 has also been linked with a potential role in the development of schizophrenia (Gauthier et al. 2010).

More recently, CNVs have been proposed to account for a substantial percentage of genetic lesions in nonsyndromic ASD cases (Beaudet 2007, Sebat et al. 2007). CNVs affecting *Shank2* and *SAPAP2* have also been identified in patients affected with ASDs or mental retardation, again suggesting a role for these families of genes in psychiatric disorders (Berkel et al. 2010, Pinto et al. 2010).

SUMMARY

Deciphering Structural and Functional Roles of Postsynaptic Scaffolding Proteins at the Synapse

The recent findings we highlight stress the dynamic and evolving view of the PSD, with emphasis here on the roles of the postsynaptic scaffolding proteins in this specialized structure. By harnessing a multitude of biochemical, molecular, electrophysiological, and behavioral methodologies, researchers in this field are methodically unraveling the precise functions subserved by individual scaffolding proteins. The application of mouse genetic engineering in recent years has especially facilitated major advancements in our knowledge of the in vivo functions carried out by the major scaffolding protein families through analysis of both loss-of-function and gain-of-function mutations. These mutant mice have collectively provided convincing confirmation of physiological functions previously demonstrated only in vitro and have led to new discoveries that have allowed us to refine and/or reinterpret the existing models (e.g., defining the functional redundancy among PSD95 family proteins; uncovering a putative calcium buffering role of Homers in neuronal and non-neuronal tissue). Despite the overt complexity of the postsynaptic compartment at excitatory synapses, the once seemingly insurmountable task of a complete molecular-genetic functional dissection of the major PSD components is emphatically feasible.

Integration of Human and Mouse Genetics to Elucidate Gene Function in Health and Disease

It has sparked great interest that a growing number of genetically modified mice harboring mutations in distinct postsynaptic scaffolding proteins exhibit behavioral phenotypes that are reminiscent of specific human psychiatric disorders. In some instances the discoveries were fortuitous, whereas in other cases the mutant mice were created with foreknowledge of the gene having been implicated in disease susceptibility or causality. Although no animal model can fully recapitulate all the core features of a particular complex human psychiatric disorder, each animal model may express a subset of core features that is easily quantifiable and amenable to detailed mechanistic investigation at a level that is not possible in humans. Thus, detailed multilevel analysis of the functional consequences of gene mutations in animal models is indispensable for searching out gene function in both health and disease, as exemplified here by the work concerning the in vivo functional roles of postsynaptic scaffolding proteins.

In considering the recent work on *SAPAP3* null mice and *Shank3* mutant mice, the proposed relevance of the mutant mouse phenotypes to a human disorder has been strengthened by complementary human genetics data linking variations in the gene (or regions harboring the gene of interest) to the same disorder or related disorders in humans (e.g., *SAPAP3* and obsessive-compulsive spectrum disorders, *Shank3* and autism-spectrum disorders). Although such findings can be viewed as strongly supportive, it is crucial to point out that evidence from human genetics studies supporting the association of a particular gene variation with a human psychiatric disorder does not establish causality of that gene, but instead establishes the overrepresentation of that particular gene variation with the diseased state. Genetic-association studies in psychiatric disorders leave open the mechanism(s) by which specific genetic variations perturb gene function and the impact of these alterations on neuronal and brain circuitry function. As a concluding note, it is valuable to expand the view beyond the relatively narrow scope of postsynaptic scaffolding proteins, as a flurry of recent human genetics studies have implicated a broad spectrum of genes related to synaptic function as contributing to susceptibility in human mental health disorders (Gilman et al. 2011, Gratacos et al. 2009, Hamdan et al. 2011, Piton et al. 2011, Voineagu et al. 2011). Deciphering causal genetic variants will undoubtedly be a monumental task that will keep our attention firmly focused on the remarkable structural and functional complexity of the synapse.

FUTURE ISSUES

1. Human genetics studies are identifying disease-linked genetic variants at an overwhelming pace. Defining which genetic variants are benign and which are pathological represents a major goal in translational neuroscience.
2. Engineering genetically modified mice with disease-relevant mutations will greatly facilitate this effort.
3. In creating new genetic mouse models, researchers should consider a variety of strategies, including but not limited to the following: null alleles, knock-in alleles, alleles with specific gene CNVs, and chromosomal aberrations. The most appropriate design will depend on the unique goals of each study.
4. Delineating cell-type-specific functions of PSD scaffolding proteins in vivo using molecular genetics tools will be of exceptional value to dissecting the circuitry basis of behavior.

DISCLOSURE STATEMENT

The authors are not aware of any affiliations, memberships, funding, or financial holdings that might be perceived as affecting the objectivity of this review.

ACKNOWLEDGMENTS

We apologize to those whose research could not be appropriately cited owing to space limitations. We thank Pedro Martins and Ana Cardoso at BioVision Studio for graphic design consultation and digital artwork. We also thank Dr. Tanya Daigle and Dr. Holly Robertson for critical comments on the manuscript. G.F. acknowledges support from The Poitras Center for Affective Disorders Research, a grant from the U.S. National Institute of Mental Health (R01MH081201), a Hartwell Individual Biomedical Research Award from The Hartwell Foundation, and a Simons Foundation Autism Research Initiative (SFARI) Award. J.T.T. is supported by a Young Investigator Award from NARSAD: The Brain and Behavior Research Foundation and a Ruth L. Kirschstein National Research Service Award from the U.S. National Institute of Mental Health (F32MH084460). J.P. is funded by an Autism Speaks Translational Postdoctoral Fellowship (7649).

LITERATURE CITED

Ango F, Prezeau L, Muller T, Tu JC, Xiao B, et al. 2001. Agonist-independent activation of metabotropic glutamate receptors by the intracellular protein Homer. *Nature* 411:962–65

Aoyama S, Shirakawa O, Ono H, Hashimoto T, Kajimoto Y, Maeda K. 2003. Mutation and association analysis of the DAP-1 gene with schizophrenia. *Psychiatry Clin. Neurosci.* 57:545–47

Bangash MA, Park JM, Melnikova T, Wang D, Jeon SK, et al. 2011. Enhanced polyubiquitination of Shank3 and NMDA receptor in a mouse model of autism. *Cell* 145:758–72

Baron MK, Boeckers TM, Vaida B, Faham S, Gingery M, et al. 2006. An architectural framework that may lie at the core of the postsynaptic density. *Science* 311:531–35

Beaudet AL. 2007. Autism: highly heritable but not inherited. *Nat. Med.* 13:534–36

Beique JC, Lin DT, Kang MG, Aizawa H, Takamiya K, Huganir RL. 2006. Synapse-specific regulation of AMPA receptor function by PSD-95. *Proc. Natl. Acad. Sci. USA* 103:19535–40

Berkel S, Marshall CR, Weiss B, Howe J, Roeth R, et al. 2010. Mutations in the SHANK2 synaptic scaffolding gene in autism spectrum disorders and mental retardation. *Nat. Genet.* 42:489–91

Berrettini WH, Ferraro TN, Goldin LR, Weeks DE, Detera-Wadleigh S, et al. 1994. Chromosome 18 DNA markers and manic-depressive illness: evidence for a susceptibility gene. *Proc. Natl. Acad. Sci. USA* 91:5918–21

Bertaso F, Roussignol G, Worley P, Bockaert J, Fagni L, Ango F. 2010. Homer1a-dependent crosstalk between NMDA and metabotropic glutamate receptors in mouse neurons. *PLoS One* 5:e9755

Bhattacharyya S, Biou V, Xu W, Schluter O, Malenka RC. 2009. A critical role for PSD-95/AKAP interactions in endocytosis of synaptic AMPA receptors. *Nat. Neurosci.* 12:172–81

Bienvenu OJ, Wang Y, Shugart YY, Welch JM, Grados MA, et al. 2009. Sapap3 and pathological grooming in humans: results from the OCD collaborative genetics study. *Am. J. Med. Genet. B* 150B:710–20

Boardman L, van der Merwe L, Lochner C, Kinnear CJ, Seedat S, et al. 2011. Investigating SAPAP3 variants in the etiology of obsessive-compulsive disorder and trichotillomania in the South African white population. *Compr. Psychiatry* 52:181–87

Bockers TM, Mameza MG, Kreutz MR, Bockmann J, Weise C, et al. 2001. Synaptic scaffolding proteins in rat brain. Ankyrin repeats of the multidomain Shank protein family interact with the cytoskeletal protein alpha-fodrin. *J. Biol. Chem.* 276:40104–12

Böckers TM, Segger-Junius M, Iglauer P, Bockmann J, Gundelfinger ED, et al. 2004. Differential expression and dendritic transcript localization of Shank family members: identification of a dendritic targeting element in the 3′ untranslated region of Shank1 mRNA. *Mol. Cell Neurosci.* 26:182–90

Boeckers TM, Kreutz MR, Winter C, Zuschratter W, Smalla KH, et al. 1999a. Proline-rich synapse-associated protein-1/cortactin binding protein 1 (ProSAP1/CortBP1) is a PDZ-domain protein highly enriched in the postsynaptic density. *J. Neurosci.* 19:6506–18

Boeckers TM, Winter C, Smalla KH, Kreutz MR, Bockmann J, et al. 1999b. Proline-rich synapse-associated proteins ProSAP1 and ProSAP2 interact with synaptic proteins of the SAPAP/GKAP family. *Biochem. Biophys. Res. Commun.* 264:247–52

Bonaglia MC, Giorda R, Beri S, De Agostini C, Novara F, et al. 2011. Molecular mechanisms generating and stabilizing terminal 22q13 deletions in 44 subjects with Phelan/McDermid syndrome. *PLoS Genet.* 7:e1002173

Bozdagi O, Sakurai T, Papapetrou D, Wang X, Dickstein DL, et al. 2010. Haploinsufficiency of the autism-associated Shank3 gene leads to deficits in synaptic function, social interaction, and social communication. *Mol. Autism* 1:15

Brakeman PR, Lanahan AA, O'Brien R, Roche K, Barnes CA, et al. 1997. Homer: a protein that selectively binds metabotropic glutamate receptors. *Nature* 386:284–88

Carr DW, Stofko-Hahn RE, Fraser ID, Cone RD, Scott JD. 1992. Localization of the cAMP-dependent protein kinase to the postsynaptic densities by A-kinase anchoring proteins. Characterization of AKAP 79. *J. Biol. Chem.* 267:16816–23

Caruana G, Bernstein A. 2001. Craniofacial dysmorphogenesis including cleft palate in mice with an insertional mutation in the discs large gene. *Mol. Cell. Biol.* 21:1475–83

Chen L, Chetkovich DM, Petralia RS, Sweeney NT, Kawasaki Y, et al. 2000. Stargazin regulates synaptic targeting of AMPA receptors by two distinct mechanisms. *Nature* 408:936–43

Chen M, Wan Y, Ade K, Ting J, Feng G, Calakos N. 2011. Sapap3 deletion anomalously activates short-term endocannabinoid-mediated synaptic plasticity. *J. Neurosci.* 31:9563–73

Chen SK, Tvrdik P, Peden E, Cho S, Wu S, et al. 2010. Hematopoietic origin of pathological grooming in Hoxb8 mutant mice. *Cell* 141:775–85

Chen X, Winters C, Azzam R, Li X, Galbraith JA, et al. 2008. Organization of the core structure of the postsynaptic density. *Proc. Natl. Acad. Sci. USA* 105(11):4453–58

Cheng MC, Lu CL, Luu SU, Tsai HM, Hsu SH, et al. 2010. Genetic and functional analysis of the DLG4 gene encoding the post-synaptic density protein 95 in schizophrenia. *PLoS One* 5:e15107

Coghlan VM, Perrino BA, Howard M, Langeberg LK, Hicks JB, et al. 1995. Association of protein kinase A and protein phosphatase 2B with a common anchoring protein. *Science* 267:108–11

Colledge M, Dean RA, Scott GK, Langeberg LK, Huganir RL, Scott JD. 2000. Targeting of PKA to glutamate receptors through a MAGUK-AKAP complex. *Neuron* 27:107–19

Crane J, Fagerness J, Osiecki L, Gunnell B, Stewart SE, et al. 2011. Family-based genetic association study of DLGAP3 in Tourette syndrome. *Am. J. Med. Genet. B* 156B:108–14

Cuthbert PC, Stanford LE, Coba MP, Ainge JA, Fink AE, et al. 2007. Synapse-associated protein 102/dlgh3 couples the NMDA receptor to specific plasticity pathways and learning strategies. *J. Neurosci.* 27:2673–82

Dahl JP, Kampman KM, Oslin DW, Weller AE, Lohoff FW, et al. 2005. Association of a polymorphism in the Homer1 gene with cocaine dependence in an African American population. *Psychiatr. Genet.* 15:277–83

Dart C, Leyland ML. 2001. Targeting of an A kinase-anchoring protein, AKAP79, to an inwardly rectifying potassium channel, Kir2.1. *J. Biol. Chem.* 276:20499–505

Delahaye A, Toutain A, Aboura A, Dupont C, Tabet AC, et al. 2009. Chromosome 22q13.3 deletion syndrome with a de novo interstitial 22q13.3 cryptic deletion disrupting SHANK3. *Eur. J. Med. Genet.* 52:328–32

Dell'Acqua ML, Faux MC, Thorburn J, Thorburn A, Scott JD. 1998. Membrane-targeting sequences on AKAP79 bind phosphatidylinositol-4, 5-bisphosphate. *EMBO J.* 17:2246–60

Du Y, Weed SA, Xiong WC, Marshall TD, Parsons JT. 1998. Identification of a novel cortactin SH3 domain-binding protein and its localization to growth cones of cultured neurons. *Mol. Cell. Biol.* 18:5838–51

Durand CM, Betancur C, Boeckers TM, Bockmann J, Chaste P, et al. 2007. Mutations in the gene encoding the synaptic scaffolding protein SHANK3 are associated with autism spectrum disorders. *Nat. Genet.* 39:25–27

Ehlers MD. 2003. Activity level controls postsynaptic composition and signaling via the ubiquitin-proteasome system. *Nat. Neurosci.* 6:231–42

Ehrlich I, Klein M, Rumpel S, Malinow R. 2007. PSD-95 is required for activity-driven synapse stabilization. *Proc. Natl. Acad. Sci. USA* 104:4176–81

El-Husseini AE, Schnell E, Chetkovich DM, Nicoll RA, Bredt DS. 2000. PSD-95 involvement in maturation of excitatory synapses. *Science* 290:1364–68

Elias GM, Funke L, Stein V, Grant SG, Bredt DS, Nicoll RA. 2006. Synapse-specific and developmentally regulated targeting of AMPA receptors by a family of MAGUK scaffolding proteins. *Neuron* 52:307–20

Falley K, Schutt J, Iglauer P, Menke K, Maas C, et al. 2009. Shank1 mRNA: dendritic transport by kinesin and translational control by the 5′ untranslated region. *Traffic* 10:844–57

Feyder M, Karlsson RM, Mathur P, Lyman M, Bock R, et al. 2010. Association of mouse Dlg4 (PSD-95) gene deletion and human DLG4 gene variation with phenotypes relevant to autism spectrum disorders and Williams' syndrome. *Am. J. Psychiatry* 167:1508–17

Feyissa AM, Chandran A, Stockmeier CA, Karolewicz B. 2009. Reduced levels of NR2A and NR2B subunits of NMDA receptor and PSD-95 in the prefrontal cortex in major depression. *Prog. Neuropsychopharmacol. Biol. Psychiatry* 33:70–75

Garcia EP, Mehta S, Blair LAC, Wells DG, Shang J, et al. 1998. SAP90 binds and clusters kainate receptors causing incomplete desensitization. *Neuron* 21:727–39

Gauthier J, Champagne N, Lafreniere RG, Xiong L, Spiegelman D, et al. 2010. De novo mutations in the gene encoding the synaptic scaffolding protein SHANK3 in patients ascertained for schizophrenia. *Proc. Natl. Acad. Sci. USA* 107:7863–68

Gauthier J, Spiegelman D, Piton A, Lafreniere RG, Laurent S, et al. 2009. Novel de novo SHANK3 mutation in autistic patients. *Am. J. Med. Genet. B* 150B:421–24

Gilman SR, Iossifov I, Levy D, Ronemus M, Wigler M, Vitkup D. 2011. Rare de novo variants associated with autism implicate a large functional network of genes involved in formation and function of synapses. *Neuron* 70:898–907

Gratacos M, Costas J, de Cid R, Bayes M, Gonzalez JR, et al. 2009. Identification of new putative susceptibility genes for several psychiatric disorders by association analysis of regulatory and non-synonymous SNPs of 306 genes involved in neurotransmission and neurodevelopment. *Am. J. Med. Genet. B* 150B:808–16

Hall DD, Davare MA, Shi M, Allen ML, Weisenhaus M, et al. 2007. Critical role of cAMP-dependent protein kinase anchoring to the L-type calcium channel Cav1.2 via A-kinase anchor protein 150 in neurons. *Biochemistry* 46:1635–46

Hamdan FF, Gauthier J, Araki Y, Lin DT, Yoshizawa Y, et al. 2011. Excess of de novo deleterious mutations in genes associated with glutamatergic systems in nonsyndromic intellectual disability. *Am. J. Hum. Genet.* 88:306–16

Han W, Kim KH, Jo MJ, Lee JH, Yang J, et al. 2006. Shank2 associates with and regulates Na^+/H^+ exchanger 3. *J. Biol. Chem.* 281:1461–9

Hayashi MK, Ames HM, Hayashi Y. 2006. Tetrameric hub structure of postsynaptic scaffolding protein homer. *J. Neurosci.* 26:8492–501

Hayashi MK, Tang C, Verpelli C, Narayanan R, Stearns MH, et al. 2009. The postsynaptic density proteins Homer and Shank form a polymeric network structure. *Cell* 137:159–71

Hirao K, Hata Y, Yao I, Deguchi M, Kawabe H, et al. 2000. Three isoforms of synaptic scaffolding molecule and their characterization. Multimerization between the isoforms and their interaction with N-methyl-D-aspartate receptors and SAP90/PSD-95-associated protein. *J. Biol. Chem.* 275:2966–72

Hoshi N, Langeberg LK, Scott JD. 2005. Distinct enzyme combinations in AKAP signalling complexes permit functional diversity. *Nat. Cell Biol.* 7:1066–73

Hoshi N, Zhang JS, Omaki M, Takeuchi T, Yokoyama S, et al. 2003. AKAP150 signaling complex promotes suppression of the M-current by muscarinic agonists. *Nat. Neurosci.* 6:564–71

Hu JH, Park JM, Park S, Xiao B, Dehoff MH, et al. 2010. Homeostatic scaling requires group I mGluR activation mediated by Homer1a. *Neuron* 68:1128–42

Hung AY, Futai K, Sala C, Valtschanoff JG, Ryu J, et al. 2008. Smaller dendritic spines, weaker synaptic transmission, but enhanced spatial learning in mice lacking Shank1. *J. Neurosci.* 28:1697–708

Hung AY, Sung CC, Brito IL, Sheng M. 2010. Degradation of postsynaptic scaffold GKAP and regulation of dendritic spine morphology by the TRIM3 ubiquitin ligase in rat hippocampal neurons. *PLoS One* 5:e9842

Irie M, Hata Y, Takeuchi M, Ichtchenko K, Toyoda A, et al. 1997. Binding of neuroligins to PSD-95. *Science.* 277:1511–15

Jeffries AR, Curran S, Elmslie F, Sharma A, Wenger S, et al. 2005. Molecular and phenotypic characterization of ring chromosome 22. *Am. J. Med. Genet. A* 137:139–47

Jurado S, Biou V, Malenka RC. 2010. A calcineurin/AKAP complex is required for NMDA receptor-dependent long-term depression. *Nat. Neurosci.* 13:1053–55

Kammermeier PJ. 2008. Endogenous homer proteins regulate metabotropic glutamate receptor signaling in neurons. *J. Neurosci.* 28:8560–67

Karolewicz B, Szebeni K, Gilmore T, Maciag D, Stockmeier CA, Ordway GA. 2009. Elevated levels of NR2A and PSD-95 in the lateral amygdala in depression. *Int. J. Neuropsychopharmacol.* 12:143–53

Kawabe H, Hata Y, Takeuchi M, Ide N, Mizoguchi A, Takai Y. 1999. nArgBP2, a novel neural member of ponsin/ArgBP2/vinexin family that interacts with synapse-associated protein 90/postsynaptic density-95-associated protein (SAPAP). *J. Biol. Chem.* 274:30914–18

Kim E, Cho KO, Rothschild A, Sheng M. 1996. Heteromultimerization and NMDA receptor-clustering activity of Chapsyn-110, a member of the PSD-95 family of proteins. *Neuron* 17:103–13

Kim E, Naisbitt S, Hsueh YP, Rao A, Rothschild A, et al. 1997. GKAP, a novel synaptic protein that interacts with the guanylate kinase-like domain of the PSD-95/SAP90 family of channel clustering molecules. *J. Cell Biol.* 136:669–78

Kim E, Niethammer M, Rothschild A, Jan YN, Sheng M. 1995. Clustering of Shaker-type K^+ channels by interaction with a family of membrane-associated guanylate kinases. *Nature* 378:85–88

Kim E, Sheng M. 2004. PDZ domain proteins of synapses. *Nat. Rev. Neurosci.* 5:771–81

Klauck TM, Faux MC, Labudda K, Langeberg LK, Jaken S, Scott JD. 1996. Coordination of three signaling enzymes by AKAP79, a mammalian scaffold protein. *Science* 271:1589–92

Kornau HC, Schenker LT, Kennedy MB, Seeburg PH. 1995. Domain interaction between NMDA receptor subunits and the postsynaptic density protein PSD-95. *Science* 269:1737–40

Kristiansen LV, Beneyto M, Haroutunian V, Meador-Woodruff JH. 2006. Changes in NMDA receptor subunits and interacting PSD proteins in dorsolateral prefrontal and anterior cingulate cortex indicate abnormal regional expression in schizophrenia. *Mol. Psychiatry* 11:737–47

Kuhlendahl S, Spangenberg O, Konrad M, Kim E, Garner CC. 1998. Functional analysis of the guanylate kinase-like domain in the synapse-associated protein SAP97. *Eur. J. Biochem.* 252:305–13

Leonard AS, Davare MA, Horne MC, Garner CC, Hell JW. 1998. SAP97 is associated with the alpha-amino-3-hydroxy-5-methylisoxazole-4-propionic acid receptor GluR1 subunit. *J. Biol. Chem.* 273:19518–24

Lim S, Naisbitt S, Yoon J, Hwang JI, Suh PG, et al. 1999. Characterization of the Shank family of synaptic proteins. Multiple genes, alternative splicing, and differential expression in brain and development. *J. Biol. Chem.* 274:29510–18

Lin L, Sun W, Wikenheiser AM, Kung F, Hoffman DA. 2010. KChIP4a regulates Kv4.2 channel trafficking through PKA phosphorylation. *Mol. Cell Neurosci.* 43:315–25

Lu Y, Allen M, Halt AR, Weisenhaus M, Dallapiazza RF, et al. 2007. Age-dependent requirement of AKAP150-anchored PKA and GluR2-lacking AMPA receptors in LTP. *EMBO J.* 26:4879–90

Lu Y, Zha XM, Kim EY, Schachtele S, Dailey ME, et al. 2011. A kinase anchor protein 150 (AKAP150)-associated protein kinase a limits dendritic spine density. *J. Biol. Chem.* 286:26496–506

Magri C, Sacchetti E, Traversa M, Valsecchi P, Gardella R, et al. 2010. New copy number variations in schizophrenia. *PLoS One* 5:e13422

McGee AW, Bredt DS. 1999. Identification of an intramolecular interaction between the SH3 and guanylate kinase domains of PSD-95. *J. Biol. Chem.* 274:17431–36

McGee AW, Topinka JR, Hashimoto K, Petralia RS, Kakizawa S, et al. 2001. PSD-93 knock-out mice reveal that neuronal MAGUKs are not required for development or function of parallel fiber synapses in cerebellum. *J. Neurosci.* 21:3085–91

Migaud M, Charlesworth P, Dempster M, Webster LC, Watabe AM, et al. 1998. Enhanced long-term potentiation and impaired learning in mice with mutant postsynaptic density-95 protein. *Nature* 396:433–39

Misceo D, Rodningen OK, Baroy T, Sorte H, Mellembakken JR, et al. 2011. A translocation between Xq21.33 and 22q13.33 causes an intragenic SHANK3 deletion in a woman with Phelan-McDermid syndrome and hypergonadotropic hypogonadism. *Am. J. Med. Genet. A* 155A:403–8

Moessner R, Marshall CR, Sutcliffe JS, Skaug J, Pinto D, et al. 2007. Contribution of SHANK3 mutations to autism spectrum disorders. *Am. J. Hum. Genet.* 81:1289–97

Naisbitt S, Kim E, Tu JC, Xiao B, Sala C, et al. 1999. Shank, a novel family of postsynaptic density proteins that binds to the NMDA receptor/PSD-95/GKAP complex and cortactin. *Neuron* 23:569–82

Niethammer M, Kim E, Sheng M. 1996. Interaction between the C terminus of NMDA receptor subunits and multiple members of the PSD-95 family of membrane-associated guanylate kinases. *J. Neurosci.* 16:2157–63

Norton N, Williams HJ, Williams NM, Spurlock G, Zammit S, et al. 2003. Mutation screening of the Homer gene family and association analysis in schizophrenia. *Am. J. Med. Genet. B* 120B:18–21

Okada D, Ozawa F, Inokuchi K. 2009. Input-specific spine entry of soma-derived Vesl-1S protein conforms to synaptic tagging. *Science* 324:904–9

Oliveria SF, Dell'Acqua ML, Sather WA. 2007. AKAP79/150 anchoring of calcineurin controls neuronal L-type Ca^{2+} channel activity and nuclear signaling. *Neuron* 55:261–75

Pak DT, Yang S, Rudolph-Correia S, Kim E, Sheng M. 2001. Regulation of dendritic spine morphology by SPAR, a PSD-95-associated RapGAP. *Neuron* 31:289–303

Peca J, Feliciano C, Ting JT, Wang W, Wells MF, et al. 2011. Shank3 mutant mice display autistic-like behaviours and striatal dysfunction. *Nature* 472:437–42

Petralia RS, Sans N, Wang YX, Wenthold RJ. 2005. Ontogeny of postsynaptic density proteins at glutamatergic synapses. *Mol. Cell Neurosci.* 29:436–52

Phelan MC, Rogers RC, Saul RA, Stapleton GA, Sweet K, et al. 2001. 22q13 deletion syndrome. *Am. J. Med. Genet.* 101:91–99

Pinto D, Pagnamenta AT, Klei L, Anney R, Merico D, et al. 2010. Functional impact of global rare copy number variation in autism spectrum disorders. *Nature* 466:368–72

Piton A, Gauthier J, Hamdan FF, Lafreniere RG, Yang Y, et al. 2011. Systematic resequencing of X-chromosome synaptic genes in autism spectrum disorders and schizophrenia. *Mol. Psychiatry* 16:867–80

Preuss UW, Ridinger M, Rujescu D, Fehr C, Koller G, et al. 2010. No association of alcohol dependence with HOMER 1 and 2 genetic variants. *Am. J. Med. Genet. B* 153B:1102–9

Qualmann B, Boeckers TM, Jeromin M, Gundelfinger ED, Kessels MM. 2004. Linkage of the actin cytoskeleton to the postsynaptic density via direct interactions of Abp1 with the ProSAP/Shank family. *J. Neurosci.* 24:2481–95

Rietschel M, Mattheisen M, Frank J, Treutlein J, Degenhardt F, et al. 2010. Genome-wide association, replication, and neuroimaging study implicates HOMER1 in the etiology of major depression. *Biol. Psychiatry* 68:578–85

Robertson HR, Gibson ES, Benke TA, Dell'Acqua ML. 2009. Regulation of postsynaptic structure and function by an A-kinase anchoring protein-membrane-associated guanylate kinase scaffolding complex. *J. Neurosci.* 29:7929–43

Rosenmund C, Carr DW, Bergeson SE, Nilaver G, Scott JD, Westbrook GL. 1994. Anchoring of protein kinase A is required for modulation of AMPA/kainate receptors on hippocampal neurons. *Nature* 368:853–56

Sala C, Futai K, Yamamoto K, Worley PF, Hayashi Y, Sheng M. 2003. Inhibition of dendritic spine morphogenesis and synaptic transmission by activity-inducible protein Homer1a. *J. Neurosci.* 23:6327–37

Sala C, Piech V, Wilson NR, Passafaro M, Liu G, Sheng M. 2001. Regulation of dendritic spine morphology and synaptic function by Shank and Homer. *Neuron* 31:115–30

Sala C, Roussignol G, Meldolesi J, Fagni L. 2005. Key role of the postsynaptic density scaffold proteins Shank and Homer in the functional architecture of Ca^{2+} homeostasis at dendritic spines in hippocampal neurons. *J. Neurosci.* 25:4587–92

Sampedro MN, Bussineau CM, Cotman CW. 1981. Postsynaptic density antigens: preparation and characterization of an antiserum against postsynaptic densities. *J. Cell Biol.* 90:675–86

Sanderson JL, Dell'Acqua ML. 2011. AKAP signaling complexes in regulation of excitatory synaptic plasticity. *Neuroscientist* 17:321–36

Sato J, Shimazu D, Yamamoto N, Nishikawa T. 2008. An association analysis of synapse-associated protein 97 (SAP97) gene in schizophrenia. *J. Neural Transm.* 115:1355–65

Satoh K, Yanai H, Senda T, Kohu K, Nakamura T, et al. 1997. DAP-1, a novel protein that interacts with the guanylate kinase-like domains of hDLG and PSD-95. *Genes Cells* 2:415–24

Schnell E, Sizemore M, Karimzadegan S, Chen L, Bredt DS, Nicoll RA. 2002. Direct interactions between PSD-95 and stargazin control synaptic AMPA receptor number. *Proc. Natl. Acad. Sci. USA* 99:13902–7

Schwab SG, Hallmayer J, Lerer B, Albus M, Borrmann M, et al. 1998. Support for a chromosome 18p locus conferring susceptibility to functional psychoses in families with schizophrenia, by association and linkage analysis. *Am. J. Hum. Genet.* 63:1139–52

Sebat J, Lakshmi B, Malhotra D, Troge J, Lese-Martin C, et al. 2007. Strong association of de novo copy number mutations with autism. *Science* 316:445–49

Sheng M, Hoogenraad CC. 2007. The postsynaptic architecture of excitatory synapses: a more quantitative view. *Annu. Rev. Biochem.* 76:823–47

Sheng M, Sala C. 2001. PDZ domains and the organization of supramolecular complexes. *Annu. Rev. Neurosci.* 24:1–29

Shin DM, Dehoff M, Luo X, Kang SH, Tu J, et al. 2003. Homer 2 tunes G protein–coupled receptors stimulus intensity by regulating RGS proteins and PLCbeta GAP activities. *J. Cell Biol.* 162:293–303

Shiraishi-Yamaguchi Y, Furuichi T. 2007. The Homer family proteins. *Genome Biol.* 8:206

Shmelkov SV, Hormigo A, Jing D, Proenca CC, Bath KG, et al. 2010. Slitrk5 deficiency impairs corticostriatal circuitry and leads to obsessive-compulsive-like behaviors in mice. *Nat. Med.* 16:598–602, 1p following

Silverman JL, Turner SM, Barkan CL, Tolu SS, Saxena R, et al. 2011. Sociability and motor functions in Shank1 mutant mice. *Brain Res.* 1380:120–37

Snyder EM, Colledge M, Crozier RA, Chen WS, Scott JD, Bear MF. 2005. Role for A kinase-anchoring proteins (AKAPS) in glutamate receptor trafficking and long term synaptic depression. *J. Biol. Chem.* 280:16962–68

Spellmann I, Rujescu D, Musil R, Mayr A, Giegling I, et al. 2011. Homer-1 polymorphisms are associated with psychopathology and response to treatment in schizophrenic patients. *J. Psychiatr. Res.* 45:234–41

Sutrala SR, Goossens D, Williams NM, Heyrman L, Adolfsson R, et al. 2007. Gene copy number variation in schizophrenia. *Schizophr. Res.* 96:93–99

Szumlinski KK, Dehoff MH, Kang SH, Frys KA, Lominac KD, et al. 2004. Homer proteins regulate sensitivity to cocaine. *Neuron* 43:401–13

Szumlinski KK, Kalivas PW, Worley PF. 2006. Homer proteins: implications for neuropsychiatric disorders. *Curr. Opin. Neurobiol.* 16:251–57

Szumlinski KK, Lominac KD, Kleschen MJ, Oleson EB, Dehoff MH, et al. 2005. Behavioral and neurochemical phenotyping of Homer1 mutant mice: possible relevance to schizophrenia. *Genes Brain Behav.* 4:273–88

Takeuchi M, Hata Y, Hirao K, Toyoda A, Irie M, Takai Y. 1997. SAPAPs. A family of PSD-95/SAP90-associated proteins localized at postsynaptic density. *J. Biol. Chem.* 272:11943–51

Tappe A, Klugmann M, Luo C, Hirlinger D, Agarwal N, et al. 2006. Synaptic scaffolding protein Homer1a protects against chronic inflammatory pain. *Nat. Med.* 12:677–81

Tarpey P, Parnau J, Blow M, Woffendin H, Bignell G, et al. 2004. Mutations in the DLG3 gene cause nonsyndromic X-linked mental retardation. *Am. J. Hum. Genet.* 75:318–24

Tomita S, Fukata M, Nicoll RA, Bredt DS. 2004. Dynamic interaction of stargazin-like TARPs with cycling AMPA receptors at synapses. *Science* 303:1508–11

Toro C, Deakin JF. 2005. NMDA receptor subunit NRI and postsynaptic protein PSD-95 in hippocampus and orbitofrontal cortex in schizophrenia and mood disorder. *Schizophr. Res.* 80:323–30

Toyooka K, Iritani S, Makifuchi T, Shirakawa O, Kitamura N, et al. 2002. Selective reduction of a PDZ protein, SAP-97, in the prefrontal cortex of patients with chronic schizophrenia. *J. Neurochem.* 83:797–806

Tu JC, Xiao B, Naisbitt S, Yuan JP, Petralia RS, et al. 1999. Coupling of mGluR/Homer and PSD-95 complexes by the Shank family of postsynaptic density proteins. *Neuron* 23:583–92

Tu JC, Xiao B, Yuan JP, Lanahan AA, Leoffert K, et al. 1998. Homer binds a novel proline-rich motif and links group 1 metabotropic glutamate receptors with IP3 receptors. *Neuron* 21:717–26

Tunquist BJ, Hoshi N, Guire ES, Zhang F, Mullendorff K, et al. 2008. Loss of AKAP150 perturbs distinct neuronal processes in mice. *Proc. Natl. Acad. Sci. USA* 105:12557–62

Voineagu I, Wang X, Johnston P, Lowe JK, Tian Y, et al. 2011. Transcriptomic analysis of autistic brain reveals convergent molecular pathology. *Nature* 474:380–84

Wang X, McCoy PA, Rodriguiz RM, Pan Y, Je HS, et al. 2011. Synaptic dysfunction and abnormal behaviors in mice lacking major isoforms of Shank3. *Hum. Mol. Genet.* 20:3093–108

Weisenhaus M, Allen ML, Yang L, Lu Y, Nichols CB, et al. 2010. Mutations in AKAP5 disrupt dendritic signaling complexes and lead to electrophysiological and behavioral phenotypes in mice. *PLoS One* 5:e10325

Welch JM, Lu J, Rodriguiz RM, Trotta NC, Peca J, et al. 2007. Cortico-striatal synaptic defects and OCD-like behaviours in Sapap3-mutant mice. *Nature* 448:894–900

Welch JM, Wang D, Feng G. 2004. Differential mRNA expression and protein localization of the SAP90/PSD-95-associated proteins (SAPAPs) in the nervous system of the mouse. *J. Comp. Neurol.* 472:24–39

Willatt L, Cox J, Barber J, Cabanas ED, Collins A, et al. 2005. 3q29 microdeletion syndrome: clinical and molecular characterization of a new syndrome. *Am. J. Hum. Genet.* 77:154–60

Wilson GM, Flibotte S, Chopra V, Melnyk BL, Honer WG, Holt RA. 2006. DNA copy-number analysis in bipolar disorder and schizophrenia reveals aberrations in genes involved in glutamate signaling. *Hum. Mol. Genet.* 15:743–49

Wilson HL, Wong AC, Shaw SR, Tse WY, Stapleton GA, et al. 2003. Molecular characterisation of the 22q13 deletion syndrome supports the role of haploinsufficiency of SHANK3/PROSAP2 in the major neurological symptoms. *J. Med. Genet.* 40:575–84

Wohr M, Roullet FI, Hung AY, Sheng M, Crawley JN. 2011. Communication impairments in mice lacking Shank1: reduced levels of ultrasonic vocalizations and scent marking behavior. *PLoS One* 6:e20631

Wong W, Scott JD. 2004. AKAP signalling complexes: focal points in space and time. *Nat. Rev. Mol. Cell. Biol.* 5:959–70

Woods DF, Bryant PJ. 1991. The discs-large tumor suppressor gene of Drosophila encodes a guanylate kinase homolog localized at septate junctions. *Cell* 66:451–64

Worley PF, Zeng W, Huang G, Kim JY, Shin DM, et al. 2007. Homer proteins in Ca^{2+} signaling by excitable and non-excitable cells. *Cell Calcium* 42:363–71

Xiao B, Tu JC, Petralia RS, Yuan JP, Doan A, et al. 1998. Homer regulates the association of group 1 metabotropic glutamate receptors with multivalent complexes of homer-related, synaptic proteins. *Neuron* 21:707–16

Yao I, Hata Y, Hirao K, Deguchi M, Ide N, et al. 1999. Synamon, a novel neuronal protein interacting with synapse-associated protein 90/postsynaptic density-95-associated protein. *J. Biol. Chem.* 274:27463–66

Yao WD, Gainetdinov RR, Arbuckle MI, Sotnikova TD, Cyr M, et al. 2004. Identification of PSD-95 as a regulator of dopamine-mediated synaptic and behavioral plasticity. *Neuron* 41:625–38

Yuan JP, Kiselyov K, Shin DM, Chen J, Shcheynikov N, et al. 2003. Homer binds TRPC family channels and is required for gating of TRPC1 by IP3 receptors. *Cell* 114:777–89

Zuchner S, Wendland JR, Ashley-Koch AE, Collins AL, Tran-Viet KN, et al. 2009. Multiple rare SAPAP3 missense variants in trichotillomania and OCD. *Mol. Psychiatry* 14:6–9

RELATED RESOURCES

SynapseWeb. Kristen M. Harris, PI. **http://synapses.clm.utexas.edu/index.asp**

Genes to Cognition. **http://www.g2conline.org/**

Simons Foundation Autism Research Initiative. **https://sfari.org/**

SFARI GENE. **http://gene.sfari.org/**

Schizophrenia Research Forum Online. **http://www.schizophreniaforum.org/**

The Attention System of the Human Brain: 20 Years After

Steven E. Petersen[1] and Michael I. Posner[2]

[1]School of Medicine, Washington University in St. Louis, St. Louis, Missouri 63110; email: sep@npg.wustl.edu

[2]Department of Psychology, University of Oregon, Eugene, Oregon 97403-1227; email: mposner@uoregon.edu

Annu. Rev. Neurosci. 2012. 35:73–89

First published online as a Review in Advance on April 12, 2012

The *Annual Review of Neuroscience* is online at neuro.annualreviews.org

This article's doi:
10.1146/annurev-neuro-062111-150525

0147-006X/12/0721-0073$20.00

Keywords

alerting network, executive network, orienting network, cingulo-opercular network, frontoparietal network

Abstract

Here, we update our 1990 *Annual Review of Neuroscience* article, "The Attention System of the Human Brain." The framework presented in the original article has helped to integrate behavioral, systems, cellular, and molecular approaches to common problems in attention research. Our framework has been both elaborated and expanded in subsequent years. Research on orienting and executive functions has supported the addition of new networks of brain regions. Developmental studies have shown important changes in control systems between infancy and childhood. In some cases, evidence has supported the role of specific genetic variations, often in conjunction with experience, that account for some of the individual differences in the efficiency of attentional networks. The findings have led to increased understanding of aspects of pathology and to some new interventions.

Contents

INTRODUCTION

Twenty years ago, when neuroimaging was in its infancy, we summarized the current state of knowledge on attention in the 1990 volume of the *Annual Review of Neuroscience* (Posner & Petersen 1990). At that time, most available evidence was from behavioral studies of normal adults or patients with varying forms of brain injury. However, the ability to image brain activity with positron emission tomography seemed to hold great promise for the physiological analysis of mental processes, including attention. In our review, we were able to integrate findings of the initial imaging studies. We never imagined that the growth of cognitive neuroscience over the subsequent 20 years would make it possible to revisit our analysis, with 4,000–6,000 imaging papers on attention or cognitive control and nearly 3,500 citations of our original review.

The original review suggested three basic concepts about the attention system. The first is that the attention system is anatomically separate from processing systems, which handle incoming stimuli, make decisions, and produce outputs. We emphasized the sources of the attentional influences, not the many processing systems that could be affected by attention. The second concept is that attention utilizes a network of anatomical areas. The third is that these anatomical areas carry out different functions that can be specified in cognitive terms. The most unique aspect of our original article, which separated it from the many excellent summaries of the attention literature appearing in the *Annual Review of Neuroscience* in the years since, is the discrete anatomical basis of the attention system: divided into three networks, each representing a different set of attentional processes. We believe that these important concepts are still operative. Here, we try to update the framework of our earlier *Annual Review* article [other summaries are available in Posner (2012a,b)].

In this review, we outline some of the major advances related to our framework that have taken place in the past 20 years. First, we reintroduce the three original networks of the attention system. We examine the nature of these networks and how the ideas related to them have evolved. The second part of the article explores additions to the original conception. Two new networks are proposed with their functional descriptions, and new methods for understanding interactions between them. The third part of the article indicates how the ideas have been extended to related topics, for example, in tying genetic variations to individual differences in network efficiency and in examining the development of attention in childhood.

THE ORIGINAL NETWORKS

The three networks we described in 1990 included an alerting network, which focused on brain stem arousal systems along with right hemisphere systems related to sustained vigilance; an orienting network focused on, among other regions, parietal cortex; and an executive network, which included midline frontal/anterior cingulate cortex. Each of these networks is explored below.

Alerting

The concept of arousal goes back to the classic work of Moruzzi & Magoun (1949) on the role of the brain stem reticular system in maintaining alertness (**Figure 1**, for macaque brain). As more became known of the neuromodulatory systems of the brain stem and thalamus, it was necessary to qualify the general concept of arousal into more differentiated components. Within cognitive psychology, a major emphasis has been on producing and maintaining optimal vigilance and performance during tasks; this is the sense of alertness that we discussed in our 1990 article.

One approach to the study of alerting is to use a warning signal prior to a target event to produce a phasic change in alertness. The warning cue leads to replacing the resting state with a new state that involves preparation for detecting and responding to an expected signal. If a speeded response is required to the target, reaction time improves following a warning. This improvement is not due to the buildup of more accurate information about the target, which is not changed by the warning signal, but the warning signal does change the speed of orienting attention and thus responding to the signal.

Several other methods have been used to study tonic alertness. These include changes over the course of the day (circadian rhythm). Reaction times are usually longer in the early morning and decline over the course of the day only to rise again during the night and peak in the early morning (Posner 1975). These measures reflect other diurnal changes such as body temperature and cortisol secretion. A long established approach to tonic alertness is to use a long and usually rather boring task to measure sustained vigilance. Some of these tasks have grown out of the job of radar operators looking for near-threshold changes over long periods of time. Vigilance tasks rely heavily on mechanisms of the right cerebral cortex (Posner & Petersen 1990). Both classical lesion data and more recent imaging data confirm that tonic alertness is heavily lateralized to the right hemisphere.

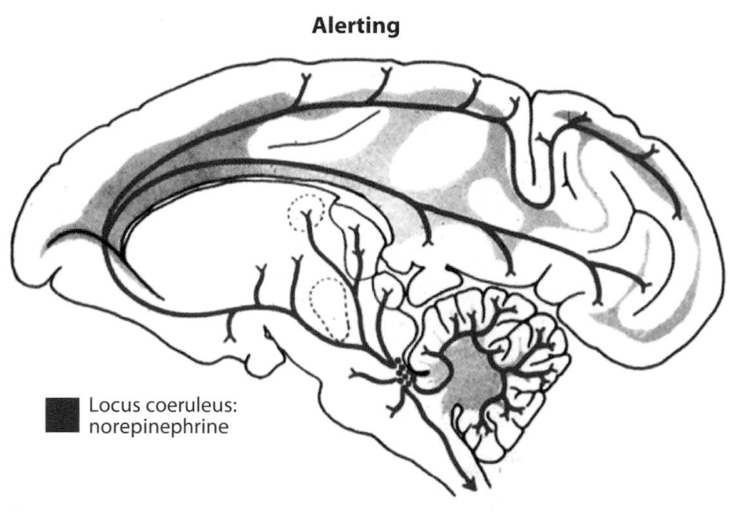

Figure 1

The locus coeruleus projections of the alerting system shown on a macaque brain. The diffuse connections interact with other, more strongly localized systems. The alerting system also includes regions of the frontal and parietal cortices (not shown). Reproduced from Aston-Jones & Cohen (2005).

Orienting

The orienting network is focused on the ability to prioritize sensory input by selecting a modality or location. Although the arguments in the original review included discussion of the pulvinar and the superior colliculus, most of our focus was on visual selection and on the parietal cortex as part of a posterior attention system (**Figure 2*a***). Consensus in the imaging literature now indicates that frontal as well as posterior areas are involved in orienting. For example, human and animal studies have implicated the frontal eye fields (FEF) in this process (Corbetta et al. 1998, Thompson et al. 2005).

In addition, parietal areas have been implicated in related forms of processing. This processing can be concrete as in the specification of directed motor or eye movements (Lindner et al. 2010) or more abstract as "movements" across a number line (Hubbard et al. 2005). In fact, the specificity of parietal regions in terms of sensory versus motor processing is a major point of contention. Nonetheless, most would agree the functions of the parietal lobe are not restricted to orienting to sensory stimuli but involve other related processes.

Executive

In our original article, the third major system was presented under the heading of target

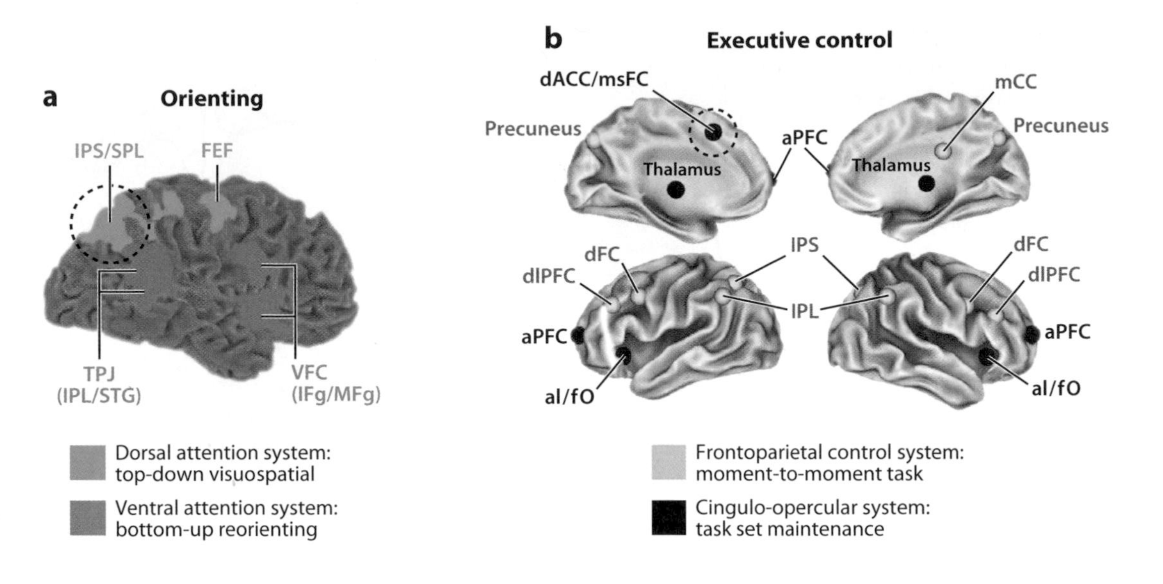

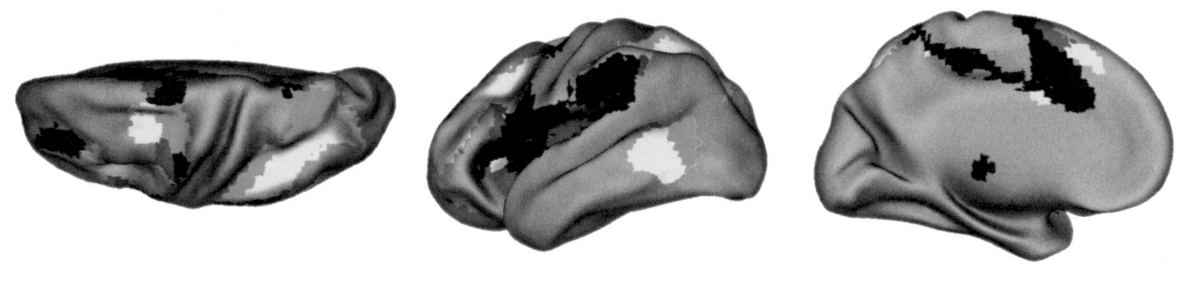

c
Grouping of regions using resting state functional connectivity MRI

Figure 2

(*a*) The dorsal and ventral orienting networks (after Corbetta & Shulman 2002). The dorsal attention network (*light green*) consists of frontal eye fields (FEF) and the intraparietal sulcus/superior parietal lobe (IPS/SPL). The ventral attention network (*teal*) consists of regions in the temporoparietal junction (TPJ) and the ventral frontal cortex (VFC). (*b*) Two networks of the executive control system. The circled region indicates the original member of the executive control system from Posner & Petersen (1990). The remaining regions come from the elaboration of the original cingulo-opercular system (*black*) and the addition of the frontoparietal system (*yellow*) (adapted from Dosenbach et al. 2007). (*c*) Resting-state correlation reflecting separate control systems. The figure illustrates three views of the brain (*left*, dorsal view; *middle*, tilted lateral view; *right*, medial view). These separable resting networks are consistent with the distinctions based on functional criteria exhibited in panels *a* and *b*: dorsal attention (*green*), ventral attention (*teal*), cingulo-opercular (*black*), frontoparietal (*yellow*) (adapted from Power et al. 2011).

detection. The main reason for this was not that target detection itself is a major attentional process, but that the moment of target detection captures awareness in a very specific way. Although it is possible to monitor for targets in many processing streams without too much difficulty, the moment of target detection produces interference across the system, slowing detection of another target (Duncan 1980). This set of processes is related to the limited capacity of the attention system, and to awareness itself, and has often been called focal attention. One might think of focal attention as the entry to the conscious state, which may involve widespread connections from the midline cortex and the anterior cingulate cortex (ACC) (**Figure 2*b***) to produce the global work space frequently associated with consciousness (Dehaene & Changeux 2011). We associated target detection and awareness of the target with the medial frontal cortex and the adjacent ACC. This brain region has been highly studied by imaging experiments partly because of its frequent activation.

Although one of us (S.P.) has vacillated significantly on this original idea over the past 20 years, it seems that the idea is still relevant. One of the reasons is that the ACC and related regions have been reliably activated when there is conflict [e.g., a requirement to withhold a dominant response to perform a subdominant response (Botvinick et al. 2001)]. The argument has been extended to include a role for these areas in the regulation of both cognition and emotion (Bush et al. 2000).

The most compelling argument for a focal attention explanation comes from the activity found in the medial frontal/anterior cingulate in such diverse operations as perception of either physical (Rainville et al. 1997) or social (Eisenberger et al. 2003) pain, processing of reward (Hampton & O'Doherty 2007), monitoring or resolution of conflict (Botvinick et al. 2001), error detection (Dehaene et al. 1994), and theory of mind (Kampe et al. 2003). These different demands all activate this region, in most cases in conjunction with the anterior insula. Some investigators advocate a separate role for the system for each of the comparisons above (e.g., as part of a pain or reward system), but, as we argue below, we support a more comprehensive view that captures more of the results, including focal attention and the regulation of processing networks. Since the original article, this network has also taken on an even more extensive role in executive control on the basis of findings showing multiple top-down control signals in these regions. This more complex functional and anatomical network is discussed in the executive control section below.

ELABORATIONS OF THE FRAMEWORK

The intervening 20 years since our original article have produced a surprising amount of support for the basic outlines of the framework described above. There has also been a significant amount of elaboration or evolution of the ideas during that timeframe. The next three sections review some of the studies deepening or expanding our understanding of the original networks.

Alerting

Our understanding of the physiology and pharmacology underlying the alerting system has changed significantly. For example, strong evidence relates the neuromodulator norepinephrine (NE) to the alerting system. A warning signal is accompanied by activity in the locus coeruleus, the source of NE (Aston-Jones & Cohen 2005). Warning-signal effects can be blocked by drugs such as guanfacine and clonodine, which decrease NE release (Marrocco & Davidson 1998). Drugs that increase NE release can also enhance the warning-signal effect. The NE pathway includes major nodes in the frontal cortex and in parietal areas relating to the dorsal but not the ventral visual pathways (Morrison & Foote 1986).

To examine the specificity of these effects to the warning signal, researchers used a cued detection task with humans, monkeys, and rats (Beane & Marrocco 2004, Marrocco & Davidson 1998) to separate information about where a target will occur (orienting) from when it will occur (alerting). To accomplish this, one of four cue conditions was presented prior to a target for a rapid response. By subtracting a double cue condition, where the participant is informed of when a target will occur but not where, from a no cue condition, they receive a specific measure of the warning influence of the signal. When the cue that indicates the target's location is subtracted from an alerting cue, the difference represents effects of orienting. Results of drug studies with humans and monkeys show that NE release influences alerting effects, whereas drugs influencing the neuromodulator acetylcholine (Ach) affect orienting but not alerting. Studies have shown that individual differences in alerting and orienting are largely uncorrelated (Fan et al.

2002) and that orienting improves to the same degree with a cue regardless of the level of alertness. These results suggest a great deal of independence between these two functions (Fernandez-Duque & Posner 1997). However, these systems usually work together in most real-world situations, when a single event often provides information on both when and where a target will occur (Fan et al. 2009).

The changes during the time between warning and target reflect a suppression of ongoing activity thought to prepare the system for a rapid response. In the central nervous system there is a negative shift in scalp-recorded EEG, known as the contingent negative variation (CNV) (Walter 1964), which often begins with the warning signal and may remain present until the target presentation. This negative potential appears to arise in part from the anterior cingulate and adjacent structures (Nagai et al. 2004) and may overlap the event-related response to the warning stimulus. The negative shift may remain present as a standing wave over the parietal area of the contralateral hemisphere (Harter & Guido 1980). If the target interval is predictable, the person may not show the CNV until just prior to target presentation.

An extensive imaging study (Sturm & Willmes 2001) showed that a largely common right hemisphere and thalamic set of areas are involved in both phasic and tonic alerting. Another imaging study, however, suggested that the warning signal effects rely more strongly on left cerebral hemisphere mechanisms (Coull et al. 2000, Fan et al. 2005). This could represent the common findings described above on hemispheric differences in which right lateralized processes often involve slower effects (tonic), whereas left hemisphere mechanisms are more likely to be involved with higher temporal (phasic) or spatial frequencies (Ivry & Robertson 1997). The exact reasons for differences in laterality found with tonic and phasic studies are still unknown.

Orienting

In a series of imaging experiments using cuing methodology in combination with event-related fMRI, Corbetta & Shulman (2002) showed that two brain systems are related to orienting to external stimuli as illustrated in **Figure 2*a***. A more dorsal system including the FEFs and the interparietal sulcus followed presentation of an arrow cue and was identified with rapid strategic control over attention. When the target was miscued, subjects had to break their focus of attention on the cued location and switch to the target location. The switch appeared to involve the temporoparietal junction (TPJ) and the ventral frontal cortex and was identified with the interrupt signal that allowed the switch to occur.

The dorsal system included the well-studied parietal regions but added a small set of frontal locations as well, particularly in the FEFs. Some have argued that covert attention shifts are slaved to the saccadic eye movement system (Rizzolatti et al. 1987), and neuroimaging studies using fMRI have shown that covert and overt shifts of attention involve similar areas (Corbetta et al. 1998). However, single-unit physiology studies in the macaque suggest important distinctions at the level of cell populations, with some cells in the FEFs active during saccades and a distinct but overlapping population of cells involved in covert shifts of attention (Schafer & Moore 2007, Thompson et al. 2005). The cells responsible for covert shifts of attention also seem to hold the location of cues during a delay interval (Armstrong et al. 2009). The two populations of cells are mixed within the FEFs and, at least to date, have not been distinguished by fMRI. However, the physiological data indicate that covert attention is distinct from the motor system governing saccades, even though they clearly interact.

As suggested by the FEF studies, it is important to be able to link the imaging and physiological results with other studies to provide more details on local computations. One strategy for doing so is to study the pharmacology of each of the attention networks. Cholinergic systems arising in the basal forebrain appear to play a critical role in orienting; lesions of the basal forebrain in monkeys interfere with orienting attention (Voytko et al. 1994). However, it appears that the site of this effect is not in the basal forebrain per se, but instead involves the superior parietal lobe. Davidson & Marrocco (2000) made injections of scopolamine directly into the lateral intraparietal area of monkeys. This area corresponds to the human superior parietal lobe and contains cells influenced by cues about spatial location. The injections have a large effect on the monkey's ability to shift attention to a target. Systemic injections of scopolamine, an anticholinergic, have a smaller effect on covert orienting of attention than do local injections in the parietal area. Cholinergic drugs do not affect the ability of a warning signal to improve the monkey's performance, so there appears to be a double dissociation, with NE involved mainly in the alerting network and Ach involved in the orienting network. These observations in the monkey have been confirmed by similar studies in the rat (Everitt & Robbins 1997). It is especially significant that comparisons in the rat studies of cholinergic and dopaminergic mechanisms have shown that only the former influence the orienting response (Everitt & Robbins 1997, Stewart et al. 2001).

The more ventral network including the TPJ (**Figure 2*a***) seemed to be more active following the target and was thus identified as part of a network responsive to sensory events. It is strongly right lateralized and lesions in this area are central to the neglect syndrome, although the interaction of TPJ with more frontal and dorsal brain areas is also critical (Shulman & Corbetta 2012). Researchers generally agree about the membership of the major nodes of the orienting network on the basis of both spatial cuing and visual search studies (Hillyard et al. 2004, Wright & Ward 2008).

Perhaps more surprising is that the brain areas involved in orienting to visual stimuli seem to overlap strongly (within fMRI resolution) with those involved with orienting to stimuli in other modalities (Driver et al. 2004). Although attention operates on sensory-specific modalities according to the incoming target, the sources of this effect appear to be common. There are also important synergies between modalities. In many cases, orienting to a location will provide priority not only to the expected modality but also to information present at the same location from other modalities (Driver et al. 2004), indicating how closely the sensory systems are integrated within the orienting network.

How can the sources of the orienting network described above influence sensory computations? Anatomically, the source of the orienting effect lies in the network of parietal, frontal, and subcortical areas mentioned above. However, the influence of attention is on the bottom-up signals arriving in sensory-specific areas: for vision, in the primary visual cortex and extrastriate areas moving forward toward the temporal lobe. That this remote influence involves synchronization between activity in the more dorsal attention areas and activity in the more ventral visual areas is an influential idea (Womelsdorf et al. 2007). The synchronization apparently leads to greater sensitivity in the visual system, allowing faster responses to visual targets and thus improved priority for processing targets.

In addition to synchronization, single-unit physiology studies conducted within ventral visual areas suggest that as items are added to a visual scene they tend to reduce the firing rate of cells responding to the target stimulus. Attention to a target seems to reduce the influence of other competing stimuli. This idea was important in the development of biased competition theory (Desimone & Duncan 1995). This theory sees attention as arising out of a winner-take-all competition within various levels of sensory and association systems. fMRI studies confirm that attention to a stimulus can occur prior to its arrival, changing the baseline neural and blood oxygen level–dependent (BOLD) response, and that the overall BOLD activity is affected in ways consistent with the biased competition theory (Desimone & Duncan 1995).

Executive Control

Our third original network has been elaborated considerably. As noted above, our original focus was on midline regions of the medial frontal cortex and anterior cingulate. We suggested that activity found during the performance of tasks was related to focal attention because trial-related activity in these regions was greater for targets than for nontargets, for conflict more than for nonconflict trials, and for errors more than for correct trials. We argued that such a system might be very useful for producing top-down regulation, thus its relationship to executive control. This role of the ACC in top-down control was based on rather slim evidence at the time but seems to still seems to be accurate and plays an important role in two prominent theories of executive control in the current literature. One theory stresses the role of the ACC in monitoring conflict and in relation to lateral frontal areas in resolving the conflict (Botvinick et al. 2001, Carter & Krug 2012). A different view arguing for two different top-down control networks is based on extensive studies of the specific aspects of the ACC during task performance and correlations with other regions at rest (**Figure 2*b*,*c***) (Dosenbach et al. 2006, 2007).

Support for two separate executive control networks arises from studies designed to discover signals related to top-down task control. Such signals might include those related to task instructions that are transient at the beginning of a task block (**Figure 3**). Transient block transition signals had been seen in earlier work (Donaldson et al. 2001, Fox et al. 2005, Konishi et al. 2001) with many different interpretations. A second type of activity is sustained across the trials of the task, putatively related to the maintenance of task parameters/top-down control (**Figure 3**). The third type of signal is related to performance feedback; an example of such feedback would be systematic differences between correct and incorrect trials (**Figure 3**).

Dosenbach et al. (2006) studied 10 different tasks (including visual and auditory words and visual objects as stimuli, with many different decision criteria, such as semantic, timing, and similarity judgments) searching for evidence of these signal types. Lateral frontal and parietal regions appeared to emphasize transient signals at the beginning of blocks, whereas medial frontal/cingulate cortex and bilateral anterior insula also showed sustained maintenance signals across task conditions. Although these experiments identified a set of regions that could be involved in top-down task level control, they provide no evidence of the relationships between regions.

Another experiment (Dosenbach et al. 2007) looked for functional correlations (at rest) between regions that showed some or all of these putative control signals, with the idea that these "functional connections" may define the systems-level relationships between the regions. Lateral frontal and parietal regions that showed primarily start-cue activity correlated well with each other (**Figure 2*c***). The midline and anterior insular regions that showed additional sustained activity also correlated well with each other (**Figure 2*c***), but these two sets of regions did not correlate strongly with each other.

These results suggested there are two separable executive control networks. Detailed evidence for this view is found in Dosenbach et al. (2008). The frontoparietal network appears to be distinct from the orienting network discussed previously, whereas the cingulo-opercular network overlaps with the original executive network. If this view is correct, there are two relatively separate executive networks. Although the best imaging evidence shows that the orienting and frontoparietal executive networks are separate in adulthood, they may have a common origin in early development (see Self-Regulation section, below).

This breakdown of executive control into two separate networks is anatomically similar to an influential idea pertaining to cognitive control (Botvinick et al. 2001, Carter & Krug 2012). However, this cognitive control view favors a single unified executive system in which lateral prefrontal cortex provides top-down control signals, guided by performance-monitoring signals generated by midline structures. Although the cognitive control view and the ideas shown in **Figure 2** are anatomically similar, several specific functional differences

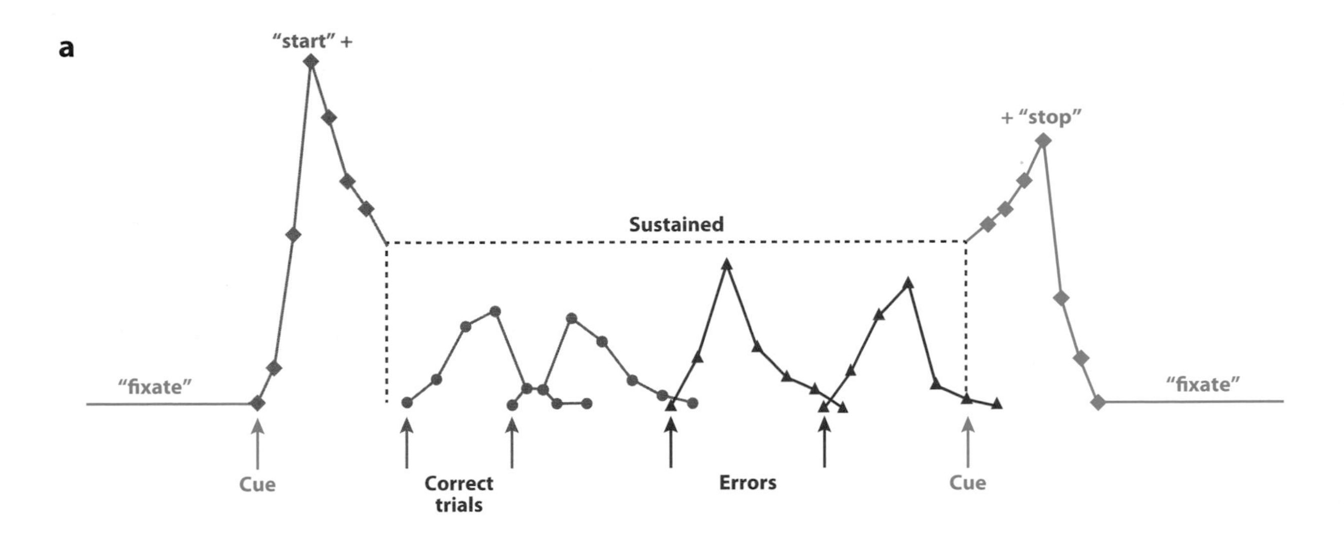

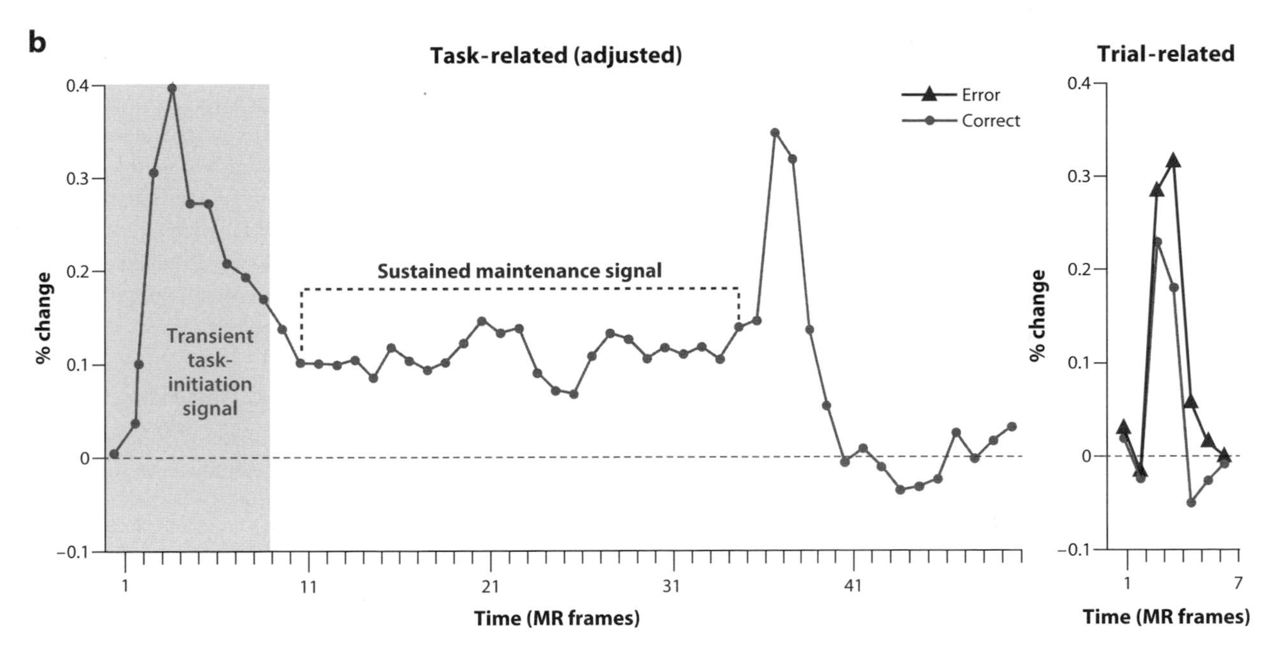

Figure 3

Executive control signals. The top panel shows three putative executive control signals: a task initiation signal in yellow, a task-maintenance signal in red, and activity related to correct (*black*) and error (*blue*) trials (adapted from Dosenbach et al. 2006). Regions showing differences in error versus correct trials are considered to be computing or receiving performance feedback. The bottom figure shows activity in the left anterior insula during a task that contains all the putative signals (plus a transient transition signal at the end of the block of trials). MR, magnetic resonance.

remain. In the dual network view (Dosenbach et al. 2008), the two executive systems act relatively independently in producing top-down control. The cingulo-opercular control system shows maintenance across trials and acts as stable background maintenance for task performance as a whole. The frontoparietal system, in contrast, showing mostly start-cue signals, is thought to relate to task switching and initiation and to adjustments within trials in real time.

Both the cognitive control view and the dual networks view explain a considerable amount of extant data, but we believe there are several reasons to choose the latter formulation.

First, lesion studies in both humans and animals seem to indicate separate aspects of control. Large lesions of the frontal midline often result in akinetic mutism in which people are capable of carrying out goal-directed activities but do not do so. On the other hand, patients with more laterally placed lesions, including those in the dorsolateral prefronal cortex (DLPFC) often exhibit perseverations with an inability to switch from one set to the other. In a compelling set of macaque experiments, Rossi et al. (2007) showed that a complete unilateral resection of the DLPFC and an interruption of the corpus callosum resulted in a unilateral inability to switch sets but an intact ability to adopt a sustained set, consistent with the human lesion data.

A second difference between the dual network and cognitive control views is concerned with the directionality of relationships. The cognitive control view requires a timing difference between the midline monitoring processes and the DLPFC implementation regions within a trial. The two-network account is tolerant of ordering effects because the two networks operate separately. Two quite different sets of data argue that cingulo-opercular involvement is often at the end of or after the trial. The first is from studies of single-unit activity in the ACC in macaques (Ito et al. 2003). During a saccade countermanding task, investigators found neurons that signaled errors and unexpected rewards after trial completion. Second, a recent human imaging study by Ploran et al. (2007) used a slow reveal task. During visual information processing, activity progressively increased with increasing visual information across several seconds in the DLPFC. This preceded late activity in the ACC and anterior insula. These results are consistent with the hypothesis that the ACC may often serve to monitor the consequences of actions, and they are inconsistent with a more rigid directionality.

The addition of two separate orienting networks and two separate executive networks raises the possibility that additional control networks will be elaborated in the future. However, for several reasons, we do not expect the number of control networks to be much larger than the number described here. The study of many complex systems, from ecosystems to protein-protein interactions, seems to indicate that these systems follow a "rule of hand" and have approximately five controlling variables (ranging from three to seven) (Gunderson & Holling 2002). For example, the maintenance of upright balanced posture appears to be controlled by at least three separate systems: vision, the vestibular system, and kinesthetic joint sensors. These systems act relatively independently and have different spatial and temporal characteristics. From this perspective, the presence of five relatively separate attention networks appears reasonable. A second argument in favor of this view is an empirical one. In a recent large-scale study of resting state networks (Power et al. 2011), with effectively all the brain represented, all the cortical networks, found by the more piecemeal approaches described above, are present.

EXTENDING THE FRAMEWORK

One of the gratifying outcomes of our original publication has been the many ways that these ideas inspired a large number of studies. We review extensions of the framework into new areas related to attention networks.

Self-Regulation

The ability to control our thoughts, feelings, and behavior in developmental psychology is called self-regulation; with adults it is often called self-control (see sidebar on Will, Self-Regulation, and Self-Control for further definitions). Neuroimaging presents strong evidence that conflict tasks such as the Stroop effect activate common areas of the anterior cingulate gyrus: the dorsal portion for more strictly cognitive tasks and the ventral area for emotion-related tasks (Botvinick et al. 2001,

WILL, SELF-REGULATION, AND SELF-CONTROL

Several names have been applied to the voluntary control of emotion and cognition. During child development, these functions are often called self-regulation. This name provides a clear contrast to the regulation that occurs through the caregiver or other external sources. In adults, the same set of voluntary functions is frequently called self-control. Regulation may also occur through nonvoluntary means, for example, by fear or by the calming aspects of drugs or therapy. In all cases, self-control or self-regulation appears to be an ability to control reflexive or otherwise dominant responses to select less dominant ones.

Conflict tasks: The Stroop effect involves the conflict between the task of naming the color of ink of conflicting color names (e.g., the word GREEN presented in RED INK). The Stroop and other conflict-related tasks have been used to measure the ability to select the less dominant response. Because the classic Stroop effect requires reading, other conflict tasks such as spatial conflict, flanker conflict, and pictorial conflict have also been used. Imaging studies with adults suggest that the conflict in these tasks have a common anatomy (Fan et al. 2003a).

Anatomy: The use of imaging has provided some evidence of a common brain network that is involved in all these senses of control. This network includes anterior cingulate (Bush et al. 2000) and anterior insula (Dosenbach et al. 2007; Sridharan et al. 2007, 2008) and also includes areas of the prefrontal cortex when inhibition of dominant responses is a strong feature (Fan et al. 2003a). The common involvement of the anterior cingulate in attention and both emotion and cognitive control has provided one basis for the argument that the executive attention network is critical to these various functions. The brain activation of conflict-related tasks such as the Stroop has also been common to studies of attention and aspects of control.

Age: Self-regulation has been a concept used mainly in developmental psychology, whereas the terms cognitive control, self-control, and willpower are usually applied to adults. There appears to be no strict dividing line. A new finding is the important role of the orienting system in providing some of the control in infants and in young children (Posner et al. 2012, Rothbart et al. 2011). Even in adults, no doubt orienting to new sensory stimuli or thoughts can be a self-control mechanism.

Future: The much broader term executive function is applied in psychology to self-control as well as the ability to solve problems, shift tasks, plan ahead, and implement goals. Although conflict resolution has been studied widely with normals, the anatomy of other functions remains to be thoroughly explored.

Bush et al. 2000). Although the cingulate anatomy is much more complex, the division into cognitive and emotion-related areas has been supported by more detailed anatomical studies (Beckmann et al. 2009).

Support for the voluntary exercise of self-regulation comes from studies that examine either the instruction to control affect or the connections involved in the exercise of that control. For example, the instruction to avoid arousal during processing of erotic events (Beauregard et al. 2001) or to ward off emotion when looking at negative pictures (Ochsner et al. 2002) produces a locus of activation in midfrontal and cingulate areas. If people are required to select an input modality, the cingulate shows functional connectivity to the selected sensory system (Crottaz-Herbette & Menon 2006) and in emotional tasks to limbic areas (Etkin et al. 2006).

Both behavioral and resting state functional data suggest substantial development of the executive attention network between infancy and childhood. A study of error detection in seven-month-old infants and adults (Berger et al. 2006) shows that both ages use the anterior cingulate area, but the usual slowing following an error does not seem present until about three years of age (Jones et al. 2003). We recently proposed (Posner et al. 2012, Rothbart et al. 2011) that during infancy control systems depend primarily on the orienting network as described previously. During later childhood and into adulthood, the time to resolve conflict correlated with parent reports of their child's ability to control his or her behavior (effortful control, EC) (Posner & Rothbart 2007, Rothbart et al. 2011). The correlation between conflict scores and parent reports of EC form one basis for the association between self-regulation and executive attention. EC is also related to the empathy that children show toward others and their ability to delay an action as well as to avoid such behaviors as lying or cheating when given the opportunity. High levels of EC and the ability to resolve conflict are related to fewer antisocial behaviors in adolescents (Rothbart 2011). These findings show that self-regulation is a psychological function crucial for child socialization, and they suggest that it can also be studied in terms of specific anatomical areas and their connections by examining the development of executive attention networks.

Differences in Network Efficiency

Although everyone has the attention networks described above, there are also individual differences in the efficiency of all brain networks. The Attention Network Test (ANT) has been used to examine the efficiency of attention networks (Fan et al. 2002). The task requires the person to press one key if the central arrow points to the left and another if it points to the right. Conflict is introduced by having flankers surrounding the target pointing in either the same (congruent) or opposite (incongruent) direction as the target. Cues presented prior to the target provide information on where or when the target will occur. There are strong individual differences in each attention network and there are surprisingly low correlations between these network scores (Fan et al. 2002), although the networks interact in more complex tasks and in everyday life (Fan et al. 2009).

Normal functions including attention are undoubtedly influenced by many genes in complex interaction with epigenetic and environmental factors. Most studies have involved various pathologies and have not centered on common human functions; hence relatively little is known about the full range of genes involved in attention networks. One strategy would be to use emerging genomic and epigenomic technologies to carry out studies of large cohorts using various attention tasks as phenotypes to determine genes that relate to performance differences. A more limited approach, based on what is known about attention networks, takes advantage of the association between different neuromodulators and attention networks to examine specific genetic alleles (e.g., related to dopamine) to examine individual performance on the appropriate network (see Green et al. 2008 for review). As one example, the ANT has been used to examine individual differences in the efficiency of executive attention. A number of polymorphisms in dopamine and serotonin genes have been associated specifically with the scores on executive attention (Green et al. 2008). This work is still just getting started, and reports are conflicting. One reason for the conflict may be that genetic variations are also influenced by environmental factors.

Genetic modulation by environmental factors is perhaps clearest for the dopamine 4 receptor gene (DRD4), which has been associated with the executive network in adult imaging studies (Fan et al. 2003b). Data at 18–20 months showed that quality of parenting interacted with the 7 repeat allele of the DRD4 gene to influence the temperamental dimensions of impulsivity, high-intensity pleasure and activity level, and all components of sensation seeking (Sheese et al. 2007). Parenting made a strong difference for children with the 7 repeat allele in moderating sensation seeking but not for those children without this allele. At 3–4 years of age, the DRD4 gene interaction with parenting was related to children's EC, suggesting that executive attention may be the mechanism for this interaction. One study found that only those children with the 7-repeat of the DRD4 showed the influence of a parent training intervention (Bakermans-Kranenburg et al. 2008), suggesting that the presence of the DRD4 7 repeat allele may make the child more susceptible to environmental influences (Bakermans-Kranenburg & Van IJzendoorn 2011, Belsky & Pluess 2009, Sheese et al. 2007). This joint influence of environment and genetics seems to continue into adulthood (Larsen et al. 2010).

Training

Because parenting and other cultural factors interact with genes to influence behavior, it should be possible to develop specific training methods that can be used to influence underlying brain networks. Two forms of training methods have been used in the literature. One involves practice of a particular attention network. Several such attention training studies have shown improved executive attention function and produced changes in attention-related brain areas (Klingberg 2011, Rueda et al. 2005). The practice of a form of meditation has been used to change the brain state in a way that improves attention, reduces stress, and also improves functional connectivity between the anterior cingulate and the striatum (Tang et al. 2007, 2009).

Evolution

The ACC is a phylogenetically old area of the brain. Comparative anatomical studies point to important differences in the evolution of cingulate connectivity between nonhuman primates and humans. Anatomical studies show great expansion of white matter, which has increased more in recent evolution than has the neocortex itself (Zilles 2005). One type of projection cell called a Von Economo neuron is found only in higher apes and a few other social species, but they are most common in humans. In the human brain, the Von Economo neurons are found only in the anterior cingulate and a related area of the anterior insula (Allman et al. 2005). This neuron is likely important in communication between the cingulate and other brain areas. The two brain areas in which Von Economo neurons are found (cingulate and anterior insula) are also shown to be in close communication during the resting state (Dosenbach et al. 2007). It is not clear, however, if the distribution of Von Economo neurons and the cingulo-opercular network are overlapping or closely juxtaposed (Power et al. 2011). Some evidence indicates that the frequency of this type of neuron increases in human development between infancy and later childhood (Allman et al. 2005). These neurons may provide the rapid and efficient connectivity needed for executive control and may help explain why self-regulation in adult humans can be so much stronger than in other organisms.

FUTURE

It has been exciting for us to see the expansion of work on networks of attention over the past 20 years. We now have the opportunity to go from genes to cells, networks, and behavior and to examine how these relationships change from infancy to old age. In development, the number of active control systems increases and their influence changes.

Although much has been learned, many questions remain unanswered. We are hopeful that the study of attention will continue to provide greater understanding of how control develops typically and in pathology (Posner 2012a, Posner et al. 2011) and will provide promising leads for translating basic research into interventions to aid children and families.

DISCLOSURE STATEMENT

The authors are not aware of any affiliations, memberships, funding, or financial holdings that might be perceived as affecting the objectivity of this review.

ACKNOWLEDGMENTS

This article was supported in part by grant HD060563 to Georgia State University subcontracted to the University of Oregon. Prof. Mary K. Rothbart made important contributions to the research and writing of this review. This article was also supported by NIH grants NS32797 and 61144 and the McDonnell Foundation.

LITERATURE CITED

Allman JM, Watson KK, Tetreault NA, Hakeem AY. 2005. Intuition and autism: a possible role for Von Economo neurons. *Trends Cogn. Sci.* 9:367–73

Armstrong KM, Chang MH, Moore T. 2009. Selection and maintenance of spatial information by frontal eye field neurons. *J. Neurosci.* 29:15621–29

Aston-Jones G, Cohen JD. 2005. An integrative theory of locus coeruleus-norepinephrine function: adaptive gain and optimal performance. *Annu. Rev. Neurosci.* 28:403–50

Bakermans-Kranenburg MJ, Van IJzendoorn MH. 2011. Differential susceptibility to rearing environment depending on dopamine-related genes: new evidence and a meta-analysis. *Dev. Psychopathol.* 23:39–52

Bakermans-Kranenburg MJ, Van IJzendoorn MH, Pijlman FT, Mesman J, Juffer F. 2008. Experimental evidence for differential susceptibility: dopamine D4 receptor polymorphism (DRD4 VNTR) moderates intervention effects on toddlers' externalizing behavior in a randomized controlled trial. *Dev. Psychol.* 44:293–300

Beane M, Marrocco RT. 2004. Norepinephrine and acetylcholine mediation of the components of reflexive attention: implications for attention deficit disorders. *Prog. Neurobiol.* 74:167–81

Beauregard M, Lévesque J, Bourgouin P. 2001. Neural correlates of conscious self-regulation of emotion. *J. Neurosci.* 21:RC165

Beckmann M, Johansen-Berg H, Rushworth MFS. 2009. Connectivity-based parcellation of human cingulate cortex and its relation to functional specialization. *J. Neurosci.* 29:1175–90

Belsky J, Pluess M. 2009. Beyond diathesis stress: differential susceptibility to environmental influences. *Psychol. Bull.* 135:885–908

Berger A, Tzur G, Posner MI. 2006. Infant brains detect arithmetic errors. *Proc. Natl. Acad. Sci. USA* 103:12649–53

Botvinick MM, Braver TS, Barch DM, Carter CS, Cohen JD. 2001. Conflict monitoring and cognitive control. *Psychol. Rev.* 108:624–52

Bush G, Luu P, Posner MI. 2000. Cognitive and emotional influences in anterior cingulate cortex. *Trends Cogn. Sci.* 4:215–22

Carter CS, Krug MK. 2012. Dynamic cognitive control and frontal-cingulate interactions. See Posner 2012b, pp. 89–98

Corbetta M, Akbudak E, Conturo TE, Snyder AZ, Ollinger JM, et al. 1998. A common network of functional areas for attention and eye movements. *Neuron* 21:761–73

Corbetta M, Shulman GL. 2002. Control of goal-directed and stimulus-driven attention in the brain. *Nat. Rev. Neurosci.* 3:201–15

Coull JT, Frith CD, Buchel C, Nobre AC. 2000. Orienting attention in time: behavioural and neuroanatomical distinction between exogenous and endogenous shifts. *Neuropsychologia* 38:808–19

Crottaz-Herbette S, Menon V. 2006. Where and when the anterior cingulate cortex modulates attentional response: combined fMRI and ERP evidence. *J. Cogn. Neurosci.* 18:766–80

Davidson MC, Marrocco RT. 2000. Local infusion of scopolamine into intraparietal cortex slows covert orienting in rhesus monkeys. *J. Neurophysiol.* 83:1536–49

Dehaene S, Changeux JP. 2011. Experimental and theoretical approaches to conscious processing. *Neuron* 70:200–27

Dehaene S, Posner MI, Tucker DM. 1994. Localization of a neural system for error detection and compensation. *Psychol. Sci.* 5:303–5

Desimone R, Duncan J. 1995. Neural mechanisms of selective visual attention. *Annu. Rev. Neurosci.* 18:193–222

Donaldson DI, Petersen SE, Ollinger JM, Buckner RL. 2001. Dissociating state and item components of recognition memory using fMRI. *Neuroimage* 13:129–42

Dosenbach NUF, Fair DA, Cohen AL, Schlaggar BL, Petersen SE. 2008. A dual-networks architecture of top-down control. *Trends Cogn. Sci.* 12:99–105

Dosenbach NUF, Fair DA, Miezin FM, Cohen AL, Wenger KK, et al. 2007. Distinct brain networks for adaptive and stable task control in humans. *Proc. Natl. Acad. Sci. USA* 104:11073–78

Dosenbach NUF, Visscher KM, Palmer ED, Miezin FM, Wenger KK, et al. 2006. A core system for the implementation of task sets. *Neuron* 50:799–812

Driver J, Eimer M, Macaluso E, van Velzen J. 2004. Neurobiology of human spatial attention: modulation, generation, and integration. See Kanwisher & Duncan 2004, pp. 267–300

Duncan J. 1980. The locus of interference in the perception of simultaneous stimuli. *Psychol. Rev.* 87:272–300

Eisenberger NI, Lieberman MD, Williams KD. 2003. Does rejection hurt? An FMRI study of social exclusion. *Science* 302:290–92

Etkin A, Egner T, Peraza DM, Kandel ER, Hirsch J. 2006. Resolving emotional conflict: a role for the rostral anterior cingulate cortex in modulating activity in the amygdala. *Neuron* 51:871–82

Everitt BJ, Robbins TW. 1997. Central cholinergic systems and cognition. *Annu. Rev. Psychol.* 48:649–84

Fan J, Flombaum JI, McCandliss BD, Thomas KM, Posner MI. 2003a. Cognitive and brain consequences of conflict. *Neuroimage* 18:42–57

Fan J, Fossella J, Sommer T, Wu Y, Posner MI. 2003b. Mapping the genetic variation of executive attention onto brain activity. *Proc. Natl. Acad. Sci. USA* 100:7406–11

Fan J, Gu X, Guise KG, Liu X, Fossella J, et al. 2009. Testing the behavioral interaction and integration of attentional networks. *Brain Cogn.* 70:209–20

Fan J, McCandliss BD, Fossella J, Flombaum JI, Posner MI. 2005. The activation of attentional networks. *Neuroimage* 26:471–79

Fan J, McCandliss BD, Sommer T, Raz A, Posner MI. 2002. Testing the efficiency and independence of attentional networks. *J. Cogn. Neurosci.* 14:340–47

Fernandez-Duque D, Posner MI. 1997. Relating the mechanisms of orienting and alerting. *Neuropsychologia* 35:477–86

Fox MD, Snyder AZ, Barch DM, Gusnard DA, Raichle ME. 2005. Transient BOLD responses at block transitions. *Neuroimage* 28:956–66

Green AE, Munafo MR, DeYoung CG, Fossella JA, Fan J, Gray JR. 2008. Using genetic data in cognitive neuroscience: from growing pains to genuine insights. *Nat. Rev. Neurosci.* 9:710–20

Gunderson LH, Holling CS. 2002. *Panarchy: Understanding Transformations in Human and Natural Systems.* Washington, DC: Island

Hampton AN, O'Doherty JP. 2007. Decoding the neural substrates of reward-related decision making with functional MRI. *Proc. Natl. Acad. Sci. USA* 104:1377–82

Harter MR, Guido W. 1980. Attention to pattern orientation: negative cortical potentials, reaction time, and the selection process. *Electroencephalogr. Clin. Neurophysiol.* 49:461–75

Hillyard SA, Di Russo F, Martinez A. 2004. The imaging of visual attention. See Kanwisher & Duncan 2004, pp. 381–90

Hubbard EM, Piazza M, Pinel P, Dehaene S. 2005. Interactions between number and space in parietal cortex. *Nat. Rev. Neurosci.* 6:435–48

Ito S, Stuphorn V, Brown JW, Schall JD. 2003. Performance monitoring by the anterior cingulate cortex during saccade countermanding. *Science* 302:120–22

Ivry R, Robertson LC. 1997. *Two Sides of Perception.* Cambridge, MA: MIT Press

Jones L, Rothbart MK, Posner MI. 2003. Development of inhibitory control in preschool children. *Dev. Sci.* 6:498–504

Kampe KK, Frith CD, Frith U. 2003. "Hey John": signals conveying communicative intention toward the self activate brain regions associated with "mentalizing," regardless of modality. *J. Neurosci.* 23:5258–63

Kanwisher N, Duncan J, eds. 2004. *Attention and Performance XX: Functional Brain Imaging of Visual Cognition*. Oxford, UK: Oxford Univ. Press

Klingberg T. 2012. Training working memory and attention. See Posner 2012b, pp. 475–86

Konishi S, Donaldson DI, Buckner RL. 2001. Transient activation during block transition. *Neuroimage* 13:364–74

Larsen H, van der Zwaluw CS, Overbeek G, Granic I, Franke B, Engels RC. 2010. A variable-number-of-tandem-repeats polymorphism in the dopamine D4 receptor gene affects social adaptation of alcohol use: investigation of a gene-environment interaction. *Psychol. Sci.* 21:1064–68

Lindner A, Iyer A, Kagan I, Andersen RA. 2010. Human posterior parietal cortex plans where to reach and what to avoid. *J. Neurosci.* 30:11715–25

Marrocco RT, Davidson MC. 1998. Neurochemistry of attention. In *The Attentive Brain*, ed. R Parasuraman, pp. 35–50. Cambridge, MA: MIT Press

Morrison JH, Foote SL. 1986. Noradrenergic and serotoninergic innervation of cortical, thalamic and tectal visual structures in Old and New World monkeys. *J. Comp. Neurol.* 243:117–28

Moruzzi G, Magoun HW. 1949. Brainstem reticular formation and activation of the EEG. *Electroencephalogr. Clin. Neurophysiol.* 1:455–73

Nagai Y, Critchley HD, Featherstone E, Fenwick PB, Trimble MR, Dolan RJ. 2004. Brain activity relating to the contingent negative variation: an fMRI investigation. *Neuroimage* 21:1232–41

Ochsner KN, Bunge SA, Gross JJ, Gabrieli JD. 2002. Rethinking feelings: an FMRI study of the cognitive regulation of emotion. *J. Cogn. Neurosci.* 14:1215–29

Ploran EJ, Nelson SM, Velanova K, Donaldson DI, Petersen SE, Wheeler ME. 2007. Evidence accumulation and the moment of recognition: dissociating perceptual recognition processes using fMRI. *J. Neurosci.* 27:11912–24

Posner MI. 1975. Psychobiology of attention. In *Handbook of Psychobiology*, ed. M Gazzaniga, C Blakemore, pp. 441–80. New York: Academic

Posner MI. 2012a. *Attention in the Social World*. New York: Oxford Univ. Press

Posner MI. 2012b. *Cognitive Neuroscience of Attention*. New York: Guilford

Posner MI, Petersen SE. 1990. The attention system of the human brain. *Annu. Rev. Neurosci.* 13:25–42

Posner MI, Rothbart MK. 2007. Research on attention networks as a model for the integration of psychological science. *Annu. Rev. Psychol.* 58:1–23

Posner MI, Rothbart MK, Sheese BE, Voelker P. 2012. Control networks and neuromodulators of early development. *Dev. Psychol.* In press

Power JD, Cohen AL, Nelson SM, Vogel AC, Church JA, et al. 2011. Functional network organization in the human brain. *Neuron* 72:665–78

Rainville P, Duncan GH, Price DD, Carrier B, Bushnell MC. 1997. Pain affect encoded in human anterior cingulate but not somatosensory cortex. *Science* 277:968–71

Rizzolatti G, Riggio L, Dascola I, Umiltá C. 1987. Reorienting attention across the horizontal and vertical meridians: evidence in favor of a premotor theory of attention. *Neuropsychologia* 25:31–40

Rossi AF, Bichot NP, Desimone R, Ungerleider LG. 2007. Top down attentional deficits in macaques with lesions of lateral prefrontal cortex. *J. Neurosci.* 27:11306–14

Rothbart MK. 2011. *Becoming Who We Are*. New York: Guilford

Rothbart MK, Sheese BE, Rueda MR, Posner MI. 2011. Developing mechanisms of self-regulation in early life. *Emot. Rev.* 3:207–13

Rueda MR, Rothbart MK, McCandliss BD, Saccomanno L, Posner MI. 2005. Training, maturation, and genetic influences on the development of executive attention. *Proc. Natl. Acad. Sci. USA* 102:14931–36

Schafer RJ, Moore T. 2007. Attention governs action in the primate frontal eye field. *Neuron* 56:541–51

Sheese BE, Voelker PM, Rothbart MK, Posner MI. 2007. Parenting quality interacts with genetic variation in dopamine receptor D4 to influence temperament in early childhood. *Dev. Psychopathol.* 19:1039–46

Shulman GL, Corbetta M. 2012. Two attentional networks: identification and function within a larger cognitive architecture. See Posner 2012b, pp. 113–27

Sridharan D, Levitin DJ, Chafe CH, Berger J, Menon V. 2007. Neural dynamics of event segmentation in music: converging evidence for dissociable ventral and dorsal networks. *Neuron* 55:521–32

Sridharan D, Levitin DJ, Menon V. 2008. A critical role for the right fronto-insular cortex in switching between central-executive and default-mode networks. *Proc. Natl. Acad. Sci. USA* 105:12569–74

Stewart C, Burke S, Marrocco R. 2001. Cholinergic modulation of covert attention in the rat. *Psychopharmacology* 155:210–18

Sturm W, Willmes K. 2001. On the functional neuroanatomy of intrinsic and phasic alertness. *Neuroimage* 14:S76–84

Tang YY, Ma Y, Fan Y, Feng H, Wang J, et al. 2009. Central and autonomic nervous system interaction is altered by short-term meditation. *Proc. Natl. Acad. Sci. USA* 106:8865–70

Tang YY, Ma Y, Wang J, Fan Y, Feng S, et al. 2007. Short-term meditation training improves attention and self-regulation. *Proc. Natl. Acad. Sci. USA* 104:17152–56

Thompson KG, Biscoe KL, Sato TR. 2005. Neuronal basis of covert spatial attention in the frontal eye field. *J. Neurosci.* 25:9479–87

Voytko ML, Olton DS, Richardson RT, Gorman LK, Tobin JR, Price DL. 1994. Basal forebrain lesions in monkeys disrupt attention but not learning and memory. *J. Neurosci.* 14:167–86

Walter G. 1964. The convergence and interaction of visual, auditory, and tactile responses in human non-specific cortex. *Ann. N. Y. Acad. Sci.* 112:320–61

Womelsdorf T, Schoffelen JM, Oostenveld R, Singer W, Desimone R, et al. 2007. Modulation of neuronal interactions through neuronal synchronization. *Science* 316:1609–12

Wright RD, Ward LM. 2008. *Orienting of Attention*. Oxford/New York: Oxford Univ. Press

Zilles K. 2005. Evolution of the human brain and comparative cyto- and receptor architecture. In *From Monkey Brain to Human Brain*, ed. S Dehaene, J-R Duhamel, MD Hauser, G Rizzolatti, pp. 41–56. Cambridge, MA: MIT Press/Bradford Books

Primary Visual Cortex: Awareness and Blindsight*

David A. Leopold

Section on Cognitive Neurophysiology and Imaging, Laboratory of Neuropsychology, National Institute of Mental Health (NIMH); Neurophysiology Imaging Facility, NIMH, National Institute of Neurological Disorders and Stroke, National Eye Institute; National Institutes of Health, Department of Health and Human Services, Bethesda, Maryland 20892; email: leopoldd@mail.nih.gov

Annu. Rev. Neurosci. 2012. 35:91–109

The *Annual Review of Neuroscience* is online at neuro.annualreviews.org

This article's doi:
10.1146/annurev-neuro-062111-150356

Keywords

V1, visual perception, blindsight, cortical lesion, consciousness, cerebral cortex, thalamus

Abstract

The primary visual cortex (V1) is the principal telencephalic recipient of visual input in humans and monkeys. It is unique among cortical areas in that its destruction results in chronic blindness. However, certain patients with V1 damage, though lacking visual awareness, exhibit visually guided behavior: blindsight. This phenomenon, together with evidence from electrophysiological, neuroimaging, and psychophysical experiments, has led to speculation that V1 activity has a special or direct role in generating conscious perception. To explore this issue, this article reviews experiments that have used two powerful paradigms—stimulus-induced perceptual suppression and chronic V1 ablation—each of which disrupts the ability to perceive salient visual stimuli. Focus is placed on recent neurophysiological, behavioral, and functional imaging studies from the nonhuman primate that shed light on V1's role in conscious awareness. In addition, anatomical pathways that relay visual information to the cortex during normal vision and in blindsight are reviewed. Although the critical role of V1 in primate vision follows naturally from its position as a bottleneck of visual signals, little evidence supports its direct contribution to visual awareness.

Contents

INTRODUCTION

Understanding the relationship between neural activity and subjective perception is one of the most fascinating and challenging goals of modern neuroscience. In the domain of vision, damage to the primary visual cortex, or V1, but not to any other cortical region, abolishes visual awareness and leads to chronic blindness. This observation, combined with data from electrophysiological and functional magnetic resonance imaging (fMRI) studies in humans and nonhuman primates, has raised speculation that neural activity in V1 may have a direct and critical role in the generation of a percept.

V1, V2, V3, V4, MT, TEO, TE: visual areas in the macaque visual cortex

Perceptual suppression: psychophysical paradigm used to induce the all or none subjective disappearance of a visual stimulus

Blindsight: residual visually capacity in the absence of awareness following damage to V1

The present article reviews experiments that shed light on this intriguing topic. We survey experiments pertaining to the visual phenomena of perceptual suppression and blindsight in an attempt to understand the role of V1 in conscious and unconscious vision. In doing so, we refer to diverse features of V1, whose anatomical connections, complex laminar organization, and electrophysiological response profile have been studied extensively in the monkey. Throughout the review, emphasis is placed on discoveries in the past decade. By necessity, several relevant topics are not discussed or are mentioned only in brief. Such topics include neural correlates of perception in V1 pertaining to paradigms other than visual suppression (reviewed in Tong 2003), perceptual correlates outside of V1, and perceptual impairments following cortical lesions in areas other than V1. We do not attempt to provide a comprehensive review of blindsight and refer the reader to recent overviews by pioneers of the field (Cowey 2010, Stoerig 2006, Weiskrantz 2009). A considerable portion of this review is devoted to describing pathways that carry retinal image information to the cortex, the details of which are important for understanding both the determinants of V1 activity during perceptual suppression and the basis for unconscious visual performance during blindsight.

EXPERIMENTAL INROADS TO THE UNCONSCIOUS

We begin by briefly describing the two featured paradigms (see **Figure 1**). The next section reviews the modulation of sensory responses in V1 during perceptual suppression, including some strikingly discrepant findings obtained from single-unit and fMRI studies. This is followed by a survey of experiments that give insight into V1-independent vision during blindsight. The final section draws upon these and other findings to evaluate the particular role of V1 in visual awareness.

Perceptual suppression can render a normally salient visual stimulus completely invisible. Stimulus paradigms that induce perceptual suppression are an important component of the psychophysicist's toolbox, as they shed light on unconscious sensory processing. Such paradigms include binocular rivalry (Blake & Logothetis 2002), motion-induced blindness (Bonneh et al. 2001), visual masking (Breitmeyer & Öğmen 2006), and various dichoptic stimulus sequences collectively termed flash suppression (Tsuchiya & Koch 2005, Wilke et al. 2003, Wolfe 1984). Psychophysical experiments have demonstrated that during perceptual suppression certain stimuli, though completely invisible, can penetrate the first stages of cortical processing. In doing so, they can generate adaptational aftereffects (Blake et al. 2006), guide manual grasping behavior (Roseboom & Arnold 2011), and recruit spatial attention (Lin & He 2009).

An example of a flash suppression stimulus sequence is shown in **Figure 1*a***. In this particular paradigm (Wilke et al. 2003), a salient target stimulus is first presented alone on the screen, often monocularly, for several hundred milliseconds. After this period, a binocular field of randomly moving dots appears in the periphery. This sequence induces the target stimulus to vanish abruptly from perception and remain entirely invisible for several seconds, provided the moving dots remain on the screen. The probability of target suppression is a function of several stimulus parameters, such as the speed at which the dots are moving. In a typical monkey neurophysiological experiment, these parameters are adjusted to induce the target to disappear on approximately 50% of the trials. Then, on the basis of the monkey's perceptual report, neural responses to an identical physical stimulus are compared when the target is subjectively visible or invisible (Leopold et al. 2003, Wilke et al. 2006, 2009). This approach allows one to assess the relationship of a given neural response to the perceptual awareness of a stimulus; the results from the visual cortex are discussed in the next section.

Blindsight refers to the ability of cortically blind patients and experimental animals to use visual information to guide behavior in the absence of visual awareness (Weiskrantz 2009). Human blindsight subjects are able to orient to and answer questions about stimuli presented to the blind part of the visual field. However, when questioned, they report being entirely unaware of the stimuli to which they are responding (Sanders et al. 1974). This situation can be somewhat perplexing for the subject. In their seminal paper, Pöppel and colleagues (Pöppel et al. 1973) asked their subject to direct his eyes to the target, to which he replied, "How can I look at something that I haven't seen?" Nonetheless, the subject was still able to carry out the task. This paradoxical phenomenon of blindsight is not simply due to low-functioning vision, but is instead due to a unique uncoupling between subjective visual perception and visually guided performance (Azzopardi & Cowey 1997). Moreover, it occurs only when damage is restricted to V1 and is generally not present when the damage extends into the extrastriate cortex (Weiskrantz 2009).

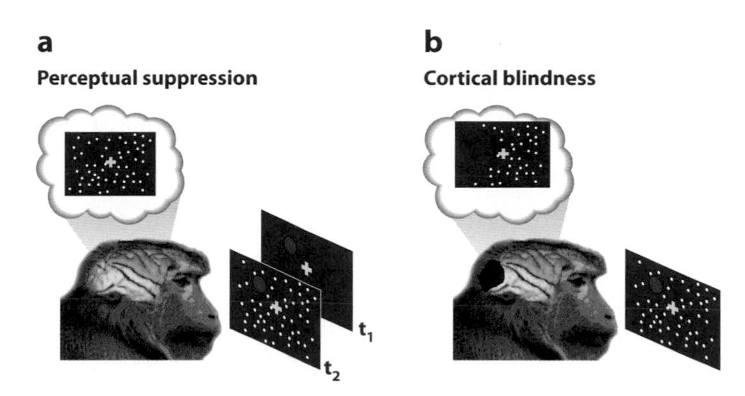

Figure 1

Paradigms to study unconscious vision in monkeys. (*a*) During perceptual suppression, a target stimulus is continuously presented on a video monitor but disappears because of a visual illusion. Depicted here is generalized flash suppression (Wilke et al. 2003), where the presentation of a bright red patch at time t_1 is followed by the appearance of dynamic surrounding white dots at time t_2, causing the red patch to disappear perceptually for up to several seconds. (*b*) Cortical blindness following V1 lesion leads to the inability to perceive stimuli in an entire region of visual space corresponding to the retinotopic position of the lesion. Following such lesions, blindsight allows for some residual behavioral responses to stimuli presented to the scotoma (blind portion of the visual field).

The blindness that follows damage to V1 in humans appears to be common among primates, but not in other mammals, which have more visual relay projections from the thalamus to other cortical areas, thus bypassing the primary visual cortex (Funk & Rosa 1998, Preuss 2007). Despite nominal blindness, the existence of some residual vision following V1 lesions in the macaque has been recognized for more than half a century (Klüver 1941). In the weeks following the surgical removal of V1,

Flash suppression: a visual stimulation paradigm in which a sequence of stimuli induces a target to undergo perceptual suppression

Visual relay: a neural pathway that receives direct or indirect retinal information and then transmits it further

LGN: lateral geniculate nucleus of the thalamus

CAMKII: calcium/calmodulin-dependent protein kinases II

macaques gradually recover the ability to use visual information to guide hand and eye movements to stimuli in the "blind" (lesion-affected) part of the visual field (Feinberg et al. 1978, Humphrey 1974, Isa & Yoshida 2009, Mohler & Wurtz 1977). While motion detection is the most consistent feature of blindsight, V1-lesioned monkeys have also been reported to discriminate simple patterns on the basis of spatial frequency, shape, texture, and color (Dineen & Keating 1981, Miller et al. 1980, Schilder et al. 1972). Some important aspects of their vision are gone forever, such as the capacity to visually recognize food, objects, or faces of familiar individuals (Humphrey 1974). Importantly, the residual vision in macaques indicates a dissociation between awareness and visually guided behavior. When visual perception was tested using both forced choice and detection tasks, macaques responded correctly to a stimulus in the blind field during the forced-choice task but then, under the same visual conditions, indicated in the detection task that no stimulus was presented (Cowey & Stoerig 1995, Moore et al. 1995). Although it is impossible to determine precisely the subjective experience of a cortically blind monkey, or human for that matter, these experiments indicate that macaques exhibit the hallmarks of blindsight and are, therefore, a good primate model for studying V1-independent vision in the human.

PERCEPTUAL SUPPRESSION OF VISUAL RESPONSES IN V1

We now focus on perceptual suppression, asking how the visibility of a stimulus affects neural responses in V1. This approach of correlating neural activity with subjective perception has been used to investigate whether V1 contributes directly to visual awareness (Tononi & Koch 2008). However, before reviewing the effect of perceptual suppression on cortical neurons, we begin with a survey of the basic anatomy and physiology of the ascending pathways carrying sensory information to V1. In humans and monkeys, nearly all visual information reaches the cortex through a primary visual pipeline that passes from the retina to the lateral geniculate nucleus (LGN) to V1. Reviewing this basic circuitry is necessary to understand how V1's basic sensory responses may interact with signals related to conscious perception. The components of this pathway are also relevant to the discussion of blindsight in a subsequent section.

Converging Visual Signals in V1

The input from the LGN to V1 consists of multiple parallel sensory pathways whose characteristic response properties originate in the retina (for a recent review, see Schiller 2010). In the macaque, LGN-projecting retinal ganglion cells have a wide range of morphologies and physiological response profiles, which are often classified into three main groups: Pα, Pβ, and Pγ. The Pβ cells, projecting almost exclusively to the parvocellular LGN layers, compose more than 80% of ganglion cells in the macaque. They have a "midget" dendritic morphology, which gives them small receptive fields for detailed form vision. Electrophysiologically, they exhibit sustained responses and typically show red/green color opponency in trichromats. The Pα neurons compose roughly 10% of the ganglion cells. Their primary target is the magnocellular layers of the LGN, though they also project to several other target structures (described below). Their "parasol" morphology translates to large, integrative receptive fields. Electrophysiologically, they tend to respond transiently and without color selectivity. The remaining ganglion cells are often grouped together as Pγ, although their morphology and physiological properties are quite diverse (Schiller & Malpeli 1977). The Pγ axons terminate in the interlaminar zones of the LGN, ventral to each magno- and parvocellular layer. The interlaminar zones are strongly associated with the koniocellular pathway, whose neurons are immunoreactive to calcium binding proteins calcium/calmodulin-dependent protein kinases II (CAMKII) or calbindin D28K (Casagrande 1994, Hendry & Yoshioka 1994), and carry blue/yellow

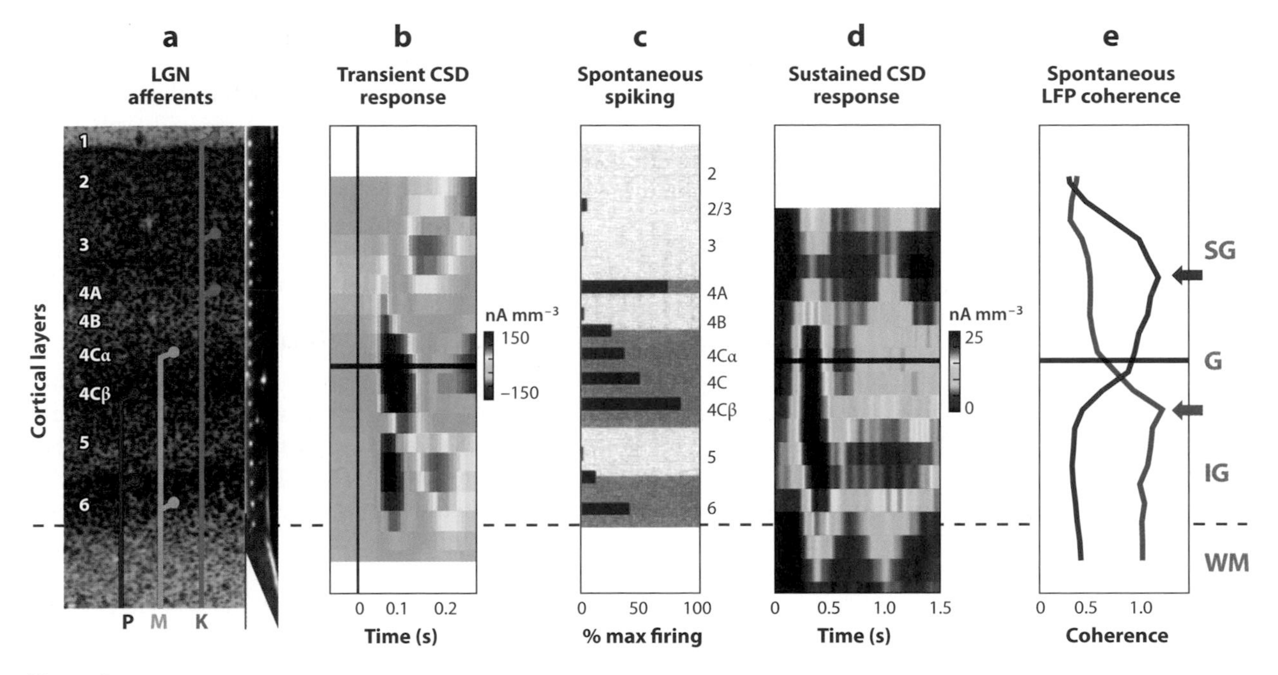

Figure 2

Sensory and spontaneous physiology across V1 layers. (*a*) The basic pathways projecting from the lateral geniculate nucleus (LGN) to the different layers of V1, including the magnocellular (M), parvocellular (P), and koniocellular (K). Adjacent is a photograph of a multicontact linear electrode array. (*b*) Current source density (CSD) response to flashed stimuli in V1. The horizontal line is drawn through the initial current sink in layer 4C (Maier et al. 2011). (*c*) Spontaneous spiking responses in different cortical layers of monkeys sitting in a dark room (Snodderly & Gur 1995). (*d*) Sustained CSD power that persists in the infragranular layers during the presentation of a simple stimulus (Maier et al. 2011). (*e*) Pattern of coherence of spontaneous high-frequency (gamma) local field potential (LFP) activity. Pairwise coherence is computed between each of two different reference positions (*blue and red arrows*) and all other laminar positions (Maier et al. 2010). Abbreviations: SG, supragranular layers; G, granular layers; IG, infragranular layers; WM, white matter.

color-opponent signals (for reviews, see Hendry & Reid 2000, Nassi & Callaway 2009). In the marmoset, injection of retrograde tracers into the koniocellular layers labels the bistratified Pγ cells in the retina (Szmajda et al. 2008).

Each ganglion cell type then relays its signals through unusually strong synapses in the LGN to V1, where the pattern of afferent projections is known in detail (for reviews, see Lund 1988, Nassi & Callaway 2009, Peters et al. 1994). Briefly, neurons from the magno- and parvocellular LGN compartments (carrying Pα and Pβ signals, respectively) project to separate subcompartments of layers 4C and 6. Koniocellular projections (carrying Pγ signals) terminate within and above layer 4A (**Figure 2*a***). The LGN projections to layer 4C are much stronger than those to layer 6. However, the intracortical projection from layer 6 to layer 4C is also prominent and may be an important factor in determining the overall strength of visual responses (Callaway 1998, Douglas & Martin 2004). Layer-6 neurons also transmit channel-specific visual signals back to the LGN in an organized fashion, with upper-tier neurons projecting to the parvocellular layers and lower-tier neurons projecting to the magnocellular layers and possibly also to the koniocellular layers (Briggs & Usrey 2009).

In addition to its LGN input, area V1 also receives afferent input from a large number of extrastriate visual cortical areas, including V2, V3, V4, MT, TEO, and TE (reviewed in Barone et al. 2000, Salin & Bullier 1995), the inferior pulvinar (Benevento & Rezak 1976), the amygdala (Freese & Amaral 2005), and the

Supragranular: laminar position within the thickness of the cerebral cortex above the middle, granular layer

GABA: gamma aminobutyric acid

Infragranular: laminar position within the thickness of the cerebral cortex below the middle, granular layer

claustrum (Baizer et al. 1997). Aside from the claustrum, which sends its densest projections to layer 4, each of these structures projects primarily to the supragranular layers. In fact, ventral stream extrastriate cortical areas V4, TEO, and TE project exclusively to layer 1. This fact is important for understanding perceptual modulation in V1, suggesting that extrastriate modulation of V1 activity affects synaptic activity in the supragranular layers.

Before we turn to how perceptual suppression affects V1 responses, we briefly review some basic features of V1 electrophysiology, including its laminar response profile and the contribution of different inputs. The responses of a given V1 neuron will be shaped to different extents by the LGN afferents, feedback from other cortical areas, input from subcortical areas, and a very large number of synaptic inputs from within V1 (Douglas & Martin 2004). To study the contribution of the feedforward pathway, it is possible to isolate and measure directly spikes arriving into V1 at the LGN terminals, provided V1 neurons are first inactivated [for example, using the gamma aminobutyric acid (GABA) agonist muscimol]. For example, this approach was used in one study to demonstrate the laminar segregation of LGN inputs on the basis of their chromatic selectivity (Chatterjee & Callaway 2003). The primary synaptic influence of these spiking afferents can be determined using current source density analysis, which computes the flow of extracellular ionic currents thought to derive from synchronized postsynaptic potentials (Schroeder et al. 1991). Following an abruptly flashed stimulus, a current sink is induced with a short latency in layer 4C, followed tens of milliseconds later by current sinks in the supragranular and infragranular layers (see **Figure 2*b***). This characteristic spatiotemporal evolution of excitatory synaptic activity from the middle layers toward the laminae above and below is thought to reflect feedforward processing of visual information through the cortical microcircuitry (Mitzdorf 1985), and it is consistent with the laminar distribution of spiking-response latencies (Nowak et al. 1995). In addition to evoked responses, spontaneous activity is also strongly influenced by LGN afferents, even in darkness. Ongoing spiking activity is markedly higher in the LGN-recipient layers compared with other layers (Snodderly & Gur 1995) (**Figure 2*c***), as is high-frequency ("gamma") local field potential (LFP) power (Maier et al. 2010).

Recent work has revealed other basic measures of V1 activity that are not as obviously derived from the pattern of LGN inputs. One study using a variant of current source density analysis found that the sustained response to a stimulus was localized in the infragranular layers, roughly 500 μm below the initial transient sink (Maier et al. 2011) (**Figure 2*d***). Another study revealed two distinct laminar zones of LFP signal coherence, with a boundary between them near the bottom of layer 4C (Maier et al. 2010) (**Figure 2*e***). The extent to which these latter findings can be explained by the LGN input, reverberation within the V1 microcircuit, or corticocortical feedback remains to be determined.

Given the multiple anatomical inputs impinging on V1 and its physiological response profile that appears largely, but not entirely, determined by its LGN afferents, we pose the following question: Does perceptual suppression affect responses to visual stimuli in V1? Obtaining a simple answer to this seemingly straightforward question has proved to be much more difficult than anticipated.

Modulation of Visual Responses During Perceptual Suppression

Using a range of psychophysical tools, including those mentioned above, researchers have investigated the neural basis of perceptual suppression in both macaques and humans. Single-unit and fMRI studies largely agree that perceptual suppression modulates neural responses to stimuli throughout the visual cortex, particularly at the highest stages of the cortical hierarchy (Fisch et al. 2009, Kreiman et al. 2002, Sheinberg & Logothetis 1997, Tong et al. 1998). At intermediate stages, such as areas MT and V4, the correlates of

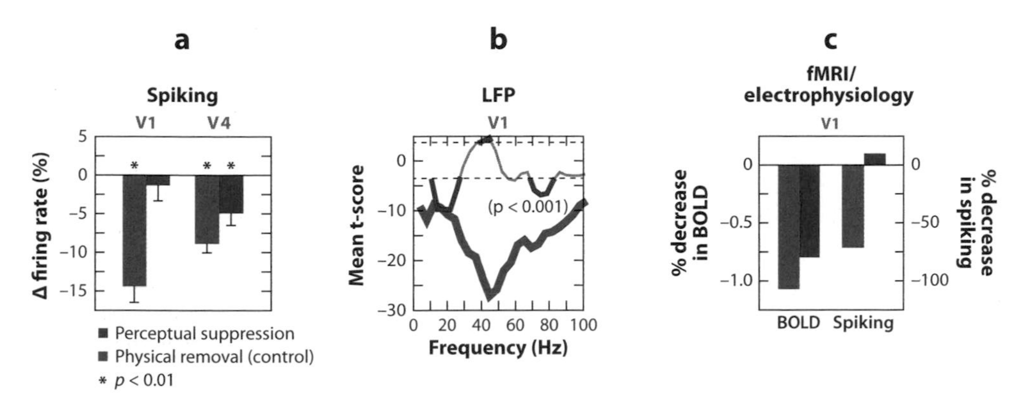

Figure 3

Neural correlates of perceptual suppression in V1. (*a*) Spiking modulation in V1 and V4 associated with perceptual suppression versus physical removal of a stimulus (Wilke et al. 2006). (*b*) Local field potential (LFP) modulation in V1 associated with perceptual suppression versus physical removal of a stimulus (Maier et al. 2008). (*c*) Comparison of the effects of perceptual suppression on the blood oxygenation level–dependent (BOLD) versus spiking signals in V1 (Maier et al. 2008).

perception are mixed throughout the population of neurons (Logothetis & Schall 1989, Wilke et al. 2006), and individual cells change their sensitivity to perceptual suppression in response to the structural details of the inducing stimulus (Maier et al. 2007).

Within V1, monkey electrophysiology and human fMRI studies have found nearly opposite results during perceptual suppression. Single-unit experiments in the macaque have consistently found that the visibility or invisibility of a stimulus has minimal, if any, effect on the firing of V1 neurons (Gail et al. 2004, Keliris et al. 2010, Leopold et al. 2005, Leopold & Logothetis 1996, Libedinsky et al. 2009, Wilke et al. 2006), in agreement with theoretical work suggesting that activity in V1 does not contribute directly to visual awareness (Crick & Koch 1995) (but for a different perspective, see Tong 2003). Compared with single-cell responses, the local LFP signal is more modulated (Gail et al. 2004, Maier et al. 2008, Wilke et al. 2006). However, this change is small relative to control trials in which the same stimulus is physically removed (**Figure 3*b***). At the same time, human fMRI experiments report strongly diminished responses in V1 during perceptual suppression resembling the physical control condition (Haynes & Rees 2005, Lee et al. 2005, Polonsky et al. 2000, Tong & Engel 2001, Wunderlich et al. 2005). As a result, the same paradigms used to argue against the role of V1 in awareness based on monkey electrophysiology have been used to argue for its role in awareness based on human fMRI.

To investigate the basis of this apparent discrepancy, a recent study combined fMRI and electrophysiological methods in V1 in monkeys experiencing flash suppression (Maier et al. 2008). During conventional visual stimulation, fMRI blood oxygenation level–dependent (BOLD) and electrophysiological responses, including spiking and LFP, were in good agreement. However, during perceptual suppression, the signals diverged markedly, even though they were measured from the same patch of tissue. When the stimulus was subjectively invisible, fMRI responses dropped to levels near to that of a control condition in which the stimulus was physically removed. By contrast, spiking responses were at the same high levels during visible and invisible periods, again indicating that neural spiking rates in V1 are unaffected by perceptual suppression (**Figure 3*c***). Responses of the LFP showed some significant perceptual modulation but proportionally much less than those of the BOLD signal. Thus, the BOLD and spiking

SC: superior colliculus

signals were fundamentally different in their responses (Logothetis 2002), and the level of the discrepancy was strongly dependent on perceptual visibility. This latter finding may be related to signal discrepancies in V1 associated with other cognitive variables, such as hemodynamic-response modulation observed in the absence of single-unit modulation during spatial attention (Posner & Gilbert 1999, Watanabe et al. 2011) and that associated with the expectation of an impending visual stimulus during a behavioral task that involves predictable stimulus presentation (Sirotin & Das 2009).

Why do BOLD signals show decreased responses in V1 to perceptually suppressed stimuli, whereas spiking responses do not? One possibility is that signals reaching V1 elicit synaptic activity that causes a hemodynamic response but is never translated into changes in the rate of action potentials. Initial results from one study indicate that synaptic activity in the supragranular layers, but not in the deeper layers, drops significantly during perceptual suppression (Leopold et al. 2008). Because cortical areas V4 and TE send their projections exclusively to the supragranular layers of V1, the reported activity changes could reflect feedback from extrastriate areas, where synaptic activity modulates with perceptual suppression. It is tempting to speculate that such synaptic modulation affects V1 BOLD responses, but why such modulation would have virtually no effect on neuronal spiking remains a puzzle. This issue warrants further investigation.

Thus, single-unit modulation during perceptual suppression provides no evidence in support of V1 playing a direct role in visual awareness. In the next section, we explore the same point from a different perspective, reviewing the neural basis of unconscious vision following damage to V1.

BLINDSIGHT: RESIDUAL VISION FOLLOWING V1 DAMAGE

The phenomenology of blindsight has two principal features. The first is blindness, or the loss of visual awareness associated with V1 damage. The second is the capacity of blind individuals to use visual signals to guide behavioral responses. Here, we address the second of these features, leaving the loss of visual awareness as a topic for the final section. Understanding the basis of residual vision during blindsight, including its unconscious nature, requires knowledge of anatomical connections. We begin by describing neural projections to the extrastriate cortex that are thought to underlie blindsight behavior (Weiskrantz 2009).

Anatomical Pathways to the Extrastriate Cortex

All retinal image information reaching the cerebral cortex ascends through synapses in the dorsal thalamus, either in the LGN or the pulvinar (see above for a review of the retinal projections to the LGN). In addition, a very small number of ganglion cells, primarily Pγ and Pα, target the inferior pulvinar (Cowey et al. 1994, O'Brien et al. 2001) as well as several other projection targets located in the forebrain and situated at distinct positions along the neuraxis (see **Figure 4*a***). Approximately one-tenth of all ganglion cells send descending projections to the superior colliculus (SC) in the midbrain. Similar to the pulvinar, the SC receives primarily Pγ and Pα inputs (Perry et al. 1984, Perry & Cowey 1984). Sparser projections terminate in the pregeniculate nucleus and in several nuclei in the hypothalamus and pretectum (Stoerig & Cowey 1997). Some ganglion cells are thought to send collateral projections to multiple targets, such as to both the LGN and the SC (Crook et al. 2008). In addition to their direct retinal input, both the LGN and the pulvinar also receive projections from the SC, suggesting a potential midbrain relay to each of the two structures (Harting et al. 1991, May 2006). There are thus at least four potential pathways by which retinal information can reach the dorsal thalamus (**Figure 4*b***).

In addition, both the LGN and pulvinar project to both V1 and the extrastriate visual cortex. Of particular interest for blindsight are

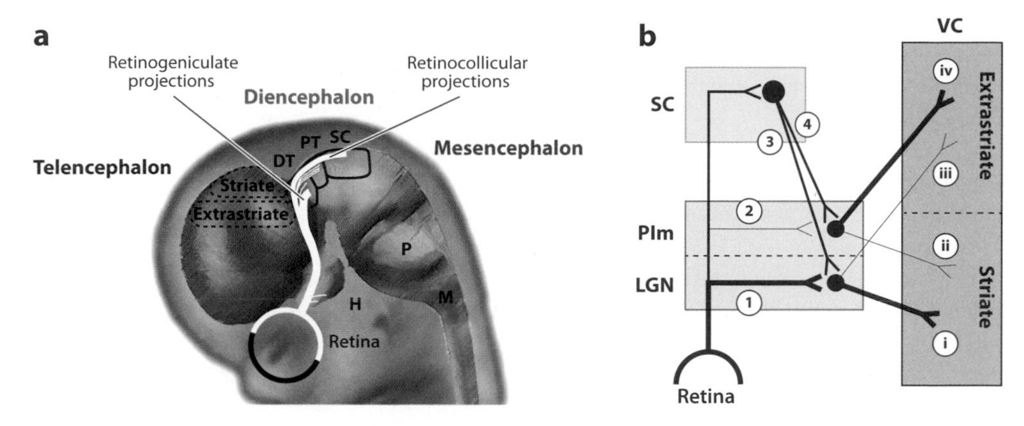

Figure 4

Visual pathways through the dorsal thalamus to the cortex. (*a*) Targets of retinal ganglion cells in the diencephalon and mesencephalon. Projections are depicted on an embryonic brain to emphasize the relative positions of the retinal projection targets with respect to the neuraxis. Note that this depiction is for schematic purposes only, as the neural connections have not been formed at this stage of development. The strongest projections are to the LGN, followed by the superficial layers of the SC. (*b*) Schematic illustration of pathways to the cortex. There are two direct pathways from the retina to the dorsal thalamus, a retinogeniculate pathway (1) and a retinopulvinar pathway (2), as well as two indirect pathways that pass through the midbrain, the retinocolliculogeniculate pathway (3) and the retinocolliculopulvinar pathway (4). Both the LGN and the inferior pulvinar project to both V1 and the extrastriate visual cortex (i–iv), with the LGN projecting predominantly to V1 (i) and the inferior pulvinar projecting predominantly to the extrastriate cortex (iv). Of particular interest for undertanding blindsight are the direct extrastriate projections (iii, iv). Abbreviations: DT, dorsal thalamus; H, hypothalamus; LGN, lateral geniculate nucleus; M, medulla oblongata; P, pons; PIm, medial division of the inferior pulvinar; PT, pretectum; SC, superior colliculus; VC, visual cortex.

the direct pathways from the thalamus to the extrastriate cortex (**Figure 4*b***). These projections have been extensively investigated using retrograde tracers injected into the extrastriate cortex, which leads to dense labeling in the pulvinar and much sparser labeling in the LGN. The LGN labeling, though sparse, has been observed in many experiments (reviewed in Rodman et al. 2001, Sincich et al. 2004). Much of the dense labeling in the pulvinar can be attributed to its role as a corticocortical relay (Sherman 2005, Shipp 2003). However, some of the retrogradely labeled pulvinar neurons, and all the labeled LGN neurons, are candidates for relaying visual information from either the SC or the retina to the extrastriate cortex.

A closer examination of these pathways reveals that the extrastriate-projecting neurons in the LGN are most commonly found in the interlaminar zones. Injection of retrograde tracers into the dorsal stream (MT) or ventral stream (V4) portions of the extrastriate cortex reveal that more than half of extrastriate-projecting neurons label positively for CAMKII and calbindin, suggesting that these neurons are part of the koniocellular pathway (Rodman et al. 2001, Sincich et al. 2004). However, unlike the konicellular neurons that project to the superficial layers of V1, neurons sending projections to the extrastriate cortex have large cell bodies with a multipolar morphology, suggesting that the term koniocellular, indicating very small cells, may be inappropriate. Strangely, the LGN projections to the extrastriate cortex terminate neither in layer 4, which is characteristic of feedforward thalamic projections, nor in the supragranular layers, which is characteristic of modulatory thalamic connections (Jones 1998). Instead, the inputs are primarily directed to layer 5, where neurons project to the thalamus, striatum, and midbrain (Benevento & Yoshida 1981). It is interesting to speculate that this laminar pattern of LGN input to extrastriate cortical areas is related to the unconscious nature of the visual signals used during blindsight.

Establishing that a putative pathway actually relays visual information to the cortex is more challenging. Given the sheer number of retinal projections to the LGN, it seems likely that extrastriate-projecting LGN neurons receive a direct retinal input and send it to the extrastriate cortex. This possibility is supported by the finding, in marmosets, of presynaptic retinal afferents synapsing on MT-projecting neurons in the koniocellular layers (Warner et al. 2010). Retinal afferents were also found to synapse on MT-projecting neurons in the histochemically defined PIm subregion of the inferior pulvinar (Warner et al. 2010).

Establishing visual pathways through the SC is even more difficult, because two synapses must act as relays. As reviewed above, Pα and Pγ ganglion cells project to the superficial layers of the SC. Within the superficial layers, a subset of neurons sends projections to the LGN and a different subset sends to the inferior pulvinar (May 2006). In the latter, initial anatomical findings in the owl monkey did not show sufficient spatial overlap between SC terminals and MT-projecting cell bodies to support such a relay (Stepniewska et al. 1999). However, recent experiments in the macaque using disynaptic tracing with a rabies virus (Lyon et al. 2010) and electrophysiological identification of neural connections with antidromic and orthodromic stimulation (Berman & Wurtz 2010) argue strongly that such a relay does exist. In fact, both recent studies identified two distinct SC relays through the pulvinar to area MT: PIm, which is the same subdivision that receives direct retinal afferents, and another relay localized in the region of the inferior pulvinar immediately adjacent to the LGN. There is also circumstantial anatomical evidence supporting a relay from the SC to the extrastriate cortex through the LGN. The SC terminals are found primarily in the interlaminar zones, which, as discussed above, is similar to the distribution of most of the extrastriate-projecting neurons (Benevento & Yoshida 1981, Stepniewska et al. 1999). This pattern is consistent with the projection pattern observed in a wide range of mammals (Harting et al. 1991). However, one study found that the laminar pattern of disynaptic labeling in the SC following extrastriate injections in areas MT and V3 was more consistent with the pulvinar route than with the LGN route, suggesting that the colliculopulvinar pathway is more prominent than the colliculogeniculate pathway, at least to certain extrastriate areas (Lyon et al. 2010).

On the basis of these neuroanatomical and neurophysiological studies, each of the four potential pathways carrying visual information from the retina to the extrastriate cortex (retina-LGN-extrastriate, retina-pulvinar-extrastriate, retina-SC-LGN-extrastriate, and retina-SC-pulvinar-extrastriate) is a viable candidate to bypass V1. It is important to point out, however, that these results were established in intact animals. Following V1 lesions, a number of significant changes to the visual system occur at many levels. This is addressed in the next section.

Changes to the Visual System Following a V1 Lesion

In the weeks following an ablation restricted to V1, massive retrograde degeneration decimates the portion of the LGN corresponding to the extent of the lesion. The outcome is a nearly complete loss of magnocellular and parvocellular neurons in the affected region of the LGN (Mihailović et al. 1971). Then, over a period of months and years, the degeneration cascades to the retina and kills half of the Pβ ganglion cells (Cowey et al. 1989, Weller & Kaas 1989). Neither the interlaminar regions of the LGN, where some neurons project directly to multiple regions of the extrastriate cortex, nor the retinal projections to the SC (Dineen et al. 1982) degenerate nearly as much (Cowey 2004). Moreover, the extrastriate-projecting neurons that survive within the LGN are much larger than normal (Hendrickson & Dineen 1982) and stain positively for calbindin D28K (Rodman et al.

2001), suggesting that the koniocellular system may be strengthened following the lesion. One study found that some of the retinal input to the remaining geniculocortical neurons projecting to V4 was mediated by GABA-ergic interneurons and that a portion of the ganglion cells stained positively for GABA (Kisvárday et al. 1991). It was subsequently shown that a small proportion of retinogeniculate neurons are GABA-positive in the normal monkey optic nerve and optic tract (Wilson et al. 1996). These findings suggest that transmission of visual signals through the LGN to the extrastriate cortex is fundamentally altered following V1 damage. As the system recovers, significant changes occur to the types of viable relay neurons, the distribution of retinal and collicular inputs, and, quite possibly, the balance of excitation and inhibition.

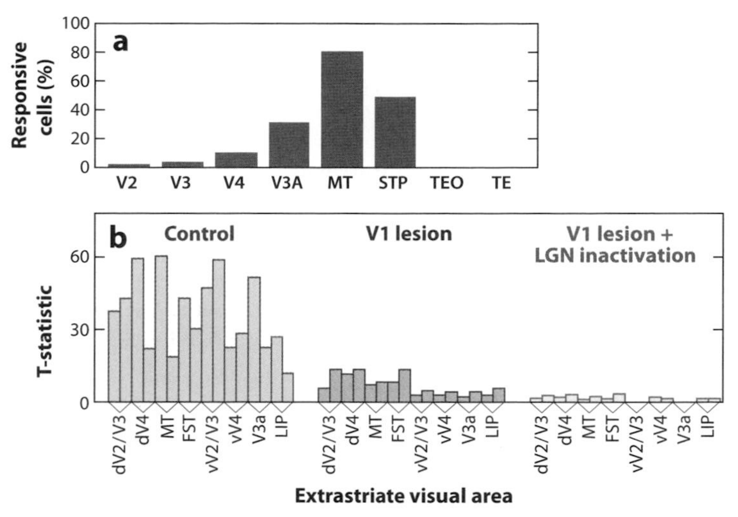

Figure 5

Extrastriate visual activation following V1 lesion and lateral geniculate nucleus (LGN) inactivation. (*a*) Residual responses of single units in multiple extrastriate visual areas following the destruction or cooling of V1 in the macaque (Bullier et al. 1994). (*b*) Functional magnetic resonance imaging (fMRI) responses in a range of extrastriate cortical areas in a normal hemisphere (*blue*), following V1 damage (*red*), and following V1 damage combined with acute inactivation of the LGN (*green*) (Schmid et al. 2010).

Extrastriate Visual Responses Without V1

Electrophysiological experiments in anesthetized macaques first demonstrated that neurons in the superior temporal polysensory area and area MT continue to respond to visual stimuli even after V1 is chronically removed, reversibly cooled, or acutely ablated (Bruce et al. 1986, Girard et al. 1992, Rodman et al. 1989). By contrast, neurons in the inferotemporal cortex are unresponsive to visual stimuli following ablation of V1 (Rocha-Miranda et al. 1975). Subsequent single-unit studies in the macaque showed that the magnitude of residual responses in extrastriate cortical areas differed between dorsal and ventral stream pathways: Dorsal areas showed a higher fraction of neurons with residual stimulus responses (summarized in Bullier et al. 1994) (see **Figure 5*a***). Studies of residual MT responses in other primates have yielded mixed results (Collins et al. 2003, Kaas & Krubitzer 1992, Rosa et al. 2000); the basis of the discrepancy is presently unknown.

Functional imaging has the advantage of simultaneously monitoring neural responses in multiple areas. It also has certain disadvantages, such as poor temporal resolution and the uncertain origin of the blood-based response. Positron emission tomography (Barbur et al. 1993) and fMRI studies (Baseler et al. 1999, Bridge et al. 2010, Goebel et al. 2001) of human blindsight patients have shown responses in the extrastriate visual cortex, particularly in area MT, to stimuli presented to the blind field. Recent fMRI studies in macaques have also shown extrastriate activity in the months following surgical ablation to V1. One study in anesthetized animals used retinotopic mapping to demonstrate preserved responses in regions V2 and V3 corresponding to the blind field (Schmid et al. 2009). Another study in awake animals showed responses in several extrastriate areas to a small stimulus confined entirely to the blind field (Schmid et al. 2010). In that study, V1-independent responses reached on average 20% of the response strength compared with the control condition (**Figure 5*b***). There was a pronounced dorsoventral asymmetry within the early extrastriate cortex: The dorsal components of areas V2/V3 and V4 showed notably higher residual activity, in agreement with a previous human study (Baseler et al. 1999).

Which Pathways Support Blindsight?

A difficult and sometimes frustrating feature of blindsight is that the experimental evidence fails to converge on a single pathway. There are at least three distinct challenges to this search. The first challenge is the biological complexity of the brain, including its parallel and redundant projections and the imperfect segregation of pathways. The second challenge is the inherent plasticity of the brain, raising the specter that the various candidate pathways change in their relative strengths over time. The third challenge is the imperfect and indirect nature of much of the evidence as it pertains to the pathways that support blindsight.

To take a concrete example, consider an electrophysiological study by Bender (1988) that used lesions to investigate potential sources of visual input into the macaque inferior pulvinar. In that study, visual responses were recorded in the pulvinar of animals that were intact or had experienced unilateral ablation of either the SC or V1. Bender found that, whereas SC ablation had minimal effects, V1 ablation completely abolished visual responses in the pulvinar. This finding suggests that SC inputs alone are unable to drive responses in the pulvinar, which would seem to refute any hypothesis of blindsight based on the colliculopulvinar pathway. However, in reference to the challenges mentioned above, rejecting the pulvinar contribution to blindsight given this finding alone would be unwise. First, regarding the biological complexity of the pathways, Bender's recordings may not have adequately sampled neurons from the two subregions of the pulvinar now suspected to be the critical visual relays (Berman & Wurtz 2010, Lyon et al. 2010). Second, regarding the inherent plasticity of the system, Bender found that, after several weeks, a few neurons in the inferior pulvinar did start to show modest visual responses. Third, regarding the indirect nature of experimental evidence, the demonstration of a physiological pathway in the anesthetized animal may or may not be related to residual visual performance in blindsight.

With these caveats in mind, a recent study points strongly to the LGN as being a critical relay in blindsight (Schmid et al. 2010). As mentioned above, following V1 ablation in macaques, fMRI responses to small stimuli in the blind field were observed in multiple extrastriate areas. Behaviorally, the monkeys were also able to respond to visual stimuli well above chance. However, following the additional pharmacological inactivation of the LGN, the residual extrastriate fMRI responses (**Figure 5*b***) as well as the monkey's behavioral performance were abolished, indicating that the LGN is critical for V1-independent vision. This result is consistent with two previous findings in macaques, one demonstrating that inactivation of all the LGN layers temporarily blocked visual responses in cortical area MT (Maunsell et al. 1990), and the other, that chemical lesions to all the LGN layers permanently abolished visual detection, with no recovery even after several months (Schiller et al. 1990). The Schmid (2010) findings also challenge explanations of blindsight that do not include the LGN. Whether the sparse direct projections from the LGN to the extrastriate cortex could support this form of residual vision has been addressed by Cowey (2010), who noted that, although the absolute number of such neurons is unknown, they are probably at least as numerous as all the retinal ganglion cells in the rat, a species that is clearly capable of visually guided behavior.

Finally, any reading of the literature makes it difficult to escape the conclusion that the SC must also be involved in blindsight. Ablation of the SC during blindsight abolishes visual performance mediated by eye movements (Kato et al. 2011, Mohler & Wurtz 1977) and visually guided reaching (Solomon et al. 1981), and it obliterates responses in the extrastriate cortex (Bruce et al. 1986, Rodman et al. 1990). The dependence on the SC has generally been interpreted as evidence for the importance of the colliculopulvinar pathway, although it is also consistent with mediation through the colliculogeniculate pathway (Rodman et al. 1990). These findings, combined with the recent results from Schmid et al. (2010), raise the possibility that retinal information reaches the extrastriate visual cortex following V1 lesions via a colliculogeniculate pathway. Whether this pathway is the ultimate answer to the blindsight puzzle, or whether the challenges outlined above will continue to keep the answer out of reach, remains to be seen.

WHAT IS THE ROLE OF V1 IN CONSCIOUS PERCEPTION?

In closing, let us consider how these and other findings illuminate the specific contribution of V1 to visual awareness. As this line of inquiry runs the danger of becoming too abstract, we formulate our question in terms of a dichotomy, which may, admittedly, also be a false one: Is V1 an essential and inseparable component of the neural processes that generate perceptual awareness, or is V1 primarily a conduit for retinal image information, receiving, processing, and passing it along to higher "perceptual" centers? Within this framework, we conclude that there is insufficient evidence to support the former proposition and that the latter is probably closer to the truth.

First, the neurophysiological results do not provide much support for the view that V1 activity is a direct contributor to visual awareness. Although V1 neural activity correlates with some aspects of perception (reviewed in Tong 2003), firing rates in V1 are only minimally affected when a stimulus is rendered completely invisible. In general, the responses of V1 neurons are much more closely tied to the sensory afferents arriving from the LGN than to perception-sensitive responses characteristic of some extrastriate visual areas.

Second, the blindness produced by V1 damage and unconscious vision supported by V1-bypassing pathways does not imply that V1 has a generative role in perception. Although damage to V1 disrupts many pathways that could contribute to visual awareness, including, for example, feedback to V1 from the extrastriate cortex (Lamme 2001), a more conservative explanation for blindness is the deafferantation of the extrastriate cortex and, possibly, the pulvinar from V1's principal feedforward visual projections. Deprived of all visual information, neither telencephalic nor higher-order thalamic centers can contribute to visual awareness. The fact that vision after V1 damage in blindsight is unconscious is not a compelling argument that V1 activity contributes directly to visual awareness. Residual visual pathways, beyond being sparse in their projections, differ from the geniculostriate pathways in many ways. They are composed mainly of Pγ and Pα channels and may involve a relay in the SC. They may draw on a special category of hypertrophic koniocellular LGN cells that project to layer 5 of the extrastriate cortex, or they may be relayed through the pulvinar exclusively to dorsal stream extrastriate cortical areas. These and many other features may help explain why the visual signals carried to the extrastriate cortex through these residual visual channels fail to reach consciousness. However, none of these explanations points to a special role for V1 in the generation of visual awareness.

Third, it is not strictly correct to say that V1 damage always leads to blindness, as pointed out frequently in the literature on human blindsight (Ffytche & Zeki 2011). In addition to the difficult task of determining what exactly blindsight patients subjectively perceive, at least two findings demonstrate that they can experience vivid visual percepts in the region of visual space corresponding to the V1 lesion. First, blindsight subject D.B. experienced "prime sight": D.B. could consciously see an afterimage generated by a visual stimulus in the blind field but, strangely, not the adapting stimulus that generated it (Weiskrantz et al. 2002). Second, when transcranial magnetic stimulation was applied bilaterally over area MT, blindsight subject G.Y. experienced perceptually visible phosphenes that traveled into his blind field (Silvanto et al. 2008). In both paradigms, the subjects were able to perceive color in the blind field when chromatic visual stimuli were applied. The bases of these phenomena are unknown, as is their generality. However, they do argue that visual awareness can occur in a region of space corresponding to a V1 lesion. Further evidence for the possibility of V1-independent visual awareness comes from humans, and quite possibly monkeys, whose vision is largely intact if their V1 damage is acquired in infancy (for a recent review, see Silvanto & Rees 2011). Based on studies in marmosets, Bourne and colleagues recently speculated that near-normal vision in the adult following V1 damage in infancy may be due to the abnormal retention of a prominent retinopulvinar pathway to area MT that, under normal conditions, is expressed only transiently in development (Bourne & Rosa 2006, Warner & Bourne 2012). Clearly this topic deserves further investigation.

In summary, the data accumulated from a wide range of anatomical, physiological, and behavioral studies in monkeys and humans paint a picture of V1 as a critical component of primate vision. Its importance, however, stems not from a direct contribution to visual awareness, but rather from its role as a highly adapted cortical lens through which the cerebral hemispheres, including the extrastriate visual cortex and other structures thought to participate directly in perception, receive visual information about the world.

DISCLOSURE STATEMENT

The author is not aware of any affiliations, memberships, funding, or financial holding that might be perceived as affecting the objectivity of this review.

ACKNOWLEDGMENTS

Thanks go to Drs. A. Maier, L. Ungerleider, Y. Chudasama, R. Wurtz, M. Schmid, and M. Mishkin for comments on the manuscript. This work was supported by the Intramural Research Programs of the National Institute of Mental Health, National Institute for Neurological Disorders and Stroke, and the National Eye Institute.

LITERATURE CITED

Azzopardi P, Cowey A. 1997. Is blindsight like normal, near-threshold vision? *Proc. Natl. Acad. Sci. USA* 94:14190–94

Baizer JS, Lock TM, Youakim M. 1997. Projections from the claustrum to the prelunate gyrus in the monkey. *Exp. Brain Res.* 113:564–68

Barbur JL, Watson JD, Frackowiak RS, Zeki S. 1993. Conscious visual perception without V1. *Brain* 116(Pt. 6):1293–302

Barone P, Batardiere A, Knoblauch K, Kennedy H. 2000. Laminar distribution of neurons in extrastriate areas projecting to visual areas V1 and V4 correlates with the hierarchical rank and indicates the operation of a distance rule. *J. Neurosci.* 20:3263–81

Baseler HA, Morland AB, Wandell BA. 1999. Topographic organization of human visual areas in the absence of input from primary cortex. *J. Neurosci.* 19:2619–27

Bender DB. 1988. Electrophysiological and behavioral experiments on the primate pulvinar. *Prog. Brain Res.* 75:55–65

Benevento LA, Rezak M. 1976. The cortical projections of the inferior pulvinar and adjacent lateral pulvinar in the rhesus monkey (*Macaca mulatta*): an autoradiographic study. *Brain Res.* 108:1–24

Benevento LA, Yoshida K. 1981. The afferent and efferent organization of the lateral geniculo-prestriate pathways in the macaque monkey. *J. Comp. Neurol.* 203:455–74

Berman RA, Wurtz RH. 2010. Functional identification of a pulvinar path from superior colliculus to cortical area MT. *J. Neurosci.* 30:6342–54

Blake R, Logothetis NK. 2002. Visual competition. *Nat. Rev. Neurosci.* 3:13–21

Blake R, Tadin D, Sobel KV, Raissian TA, Chong SC. 2006. Strength of early visual adaptation depends on visual awareness. *Proc. Natl. Acad. Sci. USA* 103:4783–88

Bonneh YS, Cooperman A, Sagi D. 2001. Motion-induced blindness in normal observers. *Nature* 411:798–801

Bourne JA, Rosa MG. 2006. Hierarchical development of the primate visual cortex, as revealed by neurofilament immunoreactivity: early maturation of the middle temporal area (MT). *Cereb. Cortex* 16:405–14

Breitmeyer BG, Öğmen H. 2006. *Visual Masking*. New York: Oxford Univ. Press

Bridge H, Hicks SL, Xie J, Okell TW, Mannan S, et al. 2010. Visual activation of extra-striate cortex in the absence of V1 activation. *Neuropsychologia* 48:4148–54

Briggs F, Usrey WM. 2009. Parallel processing in the corticogeniculate pathway of the macaque monkey. *Neuron* 62:135–46

Bruce CJ, Desimone R, Gross CG. 1986. Both striate cortex and superior colliculus contribute to visual properties of neurons in superior temporal polysensory area of macaque monkey. *J. Neurophysiol.* 55:1057–75

Bullier J, Girard P, Salin PA. 1994. The role of area 17 in the transfer of information to extrastriate visual cortex. *Cereb. Cortex* 10:1–30

Callaway E. 1998. Local circuits in primary visual cortex of the macaque monkey. *Annu. Rev. Neurosci.* 21:47–74

Casagrande VA. 1994. A third parallel visual pathway to primate area V1. *Trends Neurosci.* 17:305–10

Chatterjee S, Callaway EM. 2003. Parallel colour-opponent pathways to primary visual cortex. *Nature* 426:668–71

Collins CE, Lyon DC, Kaas JH. 2003. Responses of neurons in the middle temporal visual area after long-standing lesions of the primary visual cortex in adult new world monkeys. *J. Neurosci.* 23:2251–64

Cowey A. 2004. The 30th Sir Frederick Bartlett lecture. Fact, artefact, and myth about blindsight. *Q. J. Exp. Psychol. A* 57:577–609

Cowey A. 2010. The blindsight saga. *Exp. Brain Res.* 200:3–24

Cowey A, Stoerig P, Bannister M. 1994. Retinal ganglion cells labelled from the pulvinar nucleus in macaque monkeys. *Neuroscience* 61:691–705

Cowey A, Stoerig P, Perry VH. 1989. Transneuronal retrograde degeneration of retinal ganglion cells after damage to striate cortex in macaque monkeys: selective loss of P beta cells. *Neuroscience* 29:65–80

Cowey A, Stoerig P. 1995. Blindsight in monkeys. *Nature* 373:247–49

Crick F, Koch C. 1995. Are we aware of neural activity in primary visual cortex? *Nature* 375:121–23

Crook JD, Peterson BB, Packer OS, Robinson FR, Troy JB, Dacey DM. 2008. Y-cell receptive field and collicular projection of parasol ganglion cells in macaque monkey retina. *J. Neurosci.* 28:11277–91

Dineen J, Hendrickson A, Keating EG. 1982. Alterations of retinal inputs following striate cortex removal in adult monkey. *Exp. Brain Res.* 47:446–56

Dineen J, Keating EG. 1981. The primate visual system after bilateral removal of striate cortex. Survival of complex pattern vision. *Exp. Brain Res.* 41:338–45

Douglas RJ, Martin KAC. 2004. Neuronal circuits of the neocortex. *Annu. Rev. Neurosci.* 27:419–51

Feinberg TE, Pasik T, Pasik P. 1978. Extrageniculostriate vision in the monkey. VI. Visually guided accurate reaching behavior. *Brain Res.* 152:422–28

Ffytche DH, Zeki S. 2011. The primary visual cortex, and feedback to it, are not necessary for conscious vision. *Brain* 134:247–57

Fisch L, Privman E, Ramot M, Harel M, Nir Y, et al. 2009. Neural "ignition": enhanced activation linked to perceptual awareness in human ventral stream visual cortex. *Neuron* 64:562–74

Freese JL, Amaral DG. 2005. The organization of projections from the amygdala to visual cortical areas TE and V1 in the macaque monkey. *J. Comp. Neurol.* 486:295–317

Funk AP, Rosa MG. 1998. Visual responses of neurones in the second visual area of flying foxes (*Pteropus poliocephalus*) after lesions of striate cortex. *J. Physiol.* 513(Pt. 2):507–19

Gail A, Brinksmeyer HJ, Eckhorn R. 2004. Perception-related modulations of local field potential power and coherence in primary visual cortex of awake monkey during binocular rivalry. *Cereb. Cortex* 14:300–13

Girard P, Salin PA, Bullier J. 1992. Response selectivity of neurons in area MT of the macaque monkey during reversible inactivation of area V1. *J. Neurophysiol.* 67:1437–46

Goebel R, Muckli L, Zanella FE, Singer W, Stoerig P. 2001. Sustained extrastriate cortical activation without visual awareness revealed by fMRI studies of hemianopic patients. *Vis. Res.* 41:1459–74

Harting JK, Huerta MF, Hashikawa T, Van Lieshout DP. 1991. Projection of the mammalian superior colliculus upon the dorsal lateral geniculate nucleus: organization of tectogeniculate pathways in nineteen species. *J. Comp. Neurol.* 304:275–306

Haynes J-D, Rees G. 2005. Predicting the orientation of invisible stimuli from activity in human primary visual cortex. *Nat. Neurosci.* 8:686–91

Hendrickson A, Dineen JT. 1982. Hypertrophy of neurons in dorsal lateral geniculate nucleus following striate cortex lesions in infant monkeys. *Neurosci. Lett.* 30:217–22

Hendry SH, Yoshioka T. 1994. A neurochemically distinct third channel in the macaque dorsal lateral geniculate nucleus. *Science* 264:575–77

Hendry S, Reid R. 2000. The koniocellular pathway in primate vision. *Annu. Rev. Neurosci.* 23:127–53

Humphrey NK. 1974. Vision in a monkey without striate cortex: a case study. *Perception* 3:241–55

Isa T, Yoshida M. 2009. Saccade control after V1 lesion revisited. *Curr. Opin. Neurobiol.* 19:608–14

Jones EG. 1998. Viewpoint: the core and matrix of thalamic organization. *Neuroscience* 85:331–45

Kaas JH, Krubitzer LA. 1992. Area 17 lesions deactivate area MT in owl monkeys. *Vis. Neurosci.* 9:399–407

Kato R, Takaura K, Ikeda T, Yoshida M, Isa T. 2011. Contribution of the retino-tectal pathway to visually guided saccades after lesion of the primary visual cortex in monkeys. *Eur. J. Neurosci.* 33:1952–60

Keliris GA, Logothetis NK, Tolias AS. 2010. The role of the primary visual cortex in perceptual suppression of salient visual stimuli. *J. Neurosci.* 30:12353–65

Kisvárday ZF, Cowey A, Stoerig P, Somogyi P. 1991. Direct and indirect retinal input into degenerated dorsal lateral geniculate nucleus after striate cortical removal in monkey: implications for residual vision. *Exp. Brain Res.* 86:271–92

Klüver H. 1941. Visual functions after removal of the occipital lobes. *J. Psychol. Interdiscip. Appl.* 11:23–45

Kreiman G, Fried I, Koch C. 2002. Single-neuron correlates of subjective vision in the human medial temporal lobe. *Proc. Natl. Acad. Sci. USA* 99:8378–83

Lamme VA. 2001. Blindsight: the role of feedforward and feedback corticocortical connections. *Acta Psychol.* 107:209–28

Lee S-H, Blake R, Heeger DJ. 2005. Traveling waves of activity in primary visual cortex during binocular rivalry. *Nat. Neurosci.* 8:22–23

Leopold D, Maier A, Wilke M, Logothetis N. 2005. Binocular rivalry and the illusion of monocular vision. In *Binocular Rivalry*, ed. D Alais, R Blake, pp. 231–58. Cambridge, MA: MIT Press

Leopold DA, Aura CJ, Maier A. 2008. *Laminar analysis of local field and current source density during physical and perceptual events in monkey V1*. Presented at Annu. Meet. Soc. Neurosci., 38th, Washington, DC

Leopold DA, Logothetis NK. 1996. Activity changes in early visual cortex reflect monkeys' percepts during binocular rivalry. *Nature* 379:549–53

Leopold DA, Maier A, Logothetis NK. 2003. Measuring subjective visual perception in the nonhuman primate. *J. Conscious. Stud.* 10:115–30

Libedinsky C, Savage T, Livingstone M. 2009. Perceptual and physiological evidence for a role for early visual areas in motion-induced blindness. *J. Vis.* 9:14.1–10

Lin Z, He S. 2009. Seeing the invisible: the scope and limits of unconscious processing in binocular rivalry. *Prog. Neurobiol.* 87:195–211

Logothetis NK. 2002. The neural basis of the blood-oxygen-level-dependent functional magnetic resonance imaging signal. *Philos. Trans. R. Soc. Lond. Ser. B* 357:1003–37

Logothetis NK, Schall JD. 1989. Neuronal correlates of subjective visual perception. *Science* 245:761–63

Lund J. 1988. Anatomical organization of macaque monkey striate visual cortex. *Annu. Rev. Neurosci.* 11:253–88

Lyon DC, Nassi JJ, Callaway EM. 2010. A disynaptic relay from superior colliculus to dorsal stream visual cortex in macaque monkey. *Neuron* 65:270–79

Maier A, Adams GK, Aura C, Leopold DA. 2010. Distinct superficial and deep laminar domains of activity in the visual cortex during rest and stimulation. *Front. Syst. Neurosci.* 4:31

Maier A, Aura CJ, Leopold DA. 2011. Infragranular sources of sustained local field potential responses in macaque primary visual cortex. *J. Neurosci.* 31:1971–80

Maier A, Logothetis NK, Leopold DA. 2007. Context-dependent perceptual modulation of single neurons in primate visual cortex. *Proc. Natl. Acad. Sci. USA* 104:5620–25

Maier A, Wilke M, Aura C, Zhu C, Ye FQ, Leopold DA. 2008. Divergence of fMRI and neural signals in V1 during perceptual suppression in the awake monkey. *Nat. Neurosci.* 11:1193–200

Maunsell JH, Nealey TA, DePriest DD. 1990. Magnocellular and parvocellular contributions to responses in the middle temporal visual area (MT) of the macaque monkey. *J. Neurosci.* 10:3323–34

May PJ. 2006. The mammalian superior colliculus: laminar structure and connections. *Prog. Brain Res.* 151:321–78

Mihailović LT, Cupić D, Dekleva N. 1971. Changes in the numbers of neurons and glial cells in the lateral geniculate nucleus of the monkey during retrograde cell degeneration. *J. Comp. Neurol.* 142:223–29

Miller M, Pasik P, Pasik T. 1980. Extrageniculostriate vision in the monkey. VII. Contrast sensitivity functions. *J. Neurophysiol.* 43:1510–26

Mitzdorf U. 1985. Current source–density method and application in cat cerebral cortex: investigation of evoked potentials and EEG phenomena. *Physiol. Rev.* 65:37–100

Mohler CW, Wurtz RH. 1977. Role of striate cortex and superior colliculus in visual guidance of saccadic eye movements in monkeys. *J. Neurophysiol.* 40:74–94

Moore T, Rodman HR, Repp AB, Gross CG. 1995. Localization of visual stimuli after striate cortex damage in monkeys: parallels with human blindsight. *Proc. Natl. Acad. Sci. USA* 92:8215–18

Nassi JJ, Callaway EM. 2009. Parallel processing strategies of the primate visual system. *Nat. Rev. Neurosci.* 10:360–72

Nowak LG, Munk MH, Girard P, Bullier J. 1995. Visual latencies in areas V1 and V2 of the macaque monkey. *Vis. Neurosci.* 12:371–84

O'Brien BJ, Abel PL, Olavarria JF. 2001. The retinal input to calbindin-D28k-defined subdivisions in macaque inferior pulvinar. *Neurosci. Lett.* 312:145–48

Perry VH, Cowey A. 1984. Retinal ganglion cells that project to the superior colliculus and pretectum in the macaque monkey. *Neuroscience* 12:1125–37

Perry VH, Oehler R, Cowey A. 1984. Retinal ganglion cells that project to the dorsal lateral geniculate nucleus in the macaque monkey. *Neuroscience* 12:1101–23

Peters A, Payne BR, Budd J. 1994. A numerical analysis of the geniculocortical input to striate cortex in the monkey. *Cereb. Cortex* 4:215–29

Polonsky A, Blake R, Braun J, Heeger DJ. 2000. Neuronal activity in human primary visual cortex correlates with perception during binocular rivalry. *Nat. Neurosci.* 3:1153–59

Posner MI, Gilbert CD. 1999. Attention and primary visual cortex. *Proc. Natl. Acad. Sci. USA* 96:2585–87

Pöppel E, Held R, Frost D. 1973. Residual visual function after brain wounds involving the central visual pathways in man. *Nature* 243:295–96

Preuss T. 2007. Evolutionary specializations of primate brain systems. In *Primate Origins: Adaptations and Evolution*, ed. MJ Ravosa, M Dagosto, pp. 625–75. New York: Springer

Rocha-Miranda CE, Bender DB, Gross CG, Mishkin M. 1975. Visual activation of neurons in inferotemporal cortex depends on striate cortex and forebrain commissures. *J. Neurophysiol.* 38:475–91

Rodman HR, Gross CG, Albright TD. 1989. Afferent basis of visual response properties in area MT of the macaque. I. Effects of striate cortex removal. *J. Neurosci.* 9:2033–50

Rodman HR, Gross CG, Albright TD. 1990. Afferent basis of visual response properties in area MT of the macaque. II. Effects of superior colliculus removal. *J. Neurosci.* 10:1154–64

Rodman HR, Sorenson KM, Shim AJ, Hexter DP. 2001. Calbindin immunoreactivity in the geniculo-extrastriate system of the macaque: implications for heterogeneity in the koniocellular pathway and recovery from cortical damage. *J. Comp. Neurol.* 431:168–81

Rosa MG, Tweedale R, Elston GN. 2000. Visual responses of neurons in the middle temporal area of new world monkeys after lesions of striate cortex. *J. Neurosci.* 20:5552–63

Roseboom W, Arnold DH. 2011. Learning to reach for "invisible" visual input. *Curr. Biol.* 21:R493–94

Salin PA, Bullier J. 1995. Corticocortical connections in the visual system: structure and function. *Physiol. Rev.* 75:107–54

Sanders MD, Warrington EK, Marshall J, Wieskrantz L. 1974. "Blindsight": vision in a field defect. *Lancet* 1:707–8

Schilder P, Pasik P, Pasik T. 1972. Extrageniculostriate vision in the monkey. 3. Circle VS triangle and "red VS green" discrimination. *Exp. Brain Res.* 14:436–48

Schiller PH. 2010. Parallel information processing channels created in the retina. *Proc. Natl. Acad. Sci. USA* 107:17087–94

Schiller PH, Logothetis NK, Charles ER. 1990. Functions of the colour-opponent and broad-band channels of the visual system. *Nature* 343:68–70

Schiller PH, Malpeli JG. 1977. Properties and tectal projections of monkey retinal ganglion cells. *J. Neurophysiol.* 40:428–45

Schmid MC, Mrowka SW, Turchi J, Saunders RC, Wilke M, et al. 2010. Blindsight depends on the lateral geniculate nucleus. *Nature* 466:373–77

Schmid MC, Panagiotaropoulos T, Augath MA, Logothetis NK, Smirnakis SM. 2009. Visually driven activation in macaque areas V2 and V3 without input from the primary visual cortex. *PLoS ONE* 4:e5527

Schroeder CE, Tenke CE, Givre SJ, Arezzo JC, Vaughan HG. 1991. Striate cortical contribution to the surface-recorded pattern-reversal VEP in the alert monkey. *Vis. Res.* 31:1143–57

Sheinberg DL, Logothetis NK. 1997. The role of temporal cortical areas in perceptual organization. *Proc. Natl. Acad. Sci. USA* 94:3408–13

Sherman SM. 2005. Thalamic relays and cortical functioning. *Prog. Brain Res.* 149:107–26

Shipp S. 2003. The functional logic of cortico-pulvinar connections. *Philos. Trans. R. Soc. Lond. Ser. B* 358:1605–24

Silvanto J, Cowey A, Walsh V. 2008. Inducing conscious perception of colour in blindsight. *Curr. Biol.* 18:R950–51

Silvanto J, Rees G. 2011. What does neural plasticity tell us about role of primary visual cortex (V1) in visual awareness? *Front. Psychol.* 2:6

Sincich LC, Park KF, Wohlgemuth MJ, Horton JC. 2004. Bypassing V1: a direct geniculate input to area MT. *Nat. Neurosci.* 7:1123–28

Sirotin YB, Das A. 2009. Anticipatory haemodynamic signals in sensory cortex not predicted by local neuronal activity. *Nature* 457:475–79

Snodderly DM, Gur M. 1995. Organization of striate cortex of alert, trained monkeys (*Macaca fascicularis*): ongoing activity, stimulus selectivity, and widths of receptive field activating regions. *J. Neurophysiol.* 74:2100–25

Solomon SJ, Pasik T, Pasik P. 1981. Extrageniculostriate vision in the monkey. VIII. Critical structures for spatial localization. *Exp. Brain Res.* 44:259–70

Stepniewska I, Qi HX, Kaas JH. 1999. Do superior colliculus projection zones in the inferior pulvinar project to MT in primates? *Eur. J. Neurosci.* 11:469–80

Stoerig P. 2006. Blindsight, conscious vision, and the role of primary visual cortex. *Prog. Brain Res.* 155:217–34

Stoerig P, Cowey A. 1997. Blindsight in man and monkey. *Brain* 120(Pt. 3):535–59

Szmajda BA, Grünert U, Martin PR. 2008. Retinal ganglion cell inputs to the koniocellular pathway. *J. Comp. Neurol.* 510:251–68

Tong F. 2003. Primary visual cortex and visual awareness. *Nat. Rev. Neurosci.* 4:219–29

Tong F, Engel SA. 2001. Interocular rivalry revealed in the human cortical blind-spot representation. *Nature* 411:195–99

Tong F, Nakayama K, Vaughan JT, Kanwisher N. 1998. Binocular rivalry and visual awareness in human extrastriate cortex. *Neuron* 21:753–59

Tononi G, Koch C. 2008. The neural correlates of consciousness: an update. *Ann. N. Y. Acad. Sci.* 1124:239–61

Tsuchiya N, Koch C. 2005. Continuous flash suppression reduces negative afterimages. *Nat. Neurosci.* 8:1096–101

Warner CE, Bourne JA. 2012. *Lesions of the primate striate cortex (V1) during infancy and in adulthood differentially alter the connectivity of the middle temporal area (MT) with visual thalamic nuclei*. Presented at Annu. Meet. Austral. Neurosci. Soc., 32nd, Queensland, Aust.

Warner CE, Goldshmit Y, Bourne JA. 2010. Retinal afferents synapse with relay cells targeting the middle temporal area in the pulvinar and lateral geniculate nuclei. *Front. Neuroanat.* 4:8

Watanabe M, Cheng K, Murayama Y, Ueno K, Asamizuya T, et al. 2011. Attention but not awareness modulates the BOLD signal in the human V1 during binocular suppression. *Science* 334:829–31

Weiskrantz L. 2009. *Blindsight*. Oxford: Oxford Univ. Press

Weiskrantz L, Cowey A, Hodinott-Hill I. 2002. Prime-sight in a blindsight subject. *Nat. Neurosci.* 5:101–2

Weller RE, Kaas JH. 1989. Parameters affecting the loss of ganglion cells of the retina following ablations of striate cortex in primates. *Vis. Neurosci.* 3:327–49

Wilke M, Logothetis NK, Leopold DA. 2003. Generalized flash suppression of salient visual targets. *Neuron* 39:1043–52

Wilke M, Logothetis NK, Leopold DA. 2006. Local field potential reflects perceptual suppression in monkey visual cortex. *Proc. Natl. Acad. Sci. USA* 103:17507–12

Wilke M, Mueller K-M, Leopold DA. 2009. Neural activity in the visual thalamus reflects perceptual suppression. *Proc. Natl. Acad. Sci.* 106:9465–70

Wilson JR, Cowey A, Somogy P. 1996. GABA immunopositive axons in the optic nerve and optic tract of macaque monkeys. *Vis. Res.* 36:1357–63

Wolfe JM. 1984. Reversing ocular dominance and suppression in a single flash. *Vis. Res.* 24:471–78

Wunderlich K, Schneider KA, Kastner S. 2005. Neural correlates of binocular rivalry in the human lateral geniculate nucleus. *Nat. Neurosci.* 8:1595–602

Evolution of Synapse Complexity and Diversity

Richard D. Emes[1] and Seth G.N. Grant[2]

[1]School of Veterinary Medicine and Science, University of Nottingham, Leicestershire LE12 5RD, United Kingdom; email: richard.emes@nottingham.ac.uk

[2]School of Molecular and Clinical Medicine, Edinburgh University, Edinburgh EH16 4SB, United Kingdom: email: seth.grant@ed.ac.uk

Annu. Rev. Neurosci. 2012. 35:111–31

The *Annual Review of Neuroscience* is online at neuro.annualreviews.org

This article's doi:
10.1146/annurev-neuro-062111-150433

0147-006X/12/0721-0111$20.00

Keywords

postsynaptic density, proteome, genome, protosynapse

Abstract

Proteomic studies of the composition of mammalian synapses have revealed a high degree of complexity. The postsynaptic and presynaptic terminals are molecular systems with highly organized protein networks producing emergent physiological and behavioral properties. The major classes of synapse proteins and their respective functions in intercellular communication and adaptive responses evolved in prokaryotes and eukaryotes prior to the origins of neurons in metazoa. In eukaryotes, the organization of individual proteins into multiprotein complexes comprising scaffold proteins, receptors, and signaling enzymes formed the precursor to the core adaptive machinery of the metazoan postsynaptic terminal. Multiplicative increases in the complexity of this protosynapse machinery secondary to genome duplications drove synaptic, neuronal, and behavioral novelty in vertebrates. Natural selection has constrained diversification in mammalian postsynaptic mechanisms and the repertoire of adaptive and innate behaviors. The evolution and organization of synapse proteomes underlie the origins and complexity of nervous systems and behavior.

Contents

INTRODUCTION

The hallmark of the brain of humans and that of most other species is its high degree of complexity and diversity of neuronal morphology. Comparisons of the neuroanatomy of organisms in different phyla point to the origin of functional neuronal circuits ~600 Mya in the gelatinous Ctenophora and Cnidaria (reviewed in Lichtneckert & Reichert 2009). Despite the apparent anatomical simplicity of these organisms, their neurons possessed chemical (symmetrical and asymmetrical) and electrical synapses, as well as chemical and peptidergic neurotransmitters. Considering that morphological synapses are absent in Porifera (sponges) or earlier multicellular organisms, it is difficult to envision a set of evolutionary pressures that would have selected for the evolution of all the specialized molecular components that construct a synapse in a single step. As with the evolution of the vertebrate eye (Dawkins 1986, 1994; Nilsson & Pelger 1994), it is far more likely that the many molecular components of synapses had already existed in those earlier organisms that lacked neurons and that these components were reorganized into the visibly distinct structure we know as the neuronal synapse.

In this review we focus on the molecular evolution of synapse proteins and their organization. Akin to the neuroanatomical staining methods that exposed the striking cellular complexity of the nervous system discovered by anatomists in the nineteenth century, neuroproteomic methods in the twenty-first century uncovered a far higher degree of molecular complexity in the protein composition of synapses than had been expected from earlier electrophysiological, biochemical, or genetic studies. It is from these proteomic data that investigators can systematically explore the molecular evolution of all the individual proteins as well as their organization into networks and supramolecular structures. We first introduce the composition of synapse proteomes and genome evolution and then review the evolution of synapses.

SYNAPSE PROTEOMICS AS A BASIS FOR STUDYING SYNAPSE EVOLUTION

Because one can readily measure gross neuroanatomy, cell number, and neuronal shape, there is an extensive literature on the evolution of the nervous system based on these measurements (reviewed in Kaas 2009). By contrast, there is a paucity of studies on synapse evolution, presumably because microscopic and electrophysiological measurements are difficult to obtain and compare between species. This has been a major shortcoming in evolutionary neuroscience since synapses were identified, more than one century ago, as the basis for neuronal connections and information transfer within neuronal circuits.

In the late 1980s and 1990s, the application of complementary DNA (cDNA) cloning methods identified the genes for a number of ion channel subunits and a relatively small number of other synaptic proteins. At this point genome sequence data were sparse, and thus the scope of enquiry into the range of species and ultimately the ancestry of the genes could not be readily determined. Moreover, these were mostly single gene studies and did not explain the evolution of the synapse itself: The synapse is a macromolecular subcellular structure that is assembled from the protein products of many genes, and it is necessary to examine the evolution of its composite sets of proteins.

The year 2000 was a significant turning point for two reasons. First, proteomic methods were applied to the study of synapses, which revealed unexpectedly large numbers of proteins (Husi et al. 2000). Proteomics has discovered more synapse proteins than has any other approach and has provided the necessary starting information for molecular studies of synapse evolution. Second, the draft sequence of the human genome was released, and the ensuing explosion of genome sequencing from many organisms provided essential data for systematic studies of the phylogeny of individual genes, sets of genes, and whole organisms. The combination of synapse proteomics and genomics has been at the center of the first systematic studies of synapse evolution.

MOLECULAR COMPLEXITY IN THE SYNAPSE PROTEOME

Catalogs of proteins found within mammalian synapses have been derived from mass spectrometry profiling of whole synapses, pre- and postsynaptic fractions, and subcellular structures including synaptic vesicles and signaling complexes (reviewed in Bayes & Grant 2009). In this review we separate discussion of the presynaptic proteome (PreSP) from the postsynaptic proteome (PSP) because most evolutionary research has been performed on the PSP. This is also a useful distinction with regard to function because these proteomes have separate origins in unicellular organisms: The presynaptic release machinery is composed largely of the vesicular release machinery used by unicellular organisms to release chemicals or output information into their environment, whereas the postsynaptic machinery is the point on the cell surface at which information from the environment is received, sensed, or input to the cell.

Presynaptic proteome (PreSP): complete set of identified proteins in the presynaptic terminal of the synapse

Postsynaptic proteome (PSP): complete set of identified proteins in the postsynaptic terminal of the synapse

The mammalian PreSP comprises hundreds of proteins centered around the vesicular release of the neurotransmitter, which occurs in response to the invasion of the action potential into the presynaptic terminal. The ternary complexes formed by synaptobrevin, synaptotagmin, and SNAP25 in mammals form a core structure derived from invertebrate and eukaryotic proteins. Purification of the rodent PreSP coupled with prediction of interacting partners identified 117 core proteins including 32 proteins involved in trafficking [including adapter protein (AP)] complex, syntaxins, synapsins, and synaptotagmins), 22 signaling molecules (including G proteins and 14-3-3 proteins), and 23 cytoskeletal proteins (including actins, septins, and tubulins) (Abul-Husn et al. 2009). Proteomic analysis of the presynaptic active zone with docked synaptic vesicles identified 240 proteins including many plasma membrane and synaptic vesicle proteins (Morciano et al. 2009). Although much is known about the function of many of the individual proteins in neurotransmitter release, by contrast with the PSP relatively little is known about the organization of presynaptic molecular networks and how their complexity is functionally integrated. The PreSP appears to be less complex than the PSP, perhaps because its function is simpler—to reliably release signals when instructed by the arriving action potential—whereas the PSP needs to decode a wide variety of signals in the patterns of neurotransmitter release and other signals from the extracellular environment.

More than 1000 proteins have been identified in the PSP of mammalian brain excitatory synapses (Bayes et al. 2010; Cheng et al. 2006; Collins et al. 2006; Dosemeci et al. 2006, 2007; Fernandez et al. 2009; Hahn et al. 2009; Jordan et al. 2004; Peng et al. 2004; Satoh et al. 2002; Selimi et al. 2009; Trinidad et al. 2005,

Postsynaptic density (PSD): a dense structure observed beneath the postsynaptic membrane of vertebrate synapses using electron microscopy

MASC: MAGUK-associated signaling complex containing ionotropic NMDA and metabotropic subtypes of glutamate receptors

MAGUK: membrane-associated guanylate kinase

2008; Walikonis et al. 2000; Yoshimura et al. 2004). Less than 10% of these proteins are neurotransmitter receptors, which highlights that the majority of PSP proteins are not directly involved in electrophysiological functions and instead perform a plethora of signaling and regulatory roles. It has therefore been of great importance to understand how this high numerical complexity can be simplified, understood, and represented in a logical framework. Toward this objective, tools that classify individual proteins into their respective functional types by structure, protein domain composition, and organization and interactions with other proteins have been used (Bayes et al. 2010; Emes et al. 2008; Pocklington et al. 2006a,b). These generate molecular networks that reveal an architecture or organization with several key features. From the membrane, where receptors and channels reside and information is first received by the neuron, to the most downstream of cytoplasmic signaling pathways is a hierarchy (upstream to downstream) of highly complex networks (Coba et al. 2009). Within these networks are pathways, modules or groups of functionally similar proteins, and proteins that are highly connected (hub proteins), many of which are scaffolding proteins that assemble other proteins into multiprotein signaling complexes.

This architecture suggests that activation of a neurotransmitter receptor orchestrates a multitude of intracellular proteins via protein interactions and that these are in many classes of effectors: ion channels, receptors, and structural, biosynthetic, metabolic, and signaling enzymes. This network model of signaling was supported by phosphoproteomic experiments such as those showing that the activation of the *N*-methyl-D-aspartate receptor (NMDAR) in mouse hippocampus slices produced simultaneous changes in phosphorylation of more than 130 postsynaptic density (PSD) proteins on >200 phosphorylation sites (Coba et al. 2009). Moreover, the detailed study of the phosphorylation sites and their network relationships showed seven types of phosphorylation building blocks that are used in combination on different proteins to perform regulatory roles (Coba et al. 2009). These studies emphasize that the PSP is a molecular system employing an elaborate multidimensional interrelationship of hundreds of proteins that require mathematical methods to represent their structure and function. Thus in considering the evolution of the synapse, one must draw attention not only to the origins of the proteins, their domains, and regulatory sites, but also to their organization and physical interrelationships into higher-order structures.

Examples of such structures found within the PSP are the signaling complexes, which play a central role in detecting and processing the information that arrives at the postsynaptic terminal (**Figure 1**). They have also served as more manageable sets of proteins for experimental manipulation. The prototype postsynaptic complex is known as MASC (MAGUK-associated signaling complex) comprising ~10% of all vertebrate PSP proteins. MASC can be physically isolated from the brain using affinity purification methods (Fernandez et al. 2009, Husi et al. 2000). MAGUK proteins are scaffold proteins in the membrane-associated guanylate kinase family and have no enzymatic activity but contain protein-binding domains that allow receptors and enzymes to act in close proximity (Good et al. 2011, Nourry et al. 2003). MASC contains the principal postsynaptic machinery involved in synaptic transmission and synaptic plasticity: ionotropic and metabotropic glutamate receptors, potassium channels, cell-adhesion proteins, and MAGUK and other scaffold proteins as well as their associated signaling enzymes and structural proteins (Bayes et al. 2010; Collins et al. 2005, 2006; Emes et al. 2008; Husi et al. 2000; Pocklington et al. 2006b). The functional importance of the prototypical MAGUK protein called postsynaptic density 95 (PSD-95) was shown using knockout mice, which had impairments in synaptic plasticity, learning, and other forms of behavioral adaptation (Migaud et al. 1998). Mice carrying mutations in other MASC proteins also show impairments in synaptic plasticity and adaptive behaviors, indicating

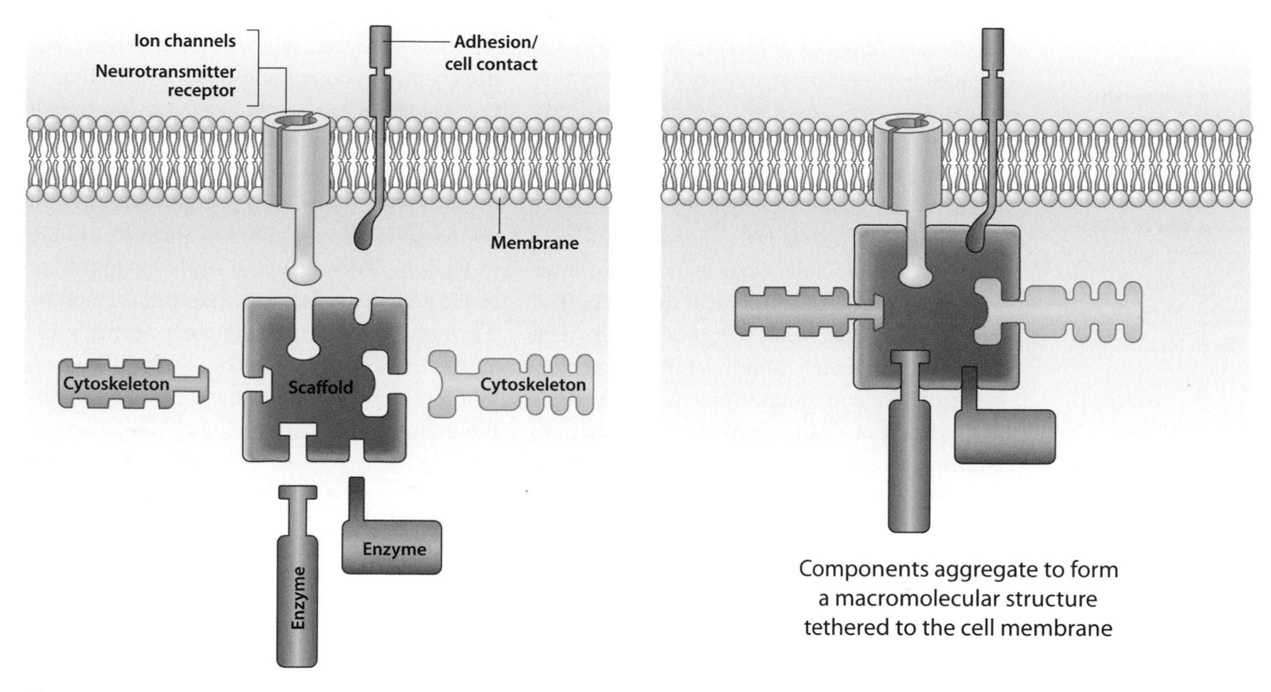

Figure 1

Organization of signaling complexes. The physical interaction of multiple types of proteins builds multiprotein complexes to allow signals to be received from the environment and communicated to intracellular biological processes and pathways. The left panel shows individual types of proteins illustrated with specific binding sites that attach to the scaffold protein, and the right panel shows them assembled into the aggregate multiprotein complex.

that these adaptive responses are an emergent property of this set of interacting proteins.

GENOME EVOLUTION AND ITS ROLE IN SYNAPSE EVOLUTION

Understanding the origins of synapses or any other aspect of brain evolution is inextricably interlinked with understanding the evolution of genomes (reviewed in Lynch 2007). Here we remind the reader of some basic principles and identify the types of genomic mutation that played key roles in synapse evolution.

The major forces affecting genome evolution are mutation, duplication, and deletion. Mutation of DNA can occur as single nucleotide events, as single nucleotide polymorphisms (SNPs) in populations of individuals, or as larger insertion or deletion (indel) events. These changes in the DNA may be silent or affect mRNA abundance, whereas if they occur in the coding sequence, they may affect the structure and hence function of the encoded protein. At the level of the gene, two major processes are associated with evolution: gain and loss. The de novo formation of a new gene from noncoding DNA is relatively rare. However, duplication of genes from existing genes or exons is more common. The formation of gene duplicates is associated with relaxation of selection pressure, allowing potential for rapid diversification. The fate of a duplicated gene will be to undertake a related role (subfunctionalization), to develop a new function (neofunctionalization), or to become nonfunctional by accumulation of deleterious mutations (pseudogenization) (Hurles 2004). More dramatic evolutionary events such as whole-genome duplication can also occur such as the two rounds of duplication in the Cambrian period (between 500 and 600 Mya) prior to the divergence of the vertebrate animals (Van de Peer et al. 2009).

Synaptome: the complete protein complement of the synapse. The synaptome is the sum of the PreSP and PSP

The comparison of genes from different organisms provides a wealth of information on which genes are shared and hence which functions we may predict for different species. The central tenet of these methods is to look for similarities between gene or protein sequences that are greater than we would expect by chance. If we detect sequence similarity, the most parsimonious explanation is to infer homology between compared sequences and predict that the genes or proteins share a common ancestor. Additional methods may compare domain composition between genes and genomes. Most protein domains are ancient and are found either at the origin of the eukaryotes or are even shared between different kingdoms (Ekman et al. 2007). However, the number of domain combinations or domain architecture dramatically increased in eukaryotes in a process termed domain accretion (Koonin et al. 2000). For example, Chothia et al. (2003) reported that vertebrate genomes contain 2.5 fold more domains per protein family compared with invertebrates. Protein domains conserved in the eukaryote PSP (present in metazoa and yeast but not detected in prokaryotes) show the incorporation of proteins containing domains involved in signal transduction, vesicle-mediated and intracellular protein transport, and ATP synthesis coupled proton transport (**Figure 2**). Molecular comparisons using these methods dominate the field of phylogenetics, which attempts to construct representations of the relationship between genes or species. This is not a trivial task, and the abundance of molecular data has revealed the complexity of gene transfer, duplication, and loss, making the drafting of a universal tree of life for all genes a complicated endeavor. However, a generally accepted relationship of species has emerged, and we show a subset of the tree of life representing species discussed within this review with some of their synaptic proteome features (**Figure 2**).

For ease of comprehension, this review describes our understanding of the origins of synapse genes and the accumulating complexity of the PSP in a step-wise manner from the most distant ancestral prokaryote organisms to a range of invertebrate and vertebrate organisms with nervous systems. We are not implying that there was a direct evolutionary trajectory from prokaryotes to vertebrate synapses by climbing the *scala naturae*. This approach is useful to identify conserved proteins and protein domains in different groups of organisms and to expand our understanding of the composition of the genetic toolkit for synapse construction. Thus one can examine the components available in the common ancestor of two groups and examine the evolution of the synaptome in different lineages.

PROKARYOTE ORIGINS OF SYNAPSE PROTEINS AND CORE PATHWAYS

Comparative genomics of the synapse has used two complementary strategies: studies of specific protein classes [e.g., glutamate receptors or scaffold proteins (Kosik 2009, Ryan et al. 2008)] or more comprehensive studies of the multiple classes of proteins comprising the range of proteins that constitute the synapse proteome (Bayes et al. 2010, Emes et al. 2008, Kosik 2009, Pocklington et al. 2006b, Ryan et al. 2008, Ryan & Grant 2009). This latter approach has tackled the problem of identifying the most ancient elements of the PSP: those conserved between vertebrates and prokaryotes (**Figure 2**). A novel method was applied to compare each mammalian PSP gene to a set of 28 bacterial and archaeal genomes (Emes & Grant 2011). By comparing species from the three superkingdoms of eukaryota, bacteria, and archaea, we identified conserved genes predating the last eukaryotic common ancestor. Using this sensitive method to identify homologs, 28.5% of genes encoding the human PSP were conserved in all superkingdoms. These represent a diverse range of 65 family types including enzymes, ribosomal proteins, and kinases. Among these, 61 genes were conserved across all the bacterial and archaeal species tested. Some of these are potentially spurious, resulting from horizontal gene

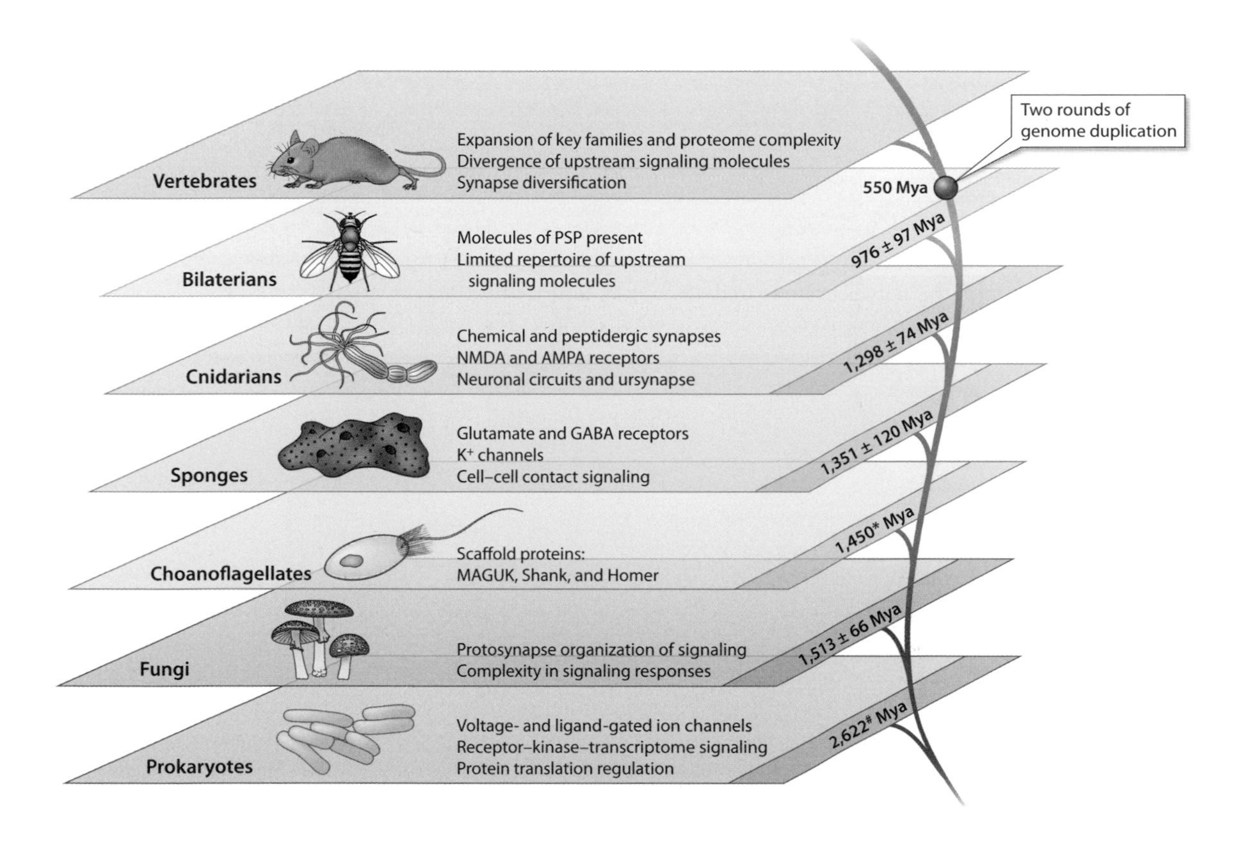

Figure 2

Cladogram of taxonomic groups and origins of PSP components. A generalized cladogram showing the groups discussed and the types of molecules and functions found in the mammalian PSP are indicated (modified from Ryan & Grant 2009). These represent the genetic toolkit for synapse construction. Dates indicate divergence time in millions of years +/− error estimates of divergence (Hedges et al. 2004). *Estimate of divergence based on midpoint of adjacent nodes (Hedges et al. 2004). #Weighted average divergence time of vertebrates and eubacteria as calculated by time tree (Hedges et al. 2006).

transfer (HGT) events between prokaryotes and eukaryotes (Koonin 2003, Lawrence & Hendrickson 2003) such as synthetase genes (Koonin et al. 2001, Wolf et al. 1999). However, ribosomal proteins, translation elongation factors, lyases, chaperone proteins, and the G protein OLA1 are monophylogenetic without obvious evidence of HGT and are therefore candidates for PSP homologs shared since the last common ancestor of prokaryotes and eukaryotes (Emes & Grant 2011). To determine the conserved biological functions underlying the detection of homologous genes, it is useful to investigate the conservation of protein domains. Owing to the functional information known about conserved domains (Finn et al. 2010, Marchler-Bauer et al. 2005, Schultz et al. 2000), we can link domains to biological processes shared between prokaryotes and eukaryotes. The domains conserved in all three superkingdoms reflected basic biological processes, such as translation, carbohydrate metabolic process, glycolysis, and tRNA aminoacylation for protein translation. Comparison of eukaryotic PSP biochemical pathways with a representative prokaryote identified ten pathways that had a significantly greater number of prokaryote-eukaryote

Prokaryote

Environment

Receptor sensor

Signal

Enzyme

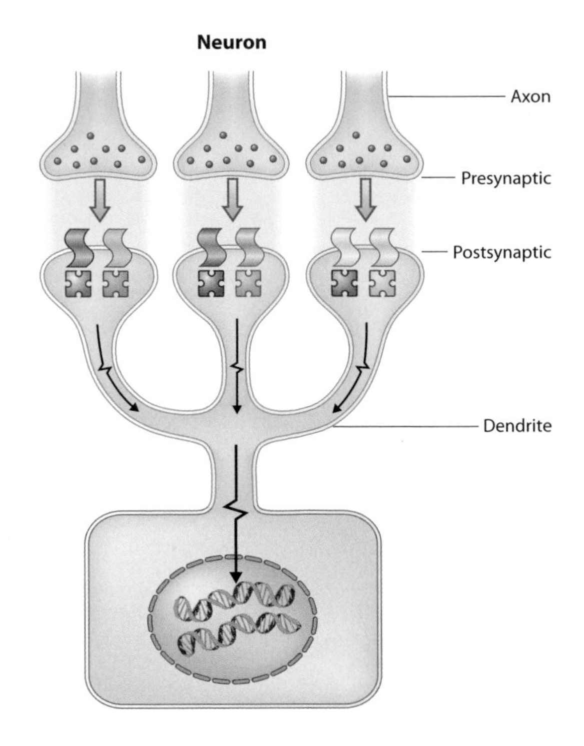

Figure 3

Sensing of the environment by prokaryotes and mammalian synapses. In the prokaryote (*left panel*) the external environment (*blue*) may be a nutrient or diffusible signal that stimulates membrane proteins (receptor sensor) that can trigger intracellular kinases (enzyme) that regulate transcription. Similar mechanisms are conserved in mammalian synapses (*right panel*); however, the sensing proteins are on the postsynaptic terminal where they respond to pulses of neurotransmitter released from the presynaptic terminal, which in turn is receiving action potentials from sensory organs. Multiple varieties of mammalian receptors and intracellular enzymes produce synaptic diversity.

homologs than expected by chance. Pathways involved in energy generation [e.g., the tricarboxylic acid (TCA) cycle] and fatty acid biosynthesis are conserved in a near complete state. In addition to the specific information about the conserved domains and proteins, these data provide evidence that some of the organization between the proteins in prokaryotes is conserved: a proportion of the homologs detected were interacting proteins prior to the divergence of prokaryotes and eukaryotes (Emes & Grant 2011).

In recent years these biosynthetic proteins and pathways have been found to be important in vertebrate and invertebrate mechanisms of synaptic plasticity. It is also important to note that there is conservation of plasticity mechanisms seen in environmental sensing proteins (**Figure 3**). The *Escherichia coli* ArcAB two-component signal transduction system is involved in responses to anaerobic environments; the chemotaxis protein CheA of the chemotactic signal transduction system and the sensor kinase ArcB (a membrane-associated protein) were all conserved. Moreover, the bacterial membrane has many voltage- and ligand-gated ion channels, which control energy production, mechanosensation, motility, and resting membrane potential (Martinac et al. 2008). Furthermore, Kralj et al. (2011) recently showed that the membrane potential of *E. coli* is dynamic and shows electrical spiking, albeit slower than in neurons. There is also evidence of prokaryote ancestry to the neurotransmitter receptors that initiate the postsynaptic response in the mammalian brain. For example, the key structural features of ligand-gated ion channels are conserved

(Bocquet et al. 2007, 2009; Nury et al. 2011), and in the case of the glutamate receptors, there is evidence of conservation of glutamate binding domains (Janovjak et al. 2011, Nakanishi et al. 1990, Sprengel et al. 2001).

This conservation from receptor-to-transcriptome signaling is interesting to consider in terms of environmental sensing and adaptive behaviors (**Figure 3**). In the case of the unicellular prokaryote, the sensing is initiated at the cell's surface, where it is in direct contact with the environment. In the case of the brain, the environment of the outside world is detected by sensory end organs (e.g., eyes, ears), which convert information into patterns of action potentials that are transmitted by nerve conduction to the synapses in the brain. These action potentials stimulate releases of pulses of neurotransmitters into the local extracellular environment where the receptors and signaling systems in the PSP are activated. The sets of synapse proteins comprising receptors and their signaling and biosynthetic pathways arose in prokaryotes, and their role in enabling the prokaryotic organisms to respond and adapt to changing environments appears to be broadly the same role they perform in the brain.

EUKARYOTE INNOVATIONS AND THE PROTOSYNAPSE

The emergence of eukaryotic cells was marked by larger and more complex genomes, linear chromosomes requiring capping with telomeres, and multiple replication origins. The increase in complexity of the transcriptome was marked by a shift from prokaryotic operons to splicing and novel RNA regulator machinery programming the proteome of complex subcellular structures, including membrane-enclosed organelles and the cytoskeleton. The vesicular machinery (which later evolved into neurotransmitter release machinery in metazoans) allowed the active movement and engulfing of material by phagocytosis, and this was likely a key step in the origins of the eukaryote: Predation of aerobic bacteria by an ancestral eukaryote cell resulted in the symbiosis of the genomes of engulfed bacteria to form the mitochondria as well as contributing expansions to the genome of the stem eukaryote (Lynch 2007). It is fascinating to consider that the vesicular mechanism that might have been responsible for eukaryotic genomic complexity, and thus the complex biology of eukaryotes, also underpinned the mechanisms of neurotransmitter vesicle recycling, which is a central function of the presynaptic terminal.

A comparison of mouse PSP orthologs with 19 eukaryote species (fungi, invertebrates, nonmammalian vertebrates, and mammals) revealed extensive conservation of the components of the synapse across the eukaryota (Emes et al. 2008). Approximately 23% of genes tested had a detectable ortholog in the yeast *Saccharomyces cerevisiae*, and this rose to ~45% having detectable orthologs in *Caenorhabditis elegans* or *Drosophila melanogaster* (these numbers should not be directly compared with those reported for the prokaryote above because here we are describing identified orthologs and hence expect to see a relatively lower percentage than that for all homologs detected by the method described above). With this finding it is evident that the genome of the common ancestor of mammals and *S. cerevisiae* that obviously lacked a nervous system or morphological synapse harbored many of the genes used to encode the constituents of the functional synapse.

As with the conserved prokaryote genes, many conserved eukaryotic genes in the PSP encode environment-sensing mechanisms driving signal transduction pathways and basic cellular functions such as protein synthesis and degradation enzymes controlling turnover of synaptic proteins (Emes et al. 2008). The detection of yeast orthologs of *NF1* (*ira2*), *PKA* (*tpk2*), *Erk2* (*fus3*), and *GNB5* (*ste4*) members of canonical pathways regulating transcription, cell morphology, and adhesion downstream of nutrient- and pheromone-sensitive GPCRs (Elion et al. 2005; Erdman & Snyder 2001; Harashima et al. 2006; Palecek et al. 2002) suggests that these components of synaptic pathways regulating protein synthesis and structural plasticity in mammals conduct analogous roles

Protosynapse: complex of synaptic proteins present in early metazoans without a defined nervous system

in unicellular responses to environmental cues (ions, nutrients) and cell-cell communication.

Although a complete analysis of PSP homologs in choanoflagellates (free living unicellular flagellates) or sponges is lacking, analysis of key PSP genes in these organisms has proved informative in understanding the organization of the networks of interacting proteins at the base of the eukaryotes. The scaffold proteins such as the MAGUKs were present in the Opisthokont common ancestor: They are detectable in the demosponge *Amphimedon queenslandica* (Sakarya et al. 2007) and in the choanoflagellate *Monosiga brevicollis* (Alie & Manuel 2010). In addition to these, sponges express GABA (γ-aminobutyric acid) receptors, K^+ channels (KIR), and metabotropic G protein–coupled glutamate receptors (MGluR). Notably absent are the NMDA ionotropic glutamate and AMPA (α-amino-3-hydroxy-5-methyl-4-isoxazolepropionic acid) receptors essential for mammalian synaptic plasticity (Richards et al. 2008); however, they are present in the cnidarian *Nematostella vecensis*, which possesses a nerve net (Sakarya et al. 2007). Additionally, multiple cnidarian species contain genes for Na_v family ion channels (Liebeskind et al. 2011).

Together these data suggest that the organization of proteins that is typical of postsynaptic multiprotein complexes—scaffold proteins, receptors, and enzymes—were present in choanoflagellates, Cnidaria and Porifera. This set of proteins in these organisms and other unicellular eukaryotes has been referred to as the protosynapse (Emes et al. 2008) because it comprises the organization of sets of proteins that confer key signaling properties on synapses (**Figure 1**). This physical organization is highly relevant in eukaryotic biology because the multiprotein complexes assembled by scaffold proteins are important for controlling the flow of cellular information in a multitude of settings (Good et al. 2011). In line with the observations on the interactomic and phosphoproteomic networks of the PSP (Coba et al. 2009, Pocklington et al. 2006b), the assembly and partitioning of proteins into complexes produce modularity and higher-order regulatory mechanisms in information processing such as amplification and forms of switching, and again these emergent properties of multiprotein signaling complexes are found in unicellular eukaryotes (Good et al. 2011).

The ancestral function of the protosynapse is thought to have provided the link between Ca^{2+} signaling and actin cytoskeleton regulation (Alie & Manuel 2010). Additionally, analysis of *M. brevicollis* has revealed the presence of multiple tyrosine kinases involved in the response to environmental stimuli (King et al. 2003, King & Carroll 2001, Manning et al. 2008, Pincus et al. 2008). This finding, again, suggests that the functional roles of the protosynapse may have been sensing and responding to a changing environment; in doing so, it can be engaged and driven by different classes of receptors and hence respond to different kinds of environmental signals. The proteomics of mammalian MASC complexes also shows that multiple types of receptors can connect to the scaffold protein complexes (Fernandez et al. 2009).

Complementary to the power of comparative genomics is the need to perform comparative proteomics on synapse proteins and complexes. In the case of humans, mice, and rats, it is clear that similar sets of proteins comprising the PSD can be isolated with an overall complexity of between 1 and 2000 proteins (Bayes et al. 2010; Cheng et al. 2006; Collins et al. 2006; Dosemeci et al. 2006, 2007; Fernandez et al. 2009; Hahn et al. 2009; Jordan et al. 2004; Peng et al. 2004; Satoh et al. 2002; Selimi et al. 2009; Trinidad et al. 2005, 2008; Walikonis et al. 2000; Yoshimura et al. 2004). The first example of the isolation of an invertebrate synapse proteome was the isolation of the synaptic MASC from *Drosophila* (fMASC) and its direct comparison with its mouse counterpart (mMASC) (Emes et al. 2008). Surprisingly, 220 fMASC proteins were identified showing that the MASC of fly and mouse [mMASC 186 proteins when isolated using a similar technique (Collins et al. 2006, Husi et al. 2000)] was of comparable size. However, although mMASC and fMASC are approximately equal in size, major differences in the types of proteins were identified. By functional annotation investigators revealed that upstream signaling/structural components (receptors, scaffolds, signal transduction molecules) accounted for ~25% of the fMASC proteome compared with >60% of the mMASC. When the composition of the fMASC was compared with yeast, 71% of fMASC genes were also found in the yeast *S. cerevisiae*, and hence only 29% appeared to be of metazoan origin. Thus the majority of downstream components were present in yeast, while the upstream signaling/structural components of fMASC and mMASC showed lineage-specific functional expansions. Thus the core functionality of the protosynapse machinery that comprises MASC evolved in unicellular eukaryotes and lineage expansion of upstream signaling molecules (such as receptors and their directly associated cytoplasmic proteins) in metazoans increased the molecular complexity of this machinery (**Figures 1** and **4**). Together the comparative proteomic and genomic studies reveal that invertebrates evolved synapses with highly complex molecular composition built around the protosynapse.

COMPLEXITY AND DIVERSITY IN METAZOAN SYNAPSES

The molecular phylogeny described above indicates that the core functionality in protein components, pathways, and organization in mammalian synapses was present prior to the first morphologically visible synapses, such as synapses in Cnidaria that evolved ~900–1400 Myr ago (**Figure 2**). The protosynapse machinery with its specialized environmental sensing capacity and regulation of transcriptome responses for adaptive changes evolved before this and was incorporated into the ursynapse. The simple nervous systems of Cnidaria were the precursors of the highly elaborate and diverse nervous systems that characterize the multitude of invertebrates and vertebrates that subsequently arose. In these anatomically complex brains, some with enormous numbers of synapses, there is considerable anatomical diversity in the postsynaptic dendritic spines. It is therefore of great interest to understand how diversity in populations of synapses arose and if this was relevant to the molecular organization of the protosynapse.

Ursynapse: last common ancestor of all morphological synapses from which all extant synapses evolved

A major event in the diversification of biological functions that characterize chordates and thus all vertebrates was the 2 complete genome duplication events ~550 Myr ago (Van de Peer et al. 2009) (**Figure 4*b***). This period corresponded with the Cambrian explosion, when there was a dramatic increase in the diversity of animal life in the fossil record, which was presumably accompanied by a diversification in nervous system complexity. Comparison of 13 vertebrate genomes (including primate, rodent, fish, chicken, and opossum genomes) showed a step-wise expansion from invertebrates in the number of conserved PSP homologs to ~85–90% conservation (Emes et al. 2008). This PSP expansion in complexity event coincides with the predicted genome duplication, where these two rounds of genome duplication are thought to have occurred prior to the divergence of the hagfish and lampreys (Holland 2009). The two rounds of genome duplication typically expanded gene families to four copies; however, some gene families have lost copies and other gene families have gained further copies by individual gene duplication events (**Figure 4*b***) (Van de Peer et al. 2009). Further evidence for the importance of gene duplication or gene retention following whole genome duplication comes from the comparison of protein domains. The number of domain types did not increase to the same extent that gene number did, which suggests that the synapse proteome expansion seen in vertebrate genomes does not represent a recruitment of proteins containing new domain types but rather the expansion of protein types already present in the PSP of early branching metazoans such as flies and worms.

Examining the expansion in the vertebrate PSP further showed that the protein families that had the greatest expansion were the upstream signaling proteins (receptors, adhesion proteins in the membrane, and their proximal

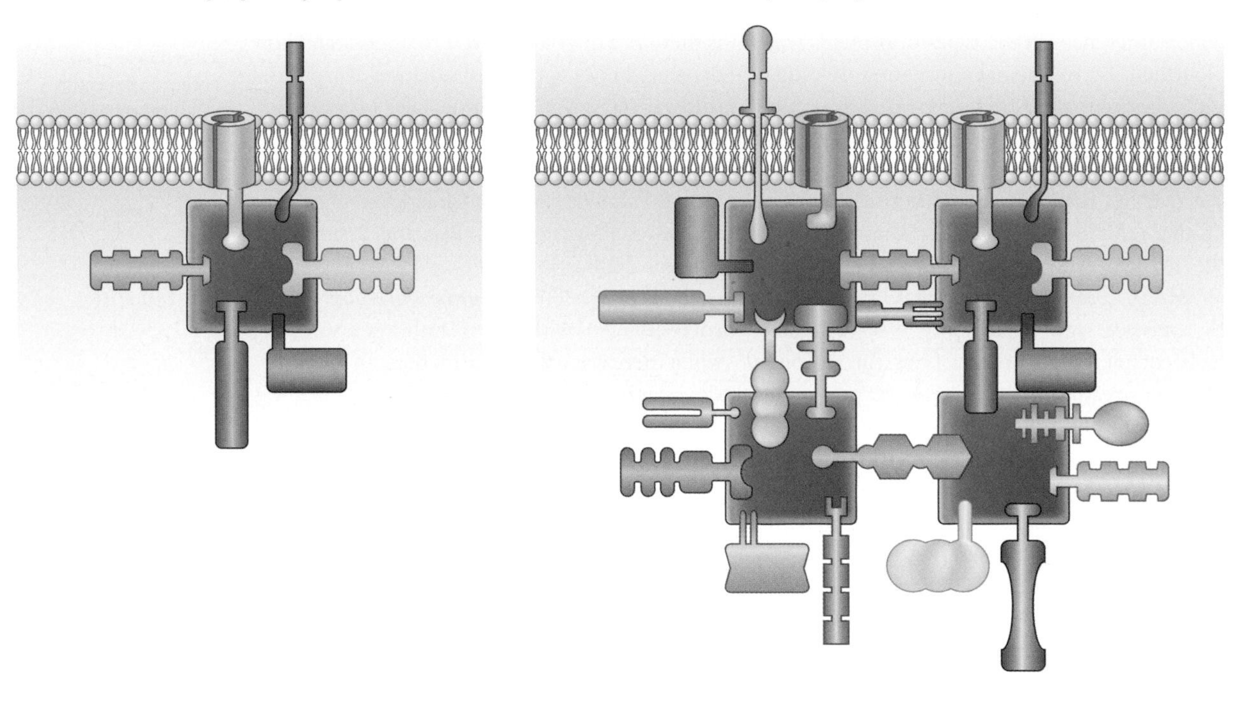

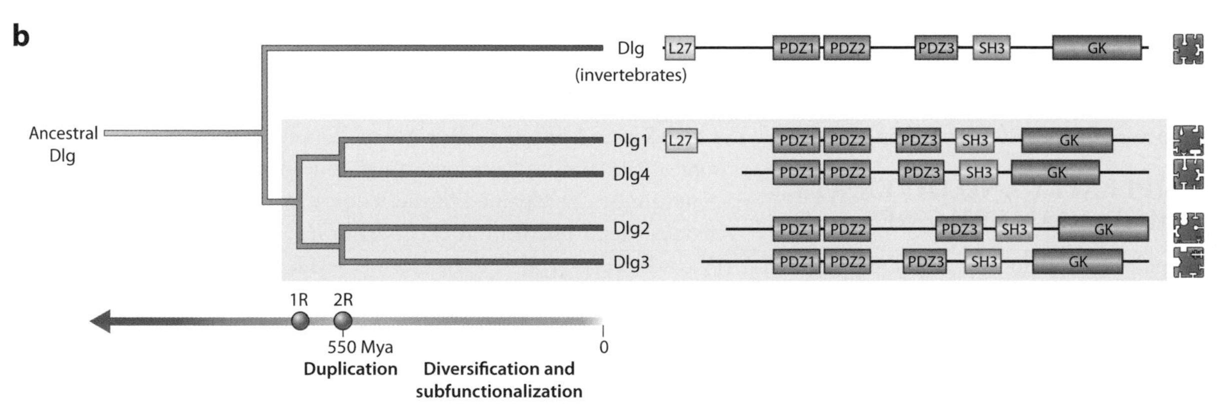

Figure 4

Genome duplication expands protosynapse complexity. (*a*) The basic components of the core signaling complexes are found in eukaryotes (*left panel*) have been multiplied by the process of gene duplication (see panel *b*), producing greater varieties of component proteins for vertebrate synapses (*right panel*). (*b*) Gene duplication of a MAGUK scaffold protein. The ancestral *Dlg* gene found in invertebrates was duplicated twice (1R and 2R) around 500–600 Mya resulting in 4 vertebrate paralogs. Following duplication, the accumulation of sequence diversity in each paralog results in functional and structural diversification of each Dlg protein. The conserved domain structure of the invertebrate Dlg and mammalian Dlg1–4 is shown (PDZ, SH3, and GK domains illustrated). The red shapes (*adjacent to each Dlg protein on right*) indicate that Dlg encodes the core scaffold proteins shown in **Figures 1**, **4*a***, and **5** with the same shape. Genome duplication similarly increased the complexity of many other PSP gene families in vertebrate lineages.

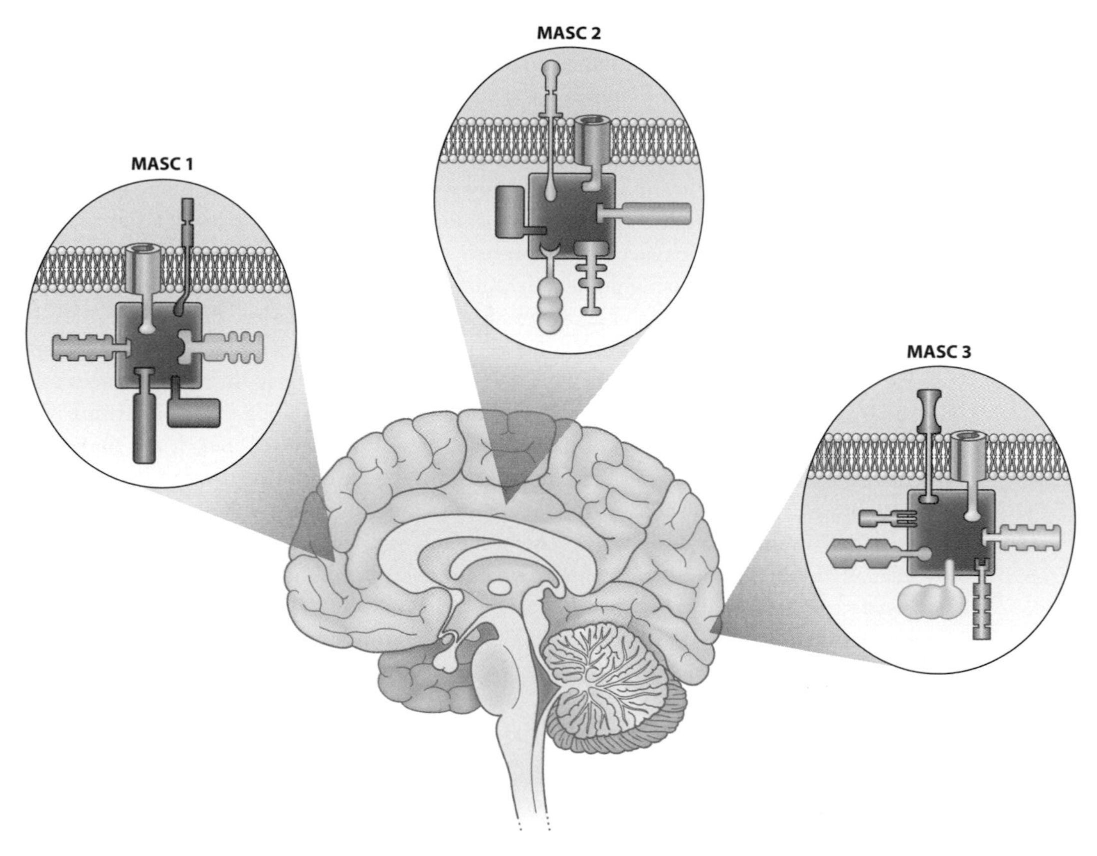

Figure 5

Synapse diversity in the mammalian brain is generated by combinations of protosynapse and MASC proteins. Three varieties of MASC complexes (labeled MASC1, 2, 3) comprising central scaffold proteins bound to receptors and adhesion protein, enzymes, and cytoskeleton are shown as in **Figure 1**. The variation in shapes of the components between the three complexes indicates that they are paralogs in expanded vertebrate gene families arising from duplication of the ancestral genes. The paralogs arising in early vertebrate evolution played a major role in diversifying neuroanatomical function.

associated proteins), and the families maintaining equal numbers of genes were those encoding the downstream cytoplasmic signaling proteins (Emes et al. 2008). This finding shows that the hierarchy of the PSP network was subject to differential expansion, and these upstream protein families were presumably retained and potentially diversified by sub- or neofunctionalization since their duplication.

The expansion in upstream proteins specifically indicates that there was adaptive advantage in the diversity of neurotransmitter receptors and adhesion proteins in the vertebrate nervous system. Obviously this diversity could generate different MASC and PSP combinations, which could ultimately be expressed in different synapse types (**Figure 5**). This mechanism of synapse diversity also produces synapses with different signaling and adhesion specificity with the potential to connect to varieties of presynaptic terminals that are also distinguished by their varying molecular compositions. A good example of how this duplication and diversification in chordate genomes resulted in signaling diversity and differential expression is in the MAGUK proteins of the Dlg family and the NMDAR, the GluN2 subunit for which directly binds to Dlg (Ryan & Grant 2009). Invertebrate genomes encode a single GluN2 and Dlg and thus assemble a single type of

MASC complex, whereas mammals have 4 genes of each, potentially producing 16 types of MASC complexes. If one considers the other proteins that are in these complexes, including the many upstream proteins that were retained with the duplication events, an astronomical number of MASC and PSP types are available to diversify the synaptic types of vertebrates.

It is now important to highlight the connection between the generation of chordate synaptic complexity from genomic duplication events and the anatomical evolution of their nervous systems. Of note is that the genomic duplication events preceded the evolution of the large and anatomically diverse nervous systems of most vertebrates. The expansions in synaptic genes provided a molecular tool kit for generating a virtually limitless number of synaptic and neuronal types that could be used to generate diversity within the brain and also between different species. Direct evidence that this mechanism of diversity has indeed produced neuronal and synaptic diversity was observed in analyzing expression patterns of PSP genes in the mouse brain (Emes et al. 2008). The expression of PSP mRNAs and proteins were examined in many regions of the mouse brain, and those proteins that showed the greatest regional and neuronal type variation were more likely to be encoded by the expanded upstream families of vertebrate genes. Moreover, the set of proteins that was most ubiquitously and uniformly expressed in all regions was that of the ancestral protosynapse, indicating that a patterning in the diversity of synapse types in the mammalian brain arose from the expansion in complexity generated by gene duplications around the protosynapse (**Figure 5**).

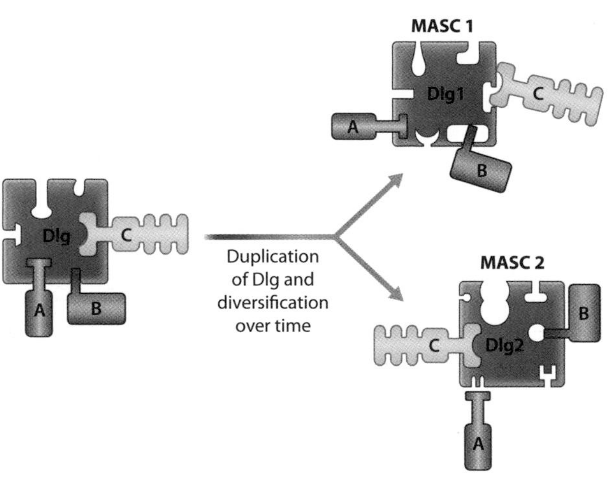

Figure 6

Duplication and diversification can lead to reorganization in signaling complexes. Duplication of scaffold proteins and diversification in their sequence modifies the organization and diversifies the molecular networks in MASC. The left-hand side shows an interaction network between Dlg and 3 binding partners (A, B, C) as in the ancestral protosynapse. As a result of gene duplication of *Dlg*, two paralogs (*Dlg1* and *Dlg2*) organize MASC1 and MASC2. With subsequent accumulated mutations in *Dlg1* and *-2*, their diversified protein binding affinity (e.g., different shapes of slots in jigsaw piece) alters the strength of interaction with binding partners so that MASC1 and MASC2 are different. The net effect is to produce diversity in MASC complexes, which can be expressed in different synapses to provide different physiological properties such as forms of synaptic plasticity.

In addition to generating a more diverse set of vertebrate synaptic genes with variation in structure and expression pattern, the organization of MASC protein networks is impacted in several important ways by genome duplication. First, as mentioned above there are multiplicative increases in the number of MASC variants produced by duplicates in its components (**Figure 4*a***). Second, a duplication in a scaffold protein that interacts with several others can lead to a rewiring of the molecular interaction network (Dreze et al. 2011), which alters the flow of signals in the network (**Figure 6**). A key organizational principle of these synaptic diversity mechanisms is that they utilize combinations of synaptic proteins in two distinct ways: the combinations of types of proteins that build up the core functionality of the protosynapse (scaffolds, receptors, enzymes, etc) and then the combinations of paralogs arising from duplication. The diversity arising from varied compositions of MASC complexes provides a way to categorize different MASCs (MASC1, MASC2, etc) and thus different classes of synapses with different physiological properties.

The NMDA receptors of the metazoan provide a prime example of the process of duplication and diversification of upstream signaling components. The NMDARs are glutamate-gated ion channels located at the surface of the postsynapse. Two rounds of duplication of the NMDA receptors in the vertebrate lineage resulted in four extant sequences. With this expansion, a dramatic change in the intracellular C-terminus of the NMDA proteins occurred. The C-terminus is the location of the phosphorylation-dependent interaction with scaffold and signaling molecules e.g., Fyn, CamKII, P85 PI3K, and PSD95. This C-terminal region (which is encoded by a single exon) is almost absent in invertebrate homologs, and hence so is the potential for multiple protein interactions in these species (Ryan et al. 2008). Thus the evolution of the intracellular portion of the NMDARs in the vertebrate lineages was likely a key stage in the link between sensing and cellular response to environmental stimuli.

INSIGHTS FROM THE HUMAN SYNAPSE

What sets humans apart from the rest of the animals, and what is the basis of human disease? The first comprehensive profiling of the human PSP and detailed study of its evolution showed that the genes of the PSP over the past 100 Myr are evolving under very strong purifying selection compared with the rest of the genome or other neuronal proteins and subcellular organelles (Bayes et al. 2010). This constraint was observed in primate and rodent lineages and shown to correlate with structural, physiological, and behavioral functions. The most conserved subset of the PSP was MASC, reinforcing its centrality in PSP function. The conserved functions of mouse and human PSP proteins were identified by systematic phenotype mapping of mutations and showed cognitive and motor functions, including learning and memory, and social functions were highly conserved. These findings again highlight the importance of adaptive behaviors as central and ancestral functions of MASC and the PSP. Another insight that arose from the proteomics of the human PSP was to identify that ∼200 genes are involved with Mendèlian diseases of which 130 were brain diseases (Bayes et al. 2010). Many of these diseases arose from gene duplication events, indicating that the cost of the evolution of paralogs was susceptibility to disease. A number of the MASC proteins in humans have been identified as mutated in patients with schizophrenia and other cognitive disorders (Fernandez et al. 2009, Kirov et al. 2011). Linking the genetic disease phenotypes to the observed constraint in vertebrate PSP evolution indicates that reduced fitness from PSP mutations is observed as strong and pervasive purifying selection.

Less is presently known about the changes in the PSP that are unique to the human lineage after it diverged from other primates around 6–8 Myr ago. Using methods to identify punctuated periods of adaptive evolution, eight genes of the PSP (*Cybrd1*, *SirpA*, *Ank2*, *Ca2*, *Cox5A*, *Pclo*, *Ndufb6*, and *Psd3*) show significant evidence of positive selection along the primate (including human) compared with the nonprimate lineage (R.D. Emes and S.G.N. Grant, manuscript in preparation). These genes may be candidates that underlie clade-specific adaptive evolution and may underpin more subtle differences in primate synapse function and hence in cognitive ability as well.

A MODEL FOR THE EVOLUTION OF THE POSTSYNAPTIC PROTEOME

Taken together, the comparative studies to date suggest a consistent model for the evolution of the PSP and its contribution to synaptic diversity and behavior. Elements of environmental sensing from surface receptor to transcriptome and core components associated with basic cellular life such as translation, energy generation, and fatty acid biosynthesis identified in bacteria and archaea were present prior to the divergence of the common ancestor of eukaryotes and prokaryotes. A number of constituents are conserved in prokaryotes as interacting proteins, suggesting that these were co-opted into the protosynapse as an interacting complex and remain in the extant PSP. In the fungi we observe homologs of the signal transduction pathways, including increased repertoire of protein kinases and the important role of scaffold proteins assembling and organizing signaling machinery. The presence of most types of synapse proteins in unicellular eukaryotes such as fungi and choanoflagellates and Porifera highlights that the evolution of the fundamental synaptic components predates the origins of identifiable neurons in metazoans. The ion channels incorporated after the cnidarian–poriferan divergence therefore would interact with a preexisting scaffold of intracellular proteins. The expanding transmembrane receptors would also plug into this preexisting network to expand rapidly the signaling complexity of the synapse (**Figure 4*a***). By comparing the predicted PSP of invertebrate metazoan species (fly, worm, bee, and mosquito) to vertebrates and by directly isolating the fly MASC component of the PSP, it is clear that the majority of protein classes were present in the invertebrates. In addition, specialization and division of labor were expanded by differential gene expression, providing combinations of proteins in different synapses (**Figure 5**). Following the divergence of the vertebrates from other deuterostomes, the driving force was of rapid expansion by duplication and diversification, particularly in the upstream signaling components such as receptors and signal transduction molecules. The expansion of the synapse proteome, therefore, predates the development of anatomically enlarged brains. This model in which the development of the synapse is a necessary step prior to the expansion and development of an enlarged nervous system has been proposed in the "synapse first" model of brain evolution (Ryan & Grant 2009).

One prediction of this model is that the synapse developed before axons and the branching network of dendrites. The increase of neuronal connectivity produced by neuronal branching rapidly increases the number of synapses and multiplies the computational power and diversity of synapses and overall signaling complexity. This may have driven the form of the nervous system we see today. Support for the theory that gene repertoire may predate the enlargement or increase in encephalization of the brain has recently emerged. Encephalization, the development of relatively excess brain size, is often measured by comparing the encephalization quotient (EQ, the log brain size versus log body size) and has been proposed as a measure of information-processing capacity or intelligence (Jerison 1977, 1985). By using high-resolution X-ray computed tomography, Rowe et al. (2011) recently showed that the particular enlargement of cerebral hemispheres and cerebellum associated with mammals occurred in a step-wise manner and was associated with expansion of key gene families. Expansion of the brain, especially the olfactory bulb and olfactory cortex, is seen in the skull of *Morganucodon oehleri*, a basal member of the mammaliaforms from the early Jurassic (~199–175 Mya). This postdates the expansion of the PSP we predict to have occurred in the ancestor of the vertebrates. A second wave of encephalization was seen with *Hadrocodium wui*, where expansion of the olfactory bulb and olfactory cortex accounts for the increase in EQ to within that seen in extant crown group mammals. Rowe et al. propose that the first wave of EQ expansion was driven by increase in olfaction and tactile sensitivity. These were then further amplified by an olfactory expansion owing to expression of the expanded olfactory receptor genome. We propose that, like the olfactory receptors, the expansion of gene repertoire in the PSP by gene duplication and expansion at the base of the vertebrates was a driver rather than a consequence of an increase in EQ. Like the exploding bubbles when uncorking champagne, mutations leading to change in brain size released the potential of the expanded PSP repertoire.

With the model that additional protein interactors plugged into an existing scaffold, why should the PSP complex expand by increasing interacting partners? The selective advantage of accumulating interacting proteins by scaffold protein binding proposes an intuitive adaptionist theory for the evolution of the synapse. The scaffold proteins provide a means to localize interacting proteins among the multifarious soup of proteins in a cell and orchestrates the flow of information in a cell (Good et al. 2011). For example, the signaling complex of mating pheromones in *S. cerevisiae* is tethered by a scaffold protein (Ste5) that acts to increase signal transduction efficiency. At low total protein concentration, the resulting colocalization will increase local concentration and hence the probability of interaction. Additionally, the scaffold proteins containing varying architectures of domains promoting protein-protein interactions (e.g., the PDZ domains of PSD-95) can act as a means to allow the rapid evolution of new pathways by changing binding specificity and hence interacting partners. This type of universal port allows different interacting proteins to plug into a preexisting network of downstream effectors.

This theory suggests that adaptive evolution by natural selection of beneficial mutations has driven the aggregation of synapse proteins and other protein complexes. This mechanism could have expanded the MASC into the even greater complexity of PSD. However, Fernandez & Lynch (2011) recently proposed a non-adaptive theory to explain the trend of protein complex development seen in eukaryotes. They suggested that the small population size seen in eukaryotes compared with the prokaryotes allows the accumulation of mildly deleterious mutations by genetic drift in key proteins (drift is less dominant in larger populations owing to stronger selection coefficients). The accumulation of these mutations in turn drives the accumulation of protein complexes to stabilize individual proteins (Fernandez & Lynch 2011). Therefore, the growth of the synapse proteome with time may simply be due to the selection pressure to maintain protein function following neutral mutations.

THE EVOLUTION OF CELL-CELL ADHESION

The essence of multicellularity is the adhesion of cells via cell-cell junctions. The most basic of these is the adherens junction containing the cadherin domains identified in sponge (Fahey & Degnan 2010) and choanoflagellate proteins (King et al. 2003, 2008). Junctions that allow cell-cell communication by passing small molecules such as gap junctions are predicted to have evolved later with cnidarian (for review, see Abedin & King 2010). Cells mixed from two species of sponge will reform as species-specific clumps (Wilson 1910) via a proteoglycan ligand for a cell-surface receptor (Dunham et al. 1983), suggesting cell-cell signaling coupled to cell adhesion is an ancient process. Cell-adhesion molecules, including cadherins, are key to synapse formation and function. These molecules are not limited to the neuronal synapse: the interaction of mammalian T-cells and antigen-presenting cells utilizes these proteins and is known as the immunological synapse (Dustin 2009, Dustin et al. 2010, Paul & Seder 1994). Moreover, presynaptic proteins such as SNARE, VAMP, and SNAP proteins are found at the immunological synapse (Griffiths et al. 2010), supporting a general model for the evolution of synaptic mechanisms in the biology of many neuronal and nonneuronal cells.

CONCLUSIONS

Studies of synapse proteomes have shed light on the organization of molecular networks and macromolecular complexes in synapses and enabled the first systematic studies on synapse evolution. These studies reveal that synapses evolved from humble beginnings in prokaryotes and the earliest forms of cellular life. The realization that the primary role of the nervous system in sensing and responding to the environment arose in the organized protein architecture of signaling complexes or protosynapses in unicellular organisms, prior to the first neurons in any multicellular organism, opens new paths to understand the origins of behavior and the evolution of the behavioral repertoire of animals. It was the organization of this molecular machinery, primarily through combinatorial use of preexisting building blocks, that was exploited to generate the remarkable synaptic diversity found in invertebrates and vertebrates. These observations suggest that to understand the function of the brain we should aim to understand the evolution in the complexity of synaptic molecular systems. How behavioral diversity arose in organisms with large and complex brains remains mysterious, and it may be that the diversity or repertoire of their adaptive behaviors was shaped by the evolutionary mechanisms discovered in the synapse. The framework of the evolution and composition of the synaptome provides a path to investigate these problems and perhaps lead to a truly unified understanding of synapse biology.

DISCLOSURE STATEMENT

The authors are not aware of any affiliations, memberships, funding, or financial holdings that might be perceived as affecting the objectivity of this review.

ACKNOWLEDGMENTS

R.D.E. was supported by a Royal Society UK Grant, RG080388, and by the School of Veterinary Medicine and Science, University of Nottingham. S.G.N.G. was supported by the Wellcome Trust Genes to Cognition Program, MRC, EU FP7 EUROSPIN, GENCODYS and SYNSYS programs.

LITERATURE CITED

Abedin M, King N. 2010. Diverse evolutionary paths to cell adhesion. *Trends Cell Biol.* 20:734–42

Abul-Husn NS, Bushlin I, Moron JA, Jenkins SL, Dolios G, et al. 2009. Systems approach to explore components and interactions in the presynapse. *Proteomics* 9:3303–15

Alie A, Manuel M. 2010. The backbone of the post-synaptic density originated in a unicellular ancestor of choanoflagellates and metazoans. *BMC Evol. Biol.* 10:34

Bayes A, Grant SG. 2009. Neuroproteomics: understanding the molecular organization and complexity of the brain. *Nat. Rev. Neurosci.* 10:635–46

Bayes A, van de Lagemaat LN, Collins MO, Croning MD, Whittle IR, et al. 2010. Characterization of the proteome, diseases and evolution of the human postsynaptic density. *Nat. Neurosci.* 14:19–21

Bocquet N, Nury H, Baaden M, Le Poupon C, Changeux JP, et al. 2009. X-ray structure of a pentameric ligand-gated ion channel in an apparently open conformation. *Nature* 457:111–14

Bocquet N, Prado de Carvalho L, Cartaud J, Neyton J, Le Poupon C, et al. 2007. A prokaryotic proton-gated ion channel from the nicotinic acetylcholine receptor family. *Nature* 445:116–19

Cheng D, Hoogenraad CC, Rush J, Ramm E, Schlager MA, et al. 2006. Relative and absolute quantification of postsynaptic density proteome isolated from rat forebrain and cerebellum. *Mol. Cell Proteomics* 5:1158–70

Chothia C, Gough J, Vogel C, Teichmann SA. 2003. Evolution of the protein repertoire. *Science* 300:1701–3

Coba MP, Pocklington AJ, Collins MO, Kopanitsa MV, Uren RT, et al. 2009. Neurotransmitters drive combinatorial multistate postsynaptic density networks. *Sci. Signal.* 2:ra19

Collins MO, Husi H, Yu L, Brandon JM, Anderson CN, et al. 2006. Molecular characterization and comparison of the components and multiprotein complexes in the postsynaptic proteome. *J. Neurochem.* 97(Suppl. 1):16–23

Collins MO, Yu L, Coba MP, Husi H, Campuzano I, et al. 2005. Proteomic analysis of in vivo phosphorylated synaptic proteins. *J. Biol. Chem.* 280:5972–82

Dawkins R. 1986. *The Blind Watchmaker*. London: Penguin

Dawkins R. 1994. Evolutionary biology. The eye in a twinkling. *Nature* 368:690–91

DeFelipe J. 2010. From the connectome to the synaptome: an epic love story. *Science* 330:1198–201

Dosemeci A, Makusky AJ, Jankowska-Stephens E, Yang X, Slotta DJ, Markey SP. 2007. Composition of the synaptic PSD-95 complex. *Mol. Cell Proteomics* 6:1749–60

Dosemeci A, Tao-Cheng JH, Vinade L, Jaffe H. 2006. Preparation of postsynaptic density fraction from hippocampal slices and proteomic analysis. *Biochem. Biophys. Res. Commun.* 339:687–94

Dreze M, Carvunis AR, Charloteaux B, Galli M, Pevzner SJ, et al. 2011. Evidence for network evolution in an Arabidopsis interactome map. *Science* 333:601–7

Dunham P, Anderson C, Rich AM, Weissmann G. 1983. Stimulus-response coupling in sponge cell aggregation: Evidence for calcium as an intracellular messenger. *Proc. Natl. Acad. Sci. USA* 80:4756–60

Dustin ML. 2009. Modular design of immunological synapses and kinapses. *Cold Spring Harb. Perspect. Biol.* 1:a002873

Dustin ML, Chakraborty AK, Shaw AS. 2010. Understanding the structure and function of the immunological synapse. *Cold Spring Harb. Perspect. Biol.* 2:a002311

Ekman D, Bjorklund AK, Elofsson A. 2007. Quantification of the elevated rate of domain rearrangements in metazoa. *J. Mol. Biol.* 372:1337–48

Elion EA, Qi M, Chen W. 2005. Signal transduction. Signaling specificity in yeast. *Science* 307:687–88

Emes RD, Grant SG. 2011. The human postsynaptic density shares conserved elements with proteomes of unicellular eukaryotes and prokaryotes. *Front. Neurosci.* 5:44

Emes RD, Pocklington AJ, Anderson CN, Bayes A, Collins MO, et al. 2008. Evolutionary expansion and anatomical specialization of synapse proteome complexity. *Nat. Neurosci.* 11:799–806

Erdman S, Snyder M. 2001. A filamentous growth response mediated by the yeast mating pathway. *Genetics* 159:919–28

Fahey B, Degnan BM. 2010. Origin of animal epithelia: insights from the sponge genome. *Evol. Dev.* 12:601–17

Fernández A, Lynch M. 2011. Non-adaptive origins of interactome complexity. *Nature* 474:502–5

Fernandez E, Collins MO, Uren RT, Kopanitsa MV, Komiyama NH, et al. 2009. Targeted tandem affinity purification of PSD-95 recovers core postsynaptic complexes and schizophrenia susceptibility proteins. *Mol. Syst. Biol.* 5:269

Finn RD, Mistry J, Tate J, Coggill P, Heger A, et al. 2010. The Pfam protein families database. *Nucleic Acids Res.* 38:D211–22

Good MC, Zalatan JG, Lim WA. 2011. Scaffold proteins: hubs for controlling the flow of cellular information. *Science* 332:680–86

Griffiths GM, Tsun A, Stinchcombe JC. 2010. The immunological synapse: a focal point for endocytosis and exocytosis. *J. Cell Biol.* 189:399–406

Hahn CG, Banerjee A, Macdonald ML, Cho DS, Kamins J, et al. 2009. The post-synaptic density of human postmortem brain tissues: an experimental study paradigm for neuropsychiatric illnesses. *PLoS One* 4:e5251

Harashima T, Anderson S, Yates JR 3rd, Heitman J. 2006. The kelch proteins Gpb1 and Gpb2 inhibit Ras activity via association with the yeast RasGAP neurofibromin homologs Ira1 and Ira2. *Mol. Cell* 22:819–30

Hedges SB, Blair JE, Venturi ML, Shoe JL. 2004. A molecular timescale of eukaryote evolution and the rise of complex multicellular life. *BMC Evol. Biol.* 4:2

Hedges SB, Dudley J, Kumar S. 2006. TimeTree: a public knowledge-base of divergence times among organisms. *Bioinformatics* 22:2971–72

Holland LZ. 2009. Chordate roots of the vertebrate nervous system: expanding the molecular toolkit. *Nat. Rev. Neurosci.* 10:736–46

Hurles M. 2004. Gene duplication: the genomic trade in spare parts. *PLoS Biol.* 2:E206

Husi H, Ward MA, Choudhary JS, Blackstock WP, Grant SG. 2000. Proteomic analysis of NMDA receptor-adhesion protein signaling complexes. *Nat. Neurosci.* 3:661–69

Janovjak H, Sandoz G, Isacoff EY. 2011. A modern ionotropic glutamate receptor with a K(+) selectivity signature sequence. *Nat. Commun.* 2:232

Jerison HJ. 1977. The theory of encephalization. *Ann. N. Y. Acad. Sci.* 299:146–60

Jerison HJ. 1985. Animal intelligence as encephalization. *Philos. Trans. R. Soc. Lond. B Biol. Sci.* 308:21–35

Jordan BA, Fernholz BD, Boussac M, Xu C, Grigorean G, et al. 2004. Identification and verification of novel rodent postsynaptic density proteins. *Mol. Cell Proteomics* 3:857–71

Kaas J, ed. 2009. *Evolutionary Neuroscience*. Oxford: Academic

King N, Carroll SB. 2001. A receptor tyrosine kinase from choanoflagellates: molecular insights into early animal evolution. *Proc. Natl. Acad. Sci. USA* 98:15032–37

King N, Hittinger CT, Carroll SB. 2003. Evolution of key cell signaling and adhesion protein families predates animal origins. *Science* 301:361–63

King N, Westbrook MJ, Young SL, Kuo A, Abedin M, et al. 2008. The genome of the choanoflagellate *Monosiga brevicollis* and the origin of metazoans. *Nature* 451:783–88

Kirov G, Pocklington AJ, Holmans P, Ivanov D, Ikeda M, et al. 2011. De novo CNV analysis implicates specific abnormalities of postsynaptic signalling complexes in the pathogenesis of schizophrenia. *Mol. Psychiatry* 17:142–53

Koonin EV. 2003. Horizontal gene transfer: the path to maturity. *Mol. Microbiol.* 50:725–27

Koonin EV, Aravind L, Kondrashov AS. 2000. The impact of comparative genomics on our understanding of evolution. *Cell* 101:573–76

Koonin EV, Makarova KS, Aravind L. 2001. Horizontal gene transfer in prokaryotes: quantification and classification. *Annu. Rev. Microbiol.* 55:709–42

Kosik KS. 2009. Exploring the early origins of the synapse by comparative genomics. *Biol. Lett.* 5:108–11

Kralj JM, Hochbaum DR, Douglass AD, Cohen AE. 2011. Electrical spiking in *Escherichia coli* probed with a fluorescent voltage-indicating protein. *Science* 333:345–48

Lawrence JG, Hendrickson H. 2003. Lateral gene transfer: When will adolescence end? *Mol. Microbiol.* 50:739–49

Lichtneckert R, Reichert H. 2009. Origin and evolution of the first nervous system. See Kaas 2009, pp. 51–78

Liebeskind BJ, Hillis DM, Zakon HH. 2011. Evolution of sodium channels predates the origin of nervous systems in animals. *Proc. Natl. Acad. Sci. USA* 108:9154–59

Lynch M. 2007. *The Origins of Genome Architecture*. Sunderland, MA: Sinauer

Manning G, Young SL, Miller WT, Zhai Y. 2008. The protist, *Monosiga brevicollis*, has a tyrosine kinase signaling network more elaborate and diverse than found in any known metazoan. *Proc. Natl. Acad. Sci. USA* 105:9674–79

Marchler-Bauer A, Anderson JB, Cherukuri PF, DeWeese-Scott C, Geer LY, et al. 2005. CDD: a Conserved Domain Database for protein classification. *Nucleic Acids Res.* 33:D192–96

Martinac B, Saimi Y, Kung C. 2008. Ion channels in microbes. *Physiol. Rev.* 88:1449–90

Migaud M, Charlesworth P, Dempster M, Webster LC, Watabe AM, et al. 1998. Enhanced long-term potentiation and impaired learning in mice with mutant postsynaptic density-95 protein. *Nature* 396:433–39

Morciano M, Beckhaus T, Karas M, Zimmermann H, Volknandt W. 2009. The proteome of the presynaptic active zone: from docked synaptic vesicles to adhesion molecules and maxi-channels. *J. Neurochem.* 108:662–75

Nakanishi N, Shneider NA, Axel R. 1990. A family of glutamate receptor genes: evidence for the formation of heteromultimeric receptors with distinct channel properties. *Neuron* 5:569–81

Nilsson DE, Pelger S. 1994. A pessimistic estimate of the time required for an eye to evolve. *Proc. Biol. Sci.* 256:53–58

Nourry C, Grant SG, Borg JP. 2003. PDZ domain proteins: plug and play! *Sci. STKE* 2003:re7

Nury H, Van Renterghem C, Weng Y, Tran A, Baaden M, et al. 2011. X-ray structures of general anaesthetics bound to a pentameric ligand-gated ion channel. *Nature* 469:428–31

Palecek SP, Parikh AS, Kron SJ. 2002. Sensing, signalling and integrating physical processes during *Saccharomyces cerevisiae* invasive and filamentous growth. *Microbiology* 148:893–907

Paul WE, Seder RA. 1994. Lymphocyte responses and cytokines. *Cell* 76:241–51

Peng J, Kim MJ, Cheng D, Duong DM, Gygi SP, Sheng M. 2004. Semiquantitative proteomic analysis of rat forebrain postsynaptic density fractions by mass spectrometry. *J. Biol. Chem.* 279:21003–11

Pincus D, Letunic I, Bork P, Lim WA. 2008. Evolution of the phospho-tyrosine signaling machinery in premetazoan lineages. *Proc. Natl. Acad. Sci. USA* 105:9680–84

Pocklington AJ, Armstrong JD, Grant SG. 2006a. Organization of brain complexity—synapse proteome form and function. *Brief Funct. Genomics Proteomics* 5:66–73

Pocklington AJ, Cumiskey M, Armstrong JD, Grant SGN. 2006b. The proteomes of neurotransmitter receptor complexes form modular networks with distributed functionality underlying plasticity and behaviour. *Mol. Syst. Biol.* 2:E1–14

Richards GS, Simionato E, Perron M, Adamska M, Vervoort M, Degnan BM. 2008. Sponge genes provide new insight into the evolutionary origin of the neurogenic circuit. *Curr. Biol.* 18:1156–61

Rowe TB, Macrini TE, Luo ZX. 2011. Fossil evidence on origin of the mammalian brain. *Science* 332:955–57

Ryan TJ, Emes RD, Grant SG, Komiyama NH. 2008. Evolution of NMDA receptor cytoplasmic interaction domains: implications for organisation of synaptic signalling complexes. *BMC Neurosci.* 9:6

Ryan TJ, Grant SG. 2009. The origin and evolution of synapses. *Nat. Rev. Neurosci.* 10:701–12
Sakarya O, Armstrong KA, Adamska M, Adamski M, Wang IF, et al. 2007. A post-synaptic scaffold at the origin of the animal kingdom. *PLoS ONE* 2:e506
Satoh K, Takeuchi M, Oda Y, Deguchi-Tawarada M, Sakamoto Y, et al. 2002. Identification of activity-regulated proteins in the postsynaptic density fraction. *Genes Cells* 7:187–97
Schultz J, Copley RR, Doerks T, Ponting CP, Bork P. 2000. SMART: a web-based tool for the study of genetically mobile domains. *Nucleic Acids Res.* 28:231–34
Selimi F, Cristea IM, Heller E, Chait BT, Heintz N. 2009. Proteomic studies of a single CNS synapse type: the parallel fiber/Purkinje cell synapse. *PLoS Biol.* 7:e83
Sprengel R, Aronoff R, Volkner M, Schmitt B, Mosbach R, Kuner T. 2001. Glutamate receptor channel signatures. *Trends Pharmacol. Sci.* 22:7–10
Trinidad JC, Thalhammer A, Specht CG, Lynn AJ, Baker PR, et al. 2008. Quantitative analysis of synaptic phosphorylation and protein expression. *Mol. Cell Proteomics* 7:684–96
Trinidad JC, Thalhammer A, Specht CG, Schoepfer R, Burlingame AL. 2005. Phosphorylation state of postsynaptic density proteins. *J. Neurochem.* 92:1306–16
Van de Peer Y, Maere S, Meyer A. 2009. The evolutionary significance of ancient genome duplications. *Nat. Rev. Genet.* 10:725–32
Walikonis RS, Jensen ON, Mann M, Provance DW Jr, Mercer JA, Kennedy MB. 2000. Identification of proteins in the postsynaptic density fraction by mass spectrometry. *J. Neurosci.* 20:4069–80
Wilson HV. 1910. Development of sponges from dissociated tissue cells. *Bull. Bur. Fish.* 30:1–30
Wolf YI, Aravind L, Grishin NV, Koonin EV. 1999. Evolution of aminoacyl-tRNA synthetases—analysis of unique domain architectures and phylogenetic trees reveals a complex history of horizontal gene transfer events. *Genome Res.* 9:689–710
Yoshimura Y, Yamauchi Y, Shinkawa T, Taoka M, Donai H, et al. 2004. Molecular constituents of the post-synaptic density fraction revealed by proteomic analysis using multidimensional liquid chromatography-tandem mass spectrometry. *J. Neurochem.* 88:759–68

RELATED RESOURCES

Bayes A, Grant SG. 2009. Neuroproteomics: understanding the molecular organization and complexity of the brain. *Nat. Rev. Neurosci.* 10:635–46

Grant SG. 2009. A general basis for cognition in the evolution of synapse signaling complexes. *Cold Spring Harb. Symp. Quant. Biol.* 74:249—57

Ryan TJ, Grant SG. 2009. The origin and evolution of synapses. *Nat. Rev. Neurosci.* 10:701–12

Genes to Cognition. **http://www.Genes2Cognition.org**. Web site for synapse proteome and related data.

Social Control of the Brain

Russell D. Fernald

Biology Department, Stanford University, Stanford, California 94305;
email: rfernald@stanford.edu

Annu. Rev. Neurosci. 2012. 35:133–51

First published online as a Review in Advance on April 12, 2012

The *Annual Review of Neuroscience* is online at neuro.annualreviews.org

This article's doi:
10.1146/annurev-neuro-062111-150520

0147-006X/12/0721-0133$20.00

Keywords

social behavior, teleost fish, control of the brain, *Astatotilapia burtoni*, immediate early genes, social behavior network

Abstract

In the course of evolution, social behavior has been a strikingly potent selective force in shaping brains to control action. Physiological, cellular, and molecular processes reflect this evolutionary force, particularly in the regulation of reproductive behavior and its neural circuitry. Typically, experimental analysis is directed at how the brain controls behavior, but the brain is also changed by behavior over evolution, during development, and through its ongoing function. Understanding how the brain is influenced by behavior offers unusual experimental challenges. General principles governing the social regulation of the brain are most evident in the control of reproductive behavior. This is most likely because reproduction is arguably the most important event in an animal's life and has been a powerful and essential selective force over evolution. Here I describe the mechanisms through which behavior changes the brain in the service of reproduction using a teleost fish model system.

Contents

INTRODUCTION

In many species, social context modulates interactions among animals, which then tune their behavior accordingly. Social interactions clearly influence the brain and circulating hormonal levels, but how does the social environment regulate the physiological, cellular, and molecular processes of the brain? Elucidating this connection is critically important because social interactions, especially those related to reproduction, are essential for the evolutionary success of all species. Social regulation of vertebrate reproduction offers a unique opportunity to understand how behavior influences the brain for two reasons. First, reproductive behaviors are often stereotypic, making them relatively easy to observe and quantify. Second, central control of reproduction is lodged in the brain-pituitary-gonadal (BPG) axis with the hypophysiotropic gonadotropin-releasing hormone (GnRH1) neurons as its final output path. Thus the neuronal circuit that produces the GnRH1 signaling peptide is the key locus for integrating the external and internal factors that control reproduction.

Understanding behavioral regulation of the brain requires an animal model in which (*a*) social interchange is essential for reproductive success; (*b*) animals can be studied in a seminatural context; (*c*) key molecular, cellular, and physiological processes are accessible; and (*d*) behavior and physiology of both individuals and groups of animals can be readily analyzed.

CHOICE OF FISH SPECIES FOR STUDY

In considering experimental systems for studies of the brain that focus on social behavior, it is important to select a system that allows a reasonable replication of the native social and environmental habitat. Although primates are an obvious choice of an experimental animal species that is closely related to humans, fish offer an interesting chance to understand how the brain controls behavior for several reasons. Fish are the largest vertebrate group, with more species than all other vertebrates combined, occupying ecological niches from freezing water to hot soda springs. Moreover, they represent more than 400 million years of vertebrate evolution, and their taxonomic dimensions exceed the distance between frogs and humans (Romer 1959). The variety of sensory modalities required for aquatic life, aside from vision, olfaction, taste, and hearing, include mechanosensory systems (e.g., lateral line), external taste buds, and numerous electroreceptor systems that have led to extensive variation in brain structures. Moreover, it is not an exaggeration to say that essentially every known kind of social system has evolved in fish species from monogamy to harems to sex-changing animals.

Given the wide variety of ecological niches exploited, appropriate species comparisons among fish offer the potential for understanding the relative contributions of environmental and social factors to brain evolution. Even more appealing, fish offer the promise of discovering mechanisms through which the brain controls social activities in ecologically meaningful contexts (Bshary et al. 2002).

Studies of animal behavior fall roughly into two groups that have quite different goals (Shettleworth 1993, Kamil 1998). Some investigators seek to identify human-like skills in other species, a thread derived broadly from a search for general cognitive abilities (Hodos & Campbell 1969). This approach has been termed the anthropocentric perspective and may, in fact, be an indirect descendent of Aristotle's *scala naturae*, a view reinforced by comparative brain research. Another camp treats cognition as strictly biological, best elucidated through observations and experiments directed at understanding adaptive modifications crafted by evolution to regulate social interactions (Robinson et al. 2008). Such adaptations typically produce cognitive skills that animals need and use in nature to navigate their social worlds. There are several ways to parse these general classes of investigation in other terms, for example, understanding aspects of human behavior using animal models versus understanding animals in meaningful biological contexts for their own sake.

Fish Have Significant Cognitive Abilities

Cognitive skills in fish species offer an opportunity to study basic skills in a tractable animal that can be kept in a seminatural environment where the whole social system can be well mimicked. The following is a brief summary of the range of cognitive capacities that could be tapped in a fish model system.

Numerosity. Counting is an important form of abstraction that led to the development of mathematics. Since the discovery in 1904 that Hans, the famous Russian trotting horse, could not really count or do mathematical calculations (or read either German or musical notation) but rather responded to his owner's cues (Candland 1993), scientists have used more rigorous methods to investigate whether and how animals can keep track of quantities. Can some fish species keep track of amounts, and when might they need to count something? Numerical abilities in a fish species appear to be handy when choosing which group to join in times of danger. "Safety in numbers" predicts that fish should select the larger group when given a choice. In fact, three-spined sticklebacks (*Gasterosteus aculeatus*), when threatened with a simulated aerial predator, chose to join larger shoals when equidistant from small shoals but made a trade-off between distance to the shoal and its numbers when given that option (Tegeder & Krause 1995). Because the number of fish in larger shoals covaries with several physical attributes of the group (e.g., area, contour, density), sophisticated experiments are required to show which feature of shoal size is actually being discriminated. These types of experiments can be done when the stimuli are presented in succession, rather than in aggregation. In one case, item-by-item presentation showed that the mosquitofish (*Gambusi holbrooki*) can distinguish small (three versus two) and large (eight versus four) shoals independent of other factors (Agrillo et al. 2009). Such numerical skills likely evolved in species in which antipredator benefits of group assemblies have evolved. We can predict that similar capacities will be found in other fish species that aggregate during predation threats.

Recognition of individuals. When an animal encounters another individual, it likely compares sensory information with a template to categorize that individual as conspecific, heterospecific, threatening, or nonthreatening. It may also recognize the individual. Recognition of individuals is a prerequisite for many behavioral interactions and has been demonstrated in a wide range of taxa. In particular, individual recognition is essential for kin recognition and hence required for any kin-selected behavior.

Individuals can be identified using multiple sensory modalities depending on the species and ecological circumstances. For fish, novel sensory systems, including electroreception, pressure reception, and polarization vision, may be important for recognizing individuals.

How many individuals would a fish need to know in some way? Sampling natural populations, Ward et al. (2005) estimated that three-spined sticklebacks living in ~20 m of a channel in a freshwater lake could meet with 900 conspecifics regularly and showed that direct experience and social cues led to relatively quick learning about the categories of individuals. However, the total number of conspecifics that stickleback's remember or whether they recognize individuals was not established (Ward et al. 2005). Bshary et al. (2002) report that some cleaner fish species can probably distinguish ~100 individuals by observing their behavior toward clients.

One of the most important realms for individual recognition is in mating pairs of fish. Noble & Curtis (1939) first described recognition between mated pairs in a cichlid fish with biparental care of the young, *Hemichromis bimaculatus*. Much later, Fricke (1973) performed elegant field experiments on fish (*Amphiprion bicinctus*) that live together among the tentacles of anemones, showing that mated individuals recognized one another on the basis of individual body color patterns rather than mutually recognizing the anemone. Moreover, after arbitrarily pairing a male and female, anemone fish could learn the identity of the new partner in 24 h and could also recognize that individual after 10 days of isolation. Given the high rates of predation on *A. bicinctus*, the ability to identify a new partner rapidly would allow animals to continue reproducing despite the loss of a familiar partner. In cichlids living in clear water habitats, individuals used primarily visual cues to recognize other individuals (Noble & Curtis 1939, Fricke 1973, Balshine-Earn & Lotem 1998).

Differences in mating systems and ecology may result in the evolution of different recognition abilities among species. For example, guppies can recognize conspecifics individually as well as distinguish among groups of conspecifics on the basis of cues about resource use and habitat (Ward et al. 2009). In contrast, sticklebacks do not recognize individuals in a social context despite prior interactions, though they do have general recognition capacities based on resource use (Ward et al. 2009) that are considered "familiarity" rather than individual recognition. This cognitive skill seems to increase the chances of grouping together and improving foraging (Ward et al. 2005, 2007). More recently, Ward et al. (2009) suggested that sticklebacks may rely on habitat information, specifically odors, to identify particular groups of individuals.

A second important function for individual recognition is to reduce the costs of contesting resources. This has been demonstrated in sea trout (*Salmo trutta*) in which familiarity with conspecifics enhanced growth (Höjesjö et al. 1998). In a twist on this skill, European minnows (*Phoxinus phoxinus*) recognized and preferred to group together in a shoal with poor competitors, although how they recognized poor competitors is unknown (Metcalfe & Thomson 1995). Evidence also indicates that kin recognition is widespread among fish species, particularly those that school (Quinn & Hara 1986, Havre & FitzGerald 1988, Olsen 1989), and that this skill requires individual recognition.

As in all laboratory experiments, individual recognition may be a consequence of artificially extended interactions among individuals. However, most data cited here were from field experiments and hence provide more convincing examples of individual recognition in fish (see also discussion below on transitive inference data: Assessing Male Fighting Abilities).

Deception

Do fish communicate honestly or can they deceive? Deception is a fundamental issue in animal communication (Maynard Smith & Harper 2003), so can fish deceive? Several authors beginning with Byrne & Whiten (1988) have distinguished between functional and intentional deception. Functional deception is widespread and includes many examples that do not require cognitive skill (e.g., mimicry, crypsis, although see Chittka & Osorio 2007), whereas intentional deception implies behavior based on intentional states (e.g., beliefs, desires). As cogently discussed by Shettleworth (1998), translating anthropocentric concepts into predictions and experiments that are testable is a serious challenge. Suitable experiments have been described for chimpanzees (Hare et al. 2000, Hare & Tomasello 2001), scrub jays (Dally et al. 2006), and ravens (Bugnyar & Heinrich 2006) with a test to demonstrate an ultimate fitness benefit of deception to the deceiver and the cost to the deceived (Hauser 1997).

Perhaps the cleanest example of deception has been described for the cleaner wrasse, *Labroides dimidiatus*. This marine cleaner fish removes ectoparasites from visiting reef fish clients. This relationship is mutual because the client gets cleaned of ectoparasites and the cleaner gets a meal, the parasite, delivered. But a problem exists: Cleaner fish prefer the client's tasty layer of mucus to its ectoparasites (Grutter & Bshary 2003). Because cleaner fish service up to 2,000 clients every day (Grutter 1997) and many of these encounters happen in the presence of observing bystanders, including future clients, does the presence of bystanders alter the cleaner's behavior? Pinto et al. (2011) used two species of client fish to ask whether being watched matters. Cleaners were tested on clients, and the introduction of a bystander led to an immediate increase in cooperation by the cleaner fish: The cleaners spent more time removing ectoparasites than eating mucus when being watched. This brief discussion of the social skills of teleost fish reveals why they are such useful model systems for understanding the neural bases of social behavior.

SOCIAL CONTROL OF THE BRAIN: WHY STUDY THE CICHLID FISH, *ASTATOTILAPIA BURTONI*?

Astatotilapia (Haplochromis) burtoni, the African cichlid fish model system we developed, has numerous social skills and, perhaps most importantly, social interactions related to the fact that social dominance tightly controls reproductive physiology. By manipulating the social system we can essentially turn on or off an animal's reproductive competence, mimicking natural changes to identify key regulatory processes. This fish model system offers several important advantages for understanding how social behavior changes the brain: (*a*) The social system of this fish can be easily replicated in the laboratory; (*b*) male status is signaled by obvious rapid color changes, making it easy to detect and quantify; (*c*) GnRH1 neurons are directly regulated by male social status and hence are causally related to behavior; (*d*) *A. burtoni* offers easy access to the brain, allowing sampling of cells and molecules of interest; and (*e*) the *A. burtoni* genome has been sequenced, enabling a class of experiments not previously possible (**http://www.genome.gov/11007951**). Taken together, these attributes make *A. burtoni* uniquely useful for studying social regulation of reproduction. Using this system, we have manipulated the social situation to produce phenotypic change in a variety of ways and have measured relevant molecular, neural, and hormonal systems.

A. burtoni males exist as one of two socially controlled, reversible phenotypes: reproductively competent dominant (D) males and reproductively incompetent nondominant (ND) males (see **Figure 1**).

D males display bright coloration, aggressively defend territories, and court females, whereas ND males display dull gray coloration, mimic females, and limit their behavior to schooling and fleeing. The major differences between these two phenotypic states of male *A. burtoni* can be summarized as follows.

Social signals regulate GnRH1 cell size, peptide level, GnRH1, and gonadotropin-releasing hormone (GnRH) mRNA receptor levels. When a D male is moved into a social system with larger (>5% in length) D males, it abruptly loses its color (<1 min) and joins other ND males and females in a school. Its GnRH1-containing neurons in the preoptic

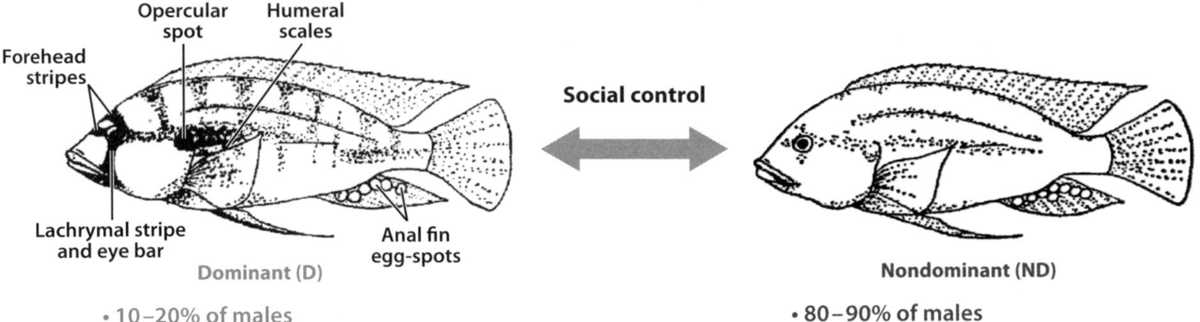

Figure 1

Socially regulated differences between dominant (D) and nondominant (ND) males. Dominant males are brightly colored yellow or blue, express complex behaviors, are reproductively competent, and comprise a small fraction of the population, whereas ND males have dull coloration, express only a few behaviors, and are reproductively incompetent.

area (POA) shrink to one-eighth their volume and produce less GnRH1 mRNA and peptide, causing hypogonadism and loss of reproductive competence (~2 weeks) (Davis & Fernald 1990, Francis et al. 1993, White et al. 2002). Similarly, androgen, estrogen, and GnRH receptor mRNA expression levels depend on social status (Au et al. 2006, Burmeister et al. 2007, Harbott et al. 2007) as do electrical properties of the GnRH1 neurons (Greenwood & Fernald 2004). Conversely, when an ND male is moved into a social system in which it is larger than other males, these changes are reversed. The male quickly (<20 s) assumes a bright territorial coloration, engages in aggressive encounters, and acquires a territory. We know that ascending animals rapidly (~20 min) activate the molecular processes related to the subsequent neural transformations (Burmeister et al. 2005).

HOW DOES SOCIAL INFORMATION INFLUENCE BEHAVIOR AND THE BRAIN?

Social living requires sophisticated cognitive abilities because successful social individuals must collect information to guide their future behavior. We have tested the abilities of *A. burtoni* to understand what they know and how they know it about their social world, and more importantly, how this knowledge affects their brain and subsequent behavior. Knowledge of how conspecifics act in key social situations would be a useful guide to predict future behavior. Consequently, in social settings, individuals gather and use information about others' behavior (Brown & Laland 2003) to regulate their own behavior. But how do animals collect social information (McGregor 2005, Brown et al. 2006), and what do they learn as a result? Here we describe several experiments in which we tested individuals' responses after social observations.

Female Mate Choice

People often assume that females choose males displaying the most exaggerated sexual traits—whether behavioral, morphological, or material, such as food and shelter. However, other factors may also contribute importantly to female mate-choice decisions. A wide range of subtle and complex external factors has been shown to influence female mate choice, suggesting sophisticated integration of cues by females. Less well understood are the physiological substrates that are likely also crucial for successful female reproductive choice. Choice of a mate by a female is very important and is dictated by a variety of factors. For example, genetic and epigenetic factors, circulating hormones, and learned behavior can contribute to a female's final mate choice (see review by Argiolas 1999). In addition to their role in solicitation behavior, hormones are intimately involved in establishing a preference for conspecifics of the opposite sex during ontogeny (e.g., Adkins-Regan 1998).

Although manipulating hormone levels experimentally can be informative, we took advantage of the naturally fluctuating levels of hormones in the female reproductive cycle to discover whether and how decisions to affiliate with males of different reproductive quality change as a function of the female's stage in the reproductive cycle. We showed in *A. burtoni* that gravid females preferentially associated with D males, whereas nongravid females showed no preference and that preference did not depend on male size (Clement et al. 2005). These data suggest that females use a hierarchy of internal and external cues in deciding on a mate.

Females who choose an inappropriate mate may pay a high cost in lost reproductive opportunity. But which cues should they use to select a mate that might be a successful reproductive partner? In general, females should be choosier about prospective mates than should males because bad mating decisions typically result in higher costs for females than for males (Trivers 1972). Consequently, studies measuring female assessment of male characteristics have produced conceptual, theoretical, and empirical hypotheses suggesting key factors that may mediate female mate choice (for reviews, see Ryan 1980, Andersson & Simmons 2006). Perhaps unsurprisingly, some evidence indicates that females may use information about male–male social interactions in their mate-choice decisions (Otter et al. 1999, Doutrelant & McGregor 2000, Mennill et al. 2002, Earley & Dugatkin 2005), but little is known about how the brain responds to this kind of information. Specifically, what are the consequences for mate choice after collecting key information? Male–male social interactions may affect female mate choice because information about a potential mate triggers changes in female reproductive physiology.

Using immediate early gene (IEG) expression as a proxy for brain activity, we asked whether social information about a preferred male influenced neural activity in females (Desjardins et al. 2010). After a gravid female *A. burtoni* chose between two socially and physically equivalent males, we staged a fight between these males. Her preferred male either won or lost. We then measured IEG expression levels in several brain nuclei including those in the vertebrate social behavior network (SBN), a collection of brain nuclei known to be important for social behavior (Newman 1999, Goodson 2005). When the female saw her preferred male win a fight, SBN brain nuclei associated with reproduction were activated, but when she saw her preferred male lose a fight, the lateral septum, a nucleus associated with anxiety, was activated instead. Thus social information alone, independent of actual social interactions, activated specific brain regions that differ significantly depending on what the female sees. These effects are seen only in gravid females, consistent with our earlier data showing that hormones are important for female mate choices.

These experiments, assessing the role of mate-choice information in the brain using a paradigm of successive presentations of mate information, suggest a method for identifying the neural consequences of social information on animals using IEG activation (Desjardins et al. 2010). IEGs are the earliest genomic responses to stimuli and require no prior activation by any other gene (Clayton 2000). This response, however, is the tip of the genetic activation iceberg because the total number of genes that comprise one neuron's inducible genomic response has been estimated from tens to hundreds (Nedivi et al. 1993); likewise, the set of rapidly inducible genes in any particular cell in the brain may be still larger (Miczek 1977), meaning that the IEG expression measured here is likely only a tiny fraction of the total gene expression. Nonetheless, this glimpse of the genetic response to social information shows not only that females attend to the information received from watching males interact but also that such information has dramatic effects on their brains in key nuclei rather than producing widespread, general arousal, generating the genetic substrate for subsequent behavioral responses.

Assessing Male Fighting Abilities

A second example of social information being used by *A. burtoni* is the process of drawing inferences through observation. Because males in this species need to be dominant and defend a territory to reproduce, individuals are innately driven to acquire a territory through fighting with an incumbent male. Males engage in vigorous aggressive fighting bouts that determine their access to a territory and subsequent mating opportunities. However, in a colony containing ~60 D males, it would take a large number of pairwise fights for a male to figure out which dominant individual is vulnerable for a territorial takeover. Plus, the cost to an individual male to engage in repeated conflicts would be substantial. The process of transitive inference (TI) could shorten this process significantly if the animals could infer their chances of winning a fight in advance. TI involves using known relationships to deduce unknown ones. For example, using $A > B$ and $B > C$ to infer $A > C$ allows an individual to acquire hierarchical information essential to logical reasoning. First described as a developmental milestone in children (Piaget 1928), TI has since been reported in nonhuman primates (Gillian 1981, McGonigle & Chalmers 1977, Rapp et al. 1996), rats (Davis 1992, Roberts & Phelps 1994), and birds (Bond et al. 2003, Steirn et al. 1995, von Fersen et al. 1991). Still, how animals acquire and represent transitive relationships and why such abilities might have evolved remain unknown.

We have shown that *A. burtoni* males can draw inferences about a hierarchy implied by pairwise fights between rival males, demonstrating TI. These fish learned the implied hierarchy vicariously as bystanders watching fights between rivals (Grosenick et al. 2007) and can use TI when trained on socially relevant stimuli. Note that they can make such inferences using indirect information alone. The key to this experiment was to show bystanders staged fights between matched animals to assure the outcomes (see Grosenick et al. 2007 for details). Testing the animals' choice robustly demonstrated TI in both the home tank and a novel tank.

As noted above, social interactions require knowledge of the environment and status of others that can be acquired indirectly by observing others' behavior. When being observed, animals can also alter their signals on the basis of who is watching. We measured how male *A. burtoni* behave when being watched in two different contexts. In the first, we showed that aggressive and courtship behaviors displayed by subordinate males depend critically on whether D males can see them; in the second, we manipulated who was watching aggressive interactions and showed that D males will change their behavior depending on audience composition. In both cases, when a more dominant individual is out of view and the audience consists of more subordinate individuals, those males signal key social information to females by displaying courtship and dominant behaviors. In contrast, when a D male is present, males cease both aggression and courtship. These data suggest that males are keenly aware of their social environment and modulate their aggressive and courtship behaviors strategically for reproductive and social advantage (Desjardins et al. 2012).

FROM SOCIAL INFORMATION TO CHANGES IN THE BRAIN

The brief description of the variety of social interactions found in fish described above offers just a glimpse of the potential mechanisms through which animals sample information from their social environments. But how does that social information change their behavior

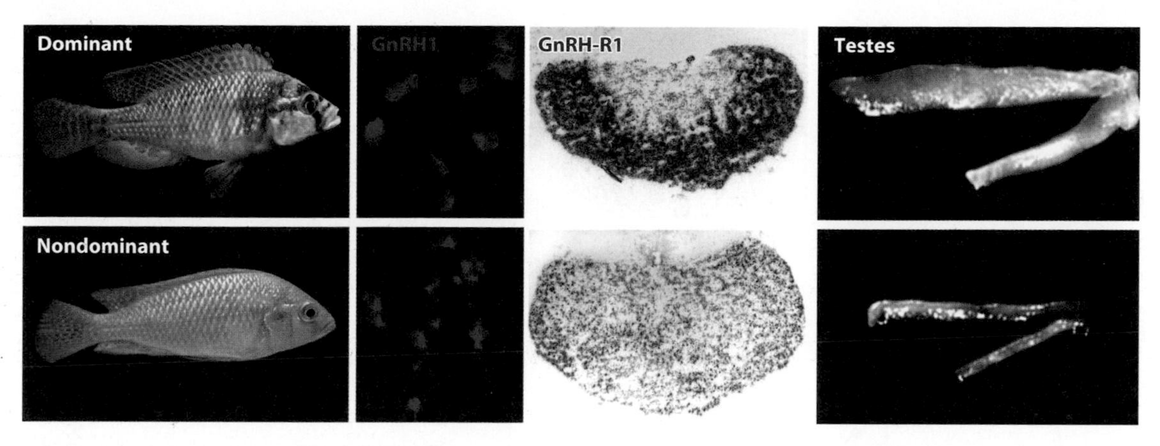

Figure 2

Social regulation of the hypothalamic-pituitary-gonadal axis in *A. burtoni*. Phenotypic characters of reproductively active dominant (D) males (*top row*) and socially suppressed ND males (*bottom row*) are shown. D males have larger GnRH1 neurons (*red*; immunohistochemical staining) in the preoptic area of the brain (Davis & Fernald 1990, White et al. 2002), higher GnRH-R1 levels (*black*, GnRH-R1 in situ hybridization; *purple*, cresyl violet counterstain) in the pituitary gland (Au et al. 2006, Flanagan et al. 2007, Maruska et al. 2011), and larger testes (Fraley & Fernald 1982, Davis & Fernald 1990, Maruska & Fernald 2011) compared with subordinate males (modified from Maruska & Fernald 2011). Comparing the dendritic morphology of GnRH1 neurons between D and ND male *A. burtoni* using confocal images provides preliminary evidence that features of the dendritic arbor morphology depend on reproductive state.

and ultimately their brain? More specifically, how does information change circuits and cells to control reproduction, and how does it ultimately reach the organs responsible for it? Although investigators have suggested potential genomic substrates for these processes in several systems (Robinson et al. 2008), the actual circuits and other parts of the responsible nervous system remain unknown.

To approach this problem, we exploit the extreme, reversible phenotypic switch between *A. burtoni* males from D to ND to understand the effects of social environment on reproduction. By combining manipulations of the social milieu with direct intervention and/or measurement of relevant neural and hormonal systems, we have discovered many socially regulated changes that provide insight into the subtle interplay among the factors responsible for causal links to reproductive function. Here I focus on the signaling pathway that begins in the brain with GnRH1 release to the pituitary and ends by controlling reproductive competence and reproduction itself. The macroscopic consequences of a change in phenotype on the reproductive axis is shown in **Figure 2**.

We measured the dendritic extent using laser confocal microscopy and found that the total dendrite length of GnRH1 neurons in D males ($n = 12$, average total length per cell = 838 μm) is dramatically greater than the total dendrite length of GnRH neurons in NDs

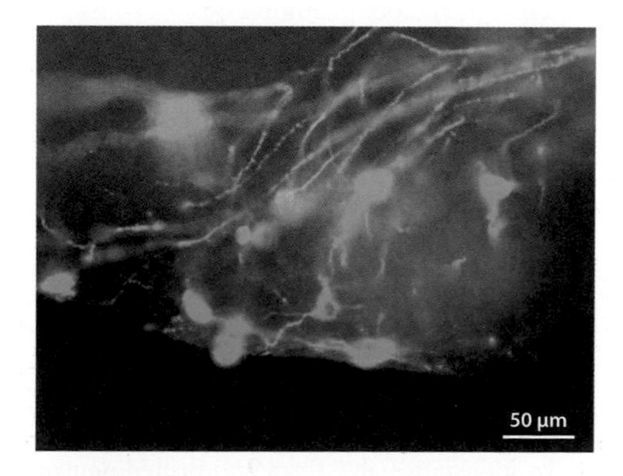

Figure 3

Individual neurons in the POA were filled with neurobiotin using a microelectrode and immunostained with antibodies to GnRH (*green*) to confirm their identity. Yellow cell is filled with neurobiotin and colabeled with GnRH antibody (A.K. Greenwood and R. Fernald, unpublished).

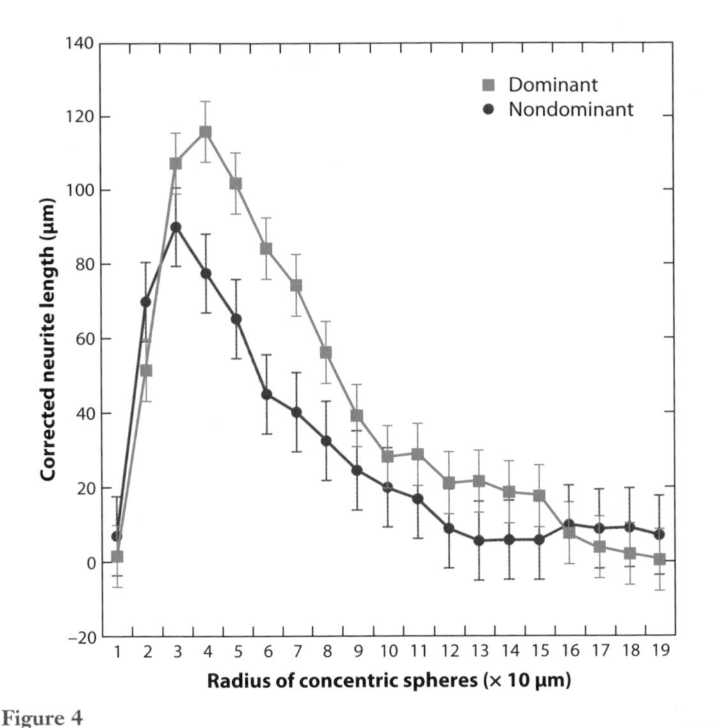

Figure 4

Neurite length measured within each shell of 10 μm concentric spheres (Sholl analysis). GnRH1 neurons from D males have a greater dendrite length beyond 30 μm from the soma (see text for discussion): D males (n = 12; average length = 838 μm), ND males (n = 8; average length = 459 μm). From M.D. Scanlon, A.K. Greenwood & R.D. Fernald, unpublished observations.

[n = 8, average total length per cell = 459 μm (Scanlon et al. 2003)] (**Figure 3**).

A Sholl (Sholl 1953) analysis of these data, which measures the dendrites in concentric circles, shows significant trends for the overall difference in dendrite arbor size between D and ND males and statistically significant differences in dendritic lengths between 30 and 80 μm. As can be seen in the graph (**Figure 4**), the difference in length emerges ~30 μm from the cell body, suggesting increased branching or extension of the dendrite ends rather than an addition of primary dendrites from the soma.

In some cases, dye from an injected neuron passed into a second uninjected neuron (e.g., unlabeled cell), suggesting that GnRH cells may be coupled, possibly by electrical synapses. Coupling by gap junctions is perhaps the most widely studied mechanism of direct electrical communication and exists in multiple regions of the mammalian brain (see Bennett & Zukin 2004 for review). Direct electrical communication between neurons can give rise to a wide repertoire of dynamic network outputs, and the particular network pattern that does arise will depend on intrinsic cellular properties that affect firing frequencies and heterogeneity across the population and at the location of the communication (e.g., soma, proximal or distal dendrite; Roberts et al. 2006).

The role of direct coupling in GnRH neurons is unknown but could be important in the production of rhythmic GnRH1 output. Fluorescence recovery after photobleaching (FRAP) (Matesic et al. 1996) and neurobiotin labeling (Hu et al. 1999) in immortalized GnRH1 neurons (GT1-7 cell line) indicated direct connections between 20% and 75% of cultured cells. In GT1-7 cells, gap junctions also contribute to synchronized pulses (Vazquez-Martinez et al. 2001), and the same group (Bose et al. 2010) recently reported the necessity of connexin 43 (Cx43) for synchrony. Moreover, gap junction protein levels appear to be regulated by at least one intracellular messenger important in GnRH secretion, cyclic adenosine monophosphate (AMP) (Matesic et al. 1996). In the rat, hypothalamic GnRH neurons express mRNA for one of the more common connexin proteins (connexin 32), but gap junctions between GnRH neurons have not been demonstrated directly in this system (Hosny & Jennes 1998).

Evidence increasingly indicates a role for gap junctions in several systems related to reproduction. For example, gap junctions have been shown to be important in the pituitary of teleost fish (Levavi-Sivan et al. 2005), and they are important in the rat spinal column motor nuclei, where they are regulated by testosterone (Coleman & Sengelaub 2002).

Most studies of electrical communication in GnRH1 neurons have focused on communication at the soma. However, two of the above studies found *in vitro* evidence for the potential involvement of neurites, some of which could be rudimentary dendrites. First, Matesic et al. (1996) reported that pharmacological agents that increase cAMP also increase connectivity, an effect possibly mediated by dendrites because these treatments increased both the number of neurites and the number of connexin 26 positive neurites. Second, Hu et al. (1999) demonstrated that interactions among neurites could account for at least some of the direct coupling observed in GT1-7 cells, but dendritic coupling of hypothalamic GnRH1 neurons remains to be studied systematically. Nonetheless, emerging evidence in mammalian GnRH neurons is consistent with our findings, namely that GnRH neurons may interact extensively at the dendrite level. Thus, the social cues that drive dendritic remodeling may result in the formation of a dendrite-mediated network of GnRH1 neurons. Modification of dendritic structure by social status is another of the socially regulated parameters in the GnRH1 neurons.

As noted above, in all vertebrates, both development and differences in reproductive state are controlled by the GnRH1 neurons in the basal forebrain (e.g., Gore 2002). In *A. burtoni* these neurons are located in the anterior parvocellular preoptic nucleus (aPPn), which is the most anterior part of the preoptic area in teleosts, a conserved vertebrate brain region (Wullimann & Mueller 2004). Other GnRH peptides exist in all vertebrates (White et al. 1998) as in *A. burtoni*. Specifically, GnRH2 and GnRH3 (White et al. 1995) are expressed in the midbrain tegmentum and the forebrain terminal nerve ganglion, respectively, but neither is found in the pituitary (Powell et al. 1995) nor do they exhibit socially induced neural plasticity in soma size or gene expression (Davis & Fernald 1990, White et al. 2002).

Social modulation of reproductive function is widespread among vertebrates with dominance status as a particularly salient factor. Examples include the suppression of ovulation (e.g., Abbott & Hearn 1978; Rood 1980; Abbott et al. 1978, 1998), control of maturation in social mammals (e.g., Vandenbergh 1973, Lombardi & Vandenbergh 1976, Bediz & Whitsett 1979, Faulkes et al. 1990), delay of first breeding in birds (e.g., Selander 1965, Wiley 1974), control of maturation in fish and other vertebrates (e.g., Sohn 1977, Fraley & Fernald 1982, McKenzie et al. 1983, Leitz 1987), changes in stress levels influencing reproduction (Fox et al. 1997, Abbott et al. 2003), and even sex change (e.g., Robertson 1972, Fricke & Fricke 1977, Cole & Robertson 1988). Thus the mechanistic discoveries from *A. burtoni* may be instructive for understanding the neural control of this process in many other species.

In vertebrates, numerous studies demonstrate activation of the reproductive axis caused by different sensory systems [e.g., olfactory (Gore et al. 2000, Rekwot et al. 2001, Murata et al. 2011), auditory (Bentley et al. 2000, Burmeister & Wilczynski 2005, Maney et al. 2007), tactile (Pfaus & Heeb 1997, Wersinger & Baum 1997), and visual (Castro et al. 2009)]. Changes in the reproductive axis are measured as changes in the number, size, or axonal densities of GnRH1-immunoreactive neurons, alterations in neuronal firing patterns, surges in circulating leutinizing hormone (LH) or steroid levels, increased testicular activity, or increases in sexual arousal and behavior. In addition to sensory channel–specific signals, contextual social interactions with multimodal sensory information such as courtship, mating, exposure to the opposite sex, parental care, and opportunities to rise in social rank are also known to influence GnRH neurons and the hypothalamic-pituitary-gonadal (HPG) axis in many vertebrates (Wu et al. 1992; Dellovade & Rissman 1994; Wersinger & Baum 1997; Rissman et al. 1997; Bakker et al. 2001; Scaggiante et al. 2004, 2006; Burmeister et al. 2005; Cameron et al. 2008; Lake et al. 2008; Mantei et al. 2008; Stevenson et al. 2008). However, despite numerous studies that demonstrate links between important social signals and activation of the reproductive axis, very few experiments examine how these social sensory signals cause changes, either directly or indirectly, within GnRH1 at the neuronal or genomic level.

One method for approaching this problem has been to use IEGs such as *egr-1*, *c-fos*, *jun*, and *arc* to identify activated neurons within reproductive and neuroendocrine circuits (Pfaus & Heeb 1997, Clayton 2000), but neuronal activation is not always associated with IEG induction. IEGs are typically not expressed in chronically activated neurons, and IEG induction is often not related to challenge-induced neuropeptide expression (Farivar et al. 2004, Hoffman & Lyo 2002, Kovacs 2008, Pfaus & Heeb 1997). Nevertheless, socially relevant reproductive stimuli are known to induce IEG expression within GnRH1 neurons across vertebrates from fishes (Burmeister & Fernald 2005) to mammals (Pfaus et al. 1994, Meredith & Fewell 2001, Gelez & Babre-Nys 2006). Burmeister et al. (2005) showed that in *A. burtoni* the perception of a social opportunity by a subordinate male who then ascends to become a D male produces a rapid (20–30 min) induction of the IEG *egr-1* (a transcription factor–encoding gene; also called *zenk*, *zif-268*, *ngfi-a*, *krox-24*, *tis8*) in the preoptic area and in GnRH1 neurons (Burmeister & Fernald 2005). This molecular response results from the recognition of a social opportunity because it is not elicited in males who are already dominant. Recent studies in *A. burtoni* also suggest that visual cues alone are not sufficient to fully suppress the reproductive axis of subordinate males and that other senses such as olfaction are likely involved (Chen & Fernald 2011, Maruska & Fernald 2012).

SOCIAL REGULATION OF GENE EXPRESSION IN THE PITUITARY: GNRH RECEPTORS AND GONADOTROPIN HORMONES

GnRH1 from the brain travels to the gonadotrope-producing cells in the anterior pituitary gland directly via neuronal projections in fish species in contrast with the specialized vascular system in other vertebrates. Once in the pituitary, GnRH1 binds to its cognate receptors on secretory cells that release LH and follicle-stimulating hormone (FSH), which stimulate steroid production and gamete development in the gonads (testes or ovaries). Multiple forms of GnRH receptors (i.e., types I, II, III) are found in mammals (Millar 2005), amphibians (Wang et al. 2001), and fishes (Robinson et al. 2001, Lethimonier et al. 2004, Moncaut et al. 2005, Flanagan et al. 2007), and they often show differential distributions, expression patterns (e.g., across season, reproductive stage, or dominance status), and varying responses to regulation by steroids, GnRH, and monoamines, all of which suggest functional specializations (Cowley et al. 1998, Levavi-Sivan et al. 2004, Au et al. 2006, Chen & Fernald 2006, Lin et al. 2010). Although we have considerable information on the signal transduction pathways and how different neurohormones and steroids influence gonadotropin synthesis and release (Bliss et al. 2010, Thackray et al. 2010), little is known about how social information modulates gonadotrope output at the pituitary.

In male *A. burtoni*, pituitary mRNA levels of *GnRH-R1*, but not *GnRH-R2*, are socially regulated such that stable D males have higher levels compared with subordinate males, and the increase during the social transition appears to occur more slowly (days) than do changes in mRNA levels of other genes, which occur within minutes to hours (Au et al. 2006, Maruska et al. 2011). However, pituitary mRNA levels of the IEG *egr-1* and of the β-subunits of LH and FSH are increased at just 30 min after social ascent, suggesting that GnRH1 release has quickly activated the pituitary gland (Maruska et al. 2011).

SOCIAL REGULATION OF GENE EXPRESSION IN THE TESTES: SPERMATOGENESIS AND STEROID PRODUCTION

Although many studies have shown how social information including mating opportunities, female presence or attractiveness, and social status can influence testicular function in terms of sperm quality (e.g., velocity, motility, number) from fishes to humans (Kilgallon & Simmons 2005, Cornwallis & Birkhead 2007, Gasparini et al. 2009, Ramm & Stockley 2009, Maruska & Fernald 2011), less is known about how social cues induce molecular changes in the testes. In *A. burtoni*, however, perception of social opportunity triggers genomic changes in mRNA levels on both rapid (minutes to hours: FSHR, androgen receptors, corticosteroid receptors) and slower [days: leutinizing hormone releasing hormone (LHR), aromatase, estrogen receptors] time scales (Maruska & Fernald 2011). During the subordinate to D male social transition, the morphological and structural changes in testicular cell composition and relative testes size take several days, whereas many molecular changes in the testes are detected more quickly (Maruska & Fernald 2011). This rapid genomic response in the most distal component of the HPG axis highlights the sensitivity and plasticity of the entire reproductive system to social information. Furthermore, the quick genomic changes in the testes raise the possibility that there may be additional and parallel signaling pathways that perhaps bypass the traditional linear cascade from brain GnRH1 release to pituitary LH/FSH release to a testicular gonadotropin receptor activation scheme.

CONCLUSIONS

Clearly, social information has a profound influence on the function of the reproductive axis in all vertebrates; however, far less is known about how this social information influences the HPG axis at the cellular and molecular levels (e.g., changes in gene expression), and some critical questions about the links between social behaviors, reproductive axis function, and the genome remain unanswered. For example, which signal pathways link reception/perception of a constellation of social cues that produce changes in gene expression along the HPG axis? It seems clear that the next step is to use the many advances in proteomics, transcriptomics, microtranscriptomics, and epigenomics, combined with comparative systems approaches including single-cell analyses, optogenetics, and transgenic methods. Determining the regulatory roles of epigenetic and small RNAs (e.g., microRNAs) in mediating socially induced changes along the reproductive axis is also an exciting area of future work (Robinson et al. 2005, 2008; Huang et al. 2011; Rajender et al. 2011) that should provide insights into our understanding of the mechanisms governing social and seasonal reproductive plasticity across taxa. The cichlid fish *A. burtoni*, with its complex and experimentally manipulable social system, the wealth of background knowledge on the social control of HPG axis function, and the recently available genomic resources will all become a valuable vertebrate model system for studying how the social environment influences genomic plasticity and function of the reproductive axis.

DISCLOSURE STATEMENT

The author is not aware of any affiliations, memberships, funding, or financial holdings that might be perceived as affecting the objectivity of this review.

ACKNOWLEDGMENTS

My thanks go to the members of my laboratory for their contributions to this work. R.D.F. was supported by NIH NS 034950, MH087930, and NSF IOS 0923588.

LITERATURE CITED

Abbott DH, Hearn JP. 1978. Physical, hormonal and behavioral aspects of sexual development in marmoset monkeys, *Callthrix jaccus*. *J. Reprod. Fertil.* 53:155–66

Abbott DH, Keverne EB, Bercovitch FB, Shively CA, Mendoza SP, et al. 2003. Are subordinates always stressed? A comparative analysis of rank differences in cortisol levels among primates. *Horm. Behav.* 43:67–82

Abbott DH, Saltzman W, Schultz-Darken NJ, Tannenbaum PL. 1998. Adaptations to subordinate status in female marmoset monkeys. *Comp. Biochem. Physiol. C* 119:261–74

Adkins-Regan E. 1998. Hormonal mechanisms of mate choice. *Am. Zool.* 38:166–78

Agrillo C, Dadda M, Serena G, Bisazza A. 2009. Use of number by fish. *PLoS One* 4:e4786

Andersson M, Simmons LW. 2006. Sexual selection and mate choice. *Trends Ecol. Evol.* 21:296–302

Argiolas A. 1999. Neuropeptides and sexual behavior. *Neurosci. Biobehav. Rev.* 23:1127–42

Au TM, Greenwood AK, Fernald RD. 2006. Differential social regulation of two pituitary gonadotropin-releasing hormone receptors. *Behav. Brain Res.* 170:342–46

Bakker J, Kelliher KR, Baum MJ. 2001. Mating induces gonadotropin-releasing hormone neuronal activation in anosmic female ferrets. *Biol. Reprod.* 64:1100–5

Balshine-Earn S, Lotem A. 1998. Individual recognition in a cooperatively breeding cichilid: evidence from video playback experiments. *Behaviour* 135:369–86

Bediz GM, Whitsett JM. 1979. Social inhibition of sexual maturation in male prairie mice. *J. Comp. Physiol. Psychol.* 93:493–500

Bennett MV, Zukin RS. 2004. Electrical coupling and neuronal synchronization in the mammalian brain. *Neuron* 41:495–511

Bentley GE, Wingfield JC, Morton ML, Ball GF. 2000. Stimulatory effects on the reproductive axis in female songbirds by conspecific and heterospecific male song. *Horm. Behav.* 37:179–89

Bliss SP, Navratil AM, Xie J, Roberson MS. 2010. GnRH signaling, the gonadotrope and endocrine control of fertility. *Front. Neuroendocrinol.* 31:322–40

Bond AB, Kamil AC, Balda RP. 2003. Social complexity and transitive inference in corvids. *Anim. Behav.* 65:479–87

Bose SK, Leclerc GM, Vazquez-Martinez R, Boockfor FR. 2010. Administration of connexin43 siRNA abolishes secretory pulse synchronization in GnRH clonal cell populations. *Mol. Cell. Endocrinol.* 314:75–83

Brown C, Laland K. 2003. Social learning in fishes: a review. *Fish Fish.* 4:280–88

Brown C, Laland K, Krause J, eds. 2006. *Fish Cognition and Behavior*. Oxford, UK: Blackwell

Byrne RW, Whiten A, eds. 1988. *Machiavellian intelligence: social complexity and the evolution of intellect in monkeys, apes, and humans*. Oxford, UK: Oxford Univ. Press

Bshary R, Wickler W, Fricke H. 2002. Fish cognition: a primate's eye view. *Anim. Cogn.* 5:1–13

Bugnyar T, Heinrich B. 2006. Pilfering ravens, *Corvus corax*, adjust their behaviour to social context and identity of competitors. *Anim. Cogn.* 9:369–76

Burmeister SS, Fernald RD. 2005. Evolutionary conservation of the egr-1 immediate-early gene response in a teleost. *J. Comp. Neurol.* 481:220–32

Burmeister SS, Jarvis ED, Fernald RD. 2005. Rapid behavioral and genomic responses to social opportunity. *PLoS Biol.* 3:e363

Burmeister SS, Kailasanath V, Fernald RD. 2007. Social dominance regulates androgen and estrogen receptor gene expression. *Horm. Behav.* 51:164–70

Burmeister SS, Wilczynski W. 2005. Social signals regulate gonadotropin-releasing hormone neurons in the green treefrog. *Brain Behav. Evol.* 65:26–32

Cameron N, Del Corpo A, Diorio J, McAllister K, Sharma S, Meaney MJ. 2008. Maternal programming of sexual behavior and hypothalamic-pituitary-gonadal function in the female rat. *PLoS One* 3:e 2210

Candland EJ. 1993. *Feral Children and Clever Animals*. New York: Oxford Univ. Press

Castro AL, Gonçalves-de-Freitas E, Volpato GL, Oliveira C. 2009. Visual communication stimulates reproduction in Nile tilapia, *Oreochromis niloticus* (L.). *Braz. J. Med. Biol. Res.* 42:368–74

Chen CC, Fernald RD. 2006. Distributions of two gonadotropin-releasing hormone receptor types in a cichlid fish suggest functional specialization. *J. Comp. Neurol.* 495:314–23

Chen CC, Fernald RD. 2011. Visual information alone changes behavior and physiology during social interactions in a cichlid fish (*Astatotilapia burtoni*). *PLoS One* 6:e20313

Chittka LA, Osorio D. 2007. Cognitive dimensions of predator responses to imperfect mimicry. *PLoS Biol.* 5:e339

Clayton DF. 2000. The genomic action potential. *Neurobiol. Learn. Mem.* 74:185–216

Clement TS, Parikh V, Schrumpf M, Fernald RD. 2005. Behavioral coping strategies in a cichlid fish: the role of social status and acute stress response in direct and displaced aggression. *Horm. Behav.* 47:336–42

Cole KS, Robertson DR. 1988. Protogyny in the caribbean reef goby, *Coryphopterus personatus*: gonad ontogeny and social influences on sex-change. *Bull. Mar. Sci.* 42:317–33

Coleman AM, Sengelaub DR. 2002. Patterns of dye coupling in lumbar motor nuclei of the rat. *J. Comp. Neurol.* 454:34–41

Cornwallis CK, Birkhead TR. 2007. Changes in sperm quality and numbers in response to experimental manipulation of male social status and female attractiveness. *Am. Nat.* 170:758–70

Cowley MA, Rao A, Wright PJ, Illing N, Millar RP, Clarke IJ. 1998. Evidence for differential regulation of multiple transcripts of the gonadotropin releasing hormone receptor in the ovine pituitary gland; effect of estrogen. *Mol. Cell. Endocrinol.* 146:141–49

Dally JM, Emery NJ, Clayton NS. 2006. Food-caching western scrub-jays keep track of who was watching when. *Science* 312:1662–65

Davis H. 1992. Transitive inference in rats (*Rattus norvegicus*). *J. Comp. Psychol.* 106:342–49

Davis MR, Fernald RD. 1990. Social control of neuronal soma size. *J. Neurobiol.* 21:1180–88

Dellovade TL, Rissman EF. 1994. Gonadotropin-releasing hormone-immunoreactive cell numbers change in response to social interactions. *Endocrinology* 134:2189–97

Desjardins JK, Klausner JQ, Fernald RD. 2010. Female genomic response to mate information. *Proc. Natl. Acad. Sci. USA* 107:21176–80

Desjardins JK, Hofmann HA, Fernald RD. 2012. Social context influences aggressive and courtship behavior in a cichlid fish. *PloS ONE.* In press

Doutrelant C, McGregor PK. 2000. Eavesdropping and mate choice in female fighting fish. *Behavior* 137:1655–69

Earley RL, Dugatkin LA. 2005. Three poeciliid pillars: fighting, mating and networking. See McGregor 2005, pp. 84–113

Farivar R, Zangenehpour S, Chaudhuri A. 2004. Cellular-resolution activity mapping of the brain using immediate-early gene expression. *Front. Biosci.* 9:104–9

Faulkes CG, Abbott DH, Jarvis JUM. 1990. Social suppression of ovarian cyclicity in captive and wild colonies of naked mole-rats, *Heterocephalus glaber*. *J. Reprod. Fertil. Suppl.* 88:559–68

Flanagan CA, Chen CC, Coetsee M, Mamputha S, Whitlock KE, et al. 2007. Expression, structure, function, and evolution of gonadotropin-releasing hormone (GnRH) receptors GnRH-R1SHS and GnRH-R2PEY in the teleost, *Astatotilapia burtoni*. *Endocrinology* 148:5060–71

Fox HE, White SA, Kao MH, Fernald RD. 1997. Stress and dominance in a social fish. *J. Neurosci.* 17:6463–69

Fraley NB, Fernald RD. 1982. Social control of developmental rate in the African cichlid, *Haplochromis burtoni*. *Z. Tierpsychol.* 60:66–82

Francis RC, Soma K, Fernald RD. 1993. Social regulation of the brain-pituitary-gonadal axis. *Proc. Natl. Acad. Sci.* 90:7794–98

Fricke HW. 1973. Individual partner recognition in fish: field studies on *Amphiprion bicinctus*. *Naturwissenschaften* 60:204–5

Fricke HW, Fricke S. 1977. Monogamy and sex change by aggressive dominance in coral reef fish. *Nature* 266:830–32

Gasparini C, Peretti AV, Pilastro A. 2009. Female presence influences sperm velocity in the guppy. *Biol. Lett.* 5:792–94

Gelez H, Fabre-Nys C. 2006. Neural pathways involved in the endocrine response of anestrous ewes to the male or its odor. *Neuroscience* 140:791–800

Gillian DJ. 1981. Reasoning in the chimpanzee: II. Transitive inference. *J. Exp. Psychol. Anim. Behav. Process.* 7:87–108

Goodson JL. 2005. The vertebrate social behavior network: evolutionary themes and variations. *Horm. Behav.* 48:11–22

Gore AC. 2002. Gonadotropin-releasing hormone (GnRH) neurons: gene expression and neuroanatomical studies. *Prog. Brain Res.* 141:193–208

Gore AC, Wersinger SR, Rissman EF. 2000. Effects of female pheromones on gonadotropin-releasing hormone gene expression and luteinizing hormone release in male wild-type and oestrogen receptor-alpha knockout mice. *J. Neuroendocrinol.* 12:1200–4

Greenwood AK, Fernald RD. 2004. Social regulation of the electrical properties of gonadotropin-releasing hormone neurons in a cichlid fish (*Astatotilapia burtoni*). *Biol. Reprod.* 71:909–18

Grosenick L, Clement TS, Fernald RD. 2007. Fish can infer social rank by observation alone. *Nature* 445:429–32

Grutter AS. 1997. Spatio-temporal variation and feeding selectivity in the diet of the cleaner fish *Labroides dimidiatus*. *Copeia* 1997:345–55

Grutter AS, Bshary R. 2003. Cleaner wrasse prefer client mucus: support for partner control mechanisms in cleaning interactions. *Proc. R. Soc. Lond. B* 270(Suppl. 2):242–44

Harbott LK, Burmeister SS, White RB, Vagell M, Fernald RD. 2007. Androgen receptors in a cichlid fish, *Astatotilapia burtoni*: structure, localization, and expression levels. *J. Comp. Neurol.* 504:57–73

Hare B, Call J, Agnetta B, Tomasello M. 2000. Chimpanzees know what conspecifics do and do not see. *Anim. Behav.* 59:771–85

Hare B, Tomasello M. 2001. Do chimpanzees know what conspecifics know? *Anim. Behav.* 61:139–51

Hauser MD. 1997. Minding the behaviour of deception. In *Machiavellian Intelligence II: Extensions and Evaluations*, ed. A Whiten, RW Byrne, pp. 112–43. Cambridge, UK: Cambridge Univ. Press

Havre N, Fitzgerald GJ. 1988. Shoaling and kin recognition in the three-spined stickleback (*Gasterosteus aculeatus*). *Biol. Behav.* 13:190–201

Hodos W, Campbell CBG. 1969. The scala naturae: why there is no theory in comparative psychology. *Psychol. Rev.* 76:337–50

Hoffman GE, Lyo D. 2002. Anatomical markers of activity in neuroendocrine systems: Are we all 'fos-ed out'? *J. Neuroendocrinol.* 14:259–68

Höjesjö J, Johnsson JI, Petersson E, Järvi T. 1998. The importance of being familiar: individual recognition and social behavior in sea trout (*Salmo trutta*). *Behav. Ecol.* 9:445–51

Hosny S, Jennes L. 1998. Identification of gap junctional connexin-32 mRNA and protein in gonadotropin-releasing hormone neurons of the female rat. *Neuroendocrinology* 67:101–8

Hu L, Olson AJ, Weiner RI, Goldsmith PC. 1999. Connexin 26 expression and extensive gap junctional coupling in cultures of GT1-7 cells secreting gonadotropin-releasing hormone. *Neuroendocrinology* 70:221–27

Huang Y, Shen XJ, Zou Q, Wang SP, Tang SM, Zhang GZ. 2011. Biological functions of microRNAs: a review. *J. Physiol. Biochem.* 67:129–39

Kamil AC. 1998. On the proper definition of cognitive ethology. In *Animal Cognition in Nature*, ed. RP Balda, IM Pepperberg, AC Kamil, pp. 1–28. San Diego, CA: Academic

Kilgallon SJ, Simmons LW. 2005. Image content influences men's semen quality. *Biol. Lett.* 1:253–55

Kovacs KJ. 2008. Measurement of immediate-early gene activation—c-fos and beyond. *J. Neuroendocrinol.* 20:665–72

Lake JI, Lange HS, O'Brien S, Sanford SE, Maney DL. 2008. Activity of the hypothalamic-pituitary-gonadal axis differs between behavioral phenotypes in female white-throated sparrows (*Zonotrichia albicollis*). *Gen. Comp. Endocrinol.* 156:426–33

Leitz T. 1987. Social control of testicular steroidogenic capacities in the Siamese fighting fish *Betta splendens*. *J. Exp. Zool.* 244:473–78

Lethimonier C, Madigou T, Muñoz-Cueto JA, Lareyre JJ, Kah O. 2004. Evolutionary aspects of GnRHs, GnRH neuronal systems and GnRH receptors in teleost fish. *Gen. Comp. Endocrinol.* 135:1–16

Levavi-Sivan B, Bloch CL, Gutnick MJ, Fleidervish IA. 2005. Electrotonic coupling in the anterior pituitary of a teleost fish. *Endocrinology* 146:1048–52

Levavi-Sivan B, Safarian H, Rosenfeld H, Elizur A, Avitan A. 2004. Regulation of gonadotropin-releasing hormone (GnRH)-receptor gene expression in tilapia: effect of GnRH and dopamine. *Biol. Reprod.* 70:1545–51

Lin CJ, Wu GC, Lee MF, Lau EL, Dufour S, Chang CF. 2010. Regulation of two forms of gonadotropin-releasing hormone receptor gene expression in the protandrous black porgy fish, *Acanthopagrus schlegeli*. *Mol. Cell Endocrinol.* 323:137–46

Lombardi JR, Vandenbergh JG. 1976. Pheromonally induced sexual maturation in females: regulation by the social environment of the male. *Science* 196:545–46

Maney DL, Goode CT, Lake JI, Lange HS, O'Brien S. 2007. Rapid neuroendocrine responses to auditory courtship signals. *Endocrinology* 148:5614–23

Mantei KE, Ramakrishnan S, Sharp PJ, Buntin JD. 2008. Courtship interactions stimulate rapid changes in GnRH synthesis in male ring doves. *Horm. Behav.* 54:669–75

Maruska KP Fernald RD. 2011. Social regulation of gene expression in the hypothalamic-pituitary-gonadal axis. *Physiology* 26:412–23

Maruska KP, Fernald RD. 2011. Plasticity of the reproductive axis caused by social status change in an African cichlid fish: II. Testicular gene expression and spermatogenesis. *Endocrinology* 152:291–302

Maruska KP, Fernald RD. 2012. Contextual chemosensory signaling in an African cichlid fish. *J. Exp. Biol.* 215:68–74

Maruska KP, Levavi-Sivan B, Biran J, Fernald RD. 2011. Plasticity of the reproductive axis caused by social status change in an African cichlid fish: I. Pituitary gonadotropins. *Endocrinology* 152:281–90

Matesic DF, Hayashi T, Trosko JE, Germak JA. 1996. Upregulation of gap junctional intercellular communication in immortalized gonadotropin-releasing hormone neurons by stimulation of the cyclic AMP pathway. *Neuroendocrinology* 64:286–97

Maynard Smith J, Harper D. 2003. *Animal Signals*. Oxford, UK: Oxford Univ. Press

McGonigle BO, Chalmers M. 1977. Are monkeys logical? *Nature* 267:694–96

McGregor PK. 2005. *Animal Communication Networks*. Cambridge, UK: Cambridge Univ. Press

McKenzie WDJ, Crews D, Kallman KD, Policansky D, Sohn JJ, et al. 1983. Age, weight, and the genetics of sexual maturation in the platyfish, *Xiphophorus maclatus*. *Copeia* 1983:770–73

Mennill DJ, Ratcliffe LM, Boag PT. 2002. Female eavesdropping on male song contests in songbirds. *Science* 296:873

Meredith M, Fewell G. 2001. Vomeronasal organ: electrical stimulation activates Fos in mating pathways and in GnRH neurons. *Brain Res.* 922:87–94

Metcalfe NB, Thomson BC. 1995. Fish recognize and prefer to shoal with poor competitors. *Proc. R. Soc. Lond. B* 259:207–10

Miczek KA. 1977. Effects of L-dopa, d-amphetamine and cocaine on intruder-evoked aggression in rats and mice. *Prog. Neuro-Pharmacol.* 1:271–77

Millar RP. 2005. GnRHs and GnRH receptors. *Anim. Reprod. Sci.* 88:5–28

Moncaut N, Somoza G, Power DM, Canario AV. 2005. Five gonadotrophin-releasing hormone receptors in a teleost fish: isolation, tissue distribution and phylogenetic relationships. *J. Mol. Endocrinol.* 34:767–79

Murata K, Wakabayashi Y, Sakamoto K, Tanaka T, Takeuchi Y, et al. 2011. Effects of brief exposure of male pheromone on multiple-unit activity at close proximity to kisspeptin neurons in the goat arcuate nucleus. *J. Reprod. Dev.* 57:197–202

Nedivi E, Hevroni D, Naot D, Israeli D, Citri Y. 1993. Numerous candidate plasticity-related genes revealed by differential cDNA cloning. *Nature* 363:718–22

Newman SW. 1999. The medial extended amygdala in male reproductive behavior. A node in the mammalian social behavior network. *Ann. N. Y. Acad. Sci.* 877:242–57

Noble GK, Curtis B. 1939. The social behavior of the jewel fish, *Hemichromis bimaculatus* (Gill). *Am. Mus. Nat. Hist.* 76:1–46

Olsen KH. 1989. Sibling recognition in juvenile Arctic charr, *Salvelinus alpirus*. *J. Fish Biol.* 34:571–81

Otter K, McGregor PK, Terry AMR, Burford FRL, Peake TM, Dabelsteen T. 1999. Do female great tits (*Parus major*) assess males by eavesdropping? A field study using interactive song playback. *Proc. Biol. Sci.* 266:1305–9

Pfaus JG, Heeb MM. 1997. Implications of immediate-early gene induction in the brain following sexual stimulation of female and male rodents. *Brain Res. Bull.* 44:397–407

Pfaus JG, Jakob A, Kleopoulos SP, Gibbs RB, Pfaff DW. 1994. Sexual stimulation induces Fos immunoreactivity within GnRH neurons of the female rat preoptic area: interaction with steroid hormones. *Neuroendocrinology* 60:283–90

Piaget J. 1928. *Judgement and Reasoning in the Child.* London: Routledge & Kegan Paul

Pinto A, Oates J, Grutter AS, Bshary R. 2011. Cleaner wrasses (*Labroides dimidiatus*) are more cooperative in the presence of an audience. *Curr. Biol.* 21:1140–44

Powell JF, Fischer WH, Park M, Craig AG, Rivier JE, et al. 1995. Primary structure of solitary form of gonadotropin-releasing hormone (GnRH) in cichlid pituitary; three forms of GnRH in brain of cichlid and pumpkinseed fish. *Regul. Pept.* 57:43–53

Quinn TP, Hara TJ. 1986. Sibling recognition and olfactory sensitivity in juvenile coho salmon (*Oncorhynchus kisutch*). *Can. J. Zool.* 64:921–25

Rajender S, Avery K, Agarwal A. 2011. Epigenetics, spermatogenesis and male infertility. *Mutat. Res.* 727:62–71

Ramm SA, Stockley P. 2009. Adaptive plasticity of mammalian sperm production in response to social experience. *Proc. Biol. Sci.* 276:745–51

Rapp PR, Kansky MT, Eichenbaum H. 1996. Learning and memory for hierarchical relationships in the monkey: effects of aging. *Behav. Neurosci.* 110:887–97

Rekwot PI, Ogwu D, Oyedipe EO, Sekoni VO. 2001. The role of pheromones and biostimulation in animal reproduction. *Anim. Reprod. Sci.* 65:157–70

Rissman EF, Li X, King JA, Millar RP. 1997. Behavioral regulation of gonadotropin-releasing hormone production. *Brain Res. Bull.* 44:459–64

Roberts CB, Best JA, Suter KJ. 2006. Dendritic processing of excitatory synaptic input in hypothalamic gonadotropin releasing-hormone neurons. *Endocrinology* 147:1545–55

Roberts WA, Phelps MT. 1994. Transitive inference in rats—a test of the spatial coding hypothesis. *Psychol. Sci.* 5:368–74

Robertson DR. 1972. Social control of sex reversal in a coral-reef fish. *Science* 177:1007–9

Robinson GE, Fernald RD, Clayton DF. 2008. Genes and social behavior. *Science* 322:896–900

Robinson GE, Grozinger CM, Whitfield CW. 2005. Sociogenomics: social life in molecular terms. *Nat. Rev. Genet.* 6:257–70

Robison RR, White RB, Illing N, Troskie BE, Morley M, et al. 2001. Gonadotropin-releasing hormone receptor in the teleost *Haplochromis burtoni*: structure, location, and function. *Endocrinology* 142:1737–43

Romer AS. 1959. *The Vertebrate Story*. Chicago, IL: Univ. Chicago Press

Rood JP. 1980. Mating relationships and breeding suppression in the dwarf mongoose. *Anim. Behav.* 23:143–50

Ryan MJ. 1980. Female mate choice in a neotropical frog. *Science* 209:523–25

Scaggiante M, Grober MS, Lorenzi V, Rasotto MB. 2004. Changes along the male reproductive axis in response to social context in a gonochoristic gobiid, *Zosterisessor ophiocephalus* (Teleostei, Gobiidae), with alternative mating tactics. *Horm. Behav.* 46:607–17

Scaggiante M, Grober MS, Lorenzi V, Rasotto MB. 2006. Variability of GnRH secretion in two goby species with socially controlled alternative male mating tactics. *Horm. Behav.* 50:107–17

Scanlon MD, Greenwood AK, Fernald RD. 2003. Dendritic plasticity in gonadotropin-releasing hormone neurons following changes in reproductive status. *Soc. Neurosci. Abstr.* No. 828.20

Selander RK. 1965. On mating systems and sexual selection. *Am. Nat.* 99:129–41

Shettleworth SJ. 1993. Varieties of learning and memory in animals. *J. Exp. Psychol. Anim. Behav. Process* 19:5–14

Shettleworth SJ. 1998. *Cognition, Evolution and Behaviour*. New York: Oxford Univ. Press

Sholl DA. 1953. Dendritic organization in the neurons of the visual and motor cortices of the cat. *J. Anat.* 87:387–406

Sohn JL. 1977. Socially induced inhibition of genetically determined maturation in the platyfish, *Xiphophorus macalatus*. *Science* 195:199–201

Steirn JN, Weaver JE, Zentall TR. 1995. Transitive inference in pigeons: simplified procedures and a test of value transfer theory. *Anim. Learn. Behav.* 23:76–82

Stevenson TJ, Bentley GE, Ubuka T, Arckens L, Hampson E, MacDougall-Shackleton SA. 2008. Effects of social cues on GnRH-I, GnRH-II, and reproductive physiology in female house sparrows (*Passer domesticus*). *Gen. Comp. Endocrinol.* 156:385–94

Tegeder RW, Krause J. 1995. Density dependence and numerosity in fright stimulated aggregation behaviour of shoaling fish. *Philos. Trans. R. Soc. Lond. B* 350:381–90

Thackray VG, Mellon PL, Coss D. 2010. Hormones in synergy: regulation of the pituitary gonadotropin genes. *Mol. Cell. Endocrinol.* 314:192–203

Trivers RL. 1972. Parental investment and sexual selection. In *Sexual Selection and the Descent of Man*, ed. B Campbell, pp. 136–79. Chicago: Aldine

Vandenbergh JG. 1973. Acceleration and inhibition of puberty in female mice by pheromones. *J. Reprod. Fertil. Suppl.* 19:411–19

Vazquez-Martinez R, Shorte SL, Boockfor FR, Frawley LS. 2001. Synchronized exocytotic bursts from gonadotropin-releasing hormone-expressing cells: dual control by intrinsic cellular pulsatility and gap junctional communication. *Endocrinology* 142:2095–101

von Fersen L, Wynne CDL, Delius JD, Staddon JER. 1991. Transitive inference formation in pigeons. *J. Exp. Psychol. Anim. Behav. Process* 17:334–41

Wang L, Bogerd J, Choi HS, Seong JY, Soh JM, et al. 2001. Three distinct types of GnRH receptor characterized in the bullfrog. *Proc. Natl. Acad. Sci. USA* 98:361–66

Ward AJW, Holbrook RI, Krause K, Hart PJB. 2005. Social recognition in sticklebacks: the role of direct experience and habitat cues. *Behav. Ecol. Sociobiol.* 57:575–83

Ward AJW, Webster MM, Hart PJB. 2007. Social recognition in wild fish populations. *Proc. R. Soc. B* 274:1071–77

Ward AJW, Webster MM, Magurran AE, Currie S, Krause J. 2009. Species and population differences in social recognition between fishes: a role for ecology. *Behav. Ecol.* 20:511–16

Wersinger SR, Baum MJ. 1997. Sexually dimorphic processing of somatosensory and chemosensory inputs to forebrain luteinizing hormone-releasing hormone neurons in mated ferrets. *Endocrinology* 138:1121–29

White RB, Eisen JA, Kasten TL, Fernald RD. 1998. Second gene for gonadotropin releasing hormone in humans. *Proc. Natl. Acad. Sci.* 95:305–9

White SA, Kasten TL, Bond CT, Adelman JP, Fernald RD. 1995. Three gonadotropin-releasing hormone genes in one organism suggest novel roles for an ancient peptide. *Proc. Natl. Acad. Sci. USA* 92:8363–67

White SA, Nguyen T, Fernald RD. 2002. Social regulation of gonadotropin-releasing hormone. *J. Exp. Biol.* 205:2567–81

Wiley RH. 1974. Effects of delayed reproduction on survival, fecundity, and the rate of population increase. *Am. Nat.* 108:705–9

Wu TJ, Segal AZ, Miller GM, Gibson MJ, Silverman AJ. 1992. FOS expression in gonadotropin-releasing hormone neurons: enhancement by steroid treatment and mating. *Endocrinology* 131:2045–50

Wullimann MF, Mueller T. 2004. Teleostean and mammalian forebrains contrasted: evidence from genes to behavior. *J. Comp. Neurol.* 475:143–62

Under Pressure: Cellular and Molecular Responses During Glaucoma, a Common Neurodegeneration with Axonopathy

Robert W. Nickells,[1] Gareth R. Howell,[2] Ileana Soto,[2] and Simon W.M. John[2,3]

[1]Department of Ophthalmology and Visual Sciences, University of Wisconsin, Madison, Wisconsin 53706; email: nickells@wisc.edu

[2]Howard Hughes Medical Institute, The Jackson Laboratory, Bar Harbor, Maine 04609; email: gareth.howell@jax.org, Ileana.SotoReyes@jax.org, simon.john@jax.org

[3]Department of Ophthalmology, Tufts University School of Medicine, Boston, Massachusetts, 02110

Annu. Rev. Neurosci. 2012. 35:153–79

First published online as a Review in Advance on April 12, 2012

The *Annual Review of Neuroscience* is online at neuro.annualreviews.org

This article's doi: 10.1146/annurev.neuro.051508.135728

0147-006X/12/0721-0153$20.00

Keywords

optic nerve head, ocular hypertension, astrocytes, retinal ganglion cells, glia

Abstract

Glaucoma is a complex neurodegenerative disorder that is expected to affect 80 million people by the end of this decade. Retinal ganglion cells (RGCs) are the most affected cell type and progressively degenerate over the course of the disease. RGC axons exit the eye and enter the optic nerve by passing through the optic nerve head (ONH). The ONH is an important site of initial damage in glaucoma. Higher intraocular pressure (IOP) is an important risk factor for glaucoma, but the molecular links between elevated IOP and axon damage in the ONH are poorly defined. In this review and focusing primarily on the ONH, we discuss recent studies that have contributed to understanding the etiology and pathogenesis of glaucoma. We also identify areas that require further investigation and focus on mechanisms identified in other neurodegenerations that may contribute to RGC dysfunction and demise in glaucoma.

Contents

INTRODUCTION

Glaucoma is an optic neuropathy that leads to the degeneration of the optic nerve and loss of retinal ganglion cell (RGC) bodies in the retina. The World Health Organization places glaucoma as the second leading cause of blindness worldwide, behind cataracts. It is estimated that nearly 80 million people will have glaucoma by the year 2020 (Quigley & Broman 2006).

Glaucoma is a collection of disease processes that result in a common pathology. Clinically, glaucoma is classified by characteristic changes of the optic nerve head (ONH), in which the neural rim of tissue comprised of RGC axons exiting the eye is reduced, leaving the optic nerve with an excavated or "cupped" appearance. In addition to changes in the ONH, glaucomatous damage is routinely defined and monitored in visual field exams by the formation of scotomas. The classic visual field defect of glaucoma often occurs first in the peripheral inferior retina and can increase to form a large defect of the entire hemisphere, which stops abruptly at the central midline. Because progression of visual field damage is slow and typically painless, patients with glaucoma commonly do not experience any problems with their vision until they have a significant level of visual loss. As many as 50% of affected individuals may go undiagnosed until late in the disease. Substantial efforts have been made to improve diagnostic technologies to detect early-onset disease and are reviewed elsewhere (Quigley 2011).

Higher intraocular pressure (IOP) at baseline is a strong risk factor for glaucoma (Quigley 1993). The relative risk conferred to both the development and progression of glaucoma is so high that elevated IOP, in combination with other characteristics, is often used to classify individuals as affected in large genetic and genomic studies (Crooks et al. 2011). The nature and complexity of IOP changes may be more relevant than the absolute IOP change. Management of IOP is currently the only standard therapy available to treat glaucoma in the clinic. Although elevated IOP is not detected in some patients, IOP lowering still benefits the vision of a significant subset of these patients (CNTGSG 1998). This underscores the need to better understand the pathology of the optic nerve and retina when subjected to IOP changes in glaucoma, which is the major focus of this review.

New data suggest that cerebrospinal fluid pressure also is an important determinant of glaucoma (Berdahl et al. 2008). The role that cerebrospinal fluid pressure plays in glaucoma likely relates to how it interacts physiologically with pressure in other fluid-filled compartments, including IOP in the globe and pressure in the arterial retinal vasculature. These pressurized, fluid-filled compartments converge at the ONH (Jonas 2011), creating a pressure gradient across the tissues of the ONH. Changes in this gradient may be a fundamental component in the initial pathology of glaucoma.

A variety of experimental models are used to study glaucoma. Each of these models has different advantages and disadvantages for assessing specific features of glaucoma, especially given the differences in ONH anatomy (Lasker/IRRF 2010). In interpreting experiments, it is important to note the specifics of the model that was used (see Howell et al. 2008, Lasker/IRRF 2010, McKinnon et al. 2009, Morrison et al. 2005, Pang & Clark 2007, Ruiz-Ederra et al. 2005). Each model can provide important insights into the pathophysiology of glaucoma, but it is now clear that, as with human glaucoma, experimental models are not mechanistically homogenous. Therefore, findings in one model should be reassessed in others. Additionally, lessons learned from other neurodegenerative diseases may be important for glaucoma and vice versa. For example, recent studies suggest similarities between Alzheimer's disease and glaucoma (Almasieh et al. 2011, Liu et al. 2011, McKinnon 2003).

Neurons have distinct functional compartments, which can be affected by distinct disease processes and execute self-autonomous degeneration pathways (compartmentalized degeneration). Whitmore et al. (2005) introduced the concept of compartmentalized degeneration for glaucoma. Their discussion of how distinct processes may affect the RGC axon, soma, or synapses provides an important framework to evaluate the etiology of the disease (Whitmore et al. 2005). The historical and current hypothesis is that an initial and critical insult damages RGC axons in the ONH as they exit the eye. In chronic glaucoma, with milder IOP-induced stress than acute glaucoma, this insult may be necessary for the degeneration of the RGC axon and other RGC compartments. Depending on the severity of insult to specific axons within a given eye, however, the importance of specific degeneration pathways may vary between different RGCs (Howell et al. 2007, Whitmore et al. 2005). In this review, it is not possible to cover in detail compartmentalized degeneration or the majority of glaucoma studies. Because the ONH is a critical structure in any discussion of the pathology of glaucoma and strong experimental evidence indicates it is a point of initial damage, the current review primarily focuses on studies of the ONH with less discussion of other tissues.

GLAUCOMA AS AN AXONOPATHY

The Laminar Region of the Optic Nerve Head

Various data point to the laminar region of the eye as the initial site of damage in glaucoma. The laminar region refers to the site where the sclera is altered to allow RGC axons to exit the eye and enter the optic nerve. Owing to these alterations, this region is the weak point in the wall of the pressurized eye. Modeling shows that pressure-generated stresses are concentrated at this weakest point, presenting an interesting paradox for axonal survival. Increased IOP increases the strain within and across the lamina. The IOP and strain changes result in an increased translaminar pressure gradient (reviewed by Burgoyne 2010). This is generally considered an important component of the disease. When considering the structure of the lamina in glaucoma, size matters. Larger optic nerves have larger laminas, which are commensurately more complicated than smaller laminas found in rodents.

The laminar region of primates is called the lamina cribrosa (LC). In humans, the LC spans an average diameter of 1.7 mm across the scleral canal, large enough to allow passage of approximately 1 million axons from the eye. The LC is comprised of approximately 10 interconnecting beams or plates of connective tissue, which form between 200 and 400 pores of varying sizes through which the axons travel (Quigley & Addicks 1981). Some pores bifurcate into branches, whereas others extend directly through the laminar region. The beams of connective tissue are principally comprised of laminin, elastin, and collagens (Morrison et al. 2005).

In addition to bundles of axons passing through the pores in the LC, the connective tissue beams provide a substrate for different cell types that provide metabolic support for the axons (Anderson & Quigley 1992). Astrocytes are the dominant cell type in the nerve head and LC. They cover the beams, line the pores, ensheathe the axon bundles, and draw nutrient support from capillaries that run within the connective tissue beams. A second cell type, termed an LC cell, has also been reported in the human LC (Hernandez et al. 1988). These cells are sometimes considered distinct from astrocytes, in that they do not express the astrocyte marker glial fibrillary acidic protein, although it is now accepted that not all subtypes of astrocytes are positive for this marker (Sofroniew & Vinters 2010). Behavioral and molecular similarities between astrocytes and LC cells (Johnson & Morrison 2009) suggest that LC cells are a subtype of astrocytes. In this discussion, we refer to them collectively as astrocytes.

In rats and mice, connective tissue beams are not present in the laminar region, presumably because the small size of the eyeball and the reduced numbers of ganglion cell axons leaving the eye (~100,000 in rats and 40,000 to 80,000 in mice) require a much smaller scleral canal. Consequently, sufficient mechanical support can be maintained with bundles of axons surrounded by astrocytes (Howell et al. 2007, May & Lütjen-Drecoll 2002, Schlamp et al. 2006). There are no connective tissue plates, but the rat laminar region has sparse collagenous beams that contain elastin fibrils (Morrison et al. 1995, Morrison et al. 2005). Because it has no collagenous plates, the rodent structure has been termed the glial lamina (Howell et al. 2007).

Patterns of Retinal Ganglion Cell Loss Suggest the Optic Nerve Head Is a Critical Site of Damage

In humans, and nonhuman primates with experimental glaucoma, optic nerve damage often appears as a classic hourglass pattern of degeneration of superior and inferior axon bundles. This pattern of degeneration in the nerve is reflected in the retina, particularly in visual field exams from humans (Quigley 2011), or from monkeys who have been trained to perform visual field tests (Harwerth et al. 1999, Harwerth et al. 2002). The hourglass pattern of degeneration corresponds to regions of the lamina with reduced connective tissue density (Quigley & Addicks 1981). This suggests that an important component of neural damage is the exposure of axon bundles to greater IOP-induced increases in strain in these areas.

The most consistent regional defects that are detected by visual field tests are arcuate scotomas, which match the retinal projections of the arcuate nerve fibers or axons (Shields 1997). Individual bundles of arcuate nerve fibers spread out from the ONH as they arch superiorly and inferiorly around the fovea, meeting at but not crossing the midline of the retina, which is anatomically called the horizontal raphe. The very sharp boundaries separating damaged from undamaged RGCs in the retina as well as the matching of RGC demise to nerve fiber projections suggest that specific bundles of RGC axons are damaged in the LC (**Figure 1**). This pattern does not closely match the trajectory of retinal blood vessels.

Due to the belief that pressure-induced distortion of the collagenous plates would mechanically collapse or damage axons and/or blood vessels, it was previously suggested that the mouse could not develop glaucoma because it lacked collagenous plates in the lamina. It is now clear that the mouse develops glaucoma with regional damage (Danias et al. 2003, Filippopoulos et al. 2006, Mabuchi et al. 2004), and the pattern of damage is similar to that in people (Howell et al. 2007, Jakobs et al. 2005, Schlamp et al. 2006), when the specific anatomy of each species is considered (**Figure 1**). Thus, collagenous plates are not required to induce glaucomatous damage.

In the mouse, the RGC axons run straight between the optic nerve and peripheral retina. Regions of RGC damage are sharply delimited and often match the path of axon bundles, appearing as pie- or fan-shaped wedges radiating from the ONH to the periphery (Howell et al. 2007, Jakobs et al. 2005, Salinas-Navarro et al. 2010, Schlamp et al. 2006, Soto et al. 2011). This pattern matches localized damage to axon bundles passing through the glial lamina (Howell et al. 2007). Because RGC axons do not remain in highly organized bundles after they pass through the lamina in rats and mice, this pattern strongly suggests that axonal damage initially occurs in the lamina. Although the mouse blood vessels run radially, the regions of damage do not match the paths of major blood vessels (Howell et al. 2007, Jakobs et al. 2005, Schlamp et al. 2006), suggesting that vascular changes alone do not influence damage. Collectively, these data implicate the ONH as a critical site of damage, which determines the topography of RGC degeneration.

RGC Axons Are Insulted at the Lamina in Glaucoma

As discussed, the pattern of RGC damage suggests that specific axon bundles are insulted in discrete regions of the lamina. Supporting this, the first morphologically detectable damage occurs in the lamina in human glaucoma (Quigley et al. 1983), in experimental glaucoma in nonhuman primates (Anderson & Hendrickson 1974, Quigley & Anderson 1976), and in an inherited glaucoma in DBA/2J mice (Howell et al. 2007). Additionally, RGC axons degenerate before their cell bodies (Howell et al. 2007, Jakobs et al. 2005, Soto et al. 2008). Because the first site of degeneration is not necessarily the location at which a neuron is insulted (reviewed in Whitmore et al. 2005), BAX-deficient DBA/2J mice were used to experimentally test if axons are directly insulted in the ONH (Howell et al. 2007). BAX is a proapoptotic molecule that is essential for the degeneration of RGC soma in mice (RGCs survive indefinitely when BAX is absent) (Libby et al. 2005, Semaan et al. 2010). Importantly, however, BAX is not required for the degeneration of the axons of these RGCs (Libby et al. 2005), consistent with the self-autonomous destruct pathways for different RGC compartments suggested by Whitmore and colleagues (Whitmore et al. 2005). Additionally, focal damage to peripheral nerves demonstrates that distal axon segments separated from the cell body by the lesion rapidly degenerate by Wallerian degeneration (Coleman & Freeman 2010, Whitmore et al. 2005), but the proximal axon segments attached to the somas remain intact. This phenomenon allows localization of a site of insult, as the proximal axon survives up to this site but the axon distal to the insult degenerates. This is also true for RGCs in BAX-deficient DBA/2J mice, which exhibited degeneration of all RGC axonal segments distal to the site of an acute crush damage to the optic nerve, although the proximal axonal segments survived up to this site (**Figure 2**). Similarly, in chronic glaucoma in these mice, axonal segments within and distal to the laminar region degenerated, whereas proximal segments attached to rescued somas remained intact (Howell et al. 2007). These results provided strong experimental evidence for an early axon insult occurring within or close to the laminar region in glaucoma (**Figure 2**).

Wallerian Degeneration Slow (*Wld*^{s}) Protects from Axon Degeneration

Further evidence of the importance of axonal injury in glaucoma was provided by two studies that utilized the Wallerian degeneration slow

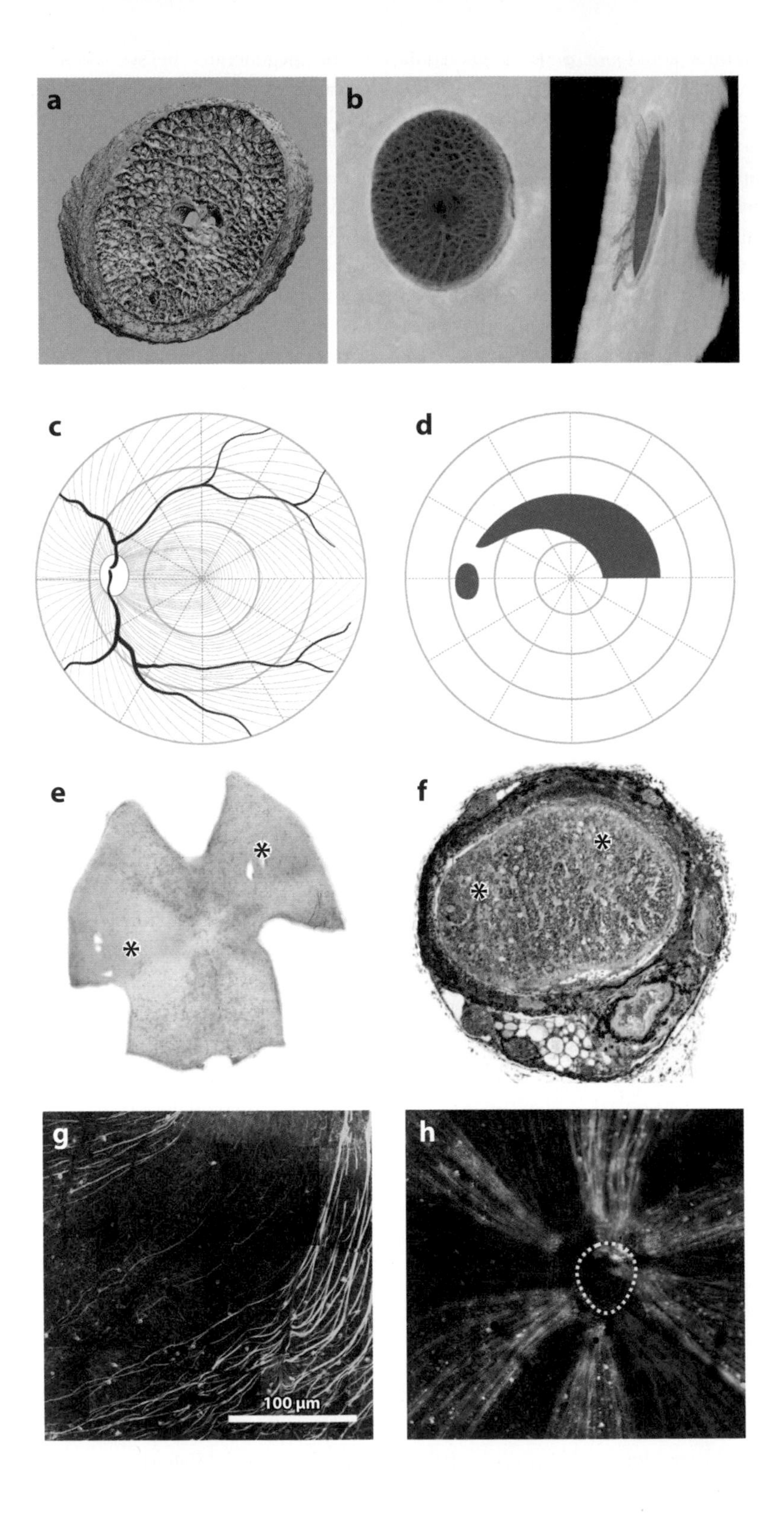
a
b
c
d
e
f
g
h
100 µm

(*Wld^s*) allele. Mice carrying the *Wld^S* allele exhibit a substantial decrease in the rate of axonal degeneration following various insults, and axons remain metabolically active even in the absence of the cell body (Beirowski et al. 2008, Deckwerth & Johnson 1994). Both glaucoma studies showed that axon degeneration was slowed or prevented by the WLDS protein (Howell et al. 2007, Beirowski et al. 2008). In the first study, DBA/2J mice with the *Wld^S* allele had significantly delayed optic nerve degeneration, nearly complete attenuation of ganglion cell soma loss, and retention of RGC activity as assessed by electrophysiological testing (Howell et al. 2007). This supports the concept that injury starts in the axon and saving the axon may prevent degeneration of the neuron in chronic glaucoma. In the second study, increased IOP was induced in transgenic rats expressing WLDS by translimbally photocoagulating episcelaral blood vessels. In this study, axons were protected but somas were apparently lost as determined by immunolabeling (Beirowski et al. 2008). No delay in soma damage was observed, despite a delay in axon degeneration.

In these studies, the different outcomes for the somas may be due to differences in the nature or magnitude of insult(s) in the models used. In this respect, differences almost certainly exist between the inherited mouse and acutely induced rat models. Unlike the DBA/2J model, this rat model produces an early acute elevation in IOP, rather than a prolonged chronic elevation, raising the possibility of a greater ischemic contribution to the pathology. Alternatively, genetic or other differences may have altered the outcome. Either way, the differences are important. They highlight the likelihood of different mechanisms having greater or lesser roles in different patients especially when the severity or abruptness of IOP elevation differs.

The WLDS protein is generated by a mutation that creates a fusion protein of the N terminus of an E4-type ubiquitin ligase (*Ube4b*) and nearly the complete sequence of nicotinimide mononucleotide adenylyltransferase (*Nmnat1*) (Coleman & Freeman 2010). The mechanism by which the WLDS protein delays axonal degeneration is not completely understood. However, it has been proposed that the activity of NMNAT1 is increased in the injured *Wld^s*

Figure 1

The optic nerve head (ONH) is a critical site of damage. (*a*) The lamina cribrosa (LC) of primates contains collagenous plates with pores of varying sizes through which axon bundles pass. The image shows digitally segmented lamina cribrosa connective tissue voxels that have been isolated from surrounding peripapillary sclera to better reveal the complex fenestrations. (*b*) Colorized 3D histomorphometric reconstruction of the normal non-human primate peripapillary sclera (*yellow*) and optic nerve head (*red central retinal vasculature, blue lamina cribrosa beams*). (*c*) In humans, arcuate nerve fiber bundles (*green*) arch superiorly and inferiorly around the fovea, meeting at but not crossing the midline of the retina. (*d*) Field view of an arcuate scotoma (inferior nerve fiber bundle defect) caused by glaucoma. The defect does not cross the midline and matches the path of axons that are bundled in the ONH, but it does not match the path of blood vessels shown in panel *c* (*red lines*). (*e–h*) Regional retinal ganglion cell (RGC) loss also occurs in glaucomatous DBA/2J mice, despite the absence of collagenous plates in the lamina. (*e,f*) In mice, RGC axons run straight between the optic nerve and peripheral retina, and fan-shaped patterns of RGC loss are observed that match axon loss in the ONH. Asterisks indicate corresponding areas of RGC loss in the retina (*e*) and ONH (*f*). (*g,h*) Regions of survival or degeneration in the retina are sharply delimited (*g*). Regions with surviving axon bundles in the lamina correspond to surviving regions in the retina, and the same is true for regions of degeneration (*h*) (compressed Z stack of central retina/ONH pseudocolored to highlight axon paths with lamina indicated as a dotted ring). As seen on the left side of the nerve, axon degeneration in the ONH occurs prior to degeneration in the retina (white-colored axons survived in the nerve fiber layer but were lost in ONH). (*a,b*) Courtesy of Drs. Claude Burgoyne (Optic Nerve Head Research Laboratory) and Crawford Downs (Ocular Biomechanics Laboratory) of the Devers Eye Institute. Adapted from Grau et al. 2006 (*a*) and Burgoyne et al. 2004 (*b*). (*c,d*) Adapted from Shields 1997. (*e,f*) Reproduced from Schlamp et al. 2006. (*g*) Reproduced from Jakobs et al. 2005. (*h*) Reproduced from Howell et al. 2007.

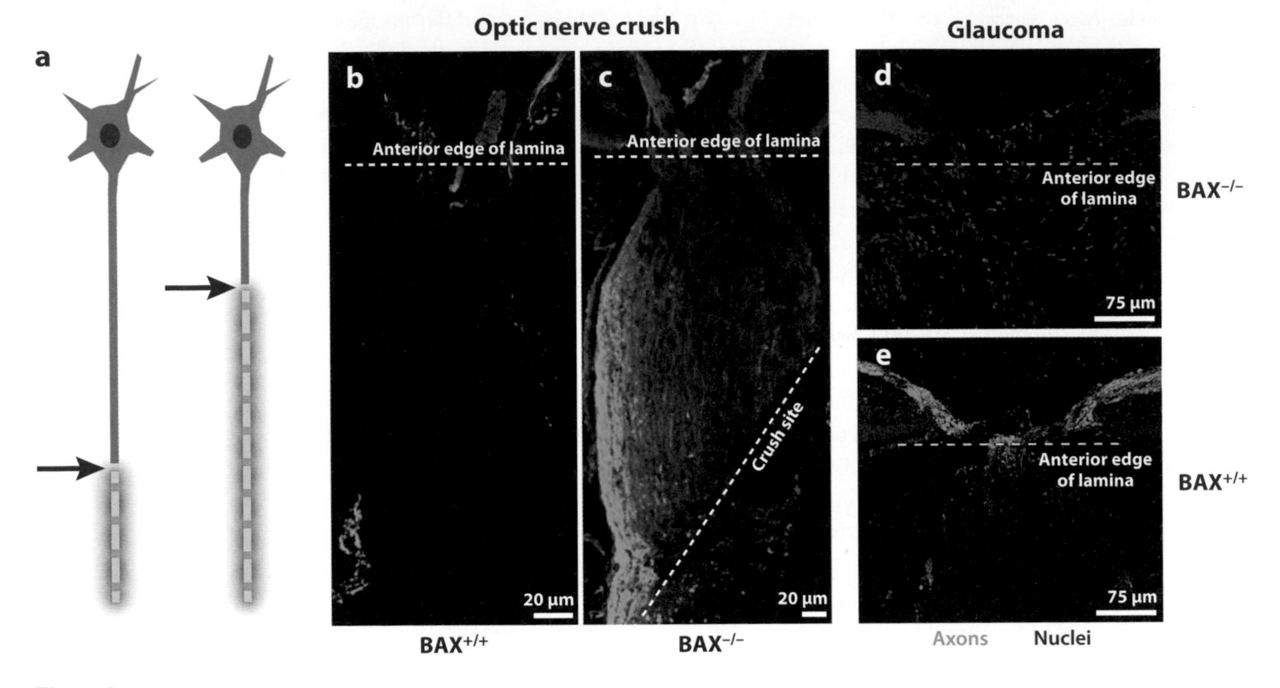

Figure 2

Axons are insulted in the lamina region in glaucoma. (*a*) For peripheral neurons, the axon segment distal to an applied transecting lesion (*black arrows*) degenerates (*dotted lines*), whereas the proximal axon segment still attached to the cell body survives (*solid lines*). (*b,c*) In BAX-deficient (*Bax*$^{-/-}$) DBA/2J mice, retinal ganglion cell (RGC) axon segments survive from the cell body to the site of optic nerve crush behind the eye (*c*), but axon segments distal to the crush site degenerate. Thus, the surviving axon segment acts as a marker for the site of insult to the axons. RGCs and axons completely degenerate in BAX-sufficient (*Bax*$^{+/+}$) DBA/2J mice (*b*). (*d,e*) RGC axons survive from the soma to the anterior edge of the lamina in BAX-deficient DBA/2J mice, indicating that the lamina is a key site of insult to axons in glaucoma (*e*). RGCs and axons completely degenerate during glaucoma in BAX-sufficient DBA/2J mice (*d*). Axons, green; nuclei, blue. Reproduced from (Howell et al. 2007).

axons (e.g., Araki et al. 2004, Babetto et al. 2010, Coleman & Freeman 2010, Sasaki et al. 2009). The protective mechanism is likely to involve NMNAT1 and the NAD biosynthesis pathway. Further understanding of the WLDS-mediated protection will add insight into the degenerative pathways involved in glaucoma.

AXON CHANGES IN GLAUCOMA

In this section, we describe efforts to understand early changes to RGCs, focusing on the axons and mechanisms of axon degeneration. When considering glaucoma, it is important to consider the path axons take from the retina through the lamina and into the optic nerve. In the retina, unmyelinated ganglion cell axons form a nerve fiber layer on the innermost surface (vitreal) of the retina. In this layer, axons are contacted by both retinal astrocytes and the end feet of Müller cells, a specialized macroglial cell type unique to the retina. As they reach the ONH, the axons abruptly turn toward the scleral canal. The axons run in distinct bundles as they exit the eye through the laminar region, where they are closely associated with astrocytes. Behind the eye, they become myelinated and are associated with oligodendrocytes.

Axon Transport Deficits Occur Early in Glaucoma

Axoplasmic transport is energy dependent, requiring the hydrolysis of ATP. In the retina proper, axonal transport is metabolically supported by both astrocytes and Müller macroglia, whereas in the ONH and lamina, astrocytes provide this function.

Alterations of axoplasmic transport occur early in glaucoma. Both anterograde and retrograde transport were compromised in the ONH of monkeys with experimental glaucoma (Anderson & Hendrickson 1974, Dandona et al. 1991, Radius & Anderson 1981), a finding that was later observed in rat models of experimental glaucoma (Chidlow et al. 2011, Pease et al. 2000, Quigley et al. 2000, Salinas-Navarro et al. 2010). Ultrastructural analysis of glaucomatous monkey ONHs also showed accumulation of organelles in both the prelaminar and postlaminar regions (Gaasterland et al. 1978). In studies of DBA/2J mice with early glaucoma, the first signs of axonal damage, which are partially characterized by the accumulation of organelles, were also detected in the glial lamina. Importantly, axon transport defects were localized and did not affect all RGCs in these mice (Howell et al. 2007, Jakobs et al. 2005). Early anterograde axonal transport defects at the level of the superior colliculus also occurred in DBA/2J mice (Crish et al. 2010).

Transport motor proteins accumulate in different locations in rat models of glaucoma. Kinesin-1 levels, which participate in anterograde transport, decline significantly in the prelaminar ONH during ocular hypertension (Munemasa et al. 2010), whereas dynein motor proteins, involved in retrograde transport, become elevated in the retrolaminar region (Martin et al. 2006). This difference could result from activated protein degrading pathways in the cell soma that can affect kinesin-1 levels but not dynein proteins, which accumulate as a consequence of lamina obstruction and/or cell soma disconnection.

Similar blockages of axonal transport at the lamina can be induced by transient elevations of IOP in animals. Interestingly, transport is reestablished, once the IOP is returned to normal (Levy 1974, Minckler et al. 1977, Quigley & Anderson 1976). These studies demonstrate that even modest IOP elevations of 4-mm Hg (Levy 1974) lead to transport blockage in as little as 24 h and that the mechanism of blockage is reversible. To date, no clear mechanism is known regarding how elevations in IOP, leading to an increase in the translaminar pressure gradient, cause disruption of axonal transport or exactly how axon transport defects contribute to axon damage. However, it is clear that axonal transport deficits precede axonal degeneration in glaucoma. There is a general agreement that disturbed axon transport is an important intrinsic event that incites RGC damage.

High-Energy Requirement in the Lamina May Contribute to Axon Damage

For various reasons, including the absence of myelin insulation on axons, the laminar region has relatively large energy demands, which may be difficult to meet during stress (see Yu-Wai-Man et al. 2011). This high metabolic need is reflected by increased levels of COX activity and a high density of voltage-gated sodium channels in the unmyelinated portion of the human ONH compared with those in the myelinated optic nerve (Barron et al. 2004). Mitochondrial concentration in the prelaminar and laminar regions of RGC axons is significantly higher than in other axonal regions (Barron et al. 2004, Minckler et al. 1977, Morgan 2004). Therefore, axon segments in the lamina are likely vulnerable to metabolic stress and changes that increase energy demand or disturb mitochondrial function (Yu-Wai-Man et al. 2011).

Changes in axonal biology due to disturbed nutrient supply or energy metabolism are suspected to underlie axonal transport defects. A recent study shows that ATP levels and axonal conduction are decreased in DBA/2J optic nerves as a result of high IOP (Baltan et al. 2010). High IOP may locally compromise the blood supply or its regulation in the lamina, contributing to metabolic stress and disruptions of axonal transport (Anderson 1996, 1999). High IOP may also directly impact mitochondrial functions by altering mitochondria (Ju et al. 2008), and mitochondrial dysfunction is reported in glaucoma patients (Abu-Amero et al. 2006).

Wallerian Degeneration and Dying Back

Two important forms of axonal degeneration, Wallerian degeneration and dying back, occur in glaucoma (Whitmore et al. 2005). Wallerian degeneration is a process whereby the distal axon rapidly degenerates after being separated from the cell body by a severe transection-type of insult. Dying back is a process in which cellular stress typically leads to disconnection of synaptic terminals and gradual and progressive degeneration of the axon toward the cell body. Both processes can occur within the same eye, depending on the severity of insult affecting individual axons (Whitmore et al. 2005). A study of DBA/2J mice showed that Wallerian degeneration occurs for a small subset of damaged axons during early glaucoma. In the same eyes, the vast majority of injured axons had very mild and focal axon swelling in the lamina, suggesting a milder injury that should impact axon transport and other functions. However, the majority of injured axons did not appear to separate completely the axon segments proximal and distal to the lesion (Howell et al. 2007). It was suggested that these more numerous and more mildly affected axons would degenerate by dying back, in agreement with data from a previous report that showed initial axon degeneration closer to the brain (Schlamp et al. 2006). More recent experiments also support a dying-back mechanism by showing early disconnection and degeneration in the brain (Crish et al. 2010).

Molecular Mechanisms of Axon Degeneration

A number of proteins and signaling pathways are known contributors to axonal degeneration and may be important for glaucoma. Activation of the death receptor-6 (DR6) gene by a cleaved N-terminal fragment of APP induces axonal degeneration through caspase-6 activation and is independent from the apoptosis pathway (Nikolaev et al. 2009). In addition to apoptosis, BAX was recently shown to contribute directly to intrinsic axon degeneration (Nikolaev et al. 2009). Although it was not required for axon degeneration in DBA/2J glaucoma, BAX deficiency did delay axon degeneration, suggesting a possible role within the axon in this glaucoma (Libby et al. 2005). Another pathway associated with apoptosis that is also independently involved in axonal degeneration is the c-Jun N-terminal kinase (JNK) pathway. By inhibiting the activation of the JNK pathway, genetic deletion of the dual leucine kinase in *Drosophila* and mouse prevents the degeneration of axons after injury (Miller et al. 2009). Some studies suggest that activation of the JNK pathway in axons after injury or disease can directly affect axonal transport (Cavalli et al. 2005, Morfini et al. 2006, Perlson et al. 2009). However, recent findings suggest that JNK3 deficiency is not protective in an inducible model of glaucoma (Quigley et al. 2011), and we have found JNK2 deficiency is not protective in DBA/2J glaucoma (R.W. Nickells, G.R. Howell, I. Soto & S.W.M. John, unpublished data). To assess fully the role of JNKs in glaucoma, researchers will need to ablate the function of JNK3 in combinations with the related kinases JNK1 or JNK2; experiments are under way. In fact, combined deficiency of both JNK2 and JNK3 strongly protects from optic nerve crush (Fernandes et al. 2012).

Recently, it was proposed that axonal degeneration is activated by the opening of the mitochondrial permeability transition pore (mPTP) after mechanical or toxic injury (Barrientos et al. 2011). Opening of the mPTP causes abnormal elevation of intramitochondrial calcium and oxidative stress that result in depolarization of the mitochondrial inner membrane, mitochondrial swelling, and subsequent uncoupling of oxidative phosphorylation. Decreases in ATP synthesis will induce metabolic failure that subsequently affects important cellular processes including axonal transport. Importantly, pharmacological or genetic inhibition of the mPTP prevents axonal degeneration in explant and cell culture, and inhibition of the JNK pathway prevents axonal degeneration and mitochondrial swelling in these cultures. As WLDS inhibited

JNK activation in axons, the protective effects of *Wld^s* may act upstream of JNK pathway activation and mPTP opening.

Similarly, decreases in ATP would also affect normal activity of the Na^+/K^+ ATPase located in axonal membranes. Loss of function of the exchanger leads to an increase in intracellular Na^+ levels, and a reversal of the Ca^{2+}/Na^+ exchanger, also yielding an increase in intracellular Ca^{2+} (Stys 2005). Importantly, this exchanger is a critical regulator of calcium homeostasis in axons (Stirling & Stys 2010). For optic nerve preparations, exposure to experimental ischemic conditions precipitates sharp elevations in intracellular calcium from both intracellular stores and external sources (Nikolaeva et al. 2005). Combined with Ca^{2+} release from both intracellular stores and Ca^{2+} influx, the deregulation of calcium homeostasis may be a principal trigger for axonal degeneration (see Whitmore et al. 2005).

The downstream consequence of Ca^{2+} overload may be the activation of the cellular degradation pathway autophagy, which has been implicated in axonal degeneration (Cheng et al. 2002, Wang et al. 2006). Mitochondrial dysfunction and mPTP opening can activate autophagy (reviewed by Rodriguez-Enriquez et al. 2004). Acute retrograde axon degeneration after neurotoxin injury or axotomy is suppressed by constitutive activation of the AKT pathway, which inhibits autophagy (Cheng et al. 2002). Additionally, suppression of autophagy by conditional deletion of the autophagy gene *Atg7* in adult mice confers a strong protection to the injured axons (Cheng et al. 2002). Further studies that directly address the involvement and activation of these pathways in glaucoma are necessary.

NONAXONAL CHANGES IN GLAUCOMA

As with all neurons, ganglion cells are made up of different but interconnected compartments. So far in this review, we have focused primarily on the axonal compartment. However, evidence also suggests that early damaging changes may directly affect other compartments including the synapse (Whitmore et al. 2005). Some of these changes may be dependent on initial damage to the axon, but other changes may be independent of axonal damage.

Signaling Between Damaged Axons and Cell Somas

The loss of axonal transport is believed to alter signaling and induce damaging processes in the retina. These signaling changes are not well understood, but various molecules that change in damaged axons, including JNKs, may be involved (see Abe & Cavalli 2008). A widely accepted mechanism is that axon damage results in a deficiency of neurotrophin support for the RGC soma. During developmental pruning of ganglion cells in the retina, RGC soma become dependent on several different neurotrophins for survival: Among these, brain-derived neurotrophic factor (BDNF) has the greatest known trophic role (Cohen-Cory et al. 1996, Cohen-Cory & Fraser 1994). BDNF is released by connected neurons in the brain. It binds to the TRKB receptor on axonal termini and is transported back to the RGC soma, where it supports the activity of the AKT kinase pathway. Bound TrkB receptors accumulated at the lamina of rats with experimental glaucoma (Pease et al. 2000), supporting the model that BDNF is unable to reach the retina in glaucoma. Application of exogenous BDNF to the eye protects RGCs in glaucoma models (Cheng et al. 2002, Di Polo et al. 1998) and after acute optic nerve damage (Mansour-Robaey et al. 1994). This protective effect of BDNF, or any neurotrophin, is only transient (Johnson et al. 2009). This may reflect either a loss of competence to respond to the trophic ligand by receptor downregulation (Cheng et al. 2002) or the activation of different signaling pathways via ligand interaction with TrkB at the soma rather than the synapse (Watson et al. 2001). Supporting this latter concept, application of exogenous BDNF to both the retina and visual cortex in cats provides a significantly greater protective effect in a model of partial optic nerve damage (Weber et al. 2010).

An alternative mechanism of damage signaling has also been proposed on the basis that blockage of retrograde transport occurs too slowly to signal damage to the soma. In a model of acute optic nerve damage in rats, changes in protein phosphorylation and gene expression occur within 30 min (Lukas et al. 2009). In vivo imaging of damaged RGC axons after acute injury demonstrates that superoxide is rapidly produced in the optic nerve and then becomes elevated in the ganglion cell somas within 24 h (Kanamori et al. 2010a). Consistent with this, scavengers of superoxide and other free radicals provide a level of protection to ganglion cells within in vitro paradigms (Kanamori et al. 2010b). These studies have been limited to acute models of ganglion cell apoptosis. It is uncertain if similar signaling mechanisms are characteristic of the less-acute levels of damage experienced in ganglion cell axons in chronic glaucoma.

Remodeling and Atrophy of Retinal Ganglion Cells

During glaucoma, RGCs do not simply degenerate but can persist in a stressed or remodeled state for prolonged periods prior to apoptotic degeneration (Jakobs et al. 2005, Libby et al. 2005, Soto et al. 2008). Studies of nonhuman primates with experimental glaucoma identified early retraction of their dendritic arbors (Weber et al. 2008). Dendritic remodeling also occurs in mouse and other models, and the somas of apparently all RGC subtypes shrink in these models (Buckingham et al. 2008, Jakobs et al. 2005, Morgan 2002, Morgan et al. 2000, Weber et al. 2008). The exact timing and molecular mechanisms of dendritic remodeling and shrinkage are not clear, but in mouse models, they appear to occur after axon injury (Jakobs et al. 2005, Soto et al. 2008). The sustained cell shrinkage may be a result of a phenomenon called the apoptotic volume decrease (Bortner & Cidlowski 2007), which is critical for apoptosis induction. In neurons, the apoptotic volume decrease appears to be regulated by a K^+ ion efflux from Kv2.1 delayed rectifier channels, which are phosphorylated at the S800 residue by activated p38 MAP Kinase (Pal et al. 2003, Redman et al. 2007). RGCs express high levels of the Kv2.1 channel (Pinto & Klumpp 1998), and p38 has been implicated in the activation of RGC death in acute models of damage (Kikuchi et al. 2000). Alternatively, inhibition of the Kv1.1 and Kv1.3 K^+ channels attenuates ganglion cell death after axotomy, suggesting that these channels may contribute to apoptotic volume decrease in these cells (Koeberle et al. 2010).

Ganglion cell dendritic processes make postsynaptic connections with amacrine and bipolar neurons in the inner plexiform layer of the retina (see Rodieck 1998). Given the dendritic remodeling in glaucoma, it is not surprising that RGC synapses are lost. Synapse loss appears to occur very early, however, and may be mediated by the complement cascade. A program of complement-mediated synapse elimination, which is activated by immature astrocytes during normal retinal development, appears to be abnormally reactivated in DBA/2J glaucoma (Stevens et al. 2007). Complement cascade molecules are induced in various animal models of glaucoma and human glaucoma (Kuehn et al. 2008, Kuehn et al. 2006, Stasi et al. 2006, Steele et al. 2006). Complement gene induction is one of the earliest events in DBA/2J glaucoma (Howell et al. 2011a, Stevens et al. 2007). Thus, glaucoma may be one of a growing number of neurodegenerative diseases in which synapse loss is one of the earliest pathologic events. In support of this, DBA/2J mice that are mutant for *C1q* have greatly reduced RGC and axon loss. However, complement molecules are also increased in the optic nerve during DBA/2J glaucoma (Howell et al. 2011a). Future experiments will clarify the timing and dependence of synapse loss in relation to axon damage, the importance of synapse loss for progression to RGC loss, and if *C1q* expression by specific cell types is necessary for synapse loss in glaucoma.

In addition to morphological atrophy, several studies have documented profound decreases in the expression of specific genes by RGCs, well in advance of completion of the apoptotic program. This was originally described for the *Thy1* gene in mouse models of acute ganglion cell damage and rat experimental glaucoma (Huang et al. 2006, Schlamp et al. 2001), and it occurred in BAX-deficient mice in which ganglion cell death was arrested (Schlamp et al. 2001). Since these initial reports, a variety of genes were found to be expressed at decreased levels prior to detectable loss of RGCs (Soto et al. 2008, Yang et al. 2007). Mechanisms underlying gene silencing in RGCs were investigated using a mouse model of acute damage. Shortly after optic nerve crush there is an exponential decay of transcript levels of several ganglion cell–specific genes, which is associated with deacetylation of H4 histones in the promoter regions of these genes. The deacetylation process appears to be mediated by nuclear translocation of histone deacetylase 3 (HDAC3), and the decrease in expression of at least some ganglion cell genes can be blocked with histone deacetylase inhibitors (Pelzel et al. 2010). A similar mechanism involving HDAC3 regulates gene silencing in the DBA/2J mouse model of glaucoma (Pelzel et al. 2012).

OTHER PATHWAYS AND CELL TYPES

In addition to RGCs, other cell types may play distinct, but key, roles in glaucoma. In this section, we discuss divergent pathways that have been recently implicated in early stages of glaucoma in the context of different cell types and whether such pathways may mediate protective or damaging responses.

Gene Expression Changes Characterize Early Stages of Glaucoma

Various studies have used gene expression arrays to identify glaucoma pathways in the ONH and retina (e.g., Howell et al. 2011a; Johnson et al. 2007, 2011; Kompass et al. 2008; Nikolskaya et al. 2009; Panagis et al. 2009; Steele et al. 2006; Yang et al. 2007). Two studies were specifically designed to identify early changes in the optic nerve. The first study, in rats, focused on eyes with less than 15% axon degeneration (Johnson et al. 2011). The most significantly upregulated biological processes (based on increased gene expression) were the cell cycle as well as the cytoskeleton and immune systems, while glucose and lipid metabolism were among the most downregulated. This study implicated astrocyte proliferation and IL-6 type cytokine gene expression, rather than astrocyte hypertrophy, as a characteristic of early pressure-induced ONH injury (Johnson et al. 2011). Microglial activation and vascular-associated gene responses were also suggested to occur early in glaucoma.

The second study addressed changes that occur prior to any detectable RGC and axon loss in DBA/2J mice (Howell et al. 2011a). This study was specifically designed to identify a group of genes (glaucoma genes) that change due to glaucoma, as opposed to other factors such as age or genotype. As it led to the development of an online community resource (Howell et al. 2011b), the study design is explained here. Individual ONHs were analyzed as separate samples. Hierarchical clustering was used to gather samples into groups based on the similarity of their gene expression profiles for this entire set of glaucoma genes. Three groups were formed on the basis of gene expression changes that occur prior to detectable glaucoma. Importantly, these groups represent three early, molecularly defined stages of glaucoma (stages 1 to 3) that are temporally ordered: Stage 1 is the earliest and closest to the no-glaucoma control. Samples from eyes at later stages of glaucoma were also studied and clustered into additional groups. Thus, the data allow consideration of new hypotheses and/or changes previously shown to occur in glaucoma in the context of whether they occur earlier or later during disease progression. Depending on the stage, thousands of genes were found to change during early glaucoma. Bioinformatics analyses identified changes in many biological processes. Some of the earliest changes affected the extracellular matrix (ECM), immune pathways, and the complement cascade. There are too many changes to discuss here. The analyses are available through an online resource (Glaucoma Discovery Platform available at **http://glaucomadb.jax.org/glaucoma**), which allows easy interrogation and visualization of the data (Howell et al. 2011b). Along with existing and future studies of genes expressed in specific cell types (e.g., Cahoy et al. 2008), these data will help to order the responses and roles of specific processes in specific cell types during glaucoma.

To allow evaluation of the timing and interdependence of events in the retina and ONH, the retinas from the same eyes were also studied (Howell et al. 2011a). Importantly, the analyses suggest that changes in the retina and optic nerve are not always interdependent: Some eyes have more extensive molecular changes in the retina than in the optic nerve, suggesting asynchrony of disease in the different tissues and that some damaging events occur directly in the retina. For the retina, one of the earliest changes is induction of the complement cascade, and genetic disruption of this cascade was highly protective against glaucoma (Howell et al. 2011a). The complement cascade may cause damage directly in the optic nerve and/or mediate synapse elimination in the retina (see Remodeling and Atrophy of Retinal Ganglion Cells, above). Activation of the endothelin system, in particular upregulation of endothelin-2, occurred early in both the ONH and retina. Endothelins are potent vasoconstrictive agents and alter axon transport. They are increased in human glaucoma and various animal models (Grieshaber et al. 2007, Yorio et al. 2002) and may result in vascular dysfunction, which could contribute to disturbed blood flow and a predisposition to transient ischemic events (Anderson 1996, 1999). The endothelin receptor antagonist, Bosentan, lessened glaucomatous decreases in the vascular lumen of the retina and protected against glaucoma in DBA/2J mice. In this model, endothelin-2 is produced by AIF1 (formerly IBA1)-expressing cells, providing a new link between microglia or monocytes and vascular abnormalities in glaucoma (Howell et al. 2011a).

Extracellular Matrix Remodeling

Various gene expression (see above) and immunolabeling studies (Johnson et al. 1996) show that an increase in ECM molecules occurs as an early response to increased IOP (Johnson & Morrison 2009). The ECM-receptor interactions pathway is upregulated in the ONH of DBA/2J mice with early glaucoma (Howell et al. 2011a). Pathway members include transmembrane proteins such as integrins and proteoglycans that are expressed by different cell types, including endothelial cells and astrocytes. These proteins directly or indirectly modulate the control of cellular activities including adhesion, migration, and proliferation. Other genes with altered expression include collagens and laminins. These molecules form the astrocyte basal lamina surrounding ONH axons and the vascular basal lamina of microcapillaries.

ECM increases may be a response to potentially damaging changes to the neurovascular unit (Del Zoppo et al. 2006), including a loss of connectivity and/or signaling between astrocytes and axons or astrocytes and endothelial cells. Alternatively, ECM changes may reflect a protective remodeling mechanism in response to the increased stress (force/area) and strain (force-induced deformation) accompanying raised IOP (Roberts et al. 2009). Focal adhesion molecules such as integrins are mechanosensors that sense changes in pressure-induced deformation and trigger increases in ECM components including tenascin C (Chiquet 1999, Sarasa-Renedo & Chiquet 2005). Tenascin C was elevated in the astrocytic lamina of glaucomatous mice (Howell et al. 2011a) and in ONHs of glaucomatous humans (Pena et al. 1999).

Support for a protective mechanism comes from biomechanical studies in primates, where the LC ECM thickens in response to increased strain (Bellezza et al. 2003; Burgoyne 2010; Burgoyne et al. 2004; Ethier 2006; Roberts et al. 2010; Sigal & Ethier 2009; Sigal et al.

2004, 2005). Although biomechanical studies are lacking for the rodent optic nerve, increased IOP is expected to exacerbate stress and strain (see Chauhan et al. 2002, Downs et al. 2011). If it continues chronically, an initially protective mechanism may become damaging. Tissue stiffening may affect cellular architecture and intercellular interactions (Ladoux et al. 2010). Basal lamina changes may impede exchange between various cell types such as astrocytes and RGC axons or between astrocytes and endothelial cells. In species with a collagenous LC, ECM changes may impede metabolic exchange with blood vessels, which are embedded in the collagenous plates (reviewed in Burgoyne 2010). Additionally, ECM remodeling in primates is eventually accompanied by substantial structural changes that are likely to create a steep increase in the energy required for axons to maintain normal transport (Burgoyne 2010). An increased demand for energy support is a major untested hypothesis for glaucoma.

Neuroinflammatory Responses and Microglia

Various gene chip (see above) as well as other studies indicate that the activation of neuroinflammatory signaling occurs very early in glaucoma. This is true in at least the majority of animal models. The course and outcome of inflammation are highly variable and depend on the nature and magnitude of inflammatory signaling as well as the tissue and cellular context. In glaucoma, it is likely that early inflammatory signaling protects against local tissue stresses (metabolic/biological stress and strain) or injury. In some contexts, ocular inflammation protects RGCs following axon injury (Leon et al. 2000, Yin et al. 2006). Following prolonged IOP elevation and/or exposure to other glaucomatous stresses, however, the initially protective response may evolve into chronic and damaging inflammation. As occurs in the central nervous system (CNS), cytokine release may ultimately promote microglial and endothelial cell activation, recruitment of monocytes and macrophages, dendritic cells, and possibly other immune cell types (Farina et al. 2007). Immune dysfunction is reported to contribute to some forms of glaucoma (Wax & Tezel 2009), with both protective and damaging roles suggested for T cells (Schwartz & London 2009). However, the roles of immunity and T cells in glaucoma remain unclear, and some studies show no detection of T cells (Ebneter et al. 2010). It remains to be determined if inflammatory processes (without classic, clinically detectable infiltrates) are primary instigators of common forms of chronic glaucoma or whether they are secondary modulators in response to tissue damage.

Inflammatory processes are likely to be controlled and mediated by both microglia and astrocytes in glaucoma. Potential roles of astrocytes are considered in more detail below. The roles of microglia in neurodegenerative disease were recently reviewed (Perry et al. 2010). Microglia are often called resident macrophages of the CNS. They have both protective and damaging functions. Microglia become reactive in animal models of experimental glaucoma (Johnson & Morrison 2009, Neufeld 1999). They may protect by phagocytosing debris and releasing trophic/survival factors. However, with chronic stimulation, they may release damaging molecules or directly phagocytose parts of the neuron (possibly, the synapses) (Rosen & Stevens 2010). ONH microglia presented both MHC class I and MHC class II surface molecules in a rat model of glaucoma (Ebneter et al. 2010). Higher densities of reactive microglia were reported in areas of axonal degeneration (Ebneter et al. 2010, Johnson et al. 2007, Taylor et al. 2011), prompting speculation that microglia contribute to a late pathological response, possibly mediating the production of scar-tissue remodeling. Other studies focused on the retina; some of these suggested early activation, with evidence for a damaging role in DBA/2J glaucoma, which was detected by inhibition of their activation using minocycline from a young age (Bosco et al. 2008). Minocycline also affects other cell types and processes. Our unpublished data (R.W. Nickells, G.R. Howell, I. Soto & S.W.M.

John) showed that minocycline administration at older ages exacerbated DBA/2J glaucoma. Thus, interpretation of these results is not simple: Inhibition may have different effects depending on disease stage.

In another study (Nakazawa et al. 2006), experimental induction of high IOP upregulated the cytokine TNF; this was followed by microglial activation, loss of oligodendrocytes, and delayed loss of RGCs. Deleting the *CD11b* gene prevented microglial activation and subsequent pathology. It was suggested that microglial activation damages oligodendrocytes resulting in RGC loss. However, oligodendrocyte loss occurs later in the inherited DBA/2J glaucoma and appears to be a secondary response to damage (Son et al. 2010). Further experiments are required to understand the roles of specific microglial processes in glaucoma. It will be important to clarify both their protective and damaging functions, with the caution that wholesale inhibition strategies may prove more damaging than beneficial (Czeh et al. 2011).

Astrocytes and Bundle Defects

Astrocytes are a major cell type in the lamina where they are intimately associated with axons. They play a major role in maintaining neuronal health (Kimelberg & Nedergaard 2011, Ransom et al. 2003, Sofroniew & Vinters 2010). They control key processes whose dysfunction may participate in glaucoma such as the control of blood flow coupled to neural activity (Pournaras et al. 2008). Given the pattern of neural damage in glaucoma, specific axon bundles are likely damaged in the lamina. However, it is not clear if astrocytes locally respond to ongoing damage to promote or limit axon injury, initiate damage to specific axon bundles, prevent the spread of damage, or a combination of all these possibilities. Importantly, it was recently shown that individual astrocytes are large and have processes spread widely throughout the mouse lamina. The processes of multiple astrocytes form individual pores or glial tubes that surround axon bundles. Similarly, a single astrocyte can contribute to the formation of multiple pores and communicate with other astrocytes throughout the lamina (Sun et al. 2009) (**Figure 3**). Thus, the pattern of damage cannot be explained by the local dysfunction or activation of individual astrocytes that are associated with a specific axon bundle. Local axon damage and astrocyte reactivity are likely restricted to only the region of the astrocyte associated with damage. However, further experiments are needed to understand the local axon injury and astrocyte response.

Astrocyte Responses Under Stressed Conditions

The roles of astrocytes in glaucoma are not clear, but we discuss a few ways in which they may impact this disease. Understanding their responses may provide novel therapeutic targets. The role of astrocytes in maintaining normal neuronal homeostasis may become more important during periods of stress, increased neuronal energy demand, or a combination of the two. How astrocytes accommodate neurons under these conditions is uncertain. Astrocytes catabolize glycogen stores and release lactate as an energy substrate to their neuronal partners (Brown & Ransom 2007, Sofroniew & Vinters 2010, Suh et al. 2007). Lactate is released from astrocytes by monocarboxylase transporters (MCT1 and MCT4), and disruption of MCT function has been associated with neuronal dysfunction. Consistent with a possible impairment of lactate shuttling to axons in early glaucoma, modest reductions in MCT1 expression levels are associated with early damage in DBA/2J glaucoma (Howell et al. 2011a; see **http://glaucomadb.jax.org**). However, recent studies suggest that, during periods of high neuronal activity, astrocytes utilize glycogen stores for their own energy demands while blocking hexokinase activity (and inhibiting glucose phosphorylation), thereby promoting direct uptake of blood glucose to neurons (DiNuzzo et al. 2010). It is conceivable, therefore, that microcirculatory changes

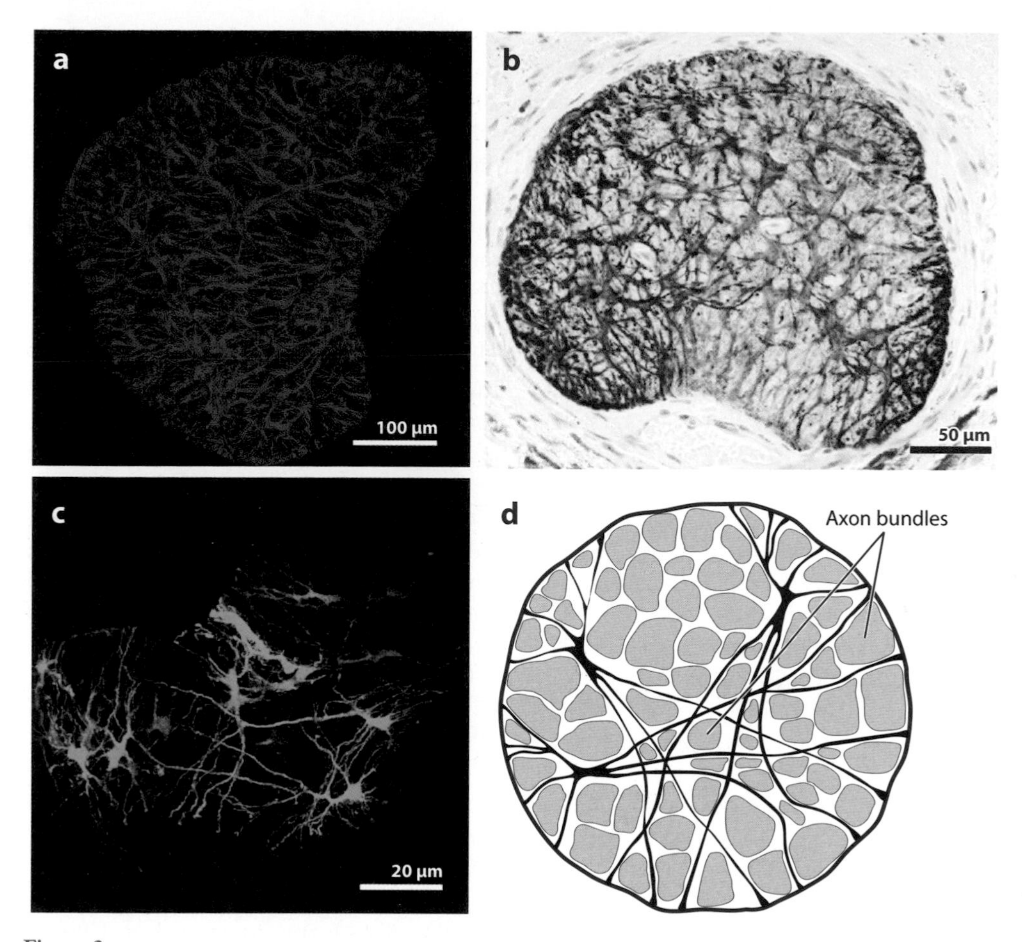

Figure 3

Astrocytes and axon bundles in the lamina. (*a,b*) The glial lamina has a network of GFAP$^+$ (glial fibrillary acidic protein) astrocytes (*a, red; b; brown*) that form pores or glial tubes through which bundles of axons pass. (*c*) Astrocytes (*green*) extend processes through the full width of the lamina, crossing numerous axon bundles to contact the pial wall via bulbous endfeet. (*d*) The functional implication of this arrangement is that a single astrocyte can contact and support multiple axon bundles and other astrocytes that are not necessarily neighbors, making the glial lamina work as a single functional unit. Given this arrangement, individual astrocytes must be in contact during glaucoma both with the remaining healthy axons and degenerating axons. Further experiments are needed to understand astrocyte to neuron interactions in glaucoma and how they modulate or restrict axon degeneration. Bars = 100 μm. (*a,c,d*) Reproduced and adapted from (Sun et al. 2010). (*b*) Reproduced and adapted from (Howell et al. 2007).

in the ONH would compromise both astrocyte and axonal physiology.

During conditions of injury or disease, astrocytes undergo a series of morphological and molecular changes that have been termed reactive astrogliosis (Sofroniew & Vinters 2010). Although reactive astrogliosis has been traditionally characterized as a harmful event, its initial purpose may be protective. Reactive astrogliosis is suggested to be a primary response of astrocytes to prevent disruption of the blood-brain barrier or to repair it. Astrocytes also regulate inflammation during injury or disease conditions in the CNS (Sofroniew 2009). In fact, inhibition of reactive astrogliosis by conditional deletion of the STAT3 protein in astrocytes results in increased inflammation, increased lesion size and demyelination, and impaired functional recovery after spinal cord injury (Herrmann et al. 2008, Okada et al. 2006). By contrast, NFκB signaling is upregulated in ONH astrocyte cultures from human glaucoma patients (Nikolskaya et al. 2009), and inactivating proinflammatory NFκB signaling in astrocytes promotes neuronal survival following spinal cord injury and retinal ischemia (Brambilla et al. 2005, Dvoriantchikova et al. 2009). Astrocytes have been implicated in proinflammatory responses in the ONH, and astrocytes subjected to stretch mount a greater inflammatory response when exposed to cytokines (Ralay Ranaivo et al. 2011). Additionally exposure to a variety of stressors, such as reactive oxygen species, can stimulate ONH astrocytes to expose MHC class II molecules, secrete TNFα, and become inducers of T cell activation (Tezel et al. 2007).

Reactive astrocytes from humans with glaucoma (Liu & Neufeld 2000) and rats with experimental glaucoma (Shareef et al. 1999) are reported to express nitric oxide synthase 2 (NOS2), leading to the suggestion that overproduction of NO damages RGC axons (Chung & David 2010). Supporting this, the NOS2 inhibitor aminoguanidine was reported to attenuate glaucoma damage in a rat model (Neufeld et al. 1999). However, similar studies in a different rat model (Pang et al. 2005) and in DBA/2J mice lacking a functional *Nos2* gene (Libby et al. 2007) failed to demonstrate a role for NO production in RGC degeneration.

Astrocytes in the ONH are also reported to increase the production of Endothelin-1 (ET-1) in response to elevation of IOP (Prasanna et al. 2005). In vitro studies have shown that ET-1 expression is elevated in ONH astrocytes in response to stretch deformation in culture, suggesting that biomechanical forces are the main inducers of this protein (Ostrow et al. 2011). Astrocytes respond to deformation via stretch-activated ion channels (SACs) that stimulate a Ca^{2+} influx. In astrocytes, this response can be partially blocked by the inhibitory peptide GsMTx4 (Bae et al. 2011, Ostrow et al. 2011), and the nonspecific SAC inhibitor streptomycin. ET-1 likely acts directly on adjacent astrocytes in an autocrine fashion by way of the ET-B receptor (Prasanna et al. 2005, Rao et al. 2007). In vitro studies using ONH astrocytes indicate that ET-1 stimulation can produce early pathogenic effects reported in the laminar region of glaucoma models, including proliferation (Murphy et al. 2010), and the production of MMPs and their inhibitors (Timps) (He et al. 2007, Rao et al. 2008). Endothelin release may also affect the vasculature.

Recent studies have demonstrated that astrocytes express phagocytic genes under normal conditions (Cahoy et al. 2008). A subpopulation of astrocytes in the myelin transition zone of the optic nerve expresses high levels of the phagocytic protein Mac-2. Researchers showed active phagocytosis of axonal membranes by these lamina and retro-laminar astrocytes occurred under normal conditions (Nguyen et al. 2011) and increased during glaucoma.

Endothelial Cells, Monocytes, and Early Damage

A single dose of gamma-radiation has been shown to protect the vast majority of treated DBA/2J mice from developing glaucoma long after their exposure to the radiation. Whole body irradiation was performed at a young age and 96% of treated eyes had no glaucoma approximately a year after treatment (Anderson et al. 2005). IOP elevation in these mice was unaltered by the radiation treatment. A study of human atomic bomb survivors in Japan suggested that a higher radiation exposure was associated with lower incidence of glaucoma (Yamada et al. 2004). The most common form of glaucoma in Japan does not involve high IOP and so the irradiation appears to protect from glaucoma within the optic nerve and/or retina in multiple species. Supporting a local effect within ocular tissues, follow up studies show that specific X-ray irradiation of only an eye protects it from glaucoma, whereas the contralateral untreated eye is not protected. Additionally, the irradiation both prevents very early axon injury in the optic nerve and prevents axon transport defects (Howell et al.

2012). Since the radiation treatment appears to completely prevent glaucomatous damage, it is likely to abrogate a critical and very early damaging mechanism in the ONH. To understand this protection, a gene expression study of ONH tissue was designed to detect very early stages/pathways that mediate glaucoma and how they are affected by the radiation treatment (Howell et al. 2012). [This study extended our previous study (Howell et al. 2011), described above and identified even earlier stages of glaucoma.] This new study found that the tissues of radiation-protected eyes experience early glaucomatous stresses, but that the molecular pathophysiology is typically stopped at a very early stage (Howell et al. 2012). Importantly, the first pathway found to change significantly in glaucoma is the leukocyte transendothelial migration pathway that is controlled by endothelial cells, and the activation of this pathway is modified by the radiation treatment (Howell et al. 2012). Consistent with an early and critical role of this pathway in glaucoma and prior to neuronal damage, monocytes enter the ONH of DBA/2J mice from the blood, but they are prevented from entering by the radiation treatment. These monocytes are CD11b+ Cd11c+ cells and express proinflammatory and damaging molecules including Endothelin-2 (EDN2) and the phagocytic marker LY6c+. The presented data support a damaging or even initiating role of these cells in glaucomatous damage (Howell et al. 2012). Our preliminary data suggest that these cells also enter the ONH in other glaucoma models (GRH, IS, SWMJ). Although further experiments are needed to test the role of these monocytes and their requirement for glaucomatous damage, these emerging data focus attention on the role of endothelial cells, the transendothelial migration pathway and monocytes in early glaucoma. It is possible that glaucomatous neural damage has a primary inflammatory component controlled by endothelial cells.

CONCLUSION

Multiple stresses and insults are likely to contribute to glaucoma. Individual cell types and RGC compartments (such as RGC soma, RGC axon, astrocytes, microglia, and endothelial cells) do not respond in isolation. Collectively understanding their responses and cross talk is likely to yield a greater insight into glaucoma and may be important for developing novel therapeutics.

DISCLOSURE STATEMENT

The authors are not aware of any affiliations, memberships, funding, or financial holdings that might be perceived as affecting the objectivity of this review.

ACKNOWLEDGMENTS

The authors are grateful for insightful discussions with Drs. Claude Burgoyne Balwantray Chauhan, Crawford Downs, Richard Libby, Albee Messing, and Donald Zack. The authors also thank K. Saidas Nair, Krish Kizhatil, Jeffrey Marchant, and Mimi de Vries for critical comments. This work was supported by grants EY012223 (R.W.N.), EY021525 (G.R.H.), and EY011721 (S.W.M.J.), as well as by the Glaucoma Foundation, American Health Assistance Foundation, and Glaucoma Research Foundations (G.R.H.). S.W.M.J. is an Investigator of the Howard Hughes Medical Institute.

LITERATURE CITED

Abe N, Cavalli V. 2008. Nerve injury signaling. *Curr. Opin. Neurobiol.* 18:276–83

Abu-Amero KK, Morales J, Bosley TM. 2006. Mitochondrial abnormalities in patients with primary open-angle glaucoma. *Invest. Ophthalmol. Vis. Sci.* 47:2533–41

Almasieh M, Zhou Y, Kelly ME, Casanova C, Di Polo A. 2011. Structural and functional neuroprotection in glaucoma: role of galantamine-mediated activation of muscarinic acetylcholine receptors. *Cell Death Dis.* 1:e27

Anderson DR. 1996. Glaucoma, capillaries and pericytes. 1. Blood flow regulation. *Ophthalmologica* 210:257–62

Anderson DR. 1999. Introductory comments on blood flow autoregulation in the optic nerve head and vascular risk factors in glaucoma. *Surv. Ophthalmol.* 43(Suppl. 1):S5–S9

Anderson DR, Hendrickson A. 1974. Effect of intraocular pressure on rapid axoplasmic transport in monkey optic nerve. *Invest. Ophthalmol.* 13:771–83

Anderson DR, Libby RT, Gould DB, Smith RS, John SWM. 2005. High-dose radiation with bone marrow transfer prevents neurodegeneration in an inherited glaucoma. *Proc. Natl. Acad. Sci. USA* 102(12):4566–71

Anderson DR, Quigley HA. 1992. The optic nerve. In *Adler's Physiology of the Eye*, ed. WM Hart Jr, pp. 616–40. St. Louis, MO: Mosby Year Book

Araki T, Sasaki Y, Milbrandt J. 2004. Increased nuclear NAD biosynthesis and SIRT1 activation prevent axonal degeneration. *Science* 305:1010–13

Babetto E, Beirowski B, Janeckova L, Brown R, Gilley J, et al. 2010. Targeting NMNAT1 to axons and synapses transforms its neuroprotective potency in vivo. *J. Neurosci.* 30:13291–304

Bae C, Sachs F, Gottlieb PA. 2011. The mechanosensitive ion channel Piezo 1 is inhibited by the peptide GsMTx4. *Biochemistry* 50:6295–300

Baltan S, Inman DM, Danilov CA, Morrison RS, Calkins DJ, Horner PJ. 2010. Metabolic vulnerability disposes retinal ganglion cell axons to dysfunction in a model of glaucomatous degeneration. *J. Neurosci.* 30:5644–52

Barrientos SA, Martinez NW, Yoo S, Jara JS, Zamorano S, et al. 2011. Axonal degeneration is mediated by the mitochondrial permeability transition pore. *J. Neurosci.* 31:966–78

Barron MJ, Griffiths P, Turnbull DM, Bates D, Nichols P. 2004. The distributions of mitochondria and sodium channels reflect the specific energy requirements and conduction properties of the human optic nerve head. *Br. J. Ophthalmol.* 88:286–90

Beirowski B, Babetto E, Coleman MP, Martin KR. 2008. The *WldS* gene delays axonal but not somatic degeneration in a rat glaucoma model. *Eur. J. Neuro.* 28:1166–79

Bellezza AJ, Rintalan CJ, Thompson HW, Downs JC, Hart RT, Burgoyne CF. 2003. Deformation of the lamina cribrosa and anterior scleral canal wall in early experimental glaucoma. *Invest. Ophthalmol. Vis. Sci.* 44:623–37

Berdahl JP, Allingham RR, Johnson DH. 2008. Cerebrospinal fluid pressure is decreased in primary open-angle glaucoma. *Ophthalmology* 115:763–68

Bortner CD, Cidlowski JA. 2007. Cell shrinkage and monovalent cation fluxes: role in apoptosis. *Arch. Biochem. Biophys.* 462:176–88

Bosco A, Inman DM, Steele MR, Wu G, Soto I, et al. 2008. Reduced retinal microglial activation and improved optic nerve integrity with minocycline treatment in the DBA/2J mouse model of glaucoma. *Invest. Ophthalmol. Vis. Sci.* 49:1437–46

Brambilla R, Bracchi-Ricard V, Hu WH, Frydel B, Bramwell A, et al. 2005. Inhibition of astroglial nuclear factor kappaB reduces inflammation and improves functional recovery after spinal cord injury. *J. Exp. Med.* 202:145–56

Brown AM, Ransom BR. 2007. Astrocyte glycogen and brain energy metabolism. *Glia* 55:1263–71

Buckingham BP, Inman DM, Lambert W, Oglesby E, Calkins DJ, et al. 2008. Progressive ganglion cell degeneration precedes neuronal loss in a mouse model of glaucoma. *J. Neurosci.* 28:2735–44

Burgoyne CF. 2010. A biomechanical paradigm for axonal insult within the optic nerve head in aging and glaucoma. *Exp. Eye Res.* 93:120–32

Burgoyne CF, Downs JC, Bellezza AJ, Hart RT. 2004. Three-dimensional reconstruction of normal and early glaucoma monkey optic nerve head connective tissues. *Invest. Ophthalmol. Vis. Sci.* 45:4388–99

Cahoy JD, Emery B, Kaushal A, Foo LC, Zamanian JL, et al. 2008. A transcriptome database for astrocytes, neurons, and oligodendrocytes: a new resource for understanding brain development and function. *J. Neurosci.* 28:264–78

Cavalli V, Kujala P, Klumperman J, Goldstein LS. 2005. Sunday Driver links axonal transport to damage signaling. *J. Cell Biol.* 168:775–87

Chauhan BC, Pan J, Archibald ML, LeVatte TL, Kelly ME, Tremblay F. 2002. Effect of intraocular pressure on optic disc topography, electrophysiology, and axonal loss in a chronic pressure-induced rat model of optic nerve damage. *Invest. Ophthalmol. Vis. Sci.* 43:2969–76

Cheng L, Sapieha P, Kittlerova P, Hauswirth WW, Di Polo A. 2002. TrkB gene transfer protects retinal ganglion cells from axotomy-induced death in vivo. *J. Neurosci.* 22:3977–86

Chidlow G, Ebneter A, Wood JP, Casson RJ. 2011. The optic nerve head is the site of axonal transport disruption, axonal cytoskeleton damage and putative axonal regeneration failure in a rat model of glaucoma. *Acta Neuropathol.* 121:737–51

Chiquet M. 1999. Regulation of extracellular matrix gene expression by mechanical stress. *Matrix Biol.* 18:417–26

Chung KK, David KK. 2010. Emerging roles of nitric oxide in neurodegeneration. *Nitric Oxide* 22:290–95

CNTGSG. 1998. Comparison of glaucomatous progression between untreated patients with normal-tension glaucoma and patients with therapeutically reduced intraocular pressures. Collaborative Normal-Tension Glaucoma Study Group. *Am. J. Ophthalmol.* 126:487–97

Cohen-Cory S, Escandón E, Fraser SE. 1996. The cellular patterns of BDNF and trkB expression suggest multiple roles for BDNF during Xenopus visual system development. *Dev. Biol.* 179:102–15

Cohen-Cory S, Fraser SE. 1994. BDNF in the development of the visual system of *Xenopus*. *Neuron* 12:747–61

Coleman MP, Freeman MR. 2010. Wallerian degeneration, WldS, and Nmnat. *Annu. Rev. Neurosci.* 33:245–67

Crish SD, Sappington RM, Inman DM, Horner PJ, Calkins DJ. 2010. Distal axonopathy with structural persistence in glaucomatous neurodegeneration. *Proc. Natl. Acad. Sci. USA* 107:5196–201

Crooks KR, Allingham RR, Qin X, Liu Y, Gibson JR, et al. 2011. Genome-wide linkage scan for primary open angle glaucoma: influences of ancestry and age at diagnosis. *PLoS One* 6:e21967

Czeh M, Gressens P, Kaindl AM. 2011. The yin and yang of microglia. *Dev. Neurosci.* 33:199–209

Dandona L, Hendrickson A, Quigley HA. 1991. Selective effects of experimental glaucoma on axonal transport by retinal ganglion cells to the dorsal lateral geniculate nucleus. *Invest. Ophthalmol. Vis. Sci.* 32:484–91

Danias J, Lee KC, Zamora MF, Chen B, Shen F, et al. 2003. Quantitative analysis of retinal ganglion cell (RGC) loss in aging DBA/2NNia glaucomatous mice: comparison with RGC loss in aging C57/BL6 mice. *Invest. Ophthalmol. Vis. Sci.* 44:5151–62

Deckwerth TL, Johnson EM Jr. 1994. Neurites can remain viable after destruction of the neuronal soma by programmed cell death (apoptosis). *Dev. Biol.* 165:63–72

Del Zoppo GJ, Milner R, Mabuchi T, Hung S, Wang X, Koziol JA. 2006. Vascular matrix adhesion and the blood-brain barrier. *Biochem. Soc. Trans.* 34:1261–66

Di Polo A, Aigner LJ, Dunn RJ, Bray GM, Aguayo AJ. 1998. Prolonged delivery of brain-derived neurotrophic factor by adenovirus-infected Müller cells temporarily rescues injured retinal ganglion cells. *Proc. Natl. Acad. Sci. USA* 95:3978–83

DiNuzzo M, Mangia S, Maraviglia B, Giove F. 2010. Changes in glucose uptake rather than lactate shuttle take center stage in subserving neuroenergetics: evidence from mathematical modeling. *J. Cereb. Blood Flow Metab.* 30:586–602

Downs JC, Roberts MD, Sigal IA. 2011. Glaucomatous cupping of the lamina cribrosa: a review of the evidence for active progressive remodeling as a mechanism. *Exp. Eye Res.* 93:133–40

Dvoriantchikova G, Barakat D, Brambilla R, Agudelo C, Hernandez E, et al. 2009. Inactivation of astroglial NF-kappa B promotes survival of retinal neurons following ischemic injury. *Eur. J. Neurosci.* 30:175–85

Ebneter A, Casson RJ, Wood JP, Chidlow G. 2010. Microglial activation in the visual pathway in experimental glaucoma: spatiotemporal characterization and correlation with axonal injury. *Invest. Ophthalmol. Vis. Sci.* 51:6448–60

Ethier CR. 2006. Scleral biomechanics and glaucoma: a connection? *Can. J. Ophthalmol.* 41:9–12, 4

Farina C, Aloisi F, Meinl E. 2007. Astrocytes are active players in cerebral innate immunity. *Trends Immunol.* 28:138–45

Fernandes KA, Harder JM, Fornarola LB, Freeman RS, Clark AF, et al. 2012. JNK2 and JNK3 are major regulators of axonal injury-induced retinal ganglion cell death. *Neurobiol. Dis.* In press

Filippopoulos T, Danias J, Chen B, Podos SM, Mittag TW. 2006. Topographic and morphologic analyses of retinal ganglion cell loss in old DBA/2NNia mice. *Invest. Ophthalmol. Vis. Sci.* 47:1968–74

Gaasterland D, Tanishima T, Kuwabara T. 1978. Axoplasmic flow during chronic experimental glaucoma I. Light and electron microscopic studies of the monkey optic nerve head during development of glaucomatous cupping. *Invest. Ophthalmol. Vis. Sci.* 17:838–46

Grau V, Downs JC, Burgoyne CF. 2006. Segmentation of trabeculated structures using an anisotropic Markov random field: application to the study of the optic nerve head in glaucoma. *IEEE Trans. Med. Imaging* 25(3):245–55

Grieshaber MC, Mozaffarieh M, Flammer J. 2007. What is the link between vascular dysregulation and glaucoma? *Surv. Ophthalmol.* 52(Suppl. 2):S144–54

Harwerth RS, Carter-Dawson L, Shen F, Smith EL III, Crawford LV. 1999. Ganglion cell losses underlying visual field defects from experimental glaucoma. *Invest. Ophthalmol. Vis. Sci.* 40:2242–50

Harwerth RS, Crawford ML, Frishman LJ, Viswanathan S, Smith EL III, Carter-Dawson L. 2002. Visual field defects and neuronal losses from experimental glaucoma. *Prog. Retin. Eye Res.* 21:91–125

He S, Prasanna G, Yorio T. 2007. Endothelin-1-mediated signaling in the expression of matrix metalloproteinases and tissue inhibitors of metalloproteinases in astrocytes. *Invest. Ophthalmol. Vis. Sci.* 48:3737–45

Hernandez MR, Igoe F, Neufeld AH. 1988. Cell culture of the human lamina cribrosa. *Invest. Ophthalmol. Vis. Sci.* 29:78–89

Herrmann JE, Imura T, Song B, Qi J, Ao Y, et al. 2008. STAT3 is a critical regulator of astrogliosis and scar formation after spinal cord injury. *J. Neurosci.* 28:7231–43

Howell GR, Libby RT, Jakobs TC, Smith RS, Phalan FC, et al. 2007. Axons of retinal ganglion cells are insulted in the optic nerve early in DBA/2J glaucoma. *J. Cell Biol.* 179:1523–37

Howell GR, Libby RT, John SW. 2008. Mouse genetic models: an ideal system for understanding glaucomatous neurodegeneration and neuroprotection. *Prog. Brain Res.* 173:303–21

Howell GR, Macalinao DG, Sousa GL, Walden M, Soto I, et al. 2011a. Molecular clustering identifies complement and endothelin induction as early events in a mouse model of glaucoma. *J. Clin. Invest.* 121:1429–44

Howell GR, Soto I, Zhu X, Ryan M, Macalinao DG, et al. 2012. Radiation treatment inhibits monocyte entry into the optic nerve head and prevents neuronal damage in a mouse model of glaucoma. *J. Clin. Invest.* 122:1246–61

Howell GR, Walton DO, King BL, Libby RT, John SWM. 2011b. Datgan, a reusable software system for facile interrogation and visualization of complex transcription profiling data. *BMC Genomics* 12:429

Huang W, Fileta J, Guo Y, Grosskreutz CL. 2006. Downregulation of Thy1 in retinal ganglion cells in experimental glaucoma. *Curr. Eye Res.* 31:265–71

Jakobs TC, Libby RT, Ben Y, John SWM, Masland RH. 2005. Retinal ganglion cell degeneration is topological but not cell type specific in DBA/2J mice. *J. Cell Biol.* 171:313–25

Johnson EC, Doser TA, Cepurna WA, Dyck JA, Jia L, et al. 2011. Cell proliferation and interleukin-6-type cytokine signaling are implicated by gene expression responses in early optic nerve head injury in rat glaucoma. *Invest. Ophthalmol. Vis. Sci.* 52:504–18

Johnson EC, Guo Y, Cepurna WO, Morrison JC. 2009. Neurotrophin roles in retinal ganglion cell survival: Lessons from rat glaucoma models. *Exp. Eye Res.* 88:808–15

Johnson EC, Jia L, Cepurna WA, Doser TA, Morrison JC. 2007. Global changes in optic nerve head gene expression after exposure to elevated intraocular pressure in a rat glaucoma model. *Invest. Ophthalmol. Vis. Sci.* 48:3161–77

Johnson EC, Morrison JC. 2009. Friend or foe? Resolving the impact of glial responses in glaucoma. *J. Glaucoma* 18:341–53

Johnson EC, Morrison JC, Farrell S, Deppmeier L, Moore CG, McGinty MR. 1996. The effect of chronically elevated intraocular pressure on the rat optic nerve head extracellular matrix. *Exp. Eye Res.* 62:663–74

Jonas JB. 2011. Role of cerebrospinal fluid pressure in the pathogenesis of glaucoma. *Acta Ophthalmol.* 89:505–14

Ju WK, Kim KY, Lindsey JD, Angert M, Duong-Polk KX, et al. 2008. Intraocular pressure elevation induces mitochondrial fission and triggers OPA1 release in glaucomatous optic nerve. *Invest. Ophthalmol. Vis. Sci.* 49:4903–11

Kanamori A, Catrinescu MM, Kanamori N, Mears KA, Beaubien R, Levin LA. 2010a. Superoxide is an associated signal for apoptosis in axonal injury. *Brain* 133:2612–25

Kanamori A, Catrinescu MM, Mahammed A, Gross Z, Levin LA. 2010b. Neuroprotection against superoxide anion radical by metallocorroles in cellular and murine models of optic neuropathy. *J. Neurochem.* 114:488–98

Kikuchi M, Tenneti L, Lipton SA. 2000. Role of p38 mitogen-activated protein kinase in axotomy-induced apoptosis of rat retinal ganglion cells. *J. Neurosci.* 20:5037–44

Kimelberg HK, Nedergaard M. 2011. Functions of astrocytes and their potential as therapeutic targets. *Neurotherapeutics* 7:338–53

Koeberle PD, Wang Y, Schlichter LC. 2010. Kv1.1 and Kv1.3 channels contribute to the degeneration of retinal ganglion cells after optic nerve transection in vivo. *Cell Death Differ.* 17:134–44

Kompass KS, Agapova OA, Li W, Kaufman PL, Rasmussen CA, Hernandez MR. 2008. Bioinformatic and statistical analysis of the optic nerve head in a primate model of ocular hypertension. *BMC Neurosci.* 9:93

Kuehn MH, Kim CY, Jiang B, Dumitrescu AV, Kwon YH. 2008. Disruption of the complement cascade delays retinal ganglion cell death following retinal ischemia-reperfusion. *Exp. Eye Res.* 87:89–95

Kuehn MH, Kim CY, Ostojic J, Bellin M, Alward WL, et al. 2006. Retinal synthesis and deposition of complement components induced by ocular hypertension. *Exp. Eye Res.* 83:620–28

Ladoux B, Anon E, Lambert M, Rabodzey A, Hersen P, et al. 2010. Strength dependence of cadherin-mediated adhesions. *Biophys. J.* 98:534–42

Lasker/IRRF. 2010. *Astrocytes and glaucomatous neurodegeneration.* **http://www.laskerfoundation.org/programs/irrf.htm**

Leon S, Yin Y, Nguyen J, Irwin N, Benowitz LI. 2000. Lens injury stimulates axon regeneration in the mature rat optic nerve. *J. Neurosci.* 20:4615–26

Levy NS. 1974. The effects of elevated intraocular pressure on slow axonal protein flow. *Invest. Ophthalmol. Vis. Sci.* 13:691–95

Libby RT, Howell GR, Pang I-H, Savinova OV, Mehalow AK, et al. 2007. Inducible nitric oxide synthase, Nos2, does not mediate optic neuropathy and retinopathy in the DBA/2J glaucoma model. *BMC Neurosci.* 8:108

Libby RT, Li Y, Savinova OV, Barter J, Smith RS, et al. 2005. Susceptibility to neurodegeneration in glaucoma is modified by Bax gene dosage. *PLoS Genet.* 1:17–26

Liu B, Neufeld AH. 2000. Expression of nitric oxide synthase-2 (NOS-2) in reactive astrocytes of the human glaucomatous optic nerve head. *Glia* 30:178–86

Liu M, Duggan J, Salt TE, Cordeiro MF. 2011. Dendritic changes in visual pathways in glaucoma and other neurodegenerative conditions. *Exp. Eye Res.* 92:244–50

Lukas TJ, Wang AL, Yuan M, Neufeld AH. 2009. Early cellular signaling responses to axonal injury. *Cell Commun. Signal.* 7:5

Mabuchi F, Aihara M, Mackey MR, Lindsey JD, Weinreb RN. 2004. Regional optic nerve damage in experimental mouse glaucoma. *Invest. Ophthalmol. Vis. Sci.* 45:4352–58

Mansour-Robaey S, Clarke DB, Wang Y-C, Bray GM, Aguayo AJ. 1994. Effects of ocular injury and administration of brain-derived neurotrophic factor on survival and regrowth of axotomized retinal ganglion cells. *Proc. Natl. Acad. Sci. USA* 91:1632–36

Martin KR, Quigley HA, Valenta D, Kielczewski J, Pease ME. 2006. Optic nerve dynein motor protein distribution changes with intraocular pressure elevation in a rat model of glaucoma. *Exp. Eye Res.* 83:255–62

May CA, Lütjen-Drecoll E. 2002. Morphology of the murine optic nerve. *Invest. Ophthalmol. Vis. Sci.* 43:2206–12

McKinnon SJ. 2003. Glaucoma: ocular Alzheimer's disease? *Front. Biosci.* 8:s1140–56

McKinnon SJ, Schlamp CL, Nickells RW. 2009. Mouse models of retinal ganglion cell death and glaucoma. *Exp. Eye Res.* 88:816–24

Miller BR, Press C, Daniels RW, Sasaki Y, Milbrandt J, DiAntonio A. 2009. A dual leucine kinase-dependent axon self-destruction program promotes Wallerian degeneration. *Nat. Neurosci.* 12:387–89

Minckler DS, Bunt AH, Johanson GW. 1977. Orthograde and retrograde axoplasmic transport during acute ocular hypertension in the monkey. *Invest. Ophthalmol. Vis. Sci.* 16:426–41

Morfini G, Pigino G, Szebenyi G, You Y, Pollema S, Brady ST. 2006. JNK mediates pathogenic effects of polyglutamine-expanded androgen receptor on fast axonal transport. *Nat. Neurosci.* 9:907–16

Morgan JE. 2002. Retinal ganglion cell shrinkage in glaucoma. *J. Glaucoma* 11:365–70

Morgan JE. 2004. Circulation and axonal transport in the optic nerve. *Eye (Lond.)* 18:1089–95

Morgan JE, Uchida H, Caprioli J. 2000. Retinal ganglion cell death in experimental glaucoma. *Br. J. Ophthalmol.* 84:303–10

Morrison JC, Farrell SK, Johnson EC, Deppmeier LMH, Moore CG, Grossmann E. 1995. Structure and composition of the rodent lamina cribrosa. *Exp. Eye Res.* 60:127–35

Morrison JC, Johnson EC, Cepurna W, Jia L. 2005. Understanding mechanisms of pressure-induced optic nerve damage. *Prog. Retin Eye Res.* 24:217–40

Munemasa Y, Kitaoka Y, Kuribayashi J, Ueno S. 2010. Modulation of mitochondria in the axon and soma of retinal ganglion cells in a rat glaucoma model. *J. Neurochem.* 115:1508–19

Murphy JA, Archibald ML, Chauhan BC. 2010. The role of endothelin-1 and its receptors in optic nerve head astrocyte proliferation. *Br. J. Ophthalmol.* 94:1233–38

Nakazawa T, Nakazawa C, Matsubara A, Noda K, Hisatomi T, et al. 2006. Tumor necrosis factor-α mediates oligodendrocyte death and delayed retinal ganglion cell loss in a mouse model of glaucoma. *J. Neurosci.* 26:12633–41

Neufeld AH. 1999. Microglia in the optic nerve head and the region of parapapillary chorioretinal atrophy in glaucoma. *Arch. Ophthalmol.* 117:1050–56

Neufeld AH, Sawada A, Becker B. 1999. Inhibition of nitric-oxide synthase 2 by aminoguanidine provides neuroprotection of retinal ganglion cells in a rat model of chronic glaucoma. *Proc. Natl. Acad. Sci. USA* 96:9944–48

Nguyen JV, Soto I, Kim KY, Bushong EA, Oglesby E, et al. 2011. Myelination transition zone astrocytes are constitutively phagocytic and have synuclein dependent reactivity in glaucoma. *Proc. Natl. Acad. Sci. USA* 108:1176–81

Nikolaev A, McLaughlin T, O'Leary DD, Tessier-Lavigne M. 2009. APP binds DR6 to trigger axon pruning and neuron death via distinct caspases. *Nature* 457:981–89

Nikolaeva MA, Mukherjee B, Stys PK. 2005. Na^{+}-dependent sources of intra-axonal Ca^{2+} release in rat optic nerve during in vitro chemical ischemia. *J. Neurosci.* 25:9960–67

Nikolskaya T, Nikolsky Y, Serebryiskaya T, Zvereva S, Sviridov E, et al. 2009. Network analysis of human glaucomatous optic nerve head astrocytes. *BMC Med. Genomics* 2:24

Okada S, Nakamura M, Katoh H, Miyao T, Shimazaki T, et al. 2006. Conditional ablation of Stat3 or Socs3 discloses a dual role for reactive astrocytes after spinal cord injury. *Nat. Med.* 12:829–34

Ostrow LW, Suchyna TM, Sachs F. 2011. Stretch induced endothelin-1 secretion by adult rat astrocytes involves calcium influx via stretch-activated ion channels (SACs). *Biochem. Biophys. Res. Commun.* 410:81–86

Pal SK, Hartnett KA, Nerbonne JM, Levitan ES, Aizenman E. 2003. Mediation of neuronal apoptosis by Kv2.1-encoded potassium channels. *J. Neurosci.* 23:4798–802

Panagis L, Zhao X, Ge Y, Ren L, Mittag TW, Danias J. 2009. Gene expression changes in areas of focal loss of retinal ganglion cells (RGC) in the retina of DBA/2J mice. *Invest. Ophthalmol. Vis. Sci.* 51:2024–34

Pang I-H, Johnson EC, Jia L, Cepurna WA, Shepard AR, et al. 2005. Evaluation of inducible nitric oxide synthase in glaucomatous optic neuropathy and pressure-induced optic nerve damage. *Invest. Ophthalmol. Vis. Sci.* 46:1313–21

Pang IH, Clark AF. 2007. Rodent models for glaucoma retinopathy and optic neuropathy. *J. Glaucoma* 16:483–505

Pease ME, McKinnon SJ, Quigley HA, Kerrigan-Baumrind LA, Zack DJ. 2000. Obstructed axonal transport of BDNF and its receptor TrkB in experimental glaucoma. *Invest. Ophthalmol. Vis. Sci.* 41:764–74

Pelzel HR, Schlamp CL, Nickells RW. 2010. Histone H4 deacetylation plays a critical role in early gene silencing during neuronal apoptosis. *BMC Neurosci.* 11:62

Pelzel HR, Schlamp CL, Waclawski M, Shaw MK, Nickells W. 2012. Silencing of $Fem1c^{R3}$ gene expression in the DBA/2J mouse precedes retinal ganglion cell death and is associated with histone deacetylase activity. *Invest. Ophthalmol. Vis. Sci.* 53:1428–35

Pena JD, Varela HJ, Ricard CS, Hernandez MR. 1999. Enhanced tenascin expression associated with reactive astrocytes in human optic nerve heads with primary open angle glaucoma. *Exp. Eye Res.* 68:29–40

Perlson E, Jeong GB, Ross JL, Dixit R, Wallace KE, et al. 2009. A switch in retrograde signaling from survival to stress in rapid-onset neurodegeneration. *J. Neurosci.* 29:9903–17

Perry VH, Nicoll JA, Holmes C. 2010. Microglia in neurodegenerative disease. *Nat. Rev. Neurol.* 6:193–201

Pinto LH, Klumpp DJ. 1998. Localization of potassium channels in the retina. *Prog. Retin. Eye Res.* 17:207–30

Pournaras CJ, Rungger-Brandle E, Riva CE, Hardarson SH, Stefansson E. 2008. Regulation of retinal blood flow in health and disease. *Prog. Retin Eye Res.* 27:284–330

Prasanna G, Hulet C, Desai D, Krishnamoorthy RR, Narayan S, et al. 2005. Effect of elevated intraocular pressure on endothelin-1 in a rat model of glaucoma. *Pharmacol. Res.* 51:41–50

Quigley HA. 1993. Open-angle glaucoma. *New Engl. J. Med.* 328:1097–106

Quigley HA. 2011. Glaucoma. *Lancet* 377:1367–77

Quigley HA, Addicks EM. 1981. Regional differences in the structure of the lamina cribrosa and their relation to glaucomatous optic nerve damage. *Arch. Ophthalmol.* 99:137–43

Quigley HA, Anderson DR. 1976. The dynamics and location of axonal transport blockage by acute intraocular pressure elevation in primate optic nerve. *Invest. Ophthalmol. Vis. Sci.* 15:606–16

Quigley HA, Broman AT. 2006. The number of people with glaucoma worldwide in 2010 and 2020. *Br. J. Ophthalmol.* 90:262–67

Quigley HA, Cone FE, Gelman SE, Yang Z, Son JL, et al. 2011. Lack of neuroprotection against experimental glaucoma in c-Jun N-terminal kinase 3 knockout mice. *Exp. Eye Res.* 92:299–305

Quigley HA, Hohman RM, Addicks EM, Massof RW, Green WR. 1983. Morphologic changes in the lamina cribrosa correlated with neural loss in open-angle glaucoma. *Am. J. Ophthalmol.* 95(5):673–91

Quigley HA, McKinnon SJ, Zack DJ, Pease ME, Kerrigan-Baumrind LA, et al. 2000. Retrograde axonal transport of BDNF in retinal ganglion cells is blocked by acute IOP elevation in rats. *Invest. Ophthalmol. Vis. Sci.* 41:3460–66

Radius RL, Anderson DR. 1981. Rapid axonal transport in primate optic nerve. *Arch. Ophthalmol.* 99:650–54

Ralay Ranaivo H, Zunich S, Choi N, Hodge J, Wainwright M. 2011. Mild stretch-induced injury increases susceptibility to interleukin-1beta-induced release of matrix metalloproteinase-9 from astrocytes. *J. Neurotrauma* 38:1757–66

Ransom B, Behar T, Nedergaard M. 2003. New roles for astrocytes (stars at last). *Trends Neurosci.* 26:520–22

Rao VR, Krishnamoorthy RR, Yorio T. 2007. Endothelin-1, endothelin A and B receptor expression and their pharmacological properties in GFAP negative human lamina cribrosa cells. *Exp. Eye Res.* 84:1115–24

Rao VR, Krishnamoorthy RR, Yorio T. 2008. Endothelin-1 mediated regulation of extracellular matrix collagens in cells of human lamina cribrosa. *Exp. Eye Res.* 86:886–94

Redman PT, He K, Hartnett KA, Jefferson BS, Hu L, et al. 2007. Apoptotic surge of potassium currents is mediated by p38 phosphorylation of Kv2.1. *Proc. Natl. Acad. Sci. USA* 104:3568–73

Roberts MD, Grau V, Grimm J, Reynaud J, Bellezza AJ, et al. 2009. Remodeling of the connective tissue microarchitecture of the lamina cribrosa in early experimental glaucoma. *Invest. Ophthalmol. Vis. Sci.* 50:681–90

Roberts MD, Liang Y, Sigal IA, Grimm J, Reynaud J, et al. 2010. Correlation between local stress and strain and lamina cribrosa connective tissue volume fraction in normal monkey eyes. *Invest. Ophthalmol. Vis. Sci.* 51:295–307

Rodieck RW. 1998. *The First Steps in Seeing*. Sunderland, MA: Sinauer

Rodriguez-Enriquez S, He L, Lemasters JJ. 2004. Role of mitochondrial permeability transition pores in mitochondrial autophagy. *Int. J. Biochem. Cell Biol.* 36:2463–72

Rosen AM, Stevens B. 2010. The role of the classical complement cascade in synapse loss during development and glaucoma. *Adv. Exp. Med. Biol.* 703:75–93

Ruiz-Ederra J, Garcia M, Hernandez M, Urcola H, Hernández-Barbáchano E, et al. 2005. The pig eye as a novel model of glaucoma. *Exp. Eye Res.* 81:561–69

Salinas-Navarro M, Alarcon-Martinez L, Valiente-Soriano FJ, Jimenez-Lopez M, Mayor-Torroglosa S, et al. 2010. Ocular hypertension impairs optic nerve axonal transport leading to progressive retinal ganglion cell degeneration. *Exp. Eye Res.* 90:168–83

Sarasa-Renedo A, Chiquet M. 2005. Mechanical signals regulating extracellular matrix gene expression in fibroblasts. *Scand. J. Med. Sci. Sports* 15:223–30

Sasaki Y, Vohra BP, Lund FE, Milbrandt J. 2009. Nicotinamide mononucleotide adenylyl transferase-mediated axonal protection requires enzymatic activity but not increased levels of neuronal nicotinamide adenine dinucleotide. *J. Neurosci.* 29:5525–35

Schlamp CL, Johnson EC, Li Y, Morrison JC, Nickells RW. 2001. Changes in Thy1 gene expression associated with damaged retinal ganglion cells. *Mol. Vis.* 7:192–201

Schlamp CL, Li Y, Dietz JA, Janssen KT, Nickells RW. 2006. Progressive ganglion cell loss and optic nerve degeneration in DBA/2J mice is variable and asymmetric. *BMC Neurosci.* 7:66

Schwartz M, London A. 2009. Immune maintenance in glaucoma: boosting the body's own neuroprotective potential. *J. Ocul. Biol. Dis. Infor.* 2:73–77

Semaan SJ, Li Y, Nickells RW. 2010. A single nucleotide polymorphism in the Bax gene promoter affects transcription and influences retinal ganglion cell death. *ASN Neuro* 2:e00032

Shareef S, Sawada A, Neufeld AH. 1999. Isoforms of nitric oxide synthase in the optic nerves of rat eyes with chronic moderately elevated intraocular pressure. *Invest. Ophthalmol. Vis. Sci.* 40:2884–91

Shields MB. 1997. *Textbook of Glaucoma*. Baltimore, MA: Williams & Wilkins

Sigal IA, Ethier CR. 2009. Biomechanics of the optic nerve head. *Exp. Eye Res.* 88:799–807

Sigal IA, Flanagan JG, Ethier CR. 2005. Factors influencing optic nerve head biomechanics. *Invest. Ophthalmol. Vis. Sci.* 46:4189–99

Sigal IA, Flanagan JG, Tertinegg I, Ethier CR. 2004. Finite element modeling of optic nerve head biomechanics. *Invest. Ophthalmol. Vis. Sci.* 45:4378–87

Sofroniew MV. 2009. Molecular dissection of reactive astrogliosis and glial scar formation. *Trends Neurosci.* 32:638–47

Sofroniew MV, Vinters HV. 2010. Astrocytes: biology and pathology. *Acta Neuropathol.* 119:7–35

Son JL, Soto I, Oglesby E, Lopez-Roca T, Pease ME, et al. 2010. Glaucomatous optic nerve injury involves early astrocyte reactivity and late oligodendrocyte loss. *Glia* 58:780–89

Soto I, Oglesby E, Buckingham BP, Son JL, Roberson EDO, et al. 2008. Retinal ganglion cells downregulate gene expression and lose their axons within the optic nerve head in a mouse glaucoma model. *J. Neurosci.* 28:548–61

Soto I, Pease ME, Son JL, Shi X, Quigley HA, Marsh-Armstrong N. 2011. Retinal ganglion cell loss in a rat ocular hypertension model is sectorial and involves early optic nerve axon loss. *Invest. Ophthalmol. Vis. Sci.* 52:434–41

Stasi K, Nagel D, Yang X, Wang RF, Ren L, et al. 2006. Complement component 1Q (C1Q) upregulation in retina of murine, primate, and human glaucomatous eyes. *Invest. Ophthalmol. Vis. Sci.* 47:1024–29

Steele MR, Inman DM, Calkins DJ, Horner PJ, Vetter ML. 2006. Microarray analysis of retinal gene expression in the DBA/2J model of glaucoma. *Invest. Ophthalmol. Vis. Sci.* 47:977–85

Stevens B, Allen NJ, Vazquez LE, Howell GR, Christopherson KS, et al. 2007. The classical complement cascade mediates CNS synapse elimination. *Cell* 131:1164–78

Stirling DP, Stys PK. 2010. Mechanisms of axonal injury: internodal nanocomplexes and calcium deregulation. *Trends Mol. Med.* 16:160–70

Stys PK. 2005. General mechanisms of axonal damage and its prevention. *J. Neurol. Sci.* 233:3–13

Suh SW, Bergher JP, Anderson CM, Treadway JL, Fosgerau K, Swanson RA. 2007. Astrocyte glycogen sustains neuronal activity during hypoglycemia: studies with the glycogen phosphorylase inhibitor CP-316, 819. *J. Pharmacol. Exp. Ther.* 321:45–50

Sun D, Lye-Barthel M, Masland RH, Jakobs TC. 2009. The morphology and spatial arrangement of astrocytes in the optic nerve head of the mouse. *J. Comp. Neurol.* 516:1–19

Sun D, Lye-Barthel M, Masland RH, Jakobs TC. 2010. Structural remodeling of fibrous astrocytes after axonal injury. *J. Neurosci.* 30(42):14008–19

Taylor S, Calder CJ, Albon J, Erichsen JT, Boulton ME, Morgan JE. 2011. Involvement of the CD200 receptor complex in microglia activation in experimental glaucoma. *Exp. Eye Res.* 92:338–43

Tezel G, Yang X, Luo C, Peng Y, Sun SL, Sun D. 2007. Mechanisms of immune system activation in glaucoma: oxidative stress-stimulated antigen presentation by the retina and optic nerve head glia. *Invest. Ophthalmol. Vis. Sci.* 48:705–14

Wang AL, Yuan M, Neufeld AH. 2006. Degeneration of neuronal cell bodies following axonal injury in WldS mice. *J. Neurosci. Res.* 84:1799–807

Watson FL, Heerssen HM, Bhattacharyya A, Klesse L, Lin MZ, Segal RA. 2001. Neurotrophins use the Erk5 pathway to mediate a retrograde survival response. *Nat. Neurosci.* 4:981–88

Wax MB, Tezel G. 2009. Immunoregulation of retinal ganglion cell fate in glaucoma. *Exp. Eye Res.* 88:825–30

Weber AJ, Harman CD, Viswanathan S. 2008. Effects of optic nerve injury, glaucoma, and neuroprotection on the survival, structure, and function of ganglion cells in the mammalian retina. *J. Physiol.* 586:4393–400

Weber AJ, Viswanathan S, Ramanathan C, Harman CD. 2010. Combined application of BDNF to the eye and brain enhances ganglion cell survival and function in the cat after optic nerve injury. *Invest. Ophthalmol. Vis. Sci.* 51:327–34

Whitmore AV, Libby RT, John SWM. 2005. Glaucoma: thinking in new ways: a role for autonomous axonal self-destruction and compartmentalised processes? *Prog. Retin. Eye Res.* 24:639–62

Yamada M, Wong FL, Fujiwara S, Akahoshi M, Suzuki G. 2004. Noncancer disease incidence in atomic bomb survivors, 1958–1998. *Radiat. Res.* 161(6):622–32

Yang Z, Quigley HA, Pease ME, Yang Y, Qian J, et al. 2007. Changes in gene expression in experimental glaucoma and optic nerve transection: the equilibrium between protective and detrimental mechanisms. *Invest. Ophthalmol. Vis. Sci.* 48:5539–48

Yin Y, Henzl MT, Lorber B, Nakazawa T, Thomas TT, et al. 2006. Oncomodulin is a macrophage-derived signal for axon regeneration in retinal ganglion cells. *Nat. Neurosci.* 9:843–52

Yorio T, Krishnamoorthy R, Prasanna G. 2002. Endothelin: Is it a contributor to glaucoma pathophysiology? *J. Glaucoma* 11:259–70

Yu-Wai-Man P, Griffiths PG, Chinnery PF. 2011. Mitochondrial optic neuropathies: disease mechanisms and therapeutic strategies. *Prog. Retin Eye Res.* 30:81–114

Early Events in Axon/Dendrite Polarization

Pei-lin Cheng[1] and Mu-ming Poo[2]

[1]Institute of Molecular Biology, Academia Sinica, Taipei 11529, Taiwan; email: plcheng@imb.sinica.edu.tw

[2]Department of Molecular and Cell Biology and Helen Wills Neuroscience Institute, University of California, Berkeley, California, 94720, USA; email: mpoo@berkeley.edu

Annu. Rev. Neurosci. 2012. 35:181–201

The *Annual Review of Neuroscience* is online at neuro.annualreviews.org

This article's doi: 10.1146/annurev-neuro-061010-113618

0147-006X/12/0721-0181$20.00

Keywords

neuronal polarization, axon determinants, cultured hippocampal neurons, morphogenesis, positive-feedback mechanisms

Abstract

Differentiation of axons and dendrites is a critical step in neuronal development. Here we review the evidence that axon/dendrite formation during neuronal polarization depends on the intrinsic cytoplasmic asymmetry inherited by the postmitotic neuron, the exposure of the neuron to extracellular chemical factors, and the action of anisotropic mechanical forces imposed by the environment. To better delineate the functions of early signals among a myriad of cellular components that were shown to influence axon/dendrite formation, we discuss their functions by distinguishing their roles as determinants, mediators, or modulators and consider selective degradation of these components as a potential mechanism for axon/dendrite polarization. Finally, we examine whether these early events of axon/dendrite formation involve local autocatalytic activation and long-range inhibition, as postulated by Alan Turing for the morphogenesis of patterned biological structure.

Contents

INTRODUCTION

The most distinct feature of neurons is their polarized morphology: a single long axon and many short and highly branched dendrites. The specific pattern of axonal and dendritic arborization is essential for neuronal functions, including the reception and integration of multiple synaptic inputs at the dendrite, as well as the initiation, conduction, and delivery of output signals via the axon. How does this polarized neuronal structure emerge during development? Which molecular mechanisms determine the axon/dendrite identity? How are the selective transport and localization of axon/dendrite components achieved? These questions directly address a central issue of developmental biology: the morphogenesis of cellular structure.

In his seminal paper on the chemical basis of morphogenesis, Alan Turing[1] (1952) laid out a conceptual scheme by which distinct patterns could emerge from an initially homogeneous system through coupled reaction-diffusion processes. Stable patterns in the distribution of substances could be created by a local autocatalytic reaction that generates a slowly diffusing activator and a fast-diffusing long-range inhibitor, which affects the activator's production. This scheme has been widely used in modeling and experimental studies of biological polarization at the tissue and cellular levels, including spontaneous regeneration of hydra (Gierer & Meinhardt 1972), polarization of nerve growth cones by extracellular guidance cues (Meinhardt 1999), and polarization and gradient sensing of chemotaxing cells (Kutscher et al. 2004). In this review, we examine the evidence indicating that the activator/inhibitor scheme proposed by Turing operates in the developing neuron to break symmetry during the neuronal polarization process.

The use of several model systems has greatly facilitated the investigation of neuronal polarization. In vivo studies of neuronal polarity formation in *Caenorhabditis elegans* have shown the importance of extracellular secreted factors, e.g., Netrin/Unc6 and Wnt, in determining polarized neurite outgrowth (Adler et al. 2006, Hilliard & Bargmann 2006, Pan et al. 2006, Prasad & Clark 2006). Recent studies of axon formation in retinal ganglion cells in zebrafish also showed that the interaction of basal processes of newborn neurons with the extracellular matrix (ECM) protein laminin is critical for directing the accumulation of early axonal markers and formation of axon growth cones (Randlett et al. 2011). The extracellular tissue environment is likely to play a critical role in controlling the orientation of the axon and dendrites when developing neurons are establishing an organized projection pattern.

[1]The year 2012 is designated by the Turing Centenary Advisory Committee (**http://www.mathcomp.leeds.ac.uk/turing2012/**) to be Alan Turing Year, inciting a year-long program of events honoring Turing's life and achievements.

The cytoplasmic asymmetry in developing neurons created by the action of polarizing cues may become amplified and stabilized by the Turing local activator and long-range inhibitor. However, in the absence of apparent extracellular cues, isolated neurons are still capable of spontaneous polarization on uniform substrates in culture (Banker & Cowan 1977), in a manner analogous to the spontaneous polarization of migrating cells in the absence of a chemoattractant (Swaney et al. 2010). If the isolated cell does not have an intrinsic cytoplasmic asymmetry, the axon/dendrite polarity may emerge by the Turing mechanism that amplifies small local fluctuation into a stable asymmetry in the cell. Much of our present understanding of the molecular mechanisms underlying neuronal polarization originated from studies on the spontaneous polarization of cultured hippocampal neurons. Several recent reviews have covered various topics related to neuronal polarization, including signal transduction events underlying axon/dendrite formation (Arimura & Kaibuchi 2007, Barnes & Polleux 2009, Hoogenraad & Bradke 2009), selective trafficking of proteins associated with neuronal polarization (Horton & Ehlers 2003, Namba et al. 2011), and regulatory mechanisms for dendrite morphogenesis (Gao 2007, Jan & Jan 2010). Rather than aiming at an exhaustive coverage of the literature on the topic of axon/dendrite development, we examine specifically early cellular events that mediate the determination of axon/dendrite identity as well as factors that are candidates for the Turing activator/inhibitor during neuronal polarization.

INTRINSIC AND EXTRINSIC FACTORS FOR NEURONAL POLARIZATION

Newborn neurons emerging from the last mitotic division may have inherited an intrinsic cytoplasmic asymmetry that influences their subsequent axon/dendrite polarization. The polarized tissue environment may also exert a polarizing action either before or after the last mitosis. Many postmitotic neurons quickly assume a bipolar morphology that is preserved during axon/dendrite differentiation in vivo (Noctor et al. 2004, Zolessi et al. 2006), suggesting that axon/dendrite identity may be determined soon after birth. Dissociated embryonic hippocampal neurons appear as spherical cells with no apparent cytoplasmic asymmetry after plating on uniform culture substrates. Multiple neuritic processes that emerge from the cell body within the first 12 hours in culture also show no apparent asymmetry. However, within one day of culturing, one of these neurites breaks the symmetry by accelerating its growth rate and eventually becomes the axon, while the others develop into dendrites. This model system is ideal for studying the role of intrinsic versus extrinsic factors in neuronal polarization and its underlying cytoplasmic signaling mechanisms.

During axon/dendrite determination in vivo, cortical principal neurons, cerebellar granules, and retinal bipolar and ganglion cells all assume bipolar morphology prior to axon-dendrite differentiation (Barnes & Polleux 2009). Determination of the axon/dendrite identity appears to occur before or during the directional (radial) neuronal migration. For cortical principal neurons and cerebellar granule cells, the leading process of the migrating cell becomes the dendrite, which arborizes extensively upon arrival at its destination; the trailing process then develops into the axon, assuming rapid growth toward its target while its soma is still undergoing radial translocation (Gao & Hatten 1993, Hatanaka & Murakami 2002, Rakic 1971). Although the last mitotic division may have endowed an asymmetry in the distribution of intrinsic determinants, it is not clear whether the axon/dendrite identity is already established during the premigratory bipolar state of the neuron or, alternatively, depends on further interaction with the substrate (e.g., radial glia) and exposure to environmental cues during migration. A recent study of axon formation in the living zebrafish retina offers new insights on this issue. Retinal ganglion cells (RGCs) are born at the apical surface of the retina and translocate their somata toward the

BDNF: brain-derived neurotrophic factor

Sema3A: Semaphorin-3A

basal surface, forming the ganglion cell layer. Using time-lapse imaging of RGCs, Randlett et al. (2011) showed that the contact of neuronal processes with basally localized ECM protein laminin1 is both necessary and sufficient for the RGC axon to polarize toward the basal lamina. Thus, regardless of whether there is intrinsic asymmetry that may act in the absence of external cues, influences from the tissue environment appear to play a dominant role in determining axon/dendrite polarity in vivo. After all, neuronal polarity must be defined with respect to its environment.

The Role of the Centrosome

The importance of intrinsic factors in setting the initial asymmetry was suggested by the finding that the centrosome and Golgi complexes were located asymmetrically in the cytoplasm of isolated hippocampal neurons, near the initiation site of the first neurite, which had a high probability of becoming the axon (de Anda et al. 2005). The centrosome is active in microtubule nucleation, and the Golgi complex provides new membrane precursors; both could facilitate neurite initiation. However, Stiess et al. (2010) demonstrated that centrosome localization is apparently not required for neurite initiation because other neurites initiated later without the localized centrosome, and the axon still formed and regenerated following laser ablation of the centrosome in cultured hippocampal neurons during the early phase of polarization.

In developing cortical neurons in vivo, the centrosome undergoes extensive translocation (Tsai & Gleeson 2005). After the last mitosis, it first sits at the base of the neuron's apical process, then translocates to the base of the trailing process (the future axon), and finally moves to the base of the leading process (the future dendrite) as the cell migrates from subventricular zone to the cortical plate. This centrosome translocation serves to transduce microtubule-based pulling forces from the leading process to the nucleus (Solecki et al. 2009). Whether the brief period of centrosome localization near the nascent axon is sufficient for axon specification is unclear, and this occurs only after the neuron has initiated neurite outgrowth. Thus, centrosomes may serve to stabilize (rather than initiate) the neurite growth by promoting polarized microtubule delivery and trafficking cellular components. Recent work also showed that impaired centrosome assembly affected neural progenitor division but not neuronal morphology (Buchman et al. 2010, Ge et al. 2010). Taken together, the evidence for asymmetric localization of the centrosome and Golgi complexes as an intrinsic determinant for axon formation remains very weak.

Extracellular Polarizing Factors

Axon/dendrite development in vivo occurs within the tissue environment that provides the exposure of the neuron to secreted extracellular factors, surrounding cell surfaces, and the ECM. Secreted factors Netrin and Wnt are critical regulators of neuronal polarity in *C. elegans* (Adler et al. 2006, Hilliard & Bargmann 2006, Montcouquiol et al. 2006, Prasad & Clark 2006). Proteins known to guide axon growth and cell migration also influence axon/dendrite formation in cultured hippocampal neurons. Undifferentiated neurites had a high probability of differentiating into axons when they were exposed to laminin (Esch et al. 2000), brain-derived neurotrophic factor (BDNF) (Shelly et al. 2007), Wnt5A (Zhang et al. 2007), NgCAM (Esch et al. 2000), or Netrin-1 (Mai et al. 2009) that was coated on substrates either in stripes or in a gradient. On the contrary, axon differentiation was prevented when the neurite contacted Semaphorin 3A (Sema3A) coated on substrate stripes (Shelly et al. 2011). In a slice overlay assay, transplanted cortical neurons exhibited polarized neurite outgrowth, with dendrites pointing toward the pial surface and the axon pointing in the opposite direction, suggesting that a gradient of Sema3A across the developing cortex may serve as an extracellular polarizing factor (Polleux et al. 1998).

Although many factors were shown to be effective in polarizing neurons in vitro, genetic deletion of any one of these factors in mice in general yielded no apparent neuronal polarity defect. An exception was reported recently for transforming growth factor β (TGF-β), a secreted morphogen found in a gradient at the ventricular zone of the developing cortex (Yi et al. 2010). Developing cortical neurons lacking the type II TGF-β receptor (TβRII) are defective in axon formation in vivo. However, the percentage of neurons showing axon formation in the knockout mice was reduced to about one-third of that found in control wild-type mice, suggesting other factors may contribute to axon formation. Furthermore, the requirement of TGF-β for axon formation is not conclusive because axons were examined only at one developmental stage; thus, the possibility remains that axon formation is retarded rather than eliminated.

Recent studies have shown that protein kinase A (PKA)–dependent phosphorylation of LKB1, a serine/threonine kinase, is essential for axon formation both in vivo and in vitro (Barnes et al. 2007, Shelly et al. 2007). Localized exposure to BDNF or local elevation of cAMP activity can promote axon differentiation by phosphorylating LKB1 at the PKA site S431, but the in vivo physiological ligand that may trigger local cAMP elevation remains elusive. The developing cortex is enriched with many secreted factors that are effective polarizing factors in vitro, including Sema3A (Lerman et al. 2007, Shelly et al. 2011), BDNF (Wong et al. 2009), TGF-β2/β3 (Yi et al. 2010), slits (Whitford et al. 2002), and ephrins (Liebl et al. 2003). Many factors likely work in concert to determine neuronal polarity, and genetic deletion of any one of them produces a weak or insignificant effect. Thus, simultaneous perturbation of multiple factors is required to dissect their contribution to neuronal polarization in vivo.

Mechanotropic Actions of the Environment

In addition to chemical factors, physical forces imposed by the environment may also be relevant for neuronal polarization and axon/dendrite development. Owing to the nonrandom organization of neural tissues, developing neurons are exposed to anisotropic physical forces, mediated by the cell's adhesion to the environment. The mechanotropic theory of nerve growth had once dominated the field of neural development (Purves & Lichtman 1985). Nearly one century ago, Ross Harrison (1910) first showed that cultured amphibian neuroblasts prefer to cling to and move along solid surfaces. Paul Weiss (1941) later showed that migrating embryonic cells and growing neurites from cultured explants tend to move along scratches on the glass surfaces or along lines of tension in the plasma clot embedding the explant, a mechanism termed contact guidance. Neurite can be experimentally initiated by applying tension to the margin of the cell body (Bray 1984, Chada et al. 1997, Zheng et al. 1991), and stretching the growth cone with a towing microelectrode is an effective stimulus for its elongation, sometimes leading to axon formation (Bray 1984; Lamoureux et al. 1989, 2002; Pfister et al. 2004). Is mechanical tension required for axon formation? Although we have no direct evidence, lowering the rest tension by neutralizing polylysine on the substrate of cultured neurons (with a polyanion) or by eliminating the tension via detaching the neurite from the substrate with laser irradiation resulted in depolymerization of microtubules and rapid neurite retraction (Dennerll et al. 1988, Lamoureux et al. 1990).

PKA: protein kinase A

Increasing in vitro evidence now indicates that mechanical forces can modulate all features of neuronal morphology, including axon growth and pruning, as well as dendrite arborization and plasticity (O'Toole et al. 2008, Previtera et al. 2010, Zheng et al. 1991). At the cellular and molecular levels, these mechanical effects must be accounted for by molecular events, such as activation of mechanosensitive ion channels. Cultured neurons showed a preference for certain features of the substrate—e.g., they anchor and turn at rigid edges (Anava et al. 2009) and prefer to extend and form more branches on softer gels (Balgude et al. 2001, Georges et al. 2006, Gunn et al. 2005)—indicating a link between the matrix compliance

PI3K: phosphoinositide 3-kinase

to the adhesion and neuronal growth behavior. Engler et al. (2006) showed that mimicking the mechanical properties of their native environment (e.g., ECM elasticity) can direct the differentiation of naïve mesenchymal stem cells into the corresponding cell lineage, with soft matrices being neurogenic. This evidence suggests that the specific anisotropic mechanical environment of the developing tissue may guide proper neuronal development. From a cellular point of view, in the presence of an anisotropic environment with a gradient of stiffness, neurite growth is pulled toward the region of higher stiffness, when surface adhesion sites are uniformly distributed. Second, in the isotropic environment, asymmetry in the distribution of the adhesion sites, either intrinsically present on the surface or induced by a gradient of extracellular factors, also results in polarized neurite growth. In the latter case, the chemical gradient is translated into a gradient of tension on the actin cytoskeleton. The mechanical action of the environment can now be divided into three steps: sensing, transduction, and response. Each can be understood in cellular and molecular terms (Schwartz 2009). Studies using new molecular and bioengineering manipulations will soon allow us to have an integrated understanding of how extrinsic physical and chemical factors act together to determine neuronal polarity and axon/dendrite development in vivo.

DETERMINANTS, MEDIATORS, AND MODULATORS OF AXON FORMATION

Axon formation could, in principle, begin with an early step of axon fate specification, followed by later steps of axon differentiation and growth, although the term specification is often used interchangeably with differentiation/growth in the literature. The distinction between axon fate specification and axon growth is difficult because axon fate cannot be expressed without axon growth, and a factor that specifies the axon fate to a neurite could do so by triggering an accelerated growth of the neurite. This notion is exemplified by the effect of local exposure to extracellular factors, e.g., laminin (Esch et al. 1999, Menager et al. 2004), BDNF (Mai et al. 2009, Shelly et al. 2007), and TGF-β (Ishihara et al. 1994, Yi et al. 2010), which can initiate axon differentiation as well as promote axon growth. Nevertheless, it is useful to define a group of factors as axon (fate) determinants that are distinct from other factors that only promote growth. Such a distinction will help us to delineate complex cytoplasmic signaling events under natural biological conditions.

Determinants of Axon Formation

We define a cytoplasmic molecule or molecular complex as an axon determinant if it satisfies the following criteria. First, the concentration or activity of this factor is changed selectively in the neurite prior to the appearance of any axonal characteristics, e.g., increased growth rate and expression of axon-specific proteins (the "early presence"). Second, preventing factor changes abolishes axon formation (the "necessity" requirement). Third, artificially altering the concentration or activity of this factor in the neurite triggers its differentiation into the axon (the "sufficiency" requirement). As shown in **Table 1**, only a few factors, including Ras, cAMP/PKA, PI3K/PIP3, and pLKB1, appear to have satisfied these criteria. These determinants are likely to be immediate cytoplasmic messengers for extracellular factors (e.g., BDNF and laminin) that polarize the neurons. In cultured hippocampal neurons, they accumulate at a higher level in one neurite of stage-two neurons prior to the polarization. Overexpression of the factor led to the formation of multiple axons, whereas downregulation resulted in no axon formation.

Early presence. Distinct accumulation of putative axon determinants has frequently been observed in one of the undifferentiated neurites of cultured hippocampal neurons, but few studies have followed the development of the same neurite over time to confirm its axonal fate. The difficulty of such time-lapse studies in living cells is that the molecule needs to be tagged, e.g., with green fluorescent protein

(GFP), which may perturb the protein's function, and overexpression of the tagged protein often leads to abnormal or even deleterious conditions. For example, expression of GFP-LKB1 in cultured hippocampal neurons resulted in the formation of multiple axons. However, under the condition of low-level expression, it is possible to map the distribution of the tagged molecule in the undifferentiated neurite and track its fate at a later time, as was done for the STE20-related adaptor (STRAD), a cofactor that complexes with LKB1. STRAD accumulation in the neurite was found to result in a high probability of axon differentiation. Thus, the LKB1/STRAD complex satisfies the criterion of "early presence" (Shelly et al. 2007). Consistently, axon differentiation of neurites in contact with stripe-coated BDNF on the substrates is also preceded by an accumulation of phosphorylated LKB1 (^{p431}LKB1) in the neurite (Shelly et al. 2007). Similarly, spontaneous GFP-shootin 1 accumulation in one neurite also resulted in a high probability of axon formation via its action on the localization of PI3K and activation of Akt (Toriyama et al. 2006).

Responsiveness to polarizing factor. Axon determinants must respond to the action of extracellular factors that promote axon formation, e.g., BDNF and laminin. Whereas laminin has been shown to be responsible for guiding retinal ganglion cell axons in vivo (Randlett et al. 2011), little evidence indicates that a single extracellular polarizing factor is responsible for triggering axon formation in other brain regions. Multiple independent axon determinants are likely to exist because diverse extracellular chemical and mechanical factors may polarize axon formation by activating different cellular effectors. Signaling cascades triggered by these effectors may eventually converge on a final common pathway that induces selective cytoskeletal rearrangement and protein trafficking required for axon formation.

Necessity and sufficiency requirements. The axon determinants are necessary and, by themselves, sufficient for axon formation. Depleting the determinant or disrupting its functions in the cell should impair axon formation, whereas elevating or activating the determinant by itself in the neurite should cause its axon to differentiate. Downregulation of the expression of a protein via RNA interference is often used to deplete the determinant, but the efficiency of RNA interference may vary; thus, the effects in preventing axon formation are in general incomplete. A residual amount of the determinant may remain owing to incomplete RNA interference. Alternatively, other determinants may function independently in mediating axon formation. To test its sufficiency, the most common approach is to apply an extracellular polarizing factor to the neurite via contact with beads or substrates coated with the polarizing factor (Menager et al. 2004, Randlett et al. 2011, Shelly et al. 2007). However, many reports showing that a factor is both necessary and sufficient for inducing axon formation have not demonstrated that the factor could be up- or downregulated in response to natural polarizing factors, thus failing the criteria of axon determinants.

Whereas experimental manipulations of a single determinant may be sufficient to influence axon formation, especially under the condition of overexpression, concerted actions of multiple determinants may be required in the natural axon formation in vivo. This notion may account for the fact that complete absence of axon formation rarely occurs in vivo when a single putative axon determinant is downregulated or eliminated by gene deletion. Cytoplasmic axon determinants that normally mediate the action of extracellular polarizing factors may also operate by initial stochastic fluctuation of its activity or distribution, followed by a local autocatalytic process that amplifies the local asymmetry, leading to spontaneous axon formation (see Amplification Mechanisms in Axon/Dendrite Formation, below).

Mediators and Modulators of Axon Formation

Within the complex signal transduction networks that are potentially involved in axon

Table 1 Molecules involving axon specification in cultured hippocampal neuron

	Molecules	Upstream regulators	Downstream effectors	Polarized distribution[a]	Rapid growth[b]	Upregulation[c]	Downregulation[c]	References
Determinants	cAMP/PKA	BDNF	LKB1	Stage 2	+	MA	NA	Cheng et al. 2011b; Shelly et al. 2007, 2010; Zheng et al. 1994
	GM1/PMGS	NGF?	TrkA, PIP3, Rac1	Stage 2	(axon)	Early polarization	NA	Da Silva et al. 2005, Pitto et al. 1998, Rodriguez et al. 2001
	PIP3, PI3K	Laminin, Ras, Shootin1	Akt	Stage 2	+	MA	NA	Jiang et al. 2005, Menager et al. 2004, Shi et al. 2003, Yoshimura et al. 2006
	pLKB1	BDNF, PKA	SAD-A/B	Stage 2/stage 3	No	MA	NA	Barnes et al. 2007, Shelly et al. 2007
Mediators	Cdc42	Rap1B	Cofilin, Par complex	Stage 2	+	MA	NA	Garvalov et al. 2007, Schwamborn & Püschel 2004
	CRMP2	GSK3β, BDNF, NT-3, Kinesin-1	Tubulin, Numb	Stage 3	+	MA	NA	Inagaki et al. 2001, Kimura et al. 2005, Nishimura et al. 2003, Yoshimura et al. 2005
	Dvl	Wnt5a	aPKC	Stage 3	+	MA	NA	Zhang et al. 2007
	GSK3β	Akt, PTEN, Dvl	MAPs	Stage 3	Inhibition	NA	MA/NA	Gärtner et al. 2006, Jiang et al. 2005, Shi et al. 2004, Yoshimura et al. 2005, Zhang et al. 2007
	ILK	Wnt5a, PI3K	Akt, GSK3β	Stage 3	—	MA	NA	Guo et al. 2007, Oinuma et al. 2007
	Par complex	TGF-β, PI3K, Cdc42	GSK3β, STEF	Stage 3	—	Unpolarized	NA	Nishimura et al. 2005, Schwamborn et al. 2007a, Shi et al. 2003, Yi et al. 2010
	pAkt	PI3K	GSK3β	Stage 3	(axon)	MA	Neuronal death	Jiang et al. 2005, Yan et al. 2006, Yoshimura et al. 2006
	Par1b/MARK2	Par complex, LKB1	GAKIN/KIF13B	No	—	NA	MA	Chen et al. 2006b, Yoshimura et al. 2010

	pSmurf1	BDNF, PKA	Par6, RhoA	Stage 3	—	MA	NA	Cheng et al. 2011a
	Rac1	PMGS	RhoA, WAVE	Stage 3	(axon)	—	NA	Kassai et al. 2008, Ng et al. 2002, Santos Da Silva et al. 2004, Tahirovic et al. 2010
	Rap1B	PI3K	Par complex, Cdc42	Stage 2	—	MA	NA	Schwamborn & Püschel 2004
	Ras	PI3K	PI3K, ILK	Stage 2	—	MA	NA	Fivaz et al. 2008, Oinuma et al. 2007, Yoshimura et al. 2006
	Rheb	PI3K	mTOR, Rap1B	Stage 3	(axon)	MA	NA	Li et al. 2008
	SAD	LKB1	—	No	—	MA	Unpolarized	Barnes et al. 2007, Kishi et al. 2005
	Tiam1/STEF	Par complex	Rac1	Stage 2	+	MA	Unpolarized	Kunda et al. 2001, Nishimura et al. 2005
Modulators	APC	—	GSK3β, Par3	Stage 2/Stage 3	—	Unpolarized/ MA	NA	Gärtner et al. 2006, Shi et al. 2004
	GAKIN/ KIF13B	—	Membrane transport (PIP3)	Stage 3	—	Unpolarized/ MA	NA	Horiguchi et al. 2006, Yoshimura et al. 2010
	pAMPK	LKB1?	Kif5 (transport of PI3K)	—	—	NA	No effect	Amato et al. 2011
	PTEN	—	PIP3, Par3	No	Inhibition	NA	MA	Jiang et al. 2005, Shi et al. 2004
	RalA	—	Par3	Stage 2	—	Unpolarized	Unpolarized	Lalli 2009
	RhoA	—	Actin/MT dynamics	No	Inhibition	Inhibited growth	Enhanced growth	Cheng et al. 2011a, Da Silva et al. 2003, Yamashita et al. 1999
	Shootin1	—	PI3K, L1-CAM	Stage 2-3	+	MA	NA	Shimada et al. 2008, Toriyama et al. 2006
	Singar	—	PI3K	No	—	No effect	MA	Mori et al. 2007
	Smurf2	—	Rap1B, Par3	No	—	NA	MA	Schwamborn et al. 2007a,b

[a]Neuronal stages at which the indicated molecules display polarized accumulation. Stage 2, accumulating at a higher level in one of the undifferentiated neurites of stage-two neurons. Stage 3, accumulating in the nascent axon of stage-three neurons. No, uniformly distributed in all processes. Note that the "stage" in the hippocampal culture model may not fully correspond to the situation in vivo.

[b]Changes in the neurite/axon outgrowth while elevating the expression or activity level of the molecules indicated.

[c]The effect of up- or down-regulation of the indicated molecules on axon formation. Abbreviations: MA, formation of multiple axons; NA, no axon formed. Unpolarized, neurite exhibits morphological characteristics of axon/dendrite but is unable to stain well with axonal/dendritic makers.

Par6: partitioning defective 6

formation (Arimura & Kaibuchi 2005, Barnes & Polleux 2009), we distinguish factors that mediate axon development ("mediators") from those that are only capable of modulating this process ("modulators"). The distribution or function of the mediator is regulated by natural axon-polarizing factors, whereas that of the modulator is not (**Figure 1**). Modulators may alter the efficacy of various steps leading to axon formation, from the receptors for polarizing signals at the plasma membrane to various cytoplasmic signal transducers and regulators of cytoskeletal structures underlying axon differentiation and growth. Modulators may be essential to maintain the integrity of signaling pathways leading to axon formation, and its up- or downregulation alone may even be sufficient to influence axon formation. Thus, the critical test for being a mediator or modulator will depend on whether the activities of those factors reported are regulated by known polarizing factors (see **Table 1**). For example, the Par3/Par6/aPKC complex is a well-known mediator of PI3K-dependent axon formation (Shi et al. 2003), which can be triggered by the polarizing factor BDNF, and Par6 phosphorylation can also be regulated by TGF-β signals in the developing cortex (Yi et al. 2010). On the other hand, we consider E3 ligase Smurf 2 to be a modulator of Rap1B levels (Schwamborn et al. 2007b), leading to changes in axon formation, because Smurf 2 itself is not regulated by polarizing factors or axon determinants.

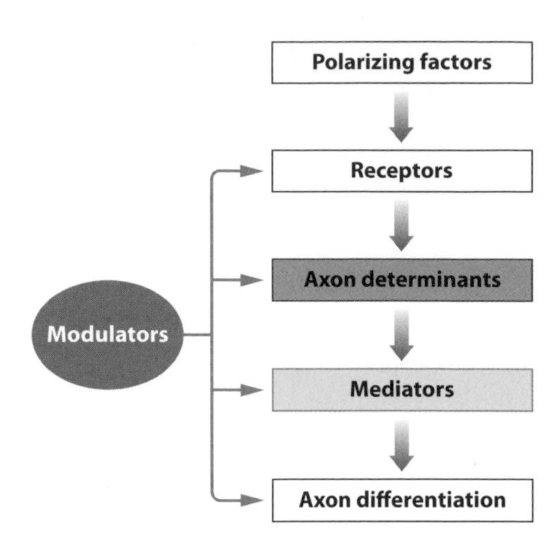

Figure 1

Definitions of determinant, mediator, and modulator of axon formation. The cytoplasmic components participating in the process of axon formation could be classified into three groups as follows: (*a*) determinant, an immediate cytoplasmic messenger that conveys the action of natural polarizing factors and exhibits early accumulation in one of the undifferentiated neurites prior to the appearance of any axonal characteristics; (*b*) mediator, a component that is regulated by natural polarizing factors and upstream determinants in the signaling pathway; and (*c*) modulator, a component that is not regulated by natural polarizing factors or axon determinants/mediators but can modulate the efficacy of various steps of the signaling pathway, leading to changes in axon formation.

Selective Protein Degradation in Neuronal Polarization

Increasing evidence indicates that regulation of the ubiquitin-proteasome system (UPS) may be involved in axon/dendrite differentiation. Treatment of cultured hippocampal neurons with general proteasome inhibitors, such as MG-132 or lactacystin, resulted in the formation of multiple axons (Yan et al. 2006, Schwamborn et al. 2007b). Because many proteins involved in axon growth are targets of the UPS, the axon-promoting effect of these inhibitors could be attributed to nonselective accumulation of proteins that caused excessive axon growth. A more interesting possibility is that normal axon/dendrite formation is mediated by UPS-dependent degradation of specific proteins that mediate or modulate axon/dendrite formation. For example, specific E3 ligases for this group of proteins may be selectively regulated during neuronal polarization in at least three ways. First, posttranslational modification (e.g., phosphorylation) alters the susceptibility of the protein for ubiquitination by the E3 ligase. Indeed, the rate of UPS-dependent degradation of dendritic Akt, which has a reduced phosphorylation status, is higher than that of axonal Akt, consistent with the asymmetrical distribution of Akt/p-Akt in polarized neurons (Yan et al. 2006). Phosphorylation of LKB1 at S431 also reduces its ubiquitination (Cheng et al. 2011a), probably owing to the formation of stable LKB1/STRAD complex (Shelly et al. 2007). Second, E3 ligase activity is regulated by its interacting proteins, such as ARF (Honda & Yasuda 1999) and F-box proteins (Kato et al. 2009). Third, the substrate specificity of E3 ligases is altered by posttranslational modification of E3 ligases themselves, as demonstrated by the phosphorylation of Smurf1 at a PKA site in response to the polarizing factor BDNF and by the change in the substrate specificity of Smurf1 in favor of reduced degradation of axon-promoting factor Par6 and increased degradation of a growth-inhibiting RhoA (Cheng et al. 2011a).

Extracellular polarizing factors may regulate the distribution and levels of growth-related proteins through their actions on the UPS. Chemotropic responses of the growth cone to extracellular guidance factors require protein degradation (Campbell & Holt 2001, Mann et al. 2003), and Netrin-1 and lysophosphatidic acid (LPA) can trigger rapid accumulation of ubiquitin-protein conjugates in the growth cone. Besides altering the properties of the substrate or the interacting proteins of the UPS, extracellular polarizing factors may also modulate the enzymatic kinetics or the substrate specificity of the ligases, leading to selective stabilization of axon/dendrite determinants or degradation of negative regulators of axon/dendrite differentiation. The recent study of the axon-polarizing effect of TGF-β signaling in neocortical neurons (Yi et al. 2010) also suggests that the recruitment of ubiquitin ligase Smurf1 via its interaction with the phosphorylated form of axon-promoting factor Par6 (a Smurf 1 substrate) at the site of TGF-β signaling may change the local cytoskeleton dynamics through Smurf1-mediated degradation of RhoA (Ozdamar et al. 2005). Thus, TGF-β signaling could induce coordinated local accumulation of axon-promoting Par6-containing complexes together with selective local downregulation of RhoA activity, leading to accelerated neurite growth required for the axon formation.

AMPLIFICATION MECHANISMS IN AXON/DENDRITE FORMATION

As noted by Ramón y Cajal (1898) more than a century ago, developing neurons in vivo resemble chemotaxing cells and thus are capable of sensing gradients of chemotactic factors to determine their direction of migration. Whereas directed migration and cell polarization work together in chemotaxing neutrophils and amoeboid cells, these cells can undergo spontaneous polarization in the absence of extracellular cues (Parent & Devreotes 1999), similar to the dissociated hippocampal neurons in culture. Studies of directed cell migration have shown that both gradient sensing and polarization in a shallow gradient of extracellular cues could be enhanced and stabilized by local autocatalytic production of a local activator and long-range cytoplasmic diffusion of an inhibitor (Parent & Devreotes 1999, Xiong et al. 2010). Spontaneous polarization could also be induced by these two processes, which stabilize cytoplasmic asymmetry originating from stochastic fluctuation of polarity determinants.

Local Autocatalytic Processes

In a nonpolarized neuron, some signaling molecules involved in neuronal polarization are found to accumulate selectively in one neurite (see Determinants of Axon Formation, above). However, such accumulation may be transient. Positive-feedback mechanisms that stabilize their localization in the neurite may represent a critical step in neuronal polarization.

PI3K/PIP3/Rho GTPase feedback loop. A positive-feedback loop mediated by phosphatidylinositol 3,4,5 trisphosphate (PIP3) and Rho GTPase is important to establish neutrophil polarity during chemotaxis (Wang et al. 2002, Weiner 2002). Production of PIP3 by activated PI3K at the leading edge of a migrating cell stimulates its own accumulation by activating Akt and Rho GTPases (Cdc42, Rac, and Rap1B), which not only regulate the actin

GSK3β: glycogen synthase kinase 3 beta

NT: neurotrophin

Trk: tropomyosin receptor kinase

cytoskeleton but also repress GSK3β as well as activate PI3K, which produces more PIP3. In cultured hippocampal neurons, PI3K activity was highly localized to the tip of the newly specified axon. Inhibitors of PI3K disrupted such asymmetrical localization and inhibited neuronal polarization. In addition, downstream effectors of Rac (Miki et al. 1998) are enriched in growing axons and play an essential role in the axon development of *Drosophila* (Ng et al. 2002, Zallen et al. 2002). Therefore, a positive feedback loop involving PI3K signaling pathways and Rho GTPase could initiate and maintain the asymmetric accumulation of PI3K signals for axon specification.

Autocrine neurotrophin loop. In many cell types, including neutrophils, platelets, and endothelial cells, chemotropic actions on the cell could be amplified by the factors secreted from the cell that acts on its own receptors in an autocrine manner. A useful demonstration is the purinergic chemotaxis in neutrophils, where high concentrations of extracellular ATP were detected in a localized area near the leading edge, together with polarized accumulation and activation of A3 adenosine receptor during cell migration (Chen et al. 2006a). Perturbation of the extracellular purinergic flux by genetic deletion of ectonucleotidases CD39/CD73 (Goepfert et al. 2001, Koszalka et al. 2004), pharmacological blockade of A3 receptor, or intravenous administration of A3 receptor agonist (Chen et al. 2006b) all lead to altered neutrophil chemotaxis. The existence of neurotrophin (NT)-induced NT release, first described for cultured PC12 and neuronal cell lines (Krüttgen et al. 1998), suggests that autocrine action of NT may provide a local positive feedback mechanism for axon formation and growth.

On the basis of the existing literature on the autocrine action of NT and the localized NT action on axon differentiation and growth (Shelly et al. 2007), we propose a self-amplifying autocrine mechanism for initiating axon differentiation and growth. Neurotrophins are highly basic molecules that tend to be associated with the cell surface after secretion, thus ideal for serving as autocrine factors. As depicted in **Figure 2**, spontaneous secretion of NT at the growth cone of undifferentiated neurites may fluctuate stochastically. One neurite that happens to secrete an amount of NT above the threshold level—capable of triggering NT secretion at a level that is more than compensating its diffusional dissipation—will result in a gradual accumulation of extracellular surface-bound NT at the growth cone, causing an accelerated axon formation. This self-amplifying NT action could be further facilitated by enhanced anterograde transport of its tropomyosin receptor kinase (Trk) receptor via transcytosis (Ascano et al. 2009) because membrane insertion of Trk receptors can also be triggered by NTs (Du et al. 2000, Haapasalo et al. 2002, Meyer-Franke et al. 1998, Zhao et al. 2009). Experimental verification of this hypothesis requires evidence that local surface accumulation of extracellular surface-bound NT occurs before or during the onset of axon differentiation and that such accumulation is both necessary and sufficient for axon formation. Recent studies have indicated that BDNF indeed satisfies these criteria as a self-amplifying autocrine for spontaneous axon formation in cultured hippocampal neurons (Cheng et al. 2011b). The examination of the distribution of BDNF secretion/TrkB expression pattern in stage-two and stage-three neurons revealed a striking reorganization of BDNF/TrkB signaling; uniformly distributed BDNF secretion/surface TrkB of stage-two neurons became accumulated at the nascent axon in stage-three neurons. These reorganizations were accompanied by spatially correlated changes in the constitutive cAMP/PKA activity toward the distal axon, similar to the redistribution of the PI3K activity (Shi et al. 2003). The coordinated localization of these cellular activities, together with the self-amplifying autocrine signaling of BDNF, may directly facilitate the axon specification and the accelerated growth of nascent axons.

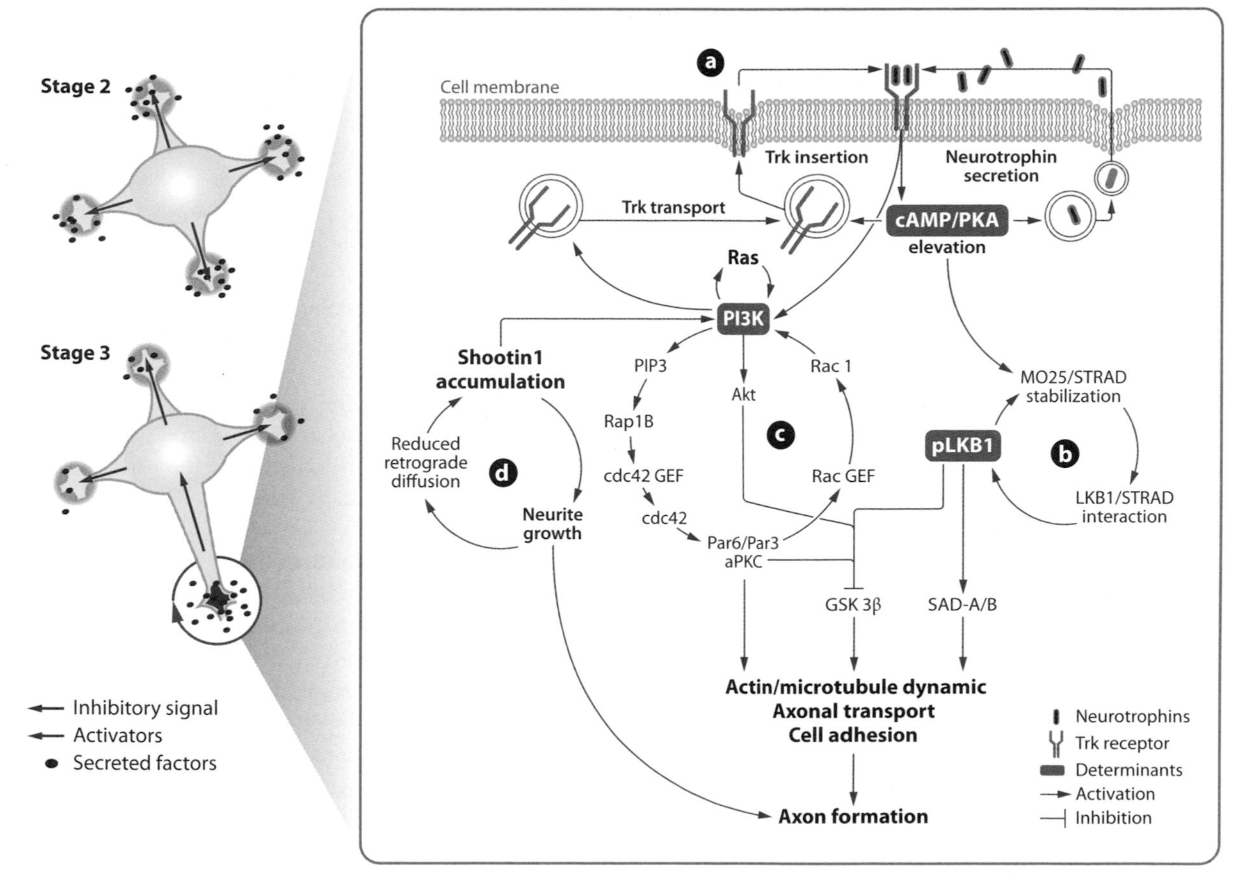

Figure 2

Local amplification mechanisms for axon formation. Prior to axon/dendrite polarization, early-stage neurons have random fluctuations within cytoplasmic axon determinants and growth-promoting activities. The initial fluctuation of a local activator in one of the neurites could be stabilized and amplified by a local autocatalytic process that generates the activator as well as by a long-range diffusible inhibitor that amplifies the local asymmetry, leading to spontaneous axon formation. The potential amplification mechanisms include (*a*) autocrine NT loop, where NT signaling triggers its own secretion and Trk receptor insertion and transport by elevating cAMP/PKA and PI3K activity, leading to further enhanced NT signaling; (*b*) PKA/LKB1/STRAD loop, where LKB1 facilitates MO25/STRAD interaction, which promotes LKB1/STRAD complex formation and in turn stabilizes LKB1 (elevation of cAMP/PKA could trigger the feedback route by increasing the level of MO25); (*c*) PI3K/PIP3/Rho GTPase feedback loop, where production of PIP3 by PI3K stimulates its own accumulation by activating Rho GTPases, which activate PI3K, producing more PIP3; and (*d*) shootin1 feedback loop, where growth cone accumulation of shootin1 induces an accelerated neurite extension, which in turn results in reduced back diffusion of shootin1 toward the soma, leading to further growth cone accumulation. Note that the "stage" in the hippocampal culture model may not fully correspond to the situation in vivo.

The autocrine action of NTs during axon formation and growth in vivo could be tested by specific downregulation of the NT synthesis and secretion in the developing neuron during its early development without affecting adjacent neurons in the surrounding tissue. Furthermore, when BDNF expression was downregulated in developing cortical neurons, the growth rate of callosal axons was reduced (Cheng et al. 2011b). Although in vivo studies have yet to unveil the role for NTs in the early stage of axon initiation, to a large extent the limitation is related to the difficulties associated with the compensatory effects of other factors or other NT family members of similar autocrine functions.

Nested positive-feedback loops. Because PI3K is a major downstream effector of BDNF (Huang & Reichardt 2003), the highly localized PI3K at the axonal tip could initially be triggered by the localized autocrine action of BDNF, and the PI3K/PIP3/Rho GTPase feedback loop thus represents another positive-feedback loop in promoting PI3K-dependent TrkB anterograde transport (Cheng et al. 2011b). Indeed, asymmetric PIP3 and Akt signaling was found to mediate chemotaxis in neuritic growth cones of *Xenopus* spinal neurons induced by a BDNF gradient (Henle et al. 2011). Other activators of PI3K, e.g., shootin1 and Singar1/2 (Mori et al. 2007, Toriyama et al. 2006), arising from asymmetric intrinsic or extrinsic factors, could also serve as the initial trigger for PI3K activation. In the latter scenario, local elevation of PI3K could then induce local accumulation of TrkB in the neurite, leading to elevated BDNF/TrkB signaling and subsequent amplification via both PI3K and cAMP/PKA pathways. These nested positive-feedback loops could thus ensure their activation for axon formation via the diverse downstream effectors, including LKB1, Par complex, and GSK3β. The initial trigger could occur as an induced or spontaneous activation of one component in these nested positive-feedback loops, and blocking the critical catalytic component in each loop (TrkB, PKA, or PI3K) could reduce (but not completely abolish) axon formation.

Neurite length–dependent feedback. Goslin & Banker (1989) first suggested a potential positive-feedback mechanism while studying the behavior of axon regeneration after transection in cultured hippocampal neurons. They found that when the remaining stump of the transected axon is 10 μm longer than other neurites of the same cell, it invariably became the axon. Conversely, when a cell's neurites were nearly equal in length, it was impossible to predict which would become the axon. The authors hypothesized that the concentration of a limiting soma-derived regulatory factor in the growth cone is a function of the neurite length: A constant anterograde transport of a growth-promoting factor together with its back diffusion to the soma will result in increasingly higher accumulation of the factor at the longest neurite and promote its growth in a positive-feedback manner. Alternatively, a constant retrograde transport of a growth-inhibiting factor together with its back diffusion to the growth cone will have a similar positive-feedback growth promotion at the longest neurite. This hypothesis predicts the existence of a growth-promoting (or -inhibiting) factor that exhibits a neurite length–dependent accumulation (or depletion) at the growth cone during the early phase of axon fate determination.

Shootin1 feedback loop. Shootin1, a novel protein, is a good example of the hypothesized neurite length–dependent growth-promoting factor. Shootin1 exhibited initial fluctuating accumulation in multiple growth cones of the unpolarized hippocampal neuron, but it soon became accumulated in a single neurite right before the neurite exhibited the accelerated growth that led to axon formation (Toriyama et al. 2006, 2010). Shootin1 is actively transported from the soma to the neuritic growth cone, and its retrograde diffusion to the soma should vary inversely with the neurite length, leading to its accumulation at the longest neurite. Inhibiting the anterograde transport of shootin1 prevented its asymmetric accumulation in neurons. Assuming the amount of available shootin1 in the soma is limited, only one axon may be allowed during neuronal polarization. Shootin1 could regulate local PI3K activity because downregulation of shootin1 by RNAi or overexpression of shootin1 could disrupt asymmetric accumulation of PI3K activity, whereas inhibiting PI3K activity had no effect on shootin1 accumulation in the growth cone (Toriyama et al. 2006). By our definition, both shootin1 and PI3K could be considered axon determinants.

PKA/LKB1/STRAD loop. In addition to the mechanism based on active transport and passive diffusion, stable accumulation of axon determinants in the nascent axon may also result from protein stability against UPS-dependent degradation. An example of stability-based accumulation is the complex of LKB1 and its cofactor STRAD. In cultured hippocampal neurons, LKB1/STRAD accumulation in one of the undifferentiated neurites is a highly reliable predictor of the axon fate of the neurite (Shelly et al. 2007). The formation of the LKB1/STRAD complex promotes LKB1 stability and PKA-dependent phosphorylation, which is critical for axon formation both in vivo and in vitro (Barnes et al. 2007, Shelly et al. 2007). Furthermore, LKB1/STRAD complex formation is promoted by the coexpression of a scaffolding protein MO25, whereas LKB1 also facilitates MO25/STRAD interaction, similar to that found in polarizing epithelial cells (Boudeau et al. 2003). Because cAMP elevation could increase the MO25 protein level (Shelly et al. 2007), local cAMP elevation may trigger a positive-feedback loop in which increased MO25 results in the initial local accumulation of a LKB1/STRAD/MO25 heterotrimeric complex, leading to enhanced LKB1/STRAD interaction and a further increase in the LKB1 stability, which in turn triggers the accumulation of more heterotrimeric complexes, allowing elevated local LKB1 phosphorylation and activation.

Long-Range Global Inhibition

Long-range cytoplasmic signaling is known to be critical for communication between axon terminals with the soma and dendrites. Axotomy of postganglionic axons resulted in regression of preganglionic synaptic inputs on the dendrites of rat sympathetic ganglion cells, and such regression can be prevented by providing nerve growth factor to the axotomized axons (Purves 1975, Purves & Njå 1978), indicating long-range axon-to-dendrite signaling. Such signaling is also demonstrated in developing *Xenopus* RGCs for conveying a retrograde spread of long-term potentiation and long-term depression of retinotectal synapses from RGC axon terminals to the bipolar cell inputs on their dendrites (Du et al. 2009). The latter axon-dendrite signaling occurred with a time course of minutes over distances of a few hundred microns and was sensitive to colchicine treatment, suggesting active microtubule-dependent transport of signals.

In cultured neurons, local contact of a neurite with a target cell (Goldberg & Schacher 1987) or laminin-coated surface (Esch et al. 1999, Menager et al. 2004) or local perfusion of a neurite with forskolin (Zheng et al. 1994) each promote local axon formation as well as growth inhibition of distant neurites of the same neuron, suggesting the existence of a long-range signal that inhibits axon formation. The nature of this inhibitory signal remains unknown. Recent studies of cyclic nucleotide signaling in developing neurons have shown that local cAMP elevation in one neurite of the unpolarized neuron may generate a long-range negative signal that would result in cAMP reduction in all other neurites of the same neuron (Shelly et al. 2010). This long-range cAMP self-suppression would ensure the formation of only one axon and multiple dendrites, but the signal that causes the distant suppression of cAMP remains to be elucidated.

CONCLUDING REMARKS

Studies based on cultured hippocampal neurons have provided us with extensive knowledge on the early events leading to axon formation. In contrast, very little is known about events that are specific to dendrite formation. Dendrite formation could be a default pathway in the absence of axon formation. Alternatively, additional dendrite-specific processes may be required to seal the dendritic fate and to promote dendrite growth. In a simple hypothesis based on the Turing model of polarization, the local activator represents the axon determinant, and the long-range inhibitor may serve as the dendrite determinant. One may further simplify the model by assuming that if the axon determinant is limited, its local accumulation in one neurite could deplete the axon determinant in other neurites without a long-range inhibitor. Although these scenarios of default dendrite determination remain to be considered, some evidence indicates that the specific processes related to dendrite growth differ from those underlying axon growth. In newly polarized, cultured hippocampal neurons, BDNF promotes axon growth but suppresses dendrite growth, whereas Sema3A had exactly the opposite effects (Shelly et al. 2011). When axons and dendrites are being formed in the developing cortex, Sema3A signaling promotes radial migration of the leading process (dendrite) of developing cortical neurons (Chen et al. 2008) and orients their dendritic arbors toward the pial surface (Polleux et al. 2000), while the trailing process (axon) is directed in the opposite direction. Thus our view on the sequence of axon/dendrite polarization, with axon formation playing the initial and dominant role, may reflect a bias due to the use of a hippocampal culture model, where the first visible polarization event is the appearance of a single rapidly growing axon. In vivo, cellular events leading to dendrite formation may function together with or even precede those underlying axon formation.

Axon/dendrite differentiation in the developing brain epitomizes the problem of morphogenesis at the cellular level: how information coded in the genome interacts with the environment to develop the three-dimensional cell structure while maintaining proper spatial relationship with other cells in the brain. To understand the morphogenesis of axonal and dendritic arbors, we need to know how expression of specific gene products could lead to the spatiotemporally controlled growth and branching of the axon and dendrites through the cell's interaction with its environment. This review examines only the initial cellular events leading to axon/dendrite polarization. New conceptual frameworks and experimental approaches need to elucidate how genetic programming and cell-cell interaction achieve the spatiotemporal control of cellular trafficking underlying the morphogenesis of axonal/dendritic arbors. Finally, the functional identity of the axon and dendrites within the neural circuit is defined largely by the pattern of synaptic connections. Given the structural plasticity and activity-dependent refinement of developing connections, maturation of axons and dendrites is complete only after the establishment of functional neural circuits.

DISCLOSURE STATEMENT

The authors are not aware of any affiliations, memberships, funding, or financial holdings that might be perceived as affecting the objectivity of this review.

LITERATURE CITED

Adler CE, Fetter RD, Bargmann CI. 2006. UNC-6/Netrin induces neuronal asymmetry and defines the site of axon formation. *Nat. Neurosci.* 9:511–18

Amato S, Liu X, Zheng B, Cantley L, Rakic P, Man HY. 2011. AMP-activated protein kinase regulates neuronal polarization by interfering with PI 3-kinase localization. *Science* 332:247–51

Anava S, Greenbaum A, Ben Jacob E, Hanein Y, Ayali A. 2009. The regulative role of neurite mechanical tension in network development. *Biophys. J.* 96:1661–70

Arimura N, Kaibuchi K. 2005. Key regulators in neuronal polarity. *Neuron* 48:881–84

Arimura N, Kaibuchi K. 2007. Neuronal polarity: from extracellular signals to intracellular mechanisms. *Nat. Rev. Neurosci.* 8:194–205

Ascano M, Richmond A, Borden P, Kuruvilla R. 2009. Axonal targeting of Trk receptors via transcytosis regulates sensitivity to neurotrophin responses. *J. Neurosci.* 29:11674–85

Balgude AP, Yu X, Szymanski A, Bellamkonda RV. 2001. Agarose gel stiffness determines rate of DRG neurite extension in 3D cultures. *Biomaterials* 22:1077–84

Banker GA, Cowan WM. 1977. Rat hippocampal neurons in dispersed cell culture. *Brain Res.* 126:397–42

Barnes AP, Lilley BN, Pan YA, Plummer LJ, Powell AW, et al. 2007. LKB1 and SAD kinases define a pathway required for the polarization of cortical neurons. *Cell* 129:549–63

Barnes AP, Polleux F. 2009. Establishment of axon-dendrite polarity in developing neurons. *Annu. Rev. Neurosci.* 32:347–81

Boudeau J, Baas AF, Deak M, Morrice NA, Kieloch A, et al. 2003. MO25alpha/beta interact with STRADalpha/beta enhancing their ability to bind, activate and localize LKB1 in the cytoplasm. *EMBO J.* 22:5102–14

Bray D. 1984. Axonal growth in response to experimentally applied mechanical tension. *Dev. Biol.* 102:379–89

Buchman JJ, Tseng HC, Zhou Y, Frank CL, Xie Z, Tsai LH. 2010. Cdk5rap2 interacts with pericentrin to maintain the neural progenitor pool in the developing neocortex. *Neuron* 66:386–402

Campbell DS, Holt CE. 2001. Chemotropic responses of retinal growth cones mediated by rapid local protein synthesis and degradation. *Neuron* 32:1013–26

Chada S, Lamoureux P, Buxbaum RE, Heidemann SR. 1997. Cytomechanics of neurite outgrowth from chick brain neurons. *J. Cell Sci.* 110(Pt. 10):1179–86

Chen G, Sima J, Jin M, Wang KY, Xue XJ, et al. 2008. Semaphorin-3A guides radial migration of cortical neurons during development. *Nat. Neurosci.* 11:36–44

Chen Y, Corriden R, Inoue Y, Yip L, Hashiguchi N, et al. 2006a. ATP release guides neutrophil chemotaxis via P2Y2 and A3 receptors. *Science* 314:1792–95

Chen YM, Wang QJ, Hu HS, Yu PC, Zhu J, et al. 2006b. Microtubule affinity-regulating kinase 2 functions downstream of the PAR-3/PAR-6/atypical PKC complex in regulating hippocampal neuronal polarity. *Proc. Natl. Acad. Sci. USA* 103:8534–39

Cheng PL, Lu H, Shelly M, Gao H, Poo MM. 2011a. Phosphorylation of E3 ligase Smurf1 switches its substrate preference in support of axon development. *Neuron* 69:231–43

Cheng PL, Song AH, Wong YH, Wang S, Zhang X, Poo MM. 2011b. Self-amplifying autocrine actions of BDNF in axon development. *Proc. Natl. Acad. Sci. USA* 108:18430–35

Da Silva JS, Hasegawa T, Miyagi T, Dotti CG, Abad-Rodriguez J. 2005. Asymmetric membrane ganglioside sialidase activity specifies axonal fate. *Nat. Neurosci.* 8:606–15

Da Silva JS, Medina M, Zuliani C, Di Nardo A, Witke W, Dotti CG. 2003. RhoA/ROCK regulation of neuritogenesis via profilin IIa-mediated control of actin stability. *J. Cell Biol.* 162:1267–79

de Anda FC, Pollarolo G, Da Silva JS, Camoletto PG, Feiguin F, Dotti CG. 2005. Centrosome localization determines neuronal polarity. *Nature* 436:704–8

Dennerll TJ, Joshi HC, Steel VL, Buxbaum RE, Heidemann SR. 1988. Tension and compression in the cytoskeleton of PC-12 neurites. II: Quantitative measurements. *J. Cell Biol.* 107:665–74

Du J, Feng L, Yang F, Lu B. 2000. Activity- and Ca(2+)-dependent modulation of surface expression of brain-derived neurotrophic factor receptors in hippocampal neurons. *J. Cell Biol.* 150:1423–34

Du JL, Wei HP, Wang ZR, Wong ST, Poo MM. 2009. Long-range retrograde spread of LTP and LTD from optic tectum to retina. *Proc. Natl. Acad. Sci. USA* 106:18890–96

Engler AJ, Sen S, Sweeney HL, Discher DE. 2006. Matrix elasticity directs stem cell lineage specification. *Cell* 126:677–89

Esch T, Lemmon V, Banker G. 1999. Local presentation of substrate molecules directs axon specification by cultured hippocampal neurons. *J. Neurosci.* 19:6417–26

Esch T, Lemmon V, Banker G. 2000. Differential effects of NgCAM and N-cadherin on the development of axons and dendrites by cultured hippocampal neurons. *J. Neurocytol.* 29:215–23

Fivaz M, Bandara S, Inoue T, Meyer T. 2008. Robust neuronal symmetry breaking by Ras-triggered local positive feedback. *Curr. Biol.* 18:44–50

Gao FB. 2007. Molecular and cellular mechanisms of dendritic morphogenesis. *Curr. Opin. Neurobiol.* 17:525–32

Gao WQ, Hatten ME. 1993. Neuronal differentiation rescued by implantation of Weaver granule cell precursors into wild-type cerebellar cortex. *Science* 260:367–69

Gärtner A, Huang X, Hall A. 2006. Neuronal polarity is regulated by glycogen synthase kinase-3 (GSK-3beta) independently of Akt/PKB serine phosphorylation. *J. Cell Sci.* 119:3927–34

Garvalov BK, Flynn KC, Neukirchen D, Meyn L, Teusch N, et al. 2007. Cdc42 regulates cofilin during the establishment of neuronal polarity. *J. Neurosci.* 27:13117–29

Ge X, Frank CL, Calderon de Anda F, Tsai LH. 2010. Hook3 interacts with PCM1 to regulate pericentriolar material assembly and the timing of neurogenesis. *Neuron* 65:191–203

Georges PC, Miller WJ, Meaney DF, Sawyer ES, Janmey PA. 2006. Matrices with compliance comparable to that of brain tissue select neuronal over glial growth in mixed cortical cultures. *Biophys. J.* 90:3012–18

Gierer A, Meinhardt H. 1972. A theory of biological pattern formation. *Kybernetik* 12:30–39

Goepfert C, Sundberg C, Sévigny J, Enjyoji K, Hoshi T, et al. 2001. Disordered cellular migration and angiogenesis in cd39-null mice. *Circulation* 104:3109–15

Goldberg DJ, Schacher S. 1987. Differential growth of the branches of a regenerating bifurcate axon is associated with differential axonal transport of organelles. *Dev. Biol.* 124:35–40

Goslin K, Banker G. 1989. Experimental observations on the development of polarity by hippocampal neurons in culture. *J. Cell Biol.* 108:1507–16

Gunn JW, Turner SD, Mann BK. 2005. Adhesive and mechanical properties of hydrogels influence neurite extension. *J. Biomed. Mater. Res. A* 72:91–97

Guo W, Jiang H, Gray V, Dedhar S, Rao Y. 2007. Role of the integrin-linked kinase (ILK) in determining neuronal polarity. *Dev. Biol.* 306:457–68

Haapasalo A, Sipola I, Larsson K, Akerman KE, Stoilov P, et al. 2002. Regulation of TRKB surface expression by brain-derived neurotrophic factor and truncated TRKB isoforms. *J. Biol. Chem.* 277:43160–67

Harrison RG. 1910. The outgrowth of the nerve fiber as a mode of protoplasmic movement. *J. Exp. Zool.* 9:787–846

Hatanaka Y, Murakami F. 2002. In vitro analysis of the origin, migratory behavior, and maturation of cortical pyramidal cells. *J. Comp. Neurol.* 454:1–14

Henle SJ, Wang G, Liang E, Wu M, Poo MM, Henley JR. 2011. Asymmetric PI(3,4,5)P3 and Akt signaling mediates chemotaxis of axonal growth cones. *J. Neurosci.* 31:7016–27

Hilliard MA, Bargmann CI. 2006. Wnt signals and frizzled activity orient anterior-posterior axon outgrowth in *C. elegans*. *Dev. Cell* 10:379–90

Honda R, Yasuda H. 1999. Association of p19(ARF) with Mdm2 inhibits ubiquitin ligase activity of Mdm2 for tumor suppressor p53. *EMBO J.* 18:22–27

Hoogenraad CC, Bradke F. 2009. Control of neuronal polarity and plasticity—a renaissance for microtubules? *Trends Cell Biol.* 19:669–76

Horiguchi K, Hanada T, Fukui Y, Chishti AH. 2006. Transport of PIP3 by GAKIN, a kinesin-3 family protein, regulates neuronal cell polarity. *J. Cell Biol.* 174:425–36

Horton AC, Ehlers MD. 2003. Neuronal polarity and trafficking. *Neuron* 40:277–95

Huang EJ, Reichardt LF. 2003. Trk receptors: roles in neuronal signal transduction. *Annu. Rev. Biochem.* 72:609–42

Inagaki N, Chihara K, Arimura N, Menager C, Kawano Y, et al. 2001. CRMP-2 induces axons in cultured hippocampal neurons. *Nat. Neurosci.* 4:781–82

Ishihara A, Saito H, Abe K. 1994. Transforming growth factor-beta 1 and -beta 2 promote neurite sprouting and elongation of cultured rat hippocampal neurons. *Brain Res.* 639:21–25

Jan YN, Jan LY. 2010. Branching out: mechanisms of dendritic arborization. *Nat. Rev. Neurosci.* 11:316–28

Jiang H, Guo W, Liang X, Rao Y. 2005. Both the establishment and the maintenance of neuronal polarity require active mechanisms: critical roles of GSK-3beta and its upstream regulators. *Cell* 120:123–35

Kassai H, Terashima T, Fukaya M, Nakao K, Sakahara M, et al. 2008. Rac1 in cortical projection neurons is selectively required for midline crossing of commissural axonal formation. *Eur. J. Neurosci.* 28:257–67

Kato M, Kito K, Ota K, Ito T. 2009. Remodeling of the SCF complex-mediated ubiquitination system by compositional alteration of incorporated F-box proteins. *Proteomics* 10:115–23

Kimura T, Watanabe H, Iwamatsu A, Kaibuchi K. 2005. Tubulin and CRMP-2 complex is transported via Kinesin-1. *J. Neurochem.* 93:1371–82

Kishi M, Pan YA, Crump JG, Sanes JR. 2005. Mammalian SAD kinases are required for neuronal polarization. *Science* 307:929–32

Koszalka P, Ozuyaman B, Huo Y, Zernecke A, Flogel U, et al. 2004. Targeted disruption of cd73/ecto-5′-nucleotidase alters thromboregulation and augments vascular inflammatory response. *Circ. Res.* 95:814–21

Krüttgen A, Möller JC, Heymach JV Jr, Shooter EM. 1998. Neurotrophins induce release of neurotrophins by the regulated secretory pathway. *Proc. Natl. Acad. Sci. USA* 95:9614–19

Kunda P, Paglini G, Quiroga S, Kosik K, Caceres A. 2001. Evidence for the involvement of Tiam1 in axon formation. *J. Neurosci.* 21:2361–72

Kutscher B, Devreotes P, Iglesias PA. 2004. Local excitation, global inhibition mechanism for gradient sensing: an interactive applet. *Sci. STKE* 2004:pl3

Lalli G. 2009. RalA and the exocyst complex influence neuronal polarity through PAR-3 and aPKC. *J. Cell Sci.* 122:1499–506

Lamoureux P, Buxbaum RE, Heidemann SR. 1989. Direct evidence that growth cones pull. *Nature* 340:159–62

Lamoureux P, Ruthel G, Buxbaum RE, Heidemann SR. 2002. Mechanical tension can specify axonal fate in hippocampal neurons. *J. Cell Biol.* 159:499–508

Lamoureux P, Steel VL, Regal C, Adgate L, Buxbaum RE, Heidemann SR. 1990. Extracellular matrix allows PC12 neurite elongation in the absence of microtubules. *J. Cell Biol.* 110:71–79

Lerman O, Ben-Zvi A, Yagil Z, Behar O. 2007. Semaphorin3A accelerates neuronal polarity in vitro and in its absence the orientation of DRG neuronal polarity in vivo is distorted. *Mol. Cell Neurosci.* 36:222–34

Li YH, Werner H, Püschel AW. 2008. Rheb and mTOR regulate neuronal polarity through Rap1B. *J. Biol. Chem.* 283:33784–92

Liebl DJ, Morris CJ, Henkemeyer M, Parada LF. 2003. mRNA expression of ephrins and Eph receptor tyrosine kinases in the neonatal and adult mouse central nervous system. *J. Neurosci. Res.* 71:7–22

Mai J, Fok L, Gao H, Zhang X, Poo MM. 2009. Axon initiation and growth cone turning on bound protein gradients. *J. Neurosci.* 29:7450–58

Mann F, Miranda E, Weinl C, Harmer E, Holt CE. 2003. B-type Eph receptors and ephrins induce growth cone collapse through distinct intracellular pathways. *J. Neurobiol.* 57:323–36

Meinhardt H. 1999. Orientation of chemotactic cells and growth cones: models and mechanisms. *J. Cell Sci.* 112(Pt. 17):2867–74

Menager C, Arimura N, Fukata Y, Kaibuchi K. 2004. PIP3 is involved in neuronal polarization and axon formation. *J. Neurochem.* 89:109–18

Meyer-Franke A, Wilkinson GA, Kruttgen A, Hu M, Munro E, et al. 1998. Depolarization and cAMP elevation rapidly recruit TrkB to the plasma membrane of CNS neurons. *Neuron* 21:681–93

Miki H, Suetsugu S, Takenawa T. 1998. WAVE, a novel WASP-family protein involved in actin reorganization induced by Rac. *EMBO J.* 17:6932–41

Montcouquiol M, Crenshaw EB 3rd, Kelley MW. 2006. Noncanonical Wnt signaling and neural polarity. *Annu. Rev. Neurosci.* 29:363–86

Mori T, Wada T, Suzuki T, Kubota Y, Inagaki N. 2007. Singar1, a novel RUN domain-containing protein, suppresses formation of surplus axons for neuronal polarity. *J. Biol. Chem.* 282:19884–93

Namba T, Nakamuta S, Funahashi Y, Kaibuchi K. 2011. The role of selective transport in neuronal polarization. *Dev. Neurobiol.* 71:445–57

Ng J, Nardine T, Harms M, Tzu J, Goldstein A, et al. 2002. Rac GTPases control axon growth, guidance and branching. *Nature* 416:442–47

Nishimura T, Fukata Y, Kato K, Yamaguchi T, Matsuura Y, et al. 2003. CRMP-2 regulates polarized Numb-mediated endocytosis for axon growth. *Nat. Cell Biol.* 5:819–26

Nishimura T, Yamaguchi T, Kato K, Yoshizawa M, Nabeshima Y, et al. 2005. PAR-6-PAR-3 mediates Cdc42-induced Rac activation through the Rac GEFs STEF/Tiam1. *Nat. Cell Biol.* 7:270–77

Noctor SC, Martinez-Cerdeño V, Ivic L, Kriegstein AR. 2004. Cortical neurons arise in symmetric and asymmetric division zones and migrate through specific phases. *Nat. Neurosci.* 7:136–44

Oinuma I, Katoh H, Negishi M. 2007. R-Ras controls axon specification upstream of glycogen synthase kinase-3beta through integrin-linked kinase. *J. Biol. Chem.* 282:303–18

O'Toole M, Lamoureux P, Miller KE. 2008. A physical model of axonal elongation: force, viscosity, and adhesions govern the mode of outgrowth. *Biophys. J.* 94:2610–20

Ozdamar B, Bose R, Barrios-Rodiles M, Wang HR, Zhang Y, Wrana JL. 2005. Regulation of the polarity protein Par6 by TGFbeta receptors controls epithelial cell plasticity. *Science* 307:1603–9

Pan, CL, Howell JE, Clark SG, Hilliard M, Cordes S, Bargmann CL, Garriga G. 2006. Multiple Wnts and frizzled receptors regulate anteriorly directed cell and growth cone migrations in Caenorhabditis elegans. *Dev. Cell* 10:367–77

Parent CA, Devreotes PN. 1999. A cell's sense of direction. *Science* 284:765–70

Pfister BJ, Iwata A, Meaney DF, Smith DH. 2004. Extreme stretch growth of integrated axons. *J. Neurosci.* 24:7978–83

Pitto M, Mutoh T, Kuriyama M, Ferraretto A, Palestini P, Masserini M. 1998. Influence of endogenous GM1 ganglioside on TrkB activity, in cultured neurons. *FEBS Lett.* 439:93–96

Polleux F, Giger RJ, Ginty DD, Kolodkin AL, Ghosh A. 1998. Patterning of cortical efferent projections by semaphorin-neuropilin interactions. *Science* 282:1904–6

Polleux F, Morrow T, Ghosh A. 2000. Semaphorin 3A is a chemoattractant for cortical apical dendrites. *Nature* 404:567–73

Prasad BC, Clark SG. 2006. Wnt signaling establishes anteroposterior neuronal polarity and requires retromer in *C. elegans. Development* 133:1757–66

Previtera ML, Langhammer CG, Langrana NA, Firestein BL. 2010. Regulation of dendrite arborization by substrate stiffness is mediated by glutamate receptors. *Ann. Biomed. Eng.* 38: 3733–43

Purves D. 1975. Functional and structural changes in mammalian sympathetic neurones following interruption of their axons. *J. Physiol.* 252:429–63

Purves D, Lichtman JW. 1985. *Principles of Neural Developement*. Sunderland, MA: Sinauer

Purves D, Njå A. 1978. Trophic maintenance of synaptic connections in autonomic ganglia. In *Neuronal Plasticity*, ed. C Cotman, pp. 27–47. New York: Raven

Rakic P. 1971. Neuron-glia relationship during granule cell migration in developing cerebellar cortex. A Golgi and electronmicroscopic study in Macacus rhesus. *J. Comp. Neurol.* 141:283–312

Ramón y Cajal S. 1898. *Histology of the Nervous System of Man and Vertebrates*. New York: Oxford Univ. Press

Randlett O, Poggi L, Zolessi FR, Harris WA. 2011. The oriented emergence of axons from retinal ganglion cells is directed by laminin contact in vivo. *Neuron* 70:266–80

Rodriguez JA, Piddini E, Hasegawa T, Miyagi T, Dotti CG. 2001. Plasma membrane ganglioside sialidase regulates axonal growth and regeneration in hippocampal neurons in culture. *J. Neurosci.* 21:8387–95

Santos Da Silva J, Schubert V, Dotti CG. 2004. RhoA, Rac1, and cdc42 intracellular distribution shift during hippocampal neuron development. *Mol. Cell Neurosci.* 27:1–7

Schwamborn JC, Khazaei MR, Püschel AW. 2007a. The interaction of mPar3 with the ubiquitin ligase Smurf2 is required for the establishment of neuronal polarity. *J. Biol. Chem.* 282:35259–68

Schwamborn JC, Müller M, Becker AH, Püschel AW. 2007b. Ubiquitination of the GTPase Rap1B by the ubiquitin ligase Smurf2 is required for the establishment of neuronal polarity. *EMBO J.* 26:1410–22

Schwamborn JC, Püschel AW. 2004. The sequential activity of the GTPases Rap1B and Cdc42 determines neuronal polarity. *Nat. Neurosci.* 7:923–29

Schwartz MA. 2009. Cell biology. The force is with us. *Science* 323:588–89

Shelly M, Cancedda L, Heilshorn S, Sumbre G, Poo MM. 2007. LKB1/STRAD promotes axon initiation during neuronal polarization. *Cell* 129:565–77

Shelly M, Cancedda L, Lim BK, Popescu AT, Cheng PL, et al. 2011. Semaphorin3A regulates neuronal polarization by suppressing axon formation and promoting dendrite growth. *Neuron* 71:433–46

Shelly M, Lim BK, Cancedda L, Heilshorn SC, Gao H, Poo MM. 2010. Local and long-range reciprocal regulation of cAMP and cGMP in axon/dendrite formation. *Science* 327:547–52

Shi SH, Cheng T, Jan LY, Jan YN. 2004. APC and GSK-3beta are involved in mPar3 targeting to the nascent axon and establishment of neuronal polarity. *Curr. Biol.* 14:2025–32

Shi SH, Jan LY, Jan YN. 2003. Hippocampal neuronal polarity specified by spatially localized mPar3/mPar6 and PI 3-kinase activity. *Cell* 112:63–75

Shimada T, Toriyama M, Uemura K, Kamiguchi H, Sugiura T, et al. 2008. Shootin1 interacts with actin retrograde flow and L1-CAM to promote axon outgrowth. *J. Cell Biol.* 181:817–29

Solecki DJ, Trivedi N, Govek EE, Kerekes RA, Gleason SS, Hatten ME. 2009. Myosin II motors and F-actin dynamics drive the coordinated movement of the centrosome and soma during CNS glial-guided neuronal migration. *Neuron* 63:63–80

Stiess M, Maghelli N, Kapitein LC, Gomis-Rüth S, Wilsch-Bräuninger M, et al. 2010. Axon extension occurs independently of centrosomal microtubule nucleation. *Science* 327:704–7

Swaney KF, Huang CH, Devreotes PN. 2010. Eukaryotic chemotaxis: a network of signaling pathways controls motility, directional sensing, and polarity. *Annu. Rev. Biophys.* 39:265–89

Tahirovic S, Hellal F, Neukirchen D, Hindges R, Garvalov BK, et al. 2010. Rac1 regulates neuronal polarization through the WAVE complex. *J. Neurosci.* 30:6930–43

Toriyama M, Sakumura Y, Shimada T, Ishii S, Inagaki N. 2010. A diffusion-based neurite length-sensing mechanism involved in neuronal symmetry breaking. *Mol. Syst. Biol.* 6:394

Toriyama M, Shimada T, Kim KB, Mitsuba M, Nomura E, et al. 2006. Shootin1: a protein involved in the organization of an asymmetric signal for neuronal polarization. *J. Cell Biol.* 175:147–57

Tsai LH, Gleeson JG. 2005. Nucleokinesis in neuronal migration. *Neuron* 46:383–88

Turing AM. 1952. The chemical basis of morphogenesis. *Proc. R. Soc. Lond. B* 237:37–72

Wang F, Herzmark P, Weiner OD, Srinivasan S, Servant G, Bourne HR. 2002. Lipid products of PI(3)Ks maintain persistent cell polarity and directed motility in neutrophils. *Nat. Cell Biol.* 4:513–18

Weiner OD. 2002. Regulation of cell polarity during eukaryotic chemotaxis: the chemotactic compass. *Curr. Opin. Cell Biol.* 14:196–202

Weiss P. 1941. Nerve patterns: the mechanics of nerve growth. *Growth* (Third Growth Symp. Suppl.) 5:163–203

Whitford KL, Marillat V, Stein E, Goodman CS, Tessier-Lavigne M, et al. 2002. Regulation of cortical dendrite development by Slit-Robo interactions. *Neuron* 33:47–61

Wong J, Webster MJ, Cassano H, Weickert CS. 2009. Changes in alternative brain-derived neurotrophic factor transcript expression in the developing human prefrontal cortex. *Eur. J. Neurosci.* 29:1311–22

Xiong Y, Huang CH, Iglesias PA, Devreotes PN. 2010. Cells navigate with a local-excitation, global-inhibition-biased excitable network. *Proc. Natl. Acad. Sci. USA* 107:17079–86

Yamashita T, Tucker KL, Barde YA. 1999. Neurotrophin binding to the p75 receptor modulates Rho activity and axonal outgrowth. *Neuron* 24:585–93

Yan D, Guo L, Wang Y. 2006. Requirement of dendritic Akt degradation by the ubiquitin-proteasome system for neuronal polarity. *J. Cell Biol.* 174:415–24

Yi JJ, Barnes AP, Hand R, Polleux F, Ehlers MD. 2010. TGF-beta signaling specifies axons during brain development. *Cell* 142:144–57

Yoshimura T, Arimura N, Kawano Y, Kawabata S, Wang S, Kaibuchi K. 2006. Ras regulates neuronal polarity via the PI3-kinase/Akt/GSK-3beta/CRMP-2 pathway. *Biochem. Biophys. Res. Commun.* 340:62–68

Yoshimura T, Kawano Y, Arimura N, Kawabata S, Kikuchi A, Kaibuchi K. 2005. GSK-3beta regulates phosphorylation of CRMP-2 and neuronal polarity. *Cell* 120:137–49

Yoshimura Y, Terabayashi T, Miki H. 2010. Par1b/MARK2 phosphorylates kinesin-like motor protein GAKIN/KIF13B to regulate axon formation. *Mol. Cell. Biol.* 30:2206–19

Zallen JA, Cohen Y, Hudson AM, Cooley L, Wieschaus E, Schejter ED. 2002. SCAR is a primary regulator of Arp2/3-dependent morphological events in *Drosophila*. *J. Cell Biol.* 156:689–701

Zhang X, Zhu J, Yang GY, Wang QJ, Qian L, et al. 2007. Dishevelled promotes axon differentiation by regulating atypical protein kinase C. *Nat. Cell Biol.* 9:743–54

Zhao L, Sheng AL, Huang SH, Yin YX, Chen B, et al. 2009. Mechanism underlying activity-dependent insertion of TrkB into the neuronal surface. *J. Cell Sci.* 122:3123–36

Zheng J, Lamoureux P, Santiago V, Dennerll T, Buxbaum RE, Heidemann SR. 1991. Tensile regulation of axonal elongation and initiation. *J. Neurosci.* 11:1117–25

Zheng JQ, Zheng Z, Poo M. 1994. Long-range signaling in growing neurons after local elevation of cyclic AMP-dependent activity. *J. Cell Biol.* 127:1693–701

Zolessi FR, Poggi L, Wilkinson CJ, Chien CB, Harris WA. 2006. Polarization and orientation of retinal ganglion cells in vivo. *Neural Dev.* 1:2

Mechanisms of Gamma Oscillations

György Buzsáki[1,2] and Xiao-Jing Wang[3]

[1]Center for Molecular and Behavioral Neuroscience, Rutgers, The State University of New Jersey, Newark, New Jersey 07102

[2]The Neuroscience Institute, New York University, School of Medicine, New York, NY 10016; email: Gyorgy.Buzsaki@nyumc.org

[3]Department of Neurobiology and Kavli Institute of Neuroscience, Yale University School of Medicine, New Haven, Connecticut 06520; email: xjwang@yale.edu

Annu. Rev. Neurosci. 2012. 35:203–25

First published online as a Review in Advance on March 20, 2012

The *Annual Review of Neuroscience* is online at neuro.annualreviews.org

This article's doi:
10.1146/annurev-neuro-062111-150444

0147-006X/12/0721-0203$20.00

Keywords

inhibitory interneurons, interneuronal network, excitatory-inhibitory loop, spike timing, dynamical cell assembly, irregular spiking, cross-frequency coupling, long-distance communication

Abstract

Gamma rhythms are commonly observed in many brain regions during both waking and sleep states, yet their functions and mechanisms remain a matter of debate. Here we review the cellular and synaptic mechanisms underlying gamma oscillations and outline empirical questions and controversial conceptual issues. Our main points are as follows: First, gamma-band rhythmogenesis is inextricably tied to perisomatic inhibition. Second, gamma oscillations are short-lived and typically emerge from the coordinated interaction of excitation and inhibition, which can be detected as local field potentials. Third, gamma rhythm typically concurs with irregular firing of single neurons, and the network frequency of gamma oscillations varies extensively depending on the underlying mechanism. To document gamma oscillations, efforts should be made to distinguish them from mere increases of gamma-band power and/or increased spiking activity. Fourth, the magnitude of gamma oscillation is modulated by slower rhythms. Such cross-frequency coupling may serve to couple active patches of cortical circuits. Because of their ubiquitous nature and strong correlation with the "operational modes" of local circuits, gamma oscillations continue to provide important clues about neuronal population dynamics in health and disease.

Contents

INTRODUCTION

The precise timing of neuronal-spike discharges is believed to be important for coding of information (O'Keefe & Recce 1993, Buzsáki & Chrobak 1995, Singer & Gray 1995, Singer 1999). The ability of various neuron types to time their action potentials with millisecond precision depends largely on the presence of fast membrane potential fluctuations (Mainen & Sejnowski 1995, Haider & McCormick 2009). In the intact brain, such high-frequency patterns are often brought about by various endogenous oscillations, the most ubiquitous of which are rhythms in the gamma-frequency range (30–90 Hz) (see Origin and Definition of Gamma Oscillation, sidebar below).

Gamma oscillations: synchronous network rhythm in 30–90 Hz that is minimally defined by an autocorrelation function and/or continuous Gabor transform

LFP: local field potential

GABA: gamma-aminobutyric acid

Numerous excellent reviews have discussed the biological processes underlying gamma oscillations (Gray 1994, Whittington et al. 2000, Laurent 2002, Traub et al. 2002, Bartos et al. 2007, Tiesinga & Sejnowski 2009, Wang 2010) as well as their role in cognitive operations (Singer & Gray 1995; Engel et al. 2001; Varela et al. 2001; Fries 2005, 2009; Wang 2010) and disease (Llinás et al. 1999, Lewis et al. 2005, Uhlhaas & Singer 2006). The present review focuses on the cellular-synaptic mechanisms of gamma oscillations, their cell-assembly-forming ability in the intact brain, and the subtypes of gamma rhythms. It also examines how gamma-reflected local-circuit operations are temporally coordinated by slower rhythms.

ARE CELL ASSEMBLIES DYNAMICALLY ORGANIZED IN GAMMA CYCLES?

To appreciate the physiological function of the gamma cycle in neural networks, we need to examine the spiking patterns of neurons at this timescale. The exact timing of neuronal spikes can be related to environmental stimuli, overt behavior, local field potential (LFP), or spiking activity of other neurons. Each of these comparisons provides a different "optimum" time window. The best prediction is obtained when information about the spike times of partner neurons are available in the 10- and 30-ms window (**Figure 1**) (Jensen & Lisman 1996, Borgers & Kopell 2003, Harris et al. 2003, Lisman 2005), i.e., the time window corresponding approximately to a gamma cycle. Neuronal assemblies, i.e., transient neuronal partnerships, can be active repeatedly in successive gamma cycles, or different assemblies can alternate in a rapid sequence.

The gamma-cycle-related lifetime of the cell assembly is closely related to several biophysical properties of neurons, including the time constant of gamma-aminobutyric acid $(GABA)_A$ and α-amino-3-hydroxy-5-methyl-4-isoxazolepropionic acid (AMPA) receptors (Johnston & Wu 1994), the membrane time constant of cortical pyramidal cells (Destexhe & Paré 1999, Leger et al. 2005), and the critical time window of spike-timing-dependent plasticity (Magee & Johnston 1997, Markram et al. 1997). Because these parameters determine the neuron's ability to integrate inputs from multiple upstream sources, a hypothesized functional role of the cell assembly is to bring together sufficient numbers of peer neurons so that their collective spiking can discharge the postsynaptic neuron (Harris et al. 2003). Consequently, from the point of view of the downstream ("reader" or "integrator") cell, ensemble activity of upstream neurons whose spikes occur within the gamma-cycle window is classified as a single event (Buzsaki 2010). Upstream neurons whose spikes fall outside this time window become part of another transient assembly.

ORIGIN AND DEFINITION OF GAMMA OSCILLATION

Berger (1929) introduced the Greek letters alpha and beta to refer to the larger amplitude rhythmic patterns below 12 Hz and the lower amplitude faster than 12-Hz patterns, respectively. Jasper & Andrews (1938) first used the term gamma waves to designate low-amplitude beta-like waves at 35–45 Hz. Other synonyms referring to this band are the 40-Hz oscillation or cognitive rhythm, both introduced by Das & Gastaut (1955). The phrase gamma oscillation became popular in the 1980s, mostly through papers by Walter Freeman (Bressler & Freeman 1980). Proper taxonomy of brain rhythms should eventually be based on mechanisms. Because mechanisms are not fully understood in most cases, the names of the brain rhythms respect historical traditions. We refer to periodic events in the 30–90-Hz band as gamma oscillations and the band above this frequency as epsilon (ε) (Freeman 2007) (also see the **Supplemental Text**: follow the **Supplemental Material link** in the online version of this article at **http://www.annualreviews.org**.).

MODELS OF GAMMA OSCILLATIONS

The similar kinetics of gamma-frequency oscillations in a variety of different brain regions and species have provided clues and constraints about the requirements of their supporting mechanisms. Gamma oscillations have been described in several areas of the neocortex (Gray et al. 1989, Murthy & Fetz 1992, Fries et al. 2001, Sirota et al. 2008), entorhinal cortex (Chrobak & Buzsáki 1998), amygdala (Halgren et al. 1977, Popescu et al. 2009), hippocampus (Buzsáki et al. 1983, Bragin et al. 1995, Whittington et al. 1995, Mann et al. 2005), striatum (Berke et al. 2004, Tort et al. 2008), olfactory bulb (Adrian 1942, Freeman 1975), and thalamus (Pinault & Deschénes 1992) as well as other areas. Common denominators of these brain regions are the presence of inhibitory interneurons and their actions through $GABA_A$ synapses. Synchronization of neurons is substantially more effective by perisomatic inhibitory postsynaptic potentials (IPSPs) than dendritic excitatory (E)PSPs (Lytton & Sejnowski 1991). From these considerations, it is reasonable to assume that a key ingredient of gamma oscillations is $GABA_A$ receptor–mediated inhibition.

I-I Model

Only three requirements are needed for gamma oscillations to emerge, as illustrated by a "stripped-down" network model consisting of only inhibitory interneurons (**Figure 2*a***) (Wang & Rinzel 1992, Whittington et al. 1995, Wang & Buzsáki 1996, Traub et al. 1996b): mutually connected inhibitory interneurons, a time constant provided by $GABA_A$ receptors, and sufficient drive to induce spiking in the interneurons. Gamma oscillations in inhibitory-inhibitory (I-I) neuron models can emerge in two different ways (see Irregular Activity of Single Neurons and Gamma Oscillations of Neuron Groups, sidebar below). When the input drive is relatively tonic, neurons can fire spikes with a well-defined periodicity (**Figure 2*a***) (Kopell & Ermentrout 2002). By contrast, when neurons receive stochastic inputs and fire spikes irregularly, sufficiently strong recurrent synaptic interactions will make the asynchronous state unstable against random fluctuations, and oscillations emerge (**Figure 2*b***) (Brunel & Hakim 1999, Brunel 2000, Brunel & Wang 2003, Geisler et al. 2005, Ardid et al. 2010, Economo & White 2012). In both cases, the emerging synchrony

I-I model: synchronization by mutual inhibition between interneurons

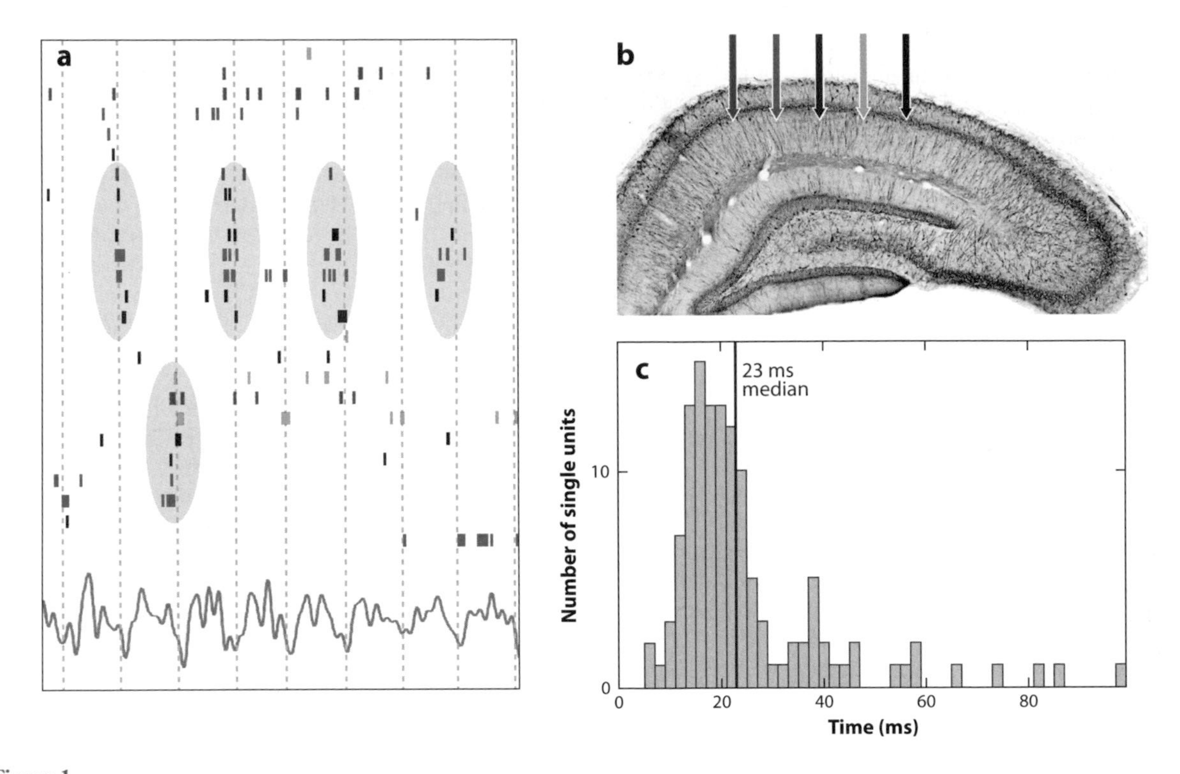

Figure 1

Dynamical cell assemblies are organized in gamma waves. (*a*) Raster plot of a subset of hippocampal pyramidal cells that were active during a 1-s period of spatial exploration on an open field out of a larger set of simultaneously recorded neurons, ordered by stochastic search over all possible orderings to highlight the temporal relationship between anatomically distributed neurons. Color-coded ticks (spikes) refer to recording locations shown in panel *b*. Vertical lines indicate troughs of theta waves (*bottom trace*). Cell-assembly organization is visible, with repeatedly synchronous firing of some subpopulations (*circled*). (*c*) Spike timing is predictable from peer activity. Distribution of timescales at which peer activity optimally improved spike-time prediction of a given cell, shown for all cells. The median optimal timescale is 23 ms (*red line*). Based on Harris et al. (2003).

is caused when a subset of the interneurons begins to discharge together and generates synchronous IPSPs in the partner neurons. In turn, the inhibited neurons will spike again with increased probability when $GABA_A$ receptor–mediated hyperpolarization has decayed, and the cycle repeats (**Figure 2*a*,*b***). Because the duration of IPSCs (inhibitory postsynaptic current) is determined by the subunit composition of the $GABA_A$ receptor (cf. Farrant & Nusser 2005), the frequency of gamma oscillations in the I-I model is determined mainly by the kinetics of the IPSPs and the net excitation of interneurons (Whittington et al. 1995, Wang & Buzsáki 1996).

In vitro experiments provided support for the sufficient role of mutual inhibition among interneurons for the generation of gamma rhythm, for instance sustained by activation of metabotropic glutamate receptors (Whittington et al. 1995). Gamma oscillations can be induced by other means as well, such as activation of muscarinic-cholinergic receptors (Fisahn et al. 1998) or kainate receptors (Fisahn et al. 2004, Hájos & Paulsen 2009). Common to all these conditions is the increased firing of synaptically coupled interneurons. When pyramidal cells and other interneuron types are added to the I-I model network, the entire network can become phase-locked to the gamma oscillations.

E-I Model

The earliest model of gamma oscillations is based on the reciprocal connections between pools of excitatory pyramidal (E) and inhibitory (I) neurons (Wilson & Cowan 1972, Freeman 1975, Leung 1982, Ermentrout & Kopell 1998, Borgers & Kopell 2003, Brunel & Wang 2003, Geisler et al. 2005). In such two-neuron pool models (**Figure 2*c***), fast excitation and the delayed feedback inhibition alternate, and with appropriate strength of excitation and inhibition, cyclic behavior may persist for a while. E-I models can also exhibit two distinct regimes, depending on whether single neurons behave periodically or highly stochastically. In the model, axon conduction and synaptic delays lead to a phase shift (~5 ms or up to 90°) between the pyramidal and interneuron spikes, and these delays determine the frequency of the gamma rhythm (Freeman 1975, Leung 1982). An appeal of the E-I model is that the delay between the timing of pyramidal cell and interneuron spikes is a prominent feature of gamma oscillations both in vivo and in vitro (**Figure 3**) (Bragin et al. 1995, Csicsvari et al. 2003, Hasenstaub et al. 2005, Mann et al. 2005, Hájos & Paulsen 2009, Tiesinga & Sejnowski 2009). In further support of the model, weakening the E-I connection by genetic knock down of AMPA receptors on fast spiking interneurons reduces the amplitude of gamma oscillations (Fuchs et al. 2007). The mainstream I-I and E-I models have been developed to explain gamma oscillations in the cortex but other gamma frequency oscillations may possibly arise from other mechanisms as well (Wang 1993, Gray & McCormick 1996, Wang 1999, Minlebaev et al. 2011).

IRREGULAR ACTIVITY OF SINGLE NEURONS AND GAMMA OSCILLATIONS OF NEURON GROUPS

A fruitful debate persists between researchers who study population gamma oscillations and ponder their functions, and researchers who study single-neuron data and observe that neuronal-spike trains are often irregular and by some measures approximate a Poisson process (Softky & Koch 1993). Recent work has offered a novel theoretical framework in which population rhythms can arise from irregularly firing neurons, thereby bridging these contrasting dynamical aspects of cortical dynamics (c.f., Wang 2010).

CELLULAR-NETWORK MECHANISMS OF GAMMA OSCILLATIONS

Perisomatic Inhibition Is Critical for Gamma Oscillations

The first support for the involvement of fast-spiking interneurons in gamma oscillations came from the correlation (spike-field coherence) between their spikes and locally recorded LFP gamma oscillations in the hippocampus of behaving rats (**Figure 3**) (Buzsáki et al. 1983). Putative fast-spiking interneurons and histologically verified parvalbumin (PV)-immunoreactive basket cells often show a broad peak in their autocorrelograms and spectrograms at gamma frequency (**Figure 3*b***), and the occurrence of their spikes follows those of the surrounding pyramidal neurons by a few milliseconds (**Figure 3*d,f***) (Bragin et al. 1995, Csicsvari et al. 2003, Mann et al. 2005, Hájos & Paulsen 2009), as in E-I models. As expected from the spike-LFP relationship (**Figure 3*a,c***) postsynaptic potentials phase-locked to the LFP gamma rhythm are present in pyramidal neurons. These gamma-correlated postsynaptic potentials in pyramidal cells reverse their polarity close to the equilibrium potential of Cl^- (**Figure 3*e,g***), indicating that the gamma-rhythm-related inhibition is mediated by $GABA_A$ receptors (Soltesz & Deschênes 1993, Whittington et al. 1995, Penttonen et al. 1998, Hasenstaub et al. 2005, Mann et al. 2005). The IPSPs paced by the PV basket cells produce coherent transmembrane fluctuations in the target pyramidal cell population (Penttonen et al. 1998, Gloveli et al. 2005, Hasenstaub et al. 2005, Mann et al. 2005, Quilichini et al. 2010) and can be detected as a strong current source in the cell-body layer (**Figure 3*d***) (Csicsvari

E-I model: synchronization by an excitatory-inhibitory loop, primarily realized by the reciprocal interaction between pyramidal neurons and interneurons

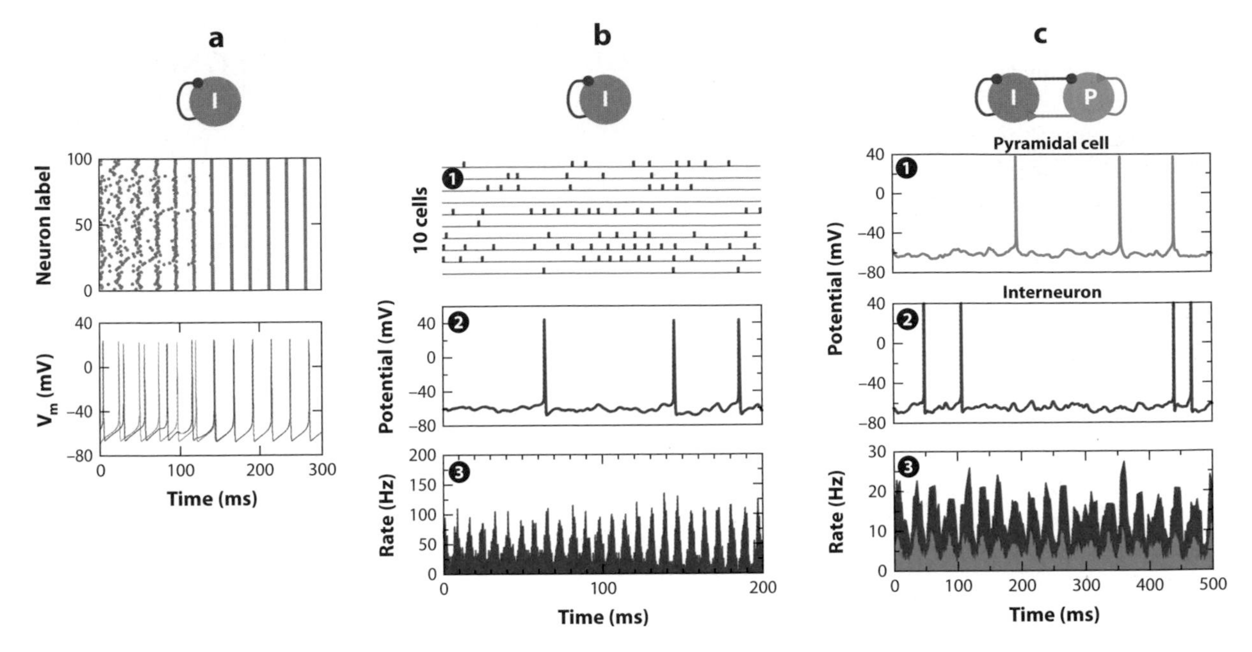

Figure 2

I-I and E-I models of gamma oscillations. (*a*) Clock-like rhythm of coupled oscillators in an interneuronal (I-I) population. (*Upper panel*) Single interneurons fire spikes periodically at ~40 Hz. Mutual inhibition via $GABA_A$ receptors quickly brings them to zero-phase synchrony; (*lower panel*) two example neurons. Adapted from Wang & Buzsáki (1996). (*b,c*) Sparsely synchronous oscillations in a neural circuit where single neuronal spiking is stochastic. Adapted from Geisler et al. (2005). (*b*) Interneuronal population in noise-dominated regime typically exhibits gamma power in the higher frequency range, in contrast to (*a*) the clock-like rhythmic case. (*c*) Reciprocally connected E-I network where pyramidal cells send fast excitation via AMPA receptors to interneurons, which in turn provide inhibition via $GABA_A$ receptors, leading to coherent oscillations in the gamma-frequency range.

Resonance: phenomenon describing a neuron or a neural circuit that is maximally responsive to an oscillatory input at a preferred frequency

et al. 2003, Mann et al. 2005). The interconnected PV-basket interneuron network with its divergent output to pyramidal cells provides an anatomical substrate for coherent timing of the pyramidal cells (**Figure 3*d***) (Kisvárday et al. 1993, Buhl et al. 1994, Sik et al. 1995). Altogether, these findings support the hypothesis that extracellularly recorded gamma waves largely correspond to synchronous IPSPs in pyramidal cells, brought about by fast-spiking interneurons (Buzsáki et al. 1983, Bragin et al. 1995, Hasenstaub et al. 2005, Freund & Katona 2007, Hájos & Paulsen 2009).

Several other findings support the critical role of fast-spiking basket neurons in gamma oscillations. Basket cells have several distinctive features among the interneuron family, including (*a*) low spike threshold (Gulyás et al. 1993), (*b*) ability to fire rapidly without fatigue (Buzsáki et al. 1983, McCormick et al. 1985, Kawaguchi & Kubota 1997), (*c*) narrow spikes conferred by a large density of KV3.1/3.2 channels (Lien & Jonas 2003), (*d*) a unique spike-conductance trajectory (Tateno & Robinson 2009), and (*e*) resonance at gamma frequency in response to stochastic excitatory conductance inputs (**Figure 4**) (Pike et al. 2000, Cardin et al. 2009, Sohal et al. 2009). Overall, these findings support the hypothesis that gamma oscillations can be induced by activation of interconnected PV interneurons by multiple means.

The involvement of other interneuron types (Freund & Buzsáki 1996, Klausberger & Somogyi 2008) in gamma generation is understood less well. Chandelier cells are likely not critical in I-I models, because they innervate only principal cells. The somatostatin-containing O-LM interneurons and Martinotti cells mainly target distal dendrites, establish few connections among themselves (Gibson et al.

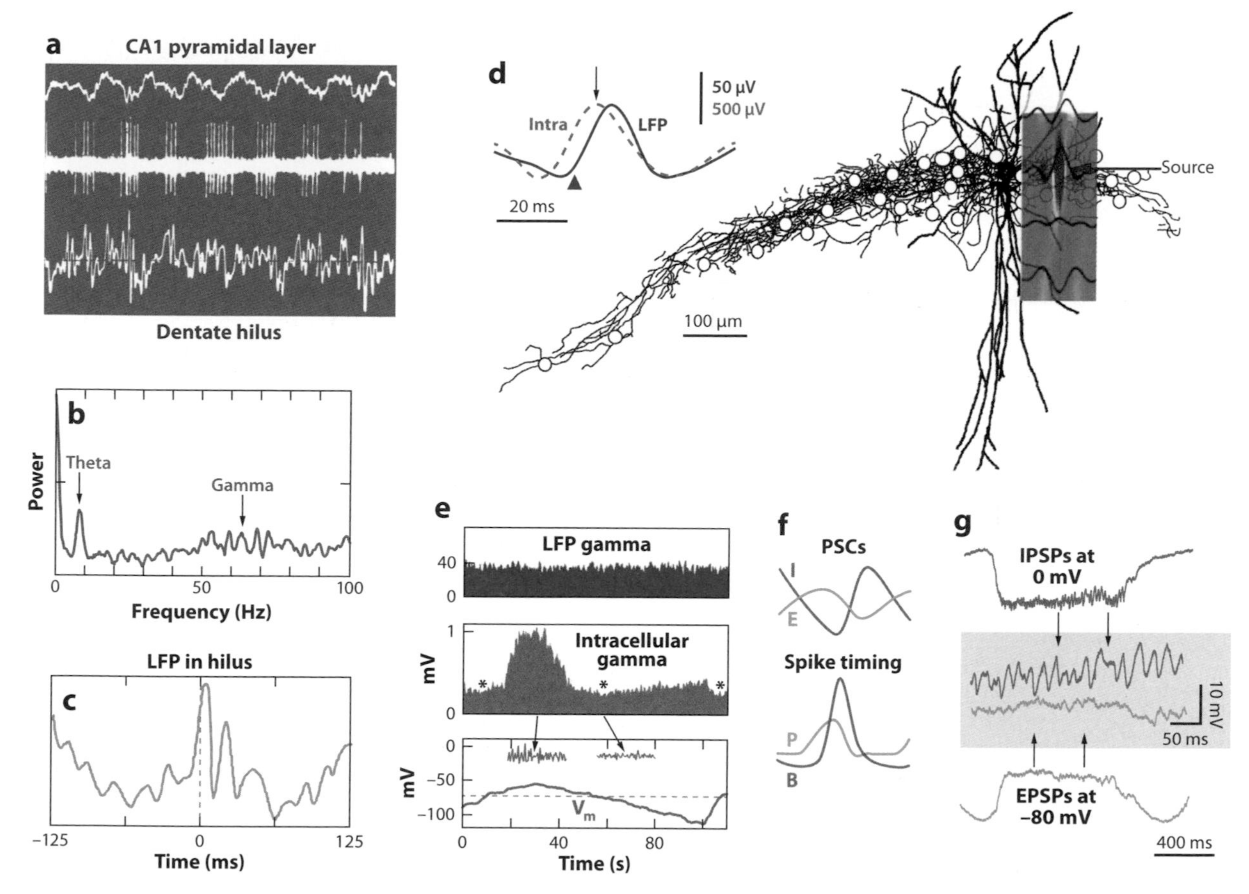

Figure 3

A critical role of parvalbumin (PV) basket cells in gamma oscillations. (*a*) Local field potential (LFP) recording from the CA1 pyramidal layer (*top*) and dentate hilus (*bottom*) and unit recording from a fast-spiking putative interneuron in the hilus (*middle trace*). Note the trains of spikes at gamma frequency, repeating periodically at theta frequency. (*b*) Power spectrum of the unit shown in panel *a*. Note the peak at theta and a broader peak at 50–80 Hz (gamma). (*c*) Spike-triggered average of the LFP in the hilus. Note the prominent phase locking of the interneuron to gamma wave phase and the cross-frequency coupling between gamma and theta waves. (*a–c*) Recordings from a behaving rat. (*d*) Camera-lucida reconstruction of the axon arbor of an immunocytochemically identified CA1 basket cell in vivo. The axon arbor outlines the CA1 pyramidal layer, showing (*circles*) putative contacts with other PV-positive neurons, (*inset*) averages of the intracellularly recorded V_m (membrane potential) and the LFP, (*triangle*) peak of the mean preferred discharge of the surrounding pyramidal cells, and (*arrow*) peak of the mean preferred discharge of the basket cell. Note the short delay between the spikes of pyramidal cells and the basket neuron. Current source density (CSD) map is superimposed on the pyramidal layer. Arrow points to current source of gamma wave (*red*). (*e*) Continuous display (110) of integrated and rectified gamma activity of the LFP and the fast intracellularly recorded V_m fluctuation (20–80 Hz; after digital removal of spikes) in a CA1 pyramidal neuron. V_m was biased by the intracellular current injection: (*dashed line*) resting membrane potential. Note the increase of the intracellular V_m gamma during both depolarization (*inset*) and hyperpolarization as well as the smallest V_m gamma power at resting membrane potential (*asterisks*) against the steady background of LFP gamma power. (*d, e*) In vivo recordings under urethane anesthesia. (*f*) Excitatory (E) and inhibitory (I) postsynaptic currents (PSCs) in a pyramidal cell, triggered by LFP gamma (*top*) and the spike timing of a pyramidal cell (P) and a basket interneuron (B) during carbachol-induced gamma oscillation in a hippocampal slice in vitro. Note that maximum discharge of the basket cell precedes the hyperpolarization of the pyramidal cell. (*g*) Intracellular recordings in a ferret prefrontal pyramidal cell in vivo illustrating the large amplitude, inhibition-dominated barrages recorded at 0 mV (*brown*) and smaller amplitude, excitation-dominated, synaptic barrages recorded at −80 mV (*tan*) for two representative UP states. Membrane potentials are expanded further (*inset*). EPSP, excitatory postsynaptic potential; IPSP, inhibitory postsynaptic potential. Reproduced with permission from (*a–c*) Buzsáki et al. (1983), (*d–e*) after Penttonen et al. (1998), (*f*) after from Mann et al. (2005), and (*g*) after Hasenstaub et al. (2005).

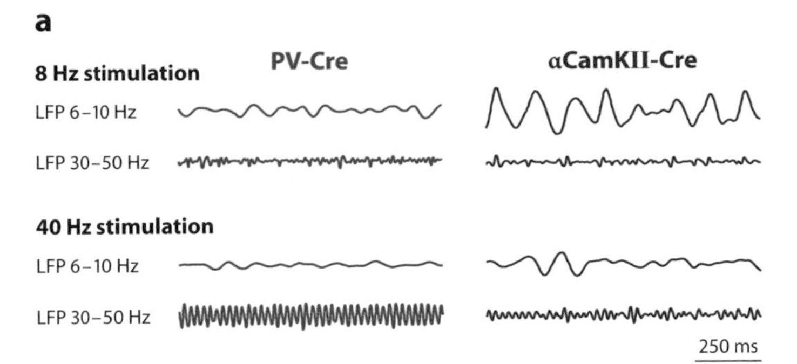

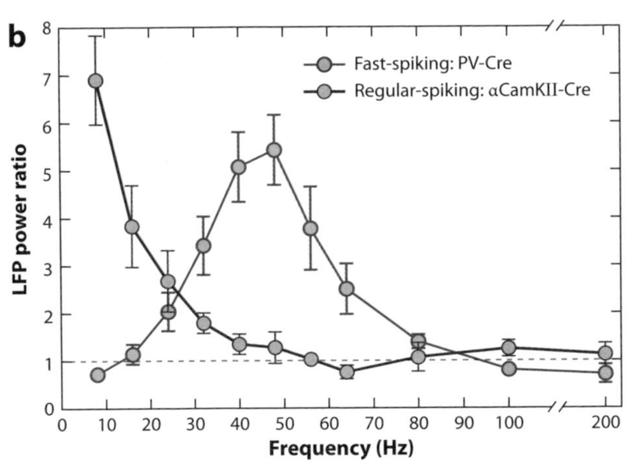

Figure 4

"Synthetic" gamma rhythm in vivo. (*a*) Local field potential (LFP) recordings in anesthetized mouse, expressing ChR2 selectively in either parvalbumin (PV) neurons (ChR2-PV-Cre) or pyramidal cells (ChR2-αCamKII-Cre). Stimulation at 8 Hz evoked rhythmic activity in the αCamKII-Cre but not the PV-Cre mouse. Conversely, stimulation at 40 Hz induced gamma oscillation in the PV-Cre but not in αCamKII-Cre mouse. (*b*) Mean LFP power ratio measured in multiple frequency bands in response to rhythmic light activation of ChR2-PV-Cre expressing neurons (*blue*) or ChR2-αCamKII-Cre expressing neurons (*purple*) at various frequencies. Reprinted from Cardin et al. (2009).

1999), and have resonance at theta, rather than gamma, frequencies (Pike et al. 2000, Gloveli et al. 2005). The postsynaptic receptor targets of CCK basket cells contain slower α2 subunits (Glickfield & Scanziani 2006, Freund & Katona 2007), and CCK interneurons are not effective in maintaining gamma oscillations (Hájos et al. 2004, Tukker et al. 2007). Hippocampal CA1 bistratified neurons showed stronger phase locking of spikes to gamma waves than did PV basket cells (Tukker et al. 2007). Their phase locking may be "inherited" from the CA3 output (Csicsvari et al. 2003), but the IPSPs they produce in the dendrites of pyramidal cells may not be faithfully transferred to the soma (Lytton & Sejnowski 1991). These other types of interneurons appear better suited to contribute to slower oscillations and, by controlling basket cells, are likely critical in establishing cross-frequency coupling (see below) between gamma and slower rhythms.

Do I-I and E-I Mechanisms Compete or Cooperate in the Brain?

Both I-I and E-I models have merits and disadvantages (Whittington et al. 2000, Tiesinga & Sejnowski 2009, Wang 2010). Because the oscillation frequency of individual neurons in the I-I model is at least partially determined by the amount of excitation, a heterogeneous input can result in a wide range of oscillation frequencies. In the face of such frequency dispersion, the population synchrony inevitably decreases. This shortcoming can be effectively compensated for by gap-junction-enhanced synchrony (Gibson et al. 1999, Hormuzdi et al. 2001, Buhl et al. 2003, Traub et al. 2004), resonant properties of basket cells, and fast and strong shunting inhibition between interneurons (Bartos et al. 2007). However, heterogeneity of neuronal firing rates may be beneficial. In networks consisting of neurons with different firing patterns and rates, gamma oscillation may function as a selection mechanism, because transient synchrony would emerge only among those neurons that are activated to approximately the same level.

In most E-I models, there is no need for I-I connections (Wilson & Cowan 1972, Whittington et al. 2000, Borgers & Kopell 2003, Brunel & Wang 2003, Geisler et al. 2005). In support of this prediction, experimentally disconnecting many I-I links in knockout mice did not strongly affect gamma power in the hippocampal CA1 region (Wulff et al. 2009). In the E-I models, the driving force of the oscillation is the activity of pyramidal cells. Note that gamma rhythms are also prominent in structures, which lack dense local E-I connections, such as the basal ganglia or ventral tegmental area (Brown et al. 2002, Berke et al. 2004, Tort et al. 2008, Fujisawa & Buzsáki 2011). E-I models require a time delay between E spikes and I spikes, since timing of the interneurons is "inherited" from the pyramidal cells. In contrast, in I-I models the spike phase of pyramidal cells largely reflects the intensity of their tonic drive. In the hippocampal CA1 region, interneurons show both phase delay or advance relative to the spikes of pyramidal cells (Bragin et al. 1995, Csicsvari et al. 2003, Tukker et al. 2007, Senior et al. 2008, Mizuseki et al. 2011). These results suggest that E-I and I-I hybrid gamma networks may work together to generate gamma frequency oscillations (Brunel & Wang 2003, Geisler et al. 2005, Tiesinga & Sejnowski 2009, Belluscio et al. 2012).

The role of recurrent excitatory (E-E) connections between principal cells in gamma models are not well-understood (Kopell et al. 2000, Whittington et al. 2000, Brunel & Wang 2003, Geisler et al. 2005). In the cortex, gamma oscillations are more prominent in the superficial, rather than the deep, layers where local recurrent connections are abundant (Chrobak & Buzsáki 1998, Quilichini et al. 2010, Buffalo et al. 2011). By contrast, the largest-amplitude gamma rhythm in the hippocampus is observed in the dentate gyrus (Buzsáki et al. 1983), even though granule cells lack recurrent excitation onto themselves. Decreasing recurrent excitatory synaptic currents in dynamic clamp studies had little effect on gamma power (Morita et al. 2008). The less critical role of E-E recurrent excitation may liberate the pyramidal cells from the timing constraints of the rhythm; therefore, they could fire spikes stochastically at various cycle phases in an input drive–dependent manner without interrupting rhythm.

LONG-RANGE SYNCHRONIZATION OF GAMMA OSCILLATIONS

Although gamma oscillations typically arise locally, patches of gamma networks can interact with each other. Synchronization of transient gamma bursts has multiple meanings, including phase-phase, phase-amplitude, and amplitude-amplitude coupling (**Figure 5**) (see Cross-Frequency Phase Coupling, sidebar below). Phase-phase synchrony between identical frequency oscillators that emerges at two (or multiple) locations can occur by phase locking (**Figure 5*b***). The magnitude of such synchrony is typically measured by phase coherence. A second form of synchrony refers to the covariation of gamma power at two (or multiple) locations, also known as amplitude or power comodulation (**Figure 5*c***). In this latter case, phase constancy between the gamma waves may or may not be present (**Figure 5*c,d***). Instead, the power (amplitude) envelopes of the gamma bursts are correlated (comodulation of power). Power-power synchrony of gamma rhythms can be effectively brought about by joint phase biasing of the power of gamma oscillations by a slower rhythm, known as cross-frequency phase-amplitude (CF_{PA}) coupling or nested oscillations (**Figure 5*c,d,e***) (Bragin et al. 1995, Schroeder & Lakatos 2009, Canolty & Knight 2010, Fell & Axmacher 2011). The third type of synchrony occurs when there is a relatively constant relationship between the gamma phase and the phase of a modulating slower rhythm (**Figure 6*e***), known as cross-frequency phase-phase (CF_{PP}), or n:m, coupling (Tass et al. 1998). Cross-frequency coupling can take place within or across structures. In practice, each relationship should be investigated with care because even stochastic signals can occasionally yield spurious coupling.

Cross-frequency phase-amplitude (CF_{PA}) coupling: phenomenon in which the amplitude of a faster oscillation is modulated by the phase of a slower rhythm

Cross-frequency phase-phase coupling (CF_{PP}): phenomenon in which the phase of a faster oscillation is coupled to multiple phases of a slower rhythm

Phase Coherence of Gamma Rhythms in Distant Networks

If multiple cell assemblies in disparate brain areas need to be synchronized, how can they be engaged in coherent gamma oscillations given the long axon conduction delays of pyramidal cells? Solid evidence for coherent gamma oscillations in distant networks is scarce; perhaps the best-established case is interhemispheric synchronization. Multiple units with similar receptive fields in the left and right primary

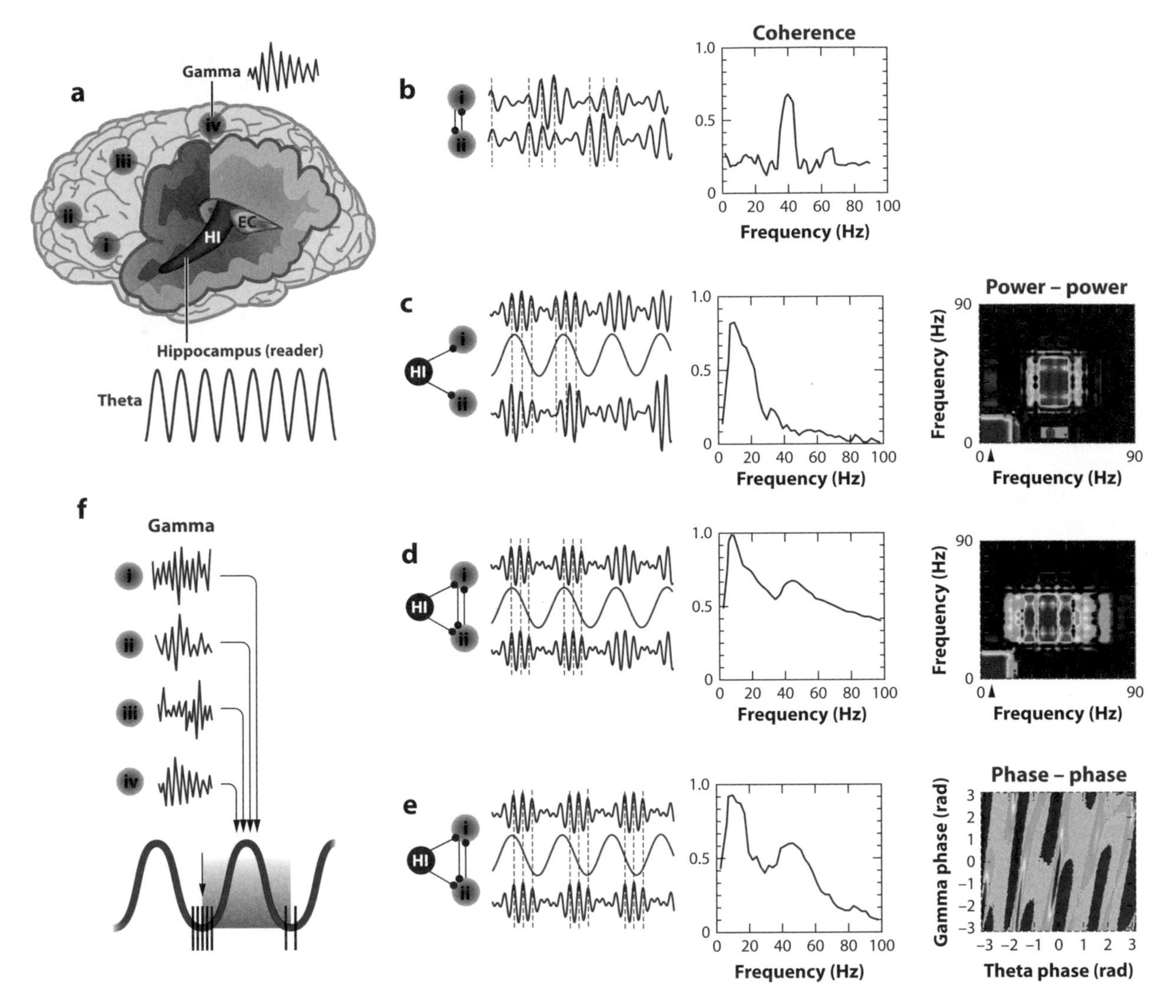

Figure 5

Oscillatory coupling mechanisms. (*a*) Schematic view of the human brain showing hot spots of transient gamma oscillations (*i–iv*) and theta oscillation in the hippocampus (HI); entorhinal cortex (EC). Oscillators of the same and different kind (e.g., theta, gamma) can influence each other in the same and different structures, thereby modulating the phase, amplitude, or both. (*b*) Phase-phase coupling of gamma oscillations between two areas. Synthetic data used for illustration purposes. Coherence spectrum (or other, more specific, phase-specific measures) between the two signals can determine the strength of phase coupling. (*c*) Cross-frequency phase-amplitude coupling. Although phase coupling between gamma waves is absent, the envelope of gamma waves at the two cortical sites is modulated by the common theta rhythm. This can be revealed by the power-power correlation (comodugram; *right*). (*d*) Gamma phase-phase coupling between two cortical sites, whose powers are modulated by the common theta rhythm. Both gamma coherence and gamma power-power coupling are high. (*e*) Cross-frequency phase-phase coupling. Phases of theta and gamma oscillations are correlated, as shown by the phase-phase plot of the two frequencies. (*f*) Hippocampal theta oscillation can modulate gamma power by its duty cycle at multiple neocortical areas so that the results of the local computations are returned to the hippocampus during the accrual ("readiness") phase of the oscillation. *a* and *f*, after Buzsáki (2010); *b–e*, after Belluscio et al. (2012).

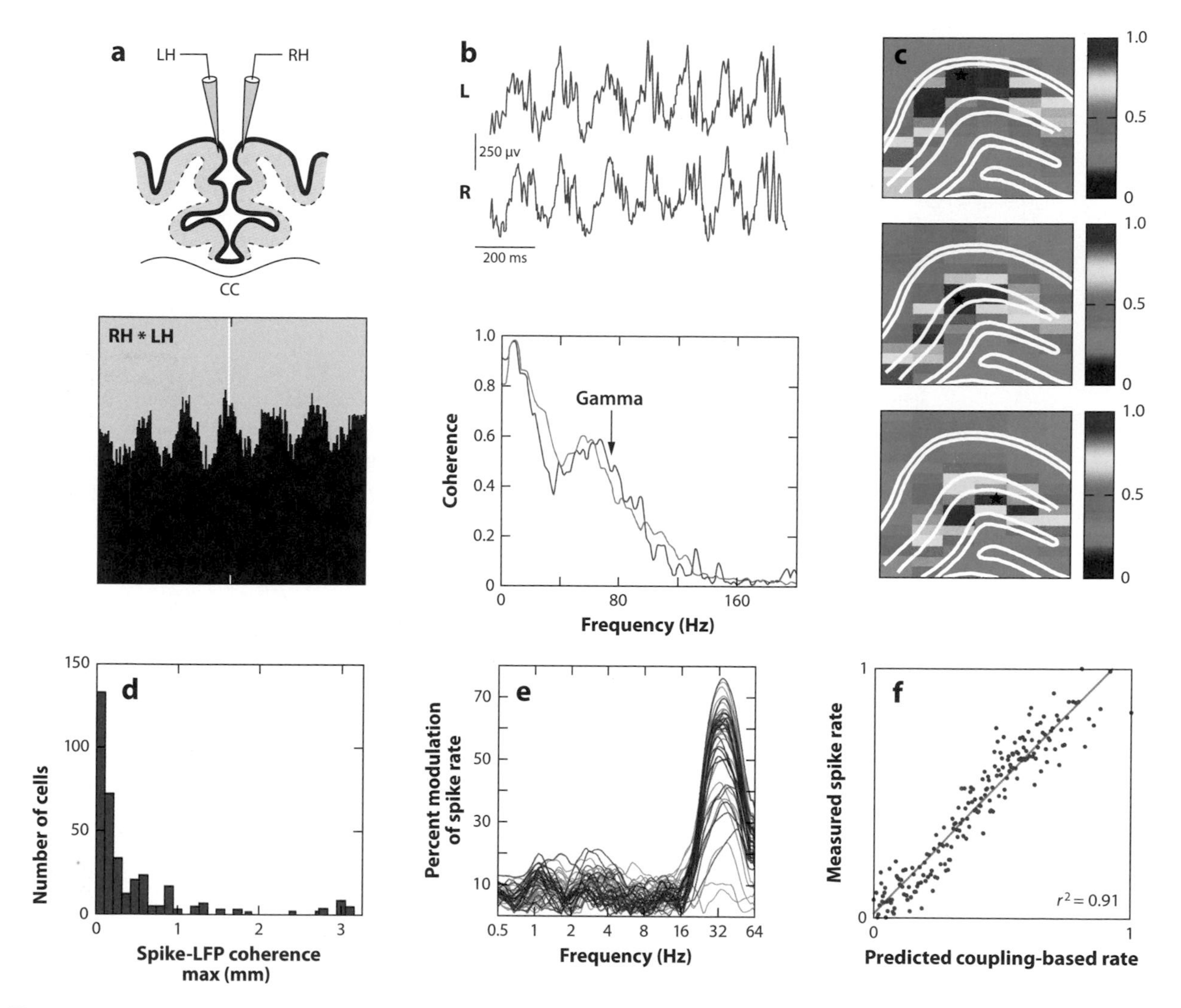

Figure 6

Long-range synchrony of gamma oscillations. (*a*) Neurons sharing receptive fields in left (LH) and right (RH) primary visual cortex of the anesthetized cat fire coherently with zero time lag at gamma frequency. (*b*) Local field potential (LFP) traces from the left (L) and right (R) hippocampal CA1 pyramidal layer of the mouse during running and coherence spectra between the traces during running (*orange*) and REM sleep (*blue*). (*c*) LFP coherence map of gamma (30–90 Hz) in the rat hippocampus during running. Coherence was calculated between the reference site (*star*) and the remaining 96 recording sites. Note the high coherence values within the same layers (*outlined by white lines*) and rapid decrease of coherence across layers. (*d*) Distribution of distances between the unit and LFP recording sites with maximum spike-LFP coherence in the gamma band. Note that, in a fraction of cases, maximum coherence is stronger at large distances between the recorded unit and the LFP. (*e*) Spike-LFP coherence in the human motor cortex. The probability of spiking correlates with frequency-specific LFP phase of the ipsilateral (*blue*) and contralateral (*green*) motor area and contralateral dorsal premotor area (*red*). (*f*) The phase-coupling-based spike rate (generated from the preferred LFP–LFP phase-coupling pattern) predicts the measured spike rate. Panels reproduced after (*a*) Engel et al. (1991), (*b*) Buzsáki et al. (2003), (*c*) Montgomery & Buzsáki (2007), (*d*) Sirota et al. (2008), and (*e,f*) Canolty et al. (2010).

visual cortex can display coherent gamma-range oscillations (**Figure 6*a***). Similarly, gamma oscillations in homologous hippocampal layers in the two hemispheres display high coherence (**Figure 6*b***). In both cases, phase synchrony is mediated by interhemispheric axon tracts, given that severing these conduits abolishes the synchrony. The high interhemispheric coherence and task-dependent inter-regional gamma synchrony (Engel et al. 1991, Roelfsema et al. 1997, Chrobak & Buzsáki 1998, Rodriguez et al. 1999, Tallon-Baudry et al. 2001, Montgomery et al. 2008) can be contrasted with the fast decrease of gamma coherence across different layers (**Figure 6*c***), owing to the noncoherent relationships among the inputs. The importance of anatomical connectivity, as opposed to physical distance, can explain the occasionally high gamma coherence between spikes and LFP at distant sites (**Figure 6*d***) and the gamma timescale covariations of firing rates of spatially distant neurons (**Figure 6*e*,*f***).

Temporal coordination between spatially separated oscillators can be established by axon collaterals of pyramidal cells (Traub et al. 1996a, Whittington et al. 2000, Bibbig et al. 2002), interleaving assemblies (Vicente et al. 2008), or long-range interneurons (Buzsáki et al. 2004). In each case, conduction delays are the primary problem because the differing delays between the different gamma inputs can destabilize the rhythm (Ermentrout & Kopell 1998), and the extra interneuron spikes brought about by the excitatory collaterals from the oscillating regions can decelerate the oscillation frequency in the target network. Reciprocal coupling between oscillators in the two hemispheres (**Figure 6*b***) can alleviate the phase-shift problem and result in 0 phase-lag synchrony, provided that the conduction delays are short enough (<4–8 ms) and that synchrony is assessed over multiple cycles (Traub et al. 2002).

Long-range interneurons may be another candidate substrate for establishing gamma synchrony (**Figure 7**) (Buzsáki et al. 2004). These interneurons distribute their axon terminals over multiple regions and layers of the cortex and even across the hemispheres (Sik et al. 1994, Gulyás et al. 2003, Tomioka et al. 2005, Jinno et al. 2006). Importantly, distally projecting axons of long-range interneurons have several-fold thicker axons and larger diameter myelin sheaths than do pyramidal cells (**Figure 7*c*,*d***), allowing for considerably faster axon conduction velocity (Jinno et al. 2006). In the I-I gamma model, replacing just 10–20% of the basket synapses with synapses of fast-conducting long-range interneurons could achieve global-phase synchrony (**Figure 7*a*,*b***) (Buzsáki et al. 2004). An obvious advantage of the hybrid basket, long-range interneuron network is that synchrony among local and distributed cell assemblies can be tuned selectively by differentially targeting the two interneuron types.

Brain-Wide Synchronization of Gamma Oscillations by Slower Rhythms

Slower temporal coordination among gamma oscillators may be achieved by modulating the gamma power by the phase of slower rhythms (**Figure 6**). Compared with faster oscillators,

Figure 7

Coupling of gamma oscillators by long-range interneurons. (*a*) Oscillations in a network with locally connected interneurons. The network is essentially asynchronous. (*Upper panel*) Spike raster of 4000 neurons; (*lower panel*) the population firing rate. (*b*) Oscillations in a network with local interneurons (B) and long-range interneurons (LR; power-law connectivity). Note clear oscillatory rhythm. (*c*) Cross-section of the axon of a long-range CA1 GABAergic interneuron projecting toward the subiculum/entorhinal cortex. In comparison, neighboring axons of pyramidal cells are also shown (*d*). Reproduced from Buzsáki et al. (2004) (*a*,*b*) and from Jinno et al. (2006) (*b*,*c*).

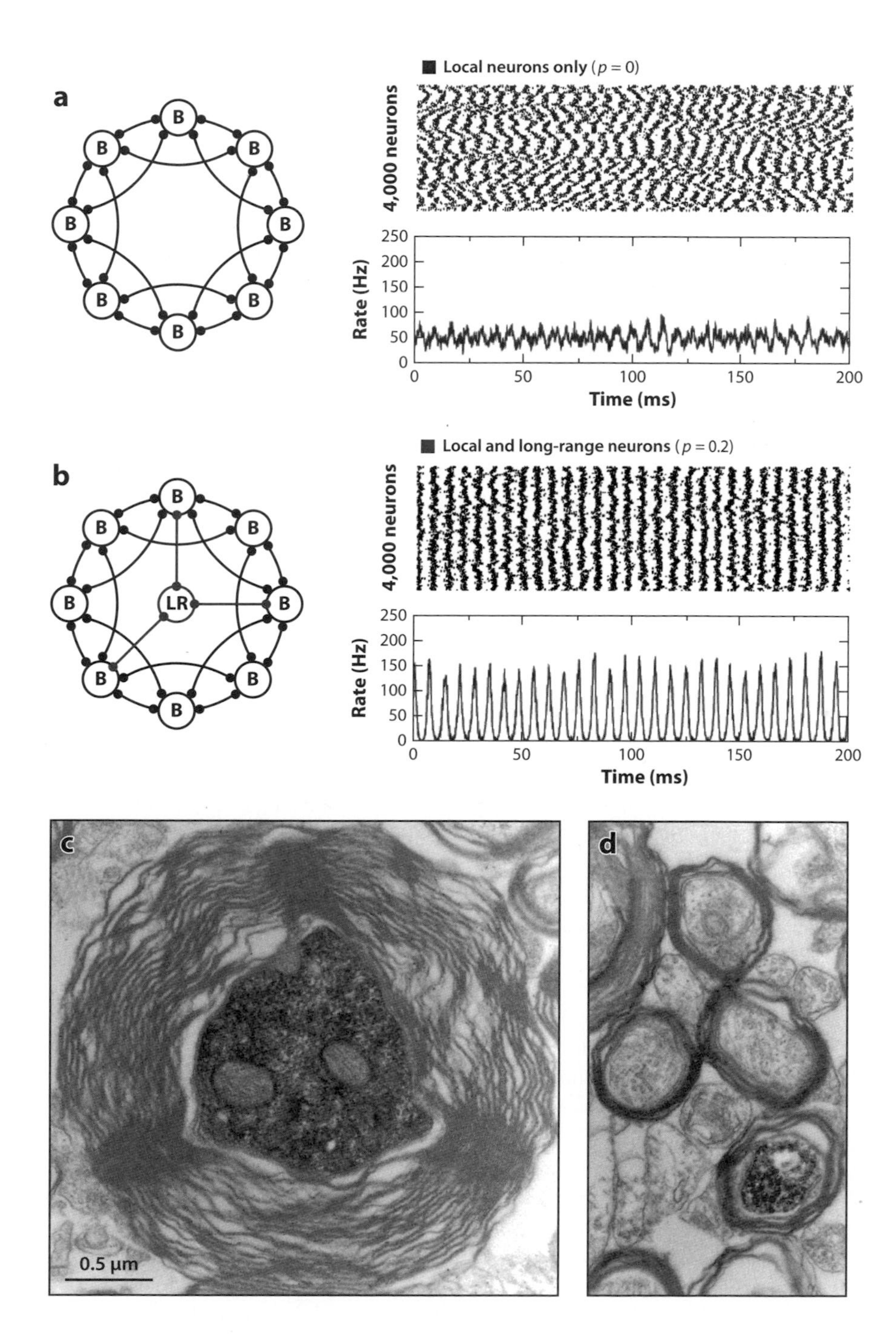
a
b
c
d
B
LR
Local neurons only ($p = 0$)
Local and long-range neurons ($p = 0.2$)
4,000 neurons
Rate (Hz)
Time (ms)
0
50
100
150
200
250
0.5 µm

slower oscillators involve more neurons in a larger volume (Von Stein & Sarnthein 2000) and are associated with larger membrane potential changes because in longer time windows spikes of many more upstream neurons can be integrated (Hasenstaub et al. 2005, Quilichini et al. 2010).

CF_{PA} coupling between gamma and other rhythms within the same and different brain regions has been well documented, including modulation by theta (**Figure 3*c***) (Buzsáki et al. 1983; Soltesz & Deschênes 1993; Bragin et al. 1995; Chrobak & Buzsáki 1998; Wang 2002; Mormann et al. 2005; Canolty et al. 2006; Demiralp et al. 2007; Tort et al. 2008, 2010; Colgin et al. 2009; Griesmayr et al. 2010), alpha (Palva et al. 2005, Cohen et al. 2009), spindle (Peyrache et al. 2011), delta (Lakatos et al. 2005), slow (Hasenstaub et al. 2005, Isomura et al. 2006), and ultraslow (Leopold et al. 2003) oscillations (Buzsáki 2006, Jensen & Colgin 2007, Schroeder & Lakatos 2009, Canolty & Knight 2010, Fell & Axmacher 2011). Because perisomatic basket cells contribute to both gamma and theta rhythms by firing theta-rhythm-paced bursts of spikes at gamma frequency, it has been hypothesized that fast-firing basket cells may play a key role in cross-frequency coupling (Buzsáki et al. 1983, Bragin et al. 1995). This is plausible because several other types of interneurons are often entrained by slower oscillations and they inhibit basket cells (Freund & Buzsáki 1996, Klausberger & Somogyi 2008). A prediction of this hypothesis is that temporal coordination by the basket cells also introduces a CF_{PP} (i.e., phase-phase or n:m) coupling relationship between theta and gamma oscillations (**Figure 5*e***). It may well be that CF_{PP} mechanisms underlie CF_{PA} coupling in most situations, but convincing demonstration of clear phase-phase coupling is hampered by the lack of adequate methods to quantify cross-frequency interactions and reliably track the true phase of nonharmonic oscillators (Tort et al. 2010, Belluscio et al. 2012).

The cross-frequency coupling of rhythms forms a multiscale timing mechanism (Buzsáki & Draguhn 2004, Jensen & Colgin 2007, Schroeder & Lakatos 2009, Canolty & Knight 2010, Fell & Axmacher 2011). Computational models have explored the potential theoretical advantages of such cross-frequency coupling (Lisman & Idiart 1995, Varela et al. 2001, Lisman 2005, Neymotin et al. 2011). The hierarchy of phase-amplitude-coupled rhythms is an effective mechanism for segmentation and linking of spike trains into cell assemblies ("letters") and assembly sequences (neural "words") (Buzsáki 2010).

Several studies have examined the relationship between cross-frequency coupling of gamma oscillations and cognitive processes. The magnitude of theta-gamma coupling in the hippocampal region varied with working memory load in patients implanted with depth electrodes (Axmacher et al. 2010). The strength of theta-gamma coupling in the hippocampus and striatum of the rat was affected by task demands (Tort et al. 2008, 2009). Similarly, the magnitude of CF_{PA} coupling between a 4-Hz oscillation and gamma power in the prefrontal cortex increased in the working memory phase of a choice task (Fujisawa & Buzsáki 2011). In an auditory task, gamma power in the frontal and temporal sites was phase-locked mainly to theta oscillations, whereas over occipital areas phase modulation was strongest by the alpha rhythm in a visual task (Voytek et al. 2010). Increased CF_{PP} coupling between alpha and beta/gamma oscillations correlates with the difficulty of arithmetic mental tasks in the human magnetoencephalogram (Palva et al. 2005), whereas in another study working memory was correlated with theta-gamma synchrony (Griesmayr et al. 2010).

Cross-frequency coupling between slow rhythms and gamma oscillations can support a "reader-initiated" mechanism for information exchange (Sirota et al. 2008). For example, the hippocampal theta rhythm can entrain local gamma oscillations in multiple cortical areas. During its duty cycle, the theta output can phase align gamma oscillations that emerge in numerous activated neocortical local circuits (**Figure 5*f***). In turn, the cell assemblies

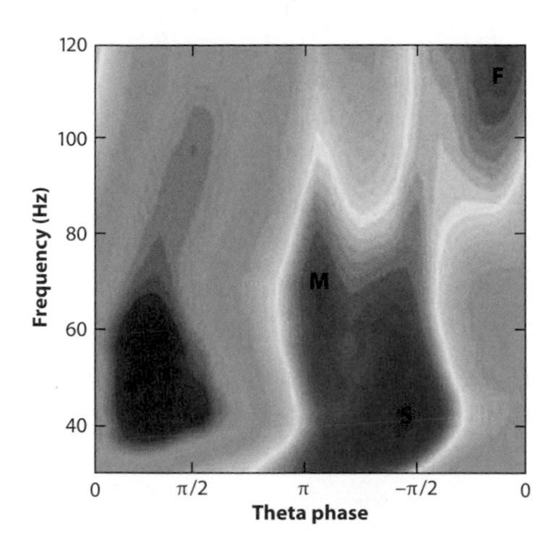

Figure 8

Multiple gamma sub-bands. Wavelet power between 30 and 150 Hz as a function of waveform-based theta cycle phases. Note the different theta-phase preference of mid-frequency (M) ($gamma_M$, 50–90 Hz, near theta peak) and slow (S) ($gamma_S$, 30–50 Hz on the descending phase of theta) gamma oscillations. Note also the dominance of fast (F) ($gamma_F$, or epsilon band, 90–150 Hz) at the trough of theta. After Belluscio et al. (2012).

WHEN GAMMA POWER IS NOT A RHYTHM

A caveat in many studies is the lack of a disciplined and quantified analysis of gamma oscillations. To identify true gamma oscillations, appropriate statistics should be applied to demonstrate periodicity (Muresan et al. 2008, Burns et al. 2011, Ray & Maunsell 2011), and additional experiments are needed to distinguish between a power increase resulting from genuine oscillations and an increase resulting from greater spiking activity (Jarvis & Mitra 2001, Crone et al. 2006, Montgomery et al. 2008, Whittingstall & Logothetis 2009, Quilichini et al. 2010, Belluscio et al. 2012, Ray & Maunsell 2011). This is especially important for higher frequencies, such as the epsilon band, but spike-afterdepolarization and -hyperpolarization components can also contribute to the gamma band power. Although spike contamination to oscillatory power can be a nuisance, by using proper analytical methods, spike power can be exploited as a proxy for the assessment of neuronal outputs even in recordings of subdural local field potentials. Studying the temporal features of such high-frequency events may provide clues about oscillatory events that modulate them, even in situations when invasive unit recordings are not an option.

associated with the transient gamma bursts can address hippocampal networks in the accrual phase of the theta cycle, corresponding to the most sensitive, plastic state (Huerta & Lisman 1995), and can combine neocortical information into a condensed hippocampal representation.

MULTIPLE GAMMA RHYTHMS

Cross-frequency coupling can assist with the separation of gamma sub-bands (Tort et al. 2008, Colgin et al. 2009). In the hippocampal CA1 region, wavelet analysis identified three distinct gamma bands: (*a*) slow gamma ($gamma_S$, 30–50 Hz) on the descending phase, (*b*) mid-frequency gamma ($gamma_M$, 50–90 Hz) near the peak, and (*c*) fast gamma ($gamma_F$, or epsilon band, 90–140 Hz) near the trough of the theta cycle (**Figure 8**) (Tort et al. 2010, Belluscio et al. 2012). Support for the different origins of gamma sub-bands is provided by their differential distribution in the different depths of the CA1 pyramidal layer and in different segments of the subiculum (Belluscio et al. 2012, Jackson et al. 2011). It is likely that the slow and mid-gamma band distinction applies to other brain regions as well (Kay 2003).

Previous works have distinguished only low and high gamma sub-bands (Csicsvari et al. 1999, Ray & Maunsell 2011) with the high sub-band defined as 60–140 Hz (Canolty et al. 2006, Colgin et al. 2009). Because power in the mid-gamma (50–90 Hz) and epsilon (90–150 Hz) bands is associated with different phases of theta oscillation (**Figure 8**) and is likely generated by different mechanisms (Belluscio et al. 2012), lumping these bands together is not justified on physiological grounds. Future studies, therefore, should distinguish sub-bands of gamma oscillations and carefully separate true and spurious gamma rhythms (see When Gamma Power is Not a Rhythm, sidebar above).

To conclude, although the word "rhythm" readily conjures up the picture of a clock, gamma rhythms occur in relatively short bursts and are quite variable in frequency, typically associated with stochastic firing of single neurons. The LFP gamma reflects largely the balancing act of excitation and inhibition, i.e., the active mode of a local circuit. Future studies on gamma oscillations will continue to inform us about the complex dynamics of brain circuits.

SUMMARY POINTS

1. Transient cell assemblies may be organized into gamma-wave cycles.
2. Perisomatic inhibition by PV basket cells is essential for gamma oscillations.
3. Gamma oscillations are short-lived and emerge from the coordinated interactions of excitation and inhibition. Thus, LFP gamma can be used to identify active operations of local circuits.
4. Network gamma oscillations may coexist with highly irregular firing of pyramidal neurons.
5. Different sub-bands of gamma oscillations can coexist or occur in isolation.
6. Long-range interneurons may be critical for gamma-phase synchrony in different brain regions
7. Cross-frequency coupling is an effective mechanism for functionally linking active cortical circuits.
8. Genuine gamma oscillations should be distinguished from mere increases of gamma-band power and/or increased spiking activity.

DISCLOSURE STATEMENT

The authors are not aware of any affiliations, memberships, funding, or financial holdings that might be perceived as affecting the objectivity of this review.

ACKNOWLEDGMENTS

We thank R. Canolty, J. Csicsvari, A. Engel, P. Fries, P. Jonas, N. Kopell, J. Maunsell, A. Peyrache, S. Ray, R.D. Traub, and M. Whittington for their comments on the manuscript as well as S. Bressler and W. Freeman for providing historical documents on gamma oscillations. Supported by the National Institutes of Health (NS034994; MH54671); the Human Frontier Science Program; the J.D. McDonnell Foundation; Zukunftskolleg, University of Konstanz, Germany (G.B.) and by the National Institutes of Health (R01 MH062349) and the Kavli Foundation (X.J.W.).

LITERATURE CITED

Adrian E. 1942. Olfactory reactions in the brain of the hedgehog. *J. Physiol.* 100:459–73

Ardid S, Wang X-J, Gomez-Cabrero D, Compte A. 2010. Reconciling coherent oscillation with modulation of irregular spiking activity in selective attention: gamma-range synchronization between sensory and executive cortical areas. *J. Neurosci.* 30(8):2856–70

Axmacher N, Henseler MM, Jensen O, Weinreich I, Elger CE, Fell J. 2010. Cross-frequency coupling supports multi-item working memory in the human hippocampus. *Proc. Natl. Acad. Sci. USA* 107:3228–33

Bartos M, Vida I, Jonas P. 2007. Synaptic mechanisms of synchronized gamma oscillations in inhibitory interneuron networks. *Nat. Rev. Neurosci.* 8:45–56

Belluscio MA, Mizuseki K, Schmidt R, Kempter R, Buzsáki G. 2012. Cross-frequency phase-phase coupling between theta and gamma oscillations in the hippocampus. *J. Neurosci.* 32(2):423–35

Berger H. 1929. Uber das Elektroenkephalogramm des Menschen. *Arch. Pyschiatr. Nervenkr.* 87:527–70

Berke JD, Okatan M, Skurski J, Eichenbaum HB. 2004. Oscillatory entrainment of striatal neurons in freely moving rats. *Neuron* 43:883–96

Bibbig A, Traub R, Whittington M. 2002. Long-range synchronization of gamma and beta oscillations and the plasticity of excitatory and inhibitory synapses: a network model. *J. Neurophysiol.* 88:1634–54

Borgers C, Kopell N. 2003. Synchronization in networks of excitatory and inhibitory neurons with sparse, random connectivity. *Neural Comput.* 15(3):509–38

Bragin A, Jando G, Nadasdy Z, Hetke J, Wise K, Buzsáki G. 1995. Gamma (40–100 Hz) oscillation in the hippocampus of the behaving rat. *J. Neurosci.* 15:47–60

Bressler SL, Freeman WJ. 1980. Frequency analysis of olfactory system EEG in cat, rabbit and rat. *Electroencephalogr. Clin. Neurophysiol.* 50:19–24

Brown P, Kupsch A, Magill PJ, Sharott A, Harnack D, Meissner W. 2002. Oscillatory local field potentials recorded from the subthalamic nucleus of the alert rat. *Exp. Neurol.* 177(2):581–85

Brunel N. 2000. Dynamics of sparsely connected networks of excitatory and inhibitory spiking neurons. *J. Comput. Neurosci.* 8(3):183–208

Brunel N, Hakim V. 1999. Fast global oscillations in networks of integrate-and-fire neurons with low firing rates. *Neural Comput.* 11:1621–71

Brunel N, Wang X-J. 2003. What determines the frequency of fast network oscillations with irregular neural discharges? I. Synaptic dynamics and excitation-inhibition balance. *J. Neurophysiol.* 90:415–30

Buffalo EA, Fries P, Landman R, Buschman TJ, Desimone R. 2011. Laminar differences in gamma and alpha coherence in the ventral stream. *Proc. Natl. Acad. Sci. USA* 108(27):11262–67

Buhl D, Harris K, Hormuzdi S, Monyer H, Buzsáki G. 2003. Selective impairment of hippocampal gamma oscillations in connexin-36 knock-out mouse in vivo. *J. Neurosci.* 23:1013–18

Buhl EH, Halasy K, Somogyi P. 1994. Diverse sources of hippocampal unitary inhibitory postsynaptic potentials and the number of synaptic release sites. *Nature* 368(6474):823–28

Burns SP, Xing D, Shapley RM. 2011. Is gamma-band activity in the local field potential of v1 cortex a "clock" or filtered noise? *J. Neurosci.* 31(26):9658–64

Buzsáki G. 2006. *Rhythms of the Brain*. New York: Oxford Univ. Press

Buzsáki G. 2010. Neural syntax: cell assemblies, synapsembles, and readers. *Neuron* 68(3):362–85

Buzsáki G, Buhl DL, Harris KD, Csicsvari J, Czéh B, Morozov A. 2003. Hippocampal network patterns of activity in the mouse. *Neuroscience* 116(1):201–11

Buzsáki G, Chrobak JJ. 1995. Temporal structure in spatially organized neuronal ensembles: a role for interneuronal networks. *Curr. Opin. Neurobiol.* 5(4):504–10

Buzsáki G, Draguhn A. 2004. Neuronal oscillations in cortical networks. *Science* 304:1926–29

Buzsáki G, Geisler C, Henze DA, Wang X-J. 2004. Interneuron diversity series: circuit complexity and axon wiring economy of cortical interneurons. *Trends Neurosci.* 27:186–93

Buzsáki G, Leung LW, Vanderwolf CH. 1983. Cellular bases of hippocampal EEG in the behaving rat. *Brain Res.* 287:139–71

Canolty R, Edwards E, Dalal S, Soltani M, Nagarajan S, et al. 2006. High gamma power is phase-locked to theta oscillations in human neocortex. *Science* 313:1626–28

Canolty RT, Ganguly K, Kennerley SW, Cadieu CF, Koepsell K, et al. 2010. Oscillatory phase coupling coordinates anatomically dispersed functional cell assemblies. *Proc. Natl. Acad. Sci. USA* 107(40):17356–61

Canolty RT, Knight RT. 2010. The functional role of cross-frequency coupling. *Trends Cogn. Sci.* 14(11):506–15

Cardin JA, Carlén M, Meletis K, Knoblich U, Zhang F, et al. 2009. Driving fast-spiking cells induces gamma rhythm and controls sensory responses. *Nature* 459:663–67

Chrobak JJ, Buzsáki G. 1998. Gamma oscillations in the entorhinal cortex of the freely behaving rat. *J. Neurosci.* 18:388–98

Cohen MX, Elger CE, Fell J. 2009. Oscillatory activity and phase-amplitude coupling in the human medial frontal cortex during decision making. *J. Cogn. Neurosci.* 21(2):390–402

Colgin LL, Denninger T, Fyhn M, Hafting T, Bonnevie T, et al. 2009. Frequency of gamma oscillations routes flow of information in the hippocampus. *Nature* 462:353–57

Crone N, Sinai A, Korzeniewska A. 2006. High-frequency gamma oscillations and human brain mapping with electrocorticography. *Prog. Brain Res.* 159:275–95

Csicsvari J, Hirase H, Czurkó A, Mamiya A, Buzsáki G. 1999. Fast network oscillations in the hippocampal CA1 region of the behaving rat. *J. Neurosci.* 19:RC20

Csicsvari J, Jamieson B, Wise K, Buzsáki G. 2003. Mechanisms of gamma oscillations in the hippocampus of the behaving rat. *Neuron* 37:311–22

Das NN, Gastaut H. 1955. Variations de l'activite electrique du cerveau, du coeur et des muscles squellettiques au cours de la meditation et de l'extase yogique. *Electroencephalogr. Clin. Neurophysiol.* Suppl. 6:211

Demiralp T, Bayraktaroglu Z, Lenz D, Junge S, Busch NA, et al. 2007. Gamma amplitudes are coupled to theta phase in human EEG during visual perception. *Int. J. Psychophysiol.* 64(1):24–30

Destexhe A, Paré D. 1999. Impact of network activity on the integrative properties of neocortical pyramidal neurons in vivo. *J. Neurophysiol.* 81(4):1531–47

Economo MN, White JA. 2011. Membrane properties and the balance between excitation and inhibition control gamma-frequency oscillations arising from feedback inhibition. *PLoS Comp. Biol.* 8(1):e1002354

Engel A, Fries P, Singer W. 2001. Dynamic predictions: oscillations and synchrony in top-down processing. *Nat. Rev. Neurosci.* 2:704–16

Engel AK, König P, Kreiter AK, Singer W. 1991. Interhemispheric synchronization of oscillatory neuronal responses in cat visual cortex. *Science* 252:1177–79

Ermentrout G, Kopell N. 1998. Fine structure of neural spiking and synchronization in the presence of conduction delays. *Proc. Natl. Acad. Sci. USA* 95:1259–64

Farrant M, Nusser Z. 2005. Variations on an inhibitory theme: phasic and tonic activation of GABA(A) receptors. *Nat. Rev. Neurosci.* 6(3):215–29

Fell J, Axmacher N. 2011. The role of phase synchronization in memory processes. *Nat. Rev. Neurosci.* 12(2):105–18

Fisahn A, Contractor A, Traub RD, Buhl EH, Heinemann SF, McBain CJ. 2004. Distinct roles for the kainate receptor subunits GluR5 and GluR6 in kainate-induced hippocampal gamma oscillations. *J. Neurosci.* 24(43):9658–68

Fisahn A, Pike FG, Buhl EH, Paulsen O. 1998. Cholinergic induction of network oscillations at 40 Hz in the hippocampus in vitro. *Nature* 394:186–89

Freeman WJ. 1975. *Mass Action in the Nervous System*. New York: Academic

Freeman WJ. 2007. Definitions of state variables and state space for brain-computer interface: Part 1. Multiple hierarchical levels of brain function. *Cogn. Neurodyn.* 1:3–14

Freund T, Buzsáki G. 1996. Interneurons of the hippocampus. *Hippocampus* 6:347–470

Freund TF, Katona I. 2007. Perisomatic inhibition. *Neuron* 56(1):33–42

Fries P. 2005. A mechanism for cognitive dynamics: neuronal communication through neuronal coherence. *Trends Cogn. Sci.* 9:474–80

Fries P. 2009. Neuronal gamma-band synchronization as a fundamental process in cortical computation. *Annu. Rev. Neurosci.* 32:209–24

Fries P, Reynolds JH, Rorie AE, Desimone R. 2001. Modulation of oscillatory neuronal synchronization by selective visual attention. *Science* 291(5508):1560–63

Fuchs EC, Zivkovic AR, Cunningham MO, Middleton S, Lebeau FE, et al. 2007. Recruitment of parvalbumin-positive interneurons determines hippocampal function and associated behavior. *Neuron* 53:591–604

Fujisawa S, Buzsáki G. 2011. A 4 Hz oscillation adaptively synchronizes prefrontal, VTA, and hippocampal activities. *Neuron* 72(1):153–65

Geisler C, Brunel N, Wang X-J. 2005. Contributions of intrinsic membrane dynamics to fast network oscillations with irregular neuronal discharges. *J. Neurophysiol.* 94(6):4344–61

Gibson J, Beierlein M, Connors B. 1999. Two networks of electrically coupled inhibitory neurons in neocortex. *Nature* 402:75–79

Glickfeld LL, Scanziani M. 2006. Distinct timing in the activity of cannabinoid-sensitive and cannabinoid-insensitive basket cells. *Nat. Neurosci.* 9(6):807–15

Gloveli T, Dugladze T, Saha S, Monyer H, Heinemann U, et al. 2005. Differential involvement of oriens/pyramidale interneurones in hippocampal network oscillations in vitro. *J. Physiol.* 562(Pt. 1):131–47

Gray CM. 1994. Synchronous oscillations in neuronal systems: mechanisms and functions. *J. Comput. Neurosci.* 1:11–38

Gray CM, McCormick DA. 1996. Chattering cells: superficial pyramidal neurons contributing to the generation of synchronous oscillations in the visual cortex. *Science* 274:109–13

Gray CM, König P, Engel A, Singer W. 1989. Oscillatory responses in cat visual cortex exhibit inter-columnar synchronization which reflects global stimulus properties. *Nature* 338:334–37

Griesmayr B, Gruber WR, Klimesch W, Sauseng P. 2010. Human frontal midline theta and its synchronization to gamma during a verbal delayed match to sample task. *Neurobiol. Learn. Mem.* 93(2):208–15

Gulyás AI, Hájos N, Katona I, Freund TF. 2003. Interneurons are the local targets of hippocampal inhibitory cells which project to the medial septum. *Eur. J. Neurosci.* 17(9):1861–72

Gulyás AI, Miles R, Sík A, Tóth K, Tamamaki N, Freund TF. 1993. Hippocampal pyramidal cells excite inhibitory neurons through a single release site. *Nature* 366(6456):683–87

Haider B, McCormick DA. 2009. Rapid neocortical dynamics: cellular and network mechanisms. *Neuron* 62(2):171–89

Hájos N, Pálhalmi J, Mann E, Németh B, Paulsen O, Freund TF. 2004. Spike timing of distinct types of GABAergic interneuron during hippocampal gamma oscillations in vitro. *J. Neurosci.* 24:9127–37

Hájos N, Paulsen O. 2009. Network mechanisms of gamma oscillations in the CA3 region of the hippocampus. *Neural. Netw.* 22:1113–19

Halgren E, Babb TL, Crandall PH. 1977. Responses of human limbic neurons to induced changes in blood gases. *Brain Res.* 132:43–63

Harris KD, Csicsvari J, Hirase H, Dragoi G, Buzski G. 2003. Organization of cell assemblies in the hippocampus. *Nature* 4 24:552–56

Hasenstaub A, Shu Y, Haider B, Kraushaar U, Duque A, McCormick D. 2005. Inhibitory postsynaptic potentials carry synchronized frequency information in active cortical networks. *Neuron* 47:423–35

Hormuzdi SG, Pais I, LeBeau FE, Towers SK, Rozov A, et al. 2001. Impaired electrical signaling disrupts gamma frequency oscillations in connexin 36-deficient mice. *Neuron* 31:487–95

Huerta PT, Lisman JE. 1995. Bidirectional synaptic plasticity induced by a single burst during cholinergic theta oscillation in CA1 in vitro. *Neuron* 15(5):1053–63

Isomura Y, Sirota A, Ozen S, Montgomery S, Mizuseki K, et al. 2006. Integration and segregation of activity in entorhinal-hippocampal subregions by neocortical slow oscillations. *Neuron* 52(5):871–82

Jackson J, Goutagny R, Williams S. 2011. Fast and slow γ rhythms are intrinsically and independently generated in the subiculum. *J. Neurosci.* 31(34):12104–17

Jarvis MR, Mitra PP. 2001. Sampling properties of the spectrum and coherency of sequences of action potentials. *Neural Comput.* 13:717–49

Jasper HH, Andrews HL. 1938. Brain potentials and voluntary muscle activity in man. *J. Neurophysiol.* 1:87–100

Jensen O, Colgin L. 2007. Cross-frequency coupling between neuronal oscillations. *Trends Cogn. Sci.* 11:267–69

Jensen O, Lisman JE. 1996. Theta/gamma networks with slow NMDA channels learn sequences and encode episodic memory: role of NMDA channels in recall. *Learn Mem.* 3(2–3):264–78

Jinno S, Klausberger T, Marton L, Dalezios Y, Roberts J, et al. 2006. Neuronal diversity in GABAergic long-range projections from the hippocampus. *J. Neurosci.* 27:8790–804

Johnston D, Wu SM-S. 1994. *Foundations of Cellular Neurophysiology*. Cambridge, MA: MIT Press

Kawaguchi Y, Kubota Y. 1997. GABAergic cell subtypes and their synaptic connections in rat frontal cortex. *Cereb Cortex* 7:476–86

Kay LM. 2003. Two species of gamma oscillations in the olfactory bulb: dependence on behavioral state and synaptic interactions. *J. Integr. Neurosci.* 2:31–44

Kisvárday ZF, Beaulieu C, Eysel UT. 1993. Network of GABAergic large basket cells in cat visual cortex (area 18): implication for lateral disinhibition. *J. Comp. Neurol.* 327(3):398–415

Klausberger T, Somogyi P. 2008. Neuronal diversity and temporal dynamics: the unity of hippocampal circuit operations. *Science* 321:53–57

Kopell N, Ermentrout GB. 2002. Mechanisms of phase-locking and frequency control in pairs of coupled neural oscillators. In *Handbook on Dynamical Systems*, ed. B Fielder, pp. 3–54. New York: Elsevier
Kopell N, Ermentrout G, Whittington M, Traub R. 2000. Gamma rhythms and beta rhythms have different synchronization properties. *Proc. Natl. Acad. Sci. USA* 97:1867–72
Lakatos P, Shah AS, Knuth KH, Ulbert I, Karmos G, Schroeder CE. 2005. An oscillatory hierarchy controlling neuronal excitability and stimulus processing in the auditory cortex. *J. Neurophysiol.* 94(3):1904–11
Laurent G. 2002. Olfactory network dynamics and the coding of multidimensional signals. *Nat. Rev. Neurosci.* 3:884–95
Leger JF, Stern EA, Aertsen A, Heck D. 2005. Synaptic integration in rat frontal cortex shaped by network activity. *J. Neurophysiol.* 93:281–93
Leopold D, Murayama Y, Logothetis N. 2003. Very slow activity fluctuations in monkey visual cortex: implications for functional brain imaging. *Cereb Cortex* 13:422–33
Leung LS. 1982. Nonlinear feedback model of neuronal populations in hippocampal CAl region. *J. Neurophysiol.* 47:845–68
Lewis D, Hashimoto T, Volk D. 2005. Cortical inhibitory neurons and schizophrenia. *Nat. Rev. Neurosci.* 6:312–24
Lien CC, Jonas P. 2003. Kv3 potassium conductance is necessary and kinetically optimized for high-frequency action potential generation in hippocampal interneurons. *J. Neurosci.* 23(6):2058–68
Lisman J. 2005. The theta/gamma discrete phase code occurring during the hippocampal phase precession may be a more general brain coding scheme. *Hippocampus* 15:913–22
Lisman JE, Idiart MA. 1995. Storage of 7 ± 2 short-term memories in oscillatory subcycles. *Science* 267:1512–15
Llinás RR, Ribary U, Jeanmonod D, Kronberg E, Mitra PP. 1999. Thalamocortical dysrhythmia: a neurological and neuropsychiatric syndrome characterized by magnetoencephalography. *Proc. Natl. Acad. Sci. USA* 96(26):15222–27
Lytton W, Sejnowski T. 1991. Simulations of cortical pyramidal neurons synchronized by inhibitory interneurons. *J. Neurophysiol.* 66:1059–79
Magee JC, Johnston D. 1997. A synaptically controlled, associative signal for Hebbian plasticity in hippocampal neurons. *Science* 275(5297):209–13
Mainen ZF, Sejnowski TJ. 1995. Reliability of spike timing in neocortical neurons. *Science* 268(5216):1503–6
Mann E, Suckling J, Hájos N, Greenfield S, Paulsen O. 2005. Perisomatic feedback inhibition underlies cholinergically induced fast network oscillations in the rat hippocampus in vitro. *Neuron* 45:105–17
Markram H, Lubke J, Frotscher M, Sakmann B. 1997. Regulation of synaptic efficacy by coincidence of postsynaptic APs and EPSPs. *Science* 275:213–15
McCormick DA, Connors BW, Lighthall JW, Prince DA. 1985. Comparative electrophysiology of pyramidal and sparsely spiny stellate neurons of the neocortex. *J. Neurophysiol.* 54(4):782–806
Minlebaev M, Colonnese M, Tsintsadze T, Sirota A, Khazipov R. 2011. Early γ oscillations synchronize developing thalamus and cortex. *Science* 334:226–29
Mizuseki K, Diba K, Pastalkova E, Buzsáki G. 2011. Hippocampal CA1 pyramidal cells form functionally distinct sublayers. *Nat. Neurosci.* 14(9):1174–81
Montgomery SM, Buzsáki G. 2007. Gamma oscillations dynamically couple hippocampal CA3 and CA1 regions during memory task performance. *Proc. Natl. Acad. Sci. USA* 104(36):14495–500
Montgomery SM, Sirota A, Buzsáki G. 2008. Theta and gamma coordination of hippocampal networks during waking and rapid eye movement sleep. *J. Neurosci.* 28:6731–41
Morita K, Kalra R, Aihara K, Robinson H. 2008. Recurrent synaptic input and the timing of gamma-frequency-modulated firing of pyramidal cells during neocortical "UP" states. *J. Neurosci.* 28:1871–81
Mormann F, Fell J, Axmacher N, Weber B, Lehnertz K, et al. 2005. Phase/amplitude reset and theta-gamma interaction in the human medial temporal lobe during a continuous word recognition memory task. *Hippocampus* 15(7):890–900
Muresan R, Jurjut O, Moca V, Singer W, Nikolic D. 2008. The oscillation score: an efficient method for estimating oscillation strength in neuronal activity. *J. Neurophysiol.* 99:1333–53
Murthy VN, Fetz EE. 1992. Coherent 25- to 35-Hz oscillations in the sensorimotor cortex of awake behaving monkeys. *Proc. Natl. Acad. Sci. USA* 89(12):5670–74

Neymotin SA, Lazarewicz MT, Sherif M, Contreras D, Finkel LH, Lytton WW. 2011. Ketamine disrupts theta modulation of gamma in a computer model of hippocampus. *J. Neurosci.* 31(32):11733–43

O'Keefe J, Recce M. 1993. Phase relationship between hippocampal place units and the EEG theta rhythm. *Hippocampus* 3:317–30

Palva J, Palva S, Kaila K. 2005. Phase synchrony among neuronal oscillations in the human cortex. *J. Neurosci.* 25:3962–72

Penttonen M, Kamondi A, Acsady L, Buzsáki G. 1998. Gamma frequency oscillation in the hippocampus of the rat: intracellular analysis in vivo. *Eur. J. Neurosci.* 10:718–28

Peyrache A, Battaglia FP, Destexhe A. 2011. Inhibition recruitment in prefrontal cortex during sleep spindles and gating of hippocampal inputs. *Proc. Natl. Acad. Sci. USA* 108:17207–12

Pike FG, Goddard RS, Suckling JM, Ganter P, Kasthuri N, Paulsen O. 2000. Distinct frequency preferences of different types of rat hippocampal neurones in response to oscillatory input currents. *J. Physiol.* 529:205–13

Pinault D, Deschénes M. 1992. Voltage-dependent 40-Hz oscillations in the rat reticular thalamic neurons in vivo. *Neuroscience* 51:245–58

Popescu AT, Popa D, Paré D. 2009. Coherent gamma oscillations couple the amygdala and striatum during learning. *Nat. Neurosci.* 12:801–7

Quilichini P, Sirota A, Buzsáki G. 2010. Intrinsic circuit organization and theta-gamma oscillation dynamics in the entorhinal cortex of the rat. *J. Neurosci.* 30(33):11128–42

Ray S, Maunsell JH. 2011. Different origins of gamma rhythm and high-gamma activity in macaque visual cortex. *PLoS Biol.* 9(4):e1000610

Rodriguez E, George N, Lachaux JP, Martinerie J, Renault B, Varela FJ. 1999. Perception's shadow: long-distance synchronization of human brain activity. *Nature* 397(6718):430–33

Roelfsema PR, Engel AK, König P, Singer W. 1997. Visuomotor integration is associated with zero time-lag synchronization among cortical areas. *Nature* 385(6612):157–61

Schroeder CE, Lakatos P. 2009. Low-frequency neuronal oscillations as instruments of sensory selection. *Trends Neurosci.* 32:9–18

Senior TJ, Huxter JR, Allen K, O'Neill J, Csicsvari J. 2008. Gamma oscillatory firing reveals distinct populations of pyramidal cells in the CA1 region of the hippocampus. *J. Neurosci.* 28(9):2274–86

Sik A, Penttonen M, Ylinen A, Buzsáki G. 1995. Hippocampal CA1 interneurons: an in vivo intracellular labeling study. *J. Neurosci.* 15(10):6651–65

Sik A, Ylinen A, Penttonen M, Buzsáki G. 1994. Inhibitory CA1-CA3-hilar region feedback in the hippocampus. *Science* 265(5179):1722–24

Singer W. 1999. Neuronal synchrony: a versatile code for the definition of relations? *Neuron* 24:49–65

Singer W, Gray C. 1995. Visual feature integration and the temporal correlation hypothesis. *Annu. Rev. Neurosci.* 18:555–86

Sirota A, Montgomery S, Fujisawa S, Isomura Y, Zugaro M, Buzsáki G. 2008. Entrainment of neocortical neurons and gamma oscillations by the hippocampal theta rhythm. *Neuron* 60:683–97

Sohal VS, Zhang F, Yizhar O, Deisseroth K. 2009. Parvalbumin neurons and gamma rhythms enhance cortical circuit performance. *Nature* 459:698–702

Soltesz I, Deschênes M. 1993. Low- and high-frequency membrane potential oscillations during theta activity in CA1 and CA3 pyramidal neurons of the rat hippocampus under ketamine-xylazine anesthesia. *J. Neurophysiol.* 70:97–116

Tallon-Baudry C, Bertrand O, Fischer C. 2001. Oscillatory synchrony between human extrastriate areas during visual short-term memory maintenance. *J. Neurosci.* 21(20):RC177

Tass P, Rosenblum MG, Weule J, Kurths J, Pikovsky A, et al. 1998. Detection of n:m phase locking from noisy data: application to magnetoencephalography. *Phys. Rev. Lett.* 81:3291–94

Tateno T, Robinson HP. 2009. Integration of broadband conductance input in rat somatosensory cortical inhibitory interneurons: an inhibition-controlled switch between intrinsic and input-driven spiking in fast-spiking cells. *J. Neurophysiol.* 101(2):1056–72

Tiesinga P, Sejnowski TJ. 2009. Cortical enlightenment: Are attentional gamma oscillations driven by ING or PING? *Neuron* 63:727–32

Tomioka R, Okamoto K, Furuta T, Fujiyama F, Iwasato T, et al. 2005. Demonstration of long-range GABAergic connections distributed throughout the mouse neocortex. *Eur. J. Neurosci.* 21(6):1587–600

Tort AB, Komorowski R, Eichenbaum H, Kopell N. 2010. Measuring phase-amplitude coupling between neuronal oscillations of different frequencies. *J. Neurophysiol.* 104(2):1195–210

Tort AB, Komorowski RW, Manns JR, Kopell NJ, Eichenbaum H. 2009. Theta-gamma coupling increases during the learning of item-context associations. *Proc. Natl. Acad. Sci. USA* 106:20942–47

Tort AB, Kramer MA, Thorn C, Gibson DJ, Kubota Y, et al. 2008. Dynamic cross-frequency couplings of local field potential oscillations in rat striatum and hippocampus during performance of a T-maze task. *Proc. Natl. Acad. Sci. USA* 105:20517–22

Traub R, Bibbig A, LeBeau F, Buhl E, Whittington M. 2004. Cellular mechanisms of neuronal population oscillations in the hippocampus in vitro. *Annu. Rev. Neurosci.* 27:247–78

Traub RD, Draguhn A, Whittington MA, Baldeweg T, Bibbig A, et al. 2002. Axonal gap junctions between principal neurons: a novel source of network oscillations, and perhaps epileptogenesis. *Rev. Neurosci.* 13(1):1–30

Traub RD, Whittington MA, Collins SB, Buzsáki G, Jefferys JGR. 1996b. Analysis of gamma rhythms in the rat hippocampus in vitro and in vivo. *J. Physiol.* 493:471–84

Traub RD, Whittington MA, Stanford IM, Jefferys JG. 1996a. A mechanism for generation of long-range synchronous fast oscillations in the cortex. *Nature* 383(6601):621–24

Tukker J, Fuentealba P, Hartwich K, Somogyi P, Klausberger T. 2007. Cell type–specific tuning of hippocampal interneuron firing during gamma oscillations in vivo. *J. Neurosci.* 27:8184–89

Uhlhaas P, Singer W. 2006. Neural synchrony in brain disorders: relevance for cognitive dysfunctions and pathophysiology. *Neuron* 52:155–68

Varela F, Lachaux J, Rodriguez E, Martinerie J. 2001. The brainweb: phase synchronization and large-scale integration. *Nat. Rev. Neurosci.* 2:229–39

Vicente R, Gollo LL, Mirasso CR, Fischer I, Pipa G. 2008. Dynamical relaying can yield zero time lag neuronal synchrony despite long conduction delays. *Proc. Natl. Acad. Sci. USA* 105:17157–62

Von Stein A, Sarnthein J. 2000. Different frequencies for different scales of cortical integration: from local gamma to long range alpha/theta synchronization. *Int. J. Psychophysiol.* 38:301–13

Voytek B, Canolty RT, Shestyuk A, Crone NE, Parvizi J, Knight RT. 2010. Shifts in gamma phase-amplitude coupling frequency from theta to alpha over posterior cortex during visual tasks. *Front. Hum. Neurosci.* 4:191

Wang X-J. 1993. Ionic basis for intrinsic 40 Hz neuronal oscillations. *NeuroReport* 5:221–24

Wang X-J. 1999. Fast burst firing and short-term synaptic plasticity: a model of neocortical chattering neurons. *Neuroscience* 89:347–62

Wang X-J. 2002. Pacemaker neurons for the theta rhythm and their synchronization in the septohippocampal reciprocal loop. *J. Neurophysiol.* 87:889–900

Wang X-J. 2010. Neurophysiological and computational principles of cortical rhythms in cognition. *Physiol. Rev.* 90(3):1195–268

Wang X-J, Buzsáki G. 1996. Gamma oscillation by synaptic inhibition in a hippocampal interneuronal network model. *J. Neurosci.* 16:6402–13

Wang X-J, Rinzel J. 1992. Alternating and synchronous rhythms in reciprocally inhibitory model neurons. *Neural Comput.* 4:84–97

Whittingstall K, Logothetis NK. 2009. Frequency-band coupling in surface EEG reflects spiking activity in monkey visual cortex. *Neuron* 64:281–89

Whittington MA, Traub RD, Jefferys JGR. 1995. Synchronized oscillations in interneuron networks driven by metabotropic glutamate receptor activation. *Nature* 373:612–15

Whittington MA, Traub RD, Kopell N, Ermentrout B, Buhl EH. 2000. Inhibition-based rhythms: experimental and mathematical observations on network dynamics. *Int. J. Psychophysiol.* 38(3):315–36

Wilson HR, Cowan JD. 1972. Excitatory and inhibitory interactions in localized populations of model neurons. *Biophys. J.* 12:1–24

Wulff P, Ponomarenko AA, Bartos M, Korotkova TM, Fuchs EC, et al. 2009. Hippocampal theta rhythm and its coupling with gamma oscillations require fast inhibition onto parvalbumin-positive interneurons. *Proc. Natl. Acad. Sci. USA* 106(9):3561–66

RELATED RESOURCES

Bartos M, Vida I, Jonas P. 2007. Synaptic mechanisms of synchronized gamma oscillations in inhibitory interneuron networks. *Nat. Rev. Neurosci.* 8:45–56

Engel A, Fries P, Singer W. 2001. Dynamic predictions: oscillations and synchrony in top-down processing. *Nat. Rev. Neurosci.* 2:704–16

György Buzsáki's Web page: **http://www.med.nyu.edu/buzsakilab/**

Uhlhaas P, Singer W. 2006. Neural synchrony in brain disorders: relevance for cognitive dysfunctions and pathophysiology. *Neuron* 52:155–68

Varela F, Lachaux J, Rodriguez E, Martinerie J. 2001. The brainweb: phase synchronization and large-scale integration. *Nat. Rev. Neurosci.* 2:229–39

Wang XJ. 2010. Neurophysiological and computational principles of cortical rhythms in cognition. *Physiol. Rev.* 90:1195–268

Whittington MA, Traub RD, Kopell N, Ermentrout B, Buhl EH. 2000. Inhibition-based rhythms: experimental and mathematical observations on network dynamics. *Int. J. Psychophysiol.* 38:315–36

Xiao-Jing Wang's Web page: **http://wang.medicine.yale.edu/**

The Restless Engram: Consolidations Never End

Yadin Dudai

Department of Neurobiology, Weizmann Institute of Science, Rehovot 76100, Israel; email: yadin.dudai@weizmann.ac.il

Annu. Rev. Neurosci. 2012. 35:227–47

First published online as a Review in Advance on March 20, 2012

The *Annual Review of Neuroscience* is online at neuro.annualreviews.org

This article's doi: 10.1146/annurev-neuro-062111-150500

0147-006X/12/0721-0227$20.00

Keywords

memory trace, long-term memory, behavioral plasticity, memory systems, episodic memory, internal representations

Abstract

Memory consolidation is the hypothetical process in which an item in memory is transformed into a long-term form. It is commonly addressed at two complementary levels of description and analysis: the cellular/synaptic level (synaptic consolidation) and the brain systems level (systems consolidation). This article focuses on selected recent advances in consolidation research, including the reconsolidation of long-term memory items, the brain mechanisms of transformation of the content and of cue-dependency of memory items over time, as well as the role of rest and sleep in consolidating and shaping memories. Taken together, the picture that emerges is of dynamic engrams that are formed, modified, and remodified over time at the systems level by using synaptic consolidation mechanisms as subroutines. This implies that, contrary to interpretations that have dominated neuroscience for a while, but similar to long-standing cognitive concepts, consolidation of at least some items in long-term memory may never really come to an end.

Contents

INTRODUCTION

Memory consolidation: hypothetical process in which a memory item is transformed into a long-term or remote form

Those who consider *In principio erat verbum* ("in the beginning there was the word") as a biblical aphorism only, philosophical connotations notwithstanding, may be gratified to discover that it applies to scientific research as well. Occasionally, scientific practice is shaped by terms whose original meaning has mutated over time. The study of memory consolidation provides an intriguing example. Since first proposed by Muller & Pilzecker (1900), the term consolidation has acquired multiple usages and meanings. It even budded off new terminology by acquiring a prefix (reconsolidation). Given the recent impressive advance of research on this topic, it seems apt to explore what memory consolidation currently means and the implications concerning our understating of memory at large.

Imaginative and resourceful as they were, Muller and Pilzecker were not the first to identify consolidation. Roman orators already knew about it (Quintillian 1C AD/1921). Though not yet so termed, consolidation entered the clinical discourse as a consequence of observations of amnesic patients (Ribot 1882). This and additional findings that preceded and coincided with the studies by Muller and Pilzecker are not reiterated here (Dudai 2004). Many impressive advances in molecular, cellular, and systems neuroscience that relate to memory mechanisms are also not discussed. Instead, the present discussion focuses on selected recent developments that have changed our view on how memories become long-term and on their subsequent fate.

CONCEPTS AND CRITERIA

Memory consolidation is the hypothetical process in which a memory item is transformed into a long-term form. It is commonly addressed at two levels of description and analysis: the cellular/synaptic level and the brain systems level. Synaptic consolidation refers to the post-encoding transformation of information into a long-term form at local nodes in the neural circuit that encodes the memory. The current central dogma of synaptic consolidation is that it involves stimulus ("teacher")-induced activation of intracellular signaling cascades, resulting in posttranslational modifications, modulation of gene expression, and synthesis of gene products that alter synaptic efficacy. Synaptic consolidation is traditionally assumed to draw to a close within hours of its initiation. The stimulus that triggers it in the local node may represent perceptually or internally driven information. Synaptic consolidation is found throughout the animal kingdom.

Systems consolidation refers to the postencoding reorganization of long-term memory (LTM) over distributed brain circuits. The process may last days to years, depending on the memory system, task, and author. The conventional taxonomy of LTM systems (Squire 2004) distinguishes between declarative memory, which is memory for facts (semantic) or events (episodic) that requires conscious awareness for retrieval, and nondeclarative memory, a collection of memory faculties that do not require conscious awareness for retrieval. Systems consolidation commonly refers to declarative memory, but may exist in nondeclarative memory as well.

"Reconsolidation" refers to a consolidation process that is initiated by reactivation of LTM. The process is assumed to transiently destabilize LTM.

How Is Consolidation Identified?

Although certain changes detected in the brain may reflect consolidation, none can so far be used as a definitive signature of consolidation. Currently, the only accepted criterion to infer consolidation is the existence of a time window of susceptibility to amnesic agents. An amnesic agent that does not exhibit time-dependent decrease in efficacy is assumed to affect maintenance or expression of memory rather than consolidation (Shema et al. 2007).

RE-CONSOLIDATION, OR IS IT?

The traditional consolidation hypothesis implied that, for any item in LTM, consolidation starts and ends just once. Accordingly, classical discussions of consolidation referred explicitly to the "fixation" of memory (Glickman 1961, McGaugh 1966). Social psychology and introspection favored a shakier engram (Bartlett 1932), but proponents of the consolidation hypothesis drew a distinction between the postulated immutability of consolidated memory items and the dynamic nature of behavior (McGaugh 1966). The view that consolidation occurs just once per item was, however, challenged by the late 1960s. Researchers reported that presentation of a reminder cue (RC) rendered a seemingly consolidated memory item labile to amnesic agents (Misanin et al. 1968). The prototypical experimental protocol goes like this: Training is followed by time to complete the postulated consolidation period. An RC, usually the conditioned stimulus (CS), is then presented to reactivate the memory. An amnesic agent is administered simultaneously or immediately afterward. LTM is then retested. Under these conditions, LTM may be blocked. No such effect is detected if retrieval is not followed by the amnesic agent or the amnesic agent is not preceded by retrieval. This reactivation-induced reopening of a consolidation-like window challenged the unidirectional memory maturation view (Spear 1973) and was termed reconsolidation (Rodriguez et al. 1993, Przybyslawski & Sara 1997).

Reservations concerning interpretations as well as paradigmatic drives diverted the exploration of reconsolidation away from mainstream memory research. Although a few groups pursued the topic (reviewed in Sara 2000), the notion lost favor, as reflected, for example, in the number of publications: Of the 27,061 papers relating to "memory" published in the psychobiology literature from 1993 to 1999, only 6 referred to "reconsolidation" (*Thomson Reuters Science Web of Knowledge*). The notion of reconsolidation was ultimately revitalized by a study that targeted an identified memory circuit in the brain (basolateral amygdala) and blocked reactivated LTM of a well-defined task (fear conditioning) with a widely used amnesic agent (the protein synthesis inhibitor anisomycin) (Nader et al. 2000). This signal paper triggered a surge of interest, data, and insights. Bibliometry

Synaptic consolidation: hypothetical process in which information is transformed into a long-term form at local nodes in the neural circuits that encode the memory

Systems consolidation: hypothetical process in which an experience-dependent internal representation is converted into a long-term form and reorganized over distributed brain circuits

Reconsolidation: postulated consolidation process initiated by reactivation of a long-term memory item in the system that already stores this item

Long-term memory (LTM): item lasting long after it is encoded; in behavioral neuroscience, "long" is conventionally considered more than one day

Declarative memory: requires conscious awareness for retrieval, usually classified into memory for facts (semantic) and memory for events (episodic)

Nondeclarative memory: can be retrieved in the absence of conscious awareness, for example, habit and skill

Engram: the physical record of a memory item in the brain; a memory trace

RC: reminder cue

CS: conditioned stimulus

US: unconditioned stimulus

again illustrates the trend: From 2001 to 2010, of the 61,950 publications on memory, 413 referred to reconsolidation (*Thomson Reuters Science Web of Knowledge*), presenting an almost 50-fold absolute increase per annum in the scientific vox populi.

Phenomena construed as reconsolidation have now been reported in many species and memory protocols. They were demonstrated mostly in synaptic consolidation but shown to occur also in systems consolidation (Debiec et al. 2002, Winocur et al. 2009). The resurrection of reconsolidation was not greeted smoothly. Reservations were raised once again concerning interpretations (McGaugh 2004). Yet, it soon became a widely accepted and stimulating observation (Dudai 2004, Nader & Hardt 2009, Alberini 2011, McKenzie & Eichenbaum 2011). The present discussion refers to only a few key questions that have gained particular attention as the field has progressed.

Boundary Conditions for Reconsolidation

Reconsolidation seems not to occur every time LTM is reactivated. Understanding the conditions under which it takes place is likely to cast light on storage and retrievability of memory in general. Among the boundary conditions for reconsolidation identified so far, two are noted here. The first relates to competition among memories that are elicited by the RC. The second relates to the role of new information upon presentation of the RC.

When multiple associations are elicited by the RC, the one that comes to dominate behavior tends to reconsolidate (Eisenberg et al. 2003). In most reconsolidation studies, the competing associations are the original CS–unconditioned stimulus (US) association and the "inhibitory" CS–US association (i.e., the outcome of experimental extinction). If one could identify exactly when to intervene with an amnesic agent in the course of retrieval/extinction training, it would be possible to favor or block one of the competing traces. This appears to depend on the task and on the kinetics of RC presentation (Eisenberg et al. 2003, Suzuki et al. 2004, Garelick & Storm 2005, Monfils et al. 2009, Perez-Cuesta & Maldonado 2009, de la Fuente et al. 2011). Yet, this approach has already been reported to allow attenuation of fear memories (Monfils et al. 2009, Schiller et al. 2010) (see below).

Another important boundary condition for reconsolidation is the requirement of novel information at the time of the reactivation session. Studying fear conditioning in the crab *Chasmagnathus*, Pedreira et al. (2004) concluded that impairing reactivated LTM by a protein synthesis inhibitor was effective only when there was a mismatch between what the animal expected and what actually occurred. Such mismatch drives learning (Rescorla & Wagner 1972). Indeed, using spatial memory and intrahippocampal infusions of a protein synthesis inhibitor in the rat, Morris et al. (2006) identified reconsolidation only when the protocol involved encoding of new information at the time of retrieval (see also Rodriguez-Ortiz et al. 2008). Similarly, Winters et al. (2009) reported that, in object recognition in the rat, the N-methyl-D-aspartate (NMDA) glutamate receptor inhibitor blocked reactivated LTM so long as salient novel contextual information was present during memory reactivation. Of relevance is also the observation that blockade of the NMDA receptor, which is critical for encoding, blocked reconsolidation, but not expression, of fear memory in the rat (Ben Mamou et al. 2006). Evidence supporting the importance of encoding in triggering reconsolidation could also be inferred from studies of human procedural (Walker et al. 2003) and declarative (Hupbach et al. 2007, Forcato et al. 2009, Kuhl et al. 2010) memory. All in all, this evidence raises the possibility that reconsolidation has to do with updating old with new information (but see Tronel et al. 2005). The possibility should also not be excluded that the two boundary conditions—trace competition and need for new information—reflect a common basic requirement, as the new information may be considered to compete with the old.

Reconsolidation as an Opportunity for Memory Enhancement

If reconsolidation updates memory, one should also be able to exploit it for reinforcing memory. Indeed, this has been demonstrated by several studies. Tronson et al. (2006) reported that, upon retrieval of long-term fear conditioning in the rat, inhibiting the activity of the enzyme protein kinase A in the amygdala impaired memory, whereas stimulating this enzyme enhanced memory. In humans, it is more practical to use sensory and verbal stimuli instead of pharmacological agents. Coccoz et al. (2011) trained volunteers to associate syllables in a distinct audiovisual context. They reactivated LTM by presenting the training context followed by one of the cue syllables, but instead of getting the opportunity to complete the test, the participants were instructed to immerse their arm in ice-cold water. A day later memory was tested, this time without interruption. The exposure to the stressor upon reactivation of the memory enhanced performance on the subsequent day. Similar results, though taxing shorter-term memory, were reported by Finn & Roediger (2011), this time using pairs of Swahili-English vocabulary words as memoranda and presenting negatively arousing pictures immediately after a cued recall test. Performance on the subsequent recall test was best for items whose initial retrieval was followed by the negative pictures.

Luckily, an arm in ice or annoying pictures are not the only ways to exploit reconsolidation for the sake of improving memory. Both schoolchildren and university students can improve their memory by practicing self-testing, because retrieval practice is a powerful mnemonic enhancer (Karpicke & Roediger 2008). This could well be the contribution of reconsolidation to success in the classroom (Roediger & Butler 2011).

Reconsolidation in the Real World

That some types of memory could be enhanced merely by testing was known before reconsolidation was implicated in the process, and the practical benefit of knowing that reconsolidation is involved is still unclear. Similarly, reconsolidation may help in understanding why episodic information becomes distorted over time (Hupbach et al. 2007, Edelson et al. 2011), but it is unlikely that this understanding could be used to remedy false memory. In contrast, in some other real-life phenomena in which reconsolidation may be involved, understanding the mechanisms may culminate in beneficial interventions. The most salient example concerns the attempt to ameliorate posttraumatic stress disorder. Two approaches are used. In one, investigators administer shortly before, during, or immediately after memory reactivation a drug that suppresses physiological manifestation of emotion. A β-blocker is the drug of choice because of its proven safety. Following this administration, patients with chronic posttraumatic stress disorder had attenuated memory for one day in human eyeblink conditioning to noise (Kindt et al. 2009), emotional enhancement of verbal information (Kroes et al. 2010), and a physiological response associated with imagery of trauma (Pitman et al. 2006). Despite these results, the clinical value of this approach is still unclear.

The other approach is nonpharmacological. Schiller et al. (2010) adapted for humans the procedure devised by Monfils et al. (2009) for the rat. Monfils et al. (2009) conditioned rats to associate tone with shock, and after 24 h, they activated the memory by the tonal CS, followed by extinction training within or after the reconsolidation window, which closes within a few hours. When tested for subsequent LTM, the rats that received extinction training within the reconsolidation window, but not afterward, displayed attenuated conditioned fear 24 h later. There was no reversal of fear as judged by spontaneous recovery, renewal (testing in a different context), reinstatement (retraining on the US only), and saving (amount of training needed for reacquisition of the task after extinction).

Schiller et al. (2010) exploited similarly the extinction-reconsolidation boundaries in humans. They trained participants to fear a visual CS by associating it with a mild shock to the wrist. A day later they presented the CS only. The participants were then trained in an extinction paradigm after 10 min or 6 h. In the 10-min group, LTM, as expressed in skin conductance response to the CS, was blocked even one year later. It now remains to be seen whether these results hold also for real-life complex recollections. It is not expected to be easy: Even in rats, higher-order associations are not blocked by blocking reconsolidation (Debiec et al. 2006), and resilient real-life traumatic memories in humans are expected to be densely associated. Nevertheless, the approach provides hope for treatment.

Can blockade of reconsolidation erase memory, or just block its expression? The tools available to assess memory erasure in reconsolidation are identical to those used in the study of extinction and consolidation. The gold standard is the lack of spontaneous recovery, reinstatement, renewal, and saving. Hence, demonstrating that the defect is a storage rather than a retrieval impairment relies on a negative finding: Memory not found, ergo memory not there. To circumvent the problem, researchers need new methods so they can identify the neuronal signature of the distinct engram (Nader & Hardt 2009).

Are Consolidation and Reconsolidation the Same?

The types of neuronal mechanisms that subserve reconsolidation are basically similar to those that subserve consolidation. First and foremost, inhibitors of macromolecular synthesis block both processes (Nader et al. 2000). Differential contributions of a spectrum of receptors, intracellular signaling, and transcription factors to reconsolidation versus consolidation have, however, been described. Examples of these differences include the obligatory involvement of brain-derived neurotrophic factor, but not the transcription factor Zif268, in consolidation and vice versa in reconsolidation of contextual fear memory in the rat hippocampus (Lee et al. 2004); the recruitment in reconsolidation of only a subset of immediate-early genes that are induced in consolidation (von Hertzen & Giese 2005); and the requirement for the interaction between specific initiation factors in the lateral amygdala in consolidation but not reconsolidation of elemental fear conditioning in the rat (Hoeffer et al. 2011). It remains to be determined whether a differential contribution to reconsolidation could be identified in mechanisms that have recently gained increased attention in consolidation research, such as additional growth factors (Chen et al. 2011), protein degradation (Lee et al. 2008), and epigenesis (Day & Sweatt 2011).

The question arises, however, whether the molecular dissociations, once found, reflect a fundamental dissociation between consolidation and reconsolidation. Differences in the contribution of specific molecular components to encoding, extinction, or reconsolidation can stem from differences in cue valence, context, or test demands (Berman & Dudai 2001, Tronson & Taylor 2007). This probably accounts for the lack in generalization of molecular signatures across reconsolidation tasks (Tronson & Taylor 2007). Hence, even if some differences are identified in the molecular signatures of consolidation and reconsolidation, the question remains whether they reflect genuine mechanistic differences that warrant proclaiming these as distinct natural kinds. The suggestion was, therefore, made that reconsolidation is use-dependent lingering consolidation, whose function is to update learned information (Dudai & Eisenberg 2004, Alberini 2005, McKenzie & Eichenbaum 2011). In that case, it might pay off to stop updating information about events that do not significantly change or such that lose their relevance. This might happen in some cases as memory ages (Milekic & Alberini 2002, Eisenberg & Dudai 2004, Inda et al. 2011).

THE ENGRAM TRANSFORMED

If reconsolidation is lingering consolidation, it brings us already into the time domain of systems consolidation. Evidence for systems consolidation stems from both human (clinical and neuropsychological) and animal research (Dudai 2004, Squire 2004, Frankland & Bontempi 2005, Wang & Morris 2010, Winocur et al. 2010, McKenzie & Eichenbaum 2011). In line with the early clinical observations that contributed to the emergence of the consolidation hypothesis (Ribot 1882, Burnham 1903), a substantial number of studies report that "global" amnesics, i.e., patients with damage in their medial temporal lobe (MTL), displayed temporally graded retrograde amnesia on declarative memory tasks. The type of memory tested, whether episodic or semantic, is highly relevant, as explained below. In addition, a substantial number of studies using animal models of amnesia confirm that the hippocampus is required for LTM for only a limited time after encoding (Squire et al. 2001; for studies with differing conclusions, see Winocur et al. 2010, Sutherland & Lehmann 2011). In addition, a substantial number of functional brain imaging studies in healthy human participants show reduced recollection-correlated activity over time in mediotemporal structures but increased activity in the neocortex (e.g., Smith & Squire 2009; see also Smith et al. 2010). Similar conclusions emerge from metabolic mapping in laboratory animals (Bontempi et al. 1999, Ross & Eichenbaum 2006).

The Standard Model of Systems Consolidation

A dominant model that attempted to explain graded retrograde amnesia was the standard consolidation theory (SCT) (McClelland et al. 1995, Squire 2004; for an influential harbinger, see Marr 1971). This model posits that the hippocampus is only a temporary repository for memory and that the neocortex stores the memory thereafter. Specifically, the model postulates that encoding, storage, and retrieval of declarative information is initially dependent on the hippocampal complex (HPC) and related MTL structures as well as neocortical areas relevant to the encoded stimuli. The hippocampal trace is probably a compressed version of the representation. Over time, the information reorganizes by replaying (see below) the hippocampal representation to the neocortex. This reinstates the corresponding neocortical memory, resulting in incremental adjustments of neocortical connections and establishment of a long-lasting, reorganized representation, while the hippocampal memory decays.

SCT: standard consolidation theory

HPC: hippocampal complex

MTT: multiple-trace theory

The Multiple-Trace and the Trace-Transformation Models

Over time, some evidence that seems incompatible with SCT has accumulated. Most significant, the effect of MTL lesions on subtypes of declarative memory is not consistent: Autobiographical episodes are the most severely affected, and the retrograde temporal gradient for this type of memory is either absent or very shallow, sparing only memories acquired several decades earlier. Driven by these observations and corresponding findings in animal models of amnesia, Nadel & Moscovitch (1997) proposed an alternative, the multiple-trace theory (MTT). MTT posits that the HPC rapidly and obligatorily encodes all episodic information. This information is sparsely encoded in distributed ensembles of HPC neurons, acts as an index for neurocortical neurons that attend the information, and binds them into a coherent representation. The resulting hippocampal-neocortical ensemble constitutes the memory trace for the episode. Because reactivation of the trace commonly occurs in an altered context, it results in newly encoded hippocampal traces, which, in turn, bind new traces in the neocortex. This results in multiple traces that share some or all the information about the initial episode. Over time, having multiple related traces facilitates the extraction of factual information into a semantic representation of the gist of the episode. This information integrates into a larger body of semantic knowledge and becomes independent of the specific episode. Contextual information about the episode, which is required for bona fide episodic recollection, continues,

Remote memory: lasts longer than a few months (in animals) to many years (in humans)

TTT: trace-transformation theory

SAM: schema assimilation model

however, to depend on the HPC as long as the memory exists. Opponents to MTT claimed that patients with well-characterized MTL lesions show intact remote, including autobiographical, memory, unless the damage exceeds the MTL (Squire & Bayley 2007). This argument has been challenged (Rosenbaum et al. 2008, Race et al. 2011). It also does not explain why functional neuroimaging identifies in healthy individuals HPC activation in retrieval of remote autobiographical memory (Gilboa et al. 2004, Viard et al. 2010). Among the open questions concerning the functional imaging data are the following: To what extent do cue-induced imagining processes (Hassabis et al. 2007), as opposed to genuine recollection, contribute to HPC activation? Does this activation reflect processes essential for, or just correlative to, retrieval?

An update of MTT, the trace-transformation theory (TTT), focuses on the proposed abstraction and transformation of HPC-neocortical episodic information into neocortical semantic representations (Winocur et al. 2010, Winocur & Moscovitch 2011). The resulting gist memories are posited to coexist and interact with those representations in which the context/episodicity is retained and that remain HPC dependent. Winocur et al. (2007) tested a TTT prediction in the rat by using context-dependent versions of two hippocampal-dependent tasks—peer-induced food preference and contextual fear conditioning. They tested the rats at short and long intervals in the training context or in a different context. According to TTT, but not according to a conservative reading of SCT (which predicts that HPC memories are reorganized in a similar form in the neocortex), the change in context is expected to affect performance at the short but not the long interval when the contextless schematic version of the memory is supposed to take over. This indeed was the case.

The Schema Assimilation Model

SCT and MTT consider systems consolidation as a gradual, lengthy process. The schema assimilation model (SAM) (Tse et al. 2007) posits that systems consolidation could be accomplished quickly if a previously established body of related knowledge, i.e., a mental schema (Bartlett 1932), is available into which the new knowledge may be assimilated. Tse et al. (2007) trained rats using hippocampal-dependent flavor-location associations. After the rats learned a set of different associations over a few weeks, a single trial learning was sufficient to consolidate rapidly the memory of a new association: Although hippocampal lesion 3 h after training disrupted subsequent LTM, a similar lesion at 48 h was ineffective, demonstrating that LTM was no longer hippocampal dependent. No such effect was seen when the rats were trained with inconsistent flavor-location-paired associates, indicating that formation of a postulated schema is a prerequisite for rapid systems consolidation. The rapid schema–dependent learning was associated with upregulation of immediate-early genes in the medial prefrontal cortex (Tse et al. 2011), whereas pharmacological intervention targeted at that area prevented the new learning as well as the recall of consolidated information. These findings are in agreement with the assertion of earlier models that initial memory is in both the HPC and the neocortex (see also Lesburgueres et al. 2011), but they are in disagreement with the assumption that the neocortex is a slow learner (on additional evidence for fast cortical learning, see Takashima et al. 2009; on sleep and consolidation, see below).

That different systems consolidation models coexist is a stimulating situation, as they provide opportunities for new hypothesis-driven experiments, which are likely to generate not only new data but also new models.

WORKING AT REST

Synaptic consolidation processes take place immediately after encoding and re-encoding. But when does systems consolidation happen? Apparently some of the action takes place when we rest and while we sleep. The contribution of rest and sleep to consolidation is one of the most fascinating frontiers in current consolidation research.

The idea that sleep enhances memory predates scientific investigation. Quintillian (1C AD/1921) turns his readers' attention to the "curious fact. . .that the interval of a single night will greatly increase the strength of the memory." It took some time for scientific research to reconfirm that this is the case (Jenkins & Dallenbach 1924). Systematic analyses of sleep and brain mechanisms followed with the development of functional brain-imaging techniques (Smith & Butler 1982, Karni et al. 1994). Ample evidence now supports the claim that memory consolidation benefits from sleep (Stickgold & Walker 2007, Diekelmann & Born 2010a; for a dissident view, see Vertes & Siegel 2005). However, questions arise regarding which (type of) memory, which (process of) consolidation, and which (mechanism of) sleep are involved.

A Reminder Concerning Sleep

Sleep is a natural, reversible physiological and mental state characterized by reduced consciousness, suspended volitional sensorimotor activity, and altered metabolism (Steriade & McCarley 2005). It involves the cyclic occurrence of phases, each conventionally defined by characteristic differences in brain activity, coordinated eye movements, and tonic muscle activity. The standard classification of sleep in primates and felines is into rapid eye movement (REM) and non-REM (NREM) stages. In humans, they alternate roughly every 90 min. NREM is further divided into substages, corresponding to the depth of sleep. NREM stage N3 (formerly stages 3 and 4), in which the deepest sleep occurs, is referred to as electroencephalogram (EEG) slow-wave sleep (SWS) based on the prevalence of EEG slow waves (below 4Hz). Other types of field-potential oscillations that characterize SWS include "spindles" (0.5–2 s, 10–15 Hz) and transient, sharp-wave "ripples" (SWR) (50–120 ms, 100–250 HZ). SWR probably reflect a transient relief of inhibition, permitting windows of opportunity for the expression of selective representations (Csicsvari et al. 1999). REM sleep is characterized by ponto-geniculo-occiptal waves and theta activity (approximately 4–7 Hz). REM and NREM also differ markedly in the level of activity of neuromodulatory systems in the brain during each of the phases (Pace-Schott & Hobosn 2002). SWS appears mostly in early sleep, whereas REM sleep occurs mostly at late sleep. Dreams, the succession of sensorimotor and affective hallucinatory experiences that occur involuntarily during sleep, are prevalent during REM but not confined to it (Nielsen 2000, Nir & Tononi 2010).

SWS: slow-wave sleep

SWR: sharp-wave "ripples"

REM: rapid eye movement

NREM: non-REM

Which Memory Systems Benefit from Consolidation in Sleep?

The evidence for the role of sleep in consolidation of acquired sensory and motor skills was initially considered more robust than that for other types of memory (Walker & Stickgold 2004). A wide spectrum of skills have been studied in this respect (Karni et al. 1994, Walker et al. 2005, Ferrara et al. 2008, Mednick et al. 2009, Wamsley et al. 2010a). It is now well established, however, that declarative memory benefits from sleep as well, though the involvement and contribution of distinct sleep stages and the underlying brain mechanisms to declarative and nondeclarative memory may differ (Diekelmann & Born 2010a,b; Walker & Stickgold 2010; also see below). A broad spectrum of tasks that involve declarative components or are considered "classical" declarative tasks have been investigated (Fenn et al. 2003, Wagner et al. 2004, Sterpenich et al. 2009, Diekelmann et al. 2011, Rauchs et al. 2011, Wilhelm et al. 2011).

Which Properties of Memory Increase the Benefit from Consolidation in Sleep?

Sleep may promote the preferential strengthening of emotional memoranda (Sterpenich et al. 2009) and of items that are expected to be subsequently retrieved (Rauchs et al. 2011, Wilhelm et al. 2011). The possibility that consolidation in sleep favors selected items gains support from multiple lines of evidence. Rudoy et al. (2009) trained awake participants to associate object locations with sound and found that only those associations that were cued during sleep with their relevant sound were strengthened. This was taken to indicate that specific associations are preferentially reactivated and strengthened during sleep. At the brain physiology level, Huber et al. (2004) reported that activity in SWS has a local component that can be triggered by a sensorimotor adaptation task that involves specific brain regions. Additional electrophysiological evidence shows that most sleep slow waves and their underlying neuronal states occur locally in the brain and, hence, are fit to process information selectively (Nir et al. 2011).

When and How in Sleep

An early report on the role of sleep in consolidation of perceptual skill suggested that REM sleep is critical (Karni et al. 1994). Furthermore, a brief nap was reported to be effective in off-line improvement of skill performance only when the nap contained both REM and SWS but not when it involved only SWS (Mednick et al. 2003). The role of REM and NREM in the effect of napping on other types of tasks that involve skill components is task dependent (Korman et al. 2007, Wamsley et al. 2010b). The possibility was also raised that, at least in some motor skills, siesta-induced improvement is not due to napping but to resting (Rieth et al. 2010). Additional studies proposed a role in skill consolidation for both REM and NREM stages (Stickgold et al. 2000). Two types of processes have been proposed: stabilization against interference and gain in performance. The suggestion was further made that stabilization benefits from the SWS stage, whereas enhancement benefits from the REM stage (Sagi 2011). However, whether skill consolidation in sleep involves enhancement in addition to stabilization remains unclear (Brawn et al. 2010).

A signal set of findings that paved the way to the exploration of the neuronal and circuit mechanisms involved in memory consolidation at large was that hippocampal place cells (Pavlides & Winson 1989) and place-cell ensembles (Wilson & McNaughton 1994), postulated to encode place representations, "replay" during sleep periods that follow performance on spatial behavioral tasks. The order of firing in the task is largely preserved in the replay (Skaggs & McNaughton 1996). Most studies reported that the replay occurred during SWS, particularly during SWR (Nadasdy et al. 1999, Lee & Wilson 2002, Diba & Buzsaki 2007, Ji & Wilson 2007). The reactivation of hippocampal maps during post-training rest/sleep periods was further reported to predict performance on hippocampal-dependent matching-to-place reward tasks (Dupret et al. 2010). SWR are associated with increased cortico-hippocampal communication (Siapas & Wilson 1998). Indeed replay in SWS was found in the neocortex (Ji & Wilson 2007, Euston et al. 2007, Payrache et al. 2009), but also in the ventral striatum (Lansink et al. 2009). The presumed "reading out" in the SWR is accompanied by compression of the replay (Nadasdy et al. 1999, Euston et al. 2007, Ji & Wilson 2007); in other words, the postulated representation is played in "fast forward" (and, as noted below, under certain circumstances in "fast backward"). The virtual speed is 15–20 times faster than in the real world (Davidson et al. 2009). Replay in REM was also reported during periods of theta modulation with a "read-out" rate close to real time (Louie & Wilson 2001).

However, most importantly, structured replay of hippocampal place cells preserving information on the distinct behavioral experience was found to also occur in the awake state. Such replay is observed time locked either to an immediate experience (Foster & Wilson 2006, Csicsvari et al. 2007, Diba & Buzsaki 2007) or to a spatially and temporally remote one (Davidson et al. 2009, Karlsson & Frank 2009). What happens in sleep may, thus, cast light on the processes and mechanisms that relate to consolidation in the awake state as well. The replay in the awake state is either forward or backward. For example, Foster & Wilson (2006) reported sequential reverse replay during awake periods immediately after a run on a track, when the rat pauses, with the reverse replay declining with familiarity, whereas Diba & Buzsaki (2007) reported forward replay at the beginning of such a run, as if in anticipation of the run, but reverse replay at the end of the run. Moreover, Dragoi & Tonegawa (2011) reported that some of the replays in aware-rest states are "preplays," i.e., sequences that match those subsequently recorded when the rats were running in a new place. The potential implications of this finding are discussed below.

A single SWR is brief, allowing replay of only a limited distance (approximately 1–2-m run), which fits routine laboratory mazes but not the real life of a wild rat. How does the brain replay realistic distances? It appears that firing sequences corresponding to long runs through a large environment are replayed in chains of shorter subsequences, with each segment corresponding to a single SWR (Davidson et al. 2009).

All in all, it has been proposed that: (*a*) Forward replay during "gaps" in the behavioral performance subserves the retrieval of path information to aid memory-guided decision making; (*b*) postexperience forward replay in both awake and sleep states is likely to subserve consolidation of acquired representations; and (*c*) reverse replay in the awake state may subserve episodic binding (Carr et al. 2011). Thus, once the episode is bound and familiar, additional fast-backward replay may not be needed (see above). Interestingly, echoing the latter proposal, human functional brain imaging in a realistic episodic task revealed immediate (within seconds) poststimulus activity in the hippocampus and in the dorsal striatum that predicted subsequent memory performance. This off-line activity may reflect episodic binding and initiation of consolidation (Ben-Yakov & Dudai 2011). Tambini et al. (2010) reported memory-related enhanced corticohippocampal functional connectivity in rest periods spanning minutes after associative encoding sessions. It is also noteworthy that reactivation of memory during waking and sleep may have different roles and outcomes concerning long-term trace stability. Hence, Diekelmann et al. (2011) reported that reactivation of object-location associations by odor cues during waking resulted in destabilization of the trace, but in SWS it resulted in fast stabilization.

The aforementioned studies potentially implicate replay in memory consolidation by way of correlation (though admittedly, only some of these studies actually correlated replay with subsequent memory). Yet interventional methods suggest a causal link as well. Girardeau et al. (2009) and Ego-Stengel & Wilson (2010) stimulated the hippocampus to selectively disrupt SWR activity in maze-trained rats. They found that disruption during post-training rest periods that included sleep impaired improvement of performance over days of training. This was taken to imply that ripple-related activity could be required for uninterrupted memory consolidation. Of further relevance, disruption of sleep continuity in the mouse by optogenetic stimulation of hypocertin/orexin neurons in the lateral hypothalamus, thereby promoting arousal, impaired later performance on novel object recognition. This was correlated with fragmentation of NREM sleep; the minimal time for uninterrupted sleep critical for consolidation on the task was estimated to be 60–120 s (Rolls et al. 2011). Although no effect on distinct representations or firing patterns was determined, these findings indicate a novel approach to the dissection of consolidation processes at large. They also strengthen the notion that sleep may be not only a correlate, but also a necessary mechanism for proper consolidation.

A few cautionary remarks are necessary. First, because replay is not unique to sleep, any unique contribution sleep provides to consolidation cannot be accounted for solely by replay. If replay in sleep has any specific contribution, it must be considered in combination with other features of sleep, such as the unique metabolic

ACSH: active consolidation in sleep hypothesis

and neuromodulatory milieu and their relevant signaling cascades (e.g., Aton et al. 2009). Thus, whatever we learn from replay in sleep could inform us about consolidation in the awake state as well.

Second, the relevance of the laboratory protocols to real life raises some issues. Many of the aforementioned protocols use task repetition and, hence, heavily tax procedures and learning sets. By contrast, realistic episodic memory is a single trial involving novelty. In this context it is worthy to reiterate that encounter with novel memoranda seems to modify the pattern of replay (Foster & Wilson 2006, Dragoi & Tonegawa 2011).

Third, and probably the most relevant question at this point in time, is whether replay is indeed specifically instrumental in consolidation. Replay may be a signature of a more global information-processing mechanism, in which case, it may be permissive but not sufficient for consolidation.

As noted above, replay is not a simple function of experience (Gupta et al. 2010, Dragoi & Tonegawa 2011). Given this, it is tempting to raise the possibility that what is played, replayed, or preplayed are combinatorial internal representations that could serve as raw material for perceiving, anticipating, reacting, recollecting, and planning. Such representations are likely to gain more visibility in sleep because of the decrease in volitional activity. Linked to a broader conceptual level, this points to the potential role of cue-invoked selection of "prerepresentations" as a Darwinian mechanism in the operation of the mind (Young 1979, Heidmann et al. 1984, Dudai 2002). Seen that way, consolidation, similar to development, perception, and retrieval, involves pruning and selecting information about the world.

How It Might Work

With the above in mind, we now consider models of how consolidation could occur in sleep. To do so, it is useful to note the postulated goals of sleep. An influential overall idea is that sleep evolved to maintain homeostasis (Crick & Mitchison 1983, Borbely & Achermann 1999, Tononi & Cirelli 2006). A specific version of this idea was developed by Tononi & Cirelli (2006). They suggest that plastic processes during wakefulness result in a net widespread increase in synaptic strength in the brain and the role of sleep is to downscale synaptic strength to a baseline level that is energetically sustainable and possibly also more useful for new learning the next day. They further propose that this function is achieved during SWS. This means that sleep plays a necessary role in sustaining memory systems, and is at least permissive yet not necessarily instrumental, let alone sufficient, for consolidation. However, as research proceeds, instrumentality may be unveiled. For example, increasing the signal-to-noise ratio of privileged representations may drive them to consolidate effectively.

A different idea is that sleep involves active processes that consolidate memory, and is hence necessary and instrumental, and possibly also sufficient, in implementing steps in consolidation. This is the "active consolidation in sleep hypothesis" (ACSH) (Diekelmann & Born 2010a). ACSH could be considered an extension of the SCT that posits that declarative memory involves initial storage in the cortico-hippocampal system (step 1), but over time, via representational replay, gets reinstated in the neocortex (step 2). ACSH adds that step 2 benefits from sleep (Diekelmann & Born 2010a). ACSH gains support from additional developments in computational models (Kali & Dayan 2004), though these models do not specify sleep per se as obligatory in implementing the stages proposed.

Diekelmann & Born (2010a) suggest how ACSH may be implemented in the brain. They draw on the sequential hypothesis of sleep proposed by Giuditta et al. (1995), among others. The sequential hypothesis proposes that information acquired during the waking period is processed first in the early sleep stages, NREM and particularly SWS. Subsequent processing occurs in the later sleep stage, REM, and information eventually emerges in a new form upon awakening. Diekelmann & Born (2010a) propose specifically that, during SWS, slow oscillations, spindles, SWR, and low cholinergic activity all coordinate to promote the reactivation and redistribution of hippocampal-dependent memories to the neocortex, thereby instantiating system consolidation. Subsequently, during REM sleep, high cholinergic and theta activity promote synaptic consolidation of the newly redistributed representations in the neocortex. Ultimately, the individual wakes up with a consolidated memory. Similar systems-synaptic sequences may take place in certain nondeclarative memories as well (Dudai 2004). This type of model is agnostic to the specific systems consolidation models discussed above.

Despite their differences, the aforementioned "homeostatic" and "active" accounts of sleep are not mutually exclusive. Whereas the former emphasizes the function of sleep in general, the latter focuses on its role in consolidation. The evolution of sleep may have been initially driven by homeostatic pressure and active consolidation became nested into it over time. Furthermore, consolidation may have evolved to comply with homeostatic needs (Fischer et al. 2005). In addition, when discussing these models, the possibility should not be neglected that we may be entrapped by an adaptationist philosophy. The mechanisms discussed may have evolved as a by-product of inherent structural and functional constraints of biological systems and not under the selective pressures we contemplate (Gould & Lewontin 1979). Analysis of this possibility, which applies to many models in biology, exceeds the scope of this discussion.

CONSOLIDATIONS INTEGRATED

Memory is the retention over time of experience-dependent internal representations or of the informational capacity to reactivate or reconstruct such representations (Dudai 2002). Consolidation is the mechanism that shifts these representations into a long-term form. In considering how this is achieved, three questions are particularly relevant. First, which level of organization of the neural system is critical for encoding the content of the distinct representation? Second, is the circuit that initially encodes the representation also the one that maintains it over time? Third, how does the system ensure that the acquired representation is updated when the world changes?

The assumption that the content of a memory item is encoded at the circuit level is not a secured given, yet is highly reasonable (Dudai 2002). Furthermore, at least in complex memory systems in the mammalian brain, the neural system that encodes the information in the first place may not be identical to the system that stores the information later on, therefore trace migration occurs (McClelland et al. 1995). Given that, an integrative broad-brush depiction of consolidation considers synaptic consolidation as the elementary mechanistic process that converts experience-dependent synaptic change into a longer-term representation. If a mismatch develops between this representation and reality, new information will modify either new or old synapses in the circuit, again by triggering synaptic consolidation. The latter, thus, functions as a subroutine activated once the external and internal cues favor off-line persistence of the change. When this change applies to information already encoded as LTM, we dub it reconsolidation.

In reality, relevant information probably pre-exists in the brain; therefore, even what we deem in the laboratory as consolidation may involve reconsolidation. In memory systems in which information migrates to other distributed brain circuits to free neuronal space and/or distill information into new forms, synaptic consolidation remains the elementary subroutine that executes the process, modifying synapses as they receive new information from other circuits that previously encoded or processed relevant information (**Figure 1**). Seen this way, synaptic consolidation is a local process indifferent to the representational semantics and activated in a similar way regardless of whether the information originated in the perceptual apparatus or in mnemonic circuits.

Consolidations all have the same computational goal—to allow the adequate level of persistence in the face of expected change (Dudai 2009). Synaptic consolidation is the term we assign to the manifestation of the process at the cellular, elementary "syntactic" level, whereas systems consolidation refers to the circuit, representational "semantic" level. Synaptic consolidation is the basic building block of systems consolidation. In simple systems, the goal of systems consolidation is achieved within the same circuit that first encoded the memory; therefore, we do not see the waves of change in which information redistributes among

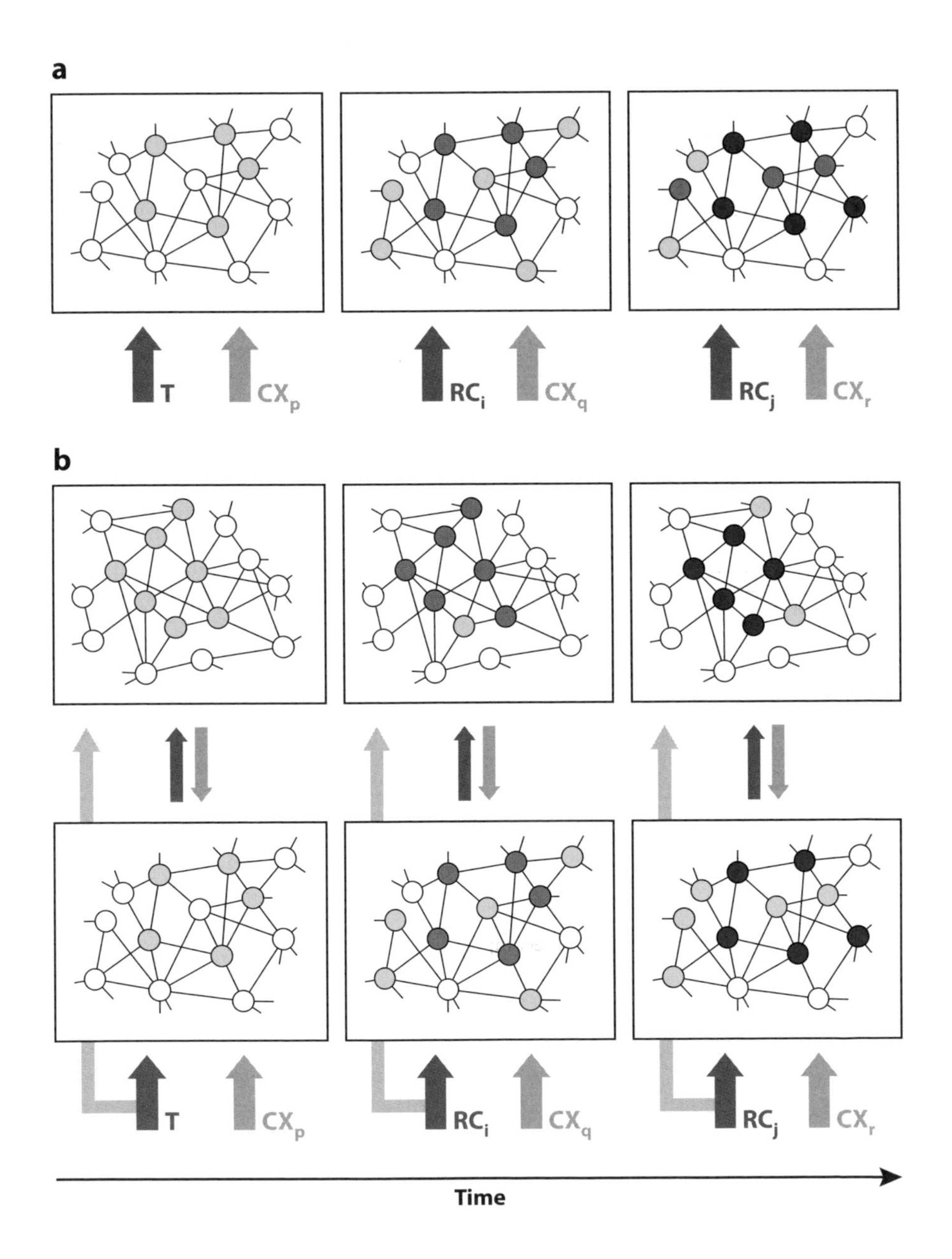

circuits. However, local migrations may still occur within the original circuit. It is all a matter of resolution.

ON THE RECONSOLIDATION OF TERMS AND IDEAS

Overall, the evidence discussed in this article suggests that consolidation of information in the behaving brain rarely stops unless one or possibly two conditions occur. Either the behavior and the context in which it is executed remain constant, there is no new information, and therefore no need to learn and update; this probably never happens even in simple systems living in boring environments, but even then, the capacity to update must remain viable. Or, alternatively, the internal representations become highly irrelevant to behavior and therefore not reactivated.

Because knowledge is always based on previous knowledge, and echoing the preamble to this chapter, it might be proper at this point to reactivate the methodology of the Vico (1710), the Italian philosopher who trusted that much can be learned about a culture from the etymology of words used. "Consolidation" is from the Latin *consolidare*, *con-* "together", *solidare* "make firm." The process that we term consolidation in memory research indeed subserves the binding together of acquired information into useful representations, but that information is evidently far from becoming solid. Shortly after the term was first introduced into memory research, emphasis was placed on the *solidare*, and as the term consolidated into the language of the science of memory, that connotation became widespread and guided research to look for stabilization mechanisms. Research in recent years has reconsolidated the connotation of the term to emphasize the inherent malleability of memories. In doing so, the neuroscience of memory reconciles with the intuitive, dynamic view of memory that dominates the cognitive sciences.

Figure 1

Schematic variants of memory consolidation. (*a*) Long-term memory (LTM) is stored in the same circuit, or in parts of the circuit, that initially encoded the memory. The "teacher" stimulus (T) triggers a set of intracellular signaling mechanisms that culminate in long-term alterations (depicted as changes in color) in the efficacy of a set of synapses that subserve encoding of the internal representation. This time-limited process, which is assumed to mature within hours, is termed synaptic consolidation and is an obligatory step in the neural registration of any type of LTM. Reactivation of the LTM by a reminder cue ($RC_{i,j}$) that is associated with new information (e.g., change in context, $CX_{q,r}$) re-triggers synaptic consolidation mechanisms in the same and in additional nodes in the circuit, resulting in synaptic alteration. This is termed reconsolidation and involves some transient destabilization of the original trace. In real life, even the initial consolidation may involve reconsolidation of previous knowledge; in which case, the differentiation between T and RC is not absolute. (*b*) LTM redistributes into new brain territories. The information is encoded first in one location (*lower panel*) and/or in parallel in both locations (*lower and upper panels*). Over time, it migrates, at least in part, from one location to another while probably undergoing metamorphosis in content and cue dependency. The potential direct input of $CX_{p,q}$ to the upstream location has been omitted for simplicity. In each of the locations, the process is executed by synaptic consolidation, whereas T/RC/CX each encodes either sensory and modulatory input (as shown in panel *a*) or information about the item already processed in LTM, manipulated in the absence or in the presence of overt retrieval. This overall process is termed systems consolidation. Hence systems consolidation recurrently recruits synaptic consolidation processes as subroutines. Systems consolidation, which matures within days (or nights) to months or even longer, traditionally deals with the transformation over time of declarative memory in the corticohippocampal system. Processes similar in nature may, however, operate in other memory and brain systems, including in distributed local circuits within the same brain region. For further details, see text. The time arrow is indicated only for the slower, horizontal axis for simplicity.

DISCLOSURE STATEMENT

The author is not aware of any affiliations, memberships, funding, or financial holdings that might be perceived as affecting the objectivity of this review.

ACKNOWLEDGMENTS

I am grateful to Aya Ben-Yakov, Bartosz Brozek, Micah Edelson, Avi Mendelsohn, Morris Moscovitch, Yuval Nir, Matthew Wilson, and Gordon Winocur for comments and discussions.

LITERATURE CITED

Alberini CM. 2005. Mechanisms of memory stabilization: Are consolidation and reconsolidation similar or distinct processes? *Trends Neurosci.* 28:51–56

Alberini CM. 2011. The role of reconsolidation and the dynamic process of long-term memory formation and storage. *Front. Behav. Neurosci.* 5:(12)1–10

Aton SJ, Seibt J, Dumoulin M, Jha SK, Steinmetz N, et al. 2009. Mechanisms of sleep-dependent consolidation of cortical plasticity. *Neuron* 61:454–66

Bartlett EC. 1932. *Remembering. A Study in Experimental and Social Psychology.* London: Cambridge Univ. Press

Ben Mamou C, Ganache K, Nader K. 2006. NMDA receptors are critical for unleashing consolidated auditory fear memories. *Nat. Neurosci.* 10:1237–39

Ben-Yakov A, Dudai Y. 2011. Constructing realistic engrams: poststimulus activity of hippocampus and dorsal striatum predicts subsequent episodic memory. *J. Neurosci.* 31:9032–42

Berman DE, Dudai Y. 2001. Memory extinction, learning anew, and learning the new: dissociations in the molecular machinery of learning in cortex. *Science* 291:2417–19

Bontempi B, Laurant-Demir C, Destrade C, Jaffard R. 1999. Time-dependent reorganization of brain circuitry underlying long-term memory storage. *Nature* 400:671–75

Borbely AA, Achermann P. 1999. Sleep homeostasis and models of sleep regulation. *J. Biol. Rhythm.* 14:559–68

Brawn TP, Fenn KM, Nusbaum HC, Margoliash D. 2010. Consolidating the effects of waking and sleep on motor-sequence learning. *J. Neurosci.* 30:13977–82

Burnham WH. 1903. Retroactive amnesia: illustrative cases and a tentative explanation. *Am. J. Psychol.* 14:382–96

Carr MF, Jadhav SP, Frank LM. 2011. Hippocampal replay in the awake state: a potential substrate for memory consolidation and retrieval. *Nat. Neurosci.* 14:147–53

Chen DY, Stern SA, Garcia-Osta A, Saunier-Rebori B, Pollonini G, et al. 2011. A critical role for IGF-II in memory consolidation and enhancement. *Nature* 469:491–97

Coccoz V, Maldionado H, Delorenzi A. 2011. The enhancement of reconsolidation with a naturalistic mild stressor improves the expression of a declarative memory in humans. *Neuroscience* 185:61–72

Crick F, Mitchison G. 1983. The function of dream sleep. *Nature* 304:111–14

Csicsvari J, Hirase H, Czurko A, Narnia A, Buzsaki G. 1999. Oscillatory coupling of hippocampal cells and interneurons in the behaving rat. *J. Neurosci.* 19:274–87

Csicsvari J, O'Neill J, Allen K, Senior T. 2007. Place-selective firing contributes to the reverse-order reactivation of CA1 pyramidal cells during sharp waves in open-field explorations. *Eur. J. Neurosci.* 26:704–16

Davidson TJ, Kloosterman F, Wilson MA. 2009. Hippocampal replay of extended experience. *Neuron* 63:497–507

Day JJ, Sweatt JD. 2011. Epigenetic mechanisms in cognition. *Neuron* 70:813–29

de la Fuente V, Freudental R, Romano A. 2011. Reconsolidation or extinction: transcription factor switch in the determination of memory course after retrieval. *J. Neurosci.* 31:5562–73

Debiec J, Doyere V, Nader K, LeDoux JE. 2006. Directly reactivated, but not indirectly reactivated, memories undergo reconsolidation in the amygdala. *Proc. Natl. Acad. Sci. USA* 103:3428–33

Debiec J, LeDoux JE, Nader K. 2002. Cellular and systems reconsolidation in the hippocampus. *Neuron* 36:527–38

Diba K, Buzsaki G. 2007. Forward and reverse hippocampal place-cell sequences during ripples. *Nature Neurosci.* 10:1241–42

Diekelmann S, Born J. 2010a. The memory function of sleep. *Nat. Rev. Neurosci.* 11:114–26

Diekelmann S, Born J. 2010b. Slow-wave sleep takes the leading role in memory reorganization. *Nat. Neurosci.* 11:218

Diekelmann S, Buchel C, Born J, Rasch B. 2011. Labile or stable: opposing consequences for memory when reactivate during waking and sleep. *Nat. Neurosci.* 14:381–86

Dragoi G, Tonegawa S. 2011. Preplay of future place cell sequences by hippocampal cellular assemblies. *Nature* 469:397–401

Dudai Y, Eisenberg M. 2004. Rites of passage of the engram: reconsolidation and the lingering consolidation hypothesis. *Neuron* 44:93–100

Dudai Y. 2002. *Memory from A to Z. Keywords, Concepts and Beyond.* Oxford: Oxford Univ. Press

Dudai Y. 2004. The neurobiology of consolidations, or, how stable is the engram? *Annu. Rev. Psychol.* 55:51–86

Dudai Y. 2009. Predicting not to predict too much: how the cellular machinery of memory anticipates the uncertain future. *Philos. Trans. R. Soc. B* 364:1255–62

Dupret D, O'Neill J, Pleydell-Bouverie B, Csicsvari J. 2010. The reorganization and reactivation of hippocampal maps predict spatial memory performance. *Nat. Neurosci.* 13:995–1002

Edelson M, Sharot T, Dolan RJ, Dudai Y. 2011. Following the crowd: brain substrates of long-term memory conformity. *Science* 333:108–11

Ego-Stengel V, Wilson MA. 2010. Disruption of ripple-associated hippocampal activity during rest impairs spatial learning in the rat. *Hippocampus* 20:1–10

Eisenberg M, Dudai Y. 2004. Reconsolidation of fresh, remote, and extinguished fear memory in Medaka: old fears don't die. *Eur. J. Neurosci.* 20:3397–403

Eisenberg M, Kobilo T, Berman DE, Dudai Y. 2003. Stability of retrieved memory: inverse correlation with trace dominance. *Science* 301:1102–4

Euston DR, Tatsuno M, McNaughton BL. 2007. Fast-forward playback of recent memory sequences in prefrontal cortex during sleep. *Science* 318:1147–50

Fenn KM, Nusbaum HC, Margoliash D. 2003. Consolidation during sleep of perceptual learning of spoken language. *Nature* 425:614–16

Ferrara M, Iaria G, Tempesta D, Curcio G, Moroni F, et al. 2008. Sleep to find your way: the role of sleep in the consolidation of memory for navigation in humans. *Hippocampus* 18:844–51

Finn B, Roediger HL III. 2011. Enhancing retention through reconsolidation: negative emotional arousal following retrieval enhances later recall. *Psychol. Sci.* 22:781–86

Fischer S, Nitschke MF, Melchert UH, Erdmann C, Born J. 2005. Motor memory consolidation in sleep shapes more efficient neuronal representations. *J. Neurosci.* 25:11248–55

Forcato C, Argibay PF, Pedreira ME, Maldonado H. 2009. Human reconsolidation does not always occur when a memory is retrieved: the relevance of the reminder structure. *Neurobiol. Learn. Mem.* 91:50–57

Foster DJ, Wilson MA. 2006. Reverse replay of behavioral sequences in hippocampal place cells during the awake state. *Nature* 440:680–83

Frankland PW, Bontempi B. 2005. The organization of recent and remote memories. *Nat. Rev. Neurosci.* 6:119–30

Garelick MG, Storm DR. 2005. The relationship between memory retrieval and memory extinction. *Proc. Natl. Acad. Sci. USA* 102:9091–92

Gilboa A, Winocur G, Grady CL, Hevenor SJ, Moscovitch M. 2004. Remembering our past: functional neuroanatomy of recollection of recent and very remote personal events. *Cereb. Cortex* 14:1214–25

Girardeau G, Benchenane K, Wiener SI, Buzsaki G, Zugaro MB. 2009. Selective suppression of hippocampal ripples impairs spatial memory. *Nat. Neurosci.* 10:1222–23

Giuditta A, Ambrosini MV, Montagnese P, Mandile P, Cotugno M, et al. 1995. The sequential hypothesis of the function of sleep. *Behav. Brain Res.* 69:157–66

Glickman SE. 1961. Preservative neural processes and consolidation of the neural trace. *Psychol. Bull.* 58:218–33

Gould SJ, Lewontin RC. 1979. The spandrels of San Marco and the Panglossian paradigm: a critique of the adaptationist programme. *Proc. R. Soc. Lond. Ser. B* 205:581–98

Gupta AS, van der Meer MAA, Touretzky DS, Redish AD. 2010. Hippocampal replay is not a simple function of experience. *Neuron* 65:695–705

Hassabis D, Kumaran D, Vann SD, Maguire EA. 2007. Patients with hippocampal amnesia cannot imagine new experiences. *Proc. Natl. Acad. Sci. USA* 104:1726–31

Heidmann A, Heidmann TM, Changeux JP. 1984. Stabilization selective de representations neuronals per resonance entre 'prerepresentations' spontanees du reseau cerebral et 'percepts' evoques par interactions avec le monde exterieur. *Compt. Rend. Acad. Sci.* 299:839–43

Hoeffer CA, Cowansage KK, Arnold EC, Banko JL, Moerke NJ, et al. 2011. Inhibition of the interactions between eukaryotic initiation factor 4E and 4G impairs long-term associative memory consolidation but not reconsolidation. *Proc. Natl. Acad. Sci. USA* 108:3383–88

Huber R, Ghilardi MF, Massimini M, Tononi G. 2004. Local sleep and learning. *Nature* 430:78–81

Hupbach A, Hardt O, Gomez R, Nadel L. 2007. Reconsolidation of episodic memories: a subtle reminder triggers integration of new information. *Learn. Mem.* 14:47–53

Inda MC, Muravieva EV, Alberini CM. 2011. Memory retrieval and passage of time: from reconsolidation and strengthening to extinction. *J. Neurosci.* 31:1635–43

Jenkins JG, Dallenbach KM. 1924. Obliviscence during sleep and waking. *Am. J. Psychol.* 35:605–12

Ji D, Wilson MA. 2007. Coordinated memory replay in the visual cortex and hippocampus during sleep. *Nat. Neurosci.* 10:100–7

Kali S, Dayan P. 2004. Off-line replay maintains declarative memories in a model of hippocampal-neocortical interactions. *Nat. Neurosci.* 7:286–94

Karlsson MP, Frank LM. 2009. Awake replay of remote experiences in the hippocampus. *Nat. Neurosci.* 12:913–18

Karni A, Tanne D, Rubenstein BS, Askenasy JJM, Sagi D. 1994. Dependence on REM sleep of overnight improvement of a perceptual skill. *Science* 265:679–82

Karpicke JD, Roediger HL III. 2008. The critical importance of retrieval for learning. *Science* 319:966–68

Kindt M, Soeter M, Vervliet B. 2009. Beyond extinction: erasing human fear responses and preventing the return of fear. *Nat. Neurosci.* 12:256–58

Korman M, Doyon J, Doljansky J, Carrier J, Dagan Y, Karni A. 2007. Daytime sleep condenses the time course of motor memory consolidation. *Nat. Neurosci.* 10:1206–13

Kroes MCW, Strange BA, Dolan RJ. 2010. β-Adrenergic blockade during memory retrieval in humans evokes sustained reduction of declarative emotional memory enhancement. *J. Neurosci.* 30:3959–63

Kuhl BA, Shah AT, DuBrow S, Wagner AD. 2010. Resistance to forgetting associated with hippocampus-mediated reactivation during new learning. 2010. *Nat. Neurosci.* 13:501–6

Lansink CS, Goltstein PM, Lankelma JV, McNaughton BL, Pennartz CMA. 2009. Hippocampus leads ventral striatum in replay of place-reward information. *PLoS Biol.* 7:e1000173

Lee AK, Wilson MA. 2002. Memory of sequential experience in the hippocampus during slow wave sleep. *Neuron* 36:1183–94

Lee JLC, Everitt BJ, Thomas KL. 2004. Independent cellular processes for hippocampal memory consolidation and reconsolidation. *Science* 304:839–43

Lee S-H, Choi J-H, Lee N, Lee H-R, Kim J-I, et al. 2008. Synaptic protein degradation underlies destabilization of retrieved fear memory. *Science* 319:1253–56

Lesburgueres E, Gobbo OL, Alaux-Cantin S, Hambucken A, Trifilieff P, Bontempi B. 2011. Early tagging of cortical networks is required for the formation of enduring associative memory. *Science* 331:924–28

Louie K, Wilson MA. 2001. Temporally structured replay of awake hippocampal ensemble activity during rapid eye movement sleep. *Neuron* 29:145–56

Marr D. 1971. Simple memory: a theory for archicortex. *Philos. Trans. R. Soc. Lond. B* 262:23–81

McClelland JL, McNaughton BL, O'Reilly RC. 1995. Why there are complementary learning systems in the hippocampus and neocortex: insights from the successes and failures of connectionist models of learning and memory. *Psychol. Rev.* 102:419–57

McGaugh JL. 1966. Time-dependent processes in memory storage. *Science* 153:1351–58

McGaugh JL. 2004. Memory reconsolidation hypothesis revived but restrained: theoretical comment on Bidenkapp and Rudy (2004). *Behav. Neurosci.* 118:1140–42

McKenzie S, Eichenbaum H. 2011. Consolidation and reconsolidation: two lives of memories? *Neuron* 71:224–33

Mednick SC, Makovski T, Cai DJ, Jiang YV. 2009. Sleep and rest facilitate implicit memory in a visual search task. *Vision Res.* 49:2557–65

Mednick SC, Nakayama K, Stickgold R. 2003. Sleep-dependent learning: A nap is as good as a night. *Nat. Neurosci.* 6:697–98

Milekic MH, Alberini CM. 2002. Temporally graded requirement for protein synthesis following memory reactivation. *Neuron* 36:521–25

Misanin JR, Miller RR, Lewis DJ. 1968. Retrograde amnesia produced by electroconvulsive shock after reactivation of consolidated memory trace. *Science* 160:554–55

Monfils M-H, Cowansage KK, Klann E, LeDoux JE. 2009. Extinction-reconsolidation boundaries: key to persistent attenuation of fear memories. *Science* 324:951–55

Morris RGM, Inglis J, Ainge JA, Olverman HJ, Tulloch J, et al. 2006. Memory reconsolidation: sensitivity of spatial memory to inhibition of protein synthesis in dorsal hippocampus during encoding and retrieval. *Neuron* 50:479–89

Muller GE, Pilzecker A. 1900. Experimentelle Beitrage zur Lehre vom Gedachtnis. *Z. Psychol.* 1:S1–300

Nadasdy Z, Hirase H, Czurko A, Csicsvari J, Buzsaki G. 1999. Reply and time compression of recurring spike sequences in the hippocampus. *J. Neurosci.* 19:9497–507

Nadel L, Moscovitch M. 1997. Memory consolidation, retrograde amnesia and the hippocampal complex. *Curr. Opin. Neurobiol.* 7:217–27

Nader K, Hardt O. 2009. A single standard for memory: the case for reconsolidation. *Nat. Rev. Neurosci.* 10:224–34

Nader K, Schafe GE, LeDoux JE. 2000. Fear memories require protein synthesis in the amygdala for reconsolidation after retrieval. *Nature* 406:722–26

Nielsen TA. 2000. A review of mentation in REM and NREM sleep: "covert" REM sleep as a possible reconciliation of two opposing models. *Behav. Brain Sci.* 23:793–1121

Nir Y, Staba RJ, Andrillon T, Vyazovskiy VV, Cirelli C, et al. 2011. Regional slow waves and spindles in human sleep. *Neuron* 70:153–69

Nir Y, Tononi G. 2010. Dreaming and the brain: from phenomenology to neurophysiology. *Trend. Cog. Sci.* 14:88–100

Pace-Schott EF, Hobson JA. 2002. The neurobiology of sleep: genetics, cellular physiology and subcortical networks. *Nat. Rev. Neurosci.* 3:591–605

Pavlides C, Winson J. 1989. Influences of hippocampal place cells firing in the awake state on the activity of these cells during subsequent sleep episodes. *J. Neurosci.* 9:2907–18

Payrache A, Khamassi M, Benchenane K, Wiener SI, Battaglia FP. 2009. Replay of rule-learning related neural patterns in the prefrontal cortex during sleep. *Nat. Neurosci.* 7:919–26

Pedreira ME, Perez-Cuesta LM, Maldonado H. 2004. Mismatch between what is expected and what actually occurs triggers memory reconsolidation or extinction. *Learn. Mem.* 11:579–85

Perez-Cuesta LM, Maldonado H. 2009. Memory reconsolidation and extinction in the crab: mutual exclusion or coexistence? *Learn. Mem.* 16:714–21

Pitman RK, Brunet A, Orr SP, Tremblay J, Nader K. 2006. A novel treatment for post-traumatic stress disorder by reconsolidation blockade with propranolol. *Neuropsychopharmacology* 3(1):S8–S9

Przybyslawski J, Sara SJ. 1997. Reconsolidation after reactivation of memory. *Behav. Brain Res.* 84:241–46

Quintillian. 1C AD/1921. *Institutio Oratoria*. London: Loeb Classical Library

Race E, Keane MM, Verfaellie M. 2011. Medial temporal lobe damage causes deficits in episodic memory and episodic thinking not attributable to deficits in narrative construction. *J. Neurosci.* 31:10262–69

Rauchs G, Feyers D, Landeau B, Bastin C, Luxen A, et al. 2011. Sleep contributes to the strengthening of some memories over others, depending on hippocampal activity at learning. *J. Neurosci.* 31:2563–68

Rescorla RA, Wagner AR. 1972. A theory of Pavlovian conditioning: variations in the effectiveness of reinforcement and non-reinforcement. In *Classical Conditioning II: Current Research and Theory*, ed. AH Black, WE Prokasy, pp. 64–99. New York: Appleton-Century-Crofts

Ribot TA. 1882/1977. *Diseases of Memory*. Washington, DC: Univ. Publ. Am.

Rieth CA, Cai DJ, McDevitt EA, Mednick SC. 2010. The role of sleep and practice in implicit and explicit motor learning. *Behav. Brain Res.* 214:470–74

Rodriguez WA, Phillips MY, Rodriguez SB, Martinez JL. 1993. Cocaine administration prior to reactivation facilitates later acquisition of an avoidance response in rats. *Psychopharmacology* 112:366–70

Rodriguez-Ortiz CJ, Garcia De La Torre P, Benavidez E, Ballesteros MA, Bermudez-Rattoni F. 2008. Intrahippocampal anisomycin infusions disrupt previously consolidated spatial memory only when memory is updated. *Neurobiol. Learn. Mem.* 89:352–59

Roediger HL, Butler AC. 2011. The critical role of retrieval practice in long-term retention. *Trends Cogn. Sci.* 15:20–27

Rolls A, Colas D, Adamantidis A, Carter M, Lanre-Amos T, et al. 2011. Optogenetic disruption of sleep continuity impairs memory consolidation. *Proc. Natl. Acad. Sci. USA* 108:13305–10

Rosenbaum RS, Moscovitch M, Foster JK, Schnyder DM, Gao F, et al. 2008. Patterns of autobiographical memory loss in medial-temporal lobe amnesic patients. *J. Cogn. Neurosci.* 20:1490–506

Ross RS, Eichenbaum H. 2006. Dynamics of hippocampal and cortical activation during consolidation of nonspatial memory. *J. Neurosci.* 26:4852–59

Rudoy JD, Voss JL, Westerberg CE, Paller KA. 2009. Strengthening individual memories by reactivating them during sleep. *Science* 326:1079

Sagi D. 2011. Perceptual learning in vision research. *Vis. Res.* 51:1552–66

Sara SJ. 2000. Retrieval and reconsolidation: toward a neurobiology of remembering. *Learn. Mem.* 7:73–84

Schiller D, Monfils MH, Raio CM, Johnson DC, LeDoux JE, Phelps EA. 2010. Preventing the return of fear in humans using reconsolidation update mechanisms. *Nature* 463:49–53

Shema R, Sacktor TC, Dudai Y. 2007. Rapid erasure of long-term memory associations in the cortex by an inhibitor of PKMzeta. *Science* 317:951–53

Siapas AG, Wilson MA. 1998. Coordinated interactions between hippocampal ripples and cortical spindles during slow-wave sleep. *Neuron* 21:1123–28

Skaggs WE, McNaughton BL. 1996. Replay of neuronal firing sequences in rat hippocampus during sleep following spatial experience. *Science* 271:1870–73

Smith C, Butler S. 1982. Paradoxical sleep at selective times following training is necessary for learning. *Physiol. Behav.* 29:469–73

Smith CN, Squire LR. 2009. Medial temporal lobe activity during retrieval of semantic memory is related to the age of the memory. *J. Neurosci.* 29:930–38

Smith JF, Alexander GE, Chen K, Husaim FT, Kim J, et al. 2010. Imaging systems level consolidation of novel associate memories: a longitudinal neuroimaging study. *Neuroimage* 50:826–36

Spear NE. 1973. Retrieval of memory in animals. *Psychol. Rev.* 80:163–94

Squire LR, Bayley PJ. 2007. The neuroscience of remote memory. *Curr. Opin. Neurobiol.* 17:185–96

Squire LR, Clark RE, Knowlton BJ. 2001. Retrograde amnesia. *Hippocampus* 11:50–55

Squire LR. 2004. Memory systems of the brain: a brief history and current perspective. *Neurobiol. Learn. Mem.* 82:171–77

Steriade MM, McCarley RW. 2005. *Brain Control of Wakefulness and Sleep*. New York: Kluver

Sterpenich V, Albouy G, Darasaud A, Schmidt C, Vendewalle G, et al. 2009. Sleep promotes the neural reorganization of remote emotional memory. *J. Neurosci.* 29:5143–52

Stickgold R, Walker MP. 2007. Sleep-dependent memory consolidation and reconsolidation. *Sleep Med.* 8:331–43

Stickgold R, Whidbee D, Schirmer B, Patel V, Hobson JA. 2000. Visual discrimination task improvement: a multi-step process occurring during sleep. *J. Cogn. Neurosci.* 12:246–54

Sutherland RJ, Lehmann H. 2011. Alternative conceptions of memory consolidation and the role of the hippocmapus at the systems level in rodents. *Curr. Opin. Neurobiol.* 21:446–51

Suzuki A, Josselyn SA, Frankland PW, Masushige S, Silva AJ, Kida S. 2004. Memory reconsolidation and extinction have distinct temporal and biochemical signatures. *J. Neurosci.* 24:4787–95

Takashima A, Nieuwenhuis ILC, Jensen O, Talamini LM, Rijpkema M, Fernandez G. 2009. Shift from hippocampal to neocortical centered retrieval network with consolidation. *J. Neurosci.* 29:10087–93

Tambini A, Ketz N, Davachi L. 2010. Enhanced brain correlations during rest are related to memory for recent experiences. *Neuron* 65:280–90

Tononi G, Cirelli C. 2006. Sleep function and synaptic homeostasis. *Sleep Med. Rev.* 10:49–62

Tronel S, Milekic MH, Alberini CM. 2005. Linking new information to a reactivated memory requires consolidation but not reconsolidation mechanisms. *PLoS Biol.* 3:1630–38

Tronson NC, Taylor JR. 2007. Molecular mechanisms of memory reconsolidation. *Nat. Rev. Neurosci.* 8:262–75

Tronson NC, Wiseman SL, Olausson P, Taylor JR. 2006. Bidirectional behavioral plasticity of memory reconsolidation depends on amygdalar protein kinase A. *Nat. Neurosci.* 9:167–69

Tse D, Langston RF, Kakeyama M, Bethus I, Spooner PA, et al. 2007. Schemas and memory consolidation. *Science* 316:76–82

Tse D, Takeuchi T, Kakeyama M, Kajii Y, Okuno H, et al. 2011. Schema-dependent gene activation and memory encoding in neocortex. *Science* 333:891–95

Vertes RP, Siegel JN. 2005. Time for the sleep community to take a critical look at the purported role of sleep in memory processing. *Sleep* 28:1228–29

Viard A, Lebreton K, Chetelat G, Desgranges B, Landeau B, et al. 2010. Patterns of hippocampal-neocortical interactions in the retrieval of episodic autobiographical memories across the entire life-span of aged adults. *Hippocampus* 20:153–65

Vico G. 1710/1988. *On the Most Ancient Wisdom of the Italians Unearthed from the Origins of the Latin Language*, transl. L.M. Palmer. Ithaca: Cornell Univ. Press

von Hertzen LSI, Giese KP. 2005. Memory reconsolidation engages only a subset of immediate-early genes induced during consolidation. *J. Neurosci.* 25:1935–42

Wagner U, Gais S, Haider H, Verleger R, Born J. 2004. Sleep inspires insight. *Nature* 427:352–55

Walker MP, Brakefield T, Hobson JA, Stickgold R. 2003. Dissociable stages of human memory consolidation and reconsolidation. *Nature* 425:616–20

Walker MP, Stickgold R, Alsop D, Gaab N, Schlaug G. 2005. Sleep-dependent motor memory plasticity in the human brain. *Neuroscience* 133:911–17

Walker MP, Stickgold R. 2004. Sleep-dependent learning and memory consolidation. *Neuron* 44:121–33

Walker MP, Stickgold R. 2010. Overnight alchemy: sleep-dependent memory evolution. *Nat. Neurosci.* 11:218

Wamsley EJ, Tucker MA, Payne JD, Stickgold R. 2010b. A brief nap is beneficial for human route-learning: the role of navigation experience and EEG spectral power. *Learn. Mem.* 17:332–36

Wamsley EJ, Perry K, Djonlagic I, Reaven LB, Stickgold R. 2010a. Cognitive replay of visuomotor learning at sleep onset: temporal dynamics and relationship to task performance. *Sleep* 33:59–68

Wang S-H, Morris RGM. 2010. Hippocampal-neocortical interactions in memory formation, consolidation, and reconsolidation. *Annu. Rev. Psychol.* 61:49–79

Wilhelm I, Diekelmann S, Molzow I, Ayoub A, Molle M, Born J. 2011. Sleep selectively enhances memory expected to be of future relevance. *J. Neurosci.* 31:1563–69

Wilson MA, McNaughton BL. 1994. Reactivation of hippocampal ensemble memories during sleep. *Science* 265:676–79

Winocur G, Frankland PW, Sekeres M, Fogel S, Moscovitch M. 2009. Changes in context-specificity during memory reconsolidation: selective effects of hippocampal lesions. *Learn. Mem.* 16:722–29

Winocur G, Moscovitch M, Bontempi B. 2010. Memory formation and long-term retention in humans and in animals: convergence towards a transformation account of hippocampal-neocortical interactions. *Neuropsychologia* 48:2339–56

Winocur G, Moscovitch M, Sekeres M. 2007. Memory consolidation or transformation: context manipulation and hippocampal representations of memory. *Nat. Neurosci.* 10:555–57

Winocur G, Moscovitch M. 2011. Memory transformation and systems consolidation. *J. Int. Neuropsychol. Soc.* 17:1–15

Winters BD, Tucci MC, DaCosta-Furtado M. 2009. Older and stronger object memories are selectively destabilized by reactivation in the presence of new information. *Learn. Mem.* 16:545–53

Young JZ. 1979. Learning as a process of selection and amplification. *J. Roy. Soc. Med.* 72:801–14

The Physiology of the Axon Initial Segment

Kevin J. Bender and Laurence O. Trussell

Oregon Hearing Research Center and Vollum Institute, Oregon Health and Science University, Portland, Oregon 97239; email: kbender@gallo.ucsf.edu, trussell@ohsu.edu

Annu. Rev. Neurosci. 2012. 35:249–65

First published online as a Review in Advance on March 20, 2012

The *Annual Review of Neuroscience* is online at neuro.annualreviews.org

This article's doi:
10.1146/annurev-neuro-062111-150339

0147-006X/12/0721-0249$20.00

Keywords

action potential, ion channel, neuromodulation, structural plasticity

Abstract

The action potential generally begins in the axon initial segment (AIS), a principle confirmed by 60 years of research; however, the most recent advances have shown that a very rich biology underlies this simple observation. The AIS has a remarkably complex molecular composition, with a wide variety of ion channels and attendant mechanisms for channel localization, and may feature membrane domains each with distinct roles in excitation. Its function may be regulated in the short term through the action of neurotransmitters, in the long term through activity- and Ca^{2+}-dependent processes. Thus, the AIS is not merely the beginning of the axon, but rather a key site in the control of neuronal excitability.

Contents

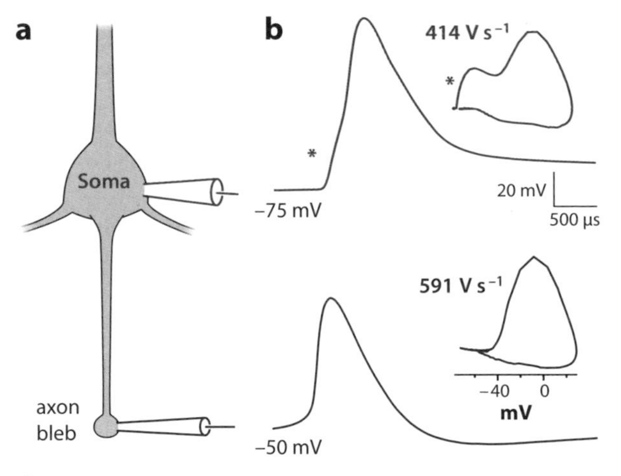

Figure 1

The axon initial segment (AIS) spike as seen in the soma and the axon. (*a*) Illustration of the recording configuration. Axon bleb is formed by the cut end of an axon in tissue slices. (*b*) Voltage traces in simultaneous recordings from the soma and bleb. The spike is triggered by a current injection into the axon bleb. Insets show dV/dt versus voltage to highlight the initial "hump" (*asterisk*) in the voltage during the rising phase of the somatic spike, corresponding to the onset of the AIS spike. Notice also the spike in the bleb initiates prior to the full somatic spike. Data provided by W. Hu and Y. Shu (Hu et al. 2009).

INTRODUCTION

The initiation of action potentials (APs) is the culmination of synaptic integration in neurons. The initial portion of the axon may seem a likely place for such initiation solely on the basis of patterns of current spread in neurons (Gesell 1940). However, it was not until the early 1950s that experimental evidence for the role of the axon in spike initiation was obtained though studies using intra- and extracellular recordings from mammalian myelinated axons and their associated cell bodies (Araki & Otani 1955; Brock et al. 1952, 1953; Coombs et al. 1957a; Fatt 1957; Freygang 1958; Fuortes et al. 1957). APs conducted antidromically and recorded in the cell body were observed to begin invariably with a small hump before the onset of the full-amplitude spike (**Figure 1**). Often, this hump occured in isolation, an observation that suggested it was a low-threshold AP that necessarily preceded a higher-threshold, full-amplitude AP. Although some controversy arose during this time as to whether this initial spike was somatic or axonal (Fatt 1957), the weight of evidence favored that the small spike began somewhere in the region between the axon hillock (the cone-shaped portion of the cell soma leading into the main body of the axon) and the first segment of myelination, a region collectively termed the axon initial segment (AIS). The small size of the hump presumably reflected passive attenuation of a local axonal spike as it spread along the axon into the soma. Orthodromic spikes, generated by synaptic stimuli or somatic current injection, also featured this initial phase of excitation, and researchers concluded that the AIS is the region at which the spike originates (Araki & Otani 1955, Coombs et al. 1957a). Initiation in the proximal axon, as opposed to the soma or dendrites, was also seen in large unmyelinated axons of lobster stretch receptors, where it was possible to record systematically at points from the soma out along the axon (Edwards & Ottoson 1958).

But why does the spike start in the AIS? Several early studies concluded that this must happen simply because the threshold voltage is significantly lower in the region where spikes first initiate (Coombs et al. 1957a, Fatt 1957); however, it remained unclear why this difference exists. Moreover, the generality of this role for the AIS was also unclear, as intracellular recordings from intact neurons were limited to a few key model preparations. Electron microscope studies in the 1960s and 1970s indicated several features common to the AIS of many neurons, including bundled microtubules and a thick undercoating of the membrane (Palay et al. 1968) (**Figure 2*a***), the latter reminiscent of what is seen in nodes of Ranvier. Striking new insights were not obtained until more modern immunohistochemical, electrophysiological, and imaging methods were brought to bear on the problem. From these, we have learned that the AIS features concentrations of specific Na^+, K^+, and Ca^{2+} channels arranged so as to determine spike generation, spike pattern, and even spike shape. The developmental, synaptic, and neuromodulatory influences on these molecular components suggest that the AIS could be a critical site for pharmacological control or could be affected in pathological states.

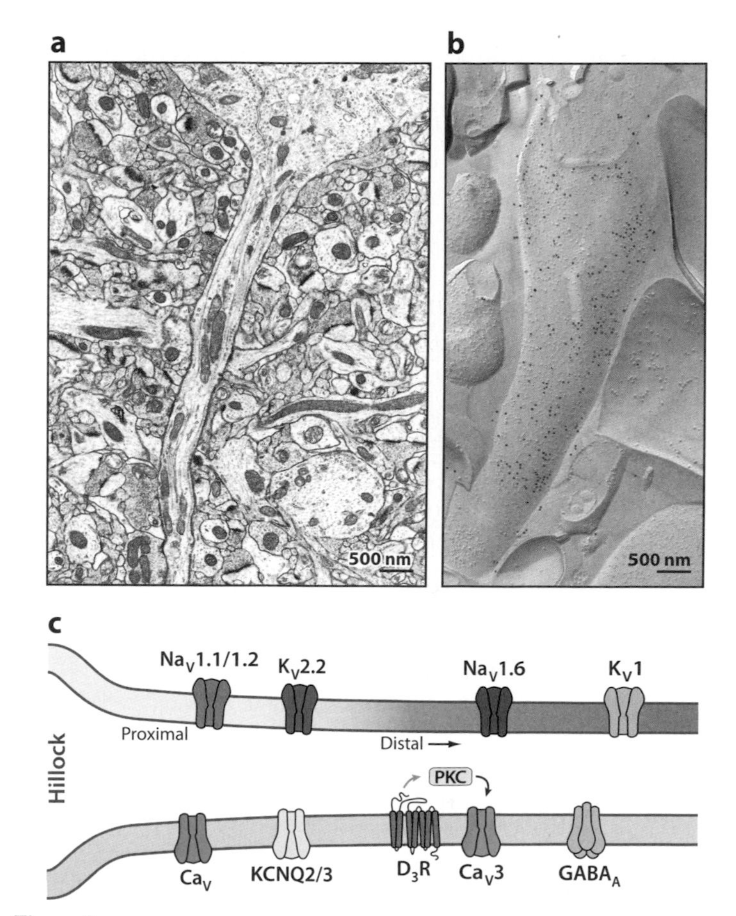

Figure 2

The structure and composition of the axon initial segment (AIS). (*a*) An electron micrograph of the AIS of a cerebral cortical pyramidal neuron (adapted from Peters et al. 1991). (*b*) Freeze-fracture immunogold labeling of $Na_V1.6$ channels in a hippocampal CA1 pyramidal cell (adapted from Lorincz & Nusser 2010). (*c*) Schematic of ion channels and receptors found in the AIS, condensed across neuron class. Top membrane illustrates differential localization of Na^+ and K^+ channels. Bottom membrane shows other channels and receptors with either no clear subcompartmental localization or incomplete immunohistochemical data (D_3R and Ca_V). Abbreviation: PKC, protein kinase C.

AXONAL SPIKE INITIATION

It may seem odd that, despite the establishment in the 1950s of the now-textbook concept of spike initiation in the axon, a tremendous effort was made 40 years later to prove this very point. In fact, this effort was vital to propel the field forward. Questions about the dogma of axonal spike initiation naturally arose when microelectrode studies concluded that dendrites, similar to axons, are active structures and can propagate APs (Llinas et al. 1968, Wong et al. 1979). Indeed, APs may even be initiated within dendrites, and in some cases, they remain there as a local spike (Schiller et al. 1997, Stuart et al. 1997). Since these efforts, the question of where spikes typically begin has been explored in a number of cell types, including principal cells and interneurons, and in myelinated and unmyelinated axons (see Defining the Axon Initial Segment, sidebar below). A wide variety of sophisticated physiological methods have been employed, including whole-cell recordings from intact axons or from swellings at cut axons (axon blebs) (**Figure 1**), excised membrane patches, cell-attached recordings of capacitative/resistive currents in axons, laser-targeted axotomy, ion and voltage imaging, and computer modeling. Using these approaches,

DEFINING THE AXON INITIAL SEGMENT

Historically, the axon initial segment (AIS) refers to the initial "nonmedullated" portion of the axon (Brock et al. 1953, Coombs et al. 1957b). Accordingly, some physiological studies identify the AIS by measuring the start site of myelination using labeling methods (Palmer & Stuart 2006, Shu et al. 2007a). Anatomical studies may also define the AIS by ultrastructural features. Other definitions and methods are needed for physiological studies of unmyelinated axons. Beyond simply referring to the proximal end of the axon, some studies define the AIS by labeling for ankyrin-G, a Na^+-channel clustering molecule (Goldberg et al. 2008, Hu et al. 2009, Kress et al. 2008). The AIS also varies widely among cell types [e.g., 20 μm in rat cerebellar Purkinje cells (Clark et al. 2005), 40–70 μm in rat layer 5 cortical pyramidal cells (Palmer & Stuart 2006)]. In ferret prefrontal cortical pyramidal cells, a 40-μm initial segment is followed by hundreds of micrometers of narrow axon before myelination begins (Shu et al. 2007a). Moreover, the presence of a morphological AIS, and the basic mechanisms of axonal ion channel clustering, varies among phyla (Rasband 2010), limiting generalized statements about sites of initiation. Clarifying what is meant by the AIS is essential to comparing conclusions among different reports.

researchers have repeatedly pinpointed the AIS as the primary region of spike initiation.

Several excellent studies have been performed in layer 5 cortical pyramidal neurons. For example, dual-patch-clamp analysis, patching the soma and the axon, has shown that the spike arises in the axon (Stuart et al. 1997). A combination of imaging and electrophysiological techniques subsequently provided unequivocal evidence that the spike arises first in the distal end of the AIS, i.e., closer to the first internode, in these cells (Palmer & Stuart 2006). Similar results have been observed in other pyramidal neurons (Hu et al. 2009, Meeks & Mennerick 2007, Popovic et al. 2011), although one study suggested that the initiation site could be beyond the AIS (Colbert & Johnston 1996).

A potential refinement of this view of spike triggering is to determine whether the site of initiation varies during different patterns of activity. Indeed, Edwards & Ottoson (1958) showed that initiation of spikes in lobster stretch receptor axons shifts toward the soma during spike bursts. In hippocampal oriens-alveus interneurons, spikes begin in the axon in response to weak stimuli, but they can shift into the somatodendritic region in response to stronger stimuli (Martina et al. 2000). In these neurons, the axon branches off from one of the cell's major dendrites. Thus, during intense activity, the relative availability of different voltage-gated ion channels along the dendrite-soma-AIS axis changes significantly, favoring more proximal initiation. Such shifts need to be confirmed experimentally and with computational approaches, as they lend greater significance to somatic ion channels during spike initiation and in the control of neuronal response properties.

Indeed, AIS spike initiation does not preclude somatic contributions to axonal output. This is most clear in Khaliq & Raman (2006), in which the excitability of cerebellar Purkinje cells was evaluated during local blockade of somatic or first-node Na^+ channels or during application of Na^+-free solutions to the initial segment. Their results (confirmed by Foust et al. 2010, Palmer et al. 2010; but see Clark et al. 2005) demonstrate that the AIS is the site of initiation and also show that somatic Na^+ channels modulate the gain of the input-output relation of Purkinje cells. This feature may be most relevant for spontaneously firing neurons and suggests that somatic excitability may be necessary to sustain high-frequency propagated spikes (Khaliq & Raman 2006). Modulation of somatic channels could, therefore, have important consequences for this interaction between AIS and somatic activity.

Na^+ Channels

Largely on theoretical grounds, it had long been thought that spikes begin in the AIS because of a reduced voltage threshold there as compared with the soma (Coombs et al. 1957a); this low threshold was thought to arise from the narrow diameter and electrical isolation of the axon combined with the properties of AIS Na^+ channels (Dodge & Cooley 1973, Mainen et al. 1995, Moore et al. 1983, Traub et al. 1994).

Numerous labeling studies confirmed that Na^+ channels are present at high abundance in the proximal axon as well as in nodes of Ranvier (Boiko et al. 2003; Duflocq et al. 2008; Lorincz & Nusser 2008, 2010; Royeck et al. 2008; Van Wart et al. 2007; Westenbroek et al. 1989). Indeed, it is now clear that the subtypes, density, and distribution of Na^+ channels are the essential issues in defining the site of spike initiation. How these and other channels are clustered and how this structure develops are reviewed elsewhere (Grubb & Burrone 2010b, Rasband 2010). Here, we explore the physiological aspects of Na^+ and other channels in the AIS.

Unexpectedly, controversy arose in physiological studies using cell-attached patches to quantify the concentration of channels. Studying layer 5 pyramidal cells, Colbert & Pan (2002) indicated that the density of channels was constant from the soma through at least 50 μm of axon. However, a later study (Kole et al. 2008) suggested that cytoskeletal elements may limit the ability of the patch technique to assess the channel number. Indeed, Na^+ imaging experiments were directly able to demonstrate the high local influx of Na^+ in the AIS (Fleidervish et al. 2010, Palmer & Stuart 2006, Bender & Trussell 2009). Moreover, within a low-Na^+ bath solution, large Na^+ currents resulted when Na^+ was locally applied in the AIS but not when it was applied to proximal dendrites (Kole et al. 2008, Palmer & Stuart 2006). Nevertheless, the study by Colbert & Pan (2002) provided important insight that the activation voltage ($V_{1/2}$) of the axonal channels was more negative than on the soma, suggesting a mechanism for reduced threshold based on channel kinetics. Research now indicates that both mechanisms hold. As discussed below, multiple subtypes of Na^+channel subunits (Na_V) are associated with the axon, and their distribution varies with cell type. $Na_V1.6$ channels are concentrated in the AIS of pyramidal cells, often toward the distal end, and have a more negative $V_{1/2}$ and less inactivation than do $Na_V1.2$ channels, which are found on the soma and the proximal axon (Hu et al. 2009; Lorincz & Nusser 2008, 2010; Schmidt-Hieber & Bischofberger 2010; Zhou & Goldin 2004). $Na_V1.2$, by contrast, are not universal components of axons; rather, their presence and biophysical properties may specifically facilitate backpropagation of APs into dendrites (Hu et al. 2009, Lorincz & Nusser 2010).

The precise assortment and location of axonal Na^+-channel subtypes varies with subtype, age, and myelination. In contrast to pyramidal cells, spinal motoneurons may employ $Na_V1.1$ as the major axonal channel (Duflocq et al. 2008), whereas, in unmyelinated retinal ganglion cell axons, $Na_V1.2$ and 1.6 are codistributed (Boiko et al. 2003). In unmyelinated hippocampal granule cells, only $Na_V1.6$ is found (Kress et al. 2010). The location-dependent changes in Na^+-channel kinetic properties in granule cells are proposed to reflect differential control by beta subunits (Schmidt-Hieber & Bischofberger 2010), a possibility yet to be tested.

Subthreshold or persistent Na^+ current (NaP) is generated in the AIS (Astman et al. 2006) and plays several roles in the triggering of axonal spikes. Kole & Stuart (2008) and Stuart & Sakmann (1995) found that Na^+ current boosts depolarizations in the AIS, thus reducing the current needed to reach threshold. The NaP of axons has also been proposed to play a role in triggering APs, and the recent study of Kole (2011) has offered a new perspective on the dynamics of spike triggering in the AIS. In pyramidal neurons, spike bursts are initiated by an afterdepolarization that is thought to be mediated by the NaP. Kole used two-photon laser axotomy to sever the axon at different lengths. Axotomy at the level of the first internode did not affect simple spikes but inhibited the afterdepolarization and abolished the ability of the neuron to generate spike bursts. Application of TTX to the first node had a similar effect. Together these data suggest that the NaP of the first node is essential to boost the voltage in the AIS to reach threshold. Thus, there is a retrograde interaction from the first node to the AIS that could help determine the mode of spike generation.

General agreement has yet to be reached regarding the precise density of functional Na^+ channels in the AIS as well as the relative changes in density from dendrite to AIS. Outside-out patch experiments suggest at least 19-fold differences between the AIS and the soma in pyramidal cells (Hu et al. 2009). Na^+ imaging and localized voltage-clamp experiments proposed differences closer to 40-fold (Kole et al. 2008). Other Na^+-imaging studies place a lower estimate of only 3-fold between the soma and the AIS (Fleidervish et al. 2010). An immuno-electron microscopy study measured a 36-fold soma-AIS difference for $Na_V1.6$ (Lorincz & Nusser 2010). This latter estimate would presumably be lower if all Na_V subunits were taken into account. Direct comparisons between physiological and anatomical methods also assume that all labeled channel proteins are functional.

Such differences in channel distribution are important to reconcile, particularly to assess the significance of differences observed among cell types. For example, immunohistochemistry, outside-out patch recording, and modeling indicate lower channel densities in unmyelinated hippocampal granule cells versus pyramidal cells (Kress et al. 2010), and only a 5-fold difference in Na_V-channel density between the AIS and the soma (Schmidt-Hieber & Bischofberger 2010). These differences may reflect compensation for the lower axial current in these small cells. Accordingly, the extent of the AIS is shorter, spike threshold is higher, and the site of initiation is closer to the soma than in pyramidal cells (Kress et al. 2010, Schmidt-Hieber & Bischofberger 2010, Schmidt-Hieber et al. 2008).

K^+ Channels

The view of the AIS has been significantly expanded by looking past Na^+ channels to explore the full spectrum of voltage- and ligand-gated channels. K_V1 channels have been localized immunohistochemically to juxtaparanodal regions as well as to the distal end of the AIS (Goldberg et al. 2008, Lorincz & Nusser 2008, Rasband et al. 1998). Greater interest in their role in the AIS emerged following the realizations that subthreshold signals may spread from the soma into the nerve terminals and that K^+ channels could potentially affect that spread (Alle & Geiger 2006, Shu et al. 2006). Inactivating K^+ channels, similar to K_V1, can affect presynaptic spike width during repetitive activity (Geiger & Jonas 2000), suggesting that channels in the AIS may have roles beyond spike triggering. Indeed, using physiological approaches, Kole et al. (2007) and Shu et al. (2007b) observed a high density of K_V1 channels in the AIS of cortical pyramidal neurons and that block of these channels with dendrotoxin broadened the axonal spike but not the somatic spike. Dendrotoxin also increased transmitter release, even when applied locally to the AIS, suggesting that the spike width set by the AIS may impact the spike width in boutons (Kole et al. 2007).

The presence of K_V1 channels has been examined using subunit specific antibodies. These studies have revealed variation in the distribution of K_V1 channels among different cell types. In layer 5 and CA3 pyramidal cells, the beginning of obvious $K_V1.1$ and 1.2 expression is immediately distal to where $Na_V1.6$ becomes prominent (Lorincz & Nusser 2008). By contrast, in mitral/tufted cells of olfactory bulb, $K_V1.1$ is absent and $K_V1.2$ has more uniform distribution. Both forms are absent in the short initial segment of cerebellar Purkinje cells. $K_V1.2$ is expressed in the distal AIS of retinal ganglion cells (Van Wart et al. 2007). Even though $K_V1.1$ and $K_V1.2$ differ in their channel kinetics (Grissmer et al. 1994), it remains unclear what specific advantages are conferred in the AIS by the presence or location of a specific K_V1 subunit.

In addition to controlling axonal spike width and passive signal propagation to terminals, axonal K_V1 channels play a role in determining excitability. Goldberg et al. (2008) identified $K_V1.1$ in the AIS of fast-spiking interneurons in the superficial layers of the barrel cortex and showed that blocking these channels removes a delay in firing onset normally seen in response to a long depolarization. Analysis with current ramps showed that K_V1-channel activation suppressed spike onset except for very fast or very long, slow ramps. Presumably, with a fast depolarization, the AP is triggered before K_V channels are activated, but with a slow ramp, K_V channels inactivate and thus no longer prevent spike generation. A similar situation was previously shown in auditory brain-stem neurons, which have very prominent K_V1 expression (Dodson et al. 2002, Ferragamo & Oertel 2002, McGinley & Oertel 2006) and which, in some cases, are localized to nodes (Oertel et al. 2008) and the AIS (Dodson et al. 2002). Ramp analysis further showed that spikes were triggered only by the very fastest stimuli, and that blocking K_V1 channels removed this property (Ferragamo & Oertel 2002, McGinley & Oertel 2006). Perhaps because of the very high current density, even prolonged stimuli could not restore spiking, thus forcing the neurons to respond quickly, or not at all, to each stimulus. Thus, K_V1 channels in the AIS appear to serve multiple functions, such as regulating spike shape and pattern as well as spike effectiveness at synaptic boutons.

KCNQ2 and KCNQ3 ($K_V7.2$, $K_V7.3$) are subunits for voltage-sensitive K^+ channels localized to the AIS by ankyrin-G, the same molecule that clusters Na^+ channels (Pan et al. 2006). These channels, which are well-known as mediators of the M current, generate a slow current that begins to activate just below spike threshold; this property, along with the channel's localization, makes them well suited for controlling the excitability of the axon. Studying hippocampal pyramidal cells, Shah et al. (2008) found that blocking KCNQ lowered spike threshold and permitted increased spontaneous AP activity. Interestingly, disruption of the association of the channel to ankyrin-G produced an effect resembling that seen with channel block, suggesting that axonal localization is key to KCNQ function.

$K_V2.2$ subunits have been localized to the AIS of principal cells of the medial nucleus of the trapezoid body where they were proposed to regulate the interval between spikes during repetitive firing (Johnston et al. 2008). By contrast, K_V3 channels have a broader distribution on the soma and the axon in these neurons (Devaux et al. 2003, Elezgarai et al. 2003), and they are largely responsible for spike repolarization (Wang et al. 1998). Steinert et al. (2008, 2011) showed that signaling by nitric oxide, induced by glutamatergic synaptic activity, reduces K_V3's contribution to AP repolarization; in this situation, $K_V2.2$ channels become the dominant regulator of spike width. Thus, levels of neuronal activity can shift the balance of AIS and non-AIS channels in determining the shape of APs.

Ca^{2+} Channels

Spike-triggered Ca^{2+} entry has been described in mammalian axons, presumably in the AIS (Callewaert et al. 1996, Luscher et al. 1996, Schiller et al. 1995), but little attention had been given to its functional significance until our study of cartwheel interneurons of the dorsal cochlear nucleus of mice (Bender & Trussell 2009). Imaging experiments revealed low Ca^{2+} signals in the soma, increasing in the AIS up to 25 μm then decreasing to low levels until boutons were reached. Parallel Ca^{2+} and Na^+ imaging suggested that these channels were codistributed with Na^+ channels, indicating that Ca^{2+} channels were localized to the AIS. This is of interest as these neurons generate a complex spike burst mediated by both Na^+ and Ca^{2+} channels. Pharmacological analysis showed that the Ca^{2+} channels were a mix of T- and R-type. Selective blockage of these channels via local antagonist application elevated spike threshold and inhibited spike burst generation. Thus, it was proposed that low-voltage-threshold Ca^{2+} current in the AIS may boost depolarizing stimuli past spike threshold, perhaps acting in concert with subthreshold or persistent Na^+ currents. Similar effects were reported for the AIS of layer 5b pyramidal neurons of barrel cortex and for Purkinje neurons.

More recently, Ca^{2+} elevations by subthreshold depolarizations as well as by spikes were reported for ferret prefrontal cortical pyramidal cells (Yu et al. 2010), but here the Ca^{2+} signals were partially blocked by P/Q, N, or R antagonists. Local blockage of these channels broadened the somatic spikes and enhanced excitability, suggesting that AIS Ca^{2+} activated Ca^{2+}-dependent K^+ channels. Clarifying the relative functional significance of the different subtypes of the AIS Ca^{2+} channels will be important, as the complement of the AIS Ca^{2+}-channel subtypes, similar to Na^+ and K^+ channels, appears to vary across neuronal class.

AXON INITIAL SEGMENT NEUROTRANSMISSION

Ionotropic Transmission

Although the majority of synapses are made onto dendrites, the AIS can also be a target for GABAergic synaptic transmission. Given their proximity to the spike-initiation site, GABA synapses can have a great influence over neuronal output. Initial segment synapses arise from two classes of interneurons: basket and chandelier cells. Basket cells, common to cortical and cerebellar structures, get their name from the unique patterning of their axonal projections. These cells form a basket of terminals primarily onto the cell body, proximal dendrites, and AIS of principal cells. In the cerebellum, multiple basket cells synapse onto a single Purkinje cell, forming the *pinceau* synapse that envelops the soma and the AIS (Sotelo & Llinas 1972). Chandelier cells, which are found in neocortex, hippocampus, and amygdala, have been termed for their axonal patterning. Their axons ramify into vertically oriented shafts, called cartridges or candles, that synapse exclusively on the initial segment of excitatory pyramidal cells. In contrast to basket cells, which target only a few cells, a single chandelier cell can innervate more than 200 pyramidal cells (Somogyi et al. 1982) and is, therefore, thought to synchronize activity over a large population of pyramidal cells (Woodruff et al. 2010).

Because both of these AIS-targeting neurons are GABAergic, researchers had long assumed that they inhibit their postsynaptic targets; however, this notion remains controversial, especially for chandelier cells. Whether chloride-permeable GABA receptors flux chloride into or out of a neuron depends on the relationship between the local membrane potential (V_m) and the chloride reversal potential (E_{Cl}), which is determined largely by chloride extrusion through the potassium-chloride cotransporter KCC2 (Blaesse et al. 2009). KCC2 expression increases in early postnatal development (Rivera et al. 1999), but before it is upregulated, GABAergic synapses can excite neurons, promoting synaptic plasticity that can instruct early activity-dependent circuit refinement (Akerman & Cline 2007). Once KCC2 expression reaches mature levels, most GABAergic synapses hyperpolarize neurons, but interestingly, KCC2 expression never increases in axons. At distal axonal sites, E_{Cl} remains above resting V_m, and chloride-permeable channels continue to enhance local membrane excitability in mature neurons (Price & Trussell 2006, Pugh & Jahr 2011, Turecek & Trussell 2001).

Similar to GABAergic signaling at distal axonal sites, the earliest physiological characterization of AIS-targeting chandelier cells showed that they depolarized cortical pyramidal cells (Szabadics et al. 2006) and interneurons in the amygdala (Woodruff et al. 2006). But in contrast to synapses some distance down the axon, chandelier synapses are in close proximity to somatic KCC2 transporters that should clear chloride from the AIS. Thus, even though the AIS lacks KCC2, this alone cannot account fully for the depolarizing nature of chandelier synapses. Subsequently, researchers determined that high AIS chloride concentrations are maintained by active chloride uptake through the sodium-potassium-chloride cotransporter NKCC1, which may be selectively expressed in the AIS. Indeed, in NKCC1-knockout mice, chandelier synapses are hyperpolarizing (Khirug et al. 2008). However, in contrast to the developmental upregulation of KCC2, NKCC1 expression is high during early development and decreases in rodents some time after postnatal day (P) 21

(Plotkin et al. 1997). Interestingly, all studies demonstrating the depolarizing effects from the chandelier cell synapse were performed in rodents younger than P24 (Khirug et al. 2008; Szabadics et al. 2006; Woodruff et al. 2009, 2011). In older animals, presumably past the age when NKCC1 expression decreases, chandelier synapses onto hippocampal pyramidal cells were largely hyperpolarizing (Canepari et al. 2010, Glickfeld et al. 2009). Whether these differences reflect a developmental progression or cell-type specificity in chandelier cell function remains unclear; no single group has examined chandelier cell function through development and across brain regions. If chandelier synapses switch polarity during development, they do so with considerable developmental lag compared with that of somatodendritic GABAergic synapses. Alternatively, chandelier synapse polarity may be circuit specific, reflecting differences in chloride extrusion mechanisms between cortical and hippocampal pyramidal cells. Furthermore, spike activity was recently found to affect chloride homeostasis by stimulating sodium-bicarbonate-chloride exchangers (Kim & Trussell 2009), thereby altering the polarity of inhibitory signaling.

G Protein–Coupled Transmission

In addition to being affected by direct GABAergic input, AIS function is also sensitive to G protein–coupled signaling cascades. In cartwheel interneurons, recently identified AIS T-type Ca^{2+} channels are regulated by dopaminergic signaling through high-affinity type 3 dopamine receptors (D_3R). Application of exogenous dopamine receptor agonists, or electrical stimulation of endogenous dopamine sources, reduced spike-evoked AIS Ca^{2+} influx by ~30%. These effects were mediated by protein kinase C (PKC), which can be enriched in the AIS (Cardell et al. 1998). This neuromodulatory pathway is remarkably specific for AIS T-type channels: Neither AIS Na^+ influx nor dendritic T-type Ca^{2+}-channel activity were affected by dopamine (Bender et al. 2010).

Though dopaminergic fibers can make direct synaptic contact with the AIS in the striatum (Freund et al. 1984), cartwheel cells do not receive dopaminergic synaptic input. Instead, dopamine modulates AIS Ca^{2+} in a paracrine fashion (Bender et al. 2010); therefore, dopamine release from a single axon could impact multiple cells. What effect does this modulation have on spike generation? AIS Ca^{2+} channels contribute to burst generation (Bender & Trussell 2009), and owing to the nonlinear nature of AP initiation, this modest reduction in AIS Ca^{2+} is amplified into a 50% loss of evoked spikes (Bender et al. 2010). Dopaminergic modulation has even greater control over spontaneous activity, effectively converting spontaneously bursting neurons into tonically simple-spiking neurons (Bender et al. 2012).

Although dopaminergic regulation of AIS Ca^{2+} channels is the first example of AIS channel neuromodulation, it will likely not be the last. KCNQ channels, which give rise to the M-current, were first discovered because of their modulation by muscarine, but whether KCNQ channels localized to the AIS are subject to cholinergic modulation remains unclear. Additional avenues for local modulation of AIS may come from metabolic or receptor-mediated control of Na_V and K_V1 channel gating (Ahn et al. 2007, Raab-Graham et al. 2006). In addition to modifying ion channel properties, neuromodulators may regulate the distribution and density of ion channels in the AIS. Indeed, channel trafficking and membrane insertion are regulated by a variety of factors, including protein kinases and accessory proteins, which can all be targets of neuromodulator signaling (Brackenbury et al. 2010, Bréchet et al. 2008, Hund et al. 2010, Laezza et al. 2007, Vacher et al. 2011). Because the precise control of the function of AIS channels has the potential to profoundly regulate neuronal signaling throughout the nervous system, further studies are clearly needed to identify the full range of targets of physiological modulation in the AIS.

PLASTICITY AND MAINTENENCE OF THE AXON INITIAL SEGMENT

Establishing the Location of the AIS in Development

The formation of the AIS is one of the first steps in establishing neuronal polarity (Rasband 2010). Typically, the axon emerges directly from the soma, with the AIS occupying the first 20–50 μm of axonal length. However, the precise location and length of the AIS can vary from neuron to neuron, and recent studies indicate that AIS location is a function of neuronal activity. Diversity in AIS position exists across as well as within neuronal classes, and it is this intraclass diversity that demonstrates that the location of the AIS has significant consequences for neuronal processing.

In the auditory brain stem, low-frequency sounds are localized along the horizontal axis in the mammalian medial superior olive and the avian nucleus laminaris, whose neurons encode interaural time differences (ITDs) in the range of tens of microseconds (Konishi 2003). In laminaris neurons, ITD discrimination is achieved, in part, by specializations in the location and length of the AIS (Kuba et al. 2006). Over the range of sound frequency encoded by laminaris, neurons that process higher-frequency sound have larger and faster unitary excitatory inputs (Slee et al. 2010). In these cells, the AIS is separated from the soma by a 40-μm stretch of axon, improving ITD sensitivity by electrically isolating the AIS from the soma. Furthermore, because synaptic inputs are filtered by the cable properties of the separating axonal length, Na^+-channel inactivation that could be produced by intense synaptic activity is minimized (Kuba et al. 2006). By contrast, synaptic inputs in lower-frequency encoding neurons are small and slow. In these neurons, the AIS abuts the soma, increasing AIS sensitivity to small synaptic inputs. But with this increased sensitivity comes the increased risk of Na^+-channel inactivation. In apparent compensation, the AIS in these neurons is quite long, minimizing the relative level of inactivation by increasing the number of Na^+ channels (Kuba et al. 2006). Interestingly, neurons in nucleus magnocellularis, which supply excitatory input to laminaris, exhibit similar frequency-dependent differences in their synaptic inputs, and again, smaller, slower unitary excitatory inputs in low-frequency encoding neurons are detected more readily by an expansion in the length of the AIS and an increase in Na^+-channel clustering (Kuba & Ohmori 2009).

Across neuronal classes, the location of the AIS can vary greatly, both within the axon and with respect to the somatodendritic compartment. In retinal ganglion cells, initial segment length varies across classes that encode different visual features (Fried et al. 2009), though the physiological relevance of these differences remains unclear. In many substantia nigra bitufted dopamine neurons, the axon emanates from a primary dendrite, in some instances more than 200 μm from the soma. This distal spike initiation site makes synapses on the parent dendritic branch privileged for both spike generation and for receiving backpropagating APs (Hausser et al. 1995), and dopaminergic signaling determines whether backpropagating spikes invade the contralateral dendrite (Gentet & Williams 2007). A similar role of axon-bearing dendrites has been proposed for alveus-oriens GABA neurons in the hippocampus (Martina et al. 2000).

Activity-Dependent Structural Plasticity of the AIS

Neuronal plasticity occurs at multiple levels and includes short- and long-term plasticity at individual synapses (Collingridge et al. 2010, Zucker & Regehr 2002), cell-wide regulation of synaptic and intrinsic excitability (Turrigiano 2011), and rewiring of neuronal connectivity (Holtmaat & Svoboda 2009, Knudsen 1999). Traditionally, the AIS has not been considered a site of such activity-dependent plasticity, but if AIS location is developmentally tuned to neuronal function, can it be retuned should activity patterns change?

In the same nucleus magnocellularis neurons in which frequency-dependent AIS tuning was described, Kuba and colleagues found that removing normal synaptic input via unilateral cochlear ablation increased the length of the AIS 1.7-fold (Kuba et al. 2010). The extra AIS length was enriched with Na^+ channels; as a result, neurons had increased whole-cell Na^+ currents, faster spikes, and lower spike thresholds. Increases in length occurred within days of cochlear ablation, reaching maximal levels within 7 days, and likely stemmed from changes in activity, rather than deprivation-induced neurodegeneration, because auditory nerves had normal synaptic function following cochlear ablation. Similar results could also be observed by immobilizing middle ear bones rather than removing the cochlea (Kuba et al. 2010).

Complementary results were obtained in cultured hippocampal neurons. Here, excitatory drive was increased, rather than decreased as in magnocellularis, either by chronically raising membrane potential by altering extracellular K^+ or by evoking APs in bursts via expression of the light-activated cation channel, channelrhodopsin-2. In response to these manipulations, AIS length did not change. Rather, the entire AIS subcompartment moved away from the soma by up to 17 μm, thus decreasing neuronal excitability. This movement occurred over days, was reversible, and involved movement of all investigated AIS components, including channels and associated scaffolding proteins (Grubb & Burrone 2010a).

Although these changes in AIS position regulate spike initiation, they can occur independent of spiking activity. In fact, the elevated K^+ levels used above to depolarize cultured neurons were sufficient to inactivate Na^+ channels and block spiking. Consistent with these results, blocking Na^+ channels with tetrodotoxin had no influence on AIS movement. Rather, movement was blocked by Ca^{2+}-channel antagonists that target T- or L-type low-voltage-gated channels (Grubb & Burrone 2010a). Thus, local Ca^{2+} signaling, either in the AIS (T types) (Bender & Trussell 2009) or in the soma (L types) (Leitch et al. 2009), likely regulates AIS structural plasticity.

Homeostatic plasticity, in which changes in excitatory drive are counterbalanced to maintain spike output within a functional limit, can be mediated through mechanisms that normalize synaptic weights or alter intrinsic excitability (Turrigiano 2011). The results described above demonstrate that the AIS is a new site for homeostatic regulation; excitability can be raised by expanding AIS length in response to a loss of synaptic input or it can be lowered by a distal movement of the AIS in response to bursting/depolarization. These changes can be mediated by relatively large alterations in cellular activity, and it is of great interest to determine whether AIS position is sensitive to more modest changes in activity. Furthermore, because AIS dysfunction is implicated in a variety of neuronal excitability disorders (Wimmer et al. 2010), its movement—or lack thereof—may be an important, but previously unappreciated, contributor.

CONCLUSIONS AND PROSPECTIVE

Many features are vital for a given neuron to carry out its characteristic function: morphology, connectivity, transmitter phenotype, synaptic plasticity. However, with respect to spike initiation, the AIS is the site most sensitive to excitatory stimuli. The remarkably rich array of ion channels distributed in the AIS thus plays a critical role in neuronal function. Accordingly, the activity-dependent regulation of the function and position of these channels should be considered one of the defining features of a neuron. This view is supported by the fact that initial segments are cell-type specific in dimensions and composition. It will be essential to assess the generality of the AIS regulatory mechanisms described above. As we learn more about control of the AIS, it should be possible to test ideas about the role of particular channels or regulatory proteins in settings where the consequences for neuronal output can be assessed.

The modern study of the AIS has relied on the most sophisticated physiological approaches available. Nevertheless, development of new approaches will be essential, particularly given the indirect nature of some methods and the uncertainties in the numerical results they produce. What are the actual densities of ion channels? Why do immunohistochemical and physiological methods yield differing results? The application of outside-out and inside-out patch measurements also are confounded by the effects on channel accessibility and stability; moreover, Na^+ imaging may give estimates different from those of other methods. It will also be necessary to determine ways to manipulate experimentally the distribution of channels or their individual properties to understand the meaning of the molecular richness of the AIS.

As we do so, these efforts will likely cast new light on neurological disorders. Degradation of the AIS is a common neuronal response to injury and anoxia (Povlishock et al. 1983, Schafer et al. 2009), and dysfunctions in AIS-channel activity may be etiological to epilepsy (Wimmer et al. 2010). Furthermore, alterations in chandelier-cell innervation have been implicated in schizophrenia and prenatal exposure to drugs of abuse (Lewis 2011, Morrow et al. 2003). A more complete understanding of AIS function, and an appreciation of its capacity for neuromodulation and plasticity, may reveal new avenues for the treatment of these disorders.

DISCLOSURE STATEMENT

The authors are not aware of any affiliations, memberships, funding, or financial holdings that might be perceived as affecting the objectivity of this review.

ACKNOWLEDGMENTS

We are grateful to Wenquin Hu, Yousheng Shu, Andrea Lorincz, and Zoltan Nusser for providing data shown in the figures and to Matthew Grubb for critically reading this manuscript. Research was supported by National Institute of Health grants DC011080 to K.J.B. and NS028901 and DC004450 to L.O.T.

LITERATURE CITED

Ahn M, Beacham D, Westenbroek RE, Scheuer T, Catterall WA. 2007. Regulation of $Na_V1.2$ channels by brain-derived neurotrophic factor, TrkB, and associated fyn kinase. *J. Neurosci.* 27:11533–42

Akerman CJ, Cline HT. 2007. Refining the roles of GABAergic signaling during neural circuit formation. *Trends Neurosci.* 30:382–89

Alle H, Geiger JR. 2006. Combined analog and action potential coding in hippocampal mossy fibers. *Science* 311:1290–93

Araki T, Otani T. 1955. Response of single motoneurons to direct stimulation in toad's spinal cord. *J. Neurophysiol.* 18:472–85

Astman N, Gutnick MJ, Fleidervish IA. 2006. Persistent sodium current in layer 5 neocortical neurons is primarily generated in the proximal axon. *J. Neurosci.* 26:3465–73

Bender KJ, Ford CP, Trussell LO. 2010. Dopaminergic modulation of axon initial segment calcium channels regulates action potential initiation. *Neuron* 68:500–11

Bender KJ, Trussell LO. 2009. Axon initial segment Ca^{2+} channels influence action potential generation and timing. *Neuron* 61:259–71

Bender KJ, Uebele VN, Renger JJ, Trussell LO. 2012. Control of firing patterns through modulation of axon initial segment T-type calcium channels. *J. Physiol.* 590:109–18

Blaesse P, Airaksinen MS, Rivera C, Kaila K. 2009. Cation-chloride cotransporters and neuronal function. *Neuron* 61:820–38

Boiko T, Van Wart A, Caldwell JH, Levinson SR, Trimmer JS, Matthews G. 2003. Functional specialization of the axon initial segment by isoform-specific sodium channel targeting. *J. Neurosci.* 23:2306–13

Brackenbury WJ, Calhoun JD, Chen C, Miyazaki H, Nukina N, et al. 2010. Functional reciprocity between Na^+ channel Nav1.6 and beta1 subunits in the coordinated regulation of excitability and neurite outgrowth. *Proc. Natl. Acad. Sci. USA* 107:2283–88

Bréchet A, Fache MP, Brachet A, Ferracci G, Baude A, et al. 2008. Protein kinase CK2 contributes to the organization of sodium channels in axonal membranes by regulating their interaction with ankyrin G. *J. Cell Biol.* 183:1101–14

Brock LG, Coombs JS, Eccles JC. 1952. The recording of potentials from motoneurones with an intracellular electrode. *J. Physiol.* 117:431–60

Brock LG, Coombs JS, Eccles JC. 1953. Intracellular recording from antidromically activated motoneurones. *J. Physiol.* 122:429–61

Callewaert G, Eilers J, Konnerth A. 1996. Axonal calcium entry during fast 'sodium' action potentials in rat cerebellar Purkinje neurones. *J. Physiol.* 495:641–47

Canepari M, Willadt S, Zecevic D, Vogt KE. 2010. Imaging inhibitory synaptic potentials using voltage sensitive dyes. *Biophys. J.* 98:2032–40

Cardell M, Landsend AS, Eidet J, Wieloch T, Blackstad TW, Ottersen OP. 1998. High resolution immunogold analysis reveals distinct subcellular compartmentation of protein kinase C gamma and delta in rat Purkinje cells. *Neuroscience* 82:709–25

Clark BA, Monsivais P, Branco T, London M, Hausser M. 2005. The site of action potential initiation in cerebellar Purkinje neurons. *Nat. Neurosci.* 8:137–39

Colbert CM, Johnston D. 1996. Axonal action-potential initiation and Na^+ channel densities in the soma and axon initial segment of subicular pyramidal neurons. *J. Neurosci.* 16:6676–86

Colbert CM, Pan E. 2002. Ion channel properties underlying axonal action potential initiation in pyramidal neurons. *Nat. Neurosci.* 5:533–38

Collingridge GL, Peineau S, Howland JG, Wang YT. 2010. Long-term depression in the CNS. *Nat. Rev. Neurosci.* 11:459–73

Coombs JS, Curtis DR, Eccles JC. 1957a. The generation of impulses in motoneurones. *J. Physiol.* 139:232–49

Coombs JS, Curtis DR, Eccles JC. 1957b. The interpretation of spike potentials of motoneurones. *J. Physiol.* 139:198–231

Devaux J, Alcaraz G, Grinspan J, Bennett V, Joho R, et al. 2003. Kv3.1b is a novel component of CNS nodes. *J. Neurosci.* 23:4509–18

Dodge FA, Cooley JW. 1973. Action potential of the motorneuron. *IBM J. Res. Dev.* 17:219–29

Dodson PD, Barker MC, Forsythe ID. 2002. Two heteromeric Kv1 potassium channels differentially regulate action potential firing. *J. Neurosci.* 22:6953–61

Duflocq A, Le Bras B, Bullier E, Couraud F, Davenne M. 2008. Nav1.1 is predominantly expressed in nodes of Ranvier and axon initial segments. *Mol. Cell Neurosci.* 39:180–92

Edwards C, Ottoson D. 1958. The site of impulse initiation in a nerve cell of a crustacean stretch receptor. *J. Physiol.* 143:138–48

Elezgarai I, Diez J, Puente N, Azkue JJ, Benitez R, et al. 2003. Subcellular localization of the voltage-dependent potassium channel Kv3.1b in postnatal and adult rat medial nucleus of the trapezoid body. *Neuroscience* 118:889–98

Fatt P. 1957. Electric potentials occurring around a neurone during its antidromic activation. *J. Neurophysiol.* 20:27–60

Ferragamo MJ, Oertel D. 2002. Octopus cells of the mammalian ventral cochlear nucleus sense the rate of depolarization. *J. Neurophysiol.* 87:2262–70

Fleidervish IA, Lasser-Ross N, Gutnick MJ, Ross WN. 2010. Na^+ imaging reveals little difference in action potential-evoked Na^+ influx between axon and soma. *Nat. Neurosci.* 13:852–60

Foust A, Popovic M, Zecevic D, McCormick DA. 2010. Action potentials initiate in the axon initial segment and propagate through axon collaterals reliably in cerebellar Purkinje neurons. *J. Neurosci.* 30:6891–902

Freund TF, Powell JF, Smith AD. 1984. Tyrosine hydroxylase-immunoreactive boutons in synaptic contact with identified striatonigral neurons, with particular reference to dendritic spines. *Neuroscience* 13:1189–215

Freygang WH Jr. 1958. An analysis of extracellular potentials from single neurons in the lateral geniculate nucleus of the cat. *J. Gen. Physiol.* 41:543–64

Fried SI, Lasker AC, Desai NJ, Eddington DK, Rizzo JF 3rd. 2009. Axonal sodium-channel bands shape the response to electric stimulation in retinal ganglion cells. *J. Neurophysiol.* 101:1972–87

Fuortes MG, Frank K, Becker MC. 1957. Steps in the production of motoneuron spikes. *J. Gen. Physiol.* 40:735–52

Geiger JR, Jonas P. 2000. Dynamic control of presynaptic Ca^{2+} inflow by fast-inactivating K^+ channels in hippocampal mossy fiber boutons. *Neuron* 28:927–39

Gentet LJ, Williams SR. 2007. Dopamine gates action potential backpropagation in midbrain dopaminergic neurons. *J. Neurosci.* 27:1892–901

Gesell R. 1940. Forces Driving the respiratory act: a fundamental concept of the integration of motor activity. *Science* 91:229–33

Glickfeld LL, Roberts JD, Somogyi P, Scanziani M. 2009. Interneurons hyperpolarize pyramidal cells along their entire somatodendritic axis. *Nat. Neurosci.* 12:21–23

Goldberg EM, Clark BD, Zagha E, Nahmani M, Erisir A, Rudy B. 2008. K^+ channels at the axon initial segment dampen near-threshold excitability of neocortical fast-spiking GABAergic interneurons. *Neuron* 58:387–400

Grissmer S, Nguyen AN, Aiyar J, Hanson DC, Mather RJ, et al. 1994. Pharmacological characterization of five cloned voltage-gated K^+ channels, types Kv1.1, 1.2, 1.3, 1.5, and 3.1, stably expressed in mammalian cell lines. *Mol. Pharmacol.* 45:1227–34

Grubb MS, Burrone J. 2010a. Activity-dependent relocation of the axon initial segment fine-tunes neuronal excitability. *Nature* 465:1070–74

Grubb MS, Burrone J. 2010b. Building and maintaining the axon initial segment. *Curr. Opin. Neurobiol.* 20:481–88

Hausser M, Stuart G, Racca C, Sakmann B. 1995. Axonal initiation and active dendritic propagation of action potentials in substantia nigra neurons. *Neuron* 15:637–47

Holtmaat A, Svoboda K. 2009. Experience-dependent structural synaptic plasticity in the mammalian brain. *Nat. Rev. Neurosci.* 10:647–58

Hu W, Tian C, Li T, Yang M, Hou H, Shu Y. 2009. Distinct contributions of Na(v)1.6 and Na(v)1.2 in action potential initiation and backpropagation. *Nat. Neurosci.* 12:996–1002

Hund TJ, Koval OM, Li J, Wright PJ, Qian L, et al. 2010. A beta(IV)-spectrin/CaMKII signaling complex is essential for membrane excitability in mice. *J. Clin. Invest.* 120:3508–19

Johnston J, Griffin SJ, Baker C, Skrzypiec A, Chernova T, Forsythe ID. 2008. Initial segment Kv2.2 channels mediate a slow delayed rectifier and maintain high frequency action potential firing in medial nucleus of the trapezoid body neurons. *J. Physiol.* 586:3493–509

Khaliq ZM, Raman IM. 2006. Relative contributions of axonal and somatic Na channels to action potential initiation in cerebellar Purkinje neurons. *J. Neurosci.* 26:1935–44

Khirug S, Yamada J, Afzalov R, Voipio J, Khiroug L, Kaila K. 2008. GABAergic depolarization of the axon initial segment in cortical principal neurons is caused by the Na-K-2Cl cotransporter NKCC1. *J. Neurosci.* 28:4635–39

Kim Y, Trussell LO. 2009. Negative shift in the glycine reversal potential mediated by a Ca^{2+}- and pH-dependent mechanism in interneurons. *J. Neurosci.* 29:11495–510

Knudsen EI. 1999. Mechanisms of experience-dependent plasticity in the auditory localization pathway of the barn owl. *J. Comp. Physiol. A* 185:305–21

Kole MH, Ilschner SU, Kampa BM, Williams SR, Ruben PC, Stuart GJ. 2008. Action potential generation requires a high sodium channel density in the axon initial segment. *Nat. Neurosci.* 11:178–86

Kole MH, Letzkus JJ, Stuart GJ. 2007. Axon initial segment Kv1 channels control axonal action potential waveform and synaptic efficacy. *Neuron* 55:633–47

Kole MH, Stuart GJ. 2008. Is action potential threshold lowest in the axon? *Nat. Neurosci.* 11:1253–55

Kole MHP. 2011. First node of Ranvier facilitates high-frequency burst encoding. *Neuron* 71:671–82

Konishi M. 2003. Coding of auditory space. *Annu. Rev. Neurosci.* 26:31–55

Kress GJ, Dowling MJ, Eisenman LN, Mennerick S. 2010. Axonal sodium channel distribution shapes the depolarized action potential threshold of dentate granule neurons. *Hippocampus* 20:558–71

Kress GJ, Dowling MJ, Meeks JP, Mennerick S. 2008. High threshold, proximal initiation, and slow conduction velocity of action potentials in dentate granule neuron mossy fibers. *J. Neurophysiol.* 100:281–91

Kuba H, Ishii TM, Ohmori H. 2006. Axonal site of spike initiation enhances auditory coincidence detection. *Nature* 444:1069–72

Kuba H, Ohmori H. 2009. Roles of axonal sodium channels in precise auditory time coding at nucleus magnocellularis of the chick. *J. Physiol.* 587:87–100

Kuba H, Oichi Y, Ohmori H. 2010. Presynaptic activity regulates Na^+ channel distribution at the axon initial segment. *Nature* 465:1075–78

Laezza F, Gerber BR, Lou JY, Kozel MA, Hartman H, et al. 2007. The FGF14(F145S) mutation disrupts the interaction of FGF14 with voltage-gated Na^+ channels and impairs neuronal excitability. *J. Neurosci.* 27:12033–44

Leitch B, Szostek A, Lin R, Shevtsova O. 2009. Subcellular distribution of L-type calcium channel subtypes in rat hippocampal neurons. *Neuroscience* 164:641–57

Lewis DA. 2011. The chandelier neuron in schizophrenia. *Dev. Neurobiol.* 71:118–27

Llinas R, Nicholson C, Freeman JA, Hillman DE. 1968. Dendritic spikes and their inhibition in alligator Purkinje cells. *Science* 160:1132–35

Lorincz A, Nusser Z. 2008. Cell-type-dependent molecular composition of the axon initial segment. *J. Neurosci.* 28:14329–40

Lorincz A, Nusser Z. 2010. Molecular identity of dendritic voltage-gated sodium channels. *Science* 328:906–9

Luscher C, Lipp P, Luscher HR, Niggli E. 1996. Control of action potential propagation by intracellular Ca^{2+} in cultured rat dorsal root ganglion cells. *J. Physiol.* 490:319–24

Mainen ZF, Joerges J, Huguenard JR, Sejnowski TJ. 1995. A model of spike initiation in neocortical pyramidal neurons. *Neuron* 15:1427–39

Martina M, Vida I, Jonas P. 2000. Distal initiation and active propagation of action potentials in interneuron dendrites. *Science* 287:295–300

McGinley MJ, Oertel D. 2006. Rate thresholds determine the precision of temporal integration in principal cells of the ventral cochlear nucleus. *Hear Res.* 216–17:52–63

Meeks JP, Mennerick S. 2007. Action potential initiation and propagation in CA3 pyramidal axons. *J. Neurophysiol.* 97:3460–72

Moore JW, Stockbridge N, Westerfield M. 1983. On the site of impulse initiation in a neurone. *J. Physiol.* 336:301–11

Morrow BA, Elsworth JD, Roth RH. 2003. Axo-axonic structures in the medial prefrontal cortex of the rat: reduction by prenatal exposure to cocaine. *J. Neurosci.* 23:5227–34

Oertel D, Shatadal S, Cao XJ. 2008. In the ventral cochlear nucleus Kv1.1 and subunits of HCN1 are colocalized at surfaces of neurons that have low-voltage-activated and hyperpolarization-activated conductances. *Neuroscience* 154:77–86

Palay SL, Sotelo C, Peters A, Orkand PM. 1968. The axon hillock and the initial segment. *J. Cell Biol.* 38:193–201

Palmer LM, Clark BA, Grundemann J, Roth A, Stuart GJ, Hausser M. 2010. Initiation of simple and complex spikes in cerebellar Purkinje cells. *J. Physiol.* 588:1709–17

Palmer LM, Stuart GJ. 2006. Site of action potential initiation in layer 5 pyramidal neurons. *J. Neurosci.* 26:1854–63

Pan Z, Kao T, Horvath Z, Lemos J, Sul JY, et al. 2006. A common ankyrin-G-based mechanism retains KCNQ and NaV channels at electrically active domains of the axon. *J. Neurosci.* 26:2599–613

Peters A, Palay SL, Webster HD. 1991. *The Fine Structure of the Nervous System*. New York: Oxford Univ. Press

Plotkin MD, Snyder EY, Hebert SC, Delpire E. 1997. Expression of the Na-K-2Cl cotransporter is developmentally regulated in postnatal rat brains: a possible mechanism underlying GABA's excitatory role in immature brain. *J. Neurobiol.* 33:781–95

Popovic MA, Foust AJ, McCormick DA, Zecevic D. 2011. The spatio-temporal characteristics of action potential initiation in layer 5 pyramidal neurons: a voltage-imaging study. *J. Physiol.* 589:4167–87

Povlishock JT, Becker DP, Cheng CL, Vaughan GW. 1983. Axonal change in minor head injury. *J. Neuropathol. Exp. Neurol.* 42:225–42

Price GD, Trussell LO. 2006. Estimate of the chloride concentration in a central glutamatergic terminal: a gramicidin perforated-patch study on the calyx of Held. *J. Neurosci.* 26:11432–36

Pugh JR, Jahr CE. 2011. Axonal GABAA receptors increase cerebellar granule cell excitability and synaptic activity. *J. Neurosci.* 31:565–74

Raab-Graham KF, Haddick PC, Jan YN, Jan LY. 2006. Activity- and mTOR-dependent suppression of Kv1.1 channel mRNA translation in dendrites. *Science* 314:144–48

Rasband MN. 2010. The axon initial segment and the maintenance of neuronal polarity. *Nat. Rev. Neurosci.* 11:552–62

Rasband MN, Trimmer JS, Schwarz TL, Levinson SR, Ellisman MH, et al. 1998. Potassium channel distribution, clustering, and function in remyelinating rat axons. *J. Neurosci.* 18:36–47

Rivera C, Voipio J, Payne JA, Ruusuvuori E, Lahtinen H, et al. 1999. The K^+/Cl^- co-transporter KCC2 renders GABA hyperpolarizing during neuronal maturation. *Nature* 397:251–55

Royeck M, Horstmann MT, Remy S, Reitze M, Yaari Y, Beck H. 2008. Role of axonal NaV1.6 sodium channels in action potential initiation of CA1 pyramidal neurons. *J. Neurophysiol.* 100:2361–80

Schafer DP, Jha S, Liu F, Akella T, McCullough LD, Rasband MN. 2009. Disruption of the axon initial segment cytoskeleton is a new mechanism for neuronal injury. *J. Neurosci.* 29:13242–54

Schiller J, Helmchen F, Sakmann B. 1995. Spatial profile of dendritic calcium transients evoked by action potentials in rat neocortical pyramidal neurones. *J. Physiol.* 487:583–600

Schiller J, Schiller Y, Stuart G, Sakmann B. 1997. Calcium action potentials restricted to distal apical dendrites of rat neocortical pyramidal neurons. *J. Physiol.* 505:605–16

Schmidt-Hieber C, Bischofberger J. 2010. Fast sodium channel gating supports localized and efficient axonal action potential initiation. *J. Neurosci.* 30:10233–42

Schmidt-Hieber C, Jonas P, Bischofberger J. 2008. Action potential initiation and propagation in hippocampal mossy fibre axons. *J. Physiol.* 586:1849–57

Shah MM, Migliore M, Valencia I, Cooper EC, Brown DA. 2008. Functional significance of axonal Kv7 channels in hippocampal pyramidal neurons. *Proc. Natl. Acad. Sci. USA* 105:7869–74

Shu Y, Duque A, Yu Y, Haider B, McCormick DA. 2007a. Properties of action-potential initiation in neocortical pyramidal cells: evidence from whole cell axon recordings. *J. Neurophysiol.* 97:746–60

Shu Y, Hasenstaub A, Duque A, Yu Y, McCormick DA. 2006. Modulation of intracortical synaptic potentials by presynaptic somatic membrane potential. *Nature* 441:761–65

Shu Y, Yu Y, Yang J, McCormick DA. 2007b. Selective control of cortical axonal spikes by a slowly inactivating K^+ current. *Proc. Natl. Acad. Sci. USA* 104:11453–58

Slee SJ, Higgs MH, Fairhall AL, Spain WJ. 2010. Tonotopic tuning in a sound localization circuit. *J. Neurophysiol.* 103:2857–75

Somogyi P, Freund TF, Cowey A. 1982. The axo-axonic interneuron in the cerebral cortex of the rat, cat and monkey. *Neuroscience* 7:2577–607

Sotelo C, Llinas R. 1972. Specialized membrane junctions between neurons in the vertebrate cerebellar cortex. *J. Cell Biol.* 53:271–89

Steinert JR, Kopp-Scheinpflug C, Baker C, Challiss RA, Mistry R, et al. 2008. Nitric oxide is a volume transmitter regulating postsynaptic excitability at a glutamatergic synapse. *Neuron* 60:642–56

Steinert JR, Robinson SW, Tong H, Haustein MD, Kopp-Scheinpflug C, Forsythe ID. 2011. Nitric oxide is an activity-dependent regulator of target neuron intrinsic excitability. *Neuron* 71:291–305

Stuart G, Sakmann B. 1995. Amplification of EPSPs by axosomatic sodium channels in neocortical pyramidal neurons. *Neuron* 15:1065–76

Stuart G, Schiller J, Sakmann B. 1997. Action potential initiation and propagation in rat neocortical pyramidal neurons. *J. Physiol.* 505:617–32

Szabadics J, Varga C, Molnar G, Olah S, Barzo P, Tamas G. 2006. Excitatory effect of GABAergic axo-axonic cells in cortical microcircuits. *Science* 311:233–35

Traub RD, Jefferys JG, Miles R, Whittington MA, Toth K. 1994. A branching dendritic model of a rodent CA3 pyramidal neurone. *J. Physiol.* 481:79–95

Turecek R, Trussell LO. 2001. Presynaptic glycine receptors enhance transmitter release at a mammalian central synapse. *Nature* 411:587–90

Turrigiano G. 2011. Too many cooks? Intrinsic and synaptic homeostatic mechanisms in cortical circuit refinement. *Annu. Rev. Neurosci.* 34:89–103

Vacher H, Yang JW, Cerda O, Autillo-Touati A, Dargent B, Trimmer JS. 2011. Cdk-mediated phosphorylation of the Kvbeta2 auxiliary subunit regulates Kv1 channel axonal targeting. *J. Cell Biol.* 192:813–24

Van Wart A, Trimmer JS, Matthews G. 2007. Polarized distribution of ion channels within microdomains of the axon initial segment. *J. Comp. Neurol.* 500:339–52

Wang LY, Gan L, Forsythe ID, Kaczmarek LK. 1998. Contribution of the Kv3.1 potassium channel to high-frequency firing in mouse auditory neurones. *J. Physiol.* 509:183–94

Westenbroek RE, Merrick DK, Catterall WA. 1989. Differential subcellular localization of the RI and RII Na^+ channel subtypes in central neurons. *Neuron* 3:695–704

Wimmer VC, Reid CA, So EY, Berkovic SF, Petrou S. 2010. Axon initial segment dysfunction in epilepsy. *J. Physiol.* 588:1829–40

Wong RK, Prince DA, Basbaum AI. 1979. Intradendritic recordings from hippocampal neurons. *Proc. Natl. Acad. Sci. USA* 76:986–90

Woodruff A, Xu Q, Anderson SA, Yuste R. 2009. Depolarizing effect of neocortical chandelier neurons. *Front. Neural Circuits* 3:15

Woodruff AR, Anderson SA, Yuste R. 2010. The enigmatic function of chandelier cells. *Front. Neurosci.* 4:201

Woodruff AR, McGarry LM, Vogels TP, Inan M, Anderson SA, Yuste R. 2011. State-dependent function of neocortical chandelier cells. *J. Neurosci.* 31:17872–86

Woodruff AR, Monyer H, Sah P. 2006. GABAergic excitation in the basolateral amygdala. *J. Neurosci.* 26:11881–87

Yu Y, Maureira C, Liu X, McCormick D. 2010. P/Q and N channels control baseline and spike-triggered calcium levels in neocortical axons and synaptic boutons. *J. Neurosci.* 30:11858–69

Zhou W, Goldin AL. 2004. Use-dependent potentiation of the Nav1.6 sodium channel. *Biophys. J.* 87:3862–72

Zucker RS, Regehr WG. 2002. Short-term synaptic plasticity. *Annu. Rev. Physiol.* 64:355–405

Attractor Dynamics of Spatially Correlated Neural Activity in the Limbic System

James J. Knierim[1] and Kechen Zhang[2]

[1]Krieger Mind/Brain Institute and Department of Neuroscience, Johns Hopkins University, Baltimore, Maryland 21218; email: jknierim@jhu.edu

[2]Department of Biomedical Engineering, Johns Hopkins University, Baltimore, Maryland 21205; email: kzhang4@jhmi.edu

Annu. Rev. Neurosci. 2012. 35:267–85

First published online as a Review in Advance on March 29, 2012

The *Annual Review of Neuroscience* is online at neuro.annualreviews.org

This article's doi:
10.1146/annurev-neuro-062111-150351

0147-006X/12/0721-0267$20.00

Keywords

hippocampus, medial entorhinal cortex, head-direction cell, place cell, grid cell

Abstract

Attractor networks are a popular computational construct used to model different brain systems. These networks allow elegant computations that are thought to represent a number of aspects of brain function. Although there is good reason to believe that the brain displays attractor dynamics, it has proven difficult to test experimentally whether any particular attractor architecture resides in any particular brain circuit. We review models and experimental evidence for three systems in the rat brain that are presumed to be components of the rat's navigational and memory system. Head-direction cells have been modeled as a ring attractor, grid cells as a plane attractor, and place cells both as a plane attractor and as a point attractor. Whereas the models have proven to be extremely useful conceptual tools, the experimental evidence in their favor, although intriguing, is still mostly circumstantial.

Contents

ATTRACTOR NETWORKS: GENERAL CONCEPTS

Attractor neural networks have occupied a prime place in the theoretical and computational literature on systems neuroscience. The elegance and explanatory power of these models are matched, unfortunately, by the ambiguity of the experimental evidence that any particular brain circuit actually incorporates attractor network architecture and dynamics. In this review, we describe the basic types of attractor networks and then evaluate the evidence for attractors in three components of the rat limbic spatial-navigation system: the head-direction (HD) circuit, the grid-cell circuit, and the place-cell circuit.

HD: head direction

An attractor is a convenient general concept for describing the stability of a dynamical system, whose state evolves in time (Milnor 1985, Amit 1989). In intuitive terms, an attractor refers to a collection of states that will eventually attract neighboring states toward that collection, as if it were a magnet. This concept is illustrated by the simplest type of attractor, the point attractor, which is a single, stable, equilibrium state of the system (**Figure 1**). The basin of attraction for an attractor refers to the collection of all neighboring states that will eventually be drawn toward that attractor.

Many types of attractors have been studied in computational neuroscience. A well-known example is a Hopfield network, which allows multiple point attractors, each corresponding to a stored memory pattern (Hopfield 1982). The model has several interesting properties, such as robustness against damage to the synaptic connections (structural stability), pattern completion (recall of a stored representation from partial or noisy inputs), and pattern separation (divergence toward different stable states). Pattern completion occurs when different initial states belong to the basin of attraction of the same attractor, whereas pattern separation occurs when similar initial states actually fall into the basins of different attractors.

A limit cycle is an attractor that represents a stable periodic oscillation. It is a stable, closed trajectory in the state space, but it does not contain any equilibrium states. Starting from different initial states, the system will eventually approach the same oscillation or the same trajectory in the state space. Examples of a limit cycle in a network are central pattern generator circuits that produce periodic rhythms (Kopell & Ermentrout 1988, Rand et al. 1988, Marder & Bucher 2001).

A relatively new type of attractor, often called a continuous attractor, allows a continuum of stable equilibrium states, or at least closely related states that approximately form a continuum. The simple examples are the one-dimensional cases of a line attractor and a ring attractor. The line attractor can represent a variable that has a minimum value and a

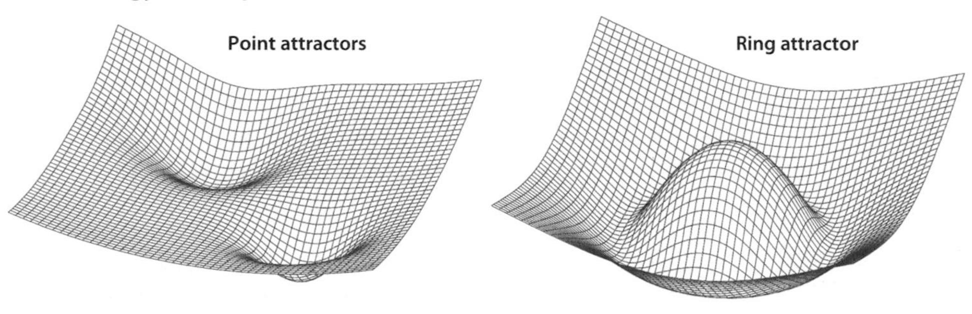

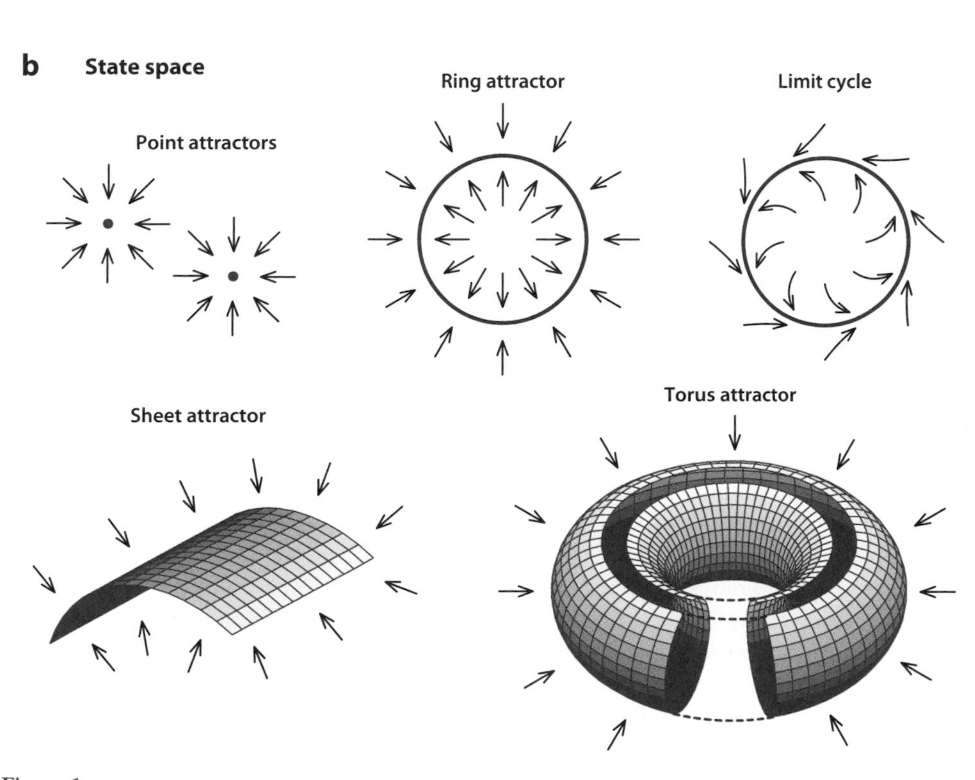

Figure 1

(*a*) In a neural network with an energy function, the state of the network goes spontaneously downhill and eventually settles into some attractor states, which correspond to local energy minima. Neural networks with symmetric connections (equal reciprocal connections between neurons) always allow an energy function (Lyapunov function) (Hopfield 1982, Cohen & Grossberg 1983, Hopfield 1984). (*b*) Schematic examples of various attractors in neural network models, with the arrows indicating how the state of a network changes. Some attractors, such as a limit cycle, cannot be described by an energy function. Connecting the opposite edges of a sheet attractor can lead to a torus attractor, which has no boundary.

maximum value, such as the orbital position of the eye in one dimension (Robinson 1989, Seung 1996, Major et al. 2004). The ring attractor is suitable for representing a variable without boundary values (Dunin-Barkowski & Osovets 1995), such as azimuth heading direction (Skaggs et al. 1995, Blair 1996, Redish et al. 1996, Zhang 1996). Both examples can be used as a velocity integrator; that is, they can be used to integrate velocity over time to produce a position signal (eye position, head direction). The continuous attractor model has also been used to model other neural systems, such as orientation tuning in visual cortex (Ben-Yishai et al. 1995, Somers et al. 1995, see also Ferster & Miller 2000), binocular rivalry (Laing & Chow 2002), and perceptual decision making (Brody et al. 2003, Machens et al. 2005, Wang 2008).

An extension of the one-dimensional line or ring attractor to two dimensions is called a plane (or sheet) attractor. This type of attractor is well-suited to representing two-dimensional variables, such as location in a room. Plane attractors have been used to account for place cells in the rat hippocampus (Samsonovich & McNaughton 1997) and view cells in the monkey hippocampus (Stringer et al. 2005). Transforming the topology of the plane into a torus allows for the periodic spatial-activity patterns demonstrated by grid cells (see below) (McNaughton et al. 2006).

These types of attractors are not mutually exclusive, in that multiple types of attractors can be embedded in the same networks. For example, continuous attractors and point attractors can be stored in the same network (Rolls et al. 2002), allowing both continuous variables, such as spatial information, and discrete variables, such as the names of individuals, to be stored.

It is important to note that attractor states refer to states of activity in the network. This does not imply that there is an anatomical correspondence between attractor states in the network and locations in the physical brain. In effect, nearby cells on a ring attractor do not need to be located next to each other anatomically. Neighboring HD cells, for example, recorded on the same electrode, can have widely different preferred firing directions (Muller et al. 1996). The connectivity between neurons, not their physical proximity, allows different topologies of attractor states.

HEAD-DIRECTION CELLS AND RING ATTRACTORS

HD cells fire whenever the rat's head is pointed in a particular direction in an allocentric coordinate frame (Ranck Jr. 1985; Taube et al. 1990a,b). These cells are sensitive to HD only in the horizontal plane (y axis); are insensitive to HD relative to the body axis; and fire continuously at a high rate whenever the rat's head is pointed in its preferred direction, with little adaptation. Rotation of salient landmarks causes a corresponding rotation of the preferred firing direction of the HD cell (Taube et al. 1990b, Taube 1998, Taube 2007), thus demonstrating that visual landmarks exert powerful control over the HD cells. However, visual input is not necessary for robust HD firing, as the cells maintain their directional selectivity (at times with a change in preferred direction) even in total darkness (Mizumori & Williams 1993, Chen et al. 1994, Goodridge et al. 1998).

Models

Most models of the HD system are variations on a basic theme developed by a number of investigators in the mid-1990s (McNaughton et al. 1991, Skaggs et al. 1995, Blair 1996, Redish et al. 1996, Zhang 1996). To explain how HD cells could fire in the same direction across multiple environments, maintain their preferred firing in the dark, yet be controlled by rotation of salient visual landmarks, McNaughton and colleagues (McNaughton et al. 1991) proposed that the HD signal was primarily updated by an angular head velocity signal. That is, if the animal is facing west, HD cells that encode that direction would be active. A head turn 90° clockwise would activate cells that respond to clockwise motion, via presumed afferents from the vestibular system. In turn, these cells would cause the cells encoding west to become silent and cause the cells that encode north to become active, via an intermediate layer of cells that encoded the combination of a particular starting direction and a particular change in direction. This conceptual model was originally implemented as a look-up table. Subsequent work incorporated the architecture of a ring attractor to implement this idea in plausible brain circuitry (Skaggs et al. 1995, Blair 1996, Redish et al. 1996, Zhang 1996, Goodridge &

Touretzky 2000, Stringer et al. 2002, Xie et al. 2002). The HD cells are conceptualized as being arrayed in a circle, with the location of each cell on the circle representing the preferred firing direction of that cell. Nearby cells are connected by strong excitatory synapses, with the strength of excitation proportional to the angular distance between the cells on the ring. Cells that are far apart on the ring are connected with inhibitory synapses. If the strength of excitation and inhibition is appropriately tuned, such an architecture exhibits attractor dynamics. From a starting condition of random excitation, stochastic fluctuations cause one part of the ring to have slightly greater excitation than other parts. The excitatory circuitry reinforces activity of other nearby cells, whereas the inhibitory connections force cells further away to be silent. The symmetric connectivity results in a localized, self-sustaining "bump" or "hill" of activity on the ring, the cells outside of which are silent.

If the location of the bump of activity was made to follow the momentary changes in the HD of the animal, such a network could form the basis of the HD signal. To accomplish this, an asymmetry in the connectivity matrix of the attractor is required to turn the ring attractor into a limit cycle and, thus, move the bump of activity around the ring. Different models added this asymmetry in different ways, but the key component was adding velocity modulation to the HD cells. With proper tuning of the weights, this asymmetry would cause the bump of activity to move in concert with the animal's HD. This circuit, dependent on self-motion cues to update the location of the activity bump, is susceptible to cumulative error. That is, if there is an error in the calculation of the angular head velocity, the bump will be offset by a certain amount from the animal's true HD. This offset would be reflected as a global shift in each HD cell's preferred firing direction. To correct for such errors, the models incorporate input from stable landmarks. Thus, the moment-by-moment updating of the activity bump is governed by the angular head velocity signals, with calibration by the stable, external landmarks to keep the bump of activity stable relative to the external world.

It is thought that the HD signal emerges as an interaction between neurons in the lateral mammillary nucleus and the dorsal tegmental nucleus of Gudden (Taube 1998, Sharp et al. 2001a), both of which contain cells that encode HD and angular head velocity (Blair et al. 1998, Stackman & Taube 1998, Bassett & Taube 2001, Sharp et al. 2001b). Thus, these nuclei contain the major components of the ring attractor models. One problem for the ring attractor models, however, is that these nuclei do not appear to contain the recurrent excitatory circuitry required for the self-sustaining activity bump. Song & Wang (2005; see also Boucheny et al. 2005) showed that a network without excitatory connections among the HD cells could drive the activity bump, as long as there was a source of tonic, excitatory drive onto the cells. The source of this tonic drive was not known. However, an intriguing possibility comes from the discovery of a cellular mechanism that enables graded, persistent activity. In these slice experiments, in which synaptic transmission was blocked, an injection of depolarizing current caused a cell to fire persistently at a tonic rate. Additional current injections caused the cell to fire at a higher rate, which persisted indefinitely. Changing the polarity of the current caused a decrement in the tonic, persistent firing rate. This type of activity has been demonstrated in a number of brain areas (Morisset & Nagy 2000; Egorov et al. 2002, 2006; Fransen et al. 2006; Winograd et al. 2008), including the anterodorsal nucleus of the thalamus (Kulkarni et al. 2011) and postsubiculum (Yoshida & Hasselmo 2009), two key components of the HD circuit. Although the firing rate of this persistent activity tends to be slower than that of many HD cells, and the time course for changing the firing rates is slower than typical head movements, these cells may still play a role in promoting the stability of the attractor states. Thus, it is conceivable that the persistent activity of HD cells derives in part from intrinsic cellular mechanisms, not recurrent excitatory circuitry, and the attractor

MEC: medial entorhinal cortex

properties may be derived from the inhibitory connections modeled by Song & Wang (2005) and Boucheny et al. (2005).

Experimental Evidence

What is the experimental evidence in favor of the ring attractor models of HD cells? With the discovery of intracellular mechanisms for persistent activity, some properties of HD cells that initially inspired the formulation of the ring attractor models can no longer be considered as strong evidence in favor of attractor dynamics in the circuit. These properties include the persistent activity of these cells, with little or no adaptation while the rat remains stationary facing the cell's preferred direction, and the continued activity of these cells in the dark. Nonetheless, the ring attractor models still provide a large amount of explanatory power to a number of known properties of the HD cell system. (*a*) The models require cells that encode both HD and angular head velocity, and such cells have been discovered in the system (McNaughton et al. 1991, Blair et al. 1998, Stackman & Taube 1998, Bassett & Taube 2001, Sharp et al. 2001b). (*b*) The HD cell tuning curves can drift relative to external landmarks, demonstrating that the signal can be purely "inside the rat's head," independent of any external sensory cue (Knierim et al. 1995, 1998; Yoganarasimha & Knierim 2005). (*c*) Strong attractor circuitry would result in a system in which individual HD cells maintain the same preferred firing directions relative to each other, even when the preferred directions drift relative to an external sensory framework. Studies that recorded multiple HD cells simultaneously have confirmed that the preferred firing directions of these cells are rigidly coupled to each other (Taube & Burton 1995, Zugaro et al. 2001, Yoganarasimha et al. 2006). (*d*) The models predict that self-motion input is the primary drive that updates the HD signal; external landmarks play a powerful, but secondary, role of correcting or preventing drift. In agreement, disruption of the vestibular system severely disrupts the HD circuit (Stackman & Taube 1997). In particular, plugging the semicircular canals causes the loss of a directional signal; nonetheless, simultaneously recorded cells in the anterodorsal nucleus of the thalamus still fire bursts in predictable sequences corresponding to clockwise and counterclockwise turns, as if the internal ring circuitry is still intact, but the update mechanism from the vestibular system causes the bump of activity to move around the ring in ways that bear no relationship to the animal's actual HD in the world (Muir et al. 2009). (*e*) External landmarks are required to keep the system calibrated relative to the external world as well as to reset the system to its "correct" orientation when the animal reenters a familiar environment. The attractor dynamics cause the correction to be either a continuous change (as the HD activity bump moves continuously to a new direction) or an abrupt shift (as the activity bump disappears at one location while reappearing at the new location). Whether the correction is continuous or abrupt depends both on the angular magnitude of the correction and on the strength of the embedded attractors (Zhang 1996, Song & Wang 2005). Consistent with the models, Knierim and colleagues (Knierim et al. 1998) showed that small angular deviations were more likely to be corrected than were large deviations, whereas Zugaro and colleagues (Zugaro et al. 2003) showed that large deviations could be corrected in a very fast, abrupt manner (80 ms).

GRID CELLS AND TWO-DIMENSIONAL CONTINUOUS ATTRACTORS

Grid cells are spatially tuned neurons that fire in multiple locations in an environment (Hafting et al. 2005). Each firing location is at the vertex of a triangular (or hexagonal) grid that tiles the entire surface of the environment. These cells, originally discovered and most prominent in the dorsocaudal medial entorhinal cortex (MEC), are also found in the presubiculum and parasubiculum (Boccara et al. 2010). In the MEC, neighboring grid cells have the same orientation and scale, differing only in phase (i.e., locational shift) (Hafting et al. 2005). In more ventral parts of the MEC, the scale of the grids increases (Hafting et al. 2005). Similar to HD cells, grid cells have a number of properties that are amenable to modeling with attractor networks. Grid cells can be controlled by salient cues and boundaries, yet they maintain their grid-like firing patterns even in total darkness (Hafting et al. 2005). Thus, a self-motion signal appears to be a key component of updating the system. Also similar to HD cells, grid cells of the same scale appear to be rigidly coupled (Fyhn et al. 2007, Hargreaves et al. 2007), although it is not known whether this rigidity extends to grid cells across all spatial scales and across hemispheres.

Attractor versus Oscillatory Interference Models

Many computational models of grid cells have now been proposed, and the large majority of them fall into two classes: oscillatory interference (OI) models and attractor models. OI models propose that the grid pattern arises from the beat frequencies that form from several oscillators with slightly different frequencies centered around the theta frequency (Blair et al. 2007, Burgess et al. 2007, Hasselmo et al. 2007, Monaco et al. 2011, Welday et al. 2011). The key requirement is that the frequency be modulated by the animal's velocity (speed and direction) (Geisler et al. 2007, Welday et al. 2011). Under appropriate conditions, the beat frequencies of the interference patterns cause a cell to reach firing threshold whenever the rat is at the vertex of the grid. In contrast, attractor models are basically two-dimensional extensions of the ring attractor models of HD cells (Fuhs & Touretzky 2006, McNaughton et al. 2006, Guanella et al. 2007, Burak & Fiete 2009). Rather than a ring, however, the neurons are arranged conceptually on a two-dimensional plane, with appropriate excitatory connections among neighbors and global inhibition to cause a two-dimensional bump of activity to form at a location on the plane. Similar to the ring attractor, each location on the plane is a stable state, and the bump can move in a continuous manner from one stable state to another, driven by asymmetric input from cells that encode linear velocity. In one mechanism, the periodically repeating pattern of grid cells comes from the toroidal architecture of the attractors (McNaughton et al. 2006). That is, just as a line attractor is made into a ring attractor by connecting the ends of the line to form a ring, a plane attractor is made into a toroidal attractor by connecting the edges of the plane to each other to form a torus (**Figure 1*b***). The simplest example of this is a square sheet of paper for which two opposing sides are joined together to form a cylinder and the ends of the cylinder are joined to form a torus (i.e., a doughnut). The movement of the bump of activity along this torus would be bounded, but if tied to the movement of the animal, it would produce a periodic, repeated pattern of activity resembling a grid cell, but with a square grid. To get a hexagonal grid, one could make the shape of the paper a rhombus (McNaughton et al. 2006) or a hexagon (in which case the resulting topology is called a twisted torus) (Guanella et al. 2007) (**Figure 2*a***). Alternatively, one could set up a center-surround excitation-inhibition connectivity scheme (e.g., a Mexican-hat function) that will naturally cause cells to fire in a hexagonal grid (the optimal packing density of circles) (Fuhs & Touretzky 2006), which is the pattern that emerges in a Turing mechanism of pattern formation (Turing 1953). Finally, it is worth mentioning that although attractor and OI models appear radically different, they actually satisfy the same general mathematical conditions for exact path integration (Issa & Zhang 2012).

OI: oscillatory interference

Experimental Evidence

Similar to HD cell models, the attractor models of grid cells are elegant and account for many properties of grid cells, but the experimental evidence that the system is organized as such an attractor is indirect and circumstantial. The

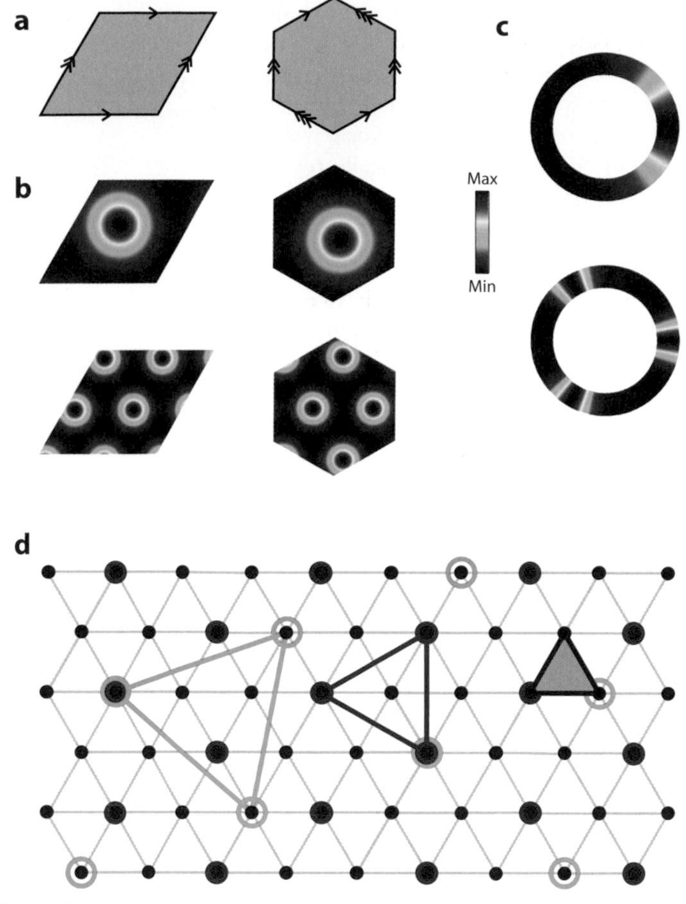

Figure 2

(*a*) Connecting the opposite edges of a sheet of cells shaped as a rhombus (*left*) or a hexagon (*right*) yields a torus or a twisted torus, respectively. Both networks can generate hexagonal firing fields if the activity bump moves according to the velocity of the animal's movement. A sheet of cells without toroidal connectivity can also support multiple activity peaks and hexagonal firing fields (Fuhs & Touretzky 2006). (*b*) Activity pattern in a torus network may have either a single peak (*top*) or multiple peaks (*bottom*), depending on the connectivity of the network. Note that these figures show the activity pattern on the attractor sheet, not the spatial firing-rate map of individual units. In all cases, recordings from a single unit would display grid-like firing fields as the bump(s) move across the attractor sheet. (*c*) Activity pattern in a ring network, showing a single peak (*top*) or multiple peaks (*bottom*). In a head-direction ring attractor model, if the speed at which the multiple bumps moves around the ring is one-third the speed of the single bump, then individual units in each attractor would show identical, single-lobed directional tuning curves. (*d*) Consistency of grid fields of different sizes may require special geometric arrangements. Let the blue dots indicate a base hexagonal grid whose spacing is given by the side length of the filled blue triangle. The spacing of a large hexagonal grid (*red circles*) consistent with the base grid will be $\sqrt{3} = 1.73$ times as large, with an orientation offset of 30°. Several larger grids are also consistent with the base grid. For instance, the spacing of the grid indicated by the green triangle is $\sqrt{7} = 2.65$ times as large.

evidence for these models is perhaps best evaluated in comparison with the evidence for the most popular alternative, the OI models, which have a number of bases of support (Hasselmo 2011). The OI models naturally account for the presence of theta phase precession in grid cells (Hafting et al. 2008) as a necessary consequence of the OI mechanism, whereas attractor models, although consistent with phase precession, do not intrinsically explain the phenomenon. OI models have some experimental support, such as the change in the resonant frequency of neurons from dorsal to ventral MEC that corresponds with the increased scale of grids along that axis (Giocomo et al. 2007). Again, attractor models can be consistent with this finding, and incorporating it into the models allows the simulated grid cells to show phase precession and scaling, but the models do not require it to produce the spatial grid patterns (Navratilova et al. 2011). Inactivation of the medial septum, which disrupts theta rhythm, abolishes the grid-like firing patterns of grid cells, in accordance with predictions of the OI theory (Brandon et al. 2011, Koenig et al. 2011). However, this result must be interpreted cautiously: The septal inactivation does more than just disrupt the rhythmic firing of the cells. It undoubtedly has other effects that may cause the disruption of grid cells. For example, the persistent cellular firing discussed above requires cholinergic input (Egorov et al. 2002), and the medial septum is a major source of cholinergic input into the MEC (Witter & Amaral 2004). Furthermore, bats have been reported to display grid cells in the absence of obvious theta oscillations (Yartsev et al. 2011), which does not disprove the OI models but presents them with a serious challenge.

Some of the attractor models naturally account for the fact that all grids within an animal appear to be of the same orientation—a necessary consequence of the toroidal architecture. The attractor map also predicts that different grid scales must come about through distinct attractor modules (McNaughton et al. 2006), and evidence suggests that this prediction is true (Barry et al. 2007). However, it is possible that OI models also require such discrete steps in grid scale to ensure that the grids stay stable relative to each other. The geometry of the grid may constrain the relative orientation and scales of grids along the dorsal-ventral axis of the MEC (**Figure 2*d***). Both models have difficulty in justifying the somewhat artificial mechanisms employed to impose the hexagonal structure on the grid. OI models have to postulate that the HD cells that provide the directional component of the velocity vector are restricted to representing only three angles, 120° apart. There is currently no strong evidence that HD cells are so organized. Toroidal attractor models have to postulate a priori that the sheet is either a rhombus or a hexagon to obtain the proper grid geometry. Although there are good arguments for why the grids should be arranged as a hexagonal lattice, on the basis of arguments of optimal packing density and natural formation of patterns from reaction-diffusion-type equations and center-surround excitation-inhibition connectivity (Fuhs & Touretzky 2006, McNaughton et al. 2006), the models still need to build this into their architecture somewhat artificially. Thus, the jury is still out on the relative merits of these classes of models. OI models have been criticized because they appear to require temporal precision greater than is observed in biological systems (Welinder et al. 2008), but newer versions have devised methods with the potential to overcome this problem (Zilli & Hasselmo 2010). Continuous attractor models are also subject to structural instability, with any small damage or disturbance capable of changing them to point attractors (Zhang 1996, Tsodyks 1999, Renart et al. 2003). As a result, the grid cells may combine aspects of both types of models. At the cellular level, OI may bootstrap the cells to fire in a grid-like manner. However, attractor dynamics may then be required to keep the system stable and to correct errors imposed by temporal noise in the oscillators. It will be of interest to see if future modeling efforts can make use of both mechanisms to create more robust models that conform closely to the biological data. (For a more detailed review of the different models of grid cells, see Giocomo et al. 2011.)

PLACE CELLS AND CONTINUOUS VERSUS POINT ATTRACTORS

The hippocampus has been modeled both as a point attractor neural network and as a continuous attractor network (Treves & Rolls 1994, Levy 1996, Samsonovich & McNaughton 1997, Battaglia & Treves 1998, Tsodyks 1999, Conklin & Eliasmith 2005, Touretzky et al. 2005, Monaco et al. 2007). In particular, the CA3 region has been the focus of such models, as its extensive system of excitatory recurrent collaterals endows it with an anatomical substrate conducive to the formation of attractor states. The discrete attractor models derive from ideas put forward by Hebb and Marr. Hebb (1949) postulated that coactive populations of neurons form, through plasticity, cell assemblies that have many of the properties now described as signatures of attractor networks. Marr (1971) formulated the idea that such a network is capable of pattern completion. That is, if a memory is stored in a recurrent network as a distributed pattern of synaptic weights, such a network is capable of reinstating the full activity pattern even if the original input pattern is incomplete, degraded, or corrupted. An attractor network is a prime candidate mechanism to perform this function. Complementary to the idea of pattern completion is pattern separation, i.e., two similar input patterns are stored as more dissimilar patterns to reduce the probability of errors in recall. The putative attractor dynamics of CA3 will promote a sigmoidal relationship between changes in the inputs to CA3 and changes in the CA3 output representation (O'Reilly & McClelland 1994). When changes are small, the CA3 attractor dynamics tend to cause CA3 to converge to the original attractor basin, such that the CA3 outputs are less different than the inputs. When changes are larger, however, the CA3 representation may form in a region of state space away from the attractor basin, making the CA3 outputs less similar than the inputs. Under these conditions, it is hypothesized that a new attractor basin will form far away from (orthogonal to) the original attractor basin. (For reviews of various computational models of the hippocampus, see Burgess 2007 and Hasselmo 2011.)

Pattern Completion and Pattern Separation

A number of experiments in recent years have tested the pattern completion/separation hypotheses in different components of the hippocampal formation. A set of studies published nearly simultaneously from three different laboratories provided evidence for the sigmoidal relationship between input and output changes in the CA3 region. Lee and colleagues (Lee et al. 2004) rotated a local reference frame (a circular track with salient cues) counter to a global reference frame (salient landmarks on the curtained walls of the environment) and found that, compared with CA1 place cells, CA3 place cells responded in a more coherent fashion to the cue-mismatch (**Figure 3**). They interpreted this as reflecting the pattern-completion functions of CA3, as the conflicting input regarding local and global cues was resolved in favor of the local cues at each location on the track, in contrast to the split representations seen in CA1. Leutgeb and colleagues (Leutgeb et al. 2004) recorded CA3 and CA1 cells in an experiment in which the rat was transported into two completely different rooms. In contrast to the findings of Lee et al. (2004), CA3 place fields under these conditions were completely independent in the two situations, even though CA1 place fields maintained some degree of similarity. Leutgeb et al. (2004) concluded that under these conditions CA3 showed pattern separation, as presumed similarities in the input representations (as reflected by the similarities in the feedforward CA1 network) were erased in CA3. The trade-off between pattern completion seen by Lee et al. (2004) and pattern separation seen by Leutgeb et al. (2004) was confirmed in a single study utilizing the immediate early gene *Arc* by Vazdarjanova & Guzowski (2004). Under conditions in which changes to the environment were small, CA3 showed a greater overlap in *Arc* expression than did CA1; conversely, when the changes to the environment were larger, CA3 representations overlapped less than those of CA1. Taken as a set, these three studies provide strong evidence for the nonlinear, attractor-like behavior of CA3 versus CA1 when environmental inputs are altered in a graded fashion (Guzowski et al. 2004).

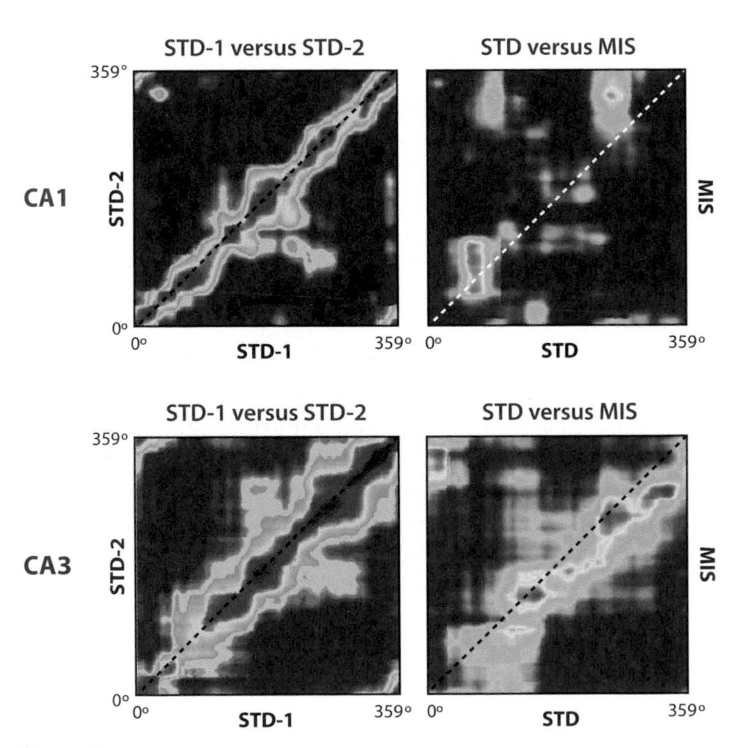

Figure 3

Spatial correlation matrices for population vectors of CA1 and CA3 place fields. Location along a standard track (session 1: STD-1; session 2: STD-2) or along a track with mismatched reference frames (MIS) is plotted along the axes. (*Left*) The red diagonal band signifies that the representations of both CA1 and CA3 were highly similar across two standard sessions. (*Right*) The band of high correlation remains in CA3 but is absent in CA1. This result demonstrates that CA3 responds more coherently to the mismatch than does CA1, consistent with pattern completion (or generalization) properties expected if CA3 incorporated attractor dynamics. Modified from Lee et al. (2004).

Nakazawa et al. (2002) tested the effects of selective deletion of the NR1 subunit of the NMDA receptor in CA3 by recording from the downstream CA1 region during a classic test of pattern completion. When a subset of the visual landmarks in a room was removed, perhaps degrading the spatial representations that provide input to CA3 and CA1, normal mice showed similar place fields between the full-cue and partial-cue environments. In contrast, mice lacking normal NMDA receptors in CA3 showed degraded place fields in CA1 in the partial-cue environments, consistent with the hypothesized role of CA3 in pattern completion.

Wills et al. (2005) recorded CA1 place fields as the geometry of an environment was gradually morphed in four stages from a circle to a square. They found that place fields in the square-like morph geometries matched the place fields of the true square, whereas place fields in the circle-like geometries matched the place fields of the true circle. The transition between square-like and circle-like place fields was sharp, which the authors interpreted as evidence for attractor dynamics in the hippocampus, presumably in the CA3 region upstream of CA1. In contrast, though employing a similar experiment, Leutgeb and colleagues (2005a) showed a gradual change in place fields as the box geometry was morphed. Place fields mostly stayed in the same locations in both the circle- and square-like environments, but the firing rate within each place field gradually changed. Insight into the discrepancy between the studies comes from Fyhn et al. (2007), who recorded from MEC grid cells and CA1 place cells when the CA1 place fields remapped. Global remapping [i.e., that seen by Wills et al. (2005)] was accompanied by shifts in the MEC grids relative to the environmental boundaries, whereas rate remapping [i.e., that seen by Leutgeb et al. (2004)] was accompanied by a stable relationship between the grids and the boundaries. It is conceivable that the Wills study caused the MEC grids to shift abruptly at the transition between the most ambiguous circle-like shape and the most ambiguous square-like shape, causing the abrupt global remapping between these two similar shapes. In contrast, because of the way the rats were pretrained and the experimental sessions were run, the grids may have remained stable in the Leutgeb experiments promoting gradual rate remapping (Leutgeb et al. 2005b).

Colgin et al. (2010) tested a similar idea by showing that CA3 and CA1 place fields displayed rate remapping when the path-integration coordinates thought to align the MEC grids were kept stable across two environments. In contrast, the place fields displayed global remapping when the path-integration coordinates between the two environments were distinctly different, which would tend to cause a shift in the grid relative to the environments. Colgin et al. (2010) interpreted these results as evidence in favor of the path-integration-based, plane attractor models of the hippocampal formation, as opposed to the single-point attractor models that were associated with notions of pattern completion and separation.

Continuous versus Point Attractor Dynamics

How does one reconcile the interpretation put forth by Colgin et al. (2010) with earlier work that interpreted the place-field data in terms of pattern completion and separation? More generally, do any of the experiments described above provide compelling, direct (rather than circumstantial) evidence in favor of attractor dynamics in the hippocampus? Because the hippocampus is an intricately connected network of multiple areas, with external input and feedback connections that can have strong influences on how the cells behave, this question has become somewhat intractable. Whereas the inputs can be turned off and the pure attractor dynamics observed in a simple computer model, this is difficult, if not impossible, to do in the biological system. Thus, even if strong attractor dynamics exist in the system, almost any experimental manipulation designed to expose those dynamics will produce results that are a hybrid between attractor dynamics and external drive onto the attractor. To test for true attractor dynamics, it is necessary to remove the external drive, as was attempted in studies of the visual cortex by Akrami et al. (2009). This is difficult to do in the hippocampus, as experimentalists do not have direct control over the highly processed sensory and cognitive representations that compose the inputs to the hippocampus (activity during sleep may help, but does not obviate, this problem). Moreover, different parts of the system may display different types of attractor dynamics. Thus, the continuous attractor dynamics supported by Colgin et al. (2010) may be present in the MEC inputs to the hippocampus. The output of the MEC attractor would then be an external input to the CA3 attractor and under most conditions would likely be a strong drive on the activity of CA3 cells. CA3 could still contain classic, point attractor dynamics, yet its place fields would also reflect the continuous attractor dynamics of its MEC inputs. If the roles of these inputs include aligning the path-integration framework with the external world and incorporating external information about the world onto this framework in support of episodic memory (perhaps via its lateral entorhinal inputs) (Knierim et al. 2006, Deshmukh & Knierim 2011), such a hybrid system may reflect properties of both the classic, discrete attractor models and the continuous attractor models.

FUTURE DIRECTIONS

Attractor models hold great sway across a number of brain systems, owing to the computational power they bring to various problems that the brain has to solve. In addition to the properties described above for limbic regions, attractor networks are useful for solving optimization problems, such as the traveling-salesman problem (Hopfield & Tank 1985). A continuous attractor network can be used to estimate signal hidden in noise, with near-optimal statistical efficiency (Pouget et al. 1998, Latham et al. 2003, Wu et al. 2008). Evidence is accumulating that the brain displays attractor dynamics in its processing; However, it is exceedingly difficult to test quantitatively any particular model of attractors in a particular brain circuit. Nonetheless, these models have proven extremely useful in providing a conceptual framework for understanding experimental data that defy simple, feedforward circuitry explanations. We finish this review by describing interesting questions for future research.

Learning the Weights of an Attractor Map

One of the major unsolved problems in attractor theory is how a neural system can learn through self-organization the synaptic-weight matrix required to form stable attractors (Kali & Dayan 2000, Stringer et al. 2005). This problem is particularly difficult with continuous attractors, which are hard to learn even in a simulation and hard to maintain because of their structural instability (any small damage or disturbance can change a continuous attractor to point attractors). The lack of anatomical topography of high-order brain systems complicates neural learning rules. Some have suggested that the weights can be learned through a developmental process in which topographically organized representations are initially present to serve as a teacher to the attractor and then disappear (McNaughton et al. 2006). The feasibility of these ideas may be tested in experiments that measure the developmental onset of place, grid, and directional signals (Martin & Berthoz 2002, Langston et al. 2010, Wills et al. 2010, Scott et al. 2011). However, there is as yet no consensus theory or model of self-organization of weights of an attractor map, and such a model would help guide interpretation and analysis of the experimental results.

Anatomical Measurement of a Connectivity Matrix

Attractor theory predicts how different neurons should be connected into a network. A main problem has been the lack of effective experimental methods for directly measuring the

a **Stored memory patterns**

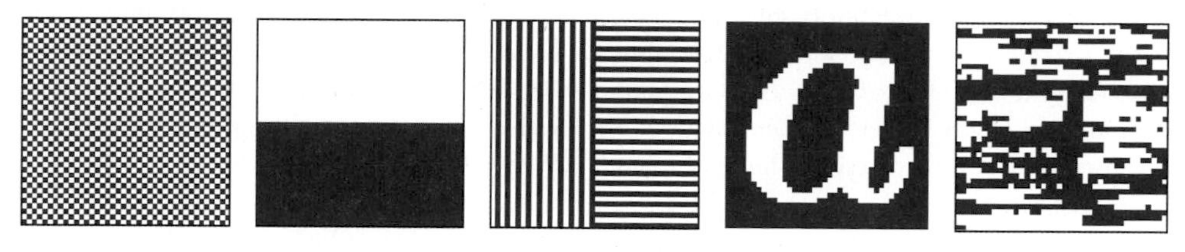

b **Examples of retrievals**

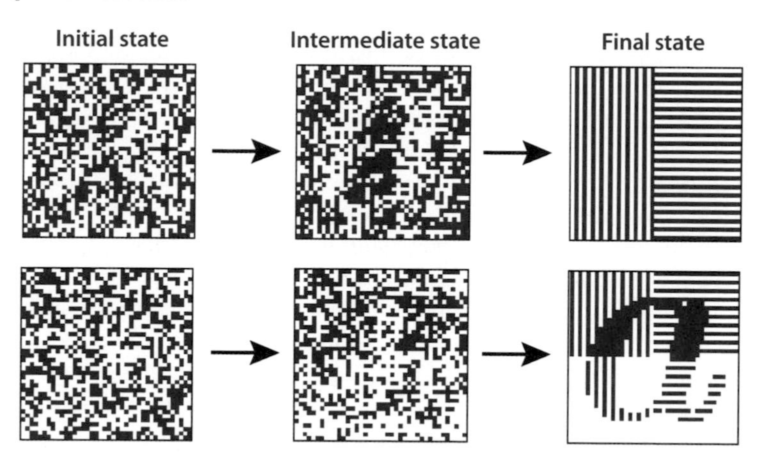

c **Examples of spurious memory patterns**

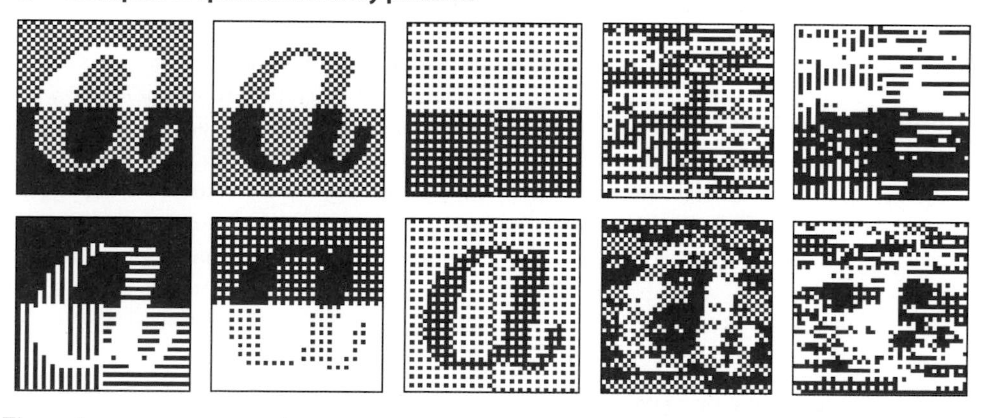

Figure 4

Spurious attractors. (*a*) Five memory patterns stored in a Hopfield network are shown as images, with each pixel corresponding to the state of a unit in the network. The last pattern is the "Loch Ness monster." (*b*) Starting from a random initial state, the network eventually settles to either a stored memory pattern or a spurious equilibrium state, which is some mixture of multiple memory patterns (sometimes with flipped polarity). (*c*) Some examples of the spurious states showing the diversity of the mixtures.

connectivity-weight matrix of a real network. The stable activity pattern on the attractor sheet may have either a single peak or multiple peaks, depending on the connectivity and exact values of the parameters. For example, the grid-cell model of Fuhs & Touretzky (2006) has multiple peaks of activity (arranged as a hexagonal grid) on the attractor sheet, whereas the model of McNaughton et al. (2006) has a single activity peak on a torus attractor (**Figure 2*b***). If one "records" from a unit in both models, the unit would appear as a grid cell, regardless of whether there is a single peak or multiple peaks in the activity pattern of the attractor. Similarly, if there were multiple peaks in an HD ring network, each individual cell could still appear as a normal HD cell (**Figure 2*c***). It is not known whether the single-peak or the multiple-peak case is closer to the real biological situation, although the two cases can imply different patterns of synaptic connectivity. The ultimate test would be direct measurement of the synaptic-connection patterns in the brain, which may directly reveal a toroidal pattern of weights in the MEC, for example. The recent advance of high-throughput connectomics may eventually provide data to address this question (Chklovskii et al. 2010, Seung 2011).

Multiple Stable States

Although rat hippocampal place cells tend to be stable for a long time in a stable environment, they are notoriously fickle in their responses to environmental or behavioral changes, displaying various forms of "remapping" and partial remapping (Knierim 2003). A theoretical explanation for the vagaries of partial remapping remains poorly understood. The ability of multiple stable states to be simultaneously active in an attractor may help explain some of these phenomena, including the existence of "split" place fields that occur in reference-frame mismatch experiments (Knierim 2002, Lee et al. 2004). One concept that may help researchers develop a theory of partial remapping is the notion that "spurious" attractors can arise in an attractor network (Amit 1989); that is, stable activity states can form that were not explicitly encoded in the network. To illustrate this concept, **Figure 4** shows a simulation of spurious attractors, in which we embedded five memory patterns in a Hopfield attractor network (**Figure 4*a***) and then ran a number of simulations in which we let the network evolve to a stable state after starting out with random input patterns. Although the system sometimes settled on one of the stored patterns (**Figure 4*b***, top row), it more often settled into a spurious, stable state (**Figure 4*b***, bottom row; **Figure 4*c***). Note that each spurious state in **Figure 4** is a mixture of several stored memory patterns. This mixture is not a simple, linear combination or superposition because different subregions of the image can combine different memory patterns in different ways, resulting in a rich variety of possible spurious states. It will be of interest to pursue whether investigation of spurious attractors may impart some insight into the sometimes seemingly capricious nature of place-cell remapping.

DISCLOSURE STATEMENT

The authors are not aware of any affiliations, memberships, funding, or financial holdings that might be perceived as affecting the objectivity of this review.

ACKNOWLEDGMENTS

The authors' work was supported by NIH grants NS039456 and MH079511.

LITERATURE CITED

Akrami A, Liu Y, Treves A, Jagadeesh B. 2009. Converging neuronal activity in inferior temporal cortex during the classification of morphed stimuli. *Cereb. Cortex* 19:760–76

Amit DJ. 1989. *Modeling Brain Function: The World of Attractor Neural Networks*. Cambridge, UK/New York: Cambridge Univ. Press.

Barry C, Hayman R, Burgess N, Jeffery KJ. 2007. Experience-dependent rescaling of entorhinal grids. *Nat. Neurosci.* 10:682–84

Bassett JP, Taube JS. 2001. Neural correlates for angular head velocity in the rat dorsal tegmental nucleus. *J. Neurosci.* 21:5740–51

Battaglia FP, Treves A. 1998. Attractor neural networks storing multiple space representations: a model for hippocampal place fields. *Phys. Rev. E* 58:7738–53

Ben-Yishai R, Bar-Or RL, Sompolinsky H. 1995. Theory of orientation tuning in visual cortex. *Proc. Natl. Acad. Sci. USA* 92:3844–48

Blair HT. 1996. A thalamocortical circuit for computing directional heading in the rat. *Adv. Neural. Inf. Process. Sys.* 8:152–58

Blair HT, Cho J, Sharp PE. 1998. Role of the lateral mammillary nucleus in the rat head direction circuit: a combined single unit recording and lesion study. *Neuron* 21:1387–97

Blair HT, Welday AC, Zhang K. 2007. Scale-invariant memory representations emerge from moire interference between grid fields that produce theta oscillations: a computational model. *J. Neurosci.* 27:3211–29

Boccara CN, Sargolini F, Thoresen VH, Solstad T, Witter MP et al. 2010. Grid cells in pre- and parasubiculum. *Nat. Neurosci.* 13:987–94

Boucheny C, Brunel N, Arleo A. 2005. A continuous attractor network model without recurrent excitation: maintenance and integration in the head direction cell system. *J. Comput. Neurosci.* 18:205–27

Brandon MP, Bogaard AR, Libby CP, Connerney MA, Gupta K, Hasselmo ME. 2011. Reduction of theta rhythm dissociates grid cell spatial periodicity from directional tuning. *Science* 332:595–99

Brody CD, Romo R, Kepecs A. 2003. Basic mechanisms for graded persistent activity: discrete attractors, continuous attractors, and dynamic representations. *Curr. Opin. Neurobiol.* 13:204–11

Burak Y, Fiete IR. 2009. Accurate path integration in continuous attractor network models of grid cells. *PLoS Comput. Biol.* 5:e1000291

Burgess N. 2007. Computational models of the spatial and mnemonic functions of the hippocampus. In *The Hippocampus Book*, ed. P Andersen, R Morris, D Amaral, T Bliss, J O'Keefe, pp. 715–49. New York: Oxford Univ. Press

Burgess N, Barry C, O'Keefe J. 2007. An oscillatory interference model of grid cell firing. *Hippocampus* 17:801–12

Chen LL, Lin LH, Barnes CA, McNaughton BL. 1994. Head-direction cells in the rat posterior cortex. II. Contributions of visual and ideothetic information to the directional firing. *Exp. Brain Res.* 101:24–34

Chklovskii DB, Vitaladevuni S, Scheffer LK. 2010. Semi-automated reconstruction of neural circuits using electron microscopy. *Curr. Opin. Neurobiol.* 20:667–75

Cohen MA, Grossberg S. 1983. Absolute stability of global pattern formation and parallel memory storage by competitive neural networks. *IEEE Trans. Sys. Man Cybern.* 13:815–26

Colgin LL, Leutgeb S, Jezek K, Leutgeb JK, Moser EI et al. 2010. Attractor-map versus autoassociation-based attractor dynamics in the hippocampal network. *J. Neurophysiol.* 104:35–50

Conklin J, Eliasmith C. 2005. A controlled attractor network model of path integration in the rat. *J. Comput. Neurosci.* 18:183–203

Deshmukh SS, Knierim JJ. 2011. Representation of non-spatial and spatial information in the lateral entorhinal cortex. *Front. Behav. Neurosci.* 5:69

Dunin-Barkowski WL, Osovets NB. 1995. Hebb-Hopfield neural networks based on one-dimensional sets of neuron states. *Neural Proc. Lett.* 2:28–31

Egorov AV, Hamam BN, Fransen E, Hasselmo ME, Alonso AA. 2002. Graded persistent activity in entorhinal cortex neurons. *Nature* 420:173–78

Egorov AV, Unsicker K, von Bohlen und Halbach O. 2006. Muscarinic control of graded persistent activity in lateral amygdala neurons. *Eur. J. Neurosci.* 13:3183–94

Ferster D, Miller KD. 2000. Neural mechanisms of orientation selectivity in the visual cortex. *Annu. Rev. Neurosci.* 23:441–71

Fransen E, Tahvildari B, Egorov AV, Hasselmo ME, Alonso AA. 2006. Mechanism of graded persistent cellular activity of entorhinal cortex layer v neurons. *Neuron* 49:735–46

Fuhs MC, Touretzky DS. 2006. A spin glass model of path integration in rat medial entorhinal cortex. *J. Neurosci.* 26:4266–76

Fyhn M, Hafting T, Treves A, Moser MB, Moser EI. 2007. Hippocampal remapping and grid realignment in entorhinal cortex. *Nature* 446:190–94

Geisler C, Robbe D, Zugaro M, Sirota A, Buzsaki G. 2007. Hippocampal place cell assemblies are speed-controlled oscillators. *Proc. Natl. Acad. Sci. USA* 104:8149–54

Giocomo LM, Zilli EA, Fransen E, Hasselmo ME. 2007. Temporal frequency of subthreshold oscillations scales with entorhinal grid cell field spacing. *Science* 315:1719–22

Giocomo LM, Moser MB, Moser EI. 2011. Computational models of grid cells. *Neuron* 71:589–603

Goodridge JP, Dudchenko PA, Worboys KA, Golob EJ, Taube JS. 1998. Cue control and head direction cells. *Behav. Neurosci.* 112:749–61

Goodridge JP, Touretzky DS. 2000. Modeling attractor deformation in the rodent head-direction system. *J. Neurophysiol.* 83:3402–10

Guanella A, Kiper D, Verschure P. 2007. A model of grid cells based on a twisted torus topology. *Int. J. Neural Syst.* 17:231–40

Guzowski JF, Knierim JJ, Moser EI. 2004. Ensemble dynamics of hippocampal regions CA3 and CA1. *Neuron* 44:581–84

Hafting T, Fyhn M, Bonnevie T, Moser MB, Moser EI. 2008. Hippocampus-independent phase precession in entorhinal grid cells. *Nature* 453:1248–52

Hafting T, Fyhn M, Molden S, Moser MB, Moser EI. 2005. Microstructure of a spatial map in the entorhinal cortex. *Nature* 436:801–6

Hargreaves EL, Yoganarasimha D, Knierim JJ. 2007. Cohesiveness of spatial and directional representations recorded from neural ensembles in the anterior thalamus, parasubiculum, medial entorhinal cortex, and hippocampus. *Hippocampus* 17:826–41

Hasselmo ME. 2011. *How We Remember: Brain Mechanisms of Episodic Memory.* Cambridge, MA: MIT Press

Hasselmo ME, Giocomo LM, Zilli EA. 2007. Grid cell firing may arise from interference of theta frequency membrane potential oscillations in single neurons. *Hippocampus* 17:1252–71

Hebb DO. 1949. *The Organization of Behavior*. New York: Wiley

Hopfield JJ. 1982. Neural networks and physical systems with emergent collective computational abilities. *Proc. Natl. Acad. Sci. USA* 79:2554–58

Hopfield JJ. 1984. Neurons with graded response have collective computational properties like those of two-state neurons. *Proc. Natl. Acad. Sci. USA* 81:3088–92

Hopfield JJ, Tank DW. 1985. "Neural" computation of decisions in optimization problems. *Biol. Cybern.* 52:141–52

Issa JB, Zhang K. 2012. Universal conditions for exact path integration in the neural systems. *Proc. Natl. Acad. Sci. USA*. In press

Kali S, Dayan P. 2000. The involvement of recurrent connections in area CA3 in establishing the properties of place fields: a model. *J. Neurosci.* 20:7463–77

Knierim JJ. 2002. Dynamic interactions between local surface cues, distal landmarks, and intrinsic circuitry in hippocampal place cells. *J. Neurosci.* 22:6254–64

Knierim JJ. 2003. Hippocampal remapping: Implications for spatial learning and navigation. In *The Neurobiology of Spatial Behaviour*, ed. KJ Jeffery, pp. 226–39. Oxford: Oxford Univ. Press

Knierim JJ, Kudrimoti HS, McNaughton BL. 1995. Place cells, head direction cells, and the learning of landmark stability. *J. Neurosci.* 15:1648–59

Knierim JJ, Kudrimoti HS, McNaughton BL. 1998. Interactions between idiothetic cues and external landmarks in the control of place cells and head direction cells. *J. Neurophysiol.* 80:425–46

Knierim JJ, Lee I, Hargreaves EL. 2006. Hippocampal place cells: parallel input streams, subregional processing, and implications for episodic memory. *Hippocampus* 16:755–64

Koenig J, Linder AN, Leutgeb JK, Leutgeb S. 2011. The spatial periodicity of grid cells is not sustained during reduced theta oscillations. *Science* 332:592–95

Kopell N, Ermentrout GB. 1988. Coupled oscillators and the design of central pattern generators. *Math. Biosci.* 90:87–109

Kulkarni M, Zhang K, Kirkwood A. 2011. Single-cell persistent activity in anterodorsal thalamus. *Neurosci. Lett.* 498:179–84

Laing CR, Chow CC. 2002. A spiking neuron model for binocular rivalry. *J. Comput. Neurosci.* 12:39–53

Langston RF, Ainge JA, Couey JJ, Canto CB, Bjerknes TL et al. 2010. Development of the spatial representation system in the rat. *Science* 328:1576–80

Latham PE, Deneve S, Pouget A. 2003. Optimal computation with attractor networks. *J. Physiol. Paris* 97:683–94

Lee I, Yoganarasimha D, Rao G, Knierim JJ. 2004. Comparison of population coherence of place cells in hippocampal subfields CA1 and CA3. *Nature* 430:456–59

Leutgeb JK, Leutgeb S, Treves A, Meyer R, Barnes CA et al. 2005a. Progressive transformation of hippocampal neuronal representations in "morphed" environments. *Neuron* 48:345–58

Leutgeb S, Leutgeb JK, Moser MB, Moser EI. 2005b. Place cells, spatial maps and the population code for memory. *Curr. Opin. Neurobiol.* 15:738–46

Leutgeb S, Leutgeb JK, Treves A, Moser MB, Moser EI. 2004. Distinct ensemble codes in hippocampal areas CA3 and CA1. *Science* 305:1295–98

Levy WB. 1996. A sequence predicting CA3 is a flexible associator that learns and uses context to solve hippocampal-like tasks. *Hippocampus* 6:579–90

Machens CK, Romo R, Brody CD. 2005. Flexible control of mutual inhibition: a neural model of two-interval discrimination. *Science* 307:1121–24

Major G, Baker R, Aksay E, Mensh B, Seung HS, Tank DW. 2004. Plasticity and tuning by visual feedback of the stability of a neural integrator. *Proc. Natl. Acad. Sci. USA* 101:7739–44

Marder E, Bucher D. 2001. Central pattern generators and the control of rhythmic movements. *Curr. Biol.* 11:R986–96

Marr D. 1971. Simple memory: a theory for archicortex. *Philos. Trans. R. Soc. Lond. Ser. B* 262:23–81

Martin PD, Berthoz A. 2002. Development of spatial firing in the hippocampus of young rats. *Hippocampus* 12:465–80

McNaughton BL, Battaglia FP, Jensen O, Moser EI, Moser MB. 2006. Path integration and the neural basis of the 'cognitive map'. *Nat. Rev. Neurosci.* 7:663–78

McNaughton BL, Chen LL, Markus EJ. 1991. "Dead reckoning", landmark learning, and the sense of direction: a neurophysiological and computational hypothesis. *J. Cogn. Neurosci.* 3:190–202

Milnor J. 1985. On the concept of attractor. *Commun. Math. Phys.* 99:177–95

Mizumori SJ, Williams JD. 1993. Directionally selective mnemonic properties of neurons in the lateral dorsal nucleus of the thalamus of rats. *J. Neurosci.* 13:4015–28

Monaco JD, Abbott LF, Kahana MJ. 2007. Lexico-semantic structure and the word-frequency effect in recognition memory. *Learn. Mem.* 14:204–13

Monaco JD, Knierim JJ, Zhang K. 2011. Sensory feedback, error correction, and remapping in a multiple oscillator model of place cell activity. *Front. Comput. Neurosci.* 5:39

Morisset V, Nagy F. 2000. Plateau potential-dependent windup of the response to primary afferent stimuli in rat dorsal horn neurons. *Eur. J. Neurosci.* 12:3087–95

Muir GM, Brown JE, Carey JP, Hirvonen TP, Della Santina CC et al. 2009. Disruption of the head direction cell signal after occlusion of the semicircular canals in the freely moving chinchilla. *J. Neurosci.* 29:14521–33

Muller RU, Ranck JB Jr, Taube JS. 1996. Head direction cells: properties and functional significance. *Curr. Opin. Neurobiol.* 6:196–206

Nakazawa K, Quirk MC, Chitwood RA, Watanabe M, Yeckel MF et al. 2002. Requirement for hippocampal CA3 NMDA receptors in associative memory recall. *Science* 297:211–18

Navratilova Z, Giocomo LM, Fellous JM, Hasselmo ME, McNaughton BL. 2011. Phase precession and variable spatial scaling in a periodic attractor map model of medial entorhinal grid cells with realistic after-spike dynamics. *Hippocampus*. doi: 10.1002/hipo.20939

O'Reilly RC, McClelland JL. 1994. Hippocampal conjunctive encoding, storage, and recall: avoiding a trade-off. *Hippocampus* 4:661–82

Pouget A, Zhang K, Deneve S, Latham PE. 1998. Statistically efficient estimation using population coding. *Neural Comput.* 10:373–401

Ranck JB Jr. 1985. Head direction cells in the deep cell layer of dorsal presubiculum in freely moving rats. In *Electrical Activity of Archicortex*, ed. G Buzsaki, CH Vanderwolf, pp. 217–20. Budapest: Akad. Kiado

Rand RH, Cohen AH, Holmes PJ. 1988. Systems of coupled oscillators as models of central pattern generators. In *Neural Control of Rhythmic Movements in Vertebrates*, ed. A Cohen, pp. 333–67. New York: Wiley

Redish AD, Elga AN, Touretzky DS. 1996. A coupled attractor model of the rodent head direction system. *Netw. Comput. Neural Syst.* 7:671–85

Renart A, Song P, Wang XJ. 2003. Robust spatial working memory through homeostatic synaptic scaling in heterogeneous cortical networks. *Neuron* 38:473–85

Robinson DA. 1989. Integrating with neurons. *Annu. Rev. Neurosci.* 12:33–45

Rolls ET, Stringer SM, Trappenberg TP. 2002. A unified model of spatial and episodic memory. *Proc. R. Soc. Lond. Ser. B* 269:1087–93

Samsonovich A, McNaughton BL. 1997. Path integration and cognitive mapping in a continuous attractor neural network model. *J. Neurosci.* 17:5900–20

Scott RC, Richard GR, Holmes GL, Lenck-Santini PP. 2011. Maturational dynamics of hippocampal place cells in immature rats. *Hippocampus* 21:347–53

Seung HS. 1996. How the brain keeps the eyes still. *Proc. Natl. Acad. Sci. USA* 93:13339–44

Seung HS. 2011. Neuroscience: towards functional connectomics. *Nature* 471:170–72

Sharp PE, Blair HT, Cho J. 2001a. The anatomical and computational basis of the rat head-direction cell signal. *Trends Neurosci.* 24:289–94

Sharp PE, Tinkelman A, Cho J. 2001b. Angular velocity and head direction signals recorded from the dorsal tegmental nucleus of gudden in the rat: implications for path integration in the head direction cell circuit. *Behav. Neurosci.* 115:571–88

Skaggs WE, Knierim JJ, Kudrimoti HS, McNaughton BL. 1995. A model of the neural basis of the rat's sense of direction. *Adv. Neural Inf. Process. Syst.* 7:173–80

Somers DC, Nelson SB, Sur M. 1995. An emergent model of orientation selectivity in cat visual cortical simple cells. *J. Neurosci.* 15:5448–65

Song P, Wang XJ. 2005. Angular path integration by moving "hill of activity": a spiking neuron model without recurrent excitation of the head-direction system. *J. Neurosci.* 25:1002–14

Stackman RW, Taube JS. 1997. Firing properties of head direction cells in the rat anterior thalamic nucleus: dependence on vestibular input. *J. Neurosci.* 17:4349–58

Stackman RW, Taube JS. 1998. Firing properties of rat lateral mammillary single units: head direction, head pitch, and angular head velocity. *J. Neurosci.* 18:9020–37

Stringer SM, Rolls ET, Trappenberg TP. 2005. Self-organizing continuous attractor network models of hippocampal spatial view cells. *Neurobiol. Learn. Mem.* 83:79–92

Stringer SM, Trappenberg TP, Rolls ET, de Araujo IE. 2002. Self-organizing continuous attractor networks and path integration: one-dimensional models of head direction cells. *Network* 13:217–42

Taube JS. 1998. Head direction cells and the neurophysiological basis for a sense of direction. *Prog. Neurobiol.* 55:225–56

Taube JS. 2007. The head direction signal: origins and sensory-motor integration. *Annu. Rev. Neurosci.* 30:181–207

Taube JS, Burton HL. 1995. Head direction cell activity monitored in a novel environment and during a cue conflict situation. *J. Neurophysiol.* 74:1953–71

Taube JS, Muller RU, Ranck JB Jr. 1990a. Head-direction cells recorded from the postsubiculum in freely moving rats. I. Description and quantitative analysis. *J. Neurosci.* 10:420–35

Taube JS, Muller RU, Ranck JB Jr. 1990b. Head-direction cells recorded from the postsubiculum in freely moving rats. II. Effects of environmental manipulations. *J. Neurosci.* 10:436–47

Touretzky DS, Weisman WE, Fuhs MC, Skaggs WE, Fenton AA, Muller RU. 2005. Deforming the hippocampal map. *Hippocampus* 15:41–55

Treves A, Rolls ET. 1994. Computational analysis of the role of the hippocampus in memory. *Hippocampus.* 4:374–91

Tsodyks M. 1999. Attractor neural network models of spatial maps in hippocampus. *Hippocampus* 9:481–89

Turing AM. 1953. The chemical basis of morphogenesis. *Bull. Math. Biol.* 52:153–97

Vazdarjanova A, Guzowski JF. 2004. Differences in hippocampal neuronal population responses to modifications of an environmental context: evidence for distinct, yet complementary, functions of CA3 and CA1 ensembles. *J. Neurosci.* 24:6489–96

Wang XJ. 2008. Decision making in recurrent neuronal circuits. *Neuron* 60:215–34

Welday AC, Shlifer G, Bloom ML, Zhang K, Blair HT. 2011. Cosine directional tuning of theta cell burst frequencies: evidence for spatial coding by oscillatory interference. *J. Neurosci.* 31:16157–76

Welinder PE, Burak Y, Fiete IR. 2008. Grid cells: the position code, neural network models of activity, and the problem of learning. *Hippocampus* 18:1283–300

Wills TJ, Cacucci F, Burgess N, O'Keefe J. 2010. Development of the hippocampal cognitive map in preweanling rats. *Science* 328:1573–76

Wills TJ, Lever C, Cacucci F, Burgess N, O'Keefe J. 2005. Attractor dynamics in the hippocampal representation of the local environment. *Science* 308:873–76

Winograd M, Destexhe A, Sanchez-Vives MV. 2008. Hyperpolarization-activated graded persistent activity in the prefrontal cortex. 105:7298–303

Witter MP, Amaral DG. 2004. Hippocampal formation. In *The Rat Nervous System*, ed. G Paxinos, pp. 635–704. Amsterdam: Elsevier. 3rd ed.

Wu S, Hamaguchi K, Amari SI. 2008. Dynamics and computation of continuous attractors. *Neural Comput.* 20:994–1025

Xie X, Hahnloser RH, Seung HS. 2002. Double-ring network model of the head-direction system. *Phys. Rev. E.* 66:041902

Yartsev MM, Witter MP, Ulanovsky N. 2011. Grid cells without theta oscillations in the entorhinal cortex of bats. *Nature* 479:103–7

Yoganarasimha D, Knierim JJ. 2005. Coupling between place cells and head direction cells during relative translations and rotations of distal landmarks. *Exp. Brain Res.* 160:344–59

Yoganarasimha D, Yu X, Knierim JJ. 2006. Head direction cell representations maintain internal coherence during conflicting proximal and distal cue rotations: comparison with hippocampal place cells. *J. Neurosci.* 26:622–31

Yoshida M, Hasselmo ME. 2009. Persistent firing supported by an intrinsic cellular mechanism in a component of the head direction system. *J. Neurosci.* 29:4945–52

Zhang K. 1996. Representation of spatial orientation by the intrinsic dynamics of the head-direction cell ensemble: a theory. *J. Neurosci.* 16:2112–26

Zilli EA, Hasselmo ME. 2010. Coupled noisy spiking neurons as velocity-controlled oscillators in a model of grid cell spatial firing. *J. Neurosci.* 30:13850–60

Zugaro MB, Arleo A, Berthoz A, Wiener SI. 2003. Rapid spatial reorientation and head direction cells. *J. Neurosci.* 23:3478–82

Zugaro MB, Berthoz A, Wiener SI. 2001. Background, but not foreground, spatial cues are taken as references for head direction responses by rat anterodorsal thalamus neurons. *J. Neurosci.* 21:RC154(1–5)

Neural Basis of Reinforcement Learning and Decision Making

Daeyeol Lee,[1,2] Hyojung Seo,[1] and Min Whan Jung[3]

[1]Department of Neurobiology, Kavli Institute for Neuroscience, Yale University School of Medicine, New Haven, Connecticut 06510; email: daeyeol.lee@yale.edu, hyojung.seo@yale.edu

[2]Department of Psychology, Yale University, New Haven, Connecticut 06520

[3]Neuroscience Laboratory, Institute for Medical Sciences, Ajou University School of Medicine, Suwon 443-721, Republic of Korea; email: min@ajou.ac.kr

Annu. Rev. Neurosci. 2012. 35:287–308

First published online as a Review in Advance on March 29, 2012

The *Annual Review of Neuroscience* is online at neuro.annualreviews.org

This article's doi:
10.1146/annurev-neuro-062111-150512

0147-006X/12/0721-0287$20.00

Keywords

prefrontal cortex, neuroeconomics, reward, striatum, uncertainty

Abstract

Reinforcement learning is an adaptive process in which an animal utilizes its previous experience to improve the outcomes of future choices. Computational theories of reinforcement learning play a central role in the newly emerging areas of neuroeconomics and decision neuroscience. In this framework, actions are chosen according to their value functions, which describe how much future reward is expected from each action. Value functions can be adjusted not only through reward and penalty, but also by the animal's knowledge of its current environment. Studies have revealed that a large proportion of the brain is involved in representing and updating value functions and using them to choose an action. However, how the nature of a behavioral task affects the neural mechanisms of reinforcement learning remains incompletely understood. Future studies should uncover the principles by which different computational elements of reinforcement learning are dynamically coordinated across the entire brain.

Contents

INTRODUCTION

Decision making refers to the process by which an organism chooses its actions and has been studied in such diverse fields as mathematics, economics, psychology, and neuroscience. Traditionally, theories of decision making have fallen into two categories. On the one hand, normative theories in economics generate well-defined criteria for identifying best choices. Such theories, including expected utility theory (von Neumann & Morgenstern 1944), deal with choices in an idealized context and often fail to account for actual choices made by humans and animals. On the other hand, descriptive psychological theories try to account for failures of normative theories by identifying a set of heuristic rules applied by decision makers. For example, prospect theory (Kahneman & Tversky 1979) can successfully account for the failures of expected utility theory in describing human decision making under uncertainty. These two complementary theoretical frameworks are essential for neurobiological studies of decision making. However, a fundamental question not commonly addressed by either approach concerns the role of learning. How do humans and animals acquire their preference for different actions and outcomes in the first place?

Our goal in this paper is to review and organize recent findings about the functions of different brain areas that underlie experience-dependent changes in choice behaviors. Reinforcement learning theory (Sutton & Barto 1998) has been widely adopted as the main theoretical framework in designing experiments as well as interpreting empirical results. Because this topic has been frequently reviewed (Dayan & Niv 2008, Bornstein & Daw 2011, Ito & Doya 2011, van der Meer & Redish 2011), we briefly summarize only the essential elements of reinforcement learning theory and focus on the following questions: First, where and how in the brain are the estimates of expected reward represented and updated by the animal's experience? Converging evidence from a number of recent studies suggests that many of these computations are carried out in multiple interconnected regions in the frontal cortex and basal ganglia. Second, how are the value signals for potential actions transformed to the final behavioral response? Competitive interactions among different pools of recurrently connected neurons are a likely mechanism for this selection process (Usher & McClelland 2001, Wang 2002), but their neuroanatomical substrates remain poorly understood (Wang 2008). Finally, how is model-based reinforcement learning implemented in the brain?

Humans and animals can acquire new knowledge about their environment without directly experiencing reward or penalty, and this knowledge can be used to influence subsequent behaviors (Tolman 1948). This is referred to as model-based reinforcement learning, whereas reinforcement learning entirely relying on experienced reward and penalty is referred to as model free. Recent studies have begun to shed some light on how these two different types of reinforcement learning are linked in the brain. We conclude with some suggestions for future research on the neurobiological mechanisms of reinforcement learning.

REINFORCEMENT LEARNING THEORIES OF DECISION MAKING

Economic Utilities and Value Functions

Economic theories of decision making focus on how numbers can be attached to alternative actions so that choices can be understood as the selection of an action that has the maximum value among all possible actions. These hypothetical quantities are often referred to as utilities and can be applied to all types of behaviors. By definition, behaviors chosen by an organism are those that maximize the organism's utility (**Figure 1*a***). Although such utilities are presumably constrained by evolution and individual experience, economic theories are largely agnostic about how they are determined. By contrast, reinforcement learning theories describe how the animal's experience alters its value functions, which in turn influence subsequent choices (**Figure 1*b***).

The goal of reinforcement learning is to maximize future rewards. Analogous to utilities in economic theories, value functions in reinforcement learning theory refer to the estimates for the sum of future rewards. However, because the animal cannot perfectly predict the future changes in its environment, value functions, unlike utilities, reflect the animal's

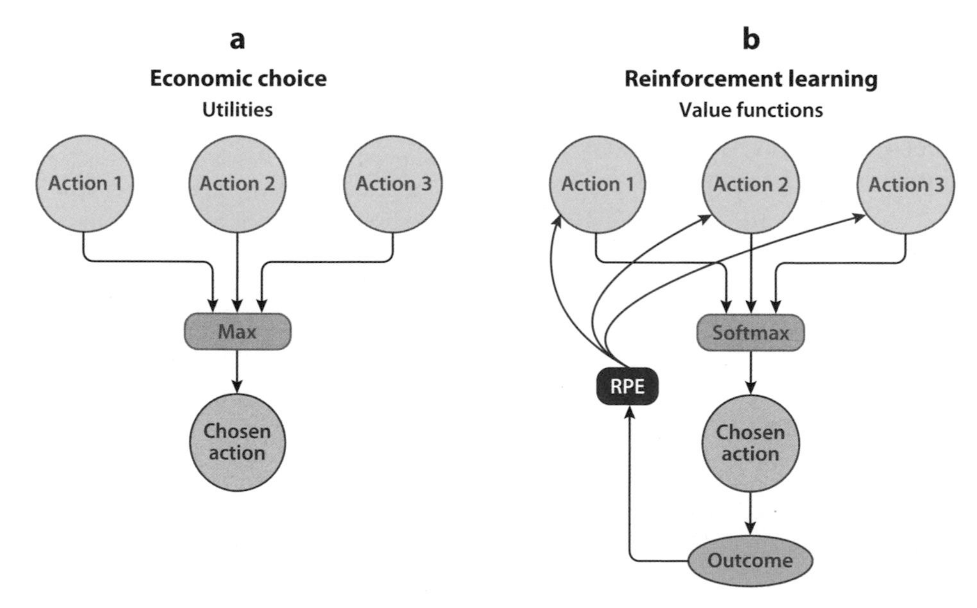

Figure 1

Economic and reinforcement learning theories of decision making. (*a*) In economic theories, decision making corresponds to the selection of an action with maximum utility. (*b*) In reinforcement learning, actions are chosen probabilistically (i.e., softmax) on the basis of their value functions. In addition, value functions are updated on the basis of the outcome (reward or penalty) resulting from the action chosen by the animal. Abbreviation: RPE, reward prediction error.

empirical estimates for its future rewards. The reinforcement learning theory utilizes two different types of value functions. First, action value function refers to the sum of future rewards expected for taking a particular action in a particular state of the environment and is often denoted by $Q(s,a)$, where s and a refer to the state of the environment and the animal's action, respectively. The term action is used formally: It can refer to not only a physical action, such as reaching for a coffee mug in a particular location with a specific limb, but also an abstract choice, such as buying a particular guitar. Second, state value function, often denoted by $V(s)$, refers to the sum of future rewards expected from a particular state of the animal's environment. If the animal always chooses only one action in a given state, then its action value function would be equal to the state value function. Otherwise, the state value function would correspond to the average of action value functions weighted by the probability of taking each action in a given state. State value functions can be used to evaluate the action outcomes, but in general, action value functions are required for selecting an action.

Model-Free versus Model-Based Reinforcement Learning

Value functions can be updated according to two different types of information. First, they can be revised according to the reward or penalty received by the animal after each action. Value functions would not change and no learning would be necessary, if choice outcomes were always perfectly predicted from the current value functions. Otherwise, value functions must be modified to reduce errors in reward predictions. The signed difference between the actual reward and the reward expected by the current value functions is referred to as a reward prediction error (Sutton & Barto 1998). In a class of reinforcement learning algorithms, referred to as simple or model-free reinforcement learning, reward prediction error is the primary source of changes in value functions. More specifically, the value function for the action chosen by the animal or the state visited by the animal is updated according to the reward prediction error, whereas the value functions for all other actions and states remain unchanged or simply decay passively (Barraclough et al. 2004, Ito & Doya 2009).

Second, value functions can be changed more flexibly. This class of reinforcement learning is referred to as model-based reinforcement learning, and the associated algorithms can update the value functions on the basis of the animal's motivational state and its knowledge of the environment without direct reward or penalty. The use of cognitive models allows the animal to adjust its value functions immediately, whenever it acquires a new piece of information about its internal state or external environment. Many lines of evidence indicate that animals as well as humans are capable of model-based reinforcement learning. For example, when an animal is satiated by a particular type of reward, the subjective value of the same food is diminished. However, if the animal were to rely entirely on simple reinforcement learning, the tendency to choose a given action would not change until it experiences the devalued reward through the same action in the same environment. Previous work has shown that rats can change their behaviors immediately according to their current motivational states following the devaluation of specific food items in a different environment. This is often used as a test for goal-directed behaviors and indicates that animals are capable of model-based reinforcement learning (Balleine & Dickinson 1998, Daw et al. 2005). Humans and animals can also simulate the consequences of potential actions that they could have chosen. This is referred to as counterfactual thinking (Roese & Olson 1995), and the information about hypothetical outcomes from unchosen actions can be incorporated into value functions when they are different from the outcomes predicted by the current value functions (Coricelli et al. 2005, Lee et al. 2005, Boorman et al. 2011). Analogous to reward prediction error, the difference between hypothetical and predicted outcomes is referred to as fictive or counterfactual reward prediction error (Lohrenz et al. 2007, Boorman et al. 2011).

Learning During Social Decision Making

During social interaction, the outcomes of actions are often jointly determined by the actions of multiple players. Such strategic situations are referred to as games (von Neumann & Morgenstern 1944). Decision makers can improve the outcomes of their choices during repeated games by applying a model-free reinforcement learning algorithm. In fact, for relatively simple games, such as two-player zero-sum games, humans and animals gradually approximate optimal strategies using model-free reinforcement learning algorithms (Mookherjee & Sopher 1994, Erev & Roth 1998, Camerer 2003, Lee et al. 2004). By contrast, players employing a model-based reinforcement learning algorithm can adjust their strategies more flexibly according to the predicted behaviors of other players. The ability to predict the beliefs and intentions of other players is often referred to as the theory of mind (Premack & Woodruff 1978), and during social decision making, this may dramatically improve the efficiency of reinforcement learning. In game theory, this is also referred to as belief learning (Camerer 2003).

In pure belief learning, the outcomes of actual choices by the decision makers do not exert any additional influence on their choices, unless such outcomes can modify the beliefs or models of the decision makers about the other players. In other words, reward or penalty does not have any separate role other than affecting the decision maker's belief about the other players. Results from studies in behavioral economics show that such pure-belief learning models do not account well for human choice behaviors (Mookherjee & Sopher 1997, Erev & Roth 1998, Feltovich 2000, Camerer 2003). Instead, human behaviors during repeated games are often consistent with hybrid learning models, such as the experience-weighted attraction model (Camerer & Ho 1999, Zhu et al. 2012). In these hybrid models, value functions are adjusted by both real and fictive reward prediction errors. Learning rates for these different reward prediction errors can be set independently. Similar to the results from human studies, hybrid models are more accurate than either model-free reinforcement learning or belief learning models at accounting for the behaviors of nonhuman primates performing a competitive game task against a computer (Lee et al. 2005, Abe & Lee 2011).

NEURAL REPRESENTATION OF VALUE FUNCTIONS

Utilities and value functions are central to economic and reinforcement learning theories of decision making, respectively. In both theories, these quantities are assumed to capture all the relevant factors influencing choices. Thus, brain areas or neurons involved in decision making are expected to harbor signals related to utilities and value functions. In fact, neural activity related to reward expectancy has been found in many different brain areas (Schultz et al. 2000, Hikosaka et al. 2006, Wallis & Kennerley 2010), including sensory cortical areas (Shuler & Bear 2006, Serences 2008, Vickery et al. 2011). The fact that signals related to reward expectancy are widespread in the brain suggests that they are likely to subserve not only reinforcement learning, but also other related cognitive processes, such as attention (Maunsell 2004, Bromberg-Martin et al. 2010a, Litt et al. 2011). Neural signals related to reward expectancy can be divided into at least two different categories, depending on whether they are related to specific actions or states. Neural signals related to action value functions would be useful in choosing a particular action, especially if such signals are observed before the execution of a motor response. Neural activity related to state value functions may play more evaluative roles. In particular, during decision making, the state value function changes from the weighted average of action values for alternative choices to the action value function for the chosen action. The latter is often referred to as the chosen value (Padoa-Schioppa & Assad 2006, Cai et al. 2011).

During decision-making experiments, choices can be made either among alternative physical movements with different spatial trajectories or among different objects regardless of the movements required to acquire them. In the reinforcement learning theory, these different options are all considered actions. Nevertheless, neural signals related to the corresponding action value functions may vary substantially according to the dimension in which choices are made. In most previous neurobiological studies, different properties of reward were linked to different physical actions. These studies have identified neural activity related to action value functions in numerous brain areas, including the posterior parietal cortex (Platt & Glimcher 1999, Dorris & Glimcher 2004, Sugrue et al. 2004, Seo et al. 2009), dorsolateral prefrontal cortex (Barraclough et al. 2004, Kim et al. 2008), premotor cortex (Pastor-Bernier & Cisek 2011), medial frontal cortex (Seo & Lee 2007, 2009; So & Stuphorn 2010; Sul et al. 2010), and striatum (Samejima et al. 2005, Lau & Glimcher 2008, Kim et al. 2009, Cai et al. 2011). Despite the limited spatial and temporal resolutions available in neuroimaging studies, metabolic activity related to action value functions has also been identified in the supplementary motor area (Wunderlich et al. 2009). In contrast, neurons in the primate orbitofrontal cortex are not sensitive to spatial locations of targets associated with specific rewards (Tremblay & Schultz 1999, Wallis & Miller 2003), suggesting that they encode action value functions related to specific objects or goals. When animals chose between two different flavors of juice, neurons in the primate orbitofrontal cortex signaled the action value functions associated with specific juice flavors rather than the directions of eye movements used to indicate the animal's choices (Padoa-Schioppa & Assad 2006).

Neurons in many different brain areas often combine value functions for alternative actions and other decision-related variables. Precisely how multiple types of signals are combined in the activity of individual neurons can, therefore, provide important clues about how such signals are computed and utilized. For example, the likelihood of choosing one of two alternative choices is determined by the difference in their action value functions; therefore, neurons encoding such signals may be closely involved in the process of action selection. During a binary-choice task, neurons in the primate posterior parietal cortex (Dorris & Glimcher 2004, Sugrue et al. 2004, Seo & Lee 2008, Seo et al. 2009), dorsolateral prefrontal cortex (Kim et al. 2008), premotor cortex (Pastor-Bernier & Cisek 2011), supplementary eye field (Seo & Lee 2009), and dorsal striatum (Cai et al. 2011) as well as in the rodent secondary motor cortex (Sul et al. 2011) and striatum (Ito & Doya 2009) encode the difference between the action value functions for two alternative actions.

Signals related to state value functions are also found in many different brain areas. During a binary-choice task, the sum or average of the action value functions for two alternative choices corresponds to the state value function before a choice is made. In the posterior parietal cortex and dorsal striatum, signals related to such state value functions and action value functions coexist (Seo & Lee 2008, Seo et al. 2009, Cai et al. 2011). Neurons encoding state value functions are also found in the ventral striatum (Cai et al. 2011), anterior cingulate cortex (Seo & Lee 2007), and amygdala (Belova et al. 2008). Neural activity related to chosen values that correspond to post-decision state value functions is also widespread in the brain and has been found in the orbitofrontal cortex (Padoa-Schioppa & Assad 2006, Sul et al. 2010), medial frontal cortex (Sul et al. 2010), dorsolateral prefrontal cortex (Kim & Lee 2011), and striatum (Lau & Glimcher 2008, Kim et al. 2009, Cai et al. 2011). Because reward prediction error corresponds to the difference between the outcome of a choice and the chosen value, neural activity related to chosen values may be utilized to compute reward prediction errors and update

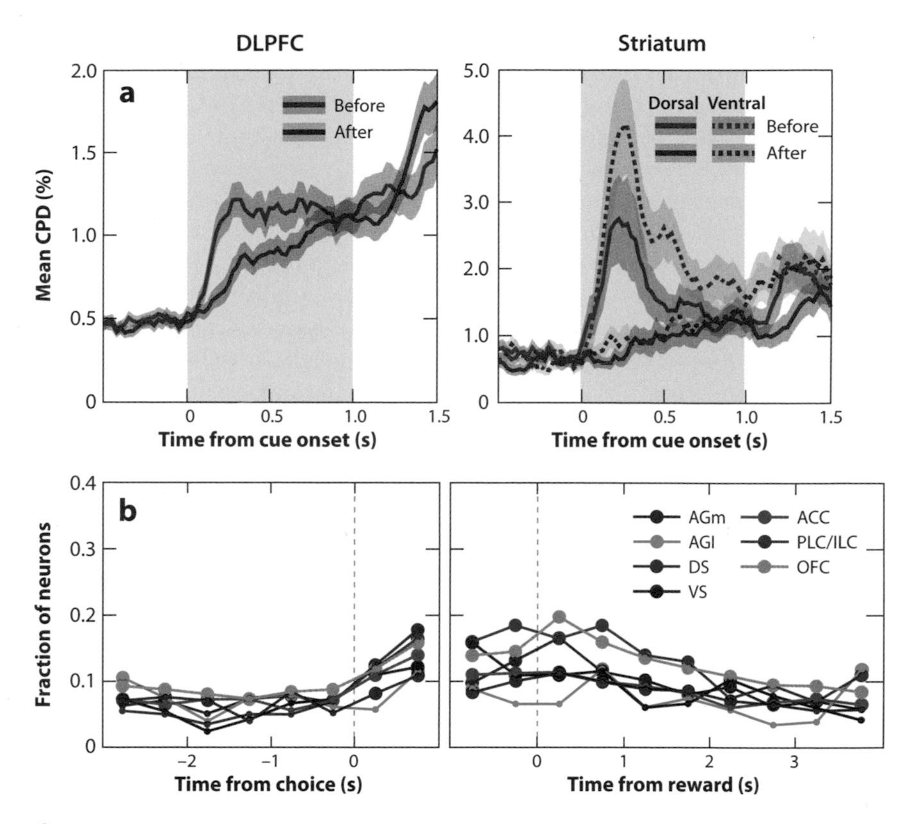

Figure 2

Time course of signals related to different state value functions during decision making. (*a*) Signals related to the state value functions before (*red*) and after (*blue*) decision making in the dorsolateral prefrontal cortex (DLPFC) (Kim et al. 2008, Kim & Lee 2011) and striatum (Cai et al. 2011) during an intertemporal choice task. These two state value functions correspond to the average of the action value functions for two options and the chosen value, respectively. During these studies, monkeys chose between a small immediate reward and a large delayed reward, and the magnitude of neural signals related to different value functions was estimated by the coefficient of partial determination (CPD). Lines correspond to the mean CPD for all the neurons recorded in each brain area; the shaded area corresponds to the standard error of the mean. (*b*) Proportion of neurons carrying chosen value signals in multiple areas of the rodent brain. Abbreviations: ACC, anterior cingulate cortex; DS, dorsal striatum; ILC, infralimbic cortex; AGl and AGm, lateral and medial agranular cortex (corresponding to the primary and secondary motor cortex, respectively); OFC, orbitofrontal cortex; PLC, prelimbic cortex; VS, ventral striatum. During these studies (Kim et al. 2009; Sul et al. 2010, 2011), the rats performed a dynamicforaging task. Large symbols indicate that the proportions are significantly ($p < 0.05$) above the chance level.

value functions. In some brain areas, such as the dorsolateral prefrontal cortex (Kim & Lee 2011) and dorsal striatum (Cai et al. 2011), activity related to chosen value signals emerges later than the signals related to the sum of the value functions for alternative actions, suggesting that action selection may take place during this delay (**Figure 2**).

NEURAL MECHANISMS OF ACTION SELECTION

During decision making, neural activity related to action value functions must be converted to the signals related to a particular action and transmitted to motor structures. Therefore, some of the brain areas encoding action-value functions or their differences are likely to be involved also in action selection. Nevertheless, the precise anatomical location playing a primary role in action selection may vary with the nature of a behavioral task. For example, compared with flexible goal-directed behaviors, actions selected by fixed stimulus-action associations or well-practiced motor sequences may rely more on the dorsolateral striatum (Knowlton et al. 1996, Hikosaka et al. 1999, Yin & Knowlton 2006). Considering that spike trains of cortical neurons are stochastic (Softky & Koch 1993), the process of action selection is likely to rely on a network of neurons temporally integrating the activity related to difference in action value functions (Soltani & Wang 2006, Krajbich et al. 2010). An analogous process has been extensively studied for action selection based on noisy sensory stimulus during perceptual decision making. For example, psychophysical performance during a two-alternative forced-choice task is well described by the so-called random-walk or drift-diffusion model in which a particular action is selected when the gradual accumulation of noisy evidence reaches a threshold for that action (Laming 1968, Roitman & Shadlen 2002, Smith & Ratcliff 2004).

Neurons in multiple brain areas involved in motor control often build up their activity gradually prior to specific movements, suggesting that these areas may be involved in action selection. Execution of voluntary movements are tightly coupled with phasic neural activity in a number of brain areas, such as the primary motor cortex (Georgopoulos et al. 1986), premotor cortex (Churchland et al. 2006), frontal eye field (Hanes & Schall 1996), supplementary eye field (Schlag & Schlag-Rey 1987), posterior parietal cortex (Andersen et al. 1987), and superior colliculus (Schiller & Stryker 1972, Wurtz & Goldberg 1972). All these structures are closely connected with motor structures in the brainstem and spinal cord. In addition, neurons in these areas display persistent activity related to the metrics of upcoming movements when the desired movement is indicated before a "go" signal, suggesting that they are also involved in motor planning and preparation (Weinrich & Wise 1982, Bruce & Goldberg 1985, Gnadt & Andersen 1988, Schall 1991, Glimcher & Sparks 1992, Smyrnis et al. 1992). Such persistent activity has often been associated with working memory (Funahashi et al. 1989, Wang 2001), but it may also subserve the temporal integration of noisy inputs (Shadlen & Newsome 2001, Roitman & Shadlen 2002, Wang 2002, Curtis & Lee 2010). In fact, neural activity in accordance with gradual evidence accumulation has been found in the same brain areas that show persistent activity related to motor planning, such as the posterior parietal cortex (Roitman & Shadlen 2002), frontal eye field (Ding & Gold 2011), and superior colliculus (Horwitz & Newsome 2001).

Computational studies have demonstrated that a network of neurons with recurrent excitation and lateral inhibition can perform temporal integration of noisy sensory inputs and produce a signal corresponding to an optimal action (Wang 2002, Lo & Wang 2006, Beck et al. 2008, Furman & Wang 2008). Most of these models have been developed to account for the pattern of activity observed in the lateral intraparietal (LIP) cortex during a perceptual decision-making task. Nevertheless, value-dependent action selection may also involve attractor dynamics in a similar network of neurons, provided that their input synapses are adjusted in a reward-dependent manner (Soltani & Wang 2006, Soltani et al. 2006). Therefore, neurons involved in the evaluation of unreliable sensory information may also contribute to value-based decision making. Consistent with this possibility, neurons in the LIP tend to change their activity according to the value of rewards expected from alternative actions (Platt & Glimcher 1999, Sugrue et al. 2004, Seo et al. 2009, Louie & Glimcher 2010, Rorie et al. 2010).

In contrast to the brain areas involved in selecting a specific physical movement, other areas may be involved in more abstract decision making. For example, the orbitofrontal cortex may play a particularly important role in making choices among different objects or goods (Padoa-Schioppa 2011). By contrast, an action selection process guided by memory and other endogenous cues rather than external sensory stimuli may rely more on the medial frontal cortex. Activity related to action value functions has been found in the supplementary and presupplementary motor areas (Sohn & Lee 2007, Wunderlich et al. 2009) as well as the supplementary eye field (Seo & Lee 2009, So & Stuphorn 2010). More importantly, neural activity related to an upcoming movement appears in the medial frontal cortex earlier than in other areas of the brain. For example, when human subjects are asked to initiate a movement voluntarily without any immediate sensory cue, scalp electroencelphalogram displays the so-called readiness potential well before movement onset, and its source has been localized to the supplementary motor area (Haggard 2008, Nachev et al. 2008). The hypothesis that internally generated voluntary movements are selected in the medial frontal cortex is also consistent with the results from single-neuron recording and neuroimaging studies. Individual neurons in the primate supplementary motor area often begin to change their activity according to an upcoming limb movement earlier than those in the premotor cortex or primary motor cortex, especially when the animal is required to produce such movements voluntarily without immediate sensory cues (Tanji & Kurata 1985, Okano & Tanji 1987). Similarly, neurons in the supplementary eye field begin to modulate their activity according to the direction of an upcoming saccade earlier than similar activity recorded in the frontal eye field and LIP (Coe et al. 2002). In a rodent performing a dynamic foraging task, signals related to the animal's choice appear in the medial motor cortex, presumably a homolog of the primate supplementary motor cortex, earlier than many other brain areas, including the primary motor cortex and basal ganglia (Sul et al. 2011). Furthermore, lesions in this area make the animal's choices less dependent on action value functions (Sul et al. 2011). Finally, analysis of BOLD activity patterns during a self-timed motor task has also identified signals related to an upcoming movement up to several seconds before the movement onset in the human supplementary motor area (Soon et al. 2008).

In summary, neural activity potentially reflecting the process of action selection has been identified in multiple regions, including areas involved in motor control, the orbitofrontal cortex, and the medial frontal cortex. Thus, future investigations should determine how these multiple areas interact cooperatively or competitively depending on the demands of specific behavioral tasks. The frame of reference in which different actions are represented varies across brain areas; therefore, how the actions encoded in one frame of reference, for example, in object space, are transformed to another, such as visual or joint space, needs to be investigated (Padoa-Schioppa 2011).

NEURAL MECHANISMS FOR UPDATING VALUE FUNCTIONS

Temporal Credit Assignment and Eligibility Trace

Reward resulting from a particular action is often revealed after a substantial delay, and an animal may carry out several other actions before collecting the reward resulting from a previous action. Therefore, it can be challenging to associate an action and its corresponding outcome correctly; this is referred to as the problem of temporal credit assignment (Sutton & Barto 1998). Not surprisingly, the loss of the ability to link specific outcomes to corresponding choices interferes with the process of updating value functions appropriately. Whereas normal animals can easily alter their preferences between two objects when the probabilities of getting rewards from the two objects are switched, humans, monkeys, and rats with lesions in the orbitofrontal cortex are impaired in such reversal learning tasks (Iversen & Mishkin 1970, Schoenbaum et al. 2002, Fellows & Farah 2003, Murray et al. 2007). These deficits may result from failures in temporal credit assignment. For example, during a probabilistic reversal learning task, in which the probabilities of rewards from different objects were dynamically and unpredictably changed, deficits produced by the lesions in the orbitofrontal cortex were due to erroneous associations between the choices and their outcomes (Walton et al. 2010).

In reinforcement learning theory, the problem of temporal credit assignment can be resolved in at least two different ways. First, a series of intermediate states can be introduced during the interval between an action and a reward, so that they can propagate the information about the reward to the value function of the correct action (Montague et al. 1996). This basic temporal difference model was initially proposed to account for the reward prediction error signals conveyed by dopamine neurons but was shown to be inconsistent with the actual temporal profiles of dopamine neuron signals (Pan et al. 2005). Second, animals can use short-term memory signals related to the states or actions they select. Such memory signals, termed eligibility traces (Sutton & Barto 1998), can facilitate action-outcome association even when the outcome is delayed. Therefore, eligibility traces can account for the temporally discontinuous shift in the phasic activity of dopamine neurons observed during classical conditioning (Pan et al. 2005). Signals related to the animal's previous choices have been observed in a number of brain areas, including the prefrontal cortex and posterior parietal cortex in monkeys (Fecteau & Munoz 2003, Barraclough et al. 2004, Genovesio et al. 2006, Seo et al. 2009, Seo & Lee 2009), as well as in many regions in the rodent frontal cortex and striatum (Kim et al. 2007, 2009; Sul et al. 2010, 2011) (**Figure 3**), and they may provide eligibility traces necessary to form associations between actions and their outcomes (Curtis & Lee 2010). In addition, neurons in many of these areas, including the orbitofrontal cortex, often encode specific conjunctions of chosen actions and their outcomes, for example, by increasing their activity when a positive outcome is obtained from a specific action (Barraclough et al. 2004, Kim et al. 2009, Roesch et al. 2009, Seo & Lee 2009, Sul et al. 2010, Abe & Lee 2011) (**Figure 3**). The brain may resolve the temporal credit assignment problem using such action-outcome conjunction signals.

Integration of Chosen Value and Reward Prediction Error

In model-free reinforcement learning, the value function for a chosen action is revised according to the reward prediction error. Therefore, signals related to the chosen value and reward prediction error must be combined in the activity of individual neurons involved in updating value functions. Signals related to reward prediction error were first identified in the midbrain dopamine neurons (Schultz 2006) but later found to exist in many other areas, including the lateral habenula (Matsumoto & Hikosaka 2007), globus pallidus (Hong & Hikosaka 2008), dorsolateral prefrontal cortex (Asaad & Eskandar 2011), anterior

Figure 3

Time course of signals related to an animal's choice, its outcome, and action-outcome conjunction in multiple brain areas of primates and rodents. (*a*) Spatial layout of the choice targets during a matching-pennies task used in single-neuron recording experiments in monkeys. (*b*) Brain regions tested during studies using monkeys (Barraclough et al. 2004, Seo & Lee 2007, Seo et al. 2009). Abbreviations: ACCd, dorsal anterior cingulate cortex; DLPFC, dorsolateral prefrontal cortex; LIP, lateral intraparietal cortex. (*c*) Fraction of neurons significantly modulating their activity according to the animal's choice (*top*), its outcome (*middle*), and choice-outcome conjunction (*bottom*) during the current (trial lag = 0) and three previous trials (trial lags = $1 \sim 3$). (*d*) Modified T-maze used in a rodent dynamic-foraging task. (*e*) Anatomical areas tested in single-neuron recording experiments in rodents (Kim et al. 2009; Sul et al. 2010, 2011). Abbreviations: ACC, anterior cingulate cortex; DS, dorsal striatum; ILC, infralimbic cortex; OFC, orbitofrontal cortex; PLC, prelimbic cortex; VS, ventral striatum. Modified with permission from Elsevier (Paxinos & Watson 1998). (*f*) Fraction of neurons significantly modulating their activity according to the animal's choice (*top*), its outcome (*middle*), and choice-outcome conjunction (*bottom*) during the current (lag = 0) and previous trials (lag = 1). Large symbols indicate that the proportions are significantly ($p < 0.05$) above the chance level.

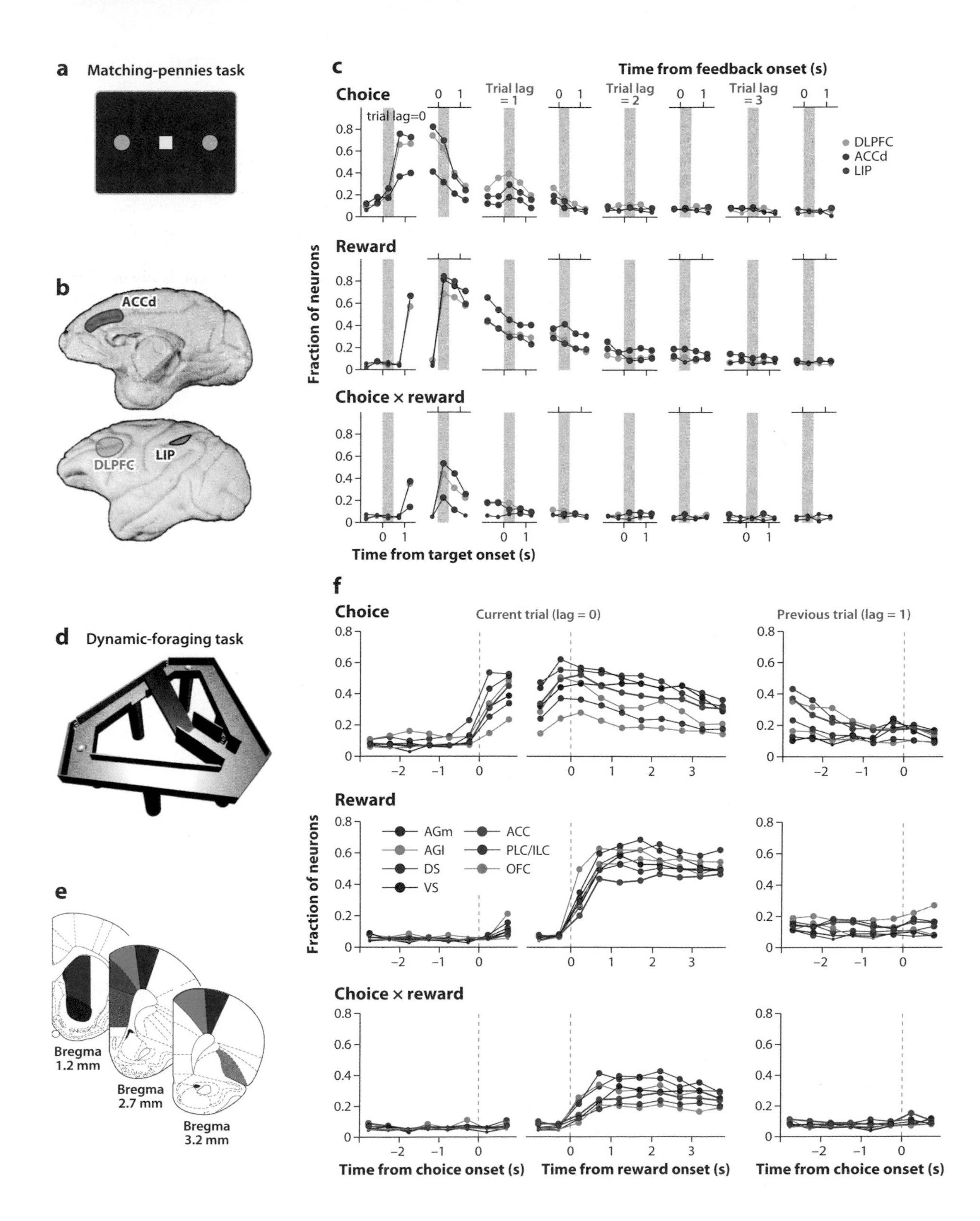
a Matching-pennies task
b
ACCd
DLPFC
LIP
c
Time from feedback onset (s)
Choice
trial lag=0
Trial lag = 1
Trial lag = 2
Trial lag = 3
DLPFC
ACCd
LIP
Reward
Choice × reward
Fraction of neurons
Time from target onset (s)
d Dynamic-foraging task
e
Bregma 1.2 mm
Bregma 2.7 mm
Bregma 3.2 mm
f
Choice
Current trial (lag = 0)
Previous trial (lag = 1)
Reward
AGm
AGl
DS
VS
ACC
PLC/ILC
OFC
Choice × reward
Fraction of neurons
Time from choice onset (s)
Time from reward onset (s)
Time from choice onset (s)

cingulate cortex (Matsumoto et al. 2007, Seo & Lee 2007), orbitofrontal cortex (Sul et al. 2010), and striatum (Kim et al. 2009, Oyama et al. 2010, Asaad & Eskandar 2011). Thus, the extraction of reward prediction error signals may be gradual and implemented through a distributed network of multiple brain areas. Dopamine neurons may then play an important role in relaying these error signals to update the value functions represented broadly in different brain areas. Signals related to chosen values are also distributed in multiple brain areas, including the medial frontal cortex, orbitofrontal cortex, and striatum (Padoa-Schioppa & Assad 2006, Lau & Glimcher 2008, Kim et al. 2009, Sul et al. 2010, Cai et al. 2011). The areas in which signals related to the chosen value and reward prediction error converge, such as the orbitofrontal cortex and striatum, may therefore play an important role in updating the value functions (Kim et al. 2009, Sul et al. 2010).

It is often hypothesized that the primary site for updating and storing action value functions is at the synapses between axons from cortical neurons and dendrites of medium spiny neurons in the striatum (Reynolds et al. 2001, Hikosaka et al. 2006, Lo & Wang 2006, Hong & Hikosaka 2011). Signals related to reward prediction error arrive at these synapses via the terminals of dopamine neurons in the substantia nigra (Levey et al. 1993, Haber et al. 2000, Schultz 2006, Haber & Knutson 2010), and multiple types of dopamine receptors in the striatum can modulate the plasticity of corticostriatal synapses according to the relative timing of presynaptic versus postsynaptic action potentials (Shen et al. 2008, Gerfen & Surmeier 2011). However, the nature of the specific information stored by these synapses remains poorly understood. In addition, whether the corticostriatal circuit carries the appropriate signals related to eligibility traces for chosen actions as well as any other necessary state information at the right time needs to be tested in future studies. Given the broad dopaminergic projections to various cortical areas (Lewis et al. 2001), value functions may be updated in many of the same cortical areas encoding the value functions at the time of decision making.

Uncertainty and Learning Rate

The learning rate, which controls the speed of learning, must be adjusted according to the uncertainty and volatility of the animal's environment. In natural environments, decision makers face many different types of uncertainty. When the probabilities of different outcomes are known, as when flipping a coin or during economic experiments, uncertainty about outcomes is referred to as risk (Kahneman & Tversky 1979) or expected uncertainty (Yu & Dayan 2005). In contrast, ambiguity (Ellsberg 1961) or unexpected uncertainty (Yu & Dayan 2005) is the term used when the exact probabilities are unknown. Ambiguity or unexpected uncertainty is high when the probabilities of different outcomes expected from a given action change frequently (Behrens et al. 2007). Such volatile environments require a large learning rate so that value functions can be modified quickly. By contrast, if the environment is largely known and stable, then the learning rate should be close to 0 so that value functions are not too easily altered by random events in the environment. Human learners can change their learning rates almost optimally when the rate of changes in reward probabilities for alternative actions is manipulated (Behrens et al. 2007). The level of volatility, and hence the learning rate, is reflected in the activity of the anterior cingulate cortex, suggesting that this region of the brain may be important for adjusting the learning rate according to the stability of the decision maker's environment (Behrens et al. 2007). Similarly, the brain areas that increase their activity during decision making under ambiguity, such as the lateral prefrontal cortex, orbitofrontal cortex, and amygdala, may also be involved in optimizing the learning rate (Hsu et al. 2005, Huettel et al. 2006). Single-neuron recording studies also indicate that the orbitofrontal cortex plays a role in evaluating the amount of uncertainty in choice outcomes (Kepecs et al. 2008, O'Neill & Schultz 2010).

NEURAL SYSTEMS FOR MODEL-FREE VERSUS MODEL-BASED REINFORCEMENT LEARNING

Model-Based Value Functions and Reward Prediction Errors

During model-based reinforcement learning, decision makers utilize their knowledge of the environment to update the estimates of outcomes expected from different actions, even without reward or penalty. A wide range of algorithms can be used to implement model-based reinforcement learning. For example, decision makers may learn the configuration and dynamics of their environment separately from the values of outcomes at different locations (Tolman 1948). Doing so enables the animal to rediscover quickly an optimal path of travel whenever the location of a desired item changes. Flexibly combining these two different types of information may rely on the prefrontal cortex (Daw et al. 2005, Pan et al. 2008, Gläscher et al. 2010) and hippocampus (Womelsdorf et al. 2010, Simon & Daw 2011). For example, activity in the human lateral prefrontal cortex increases when unexpected state transitions are observed, suggesting that this area is involved in learning the likelihood of state transitions (Gläscher et al. 2010). The hippocampus may play a role in providing information about the layout of the environment and other contextual information necessary to update the value functions. The integration of information about the behavioral context and current task demands encoded in these two areas, especially at the time of decision making, may rely on rhythmic synchronization of neural activity in the theta frequency range (Sirota et al. 2008, Benchenane et al. 2010, Hyman et al. 2010, Womelsdorf et al. 2010).

Whether and to what extent model-free and model-based forms of reinforcement learning are supported by the same brain areas remain an important area of research. Value functions estimated by model-free and model-based algorithms may be updated or represented separately in different brain areas (Daw et al. 2005), but they may also be combined to produce a unique estimate for the outcomes expected from the chosen actions. For example, when decision makers are required to combine information about reward history and social information, the reliability of the predictions based on these two different types of information is reflected separately in two different regions of the anterior cingulate cortex (Behrens et al. 2008) (**Figure 4**). By contrast, signals related to reward probability predicted by both types of information were found in the ventromedial prefrontal cortex (Behrens et al. 2008). Similarly, human ventral striatum may represent the chosen values and reward prediction errors regardless of how they are computed (Daw et al. 2011, Simon & Daw 2011, Zhu et al. 2012). This is also consistent with the finding that reward prediction error signals encoded by the midbrain dopamine neurons, as well as the neurons in the globus pallidus and lateral habenula, are in accordance with both model-free and model-based reinforcement learning (Bromberg-Martin et al. 2010b).

Hypothetical Outcomes and Mental Simulation

From the information that becomes available after completing chosen actions, decision makers can often deduce what alternative outcomes would have been possible from other actions. They can then use this information about hypothetical outcomes to update the action value functions for unchosen actions. In particular, the observed behaviors of other decision makers during social interaction are a rich source of information about such hypothetical outcomes (Camerer & Ho 1999, Camerer 2003, Lee et al. 2005, Lee 2008). Results from lesion and neuroimaging studies have demonstrated that the information about hypothetical or counterfactual outcomes may be processed in the same brain areas that are also involved in evaluating the actual outcomes

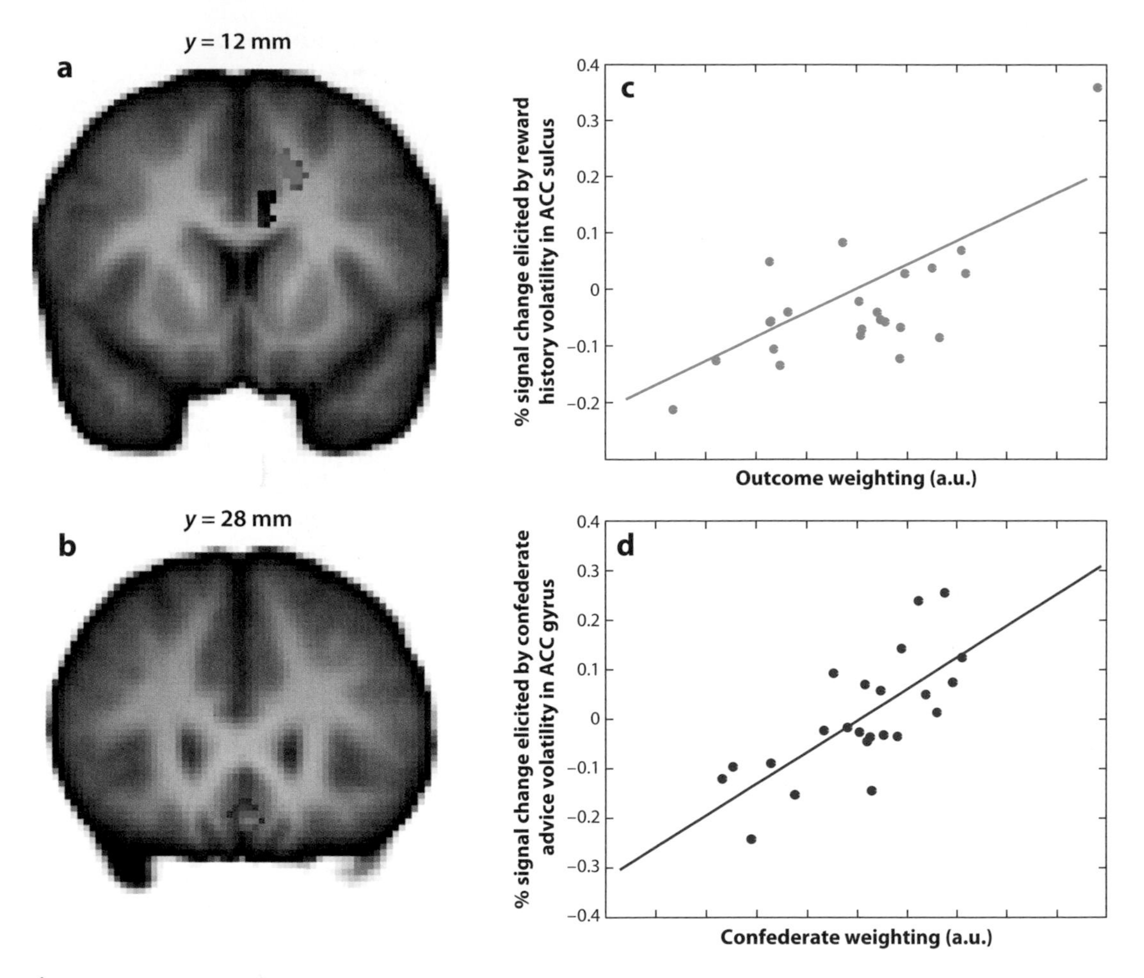

Figure 4

Areas in the human brain involved in updating model-free and model-based value functions (Behrens et al. 2008). (*a*) Regions in which the activity is correlated with the volatility in estimating the value functions based on reward history (*green*) and social information (*red*). (*b*) Activity in the ventromedital prefrontal cortex was correlated with the value functions regardless of whether they were estimated from reward history or social information. (*c*) Subjects more strongly influenced by reward history (ordinate) tended to show greater signal change in the anterior cingulate cortex (ACC) in association with reward history (abscissa; *green region* in panel *a*). (*d*) Subjects more strongly influenced by social information (ordinate) showed greater signal changes in the ACC in association with social information (abscissa; *red region* in panel *a*). Reprinted by permission from Macmillan Publisher Ltd: *Nature* 456:245–49, copyright 2008.

of chosen actions, such as the prefrontal cortex (Camille et al. 2004, Coricelli et al. 2005, Boorman et al. 2011) and striatum (Lohrenz et al. 2007). The hippocampus may also be involved in simulating the possible outcomes of future actions (Hassabis & Maguire 2007, Johnson & Redish 2007, Schacter et al. 2007, Luhmann et al. 2008). Single-neuron recording studies have also shown that neurons in the dorsal anterior cingulate cortex respond similarly to actual and hypothetical outcomes (Hayden et al. 2009). More recent neurophysiological experiments in the dorsolateral prefrontal cortex and orbitofrontal cortex further revealed that neurons in these areas tend to encode actual and hypothetical outcomes for the same action, suggesting that they may provide an important substrate for updating the action value functions for chosen and unchosen actions simultaneously (Abe & Lee 2011) (**Figure 5**).

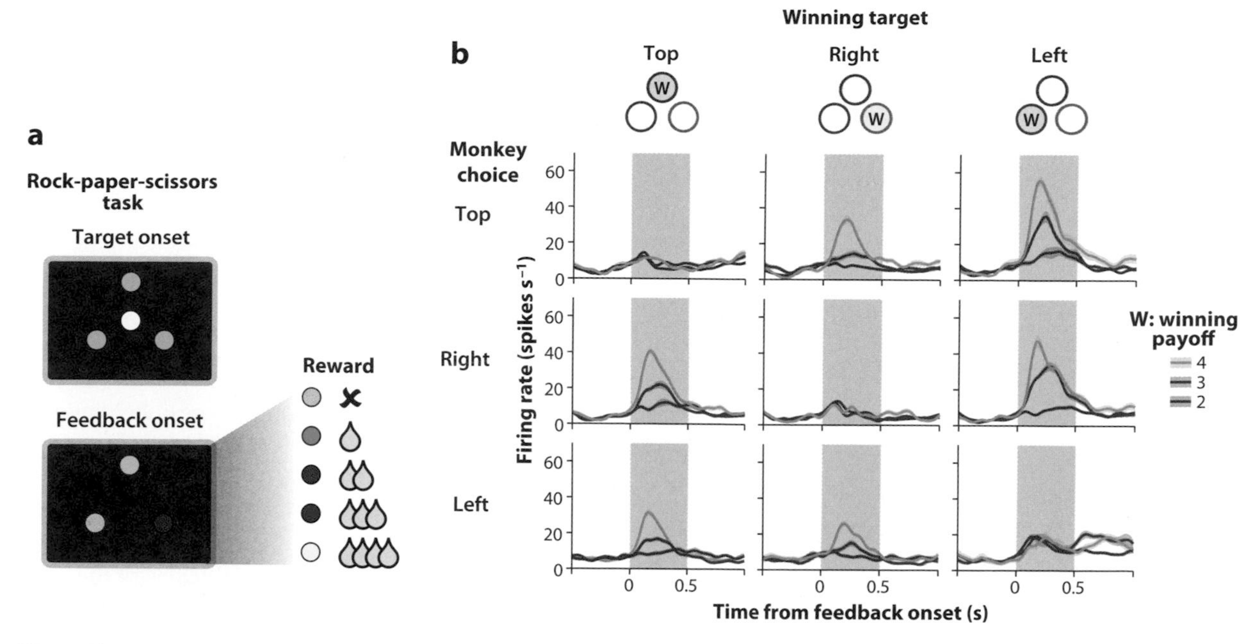

Figure 5

Neuronal activity related to hypothetical outcomes in the primate orbitofrontal cortex. (*a*) Virtual rock-paper-scissors task used for single-neuron recording studies in monkeys (Abe & Lee 2011). (*b*) An example of a neuron recorded in the orbitofrontal cortex that modulated its activity according to the magnitude of reward that was available from the unchosen winning target (indicated by "W" in the top panels). The spike density function of this neuron was estimated separately according to the position of the winning target (*columns*), the position of the target chosen by the animal (*rows*), and the magnitude of the reward available from the winning target (*colors*).

CONCLUSIONS

In recent decades, reinforcement learning theory has become a central framework in the newly emerging areas of neuroeconomics and decision neuroscience. This is hardly surprising, because unlike abstract decisions analyzed in economic theories, biological organisms seldom receive complete information about the likelihoods of different outcomes expected from alternative actions. Instead, they face the challenge of learning how to predict the outcomes of their actions by trial and error, which is the essence of reinforcement learning.

The field of reinforcement learning has yielded many different algorithms, which provide neurobiologists with exciting opportunities to test whether they can successfully account for the actual behaviors of humans and animals and for how different computational elements are implemented in the brain. In some cases, particular theoretical components closely correspond to specific brain structures, as in the case of reward prediction errors and midbrain dopamine neurons. However, in general, a relatively well-circumscribed computational step in a given algorithm is often implemented in multiple brain areas, and this relationship may change with the animal's experience and task demands. The challenge that lies ahead is, therefore, to understand whether and how the signals in different brain areas related to various components of reinforcement learning, such as action value functions and chosen value, make different contributions to the overall behaviors of the animal. An important example discussed in this article is the relationship between model-free and model-based reinforcement learning algorithms. Neural machinery of model-free reinforcement learning may be phylogenetically older, and for simple decision-making problems, it may be more robust. By contrast, for complex decision-making problems, model-based reinforcement learning algorithms can be more efficient, because they can avoid the need to relearn appropriate stimulus-action associations repeatedly by exploiting the regularities in the environment and the current information about the animal's internal state. Failures in applying appropriate reinforcement learning algorithms can lead to a variety of maladaptive behaviors observed in different mental disorders. Therefore, it would be crucial for future studies to elucidate the mechanisms that allow the brain to coordinate different types of reinforcement learning and their individual elements.

DISCLOSURE STATEMENT

The authors are not aware of any affiliations, memberships, funding, or financial holding that might be perceived as affecting the objectivity of this review.

ACKNOWLEDGMENTS

We are grateful to Soyoun Kim, Jung Hoon Sul, and Hoseok Kim for their help with the illustrations and to Jeansok Kim, Matthew Kleinman, and Tim Vickery for helpful comments on the manuscript. The research of the authors was supported by the grants from the National Institute of Drug Abuse (DA024855 and DA029330 to D.L.) and the Korea Ministry of Education, Science, and Technology (the Brain Research Center of the 21st Century Frontier Research Program, National Research Foundation grant 2011-0015618 and the Original Technology Research Program for Brain Science 2011-0019209 to M.W.J.).

LITERATURE CITED

Abe H, Lee D. 2011. Distributed coding of actual and hypothetical outcomes in the orbital and dorsolateral prefrontal cortex. *Neuron* 70:731–41

Andersen RA, Essick GK, Siegel RM. 1987. Neurons of area 7 activated by both visual stimuli and oculomotor behavior. *Exp. Brain Res.* 67:316–22

Asaad WF, Eskandar EN. 2011. Encoding of both positive and negative reward prediction errors by neurons of the primate lateral prefrontal cortex and caudate nucleus. *J. Neurosci.* 31:17772–87

Balleine BW, Dickinson A. 1998. Goal-directed instrumental action: contingency and incentive learning and their cortical substrates. *Neuropharmacology* 37:407–19

Barraclough DJ, Conroy ML, Lee D. 2004. Prefrontal cortex and decision making in a mixed-strategy game. *Nat. Neurosci.* 7:404–10

Beck JM, Ma WJ, Kiani R, Hanks T, Churchland AK, et al. 2008. Probabilistic population codes for Bayesian decision making. *Neuron* 60:1142–52

Belova MA, Paton JJ, Salzman CD. 2008. Moment-to-moment tracking of state value in the amygdala. *J. Neurosci.* 48:10023–30

Behrens TE, Woolrich MW, Walton ME, Rushworth MF. 2007. Learning the value of information in an uncertain world. *Nat. Neurosci.* 10:1214–21

Behrens TE, Hunt LT, Woolrich MW, Rushworth MF. 2008. Associative learning of social value. *Nature* 456:245–9

Benchenane K, Peyrache A, Khamassi M, Tierney PL, Gioanni Y, et al. 2010. Coherent theta oscillations and reorganization of spike timing in the hippocampal-prefrontal network upon learning. *Neuron* 66:921–36

Boorman ED, Behrens TE, Rushworth MF. 2011. Counterfactual choicer and learning in a neural network centered on human lateral frontopolar cortex. *PLoS Biol.* 9:e1001093

Bornstein AM, Daw ND. 2011. Multiplicity of control in the basal ganglia: computational roles of striatal subregions. *Curr. Opin. Neurobiol.* 21:374–80

Bromberg-Martin ES, Matsumoto M, Hikosaka O. 2010a. Dopamine in motivational control: rewarding, aversive, and alerting. *Neuron* 68:815–34

Bromberg-Martin ES, Matsumoto M, Hong S, Hikosaka O. 2010b. A pallidus-habenula-dopamine pathway signals inferred stimulus values. *J. Neurophysiol.* 104:1068–76

Bruce CJ, Goldberg ME. 1985. Primate frontal eye fields. I. Single neurons discharging before saccades. *J. Neurophysiol.* 53:603–35

Cai X, Kim S, Lee D. 2011. Heterogeneous coding of temporally discounted values in the dorsal and ventral striatum during intertemporal choice. *Neuron* 69:170–82

Camerer CF. 2003. *Behavioral Game Theory: Experiments in Strategic Interaction.* Princeton: Princeton Univ. Press

Camerer C, Ho TH. 1999. Experience-weighted attraction learning in normal form games. *Econometrica* 67:827–74

Camille N, Coricelli G, Sallet J, Pradat-Diehl P, Duhamel JR, Sirigu A. 2004. The involvement of the orbitofrontal cortex in the existence of regret. *Science* 304:1167–70

Churchland MM, Yu BM, Ryu SI, Santhnam G, Shenoy KV. 2006. Neural variability in premotor cortex provides a signature of motor preparation. *J. Neurosci.* 26:3697–712

Coe B, Tomihara K, Matsuzawa M, Hikosaka O. 2002. Visual and anticipatory bias in three cortical eye fields of the monkey during an adaptive decision-making task. *J. Neurosci.* 22:5081–90

Coricelli G, Critchley HD, Joffily M, O'Doherty JP, Sirigu A, Dolan RJ. 2005. Regret and its avoidance: a neuroimaging study of choice behavior. *Nat. Neurosci.* 8:1255–62

Curtis CE, Lee D. 2010. Beyond working memory: the role of persistent activity in decision making. *Trends Cogn. Sci.* 14:216–22

Daw ND, Gershman SJ, Seymour B, Dayan P, Dolan RJ. 2011. Model-based influences on humans' choices and striatal prediction errors. *Neuron* 69:1204–15

Daw ND, Niv Y, Dayan P. 2005. Uncertainty-based competition between prefrontal and dorsolateral striatal systems for behavioral control. *Nat. Neurosci.* 8:1704–11

Dayan P, Niv Y. 2008. Reinforcement learning: the good, the bad and the ugly. *Curr. Opin. Neurobiol.* 18:185–96

Ding L, Gold JI. 2011. Neural correlates of perceptual decision making before, during, and after decision commitment in monkey frontal eye field. *Cereb. Cortex.* doi: 10.1093/cercor/bhr178

Dorris MC, Glimcher PW. 2004. Activity in posterior parietal cortex is correlated with the relative subjective desirability of action. *Neuron* 44:365–78

Ellsberg D. 1961. Risk, ambiguity, and the Savage axioms. *Q. J. Econ.* 61:643–69

Erev I, Roth AE. 1998. Predicting how people play games: reinforcement learning in experimental games with unique, mixed strategy equilibria. *Am. Econ. Rev.* 88:848–81

Fecteau JH, Munoz DP. 2003. Exploring the consequences of the previous trial. *Nat. Rev. Neurosci.* 4: 435–43

Fellows LK, Farah MJ. 2003. Ventromedial frontal cortex mediates affective shifting in humans: evidence from a reversal learning paradigm. *Brain* 126:1830–37

Feltovich R. 2000. Reinforcement-based versus belief-based learning models in experimental asymmetric-information games. *Econometrica* 68:605–41

Funahashi S, Bruce CJ, Goldman-Rakic PS. 1989. Mnemonic coding of visual space in the monkey's dorsolateral prefrontal cortex. *J. Neurophysiol.* 61:331–49

Furman M, Wang X-J. 2008. Similarity effect and optimal control of multiple-choice decision making. *Neuron* 60:1153–68

Genovesio A, Brasted PJ, Wise SP. 2006. Representation of future and previous spatial goals by separate neural populations in prefrontal cortex. *J. Neurosci.* 26:7305–16

Georgopoulos AP, Schwartz AB, Kettner RE. 1986. Neural population coding of movement direction. *Science* 233:1416–19

Gerfen CR, Surmeier DJ. 2011. Modulation of striatal projection systems by dopamine. *Annu. Rev. Neurosci.* 34:441–66

Gläscher J, Daw N, Dayan P, O'Doherty JP. 2010. States versus rewards: dissociable neural prediction error signals underlying model-based and model-free reinforcement learning. *Neuron* 66:585–95

Glimcher PW, Sparks DL. 1992. Movement selection in advance of action in the superior colliculus. *Nature* 355:542–45

Gnadt JW, Andersen RA. 1988. Memory related motor planning activity in posterior parietal cortex of macaque. *Exp. Brain Res.* 70:216–20

Haber SN, Fudge JL, McFarland R. 2000. Striatonigrostriatal pathways in primates form an ascending spiral from the shell to the dorsolateral striatum. *J. Neurosci.* 20:2369–82

Haber SN, Knutson B. 2010. The reward circuit: linking primate anatomy and human imaging. *Neuropsychopharmacology* 35:4–26

Haggard P. 2008. Human volition: towards a neuroscience of will. *Nat. Rev. Neurosci.* 9:934–46

Hanes DP, Schall JD. 1996. Neural control of voluntary movement initiation. *Science* 274:427–30

Hassabis D, Maguire EA. 2007. Deconstructing episodic memory with construction. *Trends Cogn. Sci.* 11:299–306

Hayden BY, Pearson JM, Platt ML. 2009. Fictive reward signals in the anterior cingulate cortex. *Science* 324:948–50

Hikosaka O, Nakamura K, Nakahara H. 2006. Basal ganglia orient eyes to reward. *J. Neurophysiol.* 95:567–84

Hikoaka O, Nakahara H, Rand MK, Sakai K, Lu X, et al. 1999. Parallel neural networks for learning sequential procedures. *Trends Neurosci.* 22:464–71

Hong S, Hikosaka O. 2008. The globus pallidus sends reward-related signals to the lateral habenula. *Neuron* 60:720–29

Hong S, Hikosaka O. 2011. Dopamine-mediated learning and switching in cortico-striatal circuit explain behavioral changes in reinforcement learning. *Front. Behav. Neurosci.* 5:15

Horwitz GD, Newsome WT. 2001. Target selection for saccadic eye movements: prelude activity in the superior colliculus during a direction-discrimination task. *J. Neurophysiol.* 86:2543–58

Hsu M, Bhatt M, Adolphs R, Tranel D, Camerer CF. 2005. Neural systems responding to degrees of uncertainty in human decision-making. *Science* 310:1680–83

Huettel SA, Stowe CJ, Gordon EM, Warner BT, Platt ML. 2006. Neural signatures of economic preferences for risk and ambiguity. *Neuron* 49:765–75

Hyman JM, Zilli EA, Paley AM, Hasselmo ME. 2010. Working memory performance correlates with prefrontal-hippocampal theta interactions but not with prefrontal neuron firing rates. *Front. Integr. Neurosci.* 4:2

Ito M, Doya K. 2009. Validation of decision-making models and analysis of decision variables in the rat basal ganglia. *J. Neurosci.* 29:9861–74

Ito M, Doya K. 2011. Multiple representations and algorithms for reinforcement learning in the cortico-basal ganglia circuit. *Curr. Opin. Neurobiol.* 21:368–73

Iversen SD, Mishkin M. 1970. Perseverative interference in monkeys following selective lesions of the inferior prefrontal convexity. *Exp. Brain Res.* 11:376–86

Johnson A, Redish AD. 2007. Neural ensembles in CA3 transiently encode paths forward of the animal at a decision point. *J. Neurosci.* 27:12176–89

Kahneman D, Tversky A. 1979. Prospect theory: an analysis of decision under risk. *Econometrica* 47:263–91

Kepecs A, Uchida N, Zariwala HA, Mainen ZF. 2008. Neural correlates, computation and behavioural impact of decision confidence. *Nature* 455:227–31

Kim H, Sul JH, Huh N, Lee D, Jung MW. 2009. Role of striatum in updating values of chosen actions. *J. Neurosci.* 29:14701–12

Kim S, Hwang J, Lee D. 2008. Prefrontal coding of temporally discounted values during intertemporal choice. *Neuron* 59:161–72

Kim S, Lee D. 2011. Prefrontal cortex and impulsive decision making. *Biol. Psychiatry* 69:1140–46

Kim Y, Huh N, Lee H, Baeg E, Lee D, Jung MW. 2007. Encoding of action history in the rat ventral striatum. *J. Neurophysiol.* 98:3548–56

Knowlton BJ, Mangels JA, Squire LR. 1996. A neostriatal habit learning system in humans. *Science* 273:1399–402

Krajbich I, Armel C, Rangel A. 2010. Visual fixations and the computation and comparison of value in simple choice. *Nat. Neurosci.* 13:1292–98

Laming DRJ. 1968. *Information Theory of Choice-Reaction Times*. London: Academic

Lau B, Glimcher PW. 2008. Value representations in the primate striatum during matching behavior. *Neuron* 58:451–63

Lee D. 2008. Game theory and neural basis of social decision making. *Nat. Neurosci.* 11:404–9

Lee D, Conroy ML, McGreevy BP, Barraclough DJ. 2004. Reinforcement learning and decision making in monkeys during a competitive game. *Cogn. Brain Res.* 22:45–58

Lee D, McGreevy BP, Barraclough DJ. 2005. Learning and decision making in monkeys during a rock-paper-scissors game. *Cogn. Brain Res.* 25:416–30

Levey AI, Hersch SM, Rye DB, Sunahara RK, Niznik HB, et al. 1993. Localization of D_1 and D_2 dopamine receptors in brain with subtype-specific antibodies. *Proc. Natl. Acad. Sci. USA* 90:8861–65

Lewis DA, Melchitzky DS, Sesack SR, Whitehead RE, Auh S, Sampson A. 2001. Dopamine transporter immunoreactivity in monkey cerebral cortex: regional, laminar, and ultrastructural localization. *J. Compar. Neurol.* 432:119–36

Litt A, Plassmann H, Shiv B, Rangel A. 2011. Dissociating valuation and saliency signals during decision making. *Cereb. Cortex* 21:95–102

Lo C-C, Wang X-J. 2006. Cortico-basal ganglia circuit mechanism for a decision threshold in reaction time tasks. *Nat. Neurosci.* 9:956–63

Lohrenz T, McCabe K, Camerer CF, Montague PR. 2007. Neural signature of fictive learning signals in a sequential investment task. *Proc. Natl. Acad. Sci. USA* 104:9493–98

Louie K, Glimcher PW. 2010. Separating value from choice: delay discounting activity in the lateral intraparietal area. *J. Neurosci.* 30:5498–507

Luhmann CC, Chun MM, Yi D-J, Lee D, Wang X-J. 2008. Neural dissociation of delay and uncertainty in intertemporal choice. *J. Neurosci.* 28:14459–66

Matsumoto M, Hikosaka O. 2007. Lateral habenula as a source of negative reward signals in dopamine neurons. *Nature* 447:1111–15

Matsumoto M, Matsumoto K, Abe H, Tanaka K. 2007. Medial prefrontal cell activity signaling prediction errors of action values. *Nat. Neurosci.* 10:647–56

Maunsell JHR. 2004. Neuronal representations of cognitive state: reward or attention? *Trends Cogn. Sci.* 8:261–65

Montague PR, Dayan P, Sejnowski TJ. 1996. A framework for mesencephalic dopamine systems based on predictive Hebbian learning. *J. Neurosci.* 16:1936–47

Mookherjee D, Sopher B. 1994. Learning behavior in an experimental matching pennies game. *Games Econ. Behav.* 7:62–91

Mookherjee D, Sopher B. 1997. Learning and decision costs in experimental constant sum games. *Games Econ. Behav.* 19:97–132

Murray EA, O'Doherty JP, Schoenbaum G. 2007. What we know and do not know about the functions of the orbitofrontal cortex after 20 years of cross-species studies. *J. Neurosci.* 27:8166–69

Nachev P, Kennard C, Husain M. 2008. Functional role of the supplementary and pre-supplementary motor areas. *Nat. Rev. Neurosci.* 9:856–69

Okano K, Tanji J. 1987. Neuronal activities in the primate motor fields of the agranular frontal cortex preceding visually triggered and self-paced movement. *Exp. Brain Res.* 66:155–66

O'Neill M, Schultz W. 2010. Coding of reward risk by orbitofrontal neurons is mostly distinct from coding of reward value. *Neuron* 68:789–800

Oyama K, Hernádi I, Iijima T, Tsutsui K-I. 2010. Reward prediction error coding in dorsal striatal neurons. *J. Neurosci.* 30:11447–57

Padoa-Schioppa C. 2011. Neurobiology of economic choice: a good-based model. *Annu. Rev. Neurosci.* 34:333–59

Padoa-Schioppa C, Assad JA. 2006. Neurons in the orbitofrontal cortex encode economic value. *Nature* 441:223–26

Pan W-X, Schmidt R, Wickens JR, Hyland BI. 2005. Dopamine cells respond to predicted events during classical conditioning: evidence for eligibility traces in the reward-learning network. *J. Neurosci.* 25:6235–42

Pan X, Sawa K, Tsuda I, Tsukada M, Sakagami M. 2008. Reward prediction based on stimulus categorization in primate lateral prefrontal cortex. *Nat. Neurosci.* 11:703–12

Pastor-Bernier A, Cisek P. 2011. Neural correlates of biased competition in premotor cortex. *J. Neurosci.* 31:7083–88

Paxinos G, Watson C. 1998. *The Rat Brain in Stereotaxic Coordinates.* San Diego: Academic

Platt ML, Glimcher PW. 1999. Neural correlates of decision variables in parietal cortex. *Nature* 400:233–38

Premack D, Woodruff G. 1978. Does the chimpanzee have a theory of mind? *Behav. Brain Sci.* 4:515–26

Reynolds JNJ, Hyland BI, Wickens JR. 2001. A cellular mechanism of reward-related learning. *Nature* 413:67–70

Roesch MR, Singh T, Brown PL, Mullins SE, Schoenbaum G. 2009. Ventral striatal neurons encode the value of the chosen action in rats deciding between differently delayed or sized rewards. *J. Neurosci.* 29:13365–76

Roese NJ, Olson JM. 1995. *What Might Have Been: The Social Psychology of Counterfactual Thinking.* New York: Psychol. Press

Roitman JD, Shadlen MN. 2002. Response of neurons in the lateral intraparietal area during a combined visual discrimination reaction time task. *J. Neurosci.* 22:9475–89

Rorie AE, Gao J, McClelland JL, Newsome WT. 2010. Integration of sensory and reward information during perceptual decision-making in lateral intraparietal cortex (LIP) of the macaque monkey. *PLoS One* 5: e9308

Samejima K, Ueda Y, Doya K, Kimura M. 2005. Representation of action-specific reward values in the striatum. *Science* 310:1337–40

Schacter DL, Addis DR, Buckner RL. 2007. Remembering the past to imagine the future: the prospective brain. *Nat. Rev. Neurosci.* 8:657–61

Schall JD. 1991. Neuronal activity related to visually guided saccadic eye movements in the supplementary motor area of rhesus monkeys. *J. Neurophysiol.* 66:530–58

Schiller PH, Stryker M. 1972. Single-unit recording and stimulation in superior colliculus of the alert rhesus monkey. *J. Neurophysiol.* 35:915–24

Schlag J, Schlag-Rey M. 1987. Evidence for a supplementary eye field. *J. Neurophysiol.* 57:179–200

Schoenbaum G, Nugent SL, Saddoris MP, Setlow B. 2002. Orbitofrontal lesions in rats impair reversal but not acquisition of go, no-go odor discriminations. *Neuroreport* 13:885–90

Schultz W. 2006. Behavioral theories and the neurophysiology of reward. *Annu. Rev. Psychol.* 57:87–115

Schultz W, Tremblay L, Hollerman JR. 2000. Reward processing in primate orbitofrontal cortex and basal ganglia. *Cereb. Cortex* 10:272–84

Seo H, Barraclough DJ, Lee D. 2009. Lateral intraparietal cortex and reinforcement learning during a mixed-strategy game. *J. Neurosci.* 29:7278–89

Seo H, Lee D. 2007. Temporal filtering of reward signals in the dorsal anterior cingulate cortex during a mixed-strategy game. *J. Neurosci.* 27:8366–77

Seo H, Lee D. 2008. Cortical mechanisms for reinforcement learning in competitive games. *Philos. Trans. R. Soc. Lond. Ser. B* 363:3845–57

Seo H, Lee D. 2009. Behavioral and neural changes after gains and losses of conditioned reinforcers. *J. Neurosci.* 29:3627–41

Serences JT. 2008. Value-based modulations in human visual cortex. *Neuron* 60:1169–81

Shadlen MN, Newsome WT. 2001. Neural basis of a perceptual decision in the parietal cortex of the rhesus monkey. *J. Neurophysiol.* 86:1916–36

Shen W, Flajolet M, Greengard P, Surmeier DJ. 2008. Dichotomous dopaminergic control of striatal synaptic plasticity. *Science* 321:848–51

Simon DA, Daw ND. 2011. Neural correlates of forward planning in a spatial decision task in humans. *J. Neurosci.* 31:5526–39

Shuler MG, Bear MF. 2006. Reward timing in the primary visual cortex. *Science* 311:1606–9

Sirota A, Montgomery S, Fujisawa S, Isomura Y, Zugaro M, Buzsáki G. 2008. Entrainment of neocortical neurons and gamma oscillations by the hippocampal theta rhythm. *Neuron* 60:683–97

Smith PL, Ratcliff R. 2004. Psychology and neurobiology of simple decisions. *Trends Neurosci.* 27:161–68

Smyrnis N, Taira M, Ashe J, Georgopoulos AP. 1992. Motor cortical activity in a memorized delay task. *Exp. Brain Res.* 92:139–51

So NY, Stuphorn V. 2010. Supplementary eye field encodes option and action value for saccades with variable reward. *J. Neurophysiol.* 104:2634–53

Softky WR, Koch C. 1993. The highly irregular firing of cortical cells is inconsistent with temporal integration of random EPSPs. *J. Neurosci.* 13:334–50

Soltani A, Lee D, Wang X-J. 2006. Neural mechanism for stochastic behaviour during a competitive game. *Neural Netw.* 19:1075–90

Soltani A, Wang X-J. 2006. A biophysically based neural model of matching law behavior: melioration by stochastic synapses. *J. Neurosci.* 26:3731–44

Sohn J-W, Lee D. 2007. Order-dependent modulation of directional signals in the supplementary and pre-supplementary motor areas. *J. Neurosci.* 27:13655–66

Soon CS, Brass M, Heinze H-J, Haynes J-D. 2008. Unconscious determinants of free decisions in the human brain. *Nat. Neurosci.* 11:543–45

Sugrue LP, Corrado GS, Newsome WT. 2004. Matching behavior and the representation of value in the parietal cortex. *Science* 304:1782–87

Sul JH, Kim H, Huh N, Lee D, Jung MW. 2010. Distinct roles of rodent orbitofrontal and medial prefrontal cortex in decision making. *Neuron* 66:449–60

Sul JH, Jo S, Lee D, Jung MW. 2011. Role of rodent secondary motor cortex in value-based action selection. *Nat. Neurosci.* 14:1202–8

Sutton RS, Barto AG. 1998. *Reinforcement Learning: An Introduction*. Cambridge, MA: MIT Press

Tanji J, Kurata K. 1985. Contrasting neuronal activity in supplementary and precentral motor cortex of monkeys. I. Responses to instructions determining motor responses to forthcoming signals of different modalities. *J. Neurophysiol.* 53:129–41

Tolman EC. 1948. Cognitive maps in rats and men. *Psychol. Rev.* 55:189–208

Tremblay L, Schultz W. 1999. Relative reward preference in primate orbitofrontal cortex. *Nature* 398:704–8

Usher M, McClelland J. 2001. On the time course of perceptual choice: the leaky, competing accumulator model. *Psychol. Rev.* 108:550–92

van der Meer MAA, Redish AD. 2011. Ventral striatum: a critical look at models of learning and evaluation. *Curr. Opin. Neurobiol.* 21:387–92

Vickery TJ, Chun MM, Lee D. 2011. Ubiquity and specificity of reinforcement signals throughout the human brain. *Neuron* 72:166–77

von Neumann J, Morgenstern O. 1944. *Theory of Games and Economic Behavior*. Princeton: Princeton Univ. Press

Wallis JD, Kennerley SW. 2010. Heterogeneous reward signals in prefrontal cortex. *Curr. Opin. Neurobiol.* 20:191–98

Wallis JD, Miller EK. 2003. Neuronal activity in primate dorsolateral and orbital prefrontal cortex during performance of a reward preference task. *Eur. J. Neurosci.* 8:2069–81

Walton ME, Behrens TE, Buckley MJ, Rudebeck PH, Rushworth MF. 2010. Separable learning systems in the macaque brain and the role of orbitofrontal cortex in contingent learning. *Neuron* 65:927–39

Wang X-J. 2001. Synaptic reverberation underlying mnemonic persistent activity. *Trends Neurosci.* 24:455–63

Wang X-J. 2002. Probabilistic decision making by slow reverberation in cortical circuits. *Neuron* 36:955–68

Wang X-J. 2008. Decision making in recurrent neuronal circuits. *Neuron* 60:215–34

Weinrich M, Wise SP. 1982. The premotor cortex of the monkey. *J. Neurosci.* 2:1329–45

Womelsdorf T, Vinck M, Leung LS, Everling S. 2010. Selective theta-synchronization of choice-relevant information subserves goal-directed behavior. *Front. Hum. Neurosci.* 4:210

Wunderlich K, Rangel A, O'Doherty JP. 2009. Neural computations underlying action-based decision making in the human brain. *Proc. Natl. Acad. Sci. USA* 106:17199–204

Wurtz RH, Goldberg ME. 1972. Activity of superior colliculus in behaving monkey. 3. Cells discharging before eye movements. *J. Neurophysiol.* 35:575–86

Yin HH, Knowlton BJ. 2006. The role of the basal ganglia in habit formation. *Nat. Rev. Neurosci.* 7:464–76

Yu AJ, Dayan P. 2005. Uncertainty, neuromodulation, and attention. *Neuron* 46:681–92

Zhu L, Mathewson KE, Hsu M. 2012. Dissociable neural representations of reinforcement and belief prediction errors underlie strategic learning. *Proc. Natl. Acad. Sci. USA* 109:1419–24

Critical-Period Plasticity in the Visual Cortex

Christiaan N. Levelt[1] and Mark Hübener[2]

[1]Department of Molecular Visual Plasticity, Netherlands Institute for Neuroscience, an Institute of the Royal Netherlands Academy of Arts and Sciences, 1105BA Amsterdam, The Netherlands; email: c.levelt@nin.knaw.nl

[2]Max Planck Institute of Neurobiology, D-82152 Martinsried, Germany; email: mark@neuro.mg.de

Annu. Rev. Neurosci. 2012. 35:309–30

First published online as a Review in Advance on March 29, 2012

The *Annual Review of Neuroscience* is online at neuro.annualreviews.org

This article's doi: 10.1146/annurev-neuro-061010-113813

0147-006X/12/0721-0309$20.00

Keywords

experience-dependent plasticity, mouse, visual system, ocular dominance, monocular deprivation

Abstract

In many regions of the developing brain, neuronal circuits undergo defined phases of enhanced plasticity, termed critical periods. Work in the rodent visual cortex has led to important insights into the cellular and molecular mechanisms regulating the timing of the critical period. Although there is little doubt that the maturation of specific inhibitory circuits plays a key role in the opening of the critical period in the visual cortex, it is less clear what puts an end to it. In this review, we describe the established mechanisms and point out where more experimental work is needed. We also show that plasticity in the visual cortex is present well before, and long after, the peak of the critical period.

Contents

INTRODUCTION

In most organisms, the neuronal connections subserving sensory perception or motor output are immature at birth or hatching. They are refined over a prolonged period after birth, which, in some species, may last up to several years. During the early, prenatal phase of nervous-system wiring, molecular guidance cues play a dominant role, whereas patterned neuronal activity becomes the key player during subsequent stages of development. This patterned activity either is generated by the developing brain or, once the peripheral sensory organs are functional, may be driven by stimuli in the external world. Frequently, the impact of sensory input on nervous-system wiring is strong during a brief, well-defined phase of an animal's early life, before and after which the same stimuli have less influence. Such a phase of increased susceptibility to certain types of sensory input is called a critical period.

Importantly, the changes in neuronal connectivity imparted by sensory input during a critical period ultimately result in certain behaviors or capabilities, which the animal would not show or have if the appropriate stimuli were absent during the critical period. In fact, among the first systematic studies on critical periods are the behavioral experiments on filial imprinting in nidifugous birds conducted by Lorenz (1935). On the basis of earlier work by Heinroth (1910), Lorenz observed that, shortly after hatching in an incubator, greylag geese chicks could be imprinted on almost any moving visual object, including himself, which from then on would serve as a substitute mother for the young geese. Importantly, he realized that the time period for imprinting was short and well defined, with a clear peak between 13 and 16 h after hatching.

Since then, critical periods have been observed in many brain systems and in a large variety of species. Song learning in birds, auditory localization in barn owls, and human language acquisition are just a few of the many intensely studied examples. These studies have made clear that there are several variations to the general concept of a critical period as a well-defined, brief phase that occurs relatively shortly after birth. One important insight is that critical-period timing is not always absolute; often, susceptibility to a specific sensory input increases relatively rapidly but then declines over a much slower time course, such that the closure of a critical period is frequently a gradual process that can extend substantially into adulthood. The opening and closing of a critical period also often depend on specific features of the sensory stimulus. For example, rearing animals in the dark shifts the critical period for ocular dominance (OD) plasticity (Mower 1991) (see Ocular Dominance Plasticity, sidebar), and noise rearing can cause a similar delay in the critical period of the auditory cortex (Chang & Merzenich 2003). Likewise, the critical period for song learning in certain bird species is not always tightly coupled to the time of hatching; instead, it can depend on external triggers, such as the seasons (Nulty et al. 2010). Critical periods for specific types of sensory input may also occur more than once in the lifetime of an animal. Some song birds, e.g., canaries, go through an annual critical period, during which they are particularly sensitive to novel auditory stimuli, which they incorporate into their song repertoire (Nottebohm & Nottebohm 1978). Similarly, parental olfactory imprinting, typically limited to a short period immediately after birth, is repeated with each new offspring (Klopfer et al. 1964). These few examples indicate that the mechanisms underlying the control of critical periods are most likely complex and diverse.

The first detailed investigation of a critical period at the neuronal level was based on the original observation by Wiesel & Hubel (1963) that temporary visual deprivation of one eye of a kitten causes a dramatic change in the OD distribution among the neurons in its visual cortex (see Ocular Dominance Plasticity, sidebar). In a subsequent study, they analyzed the timing of the critical period for this type of plasticity in the cat visual cortex and found that it was maximal during a short period between four and eight weeks after birth and then declined gradually until three months after birth (Hubel & Wiesel 1970). In adult cats, even prolonged periods of monocular eye closure had no effect. This type of sensory deprivation has since become the primary model for studying critical-period regulation. Although most earlier studies addressing this question were carried out in higher mammals such as cats, ferrets, and monkeys, much of our current knowledge regarding the cellular and molecular mechanisms controlling critical periods is based on experiments carried out in mice and rats (**Figure 1**). In particular, the use of transgenic mice allowed for a detailed assessment of the molecular players involved in critical-period regulation.

OCULAR DOMINANCE PLASTICITY

In mammals, the primary visual cortex is the first station along the visual pathway where information coming from both eyes converges at the level of single neurons, and many neurons in this area can be driven by stimulation through either eye. Typically, a neuron fires more action potentials when identical visual stimuli are presented to one eye versus the other, a receptive field property termed ocular dominance (OD) (Hubel & Wiesel 1962). In the visual cortex of many higher mammals, neurons are clustered together according to the eye by which they are preferentially driven, forming alternating OD columns running across the cortex. Closing one eye early in life alters OD in the visual cortex, such that the temporarily closed eye becomes less effective in driving cortical cells, whereas the nondeprived eye gains influence. At the level of OD columns, this is reflected in an expansion of the open eye's columns at the expense of the columns associated with the closed eye. The rodent visual cortex lacks OD columns, but individual neurons in the binocular part show OD. Similar to higher mammals, a shift in the OD distribution can be readily induced by a temporary closure of one eye (Dräger 1978) (**Figure 1**). Collectively, the above-described changes are termed OD plasticity.

CRITICAL-PERIOD ONSET

A few days of monocular deprivation (MD) in juvenile mice are sufficient to induce a marked shift in OD in the visual cortex (Gordon & Stryker 1996). The effect is strongest during a relatively brief phase at the end of the fourth postnatal week, and the time between postnatal day 28 and postnatal day 32 is generally considered as the critical period for OD plasticity in the mouse visual cortex (Gordon & Stryker 1996). OD plasticity can also be induced before this period (see Precritical Period Plasticity, sidebar below) or after, but differs quantitatively and qualitatively. Soon after several laboratories had started using transgenic mouse models to investigate the molecular and cellular

Monocular deprivation (MD): temporary closure of one eye

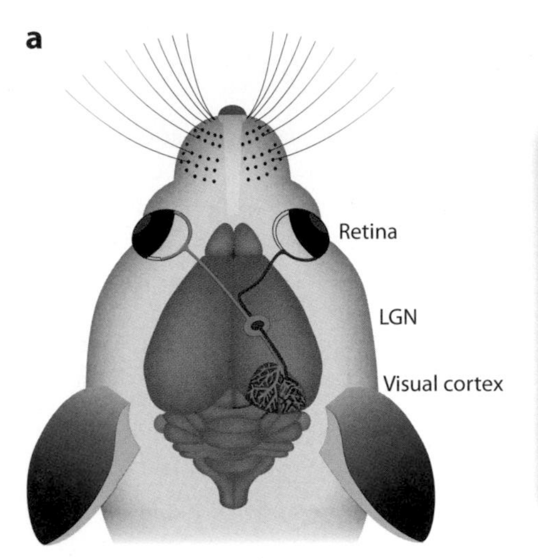

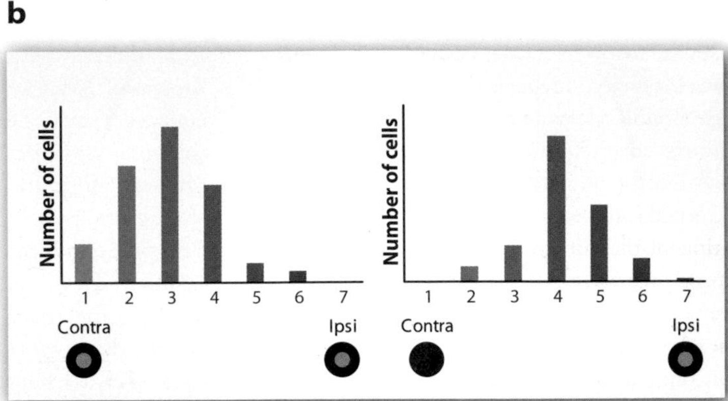

Figure 1

Ocular dominance (OD) plasticity in the mouse visual cortex. (*a*) Schematic of the mouse visual system. Most retinal ganglion cell axons (*blue*) cross the midline and terminate in the lateral geniculate nucleus (LGN) of the contralateral hemisphere. A very small number of axons from the temporal retina (*red*) do not cross at the optic chiasm; instead they terminate in the ipsilateral LGN. In the LGN, inputs from both eyes remain fully segregated. The major (medial) part of the primary visual cortex (V1) receives input from only the contralateral retina (*blue*), whereas the smaller (lateral) third of V1 also receives ipsilateral projections (*red*). In this binocular region of V1, most neurons respond to visual stimuli presented to either eye, but contralateral eye input is stronger overall. (*b*) (*Left*) On the basis of their responses to monocularly presented stimuli, neurons in the binocular part of V1 can be grouped into different OD classes, from complete contralateral (class 1) to complete ipsilateral (class 7) dominance. Neurons in class 4 are driven equally strongly by both eyes. (*Right*) Visually depriving the contralateral eye for a few days causes a shift of the OD distribution, such that the relative influence of the deprived eye becomes weaker, whereas the nondeprived eye exerts a stronger drive.

Gamma-aminobutyric acid (GABA): main inhibitory neurotransmitter in the cerebral cortex

BDNF: brain-derived neurotrophic factor

IGF-1: hormone with a molecular structure similar to that of insulin; regulates several aspects of development and has anabolic effects in adults

mechanisms underlying OD plasticity, Hensch et al. (1998) made the important discovery that the development of gamma-aminobutyric acid (GABA)ergic innervation is crucial for the onset of the critical period. In mice with reduced GABA synthesis due to genetic inactivation of the GABA-synthesizing enzyme glutamic acid decarboxylase (GAD65), the critical period for OD plasticity does not open until the animals are treated with the GABA-A receptor agonist diazepam (Fagiolini & Hensch 2000, Hensch et al. 1998). It is also possible to induce a precocious critical period in 19-day-old wild-type mice by treating them with benzodiazepines (Fagiolini & Hensch 2000).

Similarly, transgenic mice overexpressing brain-derived neurotropic factor (BDNF) during postnatal development show a premature opening of the critical period, probably because BDNF spurs the maturation of GABAergic synapses in the visual cortex (Hanover et al. 1999, Huang et al. 1999) (for an overview of signaling pathways that influence the onset or closure of the critical period, see **Figure 2**). In contrast, when wild-type mice are dark reared, BDNF expression is downregulated and the development of inhibitory innervation and critical-period onset are delayed. Also IGF-1 can accelerate the development of inhibitory innervation and the increase of visual acuity, which is believed to be a hallmark of the critical period (Ciucci et al. 2007). In addition, when mice are intracortically injected with an enzyme that removes polysialic acid from the neural cell-adhesion molecule (NCAM) of precritical-period mice, inhibitory synapses form more rapidly and OD plasticity can be induced earlier than normal (Di et al. 2007). Together, these studies indicate that several days after eye opening (around P12 in mice), all prerequisites for critical-period plasticity are in place, except for the appropriate level or type of inhibitory input.

However, it is not entirely clear why it takes more than one week after eye opening for the development of the inhibitory system to allow OD plasticity to take place. Maturation of the excitatory networks may be a further prerequisite for OD plasticity to occur. Unreliable and noisy excitatory drive may otherwise lead to a random strengthening and weakening of synaptic connections, which could slow down the formation of the adequate circuitry necessary to process sensory stimuli. Maturation of excitatory networks may also alter the tuning properties of inhibitory neurons, ultimately leading to the initiation of the critical period (Kuhlman et al. 2011).

PRECRITICAL PERIOD PLASTICITY

Despite the mounting evidence for a substantial degree of ocular dominance (OD) plasticity in the visual cortex of adult mice (see main text), only a few studies have systematically explored the degree of plasticity in very young animals. Early on, researchers determined that, although the critical period peaks around postnatal day P30, a shift in eye preference, albeit small, could already be induced at P19 (Gordon & Stryker 1996). Using Arc induction, Tagawa and colleagues (2005) demonstrated that short-term monocular deprivation starting as early as P17 caused a clear change in OD. Subsequently, intrinsic imaging revealed that removal or inactivation of the contralateral eye before natural eye opening accelerated the development of the retinotopic map of the ipsilateral eye that was apparent at P15 (Smith & Trachtenberg 2007). Together, these data indicate that at least certain aspects of binocular plasticity in the mouse visual cortex are fully functional immediately after eye opening and long before the peak of the critical period for OD plasticity.

Activity-Driven Development of Inhibitory Innervation

How visual input and neuronal activity drive the development of inhibitory innervation is not entirely clear. Activity-induced BDNF release driving GABAergic synapse formation is one possibility (Gianfranceschi et al. 2003, Hanover et al. 1999, Huang et al. 1999). Another fascinating pathway is via OTX2. This transcription factor is transcribed and translated in the retina, but it is also transported to the visual cortex where it is taken up predominantly by parvalbumin (PV)-expressing interneurons (Sugiyama et al. 2008). In the absence of OTX2, the critical period does not start, a defect that can be rescued by intracortical infusion of OTX2. OTX2 has interesting links with other cellular events associated with critical-period onset. In Xenopus embryonic development, IGF1 enhances OTX2-mediated transcription (Carron et al. 2005). This may indicate that the effects of IGF-1 on the development of cortical inhibition are mediated through OTX2. Interestingly, IGF-1 also upregulates BDNF expression (Carro et al. 2000), suggesting that IGF-1 may regulate critical-period onset through different mechanisms at various locations. Moreover, OTX2 may bind to the PSA moiety of NCAM (Joliot et al. 1991, Sugiyama et al. 2009), by which it may become sequestered instead of being transferred to PV+ interneurons. Furthermore, OTX2 appears to enter PV+ interneurons through an interaction with dense aggregates of the extracellular matrix (ECM) known as perineuronal nets that enwrap these neurons (Sugiyama et al. 2009). ECM is an important regulator of critical-period closure (see below) (Pizzorusso et al. 2002) and may thus exert its role through aiding OTX2-mediated transcription in PV+ interneurons.

The maturation of inhibition in the visual cortex also depends crucially on the integrity of a transient cell population located in the white matter of the developing cortex, the subplate neurons (Chun & Shatz 1989), which are part of an early circuit indirectly linking thalamic inputs with layer 4 neurons (Friauf et al. 1990). Ablation of subplate neurons halts the maturation of fast GABAergic transmission (Kanold & Shatz 2006) and prevents the formation of OD columns in the visual cortex of cats with normal binocular vision (Ghosh & Shatz 1992). Interestingly, MD in kittens lacking subplate neurons leads to a paradoxical OD shift, whereby OD columns of the deprived eye expand (Kanold & Shatz 2006).

Neural cell-adhesion molecule (NCAM): homophilic binding glycoprotein expressed on neurons and glia and involved in cell-cell adhesion, neurite outgrowth, and synaptic plasticity

Parvalbumin (PV): marker for specific subtypes of cortical inhibitory interneurons

OTX2: homeobox protein involved in embryonic pattern formation

ECM: extracellular matrix

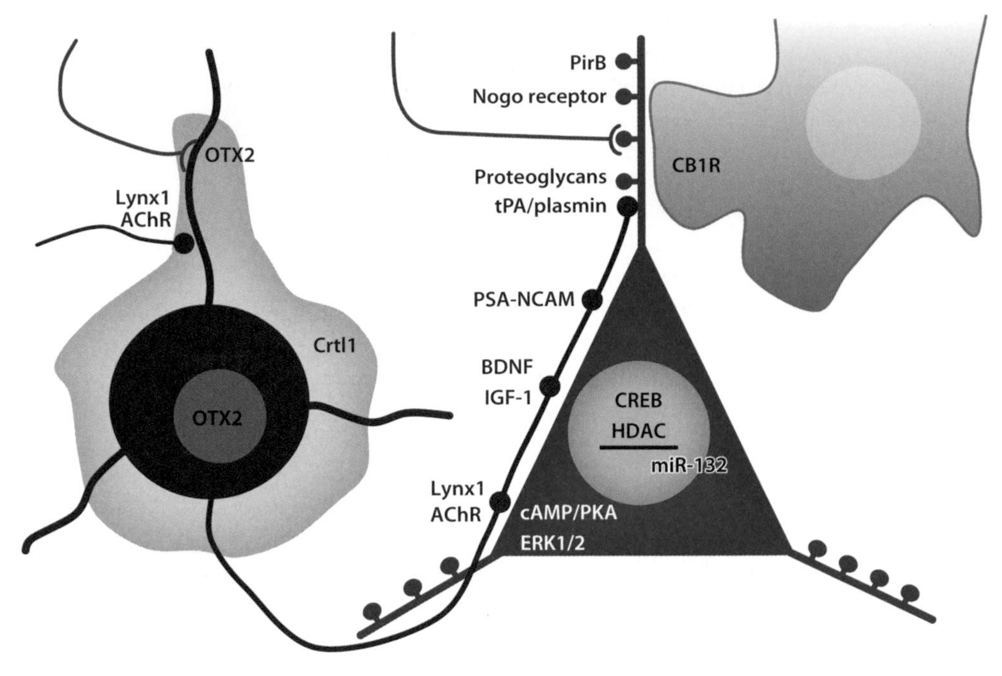

Figure 2

Cartoon showing molecules involved in regulating the critical period. Researchers have identified various signaling pathways that spur the development of inhibitory neurons (*red*) and their synapses, thereby prompting the onset of the critical period. These pathways include BDNF and IGF-1 signals and removal of the PSA moiety from the adhesion molecule NCAM. In addition, the transcription factor OTX2 is synthesized in the retina and trans-synaptically transported to the visual cortex, where it is taken up predominantly by PV-expressing interneurons, expediting their development. Researchers have uncovered other signalling pathways whose inactivation increases adult OD plasticity. Proteins involved in myelin-mediated growth inhibition such as Nogo-66 receptor and PirB limit adult OD plasticity, by inhibiting structural changes or diminishing synaptic plasticity of excitatory neurons (*blue*). Removing chondroitin sulfate from proteoglycans of the ECM enhances adult OD plasticity. The mature ECM may limit structural plasticity in the adult visual cortex. Alternatively, adult plasticity may be limited by perineuronal nets (*gray*) that are formed around PV-expressing interneurons. The development of these nets depends on the presence of Crtl1. The activity of these interneurons may also be modulated by nAChR and its inhibitor lynx1 and by the receptor for endocannabinoids, CB1R (possibly present on glia) (*green*). In addition, downregulation of CREB-mediated gene transcription appears to limit adult OD plasticity. Activating this pathway by increasing cAMP levels or inhibiting HDAC enhances plasticity, which may involve regulation of microRNA-132. Abbreviations: BDNF, brain-derived neurotropic factor; CB1R, cannabinoid receptor 1; Crtl1, cartilage link protein 1; ECM, extracellular matrix; HDAC, histone deacetylases; nAChR, nicotinic acetylcholine receptors; NCAM, neural cell-adhesion molecule; OD, ocular dominance: PirB, paired immunoglobulin-like receptor B; PSA, polysialic acid; PV, parvalbumin; tPA, tissue plasminogen activator.

PV+ Basket Cells and Critical-Period Onset

Besides uncovering an intriguing developmental mechanism, the findings regarding the role of OTX2 also strengthen the view that the development of PV+ interneurons is crucial for initiating the critical period for OD plasticity. Initial evidence for this notion came from the observation that a precocious critical period can be induced with benzodiazepine treatment of young mice. This effect is abolished when the alpha1 subunit of the pentameric GABA-A receptor is genetically rendered insensitive to benzodiazepines (Fagiolini et al. 2004). In contrast, modifying the alpha2 subunit to make it insensitive to benzodiazepines does not alter their effects on critical-period onset. The alpha1 subunit is enriched in inhibitory synapses formed by PV+ basket cells terminating on the soma and proximal dendrites of their target neurons (Klausberger et al. 2002). In contrast, the alpha2 subunits are enriched at the soma and dendrites in synapses formed by cholecystokinin-expressing basket cells (Nyiri et al. 2001) and at axoaxonic synapses formed by chandelier cells (Nusser et al. 1996). The presence of the alpha1 subunit in the GABA-A receptor decreases its decay-time constant (Brussaard et al. 1997), which may help to ensure faithful transmission of the high firing rates of PV+ interneurons, which are mostly fast spiking, to postsynaptic conductance changes. Together with the specific uptake of OTX2 by PV+ interneurons, these findings suggest that these interneurons are the dominant subtype regulating critical-period onset.

MECHANISMS UNDERLYING INHIBITION-DRIVEN CRITICAL-PERIOD ONSET

An important open question regards how the maturation of inhibitory inputs initiates the critical period. One explanation is that increased levels of inhibition modulate plasticity at excitatory synapses. Alternatively, plasticity of inhibitory output itself could alter the expression of OD. It is also possible that a dynamic interaction between these two mechanisms exists, causing a selective change in the excitatory/inhibitory balance that promotes plasticity of excitatory neurons, thereby driving their responsiveness toward the open eye (Gandhi et al. 2008).

Modulation of Excitatory Synaptic Plasticity Through Inhibition

There are various ways through which inhibitory innervation could modulate plasticity of excitatory connections in the visual cortex. One powerful mechanism is by modulating spike-timing-dependent plasticity (STDP). PV+ interneurons predominantly innervate excitatory neurons and other interneurons on the soma and proximal dendrites, where they are in the perfect position to alter backpropagation of signals into dendrites (Tsubokawa & Ross 1996). STDP depends critically on the precise timing of perisomatic inhibition through the alteration of backpropagation (Dan & Poo 2004). Indeed, blocking inhibition when pairing presynaptic activity with single postsynaptic action potentials facilitates spike-timing-dependent synaptic strengthening. Increased perisomatic inhibition is expected to make STDP more stringent (Pouille & Scanziani 2001) and may explain why increased inhibition in vivo facilitates the loss of active connections in the visual cortex (Hata et al. 1999). Additionally, a more complicated mechanism may also operate. PV+ interneurons form networks consisting of 30–50 cells that are connected through reciprocal chemical and electrical (gap junction) synapses. This arrangement makes them exquisitely sensitive to synchrony in inputs from excitatory neurons, which is translated into the synchronous firing of groups of PV+ interneurons (Galarreta & Hestrin 1999, 2002). As a result, PV+ interneurons are able to coordinate the firing of large sets of cortical excitatory neurons. This mechanism underlies the appearance of gamma-band oscillations (30–100 Hz) (Tamas et al. 2000) that reflect the coherent and rhythmic firing of large numbers of cells and are thought to be involved in the processing of sensory input (Cardin et al. 2009), attention, and learning (Fries et al. 2007). Thus, the networks of PV+ interneurons may organize the synchrony of neuronal activity in the visual cortex, which in turn influences OD plasticity through the filtering of visual information and STDP.

Inhibition Affects Expression of the Ocular Dominance Shift

Alternatively, or in addition, inhibitory inputs onto excitatory neurons may change during OD plasticity. This could alter the expression of OD plasticity and/or drive synaptic plasticity in favor of inputs from the nondeprived eye. One

V1: primary visual cortex

***N*-methyl-D-aspartate receptor (NMDAR):** an ionotropic glutamate receptor crucially involved in synaptic plasticity in the cerebral cortex

way inhibitory inputs onto excitatory cells may change is through modifications in the responsiveness of the inhibitory neurons. Recently, three studies have examined how interneurons shift their responsiveness during OD plasticity. Two of these used in vivo two-photon imaging of calcium responses in mice expressing a fluorescent protein marking interneurons (Gandhi et al. 2008, Kameyama et al. 2010). The third study employed in vivo intracellular recordings to measure the OD of individual fast-spiking interneurons (Yazaki-Sugiyama et al. 2009). These studies found that long-term MD causes interneurons, similar to excitatory neurons, to shift their responsiveness toward the nondeprived eye, refuting the idea that an OD shift after long-term MD is caused by increased inhibition of responses to the closed eye. The studies did not agree, however, on how interneurons change OD after short-term MD. Using intracellular recordings, Yazaki-Sugiyama et al. (2009) found that inhibitory neurons paradoxically show a shift toward the deprived eye, supporting a role of GABAergic inhibition early after MD in suppressing responsiveness to the nondeprived eye. This is not supported by the two other studies using calcium imaging, however. One shows that interneurons undergo a delayed shift toward the open eye (Gandhi et al. 2008), whereas the other shows that inhibitory and excitatory neurons both shift toward the open eye within the same time course (Kameyama et al. 2010). It is of importance to clarify the causes of these contradictory results to understand how interneurons alter their responsiveness during OD plasticity and what the consequences are for the responsiveness of excitatory neurons.

Inhibition of excitatory neurons can also change in response to the experience-dependent plasticity of inhibitory synapses. The most direct evidence for this notion comes from the finding that MD during the critical period increases miniature inhibitory postsynaptic current frequency in layer 4 star pyramidal neurons in slices of the binocular visual cortex (Maffei et al. 2010). Other studies have also implicated the plasticity of inhibitory synapses, albeit less directly. In vivo intracellular recordings have shown that blocking inhibitory inputs by intracellular injection of GABA antagonists increases the binocularity of neurons in the primary visual cortex (V1), whereas reducing inhibition after MD causes a reversal of the eye bias of excitatory neurons (Yazaki-Sugiyama et al. 2009). This shows that OD plasticity is accompanied by a change in the balance of excitation and inhibition at the single-cell level. Additional evidence that inhibition directly affects OD plasticity comes from the observation that, after two weeks of MD, callosal inputs from the visual cortex ipsilateral to the deprived eye suppress the responses to the deprived eye in the contralateral binocular cortex (Restani et al. 2009). Thus, callosal inputs from the deprived eye may increase their strength onto inhibitory neurons in the binocular cortex, or inhibitory synapses formed by interneurons receiving callosal inputs may strengthen after deprivation. Taken together, the evidence indicating PV+ basket cells initiate the critical period has been accumulating during the past decade and has become very convincing. The underlying mechanism, however, remains to be understood.

END OF THE CRITICAL PERIOD

Although ample evidence indicates that a single developmental event triggers the opening of the critical period, i.e., the maturation of PV+ interneuron–mediated inhibition, the mechanisms leading to the closure of the critical period seem more diverse. One pathway that can be excluded, but has long been hypothesized to be responsible, is the developmental switch in the subunits that make up the *N*-methyl-D-aspartate receptor (NMDAR). In young animals, NMDARs preferentially incorporate the NR2B subunit, which allows for more effective strengthening of nondeprived eye responses on MD than does the mature NR2A subunit (Cho et al. 2009). The switch to the mature NR2A subunit is not responsible for the closure of the critical period, however, because this occurs normally in mice deficient for NR2A (Fagiolini et al. 2003). Numerous recent studies have uncovered other interventions that do enable or enhance adult OD plasticity. It is likely that at least some of these interventions directly reverse developmental events responsible for closure of the critical period, whereas others may enhance plasticity through mechanisms unrelated to regulation of critical-period closure.

Ocular Dominance Plasticity and Inhibitors of Axon Growth

The first study describing a treatment causing the full restoration of OD plasticity in the adult rat visual cortex was inspired by the finding that dissolving the ECM using chondroitinase in the lesioned spinal cord enhances axon regrowth (Moon et al. 2001). As axonal growth and retraction are hallmarks of critical-period plasticity, Pizzorusso et al. (2002) tested whether intracortical chondroitinase injections would also allow adult OD plasticity, which indeed was the case. This suggests that the adult ECM is an important factor limiting structural plasticity, either by forming a physical barrier or by binding and presenting axon-repelling factors such as Semaphorins and Ephrins.

Dendritic spine turnover is also affected by the adult ECM. Recovery of vision in the deprived eye after long-term MD can be enhanced with chondroitinase treatment, an effect that is accompanied by a gain in spine density on pyramidal neurons (Pizzorusso et al. 2006). In mice, a protein that degrades the ECM, tissue plasminogen activator (tPA), is increased during MD and essential for OD plasticity (Mataga et al. 2002). Although inhibiting tPA/Plasmin activity in kittens does not alter the OD shift after MD, it does block recovery of vision in the deprived eye after reverse occlusion (Müller & Griesinger 1998). Plasmin activity increases spine motility in slices (Oray et al. 2004), and in tPA-deficient mice, the spine loss normally observed after three days of MD during the critical period is reduced (Mataga et al. 2004).

Interestingly, the receptors for myelin-associated growth inhibitors, Nogo-66 receptor (McGee et al. 2005) and PirB (paired immunoglobulin-like receptor B) (Syken et al. 2006, Atwal et al. 2008), are also implicated in closure of the critical period, and in mice deficient for either receptor, OD plasticity can be readily induced in adulthood. At first glance, these experiments seem to provide convincing evidence for the idea that the inhibition of structural plasticity through changes in the ECM and myelin-based factors determines the closure of the critical period. However, the ECM and the signaling molecules bound to it as well as myelin-bound growth inhibitors and their receptors (Raiker et al. 2010, Delekate et al. 2011) also affect long-term potentiation and long-term depression (LTD) (Dityatev et al. 2010, Klein 2009, Pang et al. 2004). This suggests that in the adult brain these signaling pathways have roles other than mediating growth-cone collapse and may regulate the rules of synaptic plasticity.

Inhibitory Innervation and Closure of the Critical Period

Recent evidence suggests that degrading the ECM may in fact reactivate critical-period plasticity through another pathway, namely by altering the function of PV+ interneurons. These interneurons are surrounded by dense ECM structures known as perineuronal nets, whose formation depends on cartilage link protein 1 (Crtl1) (Carulli et al. 2006). In mice deficient for this protein, OD plasticity can be readily induced during adulthood (Carulli et al. 2010). As perineuronal nets are crucial for the uptake of the transcription factor OTX2, chondroitinase treatment or Crtl1 deficiency may alter the function of PV+ interneurons owing to reduced OTX2 uptake. Although this does not exclude the possibility that regulation of structural plasticity through guidance molecules or changes in the ECM is an important factor in closing the critical period for OD plasticity, it does call for the careful interpretation of these findings. For example, reducing GABAergic input also reduces spine loss during OD plasticity

LTD: long-term depression

Cannabinoid receptor (CB1R): present on presynaptic terminals of inhibitory (as well as excitatory) neurons, providing negative feedback resulting in reduced vesicle release

cAMP responsive element binding protein (CREB): transcription factor important for long-term memory storage

by postponing the critical period (Mataga et al. 2004).

Further evidence also indicates that mature inhibitory innervation is involved in closing the critical period. After rats are monocularly deprived during the critical period, deprived eye vision can be recovered in adulthood if the animals are housed in an enriched environment. This recovery is accompanied by a reduction of GABAergic inhibition (Baroncelli et al. 2010) and can be blocked with benzodiazepine treatment, suggesting a causative relationship. More direct evidence comes from the observation that pharmacologically reducing inhibition in the visual cortex facilitates adult OD plasticity (Harauzov et al. 2010). The level of inhibition may control critical-period closure, but recent studies suggest that additional developmental changes in inhibitory synapse physiology are required. When inhibitory neurons are isolated from 12–16-day-old embryos and transplanted into the visual cortex of mice several days older, a second critical period for OD plasticity occurs 30 days after transplantation (Southwell et al. 2010). Because it seems unlikely that the transplanted interneurons decrease overall inhibition (unless they specifically innervate other interneurons), this finding suggests that the maturation state of inhibitory inputs affects the potential for OD plasticity.

Further support comes from recent findings on the role of cannabinoid receptor 1 (CB1R) in mediating LTD of inhibitory synapses and regulating critical-period plasticity. Both cholecystokinin- and PV+ interneurons respond to CB1R stimulation (Jiang et al. 2010). During the critical period, CB1R-dependent LTD of inhibitory synapses (iLTD) can be induced in layer 2/3 of the visual cortex (Jiang et al. 2010). This results in less GABA release but also reduces the potential of inhibitory synapses to undergo short-term plasticity. In adult mice, inhibitory synapses do not show CB1R-mediated changes but behave in a similar fashion as CB1R-stimulated juvenile synapses. This suggests that CB1R stimulation during development results in the maturation of inhibitory synapses, reducing their efficacy but making them more suitable for fast transmission with less depression. Interestingly, when mice are dark reared and even when adult mice are housed in the dark for a period of two weeks, inhibitory synapses become sensitive once again to CB1R stimulation. These manipulations also enable OD plasticity to be induced in adult animals. When CB1R agonists or benzodiazepines are given to animals housed in the dark, both, iLTD and OD plasticity are prevented. These data thus support the idea that closure of the critical period is due to functional changes in the speed of inhibitory synapses or their potential to undergo plasticity. However, because excitatory synapses in the pyramidal layers also show CB1R-dependent LTD (Liu et al. 2008), the observed correlation between iLTD and OD plasticity is not necessarily causal. Interestingly, astrocytes also carry CB1R, and endocannabinoids released by active neurons cause astrocytes to stimulate glutamate release from the neurons they contact (Navarrete & Araque 2010). Thus, the transplantion of immature astrocytes into the adult visual cortex of cats that activates OD plasticity (Müller & Best 1989) may also act through altered CB1R-mediated synaptic physiology.

Epigenetic Regulation of CREB-Mediated Transcription

Another pathway that has been implicated in the closure of the critical period is the epigenetic regulation of cAMP responsive element binding protein (CREB)-mediated gene transcription. MD or brief visual stimulation during the critical period rapidly upregulates CREB phosphorylation and CREB-mediated gene transcription (Pham et al. 1999, Putignano et al. 2007), but this response is much weaker in adult animals. The functional importance of CREB activity is illustrated by the finding that overexpression of dominant-negative CREB in V1 using viral vectors blocks OD plasticity (Mower et al. 2002), whereas expression of a constitutively active form in adult V1 stimulates plasticity (Pham et al. 2004). Genetic or pharmacological interference with the upstream signals also results in inhibition of OD plasticity. An important regulator of CREB activity is cAMP-dependent protein kinase A (PKA), which is activated though various G protein–coupled receptors including metabotropic glutamate receptors. Genetic inactivation of the regulatory PKA subunits RIIb or RIIa (Fischer et al. 2004, Rao et al. 2004) or pharmacological inhibition of PKA activity (Beaver et al. 2001) results in reduced OD plasticity. Interestingly, activating this signaling pathway in adult cats through various pharmacological interventions stimulates adult plasticity (Imamura et al. 1999). Together, these studies suggest that the reduction of CREB-mediated gene transcription closes the critical period. This seems to take place at several levels. Both basal cAMP levels and cAMP levels induced through type II metabotropic glutamate receptors are higher during the critical period than they are before or after, and dark rearing delays the age at which basal cAMP levels are highest (Reid et al. 1996). A recent proteomics study also showed that dark-rearing mice into adulthood increases various proteins in this pathway, including several PKA subunits, an anchor protein, and G proteins involved in cAMP production (Dahlhaus et al. 2011).

More importantly, certain epigenetic regulatory mechanisms limit CREB-mediated transcription and OD plasticity. For transcription to occur, changes in the organization of chromatin have to take place by the acetylation, methylation, phosphorylation, or SUMOylation of histones. Studies employing microarrays have found that only a small fraction (1–7%) of genes is regulated through histone acetylation (Fass et al. 2003, Van et al. 1996) and CRE-regulated genes are highly enriched in this fraction. These epigenetic mechanisms of gene regulation depend on mitogen-activated protein kinase, which is activated among others by PKA and BDNF. Epigenic gene regulation is also highly active following visual stimulation during the critical period but is strongly diminished before and after (Medini & Pizzorusso 2008). The functional implications of these findings have been tested by treating adult mice with Trichostatin, an inhibitor of enzymes involved in the deacteylation of histones. This fully restored the potential for OD plasticity to occur upon MD (Putignano et al. 2007). Recovery of the visual acuity of rats that underwent long-term MD during the critical period was also much more effective in adulthood if they were treated with Valproic acid or sodium butyrate, two drugs that reduce histone deactelyation (Silingardi et al. 2010). These findings indicate that epigenetic regulation of gene transcription is directly involved in mediating plastic changes.

An important question regards which CRE-regulated genes mediate OD plasticity. One interesting candidate is miR-132, a CRE-regulated microRNA whose visual experience-induced expression depends on epigenetic regulation. Reducing or increasing miR-132 activity inhibits OD plasticity, indicating that regulation of miR-132 is important for OD plasticity (Mellios et al. 2011, Tognini et al. 2011). This may involve the transcriptional control of the p250GAP gene, which encodes a protein involved in spine formation and stability (Impey et al. 2010). By changing inhibition, CREB-regulated gene transcription may also alter critical-period plasticity. A recent study showed that GAD65 expression in interneurons is upregulated by BDNF through CREB-mediated transcription (Sanchez-Huertas & Rico 2011). Thus, epigenetic regulation of CREB transcription may also affect inhibitory synaptic transmission, which provides at least one factor by which the different pathways may converge.

Neuromodulatory Inputs Can Increase Adult Ocular Dominance Plasticity

More recently, researchers discovered that manipulating neuromodulatory innervation in the visual cortex also increases adult visual plasticity. When rats were treated with fluoxetine, a serotonin reuptake inhibitor, vision after a prolonged period of MD was recovered more effectively than in control animals (Maya Vetencourt et al. 2008). Because serotonergic input is essential for OD plasticity to occur during the critical period (Gu & Singer 1995), its reduction may make the visual cortex less plastic with development. Also, increased cholinergic innervation enhances adult OD plasticity. In mice deficient for lynx1, an inhibitor of the nicotinic acetylcholine receptor, OD plasticity could be induced by a short period of MD in adult mice (Morishita et al. 2010). To explain these findings, both studies (Gu & Singer 1995, Morishita et al. 2010) suggest that the excitatory/inhibitory balance is altered under these conditions.

In conclusion, different cellular events inhibit adult plasticity. These include increased inhibitory input, reduced epigenetic regulation of CREB-mediated transcription, development of the ECM, signals directly or indirectly inhibiting structural plasticity, and decreased neuromodulatory activity. How reversal of any of these events can result in almost full restoration of OD plasticity in adult animals is not well understood. One possibility is that all these events impinge on the same mechanism, for example, a change in the excitation/inhibition balance. However, not all the mechanisms that reactivate plasticity in the adult mouse may be involved in closing the critical period and the restoration of plasticity may not be as complete as it seems if analyzed in more detail.

ADULT OCULAR DOMINANCE PLASTICITY

Although the critical period for OD plasticity is well defined and regulated, plasticity can still be induced to a certain degree during early adulthood. With the increased use of mice to study the mechanisms of OD plasticity this has become especially clear because mice show more adult plasticity than other species including other rodents such as rats (Antonini et al. 1999; Fagiolini et al. 1994; Pizzorusso et al. 2002; Sato & Stryker 2008, 2010). It is now clear that adult OD plasticity in mice can be measured using recordings of visually evoked potentials (Pham et al. 2004, Sawtell et al. 2003), in situ hybridization for the immediate early gene Arc (Tagawa et al. 2005), optical imaging of intrinsic signals (Heimel et al. 2007, 2010; Hofer et al. 2006; Lehmann & Löwel 2008; Sato & Stryker 2008), single-unit recordings (Antonini et al. 1999, Hofer et al. 2006) and in vivo two-photon imaging of calcium responses (Kameyama et al. 2010). However, divergent results have been obtained depending on the conditions under which OD plasticity was induced or assessed. For example, OD plasticity requires a longer period of MD in young adult versus juvenile mice. Whereas a clear shift of OD can be observed in juvenile mice within 3 days or less, adult mice require 4 (Pham et al. 2004, Tagawa et al. 2005) or 5 days (Sawtell et al. 2003) of MD for the OD shift to become significant. In other species such as cats (Blakemore et al. 1978, Jones et al. 1984), ferrets (Issa et al. 1999), and monkeys (Blakemore et al. 1978), an even longer period, ranging from 1 week to 4 weeks of MD, is necessary to induce adult plasticity.

Another factor influencing the detectability of adult plasticity in mice is the anesthetic used during the measurement of OD. In almost all studies in which an OD shift was observed in adult mice after 4–5 days of MD, halothane, urethane, or no anesthesia was used (Fischer et al. 2007; Heimel et al. 2007, 2010; Hofer et al. 2009; Pham et al. 2004; Sawtell et al. 2003; Tagawa et al. 2005). By contrast, when researchers used barbiturates or benzodiazepines an OD shift was never observed following 4–5 days of MD (Fagiolini & Hensch 2000, Pham et al. 2004) and, in most cases, not even after 7 days (Fagiolini & Hensch 2000, Heimel et al. 2007, Sato & Stryker 2008). An interesting interpretation of these findings, which is also supported by other studies, is that drugs directly acting on the GABAergic system such as barbiturates and benzodiazepines affect the readout of OD (Duffy et al. 1976, Sillito et al. 1981, Yazaki-Sugiyama et al. 2009).

In addition to the experimental conditions under which OD plasticity is induced or measured, the genetic background of the mouse strains used by the different laboratories may also affect the observed levels of adult OD plasticity. Support for the influence of genetic influence on OD plasticity during the critical period comes from a study making use of a set of recombinant inbred strains created by inbreeding offspring of C57bl/6J and DBA mice (Heimel et al. 2008). In these mouse strains, the genetic background strongly influences the magnitude of the OD shift and the extent to which this involves depression of deprived-eye responses and enhancement of open-eye responses. More recently, researchers found that even C57bl/6 mice derived from different breeders, e.g., Jackson (C57BL/6J) or Harlan (C57bl/6OlaHsd), also show such differences (Ranson et al. 2012), underscoring the importance of genetic standardization in studies of the mechanisms underlying OD plasticity.

Differences Between Juvenile and Adult Plasticity

Thus, in mice, and to a lesser extent in other species, OD plasticity can be induced during early adulthood. There are, however, clear differences between adult and juvenile OD plasticity. One such difference is that the OD shift in adults is smaller than that observed in juvenile mice. There is also a clear difference in how deprived- and nondeprived-eye responses change in the visual cortex of young and adult mice. In young mice, a rapid loss in deprived-eye responses is followed by a delayed increase in open-eye response strength (Frenkel & Bear 2004, Mrsic-Flogel et al. 2007). In contrast, in adult mice, the dominant change is an increase in open-eye response strength (Hofer et al. 2006, Sawtell et al. 2003). These results were essentially confirmed by a follow-up study directly comparing OD plasticity in juvenile and adult mice (Sato & Stryker 2008). This study also found that the ipsilateral visual cortex showed no changes following a prolonged period of MD in adults. Other studies, however, found reduced deprived-eye responses in the adult ipsilateral visual cortex. This apparent discrepancy may be due to the use of different (Hofer et al. 2006) or no (Tagawa et al. 2005) anesthetics.

The degree of adult OD plasticity also differs between cortical layers: In young adult cats, plasticity occurs only in the pyramidal layers and is absent in layer 4 (Daw et al. 1992). Whether this is also the case in rodents has not been investigated. Additionally, adult OD plasticity is less permanent than is juvenile OD plasticity. A study showed that, when mice are monocularly deprived for 13 days during the critical period, the change in visual acuity can still be observed in a behavioral task 40 days after reopening the eye. By contrast, after 18 days of MD started at p33, binocular vision was completely restored 20 days after the eye was reopened (Prusky & Douglas 2003).

Mechanisms Underlying Juvenile and Adult Ocular Dominance Plasticity

What do the differences between juvenile and adult plasticity tell us about the underlying mechanisms? As discussed above, one important change after the critical period is the reduction of structural plasticity. MD in cats (Antonini & Stryker 1996, Shatz & Stryker 1978) or monkeys (Hubel et al. 1977) causes retraction of thalamocortical projections serving the deprived eye and the growth of those serving the nondeprived eye during the critical period, but these changes are not observed in adult animals. On the postsynaptic side, structural plasticity also diminishes after the critical period. An in vivo two-photon imaging study in mice showed that dendritic-spine turnover tapers off with age (Holtmaat et al. 2006). This may be related to the reduced plasticity potential of the maturing visual cortex: While MD during the critical period results in a reduction of the spine density of the apical dendrites of layer 3 pyramidal neurons, adult mice do not exhibit this spine loss (Mataga et al. 2004). As a consequence, adult plasticity may be limited to local structural reorganization such as changes in synapse strength or formation and loss of synapses with direct neighbors (Hofer et al. 2009), whereas the growth and retraction of axons necessary to reach more distal neurons are reduced.

Various studies on adult OD plasticity support this notion. In animals in which V1 is organized into OD columns, a shift in OD in layer IV depends on the growth and retraction of the thalamocortical projection toward or away from columns serving the other eye. In contrast, the axons of pyramidal neurons in the extragranular layers project horizontally and pass through columns dominated by either eye, and local synapse turnover may be sufficient to achieve changes in OD. Accordingly, reduced axon growth and retraction could diminish plasticity in layer 4 more strongly than in the pyramidal layers. As in rodents the thalamocortical projections to layer 4 are not organized in OD columns, testing whether adult plasticity occurs in layer 4 in rodents would be interesting.

Very extensive structural rearrangements can occur in the adult visual cortex when retinal lesions are made that cause the visual cortex (the lesion projection zone) to receive no visual input. In monkeys, this leads to strongly increased dynamics of axon growth and retraction and synaptic bouton turnover (Yamahachi et al. 2009). In mice, a massive increase in spine turnover occurs in the lesion projection zone, ultimately leading to the replacement of almost all the spines (Keck et al. 2008). Interestingly, the initial change that occurs after retinal lesions is a strong reduction of inhibition (Mittmann et al. 1994). In vivo two-photon imaging has demonstrated that this occurs by structural changes of synaptic inputs and outputs of GABAergic neurons (Keck et al. 2011). For adult plasticity to occur, a reduction of inhibitory inputs caused by prolonged and/or severe reductions or changes in visual input may thus be a first requirement. Whether such a reduction of inhibition also occurs with OD plasticity is not yet clear, but an in vivo two-photon imaging study of structural changes of interneurons is compatible with this idea (Chen et al. 2011).

Altogether, these studies make clear that plasticity in V1 can still be induced after the critical period, albeit less effectively and less permanently. It remains to be investigated whether juvenile and adult plasticity involve the same underlying mechanisms. Although both forms of plasticity are NMDAR dependent (Sato & Stryker 2008, Sawtell et al. 2003), dependence on structural plasticity may differ strongly. It is also unclear whether these forms of plasticity serve the same function or whether adult plasticity is a more temporary adjustment to changes in environmental or sensory changes occurring in life.

OUTLOOK

Employing OD plasticity in the rodent visual cortex as a model system has turned out to be crucial for investigating the regulation of critical-period timing. Studies in transgenic mice and pharmacological approaches have helped to identify key players and candidate molecular mechanisms involved in controlling the opening and closing of the critical period. However, there are still gaps in our knowledge of the chain of events linking the onset of vision to the strongly enhanced susceptibility of the visual cortex to undergo a shift in eye preference several weeks later.

For example, how does visually driven activity in the retino-geniculo-cortical pathway interact with the transcription factor OTX2 (Sugiyama et al. 2008) to promote the maturation of inhibition in the visual cortex? Is transported OTX2 alone sufficient to induce this process? While there is little doubt that the maturation of inhibitory circuits and PV+ interneurons in particular are important for critical period onset (Fagiolini & Hensch 2000, Fagiolini et al. 2004, Hensch et al. 1998), how exactly these neurons take part in OD plasticity is not well understood. In part, this is due to conflicting results obtained regarding how the eye-specific responses of inhibitory interneurons change after MD (Gandhi et al. 2008, Kameyama et al. 2010, Mainardi et al. 2009, Yazaki-Sugiyama et al. 2009). Any mechanistic explanation of their role in OD plasticity has to take into account that the first change in the visual cortex after MD is a rapid drop in the deprived eye's ability to drive cortical cells, mediated by LTD (Rittenhouse et al. 1999, Yoon et al. 2009).

We are also lacking a clear picture of the mechanisms underlying closure of the critical period. Several factors limiting OD plasticity in adult animals have been identified, but it is not obvious which of these are involved in terminating the critical period and how they interact. For certain, limitations to large-scale structural rearrangements are one important factor, likely caused by changes in the ECM, but how these changes are triggered is not known.

Solving some of these questions will be greatly facilitated by recently developed tools that allow the structure and function of neurons to be imaged in vivo (Grewe & Helmchen 2009, Holtmaat & Svoboda 2009) and their activity to be manipulated with light (Fenno et al. 2011). For example, repeated two-photon imaging of specific subtypes of neurons expressing genetically encoded calcium indicators (Mank et al. 2008, Tian et al. 2009) will enable us to follow how excitatory and inhibitory neurons change their OD after MD. Performing such experiments in awake, behaving animals (Dombeck et al. 2007, Greenberg et al. 2008) will reveal how MD alters the activity levels of individual neurons in the visual cortex, which is currently not clear. This is all the more important as recent recordings from the lateral geniculate nucleus of awake mice indicate that activity in the visual pathway may change in unexpected ways following MD (Linden et al. 2009).

A more general problem for a broader understanding of critical-period regulation is that the vast majority of experiments on the underlying molecular and cellular mechanisms have been carried out in one system, the rodent visual cortex, using one specific manipulation of sensory input, MD. Studies in other species and systems have shown that some of the basic mechanisms involved in critical-period plasticity are similar to those in the visual cortex. For example, synaptic strengthening and weakening via NMDA receptors as well as structural modifications are also hallmarks of critical periods in song learning in birds (Aamodt et al. 1996, Roberts et al. 2010) and in the barn owl's auditory localization system (Feldman & Knudsen 1997, Feldman et al. 1996). However, whether the same factors that control the opening and closing of the critical period for OD plasticity in the visual cortex are also at work in these systems is not known. Thus, although unresolved questions regarding critical-period regulation in the visual cortex remain, it will also be important to broaden research into other systems (Barkat et al. 2011) to understand general principles as well as specific differences.

This is all the more important because a comprehensive understanding of critical period mechanisms may also be relevant for clinical reasons. Several neuropsychiatric disorders such as schizophrenia, Rett syndrome, and autism have been linked to alterations in cortical excitatory/inhibitory balance (Chao et al. 2010, Dani et al. 2005, Kehrer et al. 2008, Rubenstein & Merzenich 2003, Yizhar et al. 2011), which, as laid out above, plays a crucial role in critical-period regulation. It is conceivable that part of the pathology underlying these diseases is caused by errors in the duration or timing of critical periods in certain brain regions (Leblanc & Fagiolini 2011). Finally, recovery of neuronal function after trauma, stroke, or other insults to brain tissue is typically hampered by the adult brain's limited capacity for plasticity. Reopening of critical periods either by pharmacological means (Maya Vetencourt et al. 2008) or by specific deprivation or training regimes (Hofer et al. 2006, Sale et al. 2007) may promote circuit reorganization that can eventually restore brain function.

DISCLOSURE STATEMENT

The authors are not aware of any affiliations, memberships, funding, or financial holdings that might be perceived as affecting the objectivity of this review.

ACKNOWLEDGMENTS

We thank Robert Schorner for assistance with figure preparation. C.N.L. is funded by the Netherlands Organization for Scientific Research and by a grant from AgentschapNL to the NeuroBasic PharmaPhenomics consortium; M.H. is funded by the Max Planck Society. C.N.L. and M.H. receive funding from the European Community's Seventh Framework Program (FP2007–2013) under grant agreement number 223326.

LITERATURE CITED

Aamodt SM, Nordeen EJ, Nordeen KW. 1996. Blockade of NMDA receptors during song model exposure impairs song development in juvenile zebra finches. *Neurobiol. Learn. Mem.* 65(1):91–98

Antonini A, Fagiolini M, Stryker MP. 1999. Anatomical correlates of functional plasticity in mouse visual cortex. *J. Neurosci.* 19(11):4388–406

Antonini A, Stryker MP. 1996. Plasticity of geniculocortical afferents following brief or prolonged monocular occlusion in the cat. *J. Comp Neurol.* 369(1):64–82

Atwal JK, Pinkston-Gosse J, Syken J, Stawicki S, Wu Y, et al. 2008. PirB is a functional receptor for myelin inhibitors of axonal regeneration. *Science* 322(5903):967–70

Barkat TR, Polley DB, Hensch TK. 2011. A critical period for auditory thalamocortical connectivity. *Nat. Neurosci.* 14(9):1189–94

Baroncelli L, Sale A, Viegi A, Maya Vetencourt JF, De PR, et al. 2010. Experience-dependent reactivation of ocular dominance plasticity in the adult visual cortex. *Exp. Neurol.* 226(1):100–9

Beaver CJ, Ji Q, Fischer QS, Daw NW. 2001. Cyclic AMP-dependent protein kinase mediates ocular dominance shifts in cat visual cortex. *Nat. Neurosci.* 4(2):159–63

Blakemore C, Garey LJ, Vital-Durand F. 1978. The physiological effects of monocular deprivation and their reversal in the monkey's visual cortex. *J. Physiol.* 283:223–62

Brussaard AB, Kits KS, Baker RE, Willems WP, Leyting-Vermeulen JW, et al. 1997. Plasticity in fast synaptic inhibition of adult oxytocin neurons caused by switch in GABA(A) receptor subunit expression. *Neuron* 19(5):1103–14

Cardin JA, Carlen M, Meletis K, Knoblich U, Zhang F, et al. 2009. Driving fast-spiking cells induces gamma rhythm and controls sensory responses. *Nature* 459(7247):663–67

Carro E, Nunez A, Busiguina S, Torres-Aleman I. 2000. Circulating insulin-like growth factor I mediates effects of exercise on the brain. *J. Neurosci.* 20(8):2926–33

Carron C, Bourdelas A, Li HY, Boucaut JC, Shi DL. 2005. Antagonistic interaction between IGF and Wnt/JNK signaling in convergent extension in Xenopus embryo. *Mech. Dev.* 122(11):1234–47

Carulli D, Pizzorusso T, Kwok JC, Putignano E, Poli A, et al. 2010. Animals lacking link protein have attenuated perineuronal nets and persistent plasticity. *Brain* 133(Pt. 8):2331–47

Carulli D, Rhodes KE, Brown DJ, Bonnert TP, Pollack SJ, et al. 2006. Composition of perineuronal nets in the adult rat cerebellum and the cellular origin of their components. *J. Comp. Neurol.* 494(4): 559–77

Chang EF, Merzenich MM. 2003. Environmental noise retards auditory cortical development. *Science* 300(5618):498–502

Chao HT, Chen H, Samaco RC, Xue M, Chahrour M, et al. 2010. Dysfunction in GABA signalling mediates autism-like stereotypies and Rett syndrome phenotypes. *Nature* 468(7321):263–69

Chen JL, Lin WC, Cha JW, So PT, Kubota Y, Nedivi E. 2011. Structural basis for the role of inhibition in facilitating adult brain plasticity. *Nat. Neurosci.* 14(5):587–94

Cho KK, Khibnik L, Philpot BD, Bear MF. 2009. The ratio of NR2A/B NMDA receptor subunits determines the qualities of ocular dominance plasticity in visual cortex. *Proc. Natl. Acad. Sci. USA* 106(13): 5377–82

Chun JJ, Shatz CJ. 1989. Interstitial cells of the adult neocortical white matter are the remnant of the early generated subplate neuron population. *J. Comp. Neurol.* 282(4):555–69

Ciucci F, Putignano E, Baroncelli L, Landi S, Berardi N, Maffei L. 2007. Insulin-like growth factor 1 (IGF-1) mediates the effects of enriched environment (EE) on visual cortical development. *PLoS One* 2(5):e475

Dahlhaus M, Li KW, Van der Schors RC, Saiepour MH, van NP, et al. 2011. The synaptic proteome during development and plasticity of the mouse visual cortex. *Mol. Cell Proteomics* 10(5):M110

Dan Y, Poo MM. 2004. Spike timing–dependent plasticity of neural circuits. *Neuron* 44(1):23–30

Dani VS, Chang Q, Maffei A, Turrigiano GG, Jaenisch R, Nelson SB. 2005. Reduced cortical activity due to a shift in the balance between excitation and inhibition in a mouse model of Rett syndrome. *Proc. Natl. Acad. Sci. USA* 102(35):12560–65

Daw NW, Fox K, Sato H, Czepita D. 1992. Critical period for monocular deprivation in the cat visual cortex. *J. Neurophysiol.* 67(1):197–202

Delekate A, Zagrebelsky M, Kramer S, Schwab ME, Korte M. 2011. NogoA restricts synaptic plasticity in the adult hippocampus on a fast time scale. *Proc. Natl. Acad. Sci. USA* 108(6):2569–74

Di CG, Chattopadhyaya B, Kuhlman SJ, Fu Y, Belanger MC, et al. 2007. Activity-dependent PSA expression regulates inhibitory maturation and onset of critical period plasticity. *Nat. Neurosci.* 10(12): 1569–77

Dityatev A, Schachner M, Sonderegger P. 2010. The dual role of the extracellular matrix in synaptic plasticity and homeostasis. *Nat. Rev. Neurosci.* 11(11):735–46

Dombeck DA, Khabbaz AN, Collman F, Adelman TL, Tank DW. 2007. Imaging large-scale neural activity with cellular resolution in awake, mobile mice. *Neuron* 56(1):43–57

Dräger UC. 1978. Observations on monocular deprivation in mice. *J. Neurophysiol.* 41(1):28–42

Duffy FH, Burchfiel JL, Conway JL. 1976. Bicuculline reversal of deprivation amblyopia in the cat. *Nature* 260(5548):256–57

Fagiolini M, Fritschy JM, Low K, Mohler H, Rudolph U, Hensch TK. 2004. Specific GABAA circuits for visual cortical plasticity. *Science* 303(5664):1681–83

Fagiolini M, Hensch TK. 2000. Inhibitory threshold for critical-period activation in primary visual cortex. *Nature* 404(6774):183–86

Fagiolini M, Katagiri H, Miyamoto H, Mori H, Grant SG, et al. 2003. Separable features of visual cortical plasticity revealed by N-methyl-D-aspartate receptor 2A signaling. *Proc. Natl. Acad. Sci. USA* 100(5):2854–59

Fagiolini M, Pizzorusso T, Berardi N, Domenici L, Maffei L. 1994. Functional postnatal development of the rat primary visual cortex and the role of visual experience: dark rearing and monocular deprivation. *Vis. Res.* 34(6):709–20

Fass DM, Butler JE, Goodman RH. 2003. Deacetylase activity is required for cAMP activation of a subset of CREB target genes. *J. Biol. Chem.* 278(44):43014–19

Feldman DE, Brainard MS, Knudsen EI. 1996. Newly learned auditory responses mediated by NMDA receptors in the owl inferior colliculus. *Science* 271(5248):525–28

Feldman DE, Knudsen EI. 1997. An anatomical basis for visual calibration of the auditory space map in the barn owl's midbrain. *J. Neurosci.* 17(17):6820–37

Fenno L, Yizhar O, Deisseroth K. 2011. The development and application of optogenetics. *Annu. Rev. Neurosci.* 34:389–412

Fischer QS, Beaver CJ, Yang Y, Rao Y, Jakobsdottir KB, et al. 2004. Requirement for the RIIbeta isoform of PKA, but not calcium-stimulated adenylyl cyclase, in visual cortical plasticity. *J. Neurosci.* 24(41):9049–58

Fischer QS, Graves A, Evans S, Lickey ME, Pham TA. 2007. Monocular deprivation in adult mice alters visual acuity and single-unit activity. *Learn. Mem.* 14(4):277–86

Frenkel MY, Bear MF. 2004. How monocular deprivation shifts ocular dominance in visual cortex of young mice. *Neuron* 44(6):917–23

Friauf E, McConnell SK, Shatz CJ. 1990. Functional synaptic circuits in the subplate during fetal and early postnatal development of cat visual cortex. *J. Neurosci.* 10(8):2601–13

Fries P, Nikolic D, Singer W. 2007. The gamma cycle. *Trends Neurosci.* 30(7):309–16

Galarreta M, Hestrin S. 1999. A network of fast-spiking cells in the neocortex connected by electrical synapses. *Nature* 402(6757):72–75

Galarreta M, Hestrin S. 2002. Electrical and chemical synapses among parvalbumin fast-spiking GABAergic interneurons in adult mouse neocortex. *Proc. Natl. Acad. Sci. USA* 99(19):12438–43

Gandhi SP, Yanagawa Y, Stryker MP. 2008. Delayed plasticity of inhibitory neurons in developing visual cortex. *Proc. Natl. Acad. Sci. USA* 105(43):16797–802

Ghosh A, Shatz CJ. 1992. Involvement of subplate neurons in the formation of ocular dominance columns. *Science* 255(5050):1441–43

Gianfranceschi L, Siciliano R, Walls J, Morales B, Kirkwood A, et al. 2003. Visual cortex is rescued from the effects of dark rearing by overexpression of BDNF. *Proc. Natl. Acad. Sci. USA* 100(21):12486–91

Gordon JA, Stryker MP. 1996. Experience-dependent plasticity of binocular responses in the primary visual cortex of the mouse. *J. Neurosci.* 16(10):3274–86

Greenberg DS, Houweling AR, Kerr JN. 2008. Population imaging of ongoing neuronal activity in the visual cortex of awake rats. *Nat. Neurosci.* 11(7):749–51

Grewe BF, Helmchen F. 2009. Optical probing of neuronal ensemble activity. *Curr. Opin. Neurobiol.* 19(5):520–29

Gu Q, Singer W. 1995. Involvement of serotonin in developmental plasticity of kitten visual cortex. *Eur. J. Neurosci.* 7(6):1146–53

Hanover JL, Huang ZJ, Tonegawa S, Stryker MP. 1999. Brain-derived neurotrophic factor overexpression induces precocious critical period in mouse visual cortex. *J. Neurosci.* 19(22):RC40

Harauzov A, Spolidoro M, DiCristo G, De PR, Cancedda L, et al. 2010. Reducing intracortical inhibition in the adult visual cortex promotes ocular dominance plasticity. *J. Neurosci.* 30(1):361–71

Hata Y, Tsumoto T, Stryker MP. 1999. Selective pruning of more active afferents when cat visual cortex is pharmacologically inhibited. *Neuron* 22(2):375–81

Heimel JA, Hartman RJ, Hermans JM, Levelt CN. 2007. Screening mouse vision with intrinsic signal optical imaging. *Eur. J. Neurosci.* 25(3):795–804

Heimel JA, Hermans JM, Sommeijer JP, Neuro-Bsik Mouse Phenomics Consortium, Levelt CN. 2008. Genetic control of experience-dependent plasticity in the visual cortex. *Genes Brain Behav.* 7(8): 915–23

Heimel JA, Saiepour MH, Chakravarthy S, Hermans JM, Levelt CN. 2010. Contrast gain control and cortical TrkB signaling shape visual acuity. *Nat. Neurosci.* 13(5):642–48

Heinroth O. 1910. Beiträge zur Biologie, namentlich Ethologie und Psychologie der Anatiden. In *Verhandlungen des V. Internationalen Ornitologen-Kongresses, Berlin*, ed. H Schalow, pp. 589–702. Berlin: Deutsche Ornithologische Gesellschaft

Hensch TK, Fagiolini M, Mataga N, Stryker MP, Baekkeskov S, Kash SF. 1998. Local GABA circuit control of experience-dependent plasticity in developing visual cortex. *Science* 282(5393):1504–8

Hofer SB, Mrsic-Flogel TD, Bonhoeffer T, Hübener M. 2006. Prior experience enhances plasticity in adult visual cortex. *Nat. Neurosci.* 9(1):127–32

Hofer SB, Mrsic-Flogel TD, Bonhoeffer T, Hübener M. 2009. Experience leaves a lasting structural trace in cortical circuits. *Nature* 457(7227):313–17

Holtmaat A, Svoboda K. 2009. Experience-dependent structural synaptic plasticity in the mammalian brain. *Nat. Rev. Neurosci.* 10(9):647–58

Holtmaat A, Wilbrecht L, Knott GW, Welker E, Svoboda K. 2006. Experience-dependent and cell-type-specific spine growth in the neocortex. *Nature* 441(7096):979–83

Huang S, Gu Y, Quinlan EM, Kirkwood A. 2010. A refractory period for rejuvenating GABAergic synaptic transmission and ocular dominance plasticity with dark exposure. *J. Neurosci.* 30(49):16636–42

Huang ZJ, Kirkwood A, Pizzorusso T, Porciatti V, Morales B, et al. 1999. BDNF regulates the maturation of inhibition and the critical period of plasticity in mouse visual cortex. *Cell* 98(6):739–55

Hubel DH, Wiesel TN. 1962. Receptive fields, binocular interaction and functional architecture in the cat's visual cortex. *J. Physiol.* 160:106–54

Hubel DH, Wiesel TN. 1970. The period of susceptibility to the physiological effects of unilateral eye closure in kittens. *J. Physiol.* 206:419–36

Hubel DH, Wiesel TN, LeVay S. 1977. Plasticity of ocular dominance columns in monkey striate cortex. *Philos. Trans. R. Soc. Lond. Ser. B* 278(961):377–409

Imamura K, Kasamatsu T, Shirokawa T, Ohashi T. 1999. Restoration of ocular dominance plasticity mediated by adenosine 3′,5′-monophosphate in adult visual cortex. *Proc. Biol. Sci.* 266(1428):1507–16

Impey S, Davare M, Lasiek A, Fortin D, Ando H, et al. 2010. An activity-induced microRNA controls dendritic spine formation by regulating Rac1-PAK signaling. *Mol. Cell Neurosci.* 43(1):146–56

Issa NP, Trachtenberg JT, Chapman B, Zahs KR, Stryker MP. 1999. The critical period for ocular dominance plasticity in the Ferret's visual cortex. *J. Neurosci.* 19(16):6965–78

Jiang B, Huang S, De PR, Millman D, Song L, et al. 2010. The maturation of GABAergic transmission in visual cortex requires endocannabinoid-mediated LTD of inhibitory inputs during a critical period. *Neuron* 66(2):248–59

Joliot AH, Triller A, Volovitch M, Pernelle C, Prochiantz A. 1991. alpha-2,8-polysialic acid is the neuronal surface receptor of antennapedia homeobox peptide. *New Biol.* 3(11):1121–34

Jones KR, Spear PD, Tong L. 1984. Critical periods for effects of monocular deprivation: differences between striate and extrastriate cortex. *J. Neurosci.* 4(10):2543–52

Kameyama K, Sohya K, Ebina T, Fukuda A, Yanagawa Y, Tsumoto T. 2010. Difference in binocularity and ocular dominance plasticity between GABAergic and excitatory cortical neurons. *J. Neurosci.* 30(4):1551–59

Kanold PO, Shatz CJ. 2006. Subplate neurons regulate maturation of cortical inhibition and outcome of ocular dominance plasticity. *Neuron* 51(5):627–38

Keck T, Mrsic-Flogel TD, Vaz AM, Eysel UT, Bonhoeffer T, Hübener M. 2008. Massive restructuring of neuronal circuits during functional reorganization of adult visual cortex. *Nat. Neurosci.* 11(10): 1162–67

Keck T, Scheuss V, Jacobsen RI, Wierenga CJ, Eysel UT, et al. 2011. Loss of sensory input causes rapid structural changes of inhibitory neurons in adult mouse visual cortex. *Neuron* 71(5):869–82

Kehrer C, Maziashvili N, Dugladze T, Gloveli T. 2008. Altered excitatory-inhibitory balance in the NMDA-hypofunction model of schizophrenia. *Front. Mol. Neurosci.* 1:6

Klausberger T, Roberts JD, Somogyi P. 2002. Cell type– and input-specific differences in the number and subtypes of synaptic GABA(A) receptors in the hippocampus. *J. Neurosci.* 22(7):2513–21

Klein R. 2009. Bidirectional modulation of synaptic functions by Eph/ephrin signaling. *Nat. Neurosci.* 12(1):15–20

Klopfer PH, Adams DK, Klopfer MS. 1964. Maternal "imprinting" in goats. *Proc. Natl. Acad. Sci. USA* 52:911–14

Kuhlman SJ, Tring E, Trachtenberg JT. 2011. Fast-spiking interneurons have an initial orientation bias that is lost with vision. *Nat. Neurosci.* 14(9):1121–23

Leblanc JJ, Fagiolini M. 2011. Autism: a "critical period" disorder? *Neural Plast.* 2011:921680

Lehmann K, Löwel S. 2008. Age-dependent ocular dominance plasticity in adult mice. *PLoS One* 3(9):e3120

Linden ML, Heynen AJ, Haslinger RH, Bear MF. 2009. Thalamic activity that drives visual cortical plasticity. *Nat. Neurosci.* 12(4):390–92

Liu CH, Heynen AJ, Shuler MG, Bear MF. 2008. Cannabinoid receptor blockade reveals parallel plasticity mechanisms in different layers of mouse visual cortex. *Neuron* 58(3):340–45

Lorenz K. 1935. Der Kumpan in der Umwelt des Vogels. *J. Ornithol.* 83:137–213

Maffei A, Lambo ME, Turrigiano GG. 2010. Critical period for inhibitory plasticity in rodent binocular V1. *J. Neurosci.* 30(9):3304–9

Mainardi M, Landi S, Berardi N, Maffei L, Pizzorusso T. 2009. Reduced responsiveness to long-term monocular deprivation of parvalbumin neurons assessed by c-Fos staining in rat visual cortex. *PLoS One* 4(2): e4342

Mank M, Santos AF, Direnberger S, Mrsic-Flogel TD, Hofer SB, et al. 2008. A genetically encoded calcium indicator for chronic in vivo two-photon imaging. *Nat. Methods* 5(9):805–11

Mataga N, Mizuguchi Y, Hensch TK. 2004. Experience-dependent pruning of dendritic spines in visual cortex by tissue plasminogen activator. *Neuron* 44(6):1031–41

Mataga N, Nagai N, Hensch TK. 2002. Permissive proteolytic activity for visual cortical plasticity. *Proc. Natl. Acad. Sci. USA* 99(11):7717–21

Maya Vetencourt JF, Sale A, Viegi A, Baroncelli L, De PR, et al. 2008. The antidepressant fluoxetine restores plasticity in the adult visual cortex. *Science* 320(5874):385–88

McGee AW, Yang Y, Fischer QS, Daw NW, Strittmatter SM. 2005. Experience-driven plasticity of visual cortex limited by myelin and Nogo receptor. *Science* 309(5744):2222–26

Medini P, Pizzorusso T. 2008. Visual experience and plasticity of the visual cortex: a role for epigenetic mechanisms. *Front. Biosci.* 13:3000–7

Mellios N, Sugihara H, Castro J, Banerjee A, Le C, et al. 2011. miR-132, an experience-dependent microRNA, is essential for visual cortex plasticity. *Nat. Neurosci.* 14:1240–42

Mittmann T, Luhmann HJ, Schmidt-Kastner R, Eysel UT, Weigel H, Heinemann U. 1994. Lesion-induced transient suppression of inhibitory function in rat neocortex in vitro. *Neuroscience* 60(4):891–906

Moon LD, Asher RA, Rhodes KE, Fawcett JW. 2001. Regeneration of CNS axons back to their target following treatment of adult rat brain with chondroitinase ABC. *Nat. Neurosci.* 4(5):465–66

Morishita H, Miwa JM, Heintz N, Hensch TK. 2010. Lynx1, a cholinergic brake, limits plasticity in adult visual cortex. *Science* 330(6008):1238–40

Mower AF, Liao DS, Nestler EJ, Neve RL, Ramoa AS. 2002. cAMP/Ca^{2+} response element-binding protein function is essential for ocular dominance plasticity. *J. Neurosci.* 22(6):2237–45

Mower GD. 1991. The effect of dark rearing on the time course of the critical period in cat visual cortex. *Brain Res. Dev. Brain Res.* 58(2):151–58

Mrsic-Flogel TD, Hofer SB, Ohki K, Reid RC, Bonhoeffer T, Hübener M. 2007. Homeostatic regulation of eye-specific responses in visual cortex during ocular dominance plasticity. *Neuron* 54(6):961–72

Müller CM, Best J. 1989. Ocular dominance plasticity in adult cat visual cortex after transplantation of cultured astrocytes. *Nature* 342(6248):427–30

Müller CM, Griesinger CB. 1998. Tissue plasminogen activator mediates reverse occlusion plasticity in visual cortex. *Nat. Neurosci.* 1(1):47–53

Navarrete M, Araque A. 2010. Endocannabinoids potentiate synaptic transmission through stimulation of astrocytes. *Neuron* 68(1):113–26

Nottebohm F, Nottebohm ME. 1978. Relationship between song repertoire and age in the canary, *Serinus canarius*. *Z. Tierpsychol.* 46:298–305

Nulty B, Burt JM, Akcay C, Templeton CN, Campbell E, Beecher MD. 2010. Song learning in song sparrows: relative importance of autumn versus spring tutoring. *Ethology* 116:653–61

Nusser Z, Sieghart W, Benke D, Fritschy JM, Somogyi P. 1996. Differential synaptic localization of two major gamma-aminobutyric acid type A receptor alpha subunits on hippocampal pyramidal cells. *Proc. Natl. Acad. Sci. USA* 93(21):11939–44

Nyiri G, Freund TF, Somogyi P. 2001. Input-dependent synaptic targeting of alpha(2)-subunit-containing GABA(A) receptors in synapses of hippocampal pyramidal cells of the rat. *Eur. J. Neurosci.* 13(3): 428–42

Oray S, Majewska A, Sur M. 2004. Dendritic spine dynamics are regulated by monocular deprivation and extracellular matrix degradation. *Neuron* 44(6):1021–30

Pang PT, Teng HK, Zaitsev E, Woo NT, Sakata K, et al. 2004. Cleavage of proBDNF by tPA/plasmin is essential for long-term hippocampal plasticity. *Science* 306(5695):487–91

Pham TA, Graham SJ, Suzuki S, Barco A, Kandel ER, et al. 2004. A semi-persistent adult ocular dominance plasticity in visual cortex is stabilized by activated CREB. *Learn. Mem.* 11(6):738–47

Pham TA, Impey S, Storm DR, Stryker MP. 1999. CRE-mediated gene transcription in neocortical neuronal plasticity during the developmental critical period. *Neuron* 22(1):63–72

Pizzorusso T, Medini P, Berardi N, Chierzi S, Fawcett JW, Maffei L. 2002. Reactivation of ocular dominance plasticity in the adult visual cortex. *Science* 298(5596):1248–51

Pizzorusso T, Medini P, Landi S, Baldini S, Berardi N, Maffei L. 2006. Structural and functional recovery from early monocular deprivation in adult rats. *Proc. Natl. Acad. Sci. USA* 103(22):8517–22

Pouille F, Scanziani M. 2001. Enforcement of temporal fidelity in pyramidal cells by somatic feed-forward inhibition. *Science* 293(5532):1159–63

Prusky GT, Douglas RM. 2003. Developmental plasticity of mouse visual acuity. *Eur. J. Neurosci.* 17(1):167–73

Putignano E, Lonetti G, Cancedda L, Ratto G, Costa M, et al. 2007. Developmental downregulation of histone posttranslational modifications regulates visual cortical plasticity. *Neuron* 53(5):747–59

Raiker SJ, Lee H, Baldwin KT, Duan Y, Shrager P, Giger RJ. 2010. Oligodendrocyte-myelin glycoprotein and Nogo negatively regulate activity-dependent synaptic plasticity. *J. Neurosci.* 30(37):12432–45

Ranson A, Cheetham CE, Fox K, Sengpiel F. 2012. Homeostatic plasticity mechanisms are required for juvenile, but not adult, ocular dominance plasticity. *Proc. Natl. Acad. Sci. USA* 109(4):1311–16

Rao Y, Fischer QS, Yang Y, McKnight GS, LaRue A, Daw NW. 2004. Reduced ocular dominance plasticity and long-term potentiation in the developing visual cortex of protein kinase A RII alpha mutant mice. *Eur. J. Neurosci.* 20(3):837–42

Reid SN, Daw NW, Gregory DS, Flavin H. 1996. cAMP levels increased by activation of metabotropic glutamate receptors correlate with visual plasticity. *J. Neurosci.* 16(23):7619–26

Restani L, Cerri C, Pietrasanta M, Gianfranceschi L, Maffei L, Caleo M. 2009. Functional masking of deprived eye responses by callosal input during ocular dominance plasticity. *Neuron* 64(5):707–18

Rittenhouse CD, Shouval HZ, Paradiso MA, Bear MF. 1999. Monocular deprivation induces homosynaptic long-term depression in visual cortex. *Nature* 397(6717):347–50

Roberts TF, Tschida KA, Klein ME, Mooney R. 2010. Rapid spine stabilization and synaptic enhancement at the onset of behavioural learning. *Nature* 463(7283):948–52

Rubenstein JL, Merzenich MM. 2003. Model of autism: increased ratio of excitation/inhibition in key neural systems. *Genes Brain Behav.* 2(5):255–67

Sale A, Maya Vetencourt JF, Medini P, Cenni MC, Baroncelli L, et al. 2007. Environmental enrichment in adulthood promotes amblyopia recovery through a reduction of intracortical inhibition. *Nat. Neurosci.* 10(6):679–81

Sanchez-Huertas C, Rico B. 2011. CREB-dependent regulation of GAD65 transcription by BDNF/TrkB in cortical interneurons. *Cereb. Cortex* 21(4):777–88

Sato M, Stryker MP. 2008. Distinctive features of adult ocular dominance plasticity. *J. Neurosci.* 28(41):10278–86

Sato M, Stryker MP. 2010. Genomic imprinting of experience-dependent cortical plasticity by the ubiquitin ligase gene Ube3a. *Proc. Natl. Acad. Sci. USA* 107(12):5611–16

Sawtell NB, Frenkel MY, Philpot BD, Nakazawa K, Tonegawa S, Bear MF. 2003. NMDA receptor-dependent ocular dominance plasticity in adult visual cortex. *Neuron* 38(6):977–85

Shatz CJ, Stryker MP. 1978. Ocular dominance in layer IV of the cat's visual cortex and the effects of monocular deprivation. *J. Physiol.* 281:267–83

Silingardi D, Scali M, Belluomini G, Pizzorusso T. 2010. Epigenetic treatments of adult rats promote recovery from visual acuity deficits induced by long-term monocular deprivation. *Eur. J. Neurosci.* 31(12):2185–92

Sillito AM, Kemp JA, Blakemore C. 1981. The role of GABAergic inhibition in the cortical effects of monocular deprivation. *Nature* 291(5813):318–20

Smith SL, Trachtenberg JT. 2007. Experience-dependent binocular competition in the visual cortex begins at eye opening. *Nat. Neurosci.* 10(3):370–75

Southwell DG, Froemke RC, Alvarez-Buylla A, Stryker MP, Gandhi SP. 2010. Cortical plasticity induced by inhibitory neuron transplantation. *Science* 327(5969):1145–48

Sugiyama S, Di Nardo AA, Aizawa S, Matsuo I, Volovitch M, et al. 2008. Experience-dependent transfer of Otx2 homeoprotein into the visual cortex activates postnatal plasticity. *Cell* 134(3):508–20

Sugiyama S, Prochiantz A, Hensch TK. 2009. From brain formation to plasticity: insights on Otx2 homeoprotein. *Dev. Growth Differ.* 51(3):369–77

Syken J, Grandpre T, Kanold PO, Shatz CJ. 2006. PirB restricts ocular-dominance plasticity in visual cortex. *Science* 313(5794):1795–800

Tagawa Y, Kanold PO, Majdan M, Shatz CJ. 2005. Multiple periods of functional ocular dominance plasticity in mouse visual cortex. *Nat. Neurosci.* 8(3):380–88

Tamas G, Buhl EH, Lorincz A, Somogyi P. 2000. Proximally targeted GABAergic synapses and gap junctions synchronize cortical interneurons. *Nat. Neurosci.* 3(4):366–71

Tian L, Hires SA, Mao T, Huber D, Chiappe ME, et al. 2009. Imaging neural activity in worms, flies and mice with improved GCaMP calcium indicators. *Nat. Methods* 6(12):875–81

Tognini P, Putignano E, Coatti A, Pizzorusso T. 2011. Experience-dependent expression of miR-132 regulates ocular dominance plasticity. *Nat. Neurosci.* 14:1237–39

Tsubokawa H, Ross WN. 1996. IPSPs modulate spike backpropagation and associated [Ca2+]i changes in the dendrites of hippocampal CA1 pyramidal neurons. *J. Neurophysiol.* 76(5):2896–906

Van LC, Emiliani S, Verdin E. 1996. The expression of a small fraction of cellular genes is changed in response to histone hyperacetylation. *Gene Expr.* 5(4–5):245–53

Wiesel TN, Hubel DH. 1963. Single-Cell responses in striate cortex of kittens deprived of vision in one eye. *J. Neurophysiol.* 26:1003–17

Yamahachi H, Marik SA, McManus JN, Denk W, Gilbert CD. 2009. Rapid axonal sprouting and pruning accompany functional reorganization in primary visual cortex. *Neuron* 64(5):719–29

Yazaki-Sugiyama Y, Kang S, Cateau H, Fukai T, Hensch TK. 2009. Bidirectional plasticity in fast-spiking GABA circuits by visual experience. *Nature* 462(7270):218–21

Yizhar O, Fenno LE, Prigge M, Schneider F, Davidson TJ, et al. 2011. Neocortical excitation/inhibition balance in information processing and social dysfunction. *Nature* 477(7363):171–78

Yoon BJ, Smith GB, Heynen AJ, Neve RL, Bear MF. 2009. Essential role for a long-term depression mechanism in ocular dominance plasticity. *Proc. Natl. Acad. Sci. USA* 106(24):9860–65

What Is the Brain-Cancer Connection?

Lei Cao and Matthew J. During

College of Medicine, The Ohio State University, Columbus, Ohio 43210;
email: cao.76@osu.edu, matthew.during@osumc.edu

Annu. Rev. Neurosci. 2012. 35:331–45

First published online as a Review in Advance on March 29, 2012

The *Annual Review of Neuroscience* is online at neuro.annualreviews.org

This article's doi:
10.1146/annurev-neuro-062111-150546

0147-006X/12/0721-0331$20.00

Keywords

environmental enrichment, hypothalamic-sympathoneural-adipocyte (HSA) axis, BDNF, adipocyte, leptin

Abstract

A focus of much cancer research is at the molecular and cellular levels. In contrast, the effects of social interactions and psychological state are less investigated, and considered by many a "soft" science. Yet several highly rigorous studies have begun to tease out biochemical pathways by which the brain can influence the development and growth of cancer. Previous reviews have discussed the concept of stress and cancer. Here, we discuss recent work showing environments that are more complex and challenging, but not stressful per se, and that have robust effects on peripheral cancer by activating a specific neuroendocrine brain-adipocyte axis. These enriched environments lead to activation of the sympathetic innervation of fat tissue, suppression of leptin, and a reduction in cancer proliferation by inducing hypothalamic BDNF expression. We summarize this work and discuss how these data integrate into the body of literature regarding stress, the environment, and cancer.

Contents

MACROENVIRONMENT EFFECTS ON CANCER

Cancer development and growth are dependent in part on its microenvironment: the balance between factors acting to facilitate growth and induce angiogenesis and cell survival and factors inhibiting cell proliferation and promoting apoptosis (Hanahan & Weinberg 2000). This local microenvironment is influenced by systemic factors, and the cancer itself induces both local and distant changes through paracrine signaling and interactions with the immune and nervous systems (Aaronson 1991, Darnell & Posner 2006). The effect of the macroenvironment on systemic cancer; specifically, an individual's interaction with his or her physical living and social environment is much less well defined. The notion that psychosocial factors can affect cancer progression has long been suspected (Reiche et al. 2004). Clinical and epidemiological studies have recognized that specific psychosocial factors such as stress, chronic depression, and lack of social support are risk factors for the development and progression of cancer (Armaiz-Pena et al. 2009). A recent meta-analysis reveals that stress-related psychosocial factors are associated with a higher cancer incidence in premorbid healthy people, poorer survival in cancer patients, and higher cancer mortality (Chida et al. 2008). However, the reports supporting the positive relationship between stressful life events and cancer initiation have been inconsistent (Duijts et al. 2003, Lillberg et al. 2003, Price et al. 2001). In contrast, psychosocial factors such as hopelessness, denial, suppression of negative emotions, and lack of social support are found more reliably to predict progression of already-diagnosed cancers (Armaiz-Pena et al. 2009).

SOCIAL SUPPORT, LONELINESS, AND SURVIVAL

In 1988, House and colleagues wrote a review (House et al. 1988), based largely on five prospective studies published in the prior decade, indicating that social relationships had a strong influence on longevity, with an effect size similar to or greater than that of more established risk factors, including smoking, hypertension, obesity, and a sedentary lifestyle. Since that report, the field has been one of intense investigation, and in 2010, Holt-Lunstad et al. published a meta-analysis of 148 such studies (Holt-Lunstad et al. 2010) showing a mean odds ratio (OR) = 1.50, i.e., a 50% increased likelihood of survival for participants with stronger social relationships. Moreover, the effect was weakest in studies that simply discriminated between those who lived alone versus those who did not (OR = 1.13); the strongest effect occurred when a more complex assessment of social integration was determined (OR = 1.91). Overall, the data confirmed and strengthened the conclusion of the 1988 report and provided strong prospective epidemiological support of the association of social integration and all-cause mortality, including cancer. The strength of this relationship appears as robust as that of smoking. Cacioppo & Patrick (2008) have emphasized that the subjective feeling of loneliness is a better predictor than many objective measurements of social integration, for example the size of social networks. He and his collaborators have shown that loneliness is associated with a specific pattern of cerebral activation on functional magnetic resonance imaging (fMRI) (Cacioppo et al. 2009), that it activates the hypothalamic-pituitary-adrenal (HPA) axis (Cacioppo et al. 2000), and that circulating antigen-presenting cells, including dendritic cells and monocytes, are responsive to the state of loneliness with alterations in gene expression that influence immune function (Cole et al. 2011). Although many of these studies did not specifically address cancer, evidence is increasingly compelling that an individual's interaction with others, and perception of their social integration, is an important risk factor for overall morbidity and mortality, including cancer.

DEFINITION OF TERMS, PART 1

Stress and allostasis relate to general concepts associated with changes, external and/or internal, that induce an adaptive response in the organism. These concepts originated with the pioneering work of Walter Cannon and Hans Selye. Stress is most commonly considered a negative and/or excessive maladaptive state posing a risk to health and well-being. Eustress and hormesis are less established terms and relate to milder and/or briefer challenges, which induce a positive or healthy adaptive response. Allostasis is a newer term that refers to the multiple adaptive physiological responses to external change, in contrast with homeostasis, the intrinsic physiological process required to maintain a constant internal state. Allostatic load is the state in which the environment poses repeated severe, excessive, and/or prolonged challenges leading to maladaptive changes over time.

EXPERIMENTAL ANIMAL STUDIES

Animal-based studies parallel these clinical data, demonstrating that experimentally imposed stress can modulate cancer progression (Thaker & Sood 2008). Recent mechanistic studies have started to identify the specific pathways that could mediate such effects. The majority of studies on this stress-cancer relationship support the notion that psychosocial factors initiate a cascade of information-processing pathways in the central nervous system (CNS) triggering fight-or-flight stress responses by activating the sympathetic-adrenal-medullary (SAM) axis or defeat/withdraw responses produced by the HPA axis and thereby influencing multiple aspects of tumorigenesis and progression (Antoni et al. 2006, Glaser & Kiecolt-Glaser 2005). In a comprehensive review article, Antoni et al. (2006) reviewed clinical, cellular, and molecular studies to synthesize an integrated model of bio-behavioral influences on cancer pathogenesis. In this model, stress, psychosocial, and behavioral factors influence cancer-related processes by regulating neuroendocrine hormones, including glucocorticoids, epinephrine, and norepinephrine. These stress hormones can modulate the activity of multiple components of the tumor microenvironment. Direct effects include the promotion of cell proliferation, invasion, angiogenesis, metastasis, and immune evasion. Stress hormones can also indirectly influence tumorigenesis by activating oncogenic viruses and modulating immune functions, including antibody production, cytokine profiles, and cell trafficking. These pleiotropic effects of stress hormones collectively support tumor initiation and progression (Antoni et al. 2006). Most animal studies on the stress-cancer relationship impose experimental stressors associated with the negative aspects of severe stress (distress) such as restraint stress (Saul et al. 2005, Sloan et al. 2010, Steplewski et al. 1985), social isolation (Hasegawa & Saiki 2002, Palermo-Neto et al. 2003, Thaker et al. 2006), and social confrontation (Stefanski & Ben-Eliyahu 1996). Scarce attention has been turned to the positive functions of a more mild stress or challenging environment (eustress/hormesis). Failure to note the distinction between eustress and distress leads to a misunderstanding of the

HPA: hypothalamic pituitary adrenal

SAM: sympatho-adrenal-medullary (axis)

EE: environmental enrichment

nature of pathways linking environmental conditions to cancer and hampers our ability to develop effective therapies. In this review, special emphasis is placed on the effects of eustress on cancer and the underlying mechanisms.

STRESS, EUSTRESS, AND DISTRESS

"Stress" is an ambiguous term. Broadly, any actual or potential disturbance of an individual's environment (stressor) is recognized or perceived by specific brain regions. The subjective state of sensing potentially adverse changes in the environment (stress) activates the HPA axis and/or the autonomic nervous system (ANS) to help an individual mitigate a threat (Joels & Baram 2009). The real or perceived environmental demands can be appraised as threatening or benign, and the ensuing biological, behavioral, and social coping responses can be labeled as "good," "tolerable," and "toxic," depending on the degree to which an individual has control over the given stressor and has support systems and resources for handling a given stressor (McEwen & Gianaros 2010). Moreover, these physiological responses are not static and change dynamically in response to the duration of the stressor and its severity. More recently, the concept of allostasis has been introduced to describe adaptive responses to external challenges (Sterling 1988). Allostasis is defined as a dynamic regulatory process wherein homeostatic control is maintained by an active process of adaptation during exposure to physical and behavioral stressors (achieving stability through change) (McEwen & Gianaros 2010). McEwen and colleagues later introduced the term "allostatic load or overload" to refer to the consequence of allodynamic regulatory wear and tear on the body and brain promoting ill health (McEwen 1998). Allostatic load can result from either too much stress or from inefficient management of allostasis: for example, not turning off the response when no longer needed, not turning on an adequate response in the first place, or dampening the allostatic response to the recurrence of the same stressor (McEwen 1998). The concepts of allostasis and allostatic load/overload emphasize the existence of multiple interacting mediators operating in a nonlinear network influencing multiple systems such as CNS function, metabolism, cardiovascular function, neuroendocrine function, and immune function (McEwen & Gianaros 2010). In this review, we adopt the classic stress literature terms eustress (positive stress) and distress (negative stress) to describe the relevant environmental conditions, health outcomes, and disparate underlying mechanisms because the distinct activation patterns of neuroendocrine axes provide physiological support for the notion of eustress as a viable concept. Hans Selye, who first imported the concept of stress into experimental medicine, noted that departures from homeostasis could be "eustressful" or "distressful" and that health effects would vary accordingly (Milsum 1985, Selye 1974). Eustress is associated with adaptive responses and benign or beneficial effects on health, whereas maladaptive distress is associated with exposure to more severe aversive or hostile environments resulting in allostatic load and ill health (Milsum 1985, Selye 1974).

ENVIRONMENTAL ENRICHMENT AS EUSTRESS MODEL

Our research on eustress and cancer uses environmental enrichment (EE) as an experimental paradigm to explore effects of a more complex environment devoid of elements traditionally thought of as being stressful. EE refers to housing with increased space, physical activity, and social interactions that facilitate enhanced sensory, cognitive, motor, and social stimulation (Nithianantharajah & Hannan 2006). What is remarkable is just how robust and powerful an influence the EE can have on brain function. EE leads to changes in growth factor expression, enhancement of neurogenesis, and survival of cells within the CNS (Cao et al. 2004, Young et al. 1999). Moreover, EE has considerable impact on the phenotype of a variety of toxin- and genetically induced models of human neurological disease (Nithianantharajah & Hannan 2006). On this basis, we hypothesized that an EE—one optimized for cerebral health, as defined by improved learning and memory, increased neurogenesis, and reduced apoptosis and resistance to external cerebral insults—could also lead to an anticancer phenotype. We tested this hypothesis in both syngeneic models of B16 melanoma and MC38 colon cancer as well as in $APC^{min/+}$ mice, a spontaneous model with a germline mutation in APC similar to that in humans with familial adenomatous polyposis and a gene in which somatic mutations occur in 80% of human colon cancer. EE leads to a remarkable suppression of cancer growth in all three models tested, even when delayed until the tumor is well established (Cao et al. 2010). In a comprehensive set of experiments using transgenic animals, somatic gene transfer, controlled release liposomes, and osmotic minipump drug infusion, we dissected out a mechanism whereby the EE paradigm induces hypothalamic brain-derived neurotrophic factor (BDNF) as an effector immediate early gene. This action led to preferential activation of discrete sympathetic innervation of white adipose tissue (WAT), which in turn via β-adrenergic receptors on adipocytes resulted in suppression of leptin expression and release. The marked drop in serum leptin levels mediates the antiproliferative phenotype. We coined the term hypothalamic-sympathoneural-adipocyte (HSA axis to describe this brain-adipocyte axis) (**Figure 1**).

DEFINITION OF TERMS, PART TWO

Hypothalamic-sympathoneural-adipocyte (HSA) axis: The HSA axis refers to the neuroendocrine pathway linking the hypothalamus to white adipocyte tissue (WAT). The hypothalamic component is mediated via brain-derived neurotrophic factor (BDNF), which, although found throughout the hypothalamus, is highly expressed in the ventromedial (VMH), dorsomedial (DMH), and arcuate nuclei. In response to the challenges associated with an enriched environment, hypothalamic BDNF is induced as an effector immediate early gene and leads to activation of a component of the sympathetic nervous system, specifically the sympathoneural innervation of WAT. The activation of the axis therefore involves an increase in hypothalamic BDNF, WAT norepinephrine, and, via beta receptors on the adipocyte, suppression of leptin expression and release. This suppressed leptin is associated with an antiproliferative phenotype. Moreover, HSA axis activation leads to a reduction in WAT mass and a phenotypic change of white to brown-like fat within abdominal fat depots.

Environmental enrichment (EE): a more complex housing condition (compared with standard laboratory animal housing) typically including a combination of increased physical space, larger groups, objects for activity or hiding and running wheels. EE is associated with many adaptive changes in the brain and behavior, including increased hippocampal neurogenesis, reduced apoptosis, greater resistance and improved recovery to external insults and ischemia, milder phenotype in models of neurological disease, and improved motor and cognitive performance.

More recently, we have revealed a new phenotype associated with EE. We show that EE decreases adiposity, increases energy expenditure, causes resistance to diet-induced obesity, and induces a white to brown fat conversion also by activating the HSA axis (Cao et al. 2011) (**Figure 1**). The EE-induced anticancer and antiobesity phenotypes could not be accounted for by physical activity alone but could be largely reproduced by overexpression of BDNF in the hypothalamus (Cao et al. 2010, 2011). Experimental animal studies have almost invariably shown that EE has beneficial effects on an animal's well being with reduced morbidity and mortality. However, EE also moderately, but significantly, increases corticosterone levels in serum and norepinephrine levels in white fat. Both corticosterone and norepinephrine are considered signatures of stress, indicating activation of the HPA and sympathetic nervous system (SNS) respectively. It may appear paradoxical that chronic mild activation of the HPA axis could be associated with a tumor-resistant phenotype. However, both hormones are metabolic regulators catabolizing energy stores to meet immediate metabolic demands. EE increases energy utilization, so it is not surprising that both HPA and SNS are activated and that white fat is

BDNF: brain-derived neurotrophic factor

WAT: white adipose tissue

HSA: hypothalamic-sympathoneural-adipocyte (axis)

SNS: sympathetic nervous system

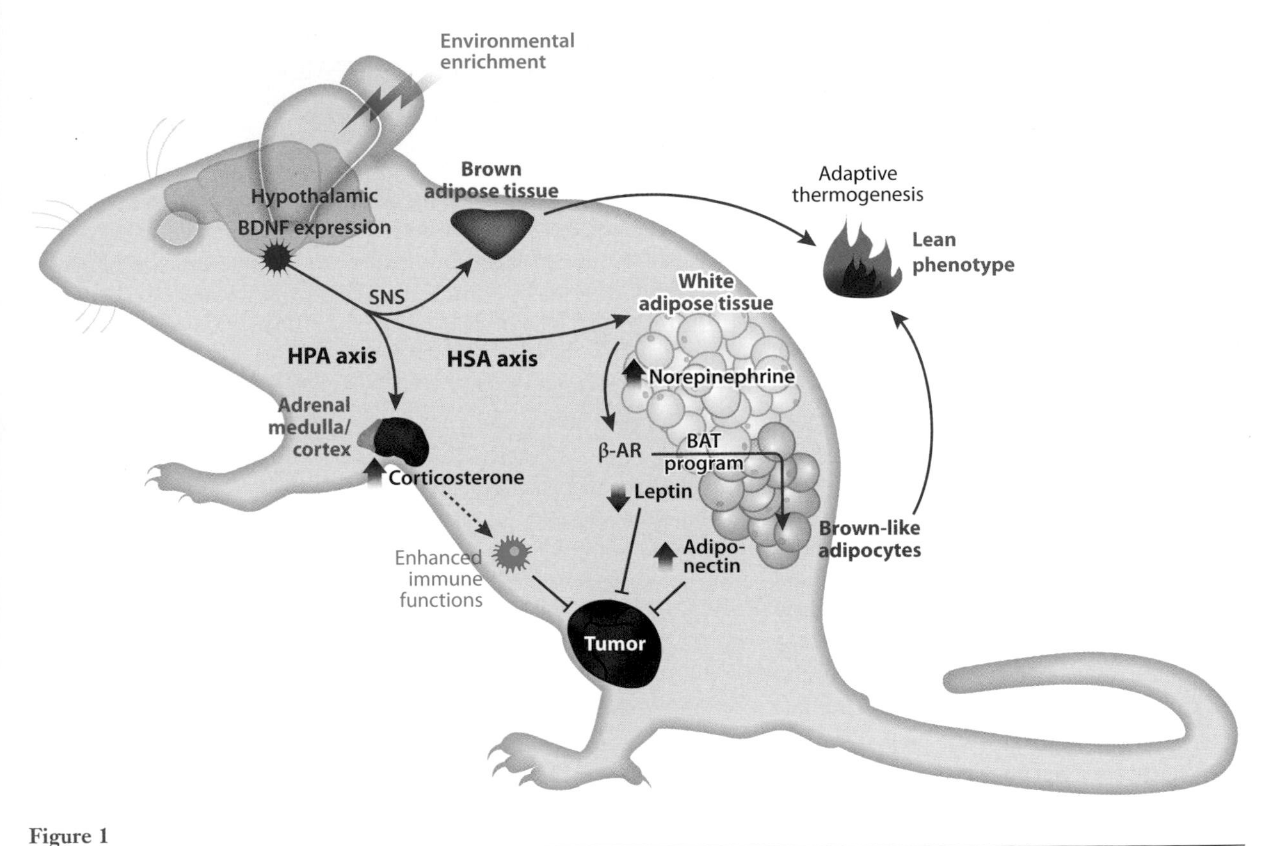

Figure 1

Mechanisms of EE-induced anticancer and antiobesity phenotypes. Abbreviations: β-AR, β-adrenergic receptors; BAT, brown adipose tissue; BDNF, brain-derived neurotrophic factor; HPA, hypothalamic-pituitary-adrenal; HSA, hypothalamic-sympathoneural-adipocyte.

mobilized to meet the demands associated with elevated energy expenditure. We believe EE can be viewed as a valuable model of eustress.

OBESITY AND CANCER

The worldwide epidemic of obesity and the global incidence of cancer are both rising. The World Health Organization recently reported that as of 2008, ~1.5 billion adults worldwide were overweight with more than 200 million men and almost 300 million women being clinically obese with a body mass index (BMI) of 30 or above (WHO 2011). Cancer is a leading cause of death worldwide, and deaths from cancer are projected to continue rising with an estimated 13.1 million deaths in 2030 (WHO 2012). Both obesity and cancer contribute to increased mortality and socioeconomic burden worldwide (IARC/WHO 2008, UICC 2008, WHO 2011). Obesity is linked to an increased risk of certain types of cancer including breast, renal, ovarian, esophagus, pancreas, prostate, hepatobiliary, colorectal, and melanoma (Reeves et al. 2007, Renehan et al. 2008, Whitlock et al. 2009). Obesity may account for 14% of all deaths from cancer, and 20% of those in women. In the United States, if both men and women maintained normal weight, 900,000 cancer deaths could be prevented (Calle et al. 2003). Although we know that weight gain can result in higher cancer rates, the converse finding of weight loss and lower cancer rates did not traditionally exist because of lack of successful weight-loss modalities (Ahn et al. 2007). Indeed, intentional weight loss to decrease cancer rates, although gaining favor, has proven difficult to achieve with diet and lifestyle therapies (Parker & Folsom 2003). Both randomized and prospective studies reveal that the long-term weight loss with intensive lifestyle intervention in diabetic patients with a BMI of 30 is ~2% at 10 years (Knowler et al. 2009) and is even less in patients with BMI above 35 (Sjostrom et al. 2007). Metabolic or bariatric surgery can provide sustained weight loss (Buchwald et al. 2004). Recent longitudinal studies on metabolic surgery have revealed that successful weight loss leads to lower cancer rates and decreased cancer mortality and helps to establish a causal association between obesity and cancer (Adams et al. 2007, Renehan 2009, Sjostrom et al. 2007). The mechanisms underlying the obesity-cancer association are multifactorial and include insulin resistance, increased growth factors and anabolic hormones, adipokines, oxidative stress, inflammation, increased bioavailable sex hormones, deterioration in immunosurveilance, and altered cellular energetics (Ashrafian et al. 2011, Longo & Fontana 2010) (**Figure 2**).

HYPOTHALAMIC-SYMPATHONEURAL-ADIPOCYTE AXIS ACTIVATION TARGETING BOTH OBESITY AND CANCER

EE induces a novel antiobesity and anticancer phenotype (Cao et al. 2009, 2010). This phenotype is not caused by exercise alone. Our mechanistic studies have begun to characterize a regulatory network encompassing multiple organ systems including the CNS, the endocrine system, and the immune systems, all of which play a role in an individual's eustress response to physical and social environments.

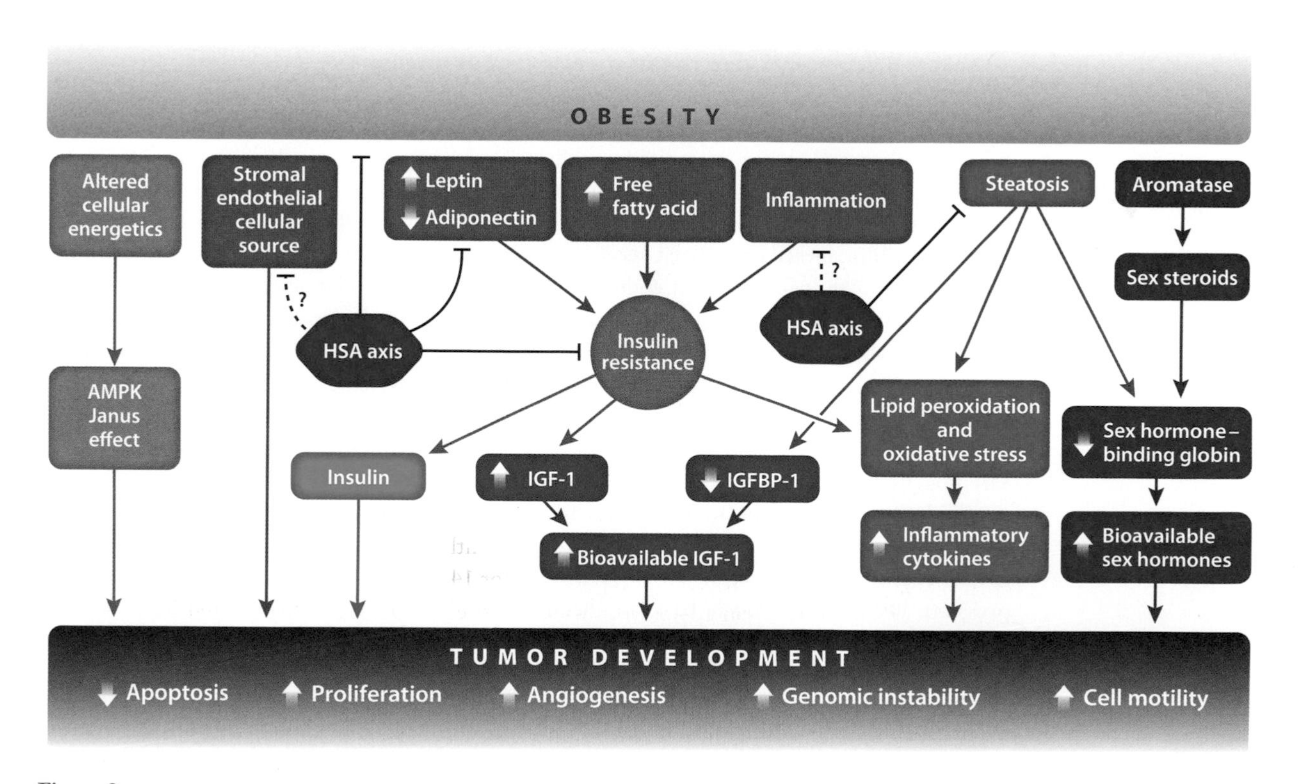

Figure 2

Mechanisms of tumor inhibition by activation of HSA axis. Abbreviations: AMPK, 5′-adenosine monophosphate-activated protein kinase; HSA, hypothalamic-sympathoneural-adipocyte.

The HSA axis is likely just one component of the body's response to eustress. Nevertheless, the HSA axis links the physical/social environment to the regulation of the fat phenotype and thereby can modulate multiple organ systems and ultimately affect cancer progression. Although highly effective in inducing an antiproliferative phenotype in nonobese animals, we predict that the effect of HSA activation is likely to be even more robust in obese individuals because of its potent modulation of fat. Indeed, three weeks of EE led to an approximate 70% decrease of melanoma mass in DIO mice, compared with a 50% decrease in mice of normal weight (Cao et al. 2010). In nonobese individuals, HSA axis activation leads to a decrease in circulating concentrations of IGF-1 and leptin and an increase in adiponectin. Our data suggest that leptin plays an essential role in the EE-induced anticancer effects.

The role of another major adipokine, adiponectin, remains to be characterized. Adipokine-independent mechanisms may also exist and are likely to be revealed in obesity models. For example, HSA activation potently alleviates obesity-associated insulin resistance, hyperinsulinemia, dislipidemia, and liver steatosis (Cao et al. 2009). Chronically elevated insulin levels have been associated with breast, colon, pancreas, and endometrial cancers (Gunter et al. 2009, Jee et al. 2005, Trevisan et al. 2001). Epidemiological studies have shown that both obesity and diabetes are risk factors for hepatocellular carcinoma (Polesel et al. 2009). The likely mechanisms of carcinogenesis include obesity-related steatosis, lipid peroxidation, and increased oxidative stress. Our data show that genetically activating the HSA axis by hypothalamic BDNF overexpression, using adeno-associated virus (AAV) vector–mediated gene transfer, completely resolves steatosis (Cao et al. 2009), suggesting potential protection against liver cancer.

In addition, an association between inflammation and cancer has been established (Coussens & Werb 2002). Obesity is associated with chronically inflamed WAT, leading to the release of proinflammatory cytokines including TNF-α, C-reactive proteins, and several interleukins such as IL-1β and IL-6 (Olefsky 2009). Our preliminary data showed no change in TNF-α or IL-6 by EE in nonobese animals. However, the effects of the HSA axis on proinflammatory cytokines may be revealed in obesity models and may contribute to the anticancer phenotype.

Several cancers are associated with altered sex steroid levels (Eliassen & Hankinson 2008). Insulin resistance and compensatory hyperinsulinemia inhibit the hepatic production of sex hormone–binding globulin (SHBG) and therefore increase the circulating concentrations of bioavailable sex hormones (Pugeat et al. 1991). WAT is considered one of the major sources of extraglandular estrogen, produced by aromatization of androgen precursors (Flototto et al. 2001). Estrogens and androgens are strong mitogens for cells and stimulate the development and progression of several tumors (Flototto et al. 2001). The HSA axis may influence SHBG by modulating WAT.

Other mechanisms implicated in the increased cancer risk and proliferation associated with obesity include cellular energetics and stromal contribution from cells originating within WAT. Obesity is associated with irregular cellular energetic and signaling pathways. Disruptive cellular energetics via the master switch 5′-adenosine monophosphate-activated protein kinase (AMPK) may lead to cancer through the Janus energy effect (Ashrafian 2006). Recent research suggests that WAT may directly mediate cancer progression by serving as a source of cells that migrate to tumors and promote neovascularization (Zhang et al. 2009). Kolonin and colleagues (Zhang et al. 2009) have shown that entry of WAT stromal and endothelial cells into the systemic circulation leads to their homing to tumor stroma and vasculature. Moreover, tumor recruitment of WAT stromal cells is sufficient to promote tumor growth (Zhang et al. 2009). Activating the HSA axis results in a marked shrinking of WAT in both obese and nonobese animals (Cao et al. 2009, 2011), which presumably affects the functional cellular source available for tumor recruitment.

How the HSA axis may affect cellular energetics and adipose stromal/vascular cell trafficking (including homing in on tumors) are fascinating topics for future research.

In a recent, comprehensive review, Hanahan & Weinberg (2011) updated the concept of cancer hallmarks and summarized the development of mechanism-based targeted therapies to treat human cancers. They note that most of the hallmark-targeting cancer drugs have been deliberately directed toward specific molecular targets, enabling particular hallmark capabilities. Such specificity should have fewer off-target effects and thus less nonspecific toxicity. However, the clinical responses have been generally transitory followed by almost-inevitable relapses. One interpretation is that the signaling pathways regulating each of the core hallmark capabilities are redundant. Consequently, some cancer cells may survive a therapeutic agent targeting one key pathway by relying on other hallmark capabilities and eventually adapt to the selective pressure imposed by the therapy. Thus the combination of mechanism-guided therapies cotargeting multiple hallmark capabilities is likely to be more effective and durable (Hanahan & Weinberg 2011).

The HSA axis has the capacity to affect multiple pathways. This axis contributes to the mechanistic understanding of the beneficial effects on health in response to a eustressful environment. Our data suggest that the brain plays a primary upstream regulatory role with BDNF as a key mediator and WAT as the principal peripheral organ responsive to the central regulator. WAT-independent mechanisms likely exist and remain to be elucidated. Nevertheless, our investigation of the HSA axis to date suggests that its activation via environmental or genetic approaches can profoundly regulate body composition and metabolism and alleviate obesity and diabetes. In contrast with metabolic surgery and caloric restriction, whose effects on cancer are associated with decreased energy intake (Albanes 1987, Ashrafian et al. 2011), HSA axis activation does not result in overt decreases in food intake in lean animals. In regulating the WAT phenotype, the HSA axis may influence several mechanisms implicated in the obesity-cancer association (**Figure 2**) and therefore represents an attractive target for combination therapy.

SI: social isolation

WHAT ARE THE MECHANISMS UNDERLYING OPPOSITE EFFECTS OF EUSTRESS VERSUS DISTRESS ON CANCER?

In contrast with EE, social isolation (SI) of naturally gregarious rodents is a widely used model for environmental deprivation in humans and can produce long-term changes including neophobia, aggression, and cognitive rigidity (Fone & Porkess 2008, Lapiz et al. 2003). SI is associated with increased tumor progression in models of breast cancer (Williams et al. 2009) and Ehrlich tumor (Palermo-Neto et al. 2008). Hasegawa & Saiki (2002) report that SI increases B16 melanoma growth implanted to male C57BL/6 mice. And the SI-enhanced tumor growth and thymus atrophy reflecting host stress responses could be abrogated by orally administering the β blocker propranolol. However, the chronic administration of corticosterone failed to modulate tumor growth despite significantly inducing thymus and spleen atrophy, suggesting SAM activation, not HPA axis activation, as the main pathway mediating SI's procancer effect (Hasegawa & Saiki 2002). Studies by Sood, Lutgendorf, and others have shown that norepinephrine and epinephrine may influence the progression of ovarian cancer by modulating the expression of vascular endothelial growth factor (VEGF) and matrix metalloproteinases (MMPs) (Lutgendorf et al. 2003, Sood et al. 2006, Thaker et al. 2007). The in vitro studies by Glaser and colleagues have shown that norepinephrine may promote other cancers such as nasopharyngeal carcinoma, myeloma, and melanoma by stimulating the secretion of VEGF and MMPs (Yang et al. 2006, 2008, 2009). These findings support the hypothesis that stress-associated activation of the SAM axis can promote tumor progression, in part, by modulating the

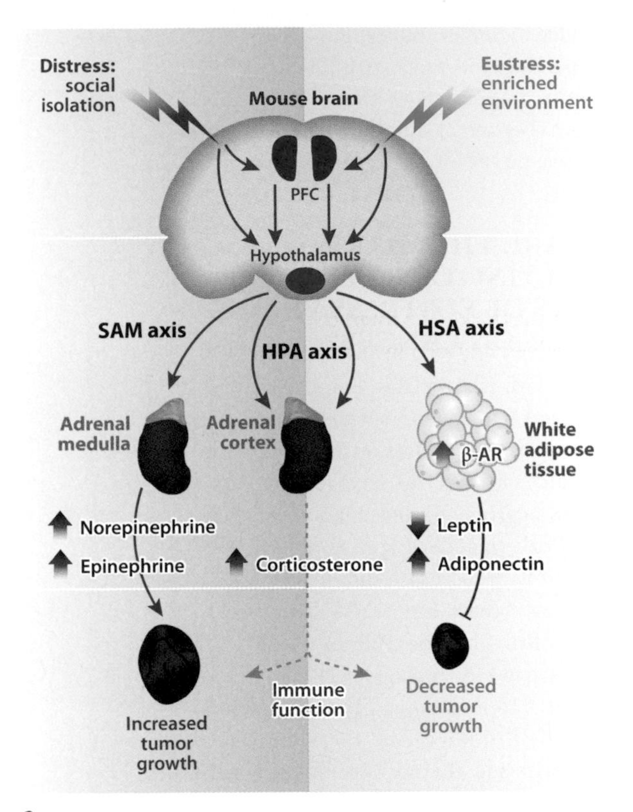

Figure 3

Mechanisms underlying the opposite effects of eustress and distress on cancer. The prefrontal cortex (PFC) may mediate the appraisal of environmental stimuli as eustressful (positive stress) or distressful (negative stress). The PFC, sitting at the top of the brain hierarchy modulating stress, is connected through multiple pathways to the hypothalamus, the gate to the periphery, which regulates the activities of three axes: hypothalamic-pituitary-adrenal (HPA), sympathetic-adrenal-medullary (SAM), and hypothalamic-sympathoneural-adipocyte (HSA). The differential activation of the three axes responding to eustressful or distressful events may lead to distinctive health outcomes (eustress associated with improved health, distress linked with impaired health) depending on the magnitude of activation of each axis, the cross talk among them, their preferential target tissues in the periphery, and their influences on the intracellular context at the target tissues. The eustressful environmental enrichment (EE) stimulates brain-derived neurotrophic factor (BDNF) in the hypothalamus, leading to HSA activation of white adipose tissue. The preferential increase of SNS tone to white fat via beta-adrenergic receptors shuts down leptin and boosts adiponectin production, resulting in tumor suppression. In contrast, the distressful social isolation may stimulate primarily the SAM axis, resulting in high-level norepinephrine in the circulation and possibly increased release in the tumor. This action would facilitate cancer growth by inducing proangiogenic factors such as vascular endothelial growth factor (VEGF). The HSA and SAM activations are interactive and could synergistically or counteractively influence tumor growth. In the EE model, the predominant HSA may override the possible weak tumor stimulation brought by mild SAM axis activation. In the SI model, the strong activation of SAM may enhance tumor growth synergistically with the weakened HSA via increased leptin production. Other abbreviation: β-AR, β-adrenergic receptors.

expression of proangiogenic and prometastatic factors. This line of research suggests that interventions targeting components of the activated SAM axis, or the use of β-adrenergic blocking agents, may represent new strategies for slowing down the progression of cancer.

Hasegawa & Saiki (2002) report that oral propranolol abrogates the accelerated B16 melanoma growth in SI. We used B16 melanoma syngeneic transplants in the same C57BL/6 strain of mice and showed remarkable tumor suppression induced by EE. Moreover, the EE-induced leptin drop and tumor inhibition could be completely blocked by oral propranolol (Cao et al. 2010). How could the same norepinephrine via beta-adrenergic receptor activation be beneficial in SI but detrimental in EE? We propose that this seemingly paradoxical situation reflects the distinct physiological responses to eustress versus distress. We hypothesize that three axes are in play: the well-known HPA and SAM axes as well as the lesser-known HSA axis (**Figure 3**). The HPA axis is activated in response to both EE and SI but is a lesser player with only a relatively minor role through its modulation of immune function. SAM activation via increased norepinephrine levels in the circulation and possibly a local increase in the solid tumor itself may facilitate melanoma growth in SI by stimulating proangiogenic factors, including VEGF, similar to the observations reported in other cancer models (Thaker et al. 2006).

The HSA response to SI is not known. On the basis of our preliminary data showing increased adiposity and a rise of leptin in SI, the HSA axis is likely suppressed in SI. In contrast, a robust activation of HSA is associated with and responsible in part for EE-induced melanoma inhibition. Notably, norepinephrine was significantly increased in WAT but not in muscle and caused no change in circulation when mice were on a normal chow diet, suggesting minimal activation of the SAM (Cao et al. 2011). Our data suggest a selective SNS regulation of WAT by EE with no significant effects on blood pressure or heart rate. The original theories of SNS activation proposed a body-wide fight-or-flight response with a general increase in sympathetic outflow (Cannon 1929). More recent studies suggest the existence of independent autonomic inputs to individual organs through distinct autonomic projections permitting fine-tuning (Anderson et al. 1987). Studies using the retrograde neuronal tracer pseudorabies virus (PRV) reveal that the SNS has a distinct organization in different body compartments. Dual tracing of retroperitoneal white fat and inguinal white fat finds no colocalization in the paraventricular nucleus of hypothalamus (PVH) or amygdala (Kreier et al. 2006). In addition, there is differential SNS drive among WAT depots as well as between WAT and brown adipose tissue (BAT) after food deprivation, cold exposure, or glucoprivation (Brito et al. 2008). The mechanisms underlying this fat pad-specific pattern of SNS drive remain to be elucidated. We speculate that specific neurons in the PVH and/or other central nuclei, including the brain stem (Foster et al. 2010, Grill 2010), regulate anatomically discrete SNS efferent pathways and may be activated by specific environmental stimulations or physiological conditions and thereby control SNS drive in a fat depot–specific manner. We are currently investigating the identity of these neurons in the EE mice.

Elucidating the effects that the differing mechanisms of eustress and distress have on cancer will likely have significant implications. For example, several studies have shown that VEGF is upregulated by norepinephrine via β-adrenergic receptors (β-AR)/cyclic adenosine monophosphate (cAMP)/protein kinase A (PKA) signaling pathways in certain cancer cells (Lutgendorf et al. 2003; Yang et al. 2006, 2009), and β blockers have been shown to abrogate the deleterious effect of distress on cancer growth (Thaker et al. 2006). Several epidemiological and clinical studies support the link between β blockers and reduced cancer risk (Algazi et al. 2004, Perron et al. 2004) or progression (De Giorgi et al. 2011). This has led some investigators to propose the use of β blockers as a therapeutic intervention for cancer (Melamed et al. 2005). However, we must keep in mind that the mediators of stress and allostasis operate in a nonlinear network. The interactions are very complex. When any one mediator is changed, compensatory changes occur in the other mediators, which depend on the time course and the magnitude of changes of each of the mediators (McEwen & Gianaros 2010). Therefore, the effects of environmental and psychosocial stress on cancer should be analyzed in the context of the relevant environmental circumstance, individual adaptation and coping strategy, and outcome on the specific cancer microenvironment biology. This is why, on one hand, the call for the use of β blockers as a treatment for certain types of cancer may be well founded; individuals are experiencing distress and maladaptive responses when their SAM axis is overactive. On the other hand, the use of β blockers may attenuate eustress-induced anticancer effects when HSA axis activation is the primary component of an individual's adaptive response to his or her environment. This notion also underscores the importance of personalized medicine, emphasizing a more patient-specific approach that considers all environmental and psychosocial factors. Ongoing mechanistic studies distinguishing eustress versus distress will provide a novel perspective for the development of genetic, psychosocial, and pharmacological interventions to alleviate the negative functions of distress and promote the positive functions of eustress for the prevention and treatment of cancer.

DISCLOSURE STATEMENT

The authors are not aware of any affiliations, memberships, funding, or financial holdings that might be perceived as affecting the objectivity of this review.

ACKNOWLEDGMENTS

We thank Eugene Y. Choi for assistance with figure preparation.

LITERATURE CITED

Aaronson SA. 1991. Growth factors and cancer. *Science* 254:1146–53

Adams TD, Gress RE, Smith SC, Halverson RC, Simper SC, et al. 2007. Long-term mortality after gastric bypass surgery. *N. Engl. J. Med.* 357:753–61

Ahn J, Schatzkin A, Lacey JV Jr, Albanes D, Ballard-Barbash R, et al. 2007. Adiposity, adult weight change, and postmenopausal breast cancer risk. *Arch. Intern. Med.* 167:2091–102

Albanes D. 1987. Total calories, body weight, and tumor incidence in mice. *Cancer Res.* 47:1987–92

Algazi M, Plu-Bureau G, Flahault A, Dondon MG, Le MG. 2004. Could treatments with beta-blockers be associated with a reduction in cancer risk?. *Rev. Epidemiol. Sante. Publique* 52:53–65

Anderson EA, Wallin BG, Mark AL. 1987. Dissociation of sympathetic nerve activity in arm and leg muscle during mental stress. *Hypertension* 9:III114–19

Antoni MH, Lutgendorf SK, Cole SW, Dhabhar FS, Sephton SE, et al. 2006. The influence of bio-behavioural factors on tumour biology: pathways and mechanisms. *Nat. Rev. Cancer* 6:240–48

Armaiz-Pena GN, Lutgendorf SK, Cole SW, Sood AK. 2009. Neuroendocrine modulation of cancer progression. *Brain Behav. Immun.* 23:10–15

Ashrafian H. 2006. Cancer's sweet tooth: the Janus effect of glucose metabolism in tumorigenesis. *Lancet* 367:618–21

Ashrafian H, Ahmed K, Rowland SP, Patel VM, Gooderham NJ, et al. 2011. Metabolic surgery and cancer: protective effects of bariatric procedures. *Cancer* 117:1788–99

Brito NA, Brito MN, Bartness TJ. 2008. Differential sympathetic drive to adipose tissues after food deprivation, cold exposure or glucoprivation. *Am. J. Physiol. Regul. Integr. Comp. Physiol.* 294:R1445–52

Buchwald H, Avidor Y, Braunwald E, Jensen MD, Pories W, et al. 2004. Bariatric surgery: a systematic review and meta-analysis. *JAMA* 292:1724–37

Cacioppo JT, Ernst JM, Burleson MH, McClintock MK, Malarkey WB, et al. 2000. Lonely traits and concomitant physiological processes: the MacArthur social neuroscience studies. *Int. J. Psychophysiol. Off. J. Int. Organ. Psychophysiol.* 35:143–54

Cacioppo JT, Norris CJ, Decety J, Monteleone G, Nusbaum H. 2009. In the eye of the beholder: individual differences in perceived social isolation predict regional brain activation to social stimuli. *J. Cogn. Neurosci.* 21:83–92

Cacioppo JT, Patrick W. 2008. *Loneliness: Human Nature and the Need for Social Connection*. New York: Norton

Calle EE, Rodriguez C, Walker-Thurmond K, Thun MJ. 2003. Overweight, obesity, and mortality from cancer in a prospectively studied cohort of US adults. *N. Engl. J. Med.* 348:1625–38

Cannon WB. 1929. Organization for physiological homeostasis. *Physiol. Rev.* 9:399–431

Cao L, Choi EY, Liu X, Martin A, Wang C, et al. 2011. White to brown fat phenotypic switch induced by genetic and environmental activation of a hypothalamic-adipocyte axis. *Cell Metab.* 14:324–38

Cao L, Jiao X, Zuzga DS, Liu Y, Fong DM, et al. 2004. VEGF links hippocampal activity with neurogenesis, learning and memory. *Nat. Genet.* 36:827–35

Cao L, Lin EJ, Cahill MC, Wang C, Liu X, During MJ. 2009. Molecular therapy of obesity and diabetes by a physiological autoregulatory approach. *Nat. Med.* 15:447–54

Cao L, Liu X, Lin EJ, Wang C, Choi EY, et al. 2010. Environmental and genetic activation of a brain-adipocyte BDNF/leptin axis causes cancer remission and inhibition. *Cell* 142:52–64

Chida Y, Hamer M, Wardle J, Steptoe A. 2008. Do stress-related psychosocial factors contribute to cancer incidence and survival? *Nat. Clin. Pract. Oncol.* 5:466–75

Cole SW, Hawkley LC, Arevalo JM, Cacioppo JT. 2011. Transcript origin analysis identifies antigen-presenting cells as primary targets of socially regulated gene expression in leukocytes. *Proc. Natl. Acad. Sci. USA* 108:3080–85

Coussens LM, Werb Z. 2002. Inflammation and cancer. *Nature* 420:860–67

Darnell RB, Posner JB. 2006. Paraneoplastic syndromes affecting the nervous system. *Semin. Oncol.* 33:270–98

De Giorgi V, Grazzini M, Gandini S, Benemei S, Lotti T, et al. 2011. Treatment with beta-blockers and reduced disease progression in patients with thick melanoma. *Arch. Intern. Med.* 171:779–81

Duijts SF, Zeegers MP, Borne BV. 2003. The association between stressful life events and breast cancer risk: a meta-analysis. *Int. J. Cancer* 107:1023–29

Eliassen AH, Hankinson SE. 2008. Endogenous hormone levels and risk of breast, endometrial and ovarian cancers: prospective studies. *Adv. Exp. Med. Biol.* 630:148–65

Flototto T, Djahansouzi S, Glaser M, Hanstein B, Niederacher D, et al. 2001. Hormones and hormone antagonists: mechanisms of action in carcinogenesis of endometrial and breast cancer. *Horm. Metab. Res.* 33:451–57

Fone KC, Porkess MV. 2008. Behavioural and neurochemical effects of post-weaning social isolation in rodents—relevance to developmental neuropsychiatric disorders. *Neurosci. Biobehav. Rev.* 32:1087–102

Foster MT, Song CK, Bartness TJ. 2010. Hypothalamic paraventricular nucleus lesion involvement in the sympathetic control of lipid mobilization. *Obesity* 18:682–89

Glaser R, Kiecolt-Glaser JK. 2005. Stress-induced immune dysfunction: implications for health. *Nat. Rev. Immunol.* 5:243–51

Grill HJ. 2010. Leptin and the systems neuroscience of meal size control. *Front. Neuroendocrinol.* 31:61–78

Gunter MJ, Hoover DR, Yu H, Wassertheil-Smoller S, Rohan TE, et al. 2009. Insulin, insulin-like growth factor-I, and risk of breast cancer in postmenopausal women. *J. Natl. Cancer Inst.* 101:48–60

Hanahan D, Weinberg RA. 2000. The hallmarks of cancer. *Cell* 100:57–70

Hanahan D, Weinberg RA. 2011. Hallmarks of cancer: the next generation. *Cell* 144:646–74

Hasegawa H, Saiki I. 2002. Psychosocial stress augments tumor development through beta-adrenergic activation in mice. *Jpn. J. Cancer Res.* 93:729–35

Holt-Lunstad J, Smith TB, Layton JB. 2010. Social relationships and mortality risk: a meta-analytic review. *PLoS Med.* 7:e 1000316

House JS, Landis KR, Umberson D. 1988. Social relationships and health. *Science* 241:540–45

Int. Agency Res. Cancer (IARC)/World Health Organ. (WHO). 2008. *The GLOBOCAN project.* **http://globocan.iarc.fr/**

Jee SH, Ohrr H, Sull JW, Yun JE, Ji M, Samet JM. 2005. Fasting serum glucose level and cancer risk in Korean men and women. *JAMA* 293:194–202

Joels M, Baram TZ. 2009. The neuro-symphony of stress. *Nat. Rev. Neurosci.* 10:459–66

Knowler WC, Fowler SE, Hamman RF, Christophi CA, Hoffman HJ, et al. 2009. 10-year follow-up of diabetes incidence and weight loss in the Diabetes Prevention Program Outcomes Study. *Lancet* 374:1677–86

Kreier F, Kap YS, Mettenleiter TC, van Heijningen C, van der Vliet J, et al. 2006. Tracing from fat tissue, liver, and pancreas: a neuroanatomical framework for the role of the brain in type 2 diabetes. *Endocrinology* 147:1140–47

Lapiz MD, Fulford A, Muchimapura S, Mason R, Parker T, Marsden CA. 2003. Influence of postweaning social isolation in the rat on brain development, conditioned behavior, and neurotransmission. *Neurosci. Behav. Physiol.* 33:13–29

Lillberg K, Verkasalo PK, Kaprio J, Teppo L, Helenius H, Koskenvuo M. 2003. Stressful life events and risk of breast cancer in 10,808 women: a cohort study. *Am. J. Epidemiol.* 157:415–23

Longo VD, Fontana L. 2010. Calorie restriction and cancer prevention: metabolic and molecular mechanisms. *Trends Pharmacol. Sci.* 31:89–98

Lutgendorf SK, Cole S, Costanzo E, Bradley S, Coffin J, et al. 2003. Stress-related mediators stimulate vascular endothelial growth factor secretion by two ovarian cancer cell lines. *Clin. Cancer Res.* 9:4514–21

McEwen BS. 1998. Protective and damaging effects of stress mediators. *N. Engl. J. Med.* 338:171–79

McEwen BS, Gianaros PJ. 2010. Central role of the brain in stress and adaptation: links to socioeconomic status, health, and disease. *Ann. N. Y. Acad. Sci.* 1186:190–222

Melamed R, Rosenne E, Shakhar K, Schwartz Y, Abudarham N, Ben-Eliyahu S. 2005. Marginating pulmonary-NK activity and resistance to experimental tumor metastasis: suppression by surgery and the prophylactic use of a beta-adrenergic antagonist and a prostaglandin synthesis inhibitor. *Brain Behav. Immun.* 19:114–26

Milsum JH. 1985. A model of the eustress system for health/illness. *Behav. Sci.* 30:179–86

Nithianantharajah J, Hannan AJ. 2006. Enriched environments, experience-dependent plasticity and disorders of the nervous system. *Nat. Rev. Neurosci.* 7:697–709

Olefsky JM. 2009. IKKepsilon: a bridge between obesity and inflammation. *Cell* 138:834–36

Palermo-Neto J, de Oliveira Massoco C, Robespierre de Souza W. 2003. Effects of physical and psychological stressors on behavior, macrophage activity, and Ehrlich tumor growth. *Brain Behav. Immun.* 17:43–54

Palermo-Neto J, Fonseca ES, Quinteiro-Filho WM, Correia CS, Sakai M. 2008. Effects of individual housing on behavior and resistance to Ehrlich tumor growth in mice. *Physiol. Behav.* 95:435–40

Parker ED, Folsom AR. 2003. Intentional weight loss and incidence of obesity-related cancers: the Iowa Women's Health Study. *Int. J. Obes. Relat. Metab. Disord.* 27:1447–52

Perron L, Bairati I, Harel F, Meyer F. 2004. Antihypertensive drug use and the risk of prostate cancer (Canada). *Cancer Causes Control* 15:535–41

Polesel J, Zucchetto A, Montella M, Dal Maso L, Crispo A, et al. 2009. The impact of obesity and diabetes mellitus on the risk of hepatocellular carcinoma. *Ann. Oncol.* 20:353–57

Price MA, Tennant CC, Butow PN, Smith RC, Kennedy SJ, et al. 2001. The role of psychosocial factors in the development of breast carcinoma: Part II. Life event stressors, social support, defense style, and emotional control and their interactions. *Cancer* 91:686–97

Pugeat M, Crave JC, Elmidani M, Nicolas MH, Garoscio-Cholet M, et al. 1991. Pathophysiology of sex hormone binding globulin (SHBG): relation to insulin. *J. Steroid. Biochem. Mol. Biol.* 40:841–9

Reeves GK, Pirie K, Beral V, Green J, Spencer E, Bull D. 2007. Cancer incidence and mortality in relation to body mass index in the Million Women Study: cohort study. *BMJ* 335:1134

Reiche EM, Nunes SO, Morimoto HK. 2004. Stress, depression, the immune system, and cancer. *Lancet Oncol.* 5:617–25

Renehan AG. 2009. Bariatric surgery, weight reduction, and cancer prevention. *Lancet Oncol.* 10:640–41

Renehan AG, Tyson M, Egger M, Heller RF, Zwahlen M. 2008. Body-mass index and incidence of cancer: a systematic review and meta-analysis of prospective observational studies. *Lancet* 371:569–78

Saul AN, Oberyszyn TM, Daugherty C, Kusewitt D, Jones S, et al. 2005. Chronic stress and susceptibility to skin cancer. *J. Natl. Cancer Inst.* 97:1760–67

Selye H. 1974. *Stress Without Distress*. Toronto: McClelland and Stewart

Sjostrom L, Narbro K, Sjostrom CD, Karason K, Larsson B, et al. 2007. Effects of bariatric surgery on mortality in Swedish obese subjects. *N. Engl. J. Med.* 357:741–52

Sloan EK, Priceman SJ, Cox BF, Yu S, Pimentel MA, et al. 2010. The sympathetic nervous system induces a metastatic switch in primary breast cancer. *Cancer Res.* 70:7042–52

Sood AK, Bhatty R, Kamat AA, Landen CN, Han L, et al. 2006. Stress hormone-mediated invasion of ovarian cancer cells. *Clin. Cancer Res.* 12:369–75

Stefanski V, Ben-Eliyahu S. 1996. Social confrontation and tumor metastasis in rats: defeat and beta-adrenergic mechanisms. *Physiol. Behav.* 60:277–82

Steplewski Z, Vogel WH, Ehya H, Poropatich C, Smith JM. 1985. Effects of restraint stress on inoculated tumor growth and immune response in rats. *Cancer Res.* 45:5128–33

Sterling P, Eyer J. 1988. Allostasis: a new paradigm to explain arousal pathology. In *Handbook of Life Stress, Cognition and Health*, ed. S Fisher, J Reason, pp. 629–49. New York: Wiley

Thaker PH, Han LY, Kamat AA, Arevalo JM, Takahashi R, et al. 2006. Chronic stress promotes tumor growth and angiogenesis in a mouse model of ovarian carcinoma. *Nat. Med.* 12:939–44

Thaker PH, Lutgendorf SK, Sood AK. 2007. The neuroendocrine impact of chronic stress on cancer. *Cell Cycle* 6:430–33

Thaker PH, Sood AK. 2008. Neuroendocrine influences on cancer biology. *Semin. Cancer Biol.* 18:164–70

Trevisan M, Liu J, Muti P, Misciagna G, Menotti A, Fucci F. 2001. Markers of insulin resistance and colorectal cancer mortality. *Cancer Epidemiol. Biomarkers Prev.* 10:937–41

Union Int. Cancer Control (UICC). 2008. *World cancer declaration*. **http://www.uicc.org/declaration**

Whitlock G, Lewington S, Sherliker P, Clarke R, Emberson J, et al. 2009. Body-mass index and cause-specific mortality in 900,000 adults: collaborative analyses of 57 prospective studies. *Lancet* 373:1083–96

Williams JB, Pang D, Delgado B, Kocherginsky M, Tretiakova M, et al. 2009. A model of gene-environment interaction reveals altered mammary gland gene expression and increased tumor growth following social isolation. *Cancer Prev. Res.* 2:850–61

World Health Organ. (WHO). 2011. *Obesity and overweight: fact sheet N°311.* **http://www.who.int/mediacentre/factsheets/fs311/en/**

World Health Organ. (WHO). 2012. *Cancer: fact sheet N°297.* **http://www.who.int/mediacentre/factsheets/fs297/en/**

Yang EV, Donovan EL, Benson DM, Glaser R. 2008. VEGF is differentially regulated in multiple myeloma-derived cell lines by norepinephrine. *Brain Behav. Immun.* 22:318–23

Yang EV, Kim SJ, Donovan EL, Chen M, Gross AC, et al. 2009. Norepinephrine upregulates VEGF, IL-8, and IL-6 expression in human melanoma tumor cell lines: implications for stress-related enhancement of tumor progression. *Brain Behav. Immun.* 23:267–75

Yang EV, Sood AK, Chen M, Li Y, Eubank TD, et al. 2006. Norepinephrine up-regulates the expression of vascular endothelial growth factor, matrix metalloproteinase (MMP)-2, and MMP-9 in nasopharyngeal carcinoma tumor cells. *Cancer Res.* 66:10357–64

Young D, Lawlor PA, Leone P, Dragunow M, During MJ. 1999. Environmental enrichment inhibits spontaneous apoptosis, prevents seizures and is neuroprotective. *Nat. Med.* 5:448–53

Zhang Y, Daquinag A, Traktuev DO, Amaya-Manzanares F, Simmons PJ, et al. 2009. White adipose tissue cells are recruited by experimental tumors and promote cancer progression in mouse models. *Cancer Res.* 69:5259–66

The Role of Organizers in Patterning the Nervous System

Clemens Kiecker and Andrew Lumsden

Medical Research Council (MRC) Center for Developmental Neurobiology, King's College, London SE1 1UL, United Kingdom; email: clemens.kiecker@kcl.ac.uk, andrew.lumsden@kcl.ac.uk

Annu. Rev. Neurosci. 2012. 35:347–67

First published online as a Review in Advance on March 29, 2012

The *Annual Review of Neuroscience* is online at neuro.annualreviews.org

This article's doi:
10.1146/annurev-neuro-062111-150543

0147-006X/12/0721-0347$20.00

Keywords

morphogen signaling, gradients, Spemann, competence, neural tube, cell lineage restriction boundaries

Abstract

The foundation for the anatomical and functional complexity of the vertebrate central nervous system is laid during embryogenesis. After Spemann's organizer and its derivatives have endowed the neural plate with a coarse pattern along its anteroposterior and mediolateral axes, this basis is progressively refined by the activity of secondary organizers within the neuroepithelium that function by releasing diffusible signaling factors. Dorsoventral patterning is mediated by two organizer regions that extend along the dorsal and ventral midlines of the entire neuraxis, whereas anteroposterior patterning is controlled by several discrete organizers. Here we review how these secondary organizers are established and how they exert their signaling functions. Organizer signals come from a surprisingly limited set of signaling factor families, indicating that the competence of target cells to respond to those signals plays an important part in neural patterning.

Contents

INTRODUCTION

Although it remains astounding, even to the experienced neurobiologist, that a structure as complex as the human brain can arise from a single cell, work in different vertebrate model organisms has started to reveal a network of tissue and genetic interactions that engineer this extraordinary feat. During embryogenesis, the primordial neuroepithelium progressively subdivides into distinct regions in a patterning process governed by small groups of cells that regulate cell fate in surrounding tissues by releasing signaling factors. These local signaling centers are called organizers to reflect their ability to confer identity on neighboring tissues in a nonautonomous fashion.

Dorsal blastopore lip: a group of dorsal mesodermal cells of the amphibian embryo where the involution of mesoderm and endoderm starts, marking the onset of gastrulation

SPEMANN'S ORGANIZER AND EARLY NEURAL PATTERNING

In 1935 Hans Spemann received the Nobel Prize in Medicine for his work with Hilde Mangold showing that transplantation of a small group of cells from the dorsal blastopore lip of a donor embryo to the ventral side of a host embryo is sufficient to induce a secondary body axis (reviewed in De Robertis & Kuroda 2004, Niehrs 2004, Stern 2001). Differently pigmented salamander embryos were used as donors and hosts, allowing for an easy distinction between cells of graft and host origin. Surprisingly, most tissues in the induced second axis were derived from the host, suggesting that the graft had induced surrounding tissue to form axial structures. Thus, Spemann named the dorsal blastopore lip the organizer, and tissues with comparable inductive activity have since been identified in all vertebrate model organisms and more recently also in some nonvertebrates (Darras et al. 2011, Meinhardt 2006, Nakamoto et al. 2011). Nowadays the term organizer is used more widely to describe groups of cells that can determine the fate of neighboring cell populations by emitting molecular signals.

The ectopic twin induced in Spemann's experiment contained a complete CNS that was properly patterned along its anteroposterior (AP, head-to-tail) and dorsoventral (DV, back-to-belly) axes, indicating that the organizer harbors both neural-inducing and neural-patterning activities. More recently, a large number of factors that are expressed in Spemann's organizer have been identified, and several were found to be secreted inhibitors of bone morphogenetic proteins (BMPs). In combination with other findings in frog and fish embryos, this led to a model whereby Spemann's organizer induces the neural plate in the dorsal ectoderm by inhibiting BMPs, whereas the ventral ectoderm forms epidermis because it remains exposed to BMPs (De Robertis & Kuroda 2004, Muñoz-Sanjuán & Brivanlou 2002). Experiments in chick embryos have since added complexity to this default model for neural induction by implicating other signaling proteins such as fibroblast growth factors (FGFs) and Wnts as additional neural inducers (Stern 2006).

During gastrulation, the organizer region stretches out and gives rise to the axial mesendoderm (AME), which comes to underlie the midline of the neural plate along its AP axis. Otto Mangold found that different AP regions of the AME induced different parts of the embryonic axis when grafted into host embryos, leading to the idea of regionally specific inductions by the organizer (Niehrs 2004). This model was challenged in the 1950s when Nieuwkoop and others proposed that the CNS is patterned by a gradient of a transformer that travels within the plane of the neural plate and induces different neural fates in a dose-dependent manner such that forebrain, midbrain, hindbrain, and spinal cord form at increasing levels of this transformer (Stern 2001). FGFs (Mason 2007), retinoic acid (Maden 2007), and Wnts all posteriorize neuroectoderm dose-dependently, but Wnts appear to be the best candidates to fulfill this role in a manner consistent with Nieuwkoop's model (Kiecker & Niehrs 2001a). Spemann's organizer also secretes inhibitors of Wnts, in addition to BMP antagonists, and these factors remain expressed in the anterior AME but are absent from the posterior AME during gastrulation (Kiecker & Niehrs 2001b). Thus, the Spemann-Mangold model of regionally specific inductions and Nieuwkoop's gradient-based model turn out to be two sides of the same coin: The anterior AME induces the forebrain by acting as a sink for posteriorizing Wnts.

AP: anteroposterior

DV: dorsoventral

Bone morphogenetic protein (BMP): subfamily of the transforming growth factor β superfamily of secreted signaling factors; initially identified by their promotion of bone and cartilage formation

Fibroblast growth factor (FGF): secreted signaling molecules that signal via tyrosine kinase receptors

Wnts: secreted lipid-modified glycoproteins that regulate multiple aspects of embryogenesis and adult homeostasis by activating several different signaling pathways

AME: axial mesendoderm

DORSOVENTRAL PATTERNING

The ectoderm that surrounds the neural plate expresses BMPs while Spemann's organizer and the extending AME express BMP inhibitors. Experiments in zebrafish embryos have suggested that this generates a gradient of BMP activity that defines mediolateral positions within the neural plate (Barth et al. 1999). Hence, a Cartesian coordinate system of two orthogonal gradients is established—a Wnt gradient along the AP axis and a BMP gradient along the mediolateral axis—and the AME defines the origin of this system by secreting BMP and Wnt antagonists (**Figure 1*a***) (Meinhardt 2006, Niehrs 2010). It is clear that such global mechanisms can establish only a crude initial pattern, and we argue below that this pattern is increasingly refined through the establishment of local (or secondary) organizers in the neuroepithelium.

Gastrulation is followed by neurulation during which the lateral folds of the neural plate roll up and fuse to form the neural tube (**Figure 1*b,c***). Thus, the initial mediolateral pattern is transposed into DV polarity: Cells that are medial in the plate end up ventral in

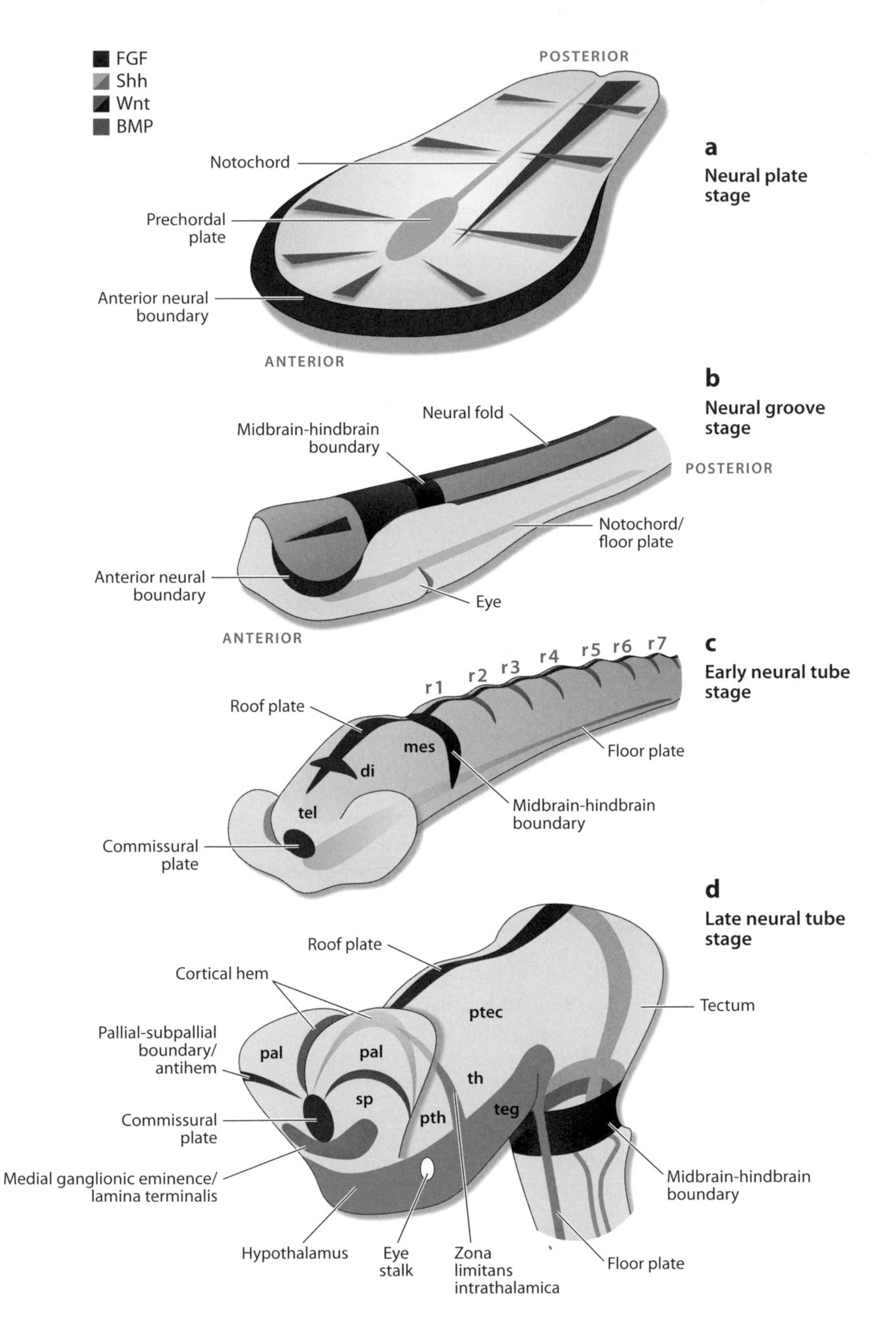

FGF
Shh
Wnt
BMP
POSTERIOR
Notochord
Prechordal plate
Anterior neural boundary
ANTERIOR
a
Neural plate stage
b
Neural groove stage
Neural fold
Midbrain-hindbrain boundary
POSTERIOR
Notochord/ floor plate
Anterior neural boundary
Eye
ANTERIOR
c
Early neural tube stage
r1
r2
r3
r4
r5
r6
r7
Roof plate
mes
di
tel
Floor plate
Midbrain-hindbrain boundary
Commissural plate
d
Late neural tube stage
Roof plate
Cortical hem
Tectum
ptec
Pallial-subpallial boundary/ antihem
pal
pal
sp
th
pth
teg
Commissural plate
Medial ganglionic eminence/ lamina terminalis
Midbrain-hindbrain boundary
Hypothalamus
Eye stalk
Zona limitans intrathalamica
Floor plate

the tube, whereas lateral neural plate cells end up dorsal.

The Notochord

During gastrulation and neurulation a large portion of the AME narrows to a thin rod, the notochord, which underlies the midline of the neural plate and later most of the neural tube (**Figure 1*a–c***). The prechordal plate, the anterior end of the AME that lies beneath the prospective anterior forebrain, is a bit wider than the notochord. The ventralmost cells of the neural tube that reside directly above the notochord form the floor plate on either side of which form motor neurons and various types of ventral interneurons.

Although not a neural structure, the notochord was one of the first tissues shown to act as a local organizer of CNS development. Microsurgical experiments in chick embryos revealed that the notochord is both necessary and sufficient to induce ventral neural identity: Sections of the spinal cord from which the notochord had been removed developed without a floor plate or motor neurons; conversely, the transplantation of pieces of notochord beneath the lateral neural plate resulted in the induction of an ectopic floor plate and motor neurons above the graft (Placzek et al. 1990, van Straaten et al. 1985).

A breakthrough in understanding this action of the notochord came with the finding that *Sonic hedgehog* (*Shh*), a vertebrate ortholog of the *Drosophila* segment polarity gene *hedgehog*, is expressed in the notochord and a bit later also in the floor plate. *Shh* encodes a secreted signaling factor and is thus a prime candidate for the inductive signal released by the notochord. Overexpression of *Shh* in mouse, zebrafish, and frog, or coculture of rat neuroectoderm with *Shh*-expressing cells result in ectopic floor plate, motor neuron, and ventral interneuron induction, indicating that Shh mimics the effect of notochord grafts (Echelard et al. 1993, Krauss et al. 1993, Roelink et al. 1994). Conversely, mice carrying a mutation in the *Shh* gene fail to form a floor plate and lack multiple ventral neural cell types (Chiang et al. 1996). Taken together, these data strongly suggest that Shh mediates the organizer function of the notochord.

Motor neurons: efferent neurons that control muscle activity

Interneurons: neurons that connect afferent and efferent neurons in multisynaptic pathways

***Shh*:** a secreted signaling factor; its active form is lipid modified and proteolytically processed

The Floor Plate

The floor plate is a strip of wedge-shaped glial cells along the ventral midline of the neural tube. Like the notochord, the floor plate expresses *Shh* and is therefore likely to contribute to the induction of ventral cell identities. Fate-mapping studies in the chick embryo in combination with detailed examinations of cellular morphologies and marker gene expression have revealed that the floor plate consists of different cell populations along both its AP and mediolateral axes (Placzek & Briscoe 2005). For example, whereas the floor plate is devoid of neural progenitors in the spinal cord and hindbrain, dopaminergic neurons are generated in the floor plate of the midbrain (Ono et al. 2007, Puelles et al. 2004).

The origin of the floor plate remains controversial. The ablation and grafting experiments described above strongly suggest that it is induced by Shh signaling from the notochord; however, others have argued that the floor plate and notochord originate from a common precursor population in Spemann's organizer

Figure 1

Main stages of neural development in a schematized amniote embryo. (*a*) Neural plate stage. Mediolateral gradients of BMP (*brown wedges*) and Shh (*green*) activity together with an anteroposterior gradient of Wnt activity (*red wedge*) establish a quasi-Cartesian coordinate system of positional information across the neural plate. (*b*) Neural groove stage. The interplay between Wnts and ANB-derived Wnt inhibitors patterns the area between the presumptive forebrain and midbrain (*red wedge*). (*c*) Early neural tube stage. (*d*) Late neural tube stage. Colors: FGF expression (*blue*), Shh expression (*green*), BMP expression (*brown*), Wnt expression (*orange/red*). Abbreviations: di, diencephalon; mes, mesencephalon; pal, pallium; ptec, pretectum; pth, prethalamus; r, rhombomere; sp, subpallium; teg, tegmentum; tel, telencephalon; th, thalamus.

and that floor plate cells are inserted into the midline of the neural plate as the AP axis of the embryo extends (Le Douarin & Halpern 2000). These views are likely not entirely incompatible: The common lineage of notochord and floor plate may endow them with shared properties, and both tissues are capable of inducing homeogenetic responses in neuroepithelium.

Morphogen: factor that is released locally, forms a concentration gradient within a tissue, and induces different cell fates dose-dependently

Ventral Neural Patterning by Shh: A Paradigm for Morphogen Signaling

The specification of multiple cell types in the ventral neural tube is arguably one of the most thoroughly studied examples of neural patterning. Considerable evidence gathered over the past 15 years indicates that Shh functions as a true morphogen in this process; i.e., it is released from a local source (notochord, floor plate) and forms a concentration gradient that specifies different cell fates in a dose-dependent fashion (Dessaud et al. 2008, Lupo et al. 2006). In mouse embryos that were genetically engineered to express fluorescently labeled Shh protein (Shh-GFP) from the *Shh* locus, a declining ventral-to-dorsal gradient of fluorescence is detectable within the ventral neural tube (Chamberlain et al. 2008).

The morphogen model is intuitively appealing because it explains a complex process pattern formation, drawing on a simple chemical activity—the diffusion of a single substance from a localized source. However, trying to understand the cellular mechanism of morphogen signaling raises a number of difficult issues. For example, how are different concentrations of a morphogen translated into distinct cell fates? In vertebrates, Shh activates the transcriptional activators Gli1 and Gli2 and antagonizes the repressor Gli3. Thus, the extracellular gradient of Shh is translated into opposing gradients of intracellular Gli1/2 and Gli3 activity along the DV axis of the neural tube (Fuccillo et al. 2006, Lei et al. 2004, Stamataki et al. 2005). These overlapping activities regulate two classes of transcriptional control genes: Class I genes such as *Pax6*, *Pax7*, and *Irx3* are repressed, and class II genes, including *Foxa2*, *Nkx2.2*, *Olig2*, *Nkx6.1*, *Dbx1*, and *Dbx2*, are induced by Shh-Gli signaling. Different thresholds of Shh signaling are required for the repression or activation of individual class I and class II genes, resulting in a nested expression pattern of these genes along the DV axis of the spinal cord. Furthermore, several class I and class II genes cross-repress each other, resulting in a sharpening of the boundaries between their expression domains (Briscoe et al. 2000). Ultimately, the combinatorial expression of class I and class II genes at a specific DV location determines which type of neural progenitor will form.

Dessaud et al. (2008) recently suggested that the duration of exposure, in addition to the extracellular concentration of Shh, determines the fate of the receiving cell. This model is likely to reflect the gradual buildup of the Shh gradient in vivo better than would a static gradient model.

Ventral Patterning in the Hindbrain and Midbrain

Early in its development, the hindbrain becomes subdivided into a series of seven to eight segments called rhombomeres (r1–8) (reviewed in Kiecker & Lumsden 2005). Nevertheless, the topological organization of neurons along the DV axis of the hindbrain is similar to that of the spinal cord. The induction of hindbrain motor neurons (which contribute to the IVth–XIIth cranial nerves) also depends on Shh signaling from the notochord/floor plate and on the nested expression of various class I and class II genes (Osumi et al. 1997, Pattyn et al. 2003, Takahashi & Osumi 2002).

The ventral part of the midbrain (and of the posterior forebrain) gives rise to the tegmentum where neural progenitors are organized in a series of morphologically visible arcs that are characterized by periodic gene expression patterns (Sanders et al. 2002). Gain- and loss-of-function experiments in chick embryos have provided evidence that this pattern is also controlled by Shh signaling (Agarwala et al. 2001, Bayly et al. 2007).

Organizers of Ventral Forebrain Development

Like the notochord, the prechordal plate expresses *Shh*, and grafts of this AME tissue to the lateral forebrain result in ectopic expression of the ventral forebrain marker *Nkx2.1* in chick embryos. However, this effect cannot be mimicked by implanting Shh-producing cells, suggesting that the inductive capacity of the prechordal plate reaches beyond mere secretion of Shh (Pera & Kessel 1997). A good candidate factor to mediate this difference is BMP7, which is expressed in the prechordal mesoderm but not in the posterior notochord and synergizes with Shh in inducing ventral forebrain identity (Dale et al. 1997). Furthermore, potent Wnt inhibitors are expressed in the prechordal plate, and these likely contribute to forebrain induction and ventralization (Kiecker & Niehrs 2001b).

The vertebrate forebrain becomes divided into the telencephalon anteriorly and the diencephalon posteriorly (**Figure 1*c***). The dorsal part of the telencephalon, the pallium, gives rise to the hippocampus and the cerebral cortex (or functionally equivalent structures in nonmammalian species), whereas the ventral part, the subpallium, gives rise to the basal ganglia. The earliest distinction between pallial and subpallial identity is established at the neural plate stage and is mediated by Shh from the prechordal plate (Gunhaga et al. 2000). However, prechordal plate-derived Shh also induces a new domain of *Shh* expression in the most ventral part of the subpallium, the presumptive medial ganglionic eminence (MGE), in a manner that is very similar to the induction of the floor plate by Shh from the notochord. This telencephalic *Shh* domain is likely to act as a secondary organizer that refines the DV pattern of the telencephalon (**Figure 1*d***) (Sousa & Fishell 2010).

The Roof Plate

Another specialized cell population, the roof plate, forms along the dorsal midline of the entire neural tube. Similar to the floor plate, the roof plate is induced by inductive signals from a nonneural tissue—in this case by BMPs and Wnts expressed in the ectoderm flanking the neural plate (Chizhikov & Millen 2004b). Neural tube closure brings the roof plate progenitors from either side of the neural plate together to form the dorsal midline of the neural tube. Identifying a specific genetic pathway for roof plate formation has been complicated by the fact that two other cell groups, the neural crest and some dorsal interneuron progenitors, also arise from the dorsal midline.

However, there is good evidence that the roof plate functions as an organizer of dorsal patterning in the spinal cord. Six populations of dorsal interneurons (dI1–6) are generated in the dorsal half of the spinal cord, and the coculture of roof plate and naïve neural plate tissue leads to the induction of at least two of those (Liem et al. 1997). Genetic ablation of the roof plate in mice results in the lack of dI1–3 and expansion of dI4–6 interneurons, confirming that the roof plate is not only sufficient but also required for dorsal spinal cord patterning (K.J. Lee et al. 2000). *Lmx1a* mutant mice in which roof plate formation is disrupted show a similar, if somewhat milder phenotype (Millonig et al. 2000). Overexpression of *Lmx1a*—which encodes a LIM homeodomain transcription factor—in the chick spinal cord results in ectopic induction of dI1 at the expense of dI2–6 interneurons in the vicinity of the electroporated cells, suggesting that a diffusible signal is secondarily induced (Chizhikov & Millen 2004a).

Identification of the signal that mediates the organizer function of the roof plate has been more problematic than identifying that of the floor plate. The roof plate expresses several Wnts and a large number of BMPs, and the genetic inactivation of individual factors often results in entirely normal spinal cord development, probably owing to compensation by other family members. The BMP-type factor GDF7 is an exception because *Gdf7*$^{-/-}$ mutant mice lack dI1 interneurons (Lee et al. 1998). Zebrafish mutants with defects in the BMP pathway fail to form Rohon-Beard neurons,

Boundary: an interface between two adjacent tissues that prevents intermingling between cells from either side of the boundary

indicating that BMPs are required for the formation of at least some dorsal cell types (Nguyen et al. 2000). *Wnt1/Wnt3a* double mutant mice show a severe reduction of dI1–3 and a concomitant expansion of dI4 and dI5 interneurons and are therefore phenotypically similar to roof plate–ablated mice, although less severely so (Muroyama et al. 2002). However, the reduction of dI1–3 interneurons could be due to defective proliferation rather than patterning because Wnts act as mitogens in gain-of-function experiments in chick (Megason & McMahon 2002).

Do roof plate signals act in a dose-dependent fashion, similar to Shh on the ventral side of the neural tube? Some evidence points toward graded effects by both BMPs and Wnts in the dorsal spinal cord (Liem et al. 1997, Megason & McMahon 2002, Timmer et al. 2002); however, the picture is far less conclusive than for Shh and it remains possible that qualitative as well as quantitative mechanisms are at work (i.e., individual BMPs or Wnts may specifically induce certain subpopulations of dorsal interneurons).

Dorsal Patterning in the Anterior Hindbrain

The dorsal part of the anterior hindbrain (r1) undergoes a series of complex morphological changes that result in the formation of the cerebellum. Granule cells, the most prevailing cell type in the cerebellum, are generated from the rhombic lip, a germinal zone at the interface between the roof plate and the r1 neuroepithelium (Hatten & Roussel 2011, Wingate 2001). The roof plate of r1, like that of the spinal cord, expresses several BMPs, and these are sufficient to initiate granule cell formation when added ectopically to r1 neuroepithelium (Alder et al. 1999). The locus coeruleus, the major noradrenergic nucleus of the brain, is also induced in the dorsal half of r1 before its neurons migrate ventrally to reach their final destination in the lateral floor of the IVth ventricle. Application of BMP antagonists to the anterior hindbrain of chick embryos results in the disappearance of or a dorsal shift of locus coeruleus neurons, suggesting that the role of BMPs in the induction of these neurons may be dose dependent (Vogel-Höpker & Rohrer 2002).

The Cortical Hem

In the telencephalon, the roof plate sinks between the two cortical hemispheres and gives rise to the monolayered choroid plexus medially and to the cortical hem laterally. Immediately adjacent to the hem, which expresses BMPs and Wnts, the hippocampus is induced and the cerebral cortex forms next to that. Genetic ablation of the telencephalic roof plate in mice results in a severe undergrowth of the cortical primordium, suggesting a nonautonomous effect of the hem on cortical specification (Monuki et al. 2001).

Ectopic application of BMPs to the developing chick telencephalon leads to holoprosencephaly (a failure to separate the cerebral hemispheres), but this is likely to be a result of increased cell death rather than a change in patterning (Golden et al. 1999). A requirement for BMP signaling in telencephalic patterning has been tested by genetically disrupting a BMP receptor gene in the mouse forebrain. These mice fail to differentiate the choroid plexus, but all other telencephalic subdivisions develop normally, arguing against a role for BMPs as an organizer signal (Hébert et al. 2002).

Wnt signaling specifies dorsal identity at the earliest stages of telencephalic development (Backman et al. 2005, Gunhaga et al. 2003), and the Wnt antagonists secreted by the prechordal plate likely help to set up early DV polarity by protecting the ventral forebrain from the dorsalizing activity of Wnts. At later stages, Wnt signaling is required for hippocampus differentiation (Galceran et al. 2000, S.M. Lee et al. 2000, Machon et al. 2003). The boundary between the pallium and the subpallium (PSB) begins to express Wnt inhibitors, raising the possibility that a gradient of Wnt activity is established across the pallium between the hem and the PSB (which has also been called the antihem) (Assimacopoulos et al. 2003, Frowein et al. 2002). However, there is little evidence for a later patterning function for Wnts beyond hippocampus induction (Chenn & Walsh 2002, Hirabayashi et al. 2004, Hirsch et al. 2007, Ivaniutsin et al. 2009, Machon et al. 2007, Muzio et al. 2005). The PSB also expresses several members of the epidermal growth factor family, transforming growth factor α and FGF7, but their roles in telencephalic development remain unknown (Assimacopoulos et al. 2003).

In summary, two major signaling centers organize DV patterning of the neural tube: the roof plate dorsally, which secretes BMPs and Wnts, and the floor plate ventrally, which secretes Shh. Both are induced by the same sets of molecular signals from the epidermis and the notochord, respectively. The notochord itself is an organizer that can mediate ventral neural patterning. In the telencephalon, an additional potential organizer, the PSB, is located midway along the DV axis.

ANTEROPOSTERIOR PATTERNING

In contrast with the DV axis, which is patterned by two signaling centers at opposite poles of the neural tube, the AP axis is patterned by several discrete local organizers.

The Midbrain-Hindbrain Boundary

The boundary between the midbrain and hindbrain (MHB) is characterized by a morphological constriction of the neural tube and is therefore also called the isthmus (**Figure 1*c,d***). The first experimental evidence that the MHB functions as an organizer came from microsurgical studies conducted in chick embryos: Transplantation of anterior hindbrain or posterior midbrain tissue into the posterior forebrain of a host embryo induced the formation of an ectopic isthmus and ectopic midbrain tissue around the graft (Bloch-Gallego et al. 1996, Gardner & Barald 1991, Martinez et al. 1991). Experimental rotation of the entire midbrain vesicle formed a double-posterior midbrain and induced ectopic midbrain and cerebellar structures in the posterior forebrain (Marín & Puelles 1994). Demonstrating a requirement for the MHB in the patterning of the surrounding tissues by microsurgery turned out to be less feasible because the isthmic organizer rapidly regenerates after surgical removal (Irving & Mason 1999).

MHB: midbrain-hindbrain boundary

Two signaling factors, Wnt1 and FGF8, are expressed on the anterior and posterior sides of the MHB, respectively. Both factors are required for midbrain-hindbrain development as *Wnt1* mutations in the mouse, and mutations in *Fgf8* in both mouse and zebrafish embryos result in defects in midbrain patterning and cerebellum formation (McMahon et al. 1992, Meyers et al. 1998, Picker et al. 1999, Reifers et al. 1998, Thomas et al. 1991). However, in gain-of-function experiments only FGF8 can mimic MHB organizer function (Crossley et al. 1996, Irving & Mason 2000, Lee et al. 1997, Martinez et al. 1999), whereas Wnt1 seems to promote cell growth and proliferation without affecting patterning (Panhuysen et al. 2004). Thus, FGF8 is the main organizer factor secreted from the MHB.

What determines the AP position of MHB formation? Many studies have uncovered a genetic module, including FGF8, Wnt1, and several transcription factors that stabilize MHB gene expression in a network of positive maintenance loops and mutually repressive interactions (Liu & Joyner 2001, Wurst & Bally-Cuif 2001). In particular, the homeodomain transcription factors Otx2 and Gbx2 play a central role in MHB positioning. *Otx2* is expressed in the forebrain and midbrain and *Gbx2* in the anterior hindbrain, and the interface between the two expression domains presages the position of the MHB from neural plate stages onward. Various gene targeting experiments in the mouse have demonstrated that an experimental shift of this interface always results in a concomitant repositioning of the MHB (Acampora et al. 1997, Broccoli et al. 1999, Millet 1999, Wassarman et al. 1997). Knockdown and cell transplantation experiments in the fish have revealed that the *otx/gbx* interface is regulated by Wnt8 at the

ANB: anterior neural border

ZLI: zona limitans intrathalamica

neural plate stage (Rhinn et al. 2005). Thus, the position of the MHB is directly defined by the early gradient of Wnt signaling that establishes the initial AP polarity of the neural plate. Similar to the mutual repression of class I and class II genes in the spinal cord, *Otx2* and *Gbx2* repress each other, thereby stabilizing the binary cell fate choice around the MHB (Liu & Joyner 2001, Wurst & Bally-Cuif 2001).

The Anterior Neural Boundary/Commissural Plate

Elegant cell ablation and transplantation experiments in zebrafish revealed an organizing function of the anterior border of the neural plate (ANB) (Houart et al. 1998). A Wnt inhibitor of the secreted Frizzled-related protein family is expressed in the ANB, and overexpression and depletion of this factor phenocopy the effects of ANB transplantation and removal, respectively (Houart et al. 2002). *Wnt8B* is expressed in the presumptive midbrain and posterior forebrain at the stages when ANB signaling is required and is therefore likely to be the main antagonist of the ANB. Thus, after the global gradient of Wnt activity has established general AP polarity in the neural plate, Wnts regulate AP identity in a more localized fashion in the prospective forebrain-midbrain region (**Figure 1*b***).

In the mouse, a role for Wnt inhibition from the ANB has yet to be demonstrated; however, FGF8, which is also expressed there, can mimic the anteriorizing effects of ANB in explants in vitro (Shimamura & Rubenstein 1997). Studies in both mouse and zebrafish embryos have demonstrated a need for FGFs in forebrain patterning (Meyers et al. 1998, Walshe & Mason 2003).

After neural tube closure, the ANB becomes a patch of cells at the anterior end of the neural tube that will eventually form the commissural plate (CP), a scaffold for the formation of forebrain commissures at later stages. The CP continues to express FGF8, and impressive in utero electroporation experiments in the mouse have revealed that FGF8 promotes anterior at the expense of posterior cortical fates. An ectopic source of FGF8 at the posterior end of the cortex resulted in a partial mirrored duplication of anterior cortical areas (Fukuchi-Shimogori & Grove 2001). These experiments identified the CP as an organizer of cortical patterning via its secretion of FGF8. Toyoda et al. (2010) recently showed that FGF8 acts directly and at a long range during this process, i.e., as a true morphogen.

The Zona Limitans Intrathalamica

The zona limitans intrathalamica (ZLI) is a narrow stripe of *Shh*-expressing cells in the alar plate of the diencephalon, transecting the neuraxis between the presumptive prethalamus and the thalamus (Kitamura et al. 1997, Shimamura et al. 1995, Zeltser et al. 2001). Gain- and loss-of-function experiments in chick, zebrafish, and mouse embryos have revealed that the ZLI acts as an organizer of diencephalic development by secreting Shh (Kiecker & Lumsden 2004, Scholpp et al. 2006, Vieira et al. 2005, Vue et al. 2009). At least in the thalamus, the activity of Shh appears to be dose dependent, with higher levels of signaling inducing the gamma-aminobutyric acid (GABA)-ergic rostral thalamus and lower levels inducing the glutamatergic caudal thalamus (Hashimoto-Torii et al. 2003, Vue et al. 2009; but see Jeong et al. 2011).

Fgf8 is expressed in a small patch in the dorsal ZLI and contributes to regulating the fate decision between the rostral and caudal thalamus (Kataoka & Shimogori 2008). Furthermore, a plethora of *Wnt*s shows sharp borders of expression at the ZLI, suggesting that Wnt signaling may also be involved in regulating the regionalization and/or proliferation of the diencephalic primordium (Bluske et al. 2009, Quinlan et al. 2009).

The apposition of any neural tissue anterior to the ZLI with any tissue between the ZLI and the MHB results in induction of *Shh*, indicating that planar interactions are sufficient to induce the ZLI organizer (Guinazu et al. 2007, Vieira et al. 2005). The ZLI forms at the interface between the expression domains of two classes of transcription factors: zinc finger proteins of the Fez family anteriorly and homeodomain proteins of the Irx family posteriorly (Hirata et al. 2006, Kobayashi et al. 2002, Rodríguez-Seguel et al. 2009, Scholpp et al. 2007). Expression of *Irx3* in chick is induced by Wnt signaling, suggesting that, as for *Gbx2* at the MHB, the early Wnt signal that posteriorizes the neural plate directly positions the ZLI (Braun et al. 2003).

Rhombomere Boundaries

Segmentation of the hindbrain into rhombomeres is controlled by graded retinoic acid signaling and by the reiterated and nested expression of tyrosine kinases and transcription factors, many of which are vertebrate orthologs of *Drosophila* gap and *Hox* genes (Kiecker & Lumsden 2005, Maden 2007). In zebrafish, the boundaries between rhombomeres express several Wnts (**Figure 1*c***), and the knockdown of these factors results in disorganized neurogenesis adjacent to the boundaries, suggesting that they may function as organizers, although no patterning defects have been demonstrated within the rhombomeres of such embryos (Amoyel et al. 2005, Riley et al. 2004).

COMMON FEATURES OF NEUROEPITHELIAL ORGANIZERS

As discussed above, organizers influence cell fate in surrounding tissues by secreting diffusible signaling factors that often act in a morphogen-like fashion—that is, they induce different responses in receiving cells at different distances from the source. In addition to this defining feature of organizers, several other commonalities have been observed regarding the establishment, maintenance, and signaling properties of neuroepithelial organizers.

Organizers Form Along Cell Lineage Restriction Boundaries

One of the hallmarks of hindbrain segmentation is the formation of cell lineage–restricted boundaries that prevent cells from moving between adjacent rhombomeres (Fraser et al. 1990, Jimenez-Guri et al. 2010). This finding prompted a search for boundaries in other parts of the neural tube, which led to the discovery that such boundaries often coincide with organizers (Kiecker & Lumsden 2005): Cell lineage restriction at the MHB has been demonstrated by sophisticated time-lapse imaging in zebrafish embryos and genetic fate mapping in the mouse (Langenberg & Brand 2005, Zervas et al. 2004); the ZLI is flanked by boundaries on either side (Zeltser et al. 2001); and signaling functions have been reported for rhombomere boundaries (see above). Lineage restriction has not been tested at the ANB, but it seems unlikely that cells intermingle freely across the neural-epidermal border. All three DV organizers—floor plate, PSB, and roof plate—also show some degree of lineage restriction (Awatramani et al. 2003, Fishell et al. 1993, Fraser et al. 1990, Jimenez-Guri et al. 2010).

Neurogenesis: the process by which proliferating neural progenitors exit the cell cycle and differentiate into functional neurons

The molecular mechanisms underlying boundary formation in the neural tube are not well understood. However, specific signaling pathways have been implicated: Eph-ephrin signaling is essential for segmentation in the hindbrain (Cooke et al. 2005, Kemp et al. 2009, Xu et al. 1999), and the Notch pathway appears to be involved in boundary formation at the ZLI, at rhombomere boundaries, and at the MHB (Cheng et al. 2004, Tossell et al. 2011, Zeltser et al. 2001).

Cell lineage restriction at organizers probably serves a dual function. First, boundaries tend to minimize contact between flanking cell populations, which may help to keep organizers in a straight line, facilitating the generation of a consistent diffusion gradient. Second, cells on either side of the organizer are kept in separate immiscible pools, thereby stabilizing a pattern after it has been induced.

Positive Feedback Maintains Organizers

Another feature shared by several neuroepithelial organizers is that their maintenance depends on the signal they produce. Both the floor plate and the roof plate are induced by their own signals, BMP and Shh; the ZLI depends on ongoing Shh signaling (Kiecker & Lumsden 2004, Zeltser 2005); and MHB integrity depends on FGF signaling (Sunmonu et al. 2011, Trokovic et al. 2005). Alan Turing's classical model for pattern formation postulated a chemical network of local self-enhancement and long-range inhibition, and the autoinduction of neuroepithelial organizers fits the local self-enhancement component of this model rather well (Meinhardt 2009).

Intrinsic Factors Regulate Differential Responses to Organizer Signals

FGF signaling from the MHB establishes the tectum anteriorly and the cerebellum posteriorly, and Shh from the ZLI induces prethalamic gene expression anteriorly and patterns the thalamus posteriorly. How can one signal induce such asymmetric responses on either side of an organizer? Two orthologs of the *Drosophila* competence factor *iroquois*, *Irx2* and *Irx3*, are expressed posterior to the MHB and ZLI, respectively. Ectopic expression of these factors anterior to the organizer results in a conversion of tectum into cerebellum and of prethalamus into thalamus (Kiecker & Lumsden 2004, Matsumoto et al. 2004). These effects are dependent on the organizer signals FGF and Shh, suggesting that Irx2 and Irx3 are not patterning factors themselves but that they convey a prepattern that determines the competence of different subdivisions of the neural tube to respond to secreted signals.

FGF-soaked beads induce ectopic midbrain and hindbrain structures to form from posterior forebrain tissue, but in the anterior forebrain FGFs anteriorize the pallium. Similarly, Shh from the ventral midline induces the hypothalamus marker *Nkx2.1* anteriorly, whereas it induces *Nkx6.1* posteriorly. The limit between these two regions of differential competence to respond to FGFs coincides with the ZLI, and the homeobox genes *Irx3* and *Six3* were shown to mediate posterior versus anterior competence (Kobayashi et al. 2002).

Taken together, intrinsic factors establish a prepattern in the developing CNS that regulates the cellular response to organizer signals. These factors are often induced by the earliest signals that pattern the neural plate—for example, *Irx3* is induced and *Six3* is repressed by posteriorizing Wnt signaling (Braun et al. 2003)—thereby linking early and late stages of neural patterning. This does not mean that organizer signals are merely permissive triggers that determine the timing and extent of regional specialization, the identity of which is prepatterned; they are also responsible for evoking different responses within the same field (as exemplified by the induction of GABAergic versus glutamatergic neurons by different doses of Shh within the *Irx3*-positive thalamus).

Hes Genes Prevent Neurogenesis in Organizer Regions

Organizers typically coincide with boundaries that are characterized by slower proliferation and a delay or absence of neurogenesis (Guthrie et al. 1991, Lumsden & Keynes 1989). Transcription factors of the Hes family that mediate Notch signaling are required to inhibit neurogenesis at the MHB of zebrafish and frog embryos (Geling et al. 2004, Ninkovic et al. 2005, Takada et al. 2005). All neural progenitors express *Hes* genes, but they usually become downregulated when cells undergo neurogenesis. An analysis in mouse embryos has revealed that in boundary regions *Hes* genes remain expressed and that it is this strong persistent expression that sets boundaries apart and allows organizer regions to form (Baek et al. 2006).

NEUROEPITHELIAL ORGANIZERS ALSO REGULATE PROLIFERATION, NEUROGENESIS, AND AXON GUIDANCE

Many organizer signals also function as mitogens, suggesting that growth, in addition to patterning, is modulated by organizers. For example, *Shh* mutant mice show not only patterning defects, but also a structural lack of many ventral neural tissues (Chiang et al. 1996). Wnt1 promotes growth in the MHB region (Panhuysen et al. 2004), and Wnts from the roof plate and cortical hem are known to regulate proliferation of the spinal cord and pallium (Chenn & Walsh 2002, Ivaniutsin et al. 2009, Megason & McMahon 2002, Muzio et al. 2005). FGFs from the MHB promote growth of the midbrain and cerebellum (Partanen 2007), but they also serve as survival factors in the midbrain (Basson et al. 2008). Similarly, FGFs secreted from the CP prevent apoptosis and promote growth in the telencephalon (Paek et al. 2009, Thomson et al. 2009). Contrary to the proliferative effects of FGFs, Shh, and Wnts, the BMP pathway often induces apoptosis when ectopically activated (Anderson et al. 2002, Lim et al. 2005, Liu et al. 2004).

To complicate the picture even further, some organizer factors promote cell cycle exit and neurogenesis (Fischer et al. 2011, Hirabayashi et al. 2004, Machon et al. 2007, Munji et al. 2011, Xie et al. 2011). These seemingly contradictory effects of the same classes of signals may be explained by temporal changes in the competence of the target cells (Hirsch et al. 2007); however, in some cases different members of the same protein family exert opposing effects on the balance between proliferation and differentiation (Borello et al. 2008, David 2010). Thus, by releasing growth-promoting and growth-inhibiting cues from localized sources, organizers help to mold the increasingly complex shape of the neural tube and coordinate the temporal progression of neurogenesis in defined subdivisions of the neural tube (Scholpp et al. 2009).

Once their regional identity has been established, differentiated neurons need to wire up precisely to form functional networks. Organizers also play a role at this stage of CNS formation, for example, by expressing axon guidance factors such as the chemoattractant netrin, which is secreted by the floor plate to guide commissural axons (Dickson 2002, Tessier-Lavigne & Goodman 1996). More recently, many of the classical morphogens that are secreted by organizers have been found to double as axon guidance molecules at later stages (Charron & Tessier-Lavigne 2005, Osterfield et al. 2003, Sánchez-Camancho et al. 2005, Zou & Lyuksyutova 2007). Shh from the floor plate cooperates with netrin in attracting commissural axons, whereas BMPs from the roof plate repel them (Augsburger et al. 1999, Charron et al. 2003). After these axons have crossed the midline, Shh signaling repels them via Hedgehog-interacting protein (Bourikas et al. 2005). Wnts are expressed in an AP gradient in the floor plate and guide the same axons anteriorly after they have crossed the midline (Lyuksyutova et al. 2003), whereas corticospinal axons are directed posteriorly by a repulsive interaction between Wnts and the atypical Wnt receptor Ryk (Liu et al. 2005). Thus, organizers influence CNS formation not only at early patterning stages, but also at later stages when functional circuits are established.

Commissural axons: nerve fibers that cross the midline of the nervous system

NEUROEPITHELIAL ORGANIZERS IN EVOLUTION

Vertebrates possess the most complex of all brains; even their closest relatives, the tunicates and hemichordates, have relatively simple nervous systems (Meinertzhagen et al. 2004). Sets of transcription factors that mark AP and DV subdivisions of the neural tube are conserved far beyond the chordate phylum (Irimia et al. 2010, Lowe et al. 2003, Reichert 2005, Tomer et al. 2010, Urbach & Technau 2008), and several orthologs of AP marker genes are even found along the head-to-foot axis of the coelenterate *Hydra*, suggesting an ancient origin of the genetic modules that regulate neural patterning (Technau & Steele 2011). By contrast, local organizers appear to be far less conserved: For example, an equivalent of the MHB is present in the urochordate *Ciona* (Imai et al. 2009) but not in the cephalochordate *Amphioxus* (Holland 2009). Both *Ciona* and *Amphioxus* seem to lack an equivalent of the ZLI, whereas a comparable region has been identified in the hemichordate *Saccoglossus* (C. Lowe, personal communication). Some organizers are even missing in lower vertebrates; no *hedgehog* expression or MGE-like differentiation has been found in the telencephalon of the lamprey, suggesting that this ventroanterior organizer is a gnathostome invention (Sugahara et al. 2011).

These observations indicate that local organizers are more recent innovations than the basic AP/DV patterning network and that they show some evolutionary flexibility that may provide a driving force for morphological change. This idea is supported by the recent finding that differences in forebrain morphology among cichlid fishes from Lake Malawi are correlated with subtle changes in signal strength, timing of signal production, and the position of forebrain organizers (Sylvester et al. 2010). Similarly, loss of eyesight in a cave-dwelling morph of the tetra *Astyanax* was shown to be caused by changes in the forebrain expression of *fgf8* and *shh* (Pottin et al. 2011). Thus, although the basic subdivisions of the brain are likely to have developed a long time ago, organizers are a more recent acquisition that may have been imposed on the underlying pattern and allow evolutionary adaptation to ecological niches.

CONCLUSIONS

Almost 90 years have passed since Spemann discovered the amphibian gastrula organizer; however, the organizer concept is more topical than ever, in particular in the developing vertebrate CNS where multiple organizers regulate patterning, proliferation, neurogenesis, cell death, and axon pathfinding. Neural organizers are generated by inductive signaling events between neighboring tissues, and they often form along, or are stabilized by, cell lineage restriction boundaries. The pattern induced by an organizer results in the formation of different cell populations that can potentially form further organizers at their interfaces, thereby subdividing the neuroepithelium into increasingly more specialized regions. In many ways, neural development can be regarded as a self-organizing process: Once initial polarity has been established, all the interactions necessary to form a functional CNS occur within the neuroepithelium itself.

DISCLOSURE STATEMENT

The authors are not aware of any affiliations, memberships, funding, or financial holdings that might be perceived as affecting the objectivity of this review.

ACKNOWLEDGMENTS

We apologize to the many researchers whose work we could not cite due to space constraints.

LITERATURE CITED

Acampora D, Avantaggiato V, Tuorto F, Simeone A. 1997. Genetic control of brain morphogenesis through *Otx* gene dosage requirement. *Development* 124:3639–50

Agarwala S, Sanders TA, Ragsdale CW. 2001. Sonic hedgehog control of size and shape in midbrain pattern formation. *Science* 291:2147–50

Alder J, Lee KJ, Jessell TM, Hatten ME. 1999. Generation of cerebellar granule neurons in vivo by transplantation of BMP-treated neural progenitor cells. *Nat. Neurosci.* 2:535–40

Amoyel M, Cheng YC, Jiang YJ, Wilkinson DG. 2005. Wnt1 regulates neurogenesis and mediates lateral inhibition of boundary cell specification in the zebrafish hindbrain. *Development* 132:775–85

Anderson RM, Lawrence AR, Stottmann RW, Bachiller D, Klingensmith J. 2002. Chordin and noggin promote organizing centers of forebrain development in the mouse. *Development* 129:4975–87

Assimacopoulos S, Grove EA, Ragsdale CW. 2003. Identification of a Pax6-dependent epidermal growth factor family signaling source at the lateral edge of the embryonic cerebral cortex. *J. Neurosci.* 23:6399–403

Augsburger A, Schuchardt A, Hoskins S, Dodd J, Butler S. 1999. BMPs as mediators of roof plate repulsion of commissural neurons. *Neuron* 24:127–41

Awatramani R, Soriano P, Rodriguez C, Mai JJ, Dymecki SM. 2003. Cryptic boundaries in roof plate and choroid plexus identified by intersectional gene activation. *Nat. Genet.* 35:70–75

Backman M, Machon O, Mygland L, van den Bout CJ, Zhong W, et al. 2005. Effects of canonical Wnt signaling on dorso-ventral specification of the mouse telencephalon. *Dev. Biol.* 279:155–68

Baek JH, Hatakeyama J, Sakamoto S, Ohtsuka T, Kageyama R. 2006. Persistent and high levels of *Hes1* expression regulate boundary formation in the developing central nervous system. *Development* 133:2467–76

Barth KA, Kishimoto Y, Rohr KB, Seydler C, Schulte-Merker S, Wilson SW. 1999. Bmp activity establishes a gradient of positional information throughout the entire neural plate. *Development* 126:4977–87

Basson MA, Echevarria D, Ahn CP, Sudarov A, Joyner AL, et al. 2008. Specific regions within the embryonic midbrain and cerebellum require different levels of FGF signaling during development. *Development* 135:889–98

Bayly RD, Ngo M, Aglyamova GV, Agarwala S. 2007. Regulation of ventral midbrain patterning by Hedgehog signaling. *Development* 134:2115–24

Bloch-Gallego E, Millet S, Alvarado-Mallart RM. 1996. Further observations on the susceptibility of diencephalic prosomeres to *En-2* induction and on the resulting histogenetic capabilities. *Mech. Dev.* 58:51–63

Bluske KK, Kawakami Y, Koyano-Nakagawa N, Nakagawa Y. 2009. Differential activity of Wnt/β-catenin signaling in the embryonic mouse thalamus. *Dev. Dyn.* 238:3297–309

Borello U, Cobos I, Long JE, McWhirter JR, Murre C, Rubenstein JL. 2008. FGF15 promotes neurogenesis and opposes FGF8 function during neocortical development. *Neural Dev.* 3:17

Bourikas D, Pekarik V, Baeriswyl T, Grunditz A, Sadhu R, et al. 2005. Sonic hedgehog guides commissural axons along the longitudinal axis of the spinal cord. *Nat. Neurosci.* 8:297–304

Braun MM, Etheridge A, Bernard A, Robertson CP, Roelink H. 2003. Wnt signaling is required at distinct stages of development for the induction of the posterior forebrain. *Development* 130:5579–87

Briscoe J, Pierani A, Jessell TM, Ericson J. 2000. A homeodomain protein code specifies progenitor cell identity and neuronal fate in the ventral neural tube. *Cell* 101:435–45

Broccoli V, Boncinelli E, Wurst W. 1999. The caudal limit of *Otx2* expression positions the isthmic organizer. *Nature* 401:164–68

Chamberlain CE, Jeong J, Guo C, Allen BL, McMahon AP. 2008. Notochord-derived Shh concentrates in close association with the apically positioned basal body in neural target cells and forms a dynamic gradient during neural patterning. *Development* 135:1097–106

Charron F, Stein E, Jeong J, McMahon AP, Tessier-Lavigne M. 2003. The morphogen sonic hedgehog is an axonal chemoattractant that collaborates with netrin-1 in midline axon guidance. *Cell* 113:11–23

Charron F, Tessier-Lavigne M. 2005. Novel brain wiring functions for classical morphogens: a role as graded positional cues in axon guidance. *Development* 132:2251–62

Cheng YC, Amoyel M, Qiu X, Jiang YJ, Xu Q, Wilkinson DG. 2004. Notch activation regulates the segregation and differentiation of rhombomere boundary cells in the zebrafish hindbrain. *Dev. Cell* 6:539–50

Chenn A, Walsh CA. 2002. Regulation of cerebral cortical size by control of cell cycle exit in neural precursors. *Science* 297:365–69

Chiang C, Litingtung Y, Lee E, Young KE, Corden JL, et al. 1996. Cyclopia and defective axial patterning in mice lacking *Sonic hedgehog* gene function. *Nature* 383:407–13

Chizhikov VV, Millen KJ. 2004a. Control of roof plate formation by *Lmx1a* in the developing spinal cord. *Development* 131:2693–705

Chizhikov VV, Millen KJ. 2004b. Mechanisms of roof plate formation in the vertebrate CNS. *Nat. Rev. Neurosci.* 5:808–12

Cooke JE, Kemp HA, Moens CB. 2005. EphA4 is required for cell adhesion and rhombomere-boundary formation in the zebrafish. *Curr. Biol.* 15:536–42

Crossley PH, Martinez S, Martin GR. 1996. Midbrain development induced by FGF8 in the chick embryo. *Nature* 380:66–68

Dale JK, Vesque C, Lints TJ, Sampath TK, Furley A, et al. 1997. Cooperation of BMP7 and SHH in the induction of forebrain ventral midline cells by prechordal mesoderm. *Cell* 90:257–69

Darras S, Gerhart J, Terasaki M, Kirschner M, Lowe CJ. 2011. ß-catenin specifies the endomesoderm and defines the posterior organizer of the hemichordate *Saccoglossus kowalevskii*. *Development* 138:959–70

David MD. 2010. Wnt-3a and Wnt-3 differently stimulate proliferation and neurogenesis of spinal neural precursors and promote neurite outgrowth by canonical signaling. *J. Neurosci. Res.* 88:3011–23

De Robertis EM, Kuroda H. 2004. Dorsal-ventral patterning and neural induction in *Xenopus* embryos. *Annu. Rev. Cell Dev. Biol.* 20:285–308

Dessaud E, McMahon AP, Briscoe J. 2008. Pattern formation in the vertebrate neural tube: a sonic hedgehog morphogen-regulated transcriptional network. *Development* 135:2489–503

Dickson BJ. 2002. Molecular mechanisms of axon guidance. *Science* 298:1959–64

Echelard Y, Epstein DJ, St-Jacques B, Shen L, Mohler J, et al. 1993. Sonic hedgehog, a member of a family of putative signaling molecules, is implicated in the regulation of CNS polarity. *Cell* 75:1417–30

Fischer T, Faus-Kessler T, Welzl G, Simeone A, Wurst W, Prakash N. 2011. Fgf15-mediated control of neurogenic and proneural gene expression regulates dorsal midbrain neurogenesis. *Dev. Biol.* 350:496–510

Fishell G, Mason CA, Hatten ME. 1993. Dispersion of neural progenitors within the germinal zones of the forebrain. *Nature* 362:636–38

Fraser S, Keynes R, Lumsden A. 1990. Segmentation in the chick embryo hindbrain is defined by cell lineage restrictions. *Nature* 344:431–35

Frowein J, Campbell K, Götz M. 2002. Expression of *Ngn1*, *Ngn2*, *Cash1*, *Gsh2* and *Sfrp1* in the developing chick telencephalon. *Mech. Dev.* 110:249–52

Fuccillo M, Joyner AL, Fishell G. 2006. Morphogen to mitogen: the multiple roles of hedgehog signalling in vertebrate neural development. *Nat. Rev. Neurosci.* 7:772–83

Fukuchi-Shimogori T, Grove EA. 2001. Neocortex patterning by the secreted signaling molecule FGF8. *Science* 294:1071–74

Galceran J, Miyashita-Lin EM, Devaney E, Rubenstein JL, Grosschedl R. 2000. Hippocampus development and generation of dentate gyrus granule cells is regulated by LEF1. *Development* 127:469–82

Gardner CA, Barald KF. 1991. The cellular environment controls the expression of engrailed-like protein in the cranial neuroepithelium of quail-chick chimeric embryos. *Development* 113:1037–48

Geling A, Plessy C, Rastegar S, Strähle U, Bally-Cuif L. 2004. Her5 acts as a prepattern factor that blocks *neurogenin1* and *coe2* expression upstream of Notch to inhibit neurogenesis at the midbrain-hindbrain boundary. *Development* 131:1993–2006

Golden JA, Bracilovic A, McFadden KA, Beesley JS, Rubenstein JL, Grinspan JB. 1999. Ectopic bone morphogenetic proteins 5 and 4 in the chicken forebrain lead to cyclopia and holoprosencephaly. *Proc. Natl. Acad. Sci. USA* 96:2439–44

Guinazu MF, Chambers D, Lumsden A, Kiecker C. 2007. Tissue interactions in the developing chick diencephalon. *Neural Dev.* 2:25

Gunhaga L, Jessell TM, Edlund T. 2000. Sonic hedgehog signaling at gastrula stages specifies ventral telencephalic cells in the chick embryo. *Development* 127:3283–93

Gunhaga L, Marklund M, Sjödal M, Hsieh JC, Jessell TM, Edlund T. 2003. Specification of dorsal telencephalic character by sequential Wnt and FGF signaling. *Nat. Neurosci.* 6:701–7

Guthrie S, Butcher M, Lumsden A. 1991. Patterns of cell division and interkinetic nuclear migration in the chick embryo hindbrain. *J. Neurobiol.* 22:742–54

Hashimoto-Torii K, Motoyama J, Hui CC, Kuroiwa A, Nakafuku M, Shimamura K. 2003. Differential activities of Sonic hedgehog mediated by Gli transcription factors define distinct neuronal subtypes in the dorsal thalamus. *Mech. Dev.* 120:1097–111

Hatten ME, Roussel MF. 2011. Development and cancer of the cerebellum. *Trends Neurosci.* 34:134–42

Hébert JM, Mishina Y, McConnell SK. 2002. BMP signaling is required locally to pattern the dorsal telencephalic midline. *Neuron* 35:1029–41

Hirabayashi Y, Itoh Y, Tabata H, Nakajima K, Akiyama T, et al. 2004. The Wnt/β-catenin pathway directs neuronal differentiation of cortical neural precursor cells. *Development* 131:2791–801

Hirata T, Nakazawa M, Muraoka O, Nakayama R, Suda Y, Hibi M. 2006. Zinc-finger genes *Fez* and *Fez-like* function in the establishment of diencephalon subdivisions. *Development* 133:3993–4004

Hirsch C, Campano LM, Wöhrle S, Hecht A. 2007. Canonical Wnt signaling transiently stimulates proliferation and enhances neurogenesis in neonatal neural progenitor cultures. *Exp. Cell Res.* 313:572–87

Holland LZ. 2009. Chordate roots of the vertebrate nervous system: expanding the molecular toolkit. *Nat. Rev. Neurosci.* 10:736–46

Houart C, Caneparo L, Heisenberg C, Barth K, Take-Uchi M, Wilson S. 2002. Establishment of the telencephalon during gastrulation by local antagonism of Wnt signaling. *Neuron* 35:255–65

Houart C, Westerfield M, Wilson SW. 1998. A small population of anterior cells patterns the forebrain during zebrafish gastrulation. *Nature* 391:788–92

Imai KS, Stolfi A, Levine M, Satou Y. 2009. Gene regulatory networks underlying the compartmentalization of the *Ciona* central nervous system. *Development* 136:285–93

Irimia M, Piñeiro C, Maeso I, Gómez-Skarmeta JL, Casares F, Garcia-Fernàndez J. 2010. Conserved developmental expression of *Fezf* in chordates and *Drosophila* and the origin of the zona limitans intrathalamica (ZLI) brain organizer. *Evodevo* 1:7

Irving C, Mason I. 1999. Regeneration of isthmic tissue is the result of a specific and direct interaction between rhombomere 1 and midbrain. *Development* 126:3981–89

Irving C, Mason I. 2000. Signalling by FGF8 from the isthmus patterns anterior hindbrain and establishes the anterior limit of *Hox* gene expression. *Development* 127:177–86

Ivaniutsin U, Chen Y, Mason JO, Price DJ, Pratt T. 2009. *Adenomatous polyposis coli* is required for early events in the normal growth and differentiation of the developing cerebral cortex. *Neural Dev.* 16:3

Jeong Y, Dolson DK, Waclaw RR, Matise MP, Sussel L, et al. 2011. Spatial and temporal requirements for sonic hedgehog in the regulation of thalamic interneuron identity. *Development* 138:531–41

Jimenez-Guri E, Udina F, Colas JF, Sharpe J, Padron-Barthe L, et al. 2010. Clonal analysis in mice underlines the importance of rhombomeric boundaries in cell movement restriction during hindbrain segmentation. *PLoS One* 5:e10112

Kataoka A, Shimogori T. 2008. Fgf8 controls regional identity in the developing thalamus. *Development* 135:2873–81

Kemp HA, Cooke JE, Moens CB. 2009. EphA4 and EfnB2a maintain rhombomere coherence by independently regulating intercalation of progenitor cells in the zebrafish neural keel. *Dev. Biol.* 327:313–26

Kiecker C, Lumsden A. 2004. Hedgehog signaling from the ZLI regulates diencephalic regional identity. *Nat. Neurosci.* 7:1242–49

Kiecker C, Lumsden A. 2005. Compartments and their boundaries in vertebrate brain development. *Nat. Rev. Neurosci.* 6:553–64

Kiecker C, Niehrs C. 2001a. A morphogen gradient of Wnt/β-catenin signalling regulates anteroposterior neural patterning in *Xenopus*. *Development* 128:4189–201

Kiecker C, Niehrs C. 2001b. The role of prechordal mesendoderm in neural patterning. *Curr. Opin. Neurobiol.* 11:27–33

Kitamura K, Miura H, Yanazawa M, Miyashita T, Kato K. 1997. Expression patterns of *Brx1* (*Rieg* gene), *Sonic hedgehog*, *Nkx2.2*, *Dlx1* and *Arx* during zona limitans intrathalamica and embryonic ventral lateral geniculate nuclear formation. *Mech. Dev.* 67:83–96

Kobayashi D, Kobayashi M, Matsumoto K, Ogura K, Nakafuku M, Shimamura K. 2002. Early subdivisions in the neural plate define distinct competence for inductive signals. *Development* 129:83–93

Krauss S, Concordet JP, Ingham PW. 1993. A functionally conserved homolog of the Drosophila segment polarity gene *hh* is expressed in tissues with polarizing activity in zebrafish embryos. *Cell* 75:1431–44

Langenberg T, Brand M. 2005. Lineage restriction maintains a stable organizer cell population at the zebrafish midbrain-hindbrain boundary. *Development* 132:3209–16

Le Douarin NM, Halpern ME. 2000. Discussion point. Origin and specification of the neural tube floor plate: insights from the chick and zebrafish. *Curr. Opin. Neurobiol.* 10:23–30

Lee KJ, Dietrich P, Jessell TM. 2000. Genetic ablation reveals that the roof plate is essential for dorsal interneuron specification. *Nature* 403:734–40

Lee KJ, Mendelsohn M, Jessell TM. 1998. Neuronal patterning by BMPs: a requirement for GDF7 in the generation of a discrete class of commissural interneurons in the mouse spinal cord. *Genes Dev.* 12:3394–407

Lee SM, Danielian PS, Fritzsch B, McMahon AP. 1997. Evidence that FGF8 signalling from the midbrain-hindbrain junction regulates growth and polarity in the developing midbrain. *Development* 124:959–69

Lee SM, Tole S, Grove EA, McMahon AP. 2000. A local Wnt-3a signal is required for development of the mammalian hippocampus. *Development* 127:457–67

Lei Q, Zelman AK, Kuang E, Li S, Matise MP. 2004. Transduction of graded Hedgehog signaling by a combination of Gli2 and Gli3 activator functions in the developing spinal cord. *Development* 131:3593–604

Liem KF Jr, Tremml G, Jessell TM. 1997. A role for the roof plate and its resident TGFβ-related proteins in neuronal patterning in the dorsal spinal cord. *Cell* 91:127–38

Lim Y, Cho G, Minarcik J, Golden J. 2005. Altered BMP signaling disrupts chick diencephalic development. *Mech. Dev.* 122:603–20

Liu A, Joyner AL. 2001. Early anterior/posterior patterning of the midbrain and cerebellum. *Annu. Rev. Neurosci.* 24:869–96

Liu Y, Helms AW, Johnson JE. 2004. Distinct activities of Msx1 and Msx3 in dorsal neural tube development. *Development* 131:1017–28

Liu Y, Shi J, Lu CC, Wang ZB, Lyuksyutova AI, et al. 2005. Ryk-mediated Wnt repulsion regulates posterior-directed growth of corticospinal tract. *Nat. Neurosci.* 8:1151–59

Lowe CJ, Wu M, Salic A, Evans L, Lander E, et al. 2003. Anteroposterior patterning in hemichordates and the origins of the chordate nervous system. *Cell* 113:853–65

Lumsden A, Keynes R. 1989. Segmental patterns of neuronal development in the chick hindbrain. *Nature* 337:424–28

Lupo G, Harris WA, Lewis KE. 2006. Mechanisms of ventral patterning in the vertebrate nervous system. *Nat. Rev. Neurosci.* 7:103–14

Lyuksyutova AI, Lu CC, Milanesio N, King LA, Guo N, et al. 2003. Anterior-posterior guidance of commissural axons by Wnt-frizzled signaling. *Science* 302:1984–88

Machon O, Backman M, Machonova O, Kozmik Z, Vacik T, et al. 2007. A dynamic gradient of Wnt signaling controls initiation of neurogenesis in the mammalian cortex and cellular specification in the hippocampus. *Dev. Biol.* 311:223–37

Machon O, van den Bout CJ, Backman M, Kemler R, Krauss S. 2003. Role of β-catenin in the developing cortical and hippocampal neuroepithelium. *Neuroscience* 122:129–43

Maden M. 2007. Retinoic acid in the development, regeneration and maintenance of the nervous system. *Nat. Rev. Neurosci.* 8:755–65

Marín F, Puelles L. 1994. Patterning of the embryonic avian midbrain after experimental inversions: a polarizing activity from the isthmus. *Dev. Biol.* 163:19–37

Martinez S, Crossley PH, Cobos I, Rubenstein JL, Martin GR. 1999. FGF8 induces formation of an ectopic isthmic organizer and isthmocerebellar development via a repressive effect on *Otx2* expression. *Development* 126:1189–200

Martinez S, Wassef M, Alvarado-Mallart RM. 1991. Induction of a mesencephalic phenotype in the 2-day-old chick prosencephalon is preceded by the early expression of the homeobox gene *en*. *Neuron* 6:971–81

Mason I. 2007. Initiation to end point: the multiple roles of fibroblast growth factors in neural development. *Nat. Rev. Neurosci.* 8:583–96

Matsumoto K, Nishihara S, Kamimura M, Shiraishi T, Otoguro T, et al. 2004. The prepattern transcription factor Irx2, a target of the FGF8/MAP kinase cascade, is involved in cerebellum formation. *Nat. Neurosci.* 7:605–12

McMahon AP, Joyner AL, Bradley A, McMahon JA. 1992. The midbrain-hindbrain phenotype of *Wnt-1*$^{-}$/*Wnt-1*$^{-}$ mice results from stepwise deletion of *engrailed*-expressing cells by 9.5 days postcoitum. *Cell* 69:581–95

Megason SG, McMahon AP. 2002. A mitogen gradient of dorsal midline Wnts organizes growth in the CNS. *Development* 129:2087–98

Meinertzhagen IA, Lemaire P, Okamura Y. 2004. The neurobiology of the ascidian tadpole larva: recent developments in an ancient chordate. *Annu. Rev. Neurosci.* 27:453–85

Meinhardt H. 2006. Primary body axes of vertebrates: generation of a near-Cartesian coordinate system and the role of Spemann-type organizer. *Dev. Dyn.* 235:2907–19

Meinhardt H. 2009. Models for the generation and interpretation of gradients. *Cold Spring Harb. Perspect. Biol.* 1:a001362

Meyers EN, Lewandowski M, Martin GR. 1998. An *Fgf8* mutant allelic series generated by Cre- and Flp-mediated recombination. *Nat. Genet.* 18:136–41

Millet S. 1999. A role for *Gbx2* in repression of *Otx2* and positioning the mid/hindbrain organizer. *Nature* 401:161–64

Millonig JH, Millen KJ, Hatten ME. 2000. The mouse *Dreher* gene *Lmx1a* controls formation of the roof plate in the vertebrate CNS. *Nature* 403:764–69

Monuki ES, Porter FD, Walsh CA. 2001. Patterning of the dorsal telencephalon and cerebral cortex by a roof plate-Lhx2 pathway. *Neuron* 32:591–604

Munji RN, Choe Y, Li G, Siegenthaler JA, Pleasure SJ. 2011. Wnt signaling regulates neuronal differentiation of cortical intermediate progenitors. *J. Neurosci.* 31:1676–87

Muñoz-Sanjuán I, Brivanlou AH. 2002. Neural induction, the default model and embryonic stem cells. *Nat. Rev. Neurosci.* 3:271–80

Muroyama Y, Fujihara M, Ikeya M, Kondoh H, Takada S. 2002. Wnt signaling plays an essential role in neuronal specification of the dorsal spinal cord. *Genes Dev.* 16:548–53

Muzio L, Soria JM, Pannese M, Piccolo S, Mallamaci A. 2005. A mutually stimulating loop involving emx2 and canonical wnt signalling specifically promotes expansion of occipital cortex and hippocampus. *Cereb. Cortex* 15:2021–28

Nakamoto A, Nagy LM, Shimizu T. 2011. Secondary embryonic axis formation by transplantation of D quadrant micromeres in an oligochaete annelid. *Development* 138:283–90

Nguyen VH, Trout J, Connors SA, Andermann P, Weinberg E, Mullins MC. 2000. Dorsal and intermediate neuronal cell types of the spinal cord are established by a BMP signaling pathway. *Development* 127:1209–20

Niehrs C. 2004. Regionally specific induction by the Spemann-Mangold organizer. *Nat. Rev. Genet.* 5:425–34

Niehrs C. 2010. On growth and form: a Cartesian coordinate system of Wnt and BMP signaling specifies bilaterian body axes. *Development* 137:845–57

Ninkovic J, Tallafuss A, Leucht C, Topczewski J, Tannhäuser B, et al. 2005. Inhibition of neurogenesis at the zebrafish midbrain-hindbrain boundary by the combined and dose-dependent activity of a new *hairy/E(spl)* gene pair. *Development* 132:75–88

Ono Y, Nakatani T, Sakamoto Y, Mizuhara E, Minaki Y, et al. 2007. Differences in neurogenic potential in floor plate cells along an anteroposterior location: midbrain dopaminergic neurons originate from mesencephalic floor plate cells. *Development* 134:3213–25

Osterfield M, Kirschner MW, Flanagan JG. 2003. Graded positional information: interpretation for both fate and guidance. *Cell* 113:425–28

Osumi N, Hirota A, Ohuchi H, Nakafuku M, Iimura T, et al. 1997. *Pax-6* is involved in the specification of hindbrain motor neuron subtype. *Development* 124:2961–72

Paek H, Gutin G, Hébert JM. 2009. FGF signaling is strictly required to maintain early telencephalic precursor cell survival. *Development* 136:2457–65

Panhuysen M, Vogt Weisenhorn DM, Blanquet V, Brodski C, Heinzmann U, et al. 2004. Effects of Wnt1 signaling on proliferation in the developing mid-/hindbrain region. *Mol. Cell. Neurosci.* 26:101–11

Partanen J. 2007. FGF signalling pathways in development of the midbrain and anterior hindbrain. *J. Neurochem.* 101:1185–93

Pattyn A, Vallstedt A, Dias JM, Sander M, Ericson J. 2003. Complementary roles for Nkx6 and Nkx2 class proteins in the establishment of motoneuron identity in the hindbrain. *Development* 130:4149–59

Pera EM, Kessel M. 1997. Patterning of the chick forebrain anlage by the prechordal plate. *Development* 124:4153–62

Picker A, Brennan C, Reifers F, Clarke JD, Holder N, Brand M. 1999. Requirement for the zebrafish mid-hindbrain boundary in midbrain polarisation, mapping and confinement of the retinotectal projection. *Development* 126:2967–78

Placzek M, Briscoe J. 2005. The floor plate: multiple cells, multiple signals. *Nat. Rev. Neurosci.* 6:230–40

Placzek M, Tessier-Lavigne M, Yamada T, Jessell T, Dodd J. 1990. Mesodermal control of neural cell identity: floor plate induction by the notochord. *Science* 250:985–88

Pottin K, Hinaux H, Rétaux S. 2011. Restoring eye size in *Astyanax mexicanus* blind cavefish embryos through modulation of the Shh and Fgf8 forebrain organising centres. *Development* 138:2467–76

Puelles E, Annino A, Tuorto F, Usiello A, Acampora D, et al. 2004. *Otx2* regulates the extent, identity and fate of neuronal progenitor domains in the ventral midbrain. *Development* 131:2037–48

Quinlan R, Graf M, Mason I, Lumsden A, Kiecker C. 2009. Complex and dynamic patterns of Wnt pathway gene expression in the developing chick forebrain. *Neural Dev.* 4:35

Reichert H. 2005. A tripartite organization of the urbilaterian brain: developmental genetic evidence from *Drosophila*. *Brain Res. Bull.* 66:491–94

Reifers F, Böhli H, Walsh EC, Crossley PH, Stainier DY, Brand M. 1998. *Fgf8* is mutated in zebrafish *acerebellar (ace)* mutants and is required for maintenance of midbrain-hindbrain boundary development and somitogenesis. *Development* 125:2381–95

Rhinn M, Lun K, Luz M, Werner M, Brand M. 2005. Positioning of the midbrain-hindbrain boundary organizer through global posteriorization of the neuroectoderm mediated by Wnt8 signaling. *Development* 132:1261–72

Riley BB, Chiang MY, Storch EM, Heck R, Buckles GR, Lekven AC. 2004. Rhombomere boundaries are Wnt signaling centers that regulate metameric patterning in the zebrafish hindbrain. *Dev. Dyn.* 231:278–91

Rodríguez-Seguel E, Alarcón P, Gómez-Skarmeta JL. 2009. The *Xenopus Irx* genes are essential for neural patterning and define the border between prethalamus and thalamus through mutual antagonism with the anterior repressors *Fezf* and *Arx*. *Dev. Biol.* 329:258–68

Roelink H, Augsburger A, Heemskerk J, Korzh V, Norlin S, et al. 1994. Floor plate and motor neuron induction by vhh-1, a vertebrate homolog of hedgehog expressed by the notochord. *Cell* 76:761–75

Sánchez-Camancho C, Rodríguez J, Ruiz JM, Trousse F, Bovolenta P. 2005. Morphogens as growth cone signalling molecules. *Brain Res. Brain Res. Rev.* 49:242–52

Sanders TA, Lumsden A, Ragsdale CW. 2002. Arcuate plan of chick midbrain development. *J. Neurosci.* 22:10742–50

Scholpp S, Delogu A, Gilthorpe J, Peukert D, Schindler S, Lumsden A. 2009. *Her6* regulates the neurogenetic gradient and neuronal identity in the thalamus. *Proc. Natl. Acad. Sci. USA* 106:19895–900

Scholpp S, Foucher I, Staudt N, Peukert D, Lumsden A, Houart C. 2007. *Otx1l*, *Otx2* and *Irx1b* establish and position the ZLI in the diencephalon. *Development* 134:3167–76

Scholpp S, Wolf O, Brand M, Lumsden A. 2006. Hedgehog signalling from the zona limitans intrathalamica orchestrates patterning of the zebrafish diencephalon. *Development* 133:855–64

Shimamura K, Hartigan DJ, Martinez S, Puelles L, Rubenstein JL. 1995. Longitudinal organization of the anterior neural plate and neural tube. *Development* 121:3923–33

Shimamura K, Rubenstein JL. 1997. Inductive interactions direct early regionalization of the mouse forebrain. *Development* 124:2709–18

Sousa VH, Fishell G. 2010. Sonic hedgehog functions through dynamic changes in temporal competence in the developing forebrain. *Curr. Opin. Genet. Dev.* 20:391–99

Stamataki D, Ulloa F, Tsoni SV, Mynett A, Briscoe J. 2005. A gradient of Gli activity mediates graded Sonic hedgehog signaling in the neural tube. *Genes Dev.* 19:626–41

Stern CD. 2001. Initial patterning of the central nervous system: how many organizers? *Nat. Rev. Neurosci.* 2:92–98

Stern CD. 2006. Neural induction: 10 years on since the 'default model'. *Curr. Opin. Cell Biol.* 18:692–97

Sugahara F, Aota S, Kuraku S, Murakami Y, Takio-Ogawa Y, et al. 2011. Involvement of Hedgehog and FGF signalling in the lamprey telencephalon: evolution of regionalization and dorsoventral patterning of the vertebrate forebrain. *Development* 138:1217–26

Sunmonu NA, Li K, Guo Q, Li JY. 2011. *Gbx2* and *Fgf8* are sequentially required for formation of the midbrain-hindbrain compartment boundary. *Development* 138:725–34

Sylvester JB, Rich CA, Loh YH, van Staaden MJ, Fraser GJ, Streelman JT. 2010. Brain diversity evolves via differences in patterning. *Proc. Natl. Acad. Sci. USA* 107:9718–23

Takada H, Hattori D, Kitayama A, Ueno N, Taira M. 2005. Identification of target genes for the *Xenopus* Hes-related protein XHR1, a prepattern factor specifying the midbrain-hindbrain boundary. *Dev. Biol.* 283:253–67

Takahashi M, Osumi N. 2002. *Pax6* regulates specification of ventral neurone subtypes in the hindbrain by establishing progenitor domains. *Development* 129:1327–38

Technau U, Steele RE. 2011. Evolutionary crossroads in developmental biology: Cnidaria. *Development* 138:1447–58

Tessier-Lavigne M, Goodman CS. 1996. The molecular biology of axon guidance. *Science* 274:1123–33

Thomas KR, Musci TS, Neumann PE, Capecchi MR. 1991. *Swaying* is a mutant allele of the proto-oncogene *Wnt-1*. *Cell* 67:969–76

Thomson RE, Kind PC, Graham NA, Etherson ML, Kennedy J, et al. 2009. Fgf receptor 3 activation promotes selective growth and expansion of occipitotemporal cortex. *Neural Dev.* 4:4

Timmer JR, Wang C, Niswander L. 2002. BMP signaling patterns the dorsal and intermediate neural tube via regulation of homeobox and helix-loop-helix transcription factors. *Development* 129:2459–72

Tomer R, Denes AS, Tessmar-Raible K, Arendt D. 2010. Profiling by image registration reveals common origin of annelid mushroom bodies and vertebrate pallium. *Cell* 142:800–9

Tossell K, Kiecker C, Wizenmann A, Lang E, Irving C. 2011. Notch signalling stabilises boundary formation at the midbrain-hindbrain organiser. *Development* 138:3745–57

Toyoda R, Assimacopoulos S, Wilcoxon J, Taylor A, Feldman P, et al. 2010. FGF8 acts as a classic diffusible morphogen to pattern the neocortex. *Development* 137:3439–48

Trokovic R, Jukkola T, Saarimaki J, Peltopuro P, Naserke T, et al. 2005. Fgfr1-dependent boundary cells between developing mid- and hindbrain. *Dev. Biol.* 278:428–39

Urbach R, Technau GM. 2008. Dorsoventral patterning of the brain: a comparative approach. *Adv. Exp. Med. Biol.* 628:42–56

van Straaten HW, Hekking JW, Thors F, Wiertz-Hoessels EL, Drukker J. 1985. Induction of an additional floor plate in the neural tube. *Acta Morphol. Neerl. Scand.* 23:91–97

Vieira C, Garda AL, Shimamura K, Martinez S. 2005. Thalamic development induced by Shh in the chick embryo. *Dev. Biol.* 284:351–63

Vogel-Höpker A, Rohrer H. 2002. The specification of noradrenergic locus coeruleus (LC) neurones depends on bone morphogenetic proteins (BMPs). *Development* 129:983–91

Vue TY, Bluske K, Alishani A, Yang LL, Koyano-Nakagawa N, et al. 2009. Sonic hedgehog signaling controls thalamic progenitor identity and nuclei specification in mice. *J. Neurosci.* 29:4484–97

Walshe J, Mason I. 2003. Unique and combinatorial functions of Fgf3 and Fgf8 during zebrafish forebrain development. *Development* 130:4337–49

Wassarman KM, Lewandowski M, Campbell K, Joyner AL, Rubenstein JL, et al. 1997. Specification of the anterior hindbrain and establishment of a normal mid/hindbrain organizer is dependent on *Gbx2* gene function. *Development* 124:2923–34

Wingate RJ. 2001. The rhombic lip and early cerebellar development. *Curr. Opin. Neurobiol.* 11:82–88

Wurst W, Bally-Cuif L. 2001. Neural plate patterning: upstream and downstream of the isthmic organizer. *Nat. Rev. Neurosci.* 2:99–108

Xie Z, Chen Y, Li Z, Bai G, Zhu Y, et al. 2011. Smad6 promotes neuronal differentiation in the intermediate zone of the dorsal neural tube by inhibition of the Wnt/β-catenin pathway. *Proc. Natl. Acad. Sci. USA* 108:12119–24

Xu Q, Mellitzer G, Robinson V, Wilkinson DG. 1999. In vivo cell sorting in complementary segmental domains mediated by Eph receptors and ephrins. *Nature* 399:267–71

Zeltser LM. 2005. Shh-dependent formation of the ZLI is opposed by signals from the dorsal diencephalon. *Development* 132:2023–33

Zeltser LM, Larsen CW, Lumsden A. 2001. A new developmental compartment in the forebrain regulated by *Lunatic fringe*. *Nat. Neurosci.* 4:683–84

Zervas M, Millet S, Ahn S, Joyner AL. 2004. Cell behaviors and genetic lineages of the mesencephalon and rhombomere 1. *Neuron* 43:345–57

Zou Y, Lyuksyutova AI. 2007. Morphogens as conserved axon guidance cues. *Curr. Opin. Neurobiol.* 17:22–28

The Complement System: An Unexpected Role in Synaptic Pruning During Development and Disease

Alexander H. Stephan,[1] Ben A. Barres,[1] and Beth Stevens[2]

[1]Department of Neurobiology, Stanford University School of Medicine, Stanford, California 94305-5125; email: astephan@stanford.edu, barres@stanford.edu

[2]Department of Neurology, F.M. Kirby Neurobiology Center, Children's Hospital Boston, Harvard Medical School, Boston, Massachusetts 02115; email: beth.stevens@childrens.harvard.edu

Annu. Rev. Neurosci. 2012. 35:369–89

The *Annual Review of Neuroscience* is online at neuro.annualreviews.org

This article's doi: 10.1146/annurev-neuro-061010-113810

0147-006X/12/0721-0369$20.00

Keywords

synapse elimination, C1q, C3, microglia, neuron-glia interactions, neurodegenerative disease

Abstract

An unexpected role for the classical complement cascade in the elimination of central nervous system (CNS) synapses has recently been discovered. Complement proteins are localized to developing CNS synapses during periods of active synapse elimination and are required for normal brain wiring. The function of complement proteins in the brain appears analogous to their function in the immune system: clearance of cellular material that has been tagged for elimination. Similarly, synapses tagged with complement proteins may be eliminated by microglial cells expressing complement receptors. In addition, developing astrocytes release signals that induce the expression of complement components in the CNS. In the mature brain, early synapse loss is a hallmark of several neurodegenerative diseases. Complement proteins are profoundly upregulated in many CNS diseases prior to signs of neuron loss, suggesting a reactivation of similar developmental mechanisms of complement-mediated synapse elimination potentially driving disease progression.

Contents

INTRODUCTION

The traditional view that the brain is an immune privileged organ has shifted dramatically with the growing realization that the nervous and immune systems interact on many levels in health and disease. Each system has an array of molecules and signaling pathways that have both novel and analogous functions in the other system. Among these proteins are components of the classical complement cascade, innate immune proteins traditionally associated with the rapid recognition and elimination of pathogens and harmful cellular debris.

New research reveals an unexpected role for the classical complement cascade in the developmental elimination, or pruning, of extranumerary synapses, a process critical for establishing precise synaptic circuits. Complement proteins are widely expressed in neurons and glia in the postnatal brain and are localized to subsets of developing synapses during periods of active synaptic remodeling (Stevens et al. 2007). Mice deficient in C1q, the initiating protein in the classical complement cascade, or the downstream complement protein C3 exhibit sustained defects in CNS synapse elimination and synaptic connectivity (Stevens et al. 2007). In the immune system, complement opsonizes, or tags, pathogenic microbes and unwanted cellular debris for rapid elimination by phagocytic macrophages or complement-mediated cell lysis. The surprising discovery that complement proteins are localized to developing synapses suggests that these classic immune molecules may be similarly opsonizing or tagging immature synapses for elimination during normal brain wiring.

Although synapse elimination is largely considered a developmental process, early synapse loss and dysfunction are becoming increasingly recognized as a hallmark of many CNS neurodegenerative diseases (Selkoe 2002, Mallucci 2009). Synapse degeneration associated with cognitive decline is also part of the normal aging process; however, the factors that trigger synapse loss in the aged and diseased brain remain elusive. One hypothesis is that synapse loss in CNS neurodegenerative diseases is caused by a reactivation, in the mature brain, of similar developmental mechanisms of synapse elimination. Indeed, components of the complement cascade are profoundly upregulated in Alzheimer's disease (AD), glaucoma, and other brain diseases (reviewed in Alexander et al. 2008, Rosen & Stevens 2010, Veerhuis et al. 2011) and are localized to synapses prior to signs of neuronal loss in animal models of neurodegenerative disease (Stevens et al. 2007). This notion suggests that the complement-mediated synapse degeneration process may be an early and critical event in driving the neurodegenerative process in glaucoma and other brain diseases. Because synapse loss appears long before pathology or cognitive decline/behavioral deficits in most neurodegenerative diseases, understanding how synapses are normally pruned during development could provide mechanistic insight into how to prevent abberant synapse elimination during disease.

These provocative findings have spurred the search for molecular mechanisms underlying complement-mediated synaptic pruning. Emerging evidence implicates glial cells—microglia and astrocytes—as key players in this process. Glia are a major source of complement in the developing and diseased CNS, but they also express complement receptors that facilitate phagocytosis and secrete an array of cytokines and other factors that can initiate the complement cascade. Given that the appearance of reactive glia is a common early step in the progression of most CNS neurodegenerative diseases, further study of glia in complement-mediated synapse elimination in the diseased and normal CNS is likely to provide important insight into mechanisms underlying complement cascade regulation and function.

In this review we focus on recent progress in understanding the role of complement in regulating CNS synapse elimination and regression and discuss mechanisms of complement cascade regulation and potential targets for therapeutic intervention in CNS neurodegenerative diseases and disorders.

FUNCTION OF THE COMPLEMENT CASCADE: CLUES FROM THE INNATE IMMUNE SYSTEM

In the periphery, complement is our first line of defense against infection through the rapid elimination of invading pathogens and regulation of the slower adaptive immune response. In addition, the complement system clears modified self cells, such as apoptotic cells and cellular debris, to protect against autoimmunity (reviewed in Medzhitov & Janeway 2002, Carroll 2004, Zipfel et al. 2007, Ricklin et al. 2010). The complement system is composed of a large family (~60) of circulating and membrane-associated proteins that act synergistically in a sequential cascade-like manner to execute and regulate its functions. Circulating complement proteins, most of which are produced in the liver, are inactive proteins or zymogens until they encounter a cell membrane or biological surface. Binding results in structural modifications, proteolytic cleavage, and assembly into active enzyme complexes (convertases) that can then activate downstream substrates in a cascade-like fashion (**Figure 1**). Thus, complement zymogens can be widely distributed until locally activated.

Complement is activated by three major routes: the classical, the alternative, and the lectin pathways, all of which converge on complement component C3, a central molecule in the complement system that ultimately drives complement effector functions, including the elimination of pathogens, debris, and cellular structures (**Figure 1**). The classical pathway is canonically triggered when C1q, the initiating protein of the cascade, interacts with one of its multiple ligands, which include antigen-antibody complexes, C reactive protein (CRP), serum pentraxins, polyanions (DNA, RNA, and lipopolysaccharides), apoptotic cells, and viral membranes (Kishore & Reid 2000, Gal et al. 2009, Kang et al. 2009). The antibody-independent lectin pathway is triggered by the binding of mannose binding lectin (MBL) and a group of related proteins that recognize terminal sugar moieties expressed on polysaccharides or glycoproteins on cell membranes (Fujita 2002, Degn et al. 2007). In contrast with the classical and lectin pathways, the alternative pathway is spontaneously and continuously activated in plasma and serves to amplify the cascade initiated by classical and lectin pathways. Once C3 is cleaved, opsonization with activated C3 fragments (C3b and iC3b) leads to elimination of target structures by triggering of phagocytosis through C3 receptors on phagocytic cells (i.e., C3R/Cd11b) (**Figure 1**). Moreover, robust activation of C3 can trigger the terminal activation of the complement cascade leading to cell lysis by inserting C5–C9 into the membrane to form the lytic membrane attack complex (MAC) (**Figure 1**).

In addition to opsonization, complement activation fragments, especially C3a, can mediate a multitude of functions including the recruitment and activation of circulating macrophages and effector cells (Nordahl et al. 2004, Zhou 2011). Much like C3a, C5a is also a potent neuroinflammatory anaphylatoxin that recruits cells expressing C5a receptors, such as macrophages in the periphery and microglia in the brain (Klos et al 2009, Zhou 2011). Although these active fragments are delivered nonspecifically to a cell surface, the progression and degree of activation of the complement cascade in the periphery is tightly controlled at

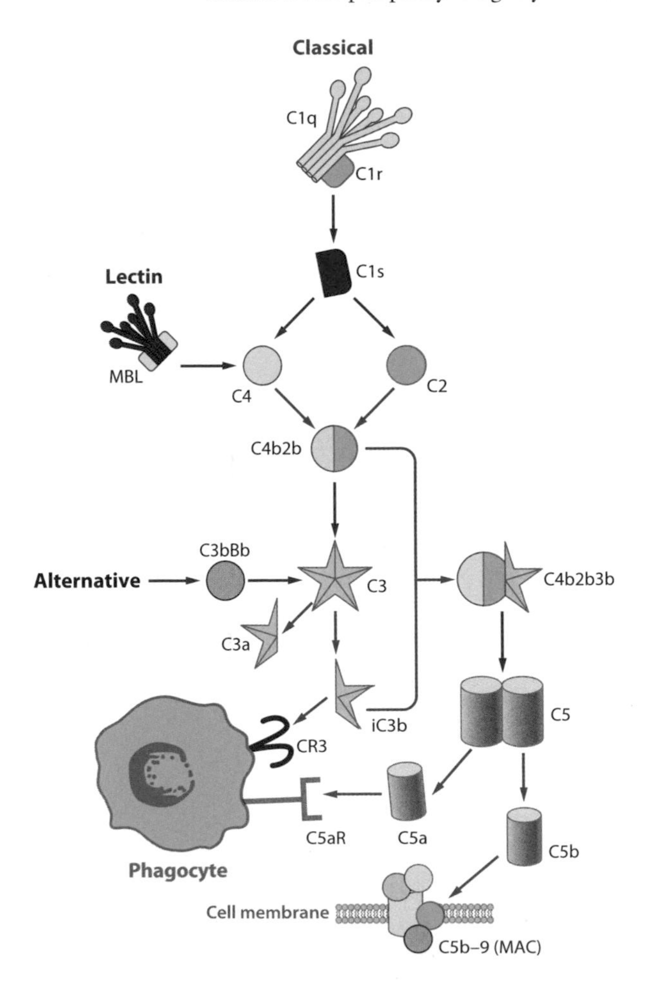

Figure 1

The complement cascade activation and function. The complement system consists of a large number of inactive components (zymogens) that are activated in a cascade-like manner to exert its biological effects in the innate immune system. Binding of complement zymogens to a membrane surface results in structural modifications, proteolytic cleavage, and the assembly into active enzyme complexes (convertases), which can then activate downstream substrates in a cascade-like fashion as shown. The complement cascade can be initiated by three major pathways: The classical pathway is induced when C1q interacts with antibodies or one of its many binding partners, such as serum pentraxins, polyanions (DNA, RNA, and lipopolysaccharides), or apoptotic cells. The C1q tail region of C1q binds proteases C1r and C1s to form the C1 complex. Binding of the C1q complex to an antibody/receptor on the cell surface induces a conformational change in the C1q molecule, which leads to activation of an autocatalytic enzymatic activity in C1r; C1r then cleaves C1s to generate the active serine protease. Once activated, C1s cleaves C4 and C2 to generate the C3 convertase, C3b2b, which in turn cleaves C3 and activates downstream cascade components. The lectin pathway is triggered by the binding of mannose binding lectin (MBL) to mannose residues on the cell surface. This activates the MBL-associated proteases mannose binding lectin serine protease 1 (MASP1) and MASP2, which then cleave C4 to generate the C4 convertase, C4b2b. The alternative pathway is spontaneously and continuously activated (via spontaneous C3 hydrolysis), which serves to amplify the cascade triggered by classical and lectin pathways. All three cascades converge on the major complement component, C3. Cleavage of C3 generates the anaphylactic peptide C3a and the opsonin C3b. Opsonization with C3b/iC3b leads to elimination of target structures by phagocytes that express C3 receptors (i.e., C3R/Cd11b). C3b later joins with C4b2a (the C3 convertase) to form the C5 convertase (C4b2a3b complex) that generates the anaphylatoxin C5a, which binds to C5a receptors (C5aR) on phagocytic/effector cells. Robust activation of complement can trigger activation of the terminal complement cascade, resulting in cell lysis through the insertion of the pore-forming C5b-C9 complex into the membrane, termed membrane attack complex (MAC).

every level by a battery of complement regulatory proteins that protect cells from aberrant elimination (Song 2006, Zipfel & Skerka 2009). Thus, a delicate balance between complement activation and inhibition is critical for proper complement function and tissue homeostasis.

COMPLEMENT EXPRESSION AND LOCALIZATION IN THE BRAIN

Complement has long been appreciated as a rapid and local immune surveillance system in the brain; however, new research has ascribed many new functions of complement in the brain that extend far beyond host defense and inflammatory processes (reviewed in Ricklin et al. 2010, Rutkowski et al. 2010b, Veerhuis et al. 2011). Although the blood brain barrier normally protects the brain from plasma-derived complement and infiltrating immune cells, many complement components can be locally produced in the brain, most often in response to injury or inflammatory signals (Zamanian et al. 2012, Veerhuis et al. 2011). Thus, local synthesis of complement is critical for local defense, neuroprotection, and homeostasis in the brain. Complement activation and its effector functions are tightly regulated by a large group of diverse molecules widely expressed throughout the organism (Zipfel & Skerka 2009), including the so-called neuroimmune regulators (NIReg), which are expressed on most nonneuronal cell types (Hoarau et al. 2011). These inhibitory molecules protect cells from uncontrolled complement-mediated damage by a range of complement inhibitor molecules, such as CD59, Complement Factor H (CFH), and Crry, which mainly interfere with C3 effector functions but also block earlier steps of complement activation. Paradoxically, inappropriate or uncontrolled complement activation can also promote inflammation, neurodegeneration, and ultimately, pathology (Sjöberg et al. 2009). This notion may be explained by the recent finding that CNS neurons, unlike most peripheral cell types, do not detectably express most of the known complement inhibitors (Cahoy et al. 2008). Thus, understanding where and when complement is normally expressed in the brain is likely to provide important insight into complement function and possible targets of intervention during development and disease.

In the CNS, complement proteins are locally synthesized by resident neurons and glial cells; however, microglia and astrocytes are the major producers of complement in the healthy and diseased CNS (Lampert-Etchells et al. 1993, Barnum 1995, Woodruff et al. 2010, Veerhuis et al. 2011). Microglia throughout the CNS express extremely high levels of C1q, as well as CR3 (CD11b) and CR5, complement receptors crucial for inducing phagocytosis of complement-coated structures and regulating cytokine signaling as well as chemotaxis, respectively (Veerhuis et al. 1999). Cultured and reactive astrocytes express high levels of C3 and other complement cascade proteins (Levi-Strauss & Mallat 1987, Cahoy et al. 2008). Complement expression by neurons has so far been described mainly in the disease context and upon injury to the CNS (Woodruff et al. 2010, Veerhuis et al. 2011). Neuronal stem cells express multiple complement receptors and differentiate and migrate in response to secreted complement. C3a–C3aR interactions were found to be a positive regulator of adult neurogenesis following ischemic injury (Rahpeymai et al. 2006, Bogestal et al. 2007, Shinjyo et al. 2009). In addition, CR2 (CD21), a receptor for activated C3 fragments, is also expressed in neural progenitor cells and regulates adult hippocampal neurogenesis (Moriyama et al. 2011).

We know that complement proteins are upregulated in neural cells following brain injury (reviewed in Veerhuis et al. 2011), but comparatively little is known about the normal function of complement proteins in the healthy brain. In situ hybridization and gene-profiling studies have unexpectedly revealed that many components of the complement system are expressed, albeit at lower levels, in the healthy brain (Stevens et al. 2007, Cahoy et al. 2008). Several components, including C1q and C3, are developmentally regulated and localized in patterns that suggest novel functions (Stevens et al. 2007, Stephan et al. 2011). Indeed, complement has been recently implicated in several nonimmune functions during the embryonic and postnatal period, including neurogenesis, migration, neuronal survival (Benard et al. 2008, Shinjyo et al. 2009, Rutkowski et al. 2010a, Benoit & Tenner 2011), and synaptic development and elimination (Stevens et al. 2007)—the focus of this review.

THE CLASSICAL COMPLEMENT CASCADE REGULATES BRAIN WIRING DURING DEVELOPMENT

The developmental expression and synaptic localization of classical complement proteins in the postnatal brain were early clues that this family of immune proteins may function in synapse development. The first two weeks of postnatal development are a period of remarkable plasticity as immature synaptic circuits are actively remodeled (reviewed in Katz & Shatz 1996, Hua & Smith 2004, Huberman et al. 2008). There is a clear spatiotemporal correlation between the appearance and association of immature astrocytes with neurons at CNS synapses during this dynamic period. Immature astrocytes secrete an array of cytokines and other molecules that promote synapse formation and synaptic plasticity (Allen & Barres 2005, Bolton & Eroglu 2009, Eroglu & Barres 2010).

A screen to determine how astrocytes influence neuronal gene expression first identified C1q as one of the few genes that were highly upregulated in developing retinal ganglion cells (RGCs) in response to an astrocyte-derived secreted factor (Stevens et al. 2007). In contrast to microglia, which continue to express C1q in the mature brain, C1q expression in retinal neurons is developmentally restricted to the early postnatal period, when RGC axons and dendrites undergo active synaptic pruning and refinement. Indeed, immunohistochemical analyses and high-resolution imaging revealed C1q and downstream complement protein C3 are localized to subsets of synapses throughout the postnatal brain and retina (Stevens et al. 2007). Given complement's well-ascribed role as opsonins in the elimination of unwanted cells, these findings suggested that the complement cascade may be tagging weak or inappropriate synapses for elimination in the developing brain.

This idea was tested in the mouse retinogeniculate system—a classical model for studying activity-dependent developmental synapse elimination (reviewed in Shatz & Sretavan 1986, Huberman 2007, Guido 2008, Hong & Chen 2011). Early in development, RGCs form transient functional synaptic connections with relay neurons in the dorsal lateral geniculate nucleus (dLGN) of the thalamus. During the first two weeks of postnatal development, there is a robust period of synaptic remodeling in which many of these transient retinogeniculate synapses are eliminated while the remaining synaptic arbors are elaborated and strengthened (Sretavan & Shatz 1984, Campbell & Shatz 1992, Hooks & Chen 2006). Whereas the role of spontaneous and experience-driven synaptic activity in developmental synaptic pruning is well established (Katz & Shatz 1996, Sanes & Lichtman 1999, Hua & Smith 2004), surprisingly little is known about the cellular and molecular mechanisms that drive the elimination of inappropriate retinogeniculate synapses.

Consistent with a role for the classical complement cascade in synaptic pruning, neuroanatomical tracings of retinogeniculate projections and electrophysiological recordings in dLGN relay neurons showed that C1q and C3 knockout (KO) mice exhibited sustained defects in synaptic refinement and elimination, as shown by their failure to segregate into eye-specific territories and by their retention of multi-innervated LGN relay neurons (Stevens et al. 2007) (**Figure 2**). Moreover, C1q KOs show an increase in the number of presynaptic boutons and exuberant excitatory connectivity in the cortex (Chu et al. 2010), suggesting complement mediates synaptic pruning and/or remodeling in other brain regions. Although C1q and C3 KOs have sustained defects in synaptic

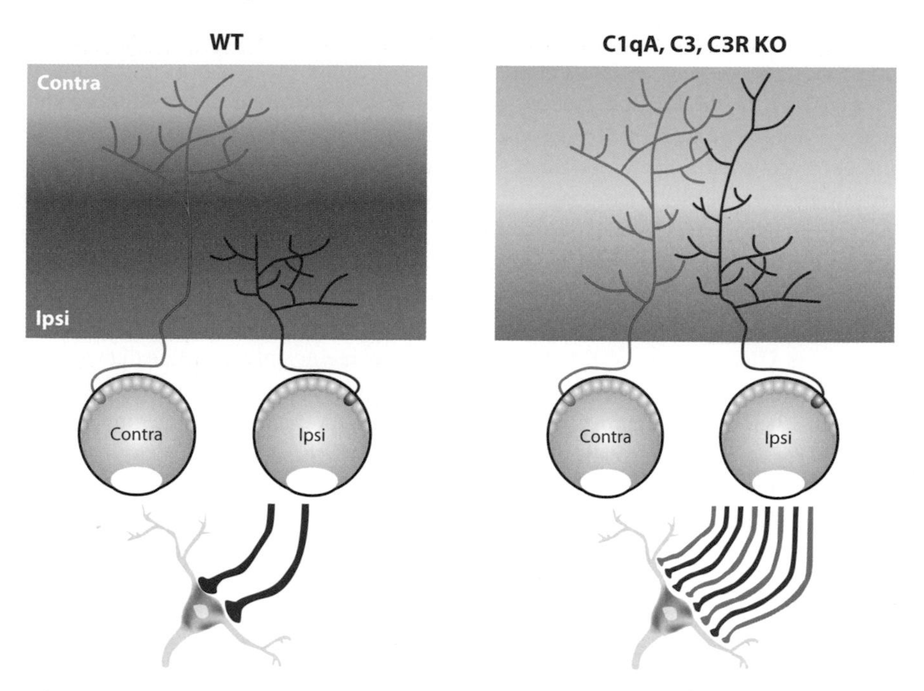

Figure 2

Classical complement cascade proteins mediate synaptic refinement in the developing retinogeniculate system. During the first postnatal week, overlapping inputs from both eyes (red versus green inputs) segregate into eye-specific territories in the dorsal lateral geniculate nucleus (dLGN) of the thalamus, resulting in the termination of ipsilateral (Ipsi) and contralateral (Contra) retinal ganglion cell (RGC) inputs in distinct nonoverlapping domains in the mature dLGN (*Left panel*). Eye-specific segregation involves the selective local pruning of overlapping parts of axonal arbors and the elaboration of the appropriate eye inputs to form the adult pattern of connections. A given relay neuron in the dLGN will ultimately receive 1–2 mature inputs from either the left or right eye (neuron, *bottom left*). Right panel: Mice deficient in classical complement cascade components, C1q and C3, and the microglia-specific complement receptor 3 (CR3) have sustained defects in eye-specific segregation compared with wild-type (WT) animals (*top, right*), depicted as increased overlap of ipsi- and contralateral RGC inputs in the dLGN (*yellow region*) and presence of binocularly innervated dLGN relay neurons (*bottom, right*).

pruning, these mice still undergo a substantial degree of synapse elimination (Stevens et al. 2007), suggesting that complement proteins cooperate with other pathways to regulate normal synaptic circuit development. Several immune-related molecules have recently been identified as mediators of synaptic refinement and plasticity in the visual system (reviewed in Boulanger 2009, Shatz 2009). These include neuronal pentraxins (e.g., NP1/2, NARP) and components of the adaptive immune system (e.g., MHC Class I (MHC-I) family of proteins and receptors) (Corriveau et al. 1998, Huh et al. 2000, Bjartmar et al. 2006, Datwani et al. 2009). Recent work suggests that C1q and MHC-I proteins colocalize at RGC synapses in the postnatal LGN (Datwani et al. 2009). Perhaps components of the complement cascade may be acting in concert with one of several of these immune-related pathways to mediate CNS synapse elimination. Together these findings raise many questions regarding the underlying cellular and molecular mechanisms.

Mechanisms of Complement-Dependent Synaptic Pruning: A Novel Role for Microglia

How are complement-tagged synapses eliminated? Emerging evidence implicates microglia as key players in developmental synaptic pruning (Ransohoff & Stevens 2011, Tremblay & Stevens 2011). In the immune system, the activated C3 fragment C3b (iC3b) opsonizes the surface of cells/debris and tags them for elimination by phagocytic macrophages that express C3 receptors (C3R/cd11b) (Carroll 2004, Gasque 2004, van Lookeren Campagne et al. 2007). Microglia, the resident phagocytes of the CNS, are the only resident brain cells to express CR3 (Tenner & Frank 1987, Guillemin & Brew 2004, Ransohoff & Perry 2009, Graeber 2010). Indeed, process-bearing activated microglia and synaptically localized C3 have been observed in the dLGN and several other postnatal brain regions, including hippocampus, cerebellum, and olfactory bulb, undergoing active synaptic remodeling (Perry et al. 1985, Dalmau et al. 1998, Fiske & Brunjes 2000, Schafer et al. 2011); until recently, however, the function of microglia in a normal brain has remained a relative mystery.

Are microglia required for the elimination of extranumerary synapses? This question was recently investigated in the mouse retinogeniculate system. Using a combination of immune electron microscopy (EM) and high-resolution imaging, microglia were found to engulf RGC presynaptic inputs during a peak pruning period in the developing dLGN (Schafer et al. 2012). Genetic or pharmacological (minocycline) disruptions in microglia-mediated engulfment during early development result in sustained functional deficits in eye-specific segregation and synaptic pruning. Futhermore, microglia-mediated engulfment of synaptic inputs was dependent on signaling between phagocytic receptor CR3, expressed by microglia and CR3 ligand, the innate immune system molecule, and complement component C3. High-resolution quantitative analyses of structural synapses revealed that adult C3 KO and CR3 KOs have significantly more structural synapses in the visual system and other brain regions, indicating that altered complement signaling early in development results in sustained defects in synaptic connectivity (Schafer et al. 2012). Recent studies have demonstrated that microglia also engulf postsynaptic elements during synaptic remodeling in the hippocampus and visual cortex, raising the question of whether complement-dependent pruning is a more global mechanism of synaptic remodeling in the CNS (Tremblay et al. 2010, Paolicelli et al. 2011). Together these new findings suggest that microglia CR3, expressed on the surface of the microglia, and C3, enriched in synaptic compartments, interact to mediate engulfment of synaptic elements undergoing active pruning and raise several fundamental questions related to the underlying mechanisms.

Which Synapses Are Eliminated?

A longstanding question in neurobiology is what determines which synapses will be eliminated during development. Synapse elimination is thought to result from competition between neighboring axons for postsynaptic territory based on differences in patterns or levels of neuronal activity (reviewed in Shatz 1990, Sanes & Lichtman 1999, Huberman et al. 2008). Based on classic studies of the neuromuscular junction, the punishment model proposes that strong synapses, which are effective in driving postsynaptic responses, actively punish and eliminate nearby weaker, less-effective synapses by inducing two postsynaptic signals: a local protective signal and a longer-range elimination, or punishment, signal (Balice-Gordon & Lichtman 1994, Jennings 1994). To date, it is not clear whether this model is correct or relevant to CNS synapses. If it is, the identity of the putative activity-dependent punishment signal remains unknown.

Could complement tag and punish synapses destined for elimination? If so, how might selectivity occur for those synapses destined to be pruned? C1q and C3 could be preferentially tagging weaker or less-active synapses for elimination by phagocytic microglia (**Figure 3**). Alternatively, C1q/C3 could bind all synapses and only those synapses that are stronger or more active would be selectively protected by local membrane-bound complement regulatory

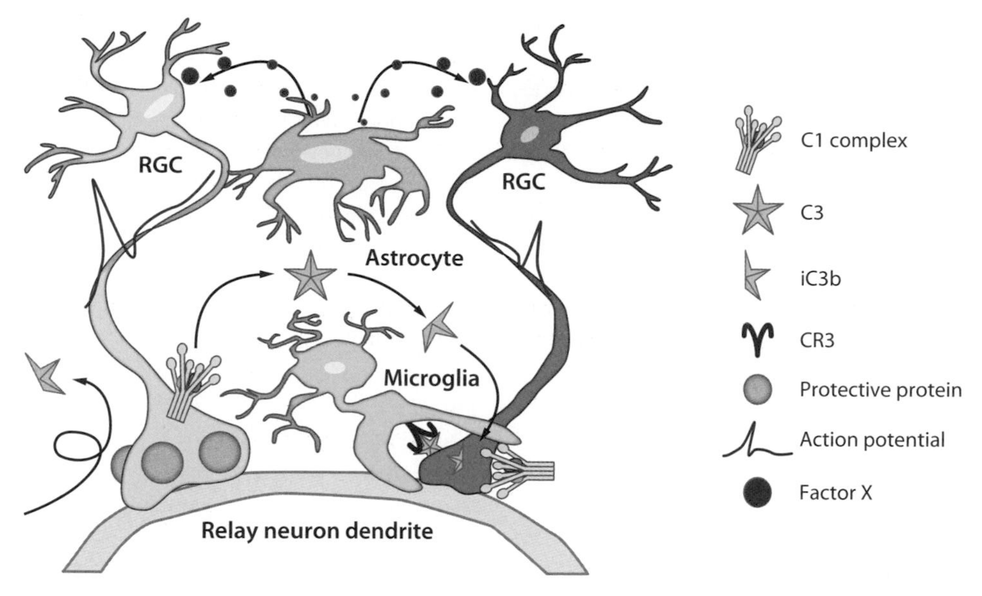

Figure 3

The complement punishment model of synapse elimination in the developing visual system. The punishment model proposes that strong synapses (*blue*), which are effective in driving postsynaptic responses, actively punish and eliminate nearby weaker, less-effective synapses (*red*) by inducing two postsynaptic signals: a local protective signal and a longer-range elimination or punishment signal (Balice-Gordon & Lichtman 1994, Jennings 1994). In the developing retina, complement is hypothesized to be upregulated in sensory neurons by unknown signals from immature astrocytes. An astrocyte-secreted factor (Factor X) upregulates C1q expression in postnatal retinal ganglion cell neurons (RGCs), which leads to deposition of C1q and the local activation of the classical complement cascade (cleavage of native C3). Synaptic deposition of activated C3 fragments (i.e., iC3b) could punish less-active retinogeniculate synapses (*red neuron, right*) of dLGN relay neurons in the thalamus (*target*). Activated microglia, which express high levels of CR3/cd11b, actively eliminate iC3b tagged synapses via CR3–C3-dependent phagocytosis. Strong synapses (*blue neuron, left*) may be protected from elimination by complement regulatory proteins or other activity-dependent protective signals).

molecules (Kim & Song 2006, Zipfel & Skerka 2009) or other activity-dependent factors. In addition, there may be a mechanism completely independent of synaptic tagging by C1q/C3 or complement regulatory molecules. For example, C3a, the anaphylatoxin and cleaved form of C3, may play a role in recruiting microglia to synaptically enriched regions (Klos et al. 2009).

In the retina, spontaneous, correlated neuronal activity from both eyes is thought to drive eye-specific segregation and retinogeniculate pruning (Penn et al. 1998, Stellwagen & Shatz 2002, Torborg & Feller 2005, Huberman 2007, Feller 2009). Although the specific properties of retinal activity that guide this process remain elusive, these findings are consistent with a model in which left- and right-eye retinal axons compete for territory on postsynaptic dLGN relay neurons. Complement and complement receptor–deficient mice have similar pruning deficits to mice in which this correlated firing has been disrupted (reviewed in Huberman 2007), suggesting that complement could act downstream of neural activity to prune inappropriate synaptic connections. Consistent with this idea, recent studies reveal that microglia-mediated pruning of RGC inputs in vivo is an activity-dependent process (Schafer et al. 2012). Indeed, microglia preferentially engulf inputs from the weaker eye, suggesting that microglia are active participants in synaptic pruning.

In vivo imaging studies in the mouse cortex have revealed that microglial dynamics and interactions with neuronal compartments change in response to neural activity and experience (Davalos et al. 2005, Nimmerjahn et al. 2005, Wake et al. 2009, Tremblay et al. 2010), but the underlying mechanisms remain elusive. Future studies will aim to address how specific synapses are eliminated by complement and microglia-dependent mechanisms and whether neuronal activity plays a role in this process.

Candidate Complement Receptors and Interacting Proteins

Although it is clear that CR3 is an important mediator of developmental synaptic engulfment, microglia have been shown to, in some instances, express the phagocytic iC3b receptor CR4 (CD11c/CD18) (Chiu et al. 2009). In addition, the newly identified iC3b/C3 receptor immunoglobulin (Ig)-superfamily member CRig (Z39Ig, VSIG4), although not yet assessed in microglia, has been shown to mediate complement-mediated phagocytosis in subpopulations of resident tissue macrophages (Helmy et al. 2006, Gorgani et al. 2008). In some instances, C1q itself can facilitate engulfment via specific C1q receptors expressed on phagocytic cells, raising the possibility that microglia may phagocytose some synapses independently of C3 or CR3 (Bobak et al. 1986, Guan et al. 1991, Nepomuceno & Tenner 1998, Tenner 1998, Eggleton et al. 2000).

The molecular characteristics of C1q enable its interaction with a wide variety of molecules (Kishore & Reid 2000, Kojouharova et al. 2010, Nayak et al. 2011). In the immune system, C1q receptors mediate the ability of C1q to activate the complement pathway and to opsonize apoptotic cells. C1q binds via its globular head domains to pathogens and cell surfaces by directly binding to certain lipids or surface proteins or to other molecules already opsonizing (coating) the pathogen or cell surface. These opsonins include IgM or IgG antibodies and the short pentraxins serum amyloid protein (SAP) and CRP, both of which acute-phase reactant proteins quickly released by the liver into the blood upon inflammation. The long pentraxin, PTX3, functions similarly. Binding of C1q to any of these molecules can initiate the complement cascade, leading to either lysis or phagocytosis, because C1q has recently been identified as crucial for promoting phagocytosis of apoptotic cells in vivo (Taylor et al. 2000, Nauta et al. 2003). Its ability to recognize a wide range of molecular patterns suggests that C1q may interact with a variety of CNS molecules in health and disease. However, the synaptic receptors that recruit C1q and, thus, complement to synapses doomed for elimination are still unknown.

Synaptic changes occur at very early stages in most neurological diseases, which may initiate an extracellular synaptic profile that attracts C1q and which in turn mediates synapse elimination. This action potentially includes the reactivation of the molecular mechanism that marks synapses for elimination in the developing CNS or the synaptic expression of disease-specific C1q-interacting molecules that activate C1q at synapses, potentially associated with the downregulation of complement inhibitors. Alternatively, detrimental structural synaptic changes, similar to the ones observed in apoptotic cells, could trigger C1q activation and subsequent complement-dependent synapse elimination in development and disease. For example, *N*-methyl-D-aspartate (NMDA) receptor stimulation induces long-term depression transiently and locally activates caspase-3 in dendrites without causing cell death (Li et al. 2010, Jo et al. 2011). These important findings raise the interesting possibility that C1q and other complement proteins preferentially tag activated apoptotic synapses to mediate their elimination just as C1q is critical for the elimination of apoptotic cells (Taylor et al. 2000).

Lastly, there is emerging evidence for other C1q homologous, secreted proteins that play important roles in synaptic plasticity and synapse formation (Yuzaki 2010). These include C1ql2, which is a synaptically localized protein (Iijima et al. 2010, Shimono et al. 2010), and other C1ql family members (C1ql1–4) that are expressed in the CNS and implicated in synapse formation or maintenance/elimination (Bolliger et al. 2011). In addition, the cerebellin family (Cbln1–4) is another family of C1q-like molecules that are secreted presynaptically and promote synapse formation and plasticity (Yuzaki 2010, Matsuda & Yuzaki 2011). Both presynaptic and postsynaptic receptors have been identified for Cbln1—neurexin-1β and GluRδ2 (Matsuda et al. 2010)—suggesting that cbln1 plays a critical synapse-organizing role during development (Martinelli & Sudhof 2011). Taken together, it seems likely that C1q-like molecules, including C1q, C1ql2, and Cbln1, are all likely to be part of the long mysterious mechanism through which activity-dependent competitive interactions between synapses lead some synapses to be maintained and others to be eliminated (Watanabe 2008).

Potential Cross Talk in Between Complement and Other Immune Pathways

Several other immune-related molecules have recently been identified as mediators of synaptic refinement and plasticity in the developing and mature brain (Boulanger 2009), including neuronal pentraxins (e.g., NP1/2, NARP) and components of the adaptive immune system (e.g., MHC-I family of proteins and receptors). It is intriguing to speculate that components of the complement pathway may be interacting with one of several of these immune-related molecules to mediate CNS synapse elimination.

Neuronal pentraxins are synaptic proteins with homology to pentraxins of the peripheral immune system, which are traditionally involved in opsonization and phagocytosis of dead cells in the immune system (Nauta 2003). Mice deficient in neuronal pentraxins, NP1 and NP2, and the receptor, NPR, have transient defects in eye-specific segregation in the dLGN (Bjartmar et al. 2006). In fact, an immune system pentraxin, the long pentraxin PTX3, which has homology to neuronal pentraxins, can enhance microglial phagocytic activity (Jeon et al. 2010). In addition, neuronal pentraxins are significantly homologous to short pentraxins such as CRP, which is a well-described binding partner of C1q. Thus neuronal pentraxins may serve as synaptic binding partners for C1q during synapse development and complement-mediated synaptic pruning.

Classical MHC-I molecules represent a large family of transmembrane immune proteins best known for their roles in the recognition and removal of foreign (non self) antigens. The MHC-I molecules and receptors were the first immune-related molecules implicated in developmental synapse elimination (Corriveau et al. 1998, Huh et al. 2000). MHC-I genes are highly expressed in brain regions undergoing activity-dependent synaptic remodeling (Corriveau et al. 1998, Huh et al. 2000, Datwani et al. 2009, McConnell et al. 2009, Shatz 2009). The MHC-I protein is enriched in synaptic compartments (i.e., dendrites), where it colocalizes with postsynaptic proteins, such as PSD95 (Corriveau et al. 1998, Huh et al. 2000, Goddard et al. 2007). Moreover, animals deficient in MHC-I molecules have defects in eye-specific segregation in the retinogeniculate pathway and ocular dominance plasticity, suggesting a role in developmental elimination of CNS synapses (Huh et al. 2000, Syken et al. 2006). MHC-I is highly expressed by activated microglia, including during normal development, which renders it possible that microglial MHC-I plays a role in synaptic pruning. In this context it is of particular interest that neuronal activity can regulate the surface expression of the other class of MHCs, MHC-II, specifically on microglia (Neumann et al. 1998, Biber et al. 2007).

Functional Consequences of Aberrant Complement Activation During Development

Insight into complement-mediated synaptic remodeling could have important implications for understanding the molecular basis of synapse loss and dysfunction in cognitive and neurodevelopmental disorders in which the balance of excitation and inhibition is altered. Consistent with this idea, mice deficient in a functional classical complement cascade component (C1qA KO mice) exhibit enhanced excitatory synaptic connectivity in the mature cortex (Chu et al. 2010).

By comparison, inappropriate complement activation during synapse development could alter neural connectivity by excessively targeting synapses for elimination. Activation of microglia and the innate inflammatory process occurs after acute seizures. Indeed, complement (C1q and C3) is chronically upregulated and activated in the adult brain during early phases of epileptogenesis in both experimental and human temporal lobe epilepsy, suggesting that aberrant complement activation could play a role in destabilizing neural networks. Similarly, maternal infection during fetal development may activate immune system genes within the fetal brain (Patterson 2011), which may alter synaptic development and lead to autism or other neuropsychiatric diseases.

Recent genome-wide association studies and analyses of postmortem human brain tissue have suggested that abnormal microglial function and/or complement cascade activation may play a role in autism and psychiatric disorders such as schizophrenia (Pardo et al. 2005, Vargas et al. 2005, Hashimoto 2008, Monji et al. 2009, Chen et al. 2010, Morgan et al. 2010, Havik et al. 2011). Thus, important future questions are whether and how microglia and/or the complement cascade underlie disruptions in neuronal connectivity associated with these psychiatric disorders.

THE ROLE OF COMPLEMENT IN NEURODEGENERATIVE DISEASES AND CNS INJURY

Does a normal mechanism of developmental synapse elimination become reactivated in and drive adult neurodegenerative diseases? The hallmark of many neurodegenerative diseases is the vast loss of neurons, induced by a wide range of molecular and cellular defects, many of which are still unknown or only partially understood. However, it has recently emerged that neuron death is preceded by aberrant synaptic functioning and massive synapse loss (Selkoe 2002, Mallucci 2009). Therefore, synaptic dysfunction and synapse loss likely directly lead to neuron death and drive disease progression. This is critical when considering how best to treat these diseases, as it will obviously be pointless to develop therapies that keep neurons alive when their synapses are dysfunctional or degenerated.

Neuroinflammation, including microglial activation, reactive gliosis, and massive and early activation of the classical complement cascade, is a cardinal feature of AD and many or most other neurodegenerative diseases (Nguyen et al. 2002, Wyss-Coray & Mucke 2002). This striking degree of neuroinflammation has, however, long been considered to be a secondary event caused by neurodegeneration. The important role of the classical complement cascade and activated microglia in eliminating synapses throughout the normal developing brain, however, raises the intriguing hypothesis that complement activation actively drives the loss of synapses early in neurodegenerative disease, which in turn drives the loss of neurons and, thus, disease progression (**Figure 4**). Below we review emerging evidence that supports this possibility. The relatively low level, or possibly total lack, of complement inhibitor expression by CNS neurons likely makes them much more vulnerable to the action of the complement cascade compared with other body cell types.

Glaucoma is one of the most common neurodegenerative diseases, characterized by the elevation of intraocular pressure, the loss of RGC neurons, and optic nerve degeneration in humans, eventually resulting in blindness (see John & Howell 2012, this volume). Glaucomatous DBA/2J mice closely resemble the human disease, and this model system revealed that synapse loss precedes neuronal loss and may contribute to disease progression (Whitmore et al. 2005, Stevens et al. 2007). Classical complement component expression is

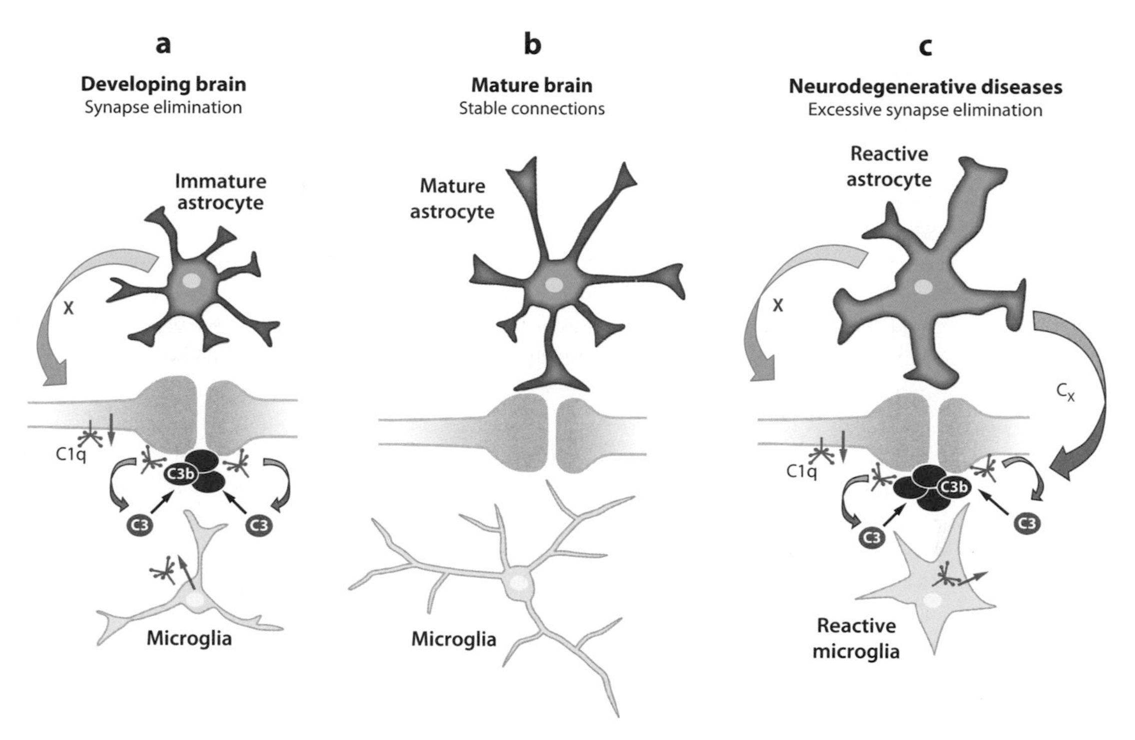

Figure 4

Complement-mediated synapse elimination during development and in neurodegenerative diseases. (*a*) In the developing brain, astrocytes induce the production of C1q in neurons through an unidentified molecular signal ("X"). Neuron and microglia-derived C1q tags weak or superfluous synapses for removal through the classical complement pathway, resulting in C3 cleavage and synaptic C3b deposition. Complement-tagged synapses are removed through phagocytosis by microglia. (*b*) In the absence of activated complement, synapses remain stable. (*c*) We propose that complement-mediated synapse elimination drives the development/progression of neurodegenerative diseases. As observed in the developing brain, reactive astrocytes release signal(s) ("X") that induce C1q production in neurons. Neuronal and microglia-derived C1q is recruited to synapses; recruitment then triggers the activation of downstream classical complement components, produced in excess by reactive astrocytes (C_X), reactive microglia, and neurons, resulting in microglia-mediated synapse elimination. Modified from Schafer & Stevens (2010).

upregulated in the murine retina during early glaucoma stages, before signs of neurodegeneration are detectable (Steele et al. 2006, Howell et al. 2011), and has also been found to be elevated in human glaucomatous retina tissue (Stasi et al. 2006). Furthermore, in the glaucomatous mouse eye, C1q immunoreactivity was shown to be upregulated in the inner plexiform layer of the retina, the synapse-rich compartment host to postsynaptic connections of RGCs. This increase in C1q immunoreactivity was temporally correlated with a decrease in synapse density, which preceded the first signs of dendrite atrophy and RGC loss (Stevens et al. 2007). Most importantly, C1q deficiency in the DBA/2J background conveys significant neuroprotection in the glaucomatous eye (Howell et al. 2011). This compelling evidence suggests that complement-mediated synapse elimination may be an early and critical event in driving the neurodegenerative process in glaucoma.

AD is the most common form of neurodegenerative dementia. A key molecular characteristic of AD is the increased generation of the amyloid-beta peptide (Aβ), a proteolytically generated derivative of the amyloid precursor protein, which accumulates in the extracellular milieu to form amyloid plaques (Glenner & Wong 1984). Many studies have established that pronounced synaptic dysfunction and synapse loss are early features of this disease in both rodents and humans (Selkoe 2002, Spires-Jones et al. 2007, Koffie et al. 2011). Gliosis, microglia activation, and an increased expression and activation of virtually all complement components occur in the AD brain (Veerhuis et al. 2011). C1q is up to 80-fold upregulated in human AD brains (Yasojima et al. 1999). C1q deficieny in a mouse model of AD causes decreased synapse loss, AD pathology, and an improvement in cognitive function, providing direct evidence for a detrimental role of the classical complement pathway in this disease (Fonseca et al. 2004). Furthermore, several complement cascade interactors have recently emerged as susceptibility genes in AD, including ApoJ/Clusterin, a complement inhibitor, and CR1 (Jun et al. 2010, Chibnik et al. 2011, Degn et al. 2011), and enhanced levels of several complement components were detected in the cerebrospinal fluid of even presymptomatic individuals that carry familial AD disease mutations (Ringman et al. 2012). Finally, Aβ oligomers cause synapse degeneration (Wilcox et al. 2011), which is of particular interest because binding of Aβ to C1q activates the classical arm of the complement cascade (Tacnet-Delorme et al. 2001, Sim et al. 2007). Oligomeric Aβ-induced synapse loss can be detected in close proximity to amyloid plaques (Spires-Jones et al. 2007, Koffie et al. 2009), which constitutes, as the authors propose, a reservoir for Aβ-oligomers. These toxic oligomers colocalize with a subset of excitatory, degenerating synapses in vivo. Given that complement components can already be detected on amyloid plaques during early AD stages in humans (Zanjani et al. 2005), when synapse loss drives neurodegeneration, the interaction of complement with oligomeric Aβ at synapses may cause the microglia-mediated loss of these structures to drive disease progression. Taken together, these findings strongly support the idea that synaptic activation of the classical complement cascade in AD may drive disease progression by synapse elimination.

Synapse loss and neuroinflammation, including complement upregulation, are also major events in many other neurodegenerative diseases, including Huntington's disease, Parkinson's disease, and multiple sclerosis. In contrast with these neurodegenerative diseases, although C1q is highly elevated in both the mouse and human diseases, it is less clear if the loss of central synapses is a crucial early component of amyotrophic lateral sclerosis (ALS), a fatal adult-onset disorder confined to the voluntary motor system. However, neither C4 deficiency (Chiu et al. 2009) nor C3 deficiency (J.W. Lewcock, unpublished observation) is protective in the mouse SOD1 ALS model, arguing against a prominent role of complement at least in murine models of ALS. However the situation may be different for spinal muscular atrophy, an often fatal autosomal-recessive disorder of infancy caused by homozygous deletion or rare missense mutations in the survival MN 1 (*SMN1*) gene, accompanied by selective loss of motor neurons within the anterior horns of the spinal cord and early reactive gliosis (Papadimitriou et al. 2010). A recent study demonstrated that prominent CNS synapse loss precedes motor neuron degeneration in a mouse model of this disease (Mentis et al. 2011). This is one of the best examples of early synapse degeneration in a mouse model of neurodegenerative disease and may open up new avenues of future research that may reveal a role for complement in this disorder.

Although we have focused on the classical complement cascade in this review, emerging evidence implicates other limbs of the complement cascade in neurological disease susceptibility and pathophysiology. In particular, the alternative complement cascade contributes to damage after brain trauma (Leinhase et al. 2006), and genetic variations of several alternative complement cascade genes, including the complement regulator Factor H, Factor B, C2, and Factor I, confer a risk for age-related macular degeneration (AMD), a common form of blindness (Klein et al. 2005). Although it is not yet clear whether loss of retinal synapses is an early component of AMD, this will be an important avenue of future investigation.

CONCLUSIONS AND PERSPECTIVES

The role of immune cells and immune pathways in the developing, adult, and diseased brain, long neglected, has emerged as an exciting area of research. Studying the role of the classical complement cascade pathway and proteins that interact with it is leading to a better understanding of some classical questions in neurobiology: How do synapses form, how are they eliminated, how does an activity-dependent competition sculpt the synaptic wiring of the developing brain, and why do synapses degenerate in neurodegenerative disease? The studies we have reviewed raise many exciting questions for future research. What are the glial signals that control activation of the complement cascade in the brain? Why are some synapses targeted by the complement cascade and not others? What are the critical synaptic receptors for C1q? Do neurons express novel molecules that control complement activation? Does the complement cascade drive synapse loss that accompanies normal aging? Might there be nonclassical roles for complement proteins in the CNS? Perhaps most importantly, will the development of therapeutic inhibitors of the classical complement cascade lead to a new therapy for neurodegenerative disorders and other neurological diseases and injuries? A better understanding of the roles of complement proteins in the CNS has the potential to broaden our understanding of neurodegenerative disease development and progression as well as open up new treatment strategies to interfere with the detrimental course common to all these diseases.

DISCLOSURE STATEMENT

B. Barres is a cofounder of Annexon Inc., a new company that will develop therapeutics for neurological diseases.

ACKNOWLEDGMENTS

We thank Drs. Arnon Rosenthal, Ryuta Koyama, Isaac Chiu, and Kenneth Colodner for helpful discussions and critical reading of the manuscript. We are grateful to Dr. Ryuta Koyama for the design and preparation of the illustrations in **Figures 1**, **2**, and **3**. We apologize to authors whose work could not be discussed because of space limitations. The work in the authors' labs reported here was supported by funding from the Swiss National Science Foundation (A.H.S.), NIDA (B.A.B.), the Smith Family Foundation (B.S.), NINDS (B.S.), and the Ellison Medical Foundation (B.S.).

LITERATURE CITED

Alexander JJ, Anderson AJ, Barnum SR, Stevens B, Tenner AJ. 2008. The complement cascade: Yin-Yang in neuroinflammation—neuro-protection and -degeneration. *J. Neurochem.* 107:1169–87

Allen NJ, Barres BA. 2005. Signaling between glia and neurons: focus on synaptic plasticity. *Curr. Opin. Neurobiol.* 15:542–48

Balice-Gordon RJ, Lichtman JW. 1994. Long-term synapse loss induced by focal blockade of postsynaptic receptors. *Nature* 372:519–24

Barnum SR. 1995. Complement biosynthesis in the central nervous system. *Crit. Rev. Oral Biol. Med.* 6:132–46

Benard M, Raoult E, Vaudry D, Leprince J, Falluel-Morel A, et al. 2008. Role of complement anaphylatoxin receptors (C3aR, C5aR) in the development of the rat cerebellum. *Mol. Immunol.* 45:3767–74

Benoit ME, Tenner AJ. 2011. Complement protein C1q-mediated neuroprotection is correlated with regulation of neuronal gene and microRNA expression. *J. Neurosci. Off. J. Soc. Neurosci.* 31:3459–69

Biber K, Neumann H, Inoue K, Boddeke HW. 2007. Neuronal 'On' and 'Off' signals control microglia. *Trends Neurosci.* 30:596–602

Bjartmar L, Huberman AD, Ullian EM, Renteria RC, Liu X, et al. 2006. Neuronal pentraxins mediate synaptic refinement in the developing visual system. *J. Neurosci. Off. J. Soc. Neurosci.* 26:6269–81

Bobak DA, Frank MM, Tenner AJ. 1986. Characterization of C1q receptor expression on human phagocytic cells: effects of PDBu and fMLP. *J. Immunol.* 136:4604–10

Bogestal YR, Barnum SR, Smith PL, Mattisson V, Pekny M, Pekna M. 2007. Signaling through C5aR is not involved in basal neurogenesis. *J. Neurosci. Res.* 85:2892–97

Bolliger MF, Martinelli DC, Sudhof TC. 2011. The cell-adhesion G protein-coupled receptor BAI3 is a high-affinity receptor for C1q-like proteins. *Proc. Natl. Acad. Sci. USA* 108:2534–39

Bolton MM, Eroglu C. 2009. Look who is weaving the neural web: glial control of synapse formation. *Curr. Opin. Neurobiol.* 19:491–97

Boulanger LM. 2009. Immune proteins in brain development and synaptic plasticity. *Neuron* 64:93–109

Cahoy JD, Emery B, Kaushal A, Foo LC, Zamanian JL, et al. 2008. A transcriptome database for astrocytes, neurons, and oligodendrocytes: a new resource for understanding brain development and function. *J. Neurosci. Off. J. Soc. Neurosci.* 28:264–78

Campbell G, Shatz CJ. 1992. Synapses formed by identified retinogeniculate axons during the segregation of eye input. *J. Neurosci. Off. J. Soc. Neurosci.* 12:1847–58

Carroll MC. 2004. The complement system in regulation of adaptive immunity. *Nat. Immunol.* 5:981–86

Chen SK, Tvrdik P, Peden E, Cho S, Wu S, et al. 2010. Hematopoietic origin of pathological grooming in Hoxb8 mutant mice. *Cell* 141:775–85

Chibnik LB, Shulman JM, Leurgans SE, Schneider JA, Wilson RS, et al. 2011. CR1 is associated with amyloid plaque burden and age-related cognitive decline. *Ann. Neurol.* 69:560–69

Chiu IM, Phatnani H, Kuligowski M, Tapia JC, Carrasco MA, et al. 2009. Activation of innate and humoral immunity in the peripheral nervous system of ALS transgenic mice. *Proc. Natl. Acad. Sci. USA* 106:20960–65

Chu Y, Jin X, Parada I, Pesic A, Stevens B, et al. 2010. Enhanced synaptic connectivity and epilepsy in C1q knockout mice. *Proc. Natl. Acad. Sci. USA* 107:7975–80

Corriveau RA, Huh GS, Shatz CJ. 1998. Regulation of class I MHC gene expression in the developing and mature CNS by neural activity. *Neuron* 21:505–20

Dalmau I, Finsen B, Zimmer J, Gonzalez B, Castellano B. 1998. Development of microglia in the postnatal rat hippocampus. *Hippocampus* 8:458–74

Datwani A, McConnell MJ, Kanold PO, Micheva KD, Busse B, et al. 2009. Classical MHCI molecules regulate retinogeniculate refinement and limit ocular dominance plasticity. *Neuron* 64:463–70

Davalos D, Grutzendler J, Yang G, Kim JV, Zuo Y, et al. 2005. ATP mediates rapid microglial response to local brain injury in vivo. *Nat. Neurosci.* 8:752–58

Degn SE, Jensenius JC, Thiel S. 2011. Disease-causing mutations in genes of the complement system. *Am. J. Hum. Genet.* 88:689–705

Degn SE, Thiel S, Jensenius JC. 2007. New perspectives on mannan-binding lectin-mediated complement activation. *Immunobiology* 212:301–11

Eggleton P, Tenner AJ, Reid KB. 2000. C1q receptors. *Clin. Exp. Immunol.* 120:406–12

Eroglu C, Barres BA. 2010. Regulation of synaptic connectivity by glia. *Nature* 468:223–31

Feller MB. 2009. Retinal waves are likely to instruct the formation of eye-specific retinogeniculate projections. *Neural Dev.* 4:24

Fiske BK, Brunjes PC. 2000. Microglial activation in the developing rat olfactory bulb. *Neuroscience* 96:807–15

Fonseca MI, Zhou J, Botto M, Tenner AJ. 2004. Absence of C1q leads to less neuropathology in transgenic mouse models of Alzheimer's disease. *J. Neurosci. Off. J. Soc. Neurosci.* 24:6457–65

Fujita T. 2002. Evolution of the lectin-complement pathway and its role in innate immunity. *Nat. Rev. Immunol.* 2:346–53

Gal P, Dobo J, Zavodszky P, Sim RB. 2009. Early complement proteases: C1r, C1s and MASPs. A structural insight into activation and functions. *Mol. Immunol.* 46:2745–52

Gasque P. 2004. Complement: a unique innate immune sensor for danger signals. *Mol. Immunol.* 41:1089–98

Glenner GG, Wong CW. 1984. Alzheimer's disease: initial report of the purification and characterization of a novel cerebrovascular amyloid protein. *Biochem. Biophys. Res. Commun.* 120:885–90

Goddard CA, Butts DA, Shatz CJ. 2007. Regulation of CNS synapses by neuronal MHC class I. *Proc. Natl. Acad. Sci. USA* 104:6828–33

Gorgani NN, He JQ, Katschke KJ Jr, Helmy KY, Xi H, et al. 2008. Complement receptor of the Ig superfamily enhances complement-mediated phagocytosis in a subpopulation of tissue resident macrophages. *J. Immunol.* 181:7902–8

Graeber MB. 2010. Changing face of microglia. *Science* 330:783–88

Guan EN, Burgess WH, Robinson SL, Goodman EB, McTigue KJ, Tenner AJ. 1991. Phagocytic cell molecules that bind the collagen-like region of C1q. Involvement in the C1q-mediated enhancement of phagocytosis. *J. Biol. Chem.* 266:20345–55

Guido W. 2008. Refinement of the retinogeniculate pathway. *J. Physiol.* 586:4357–62

Guillemin GJ, Brew BJ. 2004. Microglia, macrophages, perivascular macrophages, and pericytes: a review of function and identification. *J. Leukoc. Biol.* 75:388–97

Hashimoto K. 2008. Microglial activation in schizophrenia and minocycline treatment. *Prog. Neuropsychopharmacol. Biol. Psychiatry* 32:1758–59; author reply 60

Havik B, Le Hellard S, Rietschel M, Lybaek H, Djurovic S, et al. 2011. The complement control-related genes CSMD1 and CSMD2 associate to schizophrenia. *Biol. Psychiatry* 70:35–42

Helmy KY, Katschke KJ Jr, Gorgani NN, Kljavin NM, Elliott JM, et al. 2006. CRIg: a macrophage complement receptor required for phagocytosis of circulating pathogens. *Cell* 124:915–27

Hoarau JJ, Krejbich-Trotot P, Jaffar-Bandjee MC, Das T, Thon-Hon GV, et al. 2011. Activation and control of CNS innate immune responses in health and diseases: a balancing act finely tuned by neuroimmune regulators (NIReg). *CNS Neurol. Disord. Drug Targets* 10:25–43

Hong YK, Chen C. 2011. Wiring and rewiring of the retinogeniculate synapse. *Curr. Opin. Neurobiol.* 21:228–37

Hooks BM, Chen C. 2006. Distinct roles for spontaneous and visual activity in remodeling of the retinogeniculate synapse. *Neuron* 52:281–91

Howell GR, Macalinao DG, Sousa GL, Walden M, Soto I, et al. 2011. Molecular clustering identifies complement and endothelin induction as early events in a mouse model of glaucoma. *J. Clin. Invest.* 121:1429–44

Hua JY, Smith SJ. 2004. Neural activity and the dynamics of central nervous system development. *Nat. Neurosci.* 7:327–32

Huberman AD. 2007. Mechanisms of eye-specific visual circuit development. *Curr. Opin. Neurobiol.* 17:73–80

Huberman AD, Feller MB, Chapman B. 2008. Mechanisms underlying development of visual maps and receptive fields. *Annu. Rev. Neurosci.* 31:479–509

Huh GS, Boulanger LM, Du H, Riquelme PA, Brotz TM, Shatz CJ. 2000. Functional requirement for class I MHC in CNS development and plasticity. *Science* 290:2155–59

Iijima T, Miura E, Watanabe M, Yuzaki M. 2010. Distinct expression of C1q-like family mRNAs in mouse brain and biochemical characterization of their encoded proteins. *Eur. J. Neurosci.* 31:1606–15

Jennings C. 1994. Developmental neurobiology. Death of a synapse. *Nature* 372:498–99

Jeon H, Lee S, Lee WH, Suk K. 2010. Analysis of glial secretome: the long pentraxin PTX3 modulates phagocytic activity of microglia. *J. Neuroimmunol.* 229:63–72

Jo J, Whitcomb DJ, Olsen KM, Kerrigan TL, Lo SC, et al. 2011. Abeta(1–42) inhibition of LTP is mediated by a signaling pathway involving caspase-3, Akt1 and GSK-3beta. *Nat. Neurosci.* 14:545–47

Jun G, Naj AC, Beecham GW, Wang LS, Buros J, et al. 2010. Meta-analysis confirms CR1, CLU, and PICALM as Alzheimer disease risk loci and reveals interactions with APOE genotypes. *Arch. Neurol.* 67:1473–84

Kang YH, Tan LA, Carroll MV, Gentle ME, Sim RB. 2009. Target pattern recognition by complement proteins of the classical and alternative pathways. *Adv. Exp. Med. Biol.* 653:117–28

Katz LC, Shatz CJ. 1996. Synaptic activity and the construction of cortical circuits. *Science* 274:1133–38

Kim DD, Song WC. 2006. Membrane complement regulatory proteins. *Clin. Immunol.* 118:127–36

Kishore U, Reid KB. 2000. C1q: structure, function, and receptors. *Immunopharmacology* 49:159–70

Klein RJ, Zeiss C, Chew EY, Tsai JY, Sackler RS, et al. 2005. Complement factor H polymorphism in age-related macular degeneration. *Science* 308:385–89

Klos A, Tenner AJ, Johswich K-O, Ager RR, Reis ES, Köhl J. 2009. The role of the anaphylatoxins in health and disease. *Mol. Immunol.* 46:2753–66

Koffie RM, Hyman BT, Spires-Jones TL. 2011. Alzheimer's disease: synapses gone cold. *Mol. Neurodegener.* 6:63

Koffie RM, Meyer-Luehmann M, Hashimoto T, Adams KW, Mielke ML, et al. 2009. Oligomeric amyloid beta associates with postsynaptic densities and correlates with excitatory synapse loss near senile plaques. *Proc. Natl. Acad. Sci. USA* 106:4012–17

Kojouharova M, Reid K, Gadjeva M. 2010. New insights into the molecular mechanisms of classical complement activation. *Mol. Immunol.* 47:2154–60

Lampert-Etchells M, Pasinetti GM, Finch CE, Johnson SA. 1993. Regional localization of cells containing complement C1q and C4 mRNAs in the frontal cortex during Alzheimer's disease. *Neurodegeneration* 2:111–21

Leinhase I, Holers VM, Thurman JM, Harhausen D, Schmidt OI, et al. 2006. Reduced neuronal cell death after experimental brain injury in mice lacking a functional alternative pathway of complement activation. *BMC Neurosci.* 7:55

Levi-Strauss M, Mallat M. 1987. Primary cultures of murine astrocytes produce C3 and factor B, two components of the alternative pathway of complement activation. *J. Immunol.* 139:2361–66

Li Z, Jo J, Jia JM, Lo SC, Whitcomb DJ, et al. 2010. Caspase-3 activation via mitochondria is required for long-term depression and AMPA receptor internalization. *Cell* 141:859–71

Mallucci GR. 2009. Prion neurodegeneration: starts and stops at the synapse. *Prion* 3:195–201

Martinelli DC, Sudhof TC. 2011. Cerebellins meet neurexins (commentary on Matsuda & Yuzaki). *Eur. J. Neurosci.* 33:1445–46

Matsuda K, Miura E, Miyazaki T, Kakegawa W, Emi K, et al. 2010. Cbln1 is a ligand for an orphan glutamate receptor delta2, a bidirectional synapse organizer. *Science* 328:363–68

Matsuda K, Yuzaki M. 2011. Cbln family proteins promote synapse formation by regulating distinct neurexin signaling pathways in various brain regions. *Eur. J. Neurosci.* 33:1447–61

McConnell MJ, Huang YH, Datwani A, Shatz CJ. 2009. H2-K(b) and H2-D(b) regulate cerebellar long-term depression and limit motor learning. *Proc. Natl. Acad. Sci. USA* 106:6784–89

Medzhitov R, Janeway CA Jr. 2002. Decoding the patterns of self and nonself by the innate immune system. *Science* 296:298–300

Mentis GZ, Blivis D, Liu W, Drobac E, Crowder ME, et al. 2011. Early functional impairment of sensory-motor connectivity in a mouse model of spinal muscular atrophy. *Neuron* 69:453–67

Monji A, Kato T, Kanba S. 2009. Cytokines and schizophrenia: microglia hypothesis of schizophrenia. *Psychiatry Clin. Neurosci.* 63:257–65

Morgan JT, Chana G, Pardo CA, Achim C, Semendeferi K, et al. 2010. Microglial activation and increased microglial density observed in the dorsolateral prefrontal cortex in autism. *Biol. Psychiatry* 68:368–76

Moriyama M, Fukuhara T, Britschgi M, He Y, Narasimhan R, et al. 2011. Complement receptor 2 is expressed in neural progenitor cells and regulates adult hippocampal neurogenesis. *J. Neurosci. Off. J. Soc. Neurosci.* 31:3981–89

Nauta A. 2003. Recognition and clearance of apoptotic cells: a role for complement and pentraxins. *Trends Immunol.* 24:148–54

Nauta AJ, Bottazzi B, Mantovani A, Salvatori G, Kishore U, et al. 2003. Biochemical and functional characterization of the interaction between pentraxin 3 and C1q. *Eur. J. Immunol.* 33:465–73

Nayak A, Pedenekar L, Reid KB, Kishore U. 2011. Complement and non-complement activating functions of C1q: a prototypical innate immune molecule. *Innate Immun.* doi: 10.1177/1753425910396252

Nepomuceno RR, Tenner AJ. 1998. C1qRP, the C1q receptor that enhances phagocytosis, is detected specifically in human cells of myeloid lineage, endothelial cells, and platelets. *J. Immunol.* 160:1929–35

Neumann H, Misgeld T, Matsumuro K, Wekerle H. 1998. Neurotrophins inhibit major histocompatibility class II inducibility of microglia: involvement of the p75 neurotrophin receptor. *Proc. Natl. Acad. Sci. USA* 95:5779–84

Nguyen MD, Julien JP, Rivest S. 2002. Innate immunity: the missing link in neuroprotection and neurodegeneration? *Nat. Rev. Neurosci.* 3:216–27

Nickells RW, Howell GR, Soto I, John SWM. 2012. Under pressure: cellular and molecular responses during glaucoma, a common neurodegeneration with axonopathy. *Annu. Rev. Neurosci.* 35:153–79

Nimmerjahn A, Kirchhoff F, Helmchen F. 2005. Resting microglial cells are highly dynamic surveillants of brain parenchyma in vivo. *Science* 308:1314–18

Nordahl EA, Rydengard V, Nyberg P, Nitsche DP, Morgelin M, et al. 2004. Activation of the complement system generates antibacterial peptides. *Proc. Natl. Acad. Sci. USA* 101:16879–84

Paolicelli RC, Bolasco G, Pagani F, Maggi L, Scianni M, et al. 2011. Synaptic pruning by microglia is necessary for normal brain development. *Science* 333:1456–58

Papadimitriou D, Le Verche V, Jacquier A, Ikiz B, Przedborski S, Re DB. 2010. Inflammation in ALS and SMA: sorting out the good from the evil. *Neurobiol. Dis.* 37:493–502

Pardo CA, Vargas DL, Zimmerman AW. 2005. Immunity, neuroglia and neuroinflammation in autism. *Int. Rev. Psychiatry* 17:485–95

Patterson PH. 2011. Maternal infection and immune involvement in autism. *Trends Mol. Med.* 17:389–94

Penn AA, Riquelme PA, Feller MB, Shatz CJ. 1998. Competition in retinogeniculate patterning driven by spontaneous activity. *Science* 279:2108–12

Perry VH, Hume DA, Gordon S. 1985. Immunohistochemical localization of macrophages and microglia in the adult and developing mouse brain. *Neuroscience* 15:313–26

Rahpeymai Y, Hietala MA, Wilhelmsson U, Fotheringham A, Davies I, et al. 2006. Complement: a novel factor in basal and ischemia-induced neurogenesis. *EMBO J.* 25:1364–74

Ransohoff RM, Perry VH. 2009. Microglial physiology: unique stimuli, specialized responses. *Annu. Rev. Immunol.* 27:119–45

Ransohoff RM, Stevens B. 2011. How many cell types does it take to wire a brain? *Science* 333:1391–92

Ricklin D, Hajishengallis G, Yang K, Lambris JD. 2010. Complement: a key system for immune surveillance and homeostasis. *Nat. Immunol.* 11:785–97

Ringman JM, Schulman H, Becker C, Jones T, Bai Y, et al. 2012. Proteomic changes in cerebrospinal fluid of presymptomatic and affected persons carrying familial Alzheimer disease mutations. *Arch. Neurol.* 69(1):96–104

Rosen AM, Stevens B. 2010. The role of the classical complement cascade in synapse loss during development and glaucoma. *Adv. Exp. Med. Biol.* 703:75–93

Rutkowski MJ, Sughrue ME, Kane AJ, Ahn BJ, Fang S, Parsa AT. 2010a. The complement cascade as a mediator of tissue growth and regeneration. *Inflamm. Res.* 59:897–905

Rutkowski MJ, Sughrue ME, Kane AJ, Mills SA, Fang S, Parsa AT. 2010b. Complement and the central nervous system: emerging roles in development, protection and regeneration. *Immunol. Cell Biol.* 88:781–86

Sanes JR, Lichtman JW. 1999. Development of the vertebrate neuromuscular junction. *Annu. Rev. Neurosci.* 22:389–442

Schafer DP, Lehrman EK, Kautzman A, Koyama R, Mardinly AR, et al. 2011. Microglia shape neural circuits in the healthy, developing brain. *Soc. Neurosci.* 663.03(Abstr.)

Schafer DP, Lehrman EK, Kautzman A, Koyama R, Mardinly AR, et al. 2012. Microglia sculpt postnatal neural circuits in an activity and complement-dependent manner. *Neuron.* In press

Schafer DP, Stevens B. 2010. Synapse elimination during development and disease: immune molecules take centre stage. *Biochem. Soc. Trans.* 38:476–81

Selkoe DJ. 2002. Alzheimer's disease is a synaptic failure. *Science* 298:789–91

Shatz CJ. 1990. Impulse activity and the patterning of connections during CNS development. *Neuron* 5:745–56

Shatz CJ. 2009. MHC class I: an unexpected role in neuronal plasticity. *Neuron* 64:40–45

Shatz CJ, Sretavan DW. 1986. Interactions between retinal ganglion cells during the development of the mammalian visual system. *Annu. Rev. Neurosci.* 9:171–207

Shimono C, Manabe R, Yamada T, Fukuda S, Kawai J, et al. 2010. Identification and characterization of nCLP2, a novel C1q family protein expressed in the central nervous system. *J. Biochem.* 147:565–79

Shinjyo N, Stahlberg A, Dragunow M, Pekny M, Pekna M. 2009. Complement-derived anaphylatoxin C3a regulates in vitro differentiation and migration of neural progenitor cells. *Stem Cells* 27:2824–32

Sim RB, Kishore U, Villiers CL, Marche PN, Mitchell DA. 2007. C1q binding and complement activation by prions and amyloids. *Immunobiology* 212:355–62

Sjöberg AP, Trouw LA, Blom AM. 2009. Complement activation and inhibition: a delicate balance. *Trends Immunol.* 30:83–90

Song W-C. 2006. Complement regulatory proteins and autoimmunity. *Autoimmunity* 39:403–10

Spires-Jones TL, Meyer-Luehmann M, Osetek JD, Jones PB, Stern EA, et al. 2007. Impaired spine stability underlies plaque-related spine loss in an Alzheimer's disease mouse model. *Am. J. Pathol.* 171:1304–11

Sretavan D, Shatz CJ. 1984. Prenatal development of individual retinogeniculate axons during the period of segregation. *Nature* 308:845–48

Stasi K, Nagel D, Yang X, Wang RF, Ren L, et al. 2006. Complement component 1Q (C1Q) upregulation in retina of murine, primate, and human glaucomatous eyes. *Invest. Ophthalmol. Vis. Sci.* 47:1024–29

Steele MR, Inman DM, Calkins DJ, Horner PJ, Vetter ML. 2006. Microarray analysis of retinal gene expression in the DBA/2J model of glaucoma. *Invest. Ophthalmol. Vis. Sci.* 47:977–85

Stellwagen D, Shatz CJ. 2002. An instructive role for retinal waves in the development of retinogeniculate connectivity. *Neuron* 33:357–67

Stephan AH, Fraser D, Yabe Y, Mateos JM, Tun C, et al. 2011. Does complement component C1q-mediated synapse elimination drive cognitive decline in the aging brain? *Soc. Neurosci.* 659.14(Abstr.)

Stevens B, Allen NJ, Vazquez LE, Howell GR, Christopherson KS, et al. 2007. The classical complement cascade mediates CNS synapse elimination. *Cell* 131:1164–78

Syken J, Grandpre T, Kanold PO, Shatz CJ. 2006. PirB restricts ocular-dominance plasticity in visual cortex. *Science* 313:1795–800

Tacnet-Delorme P, Chevallier S, Arlaud GJ. 2001. Beta-amyloid fibrils activate the C1 complex of complement under physiological conditions: evidence for a binding site for A beta on the C1q globular regions. *J. Immunol.* 167:6374–81

Taylor PR, Carugati A, Fadok VA, Cook HT, Andrews M, et al. 2000. A hierarchical role for classical pathway complement proteins in the clearance of apoptotic cells in vivo. *J. Exp. Med.* 192:359–66

Tenner AJ. 1998. C1q receptors: regulating specific functions of phagocytic cells. *Immunobiology* 199:250–64

Tenner AJ, Frank MM. 1987. A sensitive specific hemolytic assay for proenzyme C1. *Complement* 4:42–52

Torborg CL, Feller MB. 2005. Spontaneous patterned retinal activity and the refinement of retinal projections. *Prog. Neurobiol.* 76:213–35

Tremblay M-È, Stevens B, Sierra A, Wake H, Bessis A, Nimmerjahn A. 2011. The role of microglia in the healthy brain. *J. Neurosci.* 31(45):16064–69

Tremblay ME, Lowery RL, Majewska AK. 2010. Microglial interactions with synapses are modulated by visual experience. *PLoS Biol.* 8:e1000527

van Lookeren Campagne M, Wiesmann C, Brown EJ. 2007. Macrophage complement receptors and pathogen clearance. *Cell Microbiol.* 9:2095–102

Vargas DL, Nascimbene C, Krishnan C, Zimmerman AW, Pardo CA. 2005. Neuroglial activation and neuroinflammation in the brain of patients with autism. *Ann. Neurol.* 57:67–81

Veerhuis R, Janssen I, De Groot CJ, Van Muiswinkel FL, Hack CE, Eikelenboom P. 1999. Cytokines associated with amyloid plaques in Alzheimer's disease brain stimulate human glial and neuronal cell cultures to secrete early complement proteins, but not C1-inhibitor. *Exp. Neurol.* 160:289–99

Veerhuis R, Nielsen HM, Tenner AJ. 2011. Complement in the brain. *Mol. Immunol.* 48(14):1592–603

Wake H, Moorhouse AJ, Jinno S, Kohsaka S, Nabekura J. 2009. Resting microglia directly monitor the functional state of synapses in vivo and determine the fate of ischemic terminals. *J. Neurosci. Off. J. Soc. Neurosci.* 29:3974–80

Watanabe M. 2008. Molecular mechanisms governing competitive synaptic wiring in cerebellar Purkinje cells. *Tohoku. J. Exp. Med.* 214:175–90

Whitmore AV, Libby RT, John SW. 2005. Glaucoma: thinking in new ways—a role for autonomous axonal self-destruction and other compartmentalised processes? *Prog. Retin. Eye Res.* 24:639–62

Wilcox KC, Lacor PN, Pitt J, Klein WL. 2011. Abeta oligomer-induced synapse degeneration in Alzheimer's disease. *Cell Mol. Neurobiol.* 31(6):939–48

Woodruff TM, Ager RR, Tenner AJ, Noakes PG, Taylor SM. 2010. The role of the complement system and the activation fragment C5a in the central nervous system. *Neuromol. Med.* 12:179–92

Wyss-Coray T, Mucke L. 2002. Inflammation in neurodegenerative disease—a double-edged sword. *Neuron* 35:419–32

Yasojima K, Schwab C, McGeer EG, McGeer PL. 1999. Up-regulated production and activation of the complement system in Alzheimer's disease brain. *Am. J. Pathol.* 154:927–36

Yuzaki M. 2010. Synapse formation and maintenance by C1q family proteins: a new class of secreted synapse organizers. *Eur. J. Neurosci.* 32:191–97

Zamanian JL, Xu L, Foo LC, Nouri N, Zhou L, et al. 2012. Genomic analysis of reactive astrogliosis. *J. Neurosci.* In press

Zanjani H, Finch CE, Kemper C, Atkinson J, McKeel D, et al. 2005. Complement activation in very early Alzheimer disease. *Alzheimer. Dis. Assoc. Disord.* 19:55–66

Zhou W. 2012. The new face of anaphylatoxins in immune regulation. *Immunobiology* 217(2):225–34

Zipfel PF, Skerka C. 2009. Complement regulators and inhibitory proteins. *Nat. Rev. Immunol.* 9:729–40

Zipfel PF, Wurzner R, Skerka C. 2007. Complement evasion of pathogens: Common strategies are shared by diverse organisms. *Mol. Immunol.* 44:3850–57

Brain Plasticity Through the Life Span: Learning to Learn and Action Video Games

Daphne Bavelier,[1,2] C. Shawn Green,[3]
Alexandre Pouget,[1,2] and Paul Schrater[4]

[1]Department of Psychology and Education Sciences, University of Geneva, 1211 Geneva 4, Switzerland

[2]Department of Brain and Cognitive Sciences, University of Rochester, Rochester, New York 14627-0268, USA; email: daphne@bcs.rochester.edu, alex@bcs.rochester.edu

[3]Department of Psychology, Eye Research Institute, University of Wisconsin, Madison, Wisconsin 53706, USA: email: csgreen2@wisc.edu

[4]Departments of Psychology and Computer Science, University of Minnesota, Minnesota 55455, USA; email: schrater@umn.edu

Annu. Rev. Neurosci. 2012. 35:391–416

The *Annual Review of Neuroscience* is online at neuro.annualreviews.org

This article's doi:
10.1146/annurev-neuro-060909-152832

0147-006X/12/0721-0391$20.00

Keywords

transfer, generalization, probabilistic inference, cognitive control, resource allocation, knowledge, hierarchy, learning rules

Abstract

The ability of the human brain to learn is exceptional. Yet, learning is typically quite specific to the exact task used during training, a limiting factor for practical applications such as rehabilitation, workforce training, or education. The possibility of identifying training regimens that have a broad enough impact to transfer to a variety of tasks is thus highly appealing. This work reviews how complex training environments such as action video game play may actually foster brain plasticity and learning. This enhanced learning capacity, termed learning to learn, is considered in light of its computational requirements and putative neural mechanisms.

Contents

INTRODUCTION

Human beings have a tremendous capacity to learn. And although there are unquestionably gradients in the ability to learn that arise as a function of intrinsic factors such as age and individual genetic propensities, nearly all humans demonstrate the ability to acquire new skills and to alter behavior given appropriate training. One common finding, however, is that the learning that emerges as a result of training is often quite specific to the trained stimuli, context, and task. Such specificity has been documented in every subfield of neuroscience that focuses on learning—e.g., motor learning (Shapiro & Schmidt 1982), expertise (Chase & Simon 1973), or memory (Godden & Baddeley 1975). In addition to being of theoretical interest, this fact is of great practical relevance in areas such as rehabilitation and education in which the end goal necessarily requires that the benefits of learning extend beyond the confines of the particular training regimen. For instance, it is of little use for a patient with damage to the motor cortex to improve on reaching movements in therapy if it does not also improve their ability to reach for a bottle of milk in their refrigerator at home (Frey et al. 2011). Similarly, it is of limited benefit if pupils trained on one type of math problem are unable to solve similar problems when presented outside the classroom. Thus, for all practical purposes, the specificity that typically accompanies learning is a curse.

A crucial issue in the field of learning, therefore, concerns training conditions that result in observable benefits for untrained skills and tasks. A handful of behavioral interventions have recently been noted to induce more general learning than that typically documented in the learning literature. These learning paradigms are usually more complex than standard laboratory manipulations, reminiscent of the "enriched environment effects" seen in animal rearing (Renner & Rosenzweig 1987), and they typically correspond to real-life experiences, such as aerobic activity (Hillman et al. 2008), athletic training (Erickson & Kramer 2009), musical training (Schellenberg 2004), mind-body training (Lutz et al. 2008, Tang & Posner 2009), working memory training (Jaeggi et al. 2008, Klingberg 2010), and, the focus of this article, action video game training (Green & Bavelier 2003, Spence & Feng 2010). Here we first review the variety of tasks on which performance is enhanced as a result of action video game experiences. We then ask why action games might produce benefits on such a wide range of tasks. In particular, rather than hypothesizing that video games teach myriad individual specific skills (i.e., one for each laboratory test during which they have been shown to produce benefit), we instead consider the possibility that what video games teach is the capability to quickly learn to perform new tasks—a capability that has been dubbed "learning to learn" (Harlow 1949, Kemp et al. 2010).

The idea that the ultimate goal of training and education might be to enable learning to learn is not a new one. On the basis of his work on pairing tasks, in which training on task A may impact training on new task B, Thorndike & Woodworth (1901) observed, "It might be that improvement in one function might fail to give in another improved ability, but succeed in giving ability to improve faster " (p. 248). Around the same time, the educational psychologist Alfred Binet [1984 (1909)] described his classroom instructional goals by saying, "[O]our first job was not to teach [the students] the things which seemed to us the most useful to them, but to teach them how to learn" (p. 111). To this end, Binet asked students to play games such as "statue," wherein the students learned to stay focused, quiet, and still for long periods of time. Although no new academic concepts or facts were taught by these games, they allowed the children to develop skills, such as attention and control, that underlie the ability to learn in school. In the 1940s, Harlow (1949) outlined what he called "the formation of learning sets" wherein animals learned general rules, such as "win-stay, lose-switch," that allowed them to quickly master new tasks that abided by this general rule. This work led Harlow to state, "The learning of primary importance to the primates . . . is the learning how to learn efficiently in the situations the animal frequently encounters" (p. 51). Despite this early interest in the topic, the factors that facilitate learning to learn and the computational principles on which it relies remain largely unexplored. Here we discuss these issues by anchoring our discussion in recent advances in the field of cognitive training through action video game play.

BENEFITS OF ACTION VIDEO GAME EXPERIENCE

An overview of the existing literature on action video game (one specific subgenre of video game) play indicates performance benefits after action game play in many domains typically thought of as distinct. This breadth is particularly notable and runs counter to the predominant pattern of results in the learning literature, wherein training facilitates the trained task but typically leads to little benefit to other, even related tasks. In the case of video game training, the transfer of learning takes place between tasks and environments that have different goals and feel. Most laboratory tasks on which avid action video game players (VGPs) are tested involve simple stimuli and highly repetitive choices, making for a rather dull environment. In sharp contrast, action video games include complex visual scenes, an enthralling feel, and a wide variety of goals at different timescales (from momentary goals such as "jump over the rock" to game-long goals such as "rescue the princess") (see **Figure 1**). Although it seems reasonable to expect VGPs to show transfer across a broad range of similar games (e.g., first-person shooters or third-person shooters), the fact that VGPs best their nonaction game–playing peers (NVGPs) on standard laboratory tests that are quite dissimilar to video games begs further investigation.

VGP: action video game player

NVGP: nonaction–video game players

Vision

Action video game play enhances the spatial and temporal resolution of vision as well as its sensitivity (Green et al. 2010). For example, when asked to determine the orientation of a T that is flanked by distracting shapes above and below, VGPs can tolerate the distractors being nearer to the T shape while still maintaining a high level of accuracy (Green & Bavelier 2007). Such capacity to resolve small details in the context of clutter, also called crowding acuity, is thought of as a limiting property of spatial vision and is often compromised in low-vision patients who complain that the small print of newspapers is unreadable because letters are unstable and mingle into one another (Legge et al. 1985). In addition, the enhancements noted in this ability as a result of action video game play occur both within and outside the typical eccentricities of game play, a finding that stands in contrast to the bulk of the perceptual learning literature, which finds that learning is typically specific to the trained retinal location.

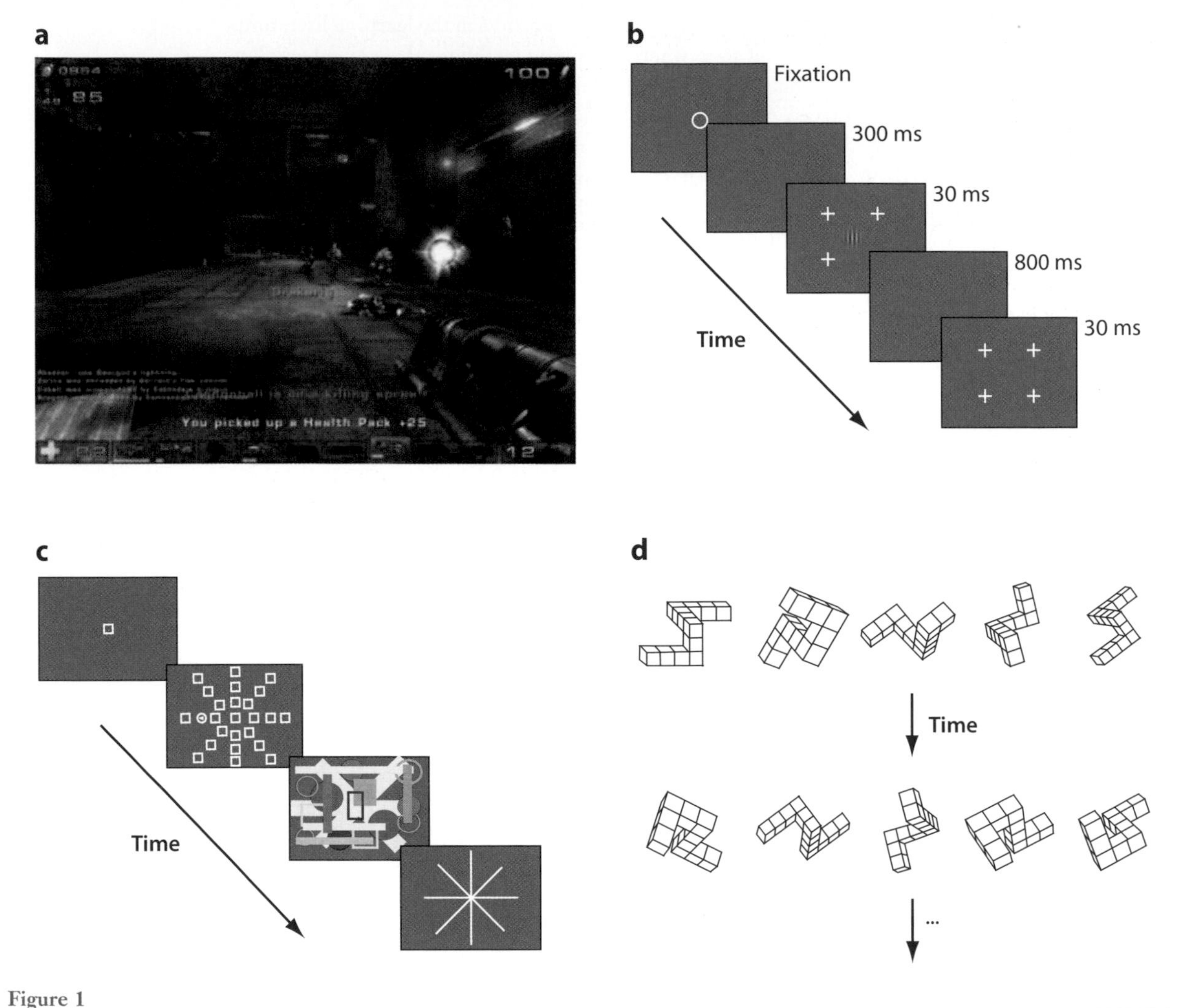

Figure 1

Action video game play (*a*) bears little resemblance in terms of stimuli and goals to perceptual, attentional, or cognitive tasks at which VGPs are found to excel in the laboratory (*b*: contrast sensitivity, *c*: visual search, *d*: mental rotation). This raises the question of why action video game play enhances such a varied set of skills.

Along with this improvement in spatial resolution, enhanced temporal resolution of vision as a result of action game experience has also been noted. VGPs show, for example, significant reductions in backward masking (wherein a display of interest is more difficult to see if it is quickly followed by another display) as compared with NVGPs. This difference suggests a change in the dynamics of the visual system: VGPs can resolve events at a higher temporal frequency (Li et al. 2010). Although backward masking, which is believed to reflect limitations on cortical processing, is improved in VGPs, this is not true of forward masking (wherein the masking display precedes the display of interest), which is believed to reflect mostly limitations of early, retinal factors. As we discuss below, these results are consistent with the view that action video game play retrains cortical networks such that each layer of the processing hierarchy makes better use of the information it receives from earlier layers, rather than changing the nature of the early sensory input (e.g., altering the optics of the eye).

Finally, a third aspect of vision found to be enhanced in VGPs is contrast sensitivity, or the ability to detect small changes in levels of gray. Although this effect appears across spatial frequencies, it is particularly pronounced at intermediate spatial frequencies (Li et al. 2009). Thus, contrary to the folk belief that screen time is bad for eyesight, action video game play appears to enhance how well one sees. Video game training, therefore, may become a powerful tool to improve eyesight in situations in which the optics of the eye are not implicated in producing the poor vision (e.g., amblyopia; Bavelier et al. 2010, Li et al. 2011).

Cognitive Functions

VGPs have also been documented to better their nongamer peers on several aspects of cognition such as visual short-term memory (Anderson et al. 2011, Boot et al. 2008), spatial cognition (Greenfield 2009), multitasking (Green & Bavelier 2006a), and some aspects of executive function (Anderson & Bavelier 2011, Chisholm & Kingstone 2011, Colzato et al. 2010, Karle et al. 2010; but see Bailey et al. 2010 for a different view). For example, whereas NVGPs were markedly slower and less accurate when asked to perform both a peripheral visual search task and a demanding central identification task concurrently [relative to performing the peripheral search task alone, VGPs showed no falloff in performance (Green & Bavelier 2006b)]. Several studies have also documented enhanced task-switching abilities in VGPs, meaning they pay less of a price for switching from one task to another (Andrews & Murphy 2006, Boot et al. 2008, Cain et al. 2012, Colzato et al. 2010, Green et al. 2012, Karle et al. 2010, Strobach et al. 2012). These results are all the more surprising given that such task-switching skills appear to be negatively affected by the extensive multitasking seen in many young adults who are heavy users of a variety of forms of technology (Ophir et al. 2009). This contrast highlights the need to assess separately the effects of different technology use on brain function.

Another cognitive domain enhanced by action games is spatial cognition. The beneficial aspect of video game play on spatial cognition was noted in the early days of video game development (McClurg & Chaille 1987, Okagaki & Frensch 1994, Subrahmanyam & Greenfield 1994). More recently, action game play has been shown to enhance mental rotation abilities (Feng et al. 2007). These results have received much attention because spatial skills are typically positively correlated with mathematical achievement in school (Halpern et al. 2007, Spelke 2005). Whether action game play may indeed foster mathematical ability is currently being investigated.

Decision Making

Benefits are also noted in decision making. In one perceptual decision-making task, for instance, VGPs and NVGPs were asked to determine whether the main flow of motion within a random dot kinematogram was to the left or to the right (**Figure 2*a***). Unlike most previous tasks used to study the effect of action gaming, in this task, participants are in control of when to terminate display presentation (by making a decision and pressing the corresponding key). It thereby provides a measure of how participants accumulate information over time in the service of decision making. Indeed, these types of task are well understood in terms of one's ability to first extract and integrate information from the environment and then to stop the integration and to select an action on the basis of the accumulated information (Palmer et al. 2005, Ratcliff & McKoon 2008). By presenting trials with variable signal-to-noise levels, the task allows the full chronometric and psychometric curves to be mapped, providing a unique description of how information about motion direction accrues over time (**Figure 2*b***). We found that action game play enhances the rate at which information accumulates over time by ~20% as compared with control participants (**Figure 2*c***). The net result is that VGPs

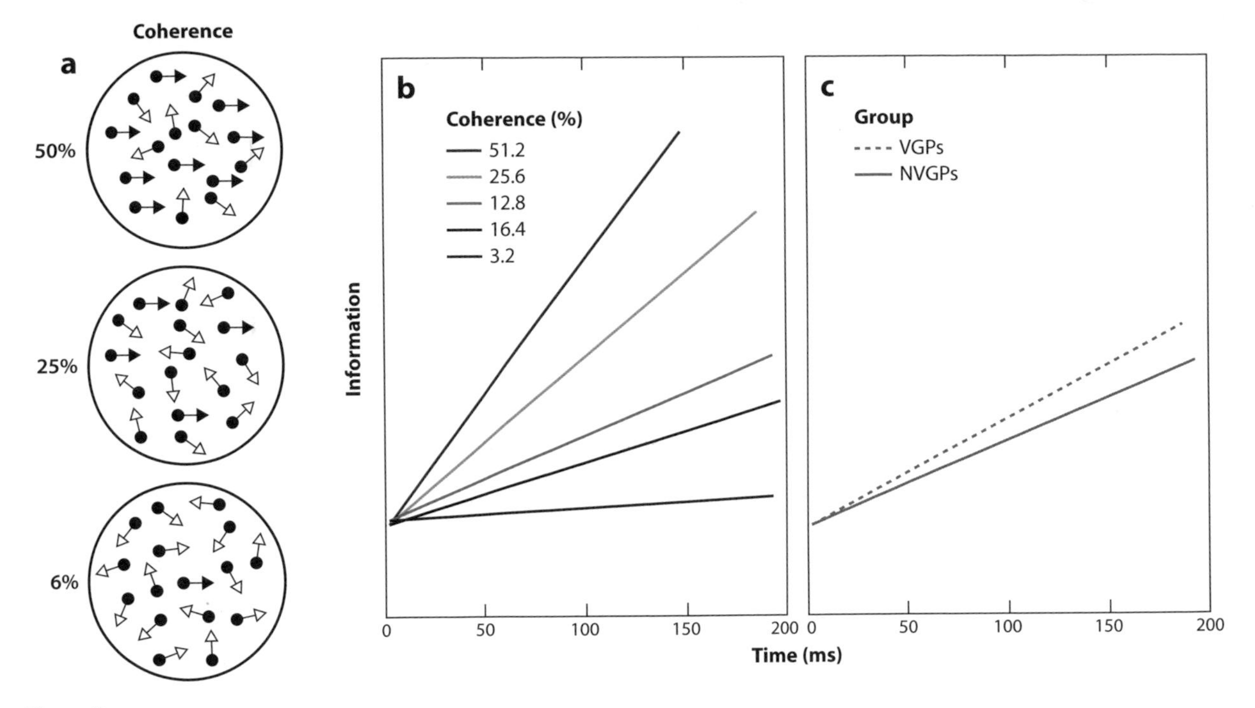

Figure 2

(*a*) In this perceptual decision-making task, participants are asked to judge whether the motion is mostly leftward or rightward. The number of dots moving coherently in one direction or the other can vary from none to many, allowing the investigator perfect control over the task difficulty. (*b*) As the level of coherence goes from low to high, the task varies from extremely difficult to easy. Accordingly, electrophysiological recordings in monkeys show that the amount of information about motion direction accumulated in the lateral intraparietal cortex (LIP), a neuronal structure associated with such decisions, increases (adapted from Beck et al. 2008, figure 3*d*). (*c*) Another way to increase the rate at which information accumulates is to enhance the decision-making characteristics of the viewer. This was shown to be the case in action video game players (VGPs) or after training on action video games: VGPs accumulated on average 20% more information per unit of time as compared with nonaction–video game players (NVGPs) (Green et al. 2010).

can make more correct decisions per units of time. This is of practical relevance as illustrated by the recruitment by the Royal Air Force of young gamers to pilot unmanned drones or by findings indicating that young laparoscopic surgeons who are gamers outperform more seasoned surgeons, executing surgery procedure faster and as accurately if not more accurately (Daily Mail 2009, McKinley et al. 2011, Rosser et al. 2007, Schlickum et al. 2009).

Reaction Time and Speed-Accuracy Trade-Off

More generally, a fourth domain of improvement concerns the ability to make quick and accurate decisions in response to changes in the environment, such as when one must brake suddenly while driving. Such improvement was noted in the decision-making tasks described above. Action video game play sped up reaction times, and this was true whether the task was in the visual or the auditory modality. Such tasks entail concurrently collection reaction times and accuracy along a wide range of task difficulties, easily revealing a speed–accuracy trade-off if present. Despite faster reaction times, accuracies were left unchanged, establishing that action game play does not result in trading speed for accuracy. A meta-analysis of more than 80 experimental conditions in which gamers and nongamers have been tested confirmed that VGPs are on average 12% faster than NVGPs, yet VGPs make no more errors (Dye et al. 2009a). This relationship held from simple decision tasks leading to reaction times as short as 250 ms, all the way to demanding, serial, visual search tasks eliciting reaction times as long as 1.5 s. The fact that the relationship was purely multiplicative, with no additive component, demonstrates that the faster reaction times seen in VGPs throughout the literature are not attributable to postdecisional factors such as faster motor execution time. Indeed, if the only difference between groups was in their ability to map a decision that had already been made into a button press on a keyboard, the difference in reaction time between groups should not depend on how long it took for the decision to be reached (i.e., would be only additive rather than multiplicative). Furthermore, the fact that there was no difference in accuracy in any of these tasks suggests that the differences also cannot be attributed to differences in criteria, or VGPs being "trigger-happy" or willing to trade reductions in accuracy for increases in speed. Whether this pattern of greater speed with matched accuracy will remain when using complex tasks with longer timescales and/or multiple task components is currently being investigated (A.F. Anderson, C.S. Green, and D. Bavelier, manuscript in preparation).

Attention

Several different facets of attention are improved following action game play. For instance, many aspects of top-down attentional control such as selective attention, divided attention, and sustained attention are enhanced in VGPs. In contrast, however, when attention is driven in a bottom-up fashion (i.e., by exogenous cueing) no differences have been found between VGPs and NVGPs, despite the fact that orienting to abrupt events is a key component of game play (Castel et al. 2005, Dye et al. 2009b, Hubert-Wallander et al. 2011). This pattern of results clearly illustrates the need for a careful investigation of those aspects of performance modified by action game play. Simply because a process is required during game play does not guarantee changes in that process. In a later section, we review in greater detail those aspects of attention and executive control that are modified by action game experience (see Resources, Knowledge, and Learning Rules in Action Video Game Play).

Causality

A key question concerns whether the effects of action video game play are causal or are instead reflective of population bias, wherein action gaming tends to attract individuals with inherently superior skills. Because our interest is in learning rather than in identifying individuals with extraordinary skills, causality is a crucial factor. The only way to establish firmly that the relationship is indeed causal is via well-controlled training studies (see Green & Bavelier 2012, box 1 for further discussion). In such studies, individuals who do not naturally play fast-paced, action video games are recruited and pretested on the task(s) of interest. A randomly selected sample of half of these subjects is then assigned to play an action game (e.g., Medal of Honor, Call of Duty, Unreal Tournament), whereas the other half is assigned to play an equally entertaining, but nonaction video game (e.g., Tetris, The Sims, Restaurant Empire). Note that in addition to controlling for simple test/retest effects (subjects may improve at posttest simply because they have experienced the task a second time) the presence of an active control condition guarantees that any effects observed in the action game training group are truly the result of action game play rather than simply a reflection of the power of an intervention per se. Indeed, individuals often feel special upon being included in an active intervention study and, as a result of receiving more attention, may perform better independently of the content of the intervention, an effect also known as the Hawthorne effect (Benson 2001).

Causality has been established through training studies, whereby both groups of participants come to the laboratory regularly to play 1–2 h per day over a period of 2–10 weeks, depending on the duration of the training (e.g., 10, 30, or 50 h of training were used in our work) (**Figure 3**). At the end of training, subjects are tested again on the same tasks of

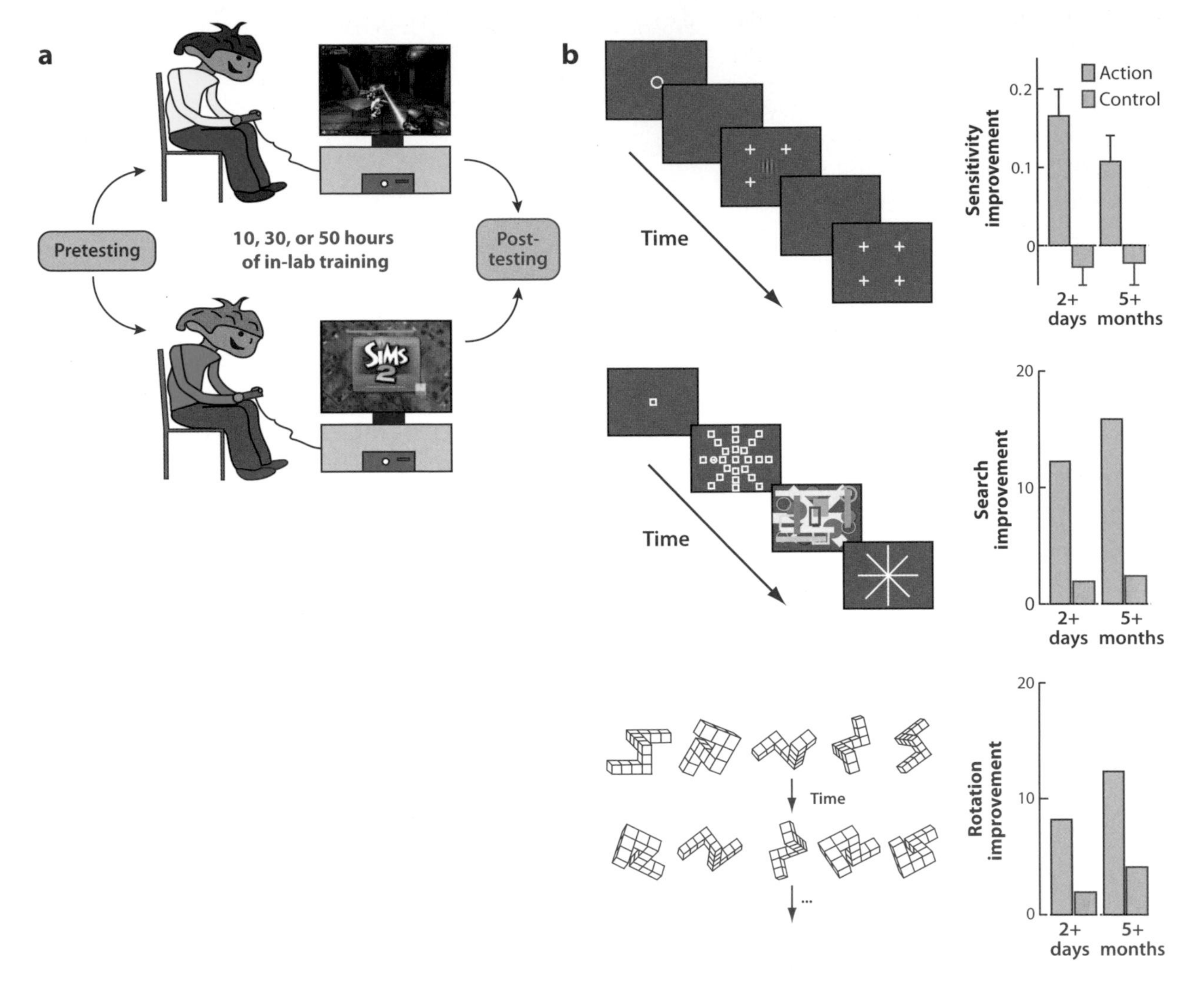

Figure 3

(*a*) Design of training studies to evaluate causality. Participants are randomly assigned to either the experimental group or a control group before being evaluated on perceptual, attentional, or cognitive laboratory tests. Both groups are then required to play commercially developed entertainment games. All experimental games are of the action genre (Unreal Tournament, Medal of Honor, Call of Duty), whereas control games are of other genres (Tetris, The Sims, Restaurant Empire). Following the end of training (and a 24-h waiting period) subjects are retested on the original measures. If the effect of action game play is truly causal, those assigned to the experimental games should display greater improvements between pre- and posttest than those assigned to the control games. (*b*) Such causal effects have been established for a variety of perceptual, attention, and cognitive tasks; some studies showed that the effects are quite long-lasting because robust effects were observed not only 1–2 days after the cessation of training (*red bars*) but also as long as 5+ months posttraining (*blue bars*) (panel *b* adapted from Feng et al. 2007, Li et al. 2009).

interest as those during pretest. However, it is important to note that in all cases we require that participants come back in the following days for posttesting, with no less than 24 h elapsing between the cessation of training and the beginning of posttesting. Indeed, our focus is on the durable learning effects of action game play and not on any transient effects of playing games (e.g., changes in arousal) that could contaminate posttest performance if assessed shortly after the end of gaming. If action game play truly has a causal effect on the skill under study, we expect greater pre- to posttest improvement in the action trainees than in the control trainees. Such causal effects have been established for vision (Green & Bavelier 2007; Li et al. 2009, 2010), some cognitive functions such as multitasking (Green et al. 2012) and mental rotation (Feng et al. 2007, Spence et al. 2009), many aspects of attention (Cohen et al. 2007; Feng et al. 2007; Green & Bavelier 2003, 2006a,b; Spence et al. 2009), and decision making (Green et al. 2010).

A COMMON MECHANISM: LEARNING TO LEARN

As illustrated above (see Benefits of Action Video Game Experience), the range of tasks seen to improve after action game play is quite atypical in the field of learning. VGPs perform better than NVGPs do on tasks neither group had previously experienced and that are, as noted earlier, quite different in nature from action game play. This finding begs the question of what exactly action video games teach that results in better performance. Rather than hypothesizing distinct mechanisms for each task improved, it seems more parsimonious to consider one common cause: learning to learn.

To understand what is meant by learning to learn, it is important first to realize that all the tasks on which VGPs have been found to improve share one fundamental computational principle: All require subjects to make a decision based on a limited amount of noisy data. Note that most everyday decisions fall into this category. Consider a radiologist reading computed tomography (CT) scans to determine the presence of bone fractures. Some scans will be clear-cut, leading to the immediate conclusion that a fracture is present. Other scans may present more subtle evidence, such that the radiologist may conclude a fracture has occurred, but his/her level of confidence will be much lower. In making these decisions, the radiologist is performing what is known as probabilistic inference. In essence, the radiologist computes the probability that each choice (fracture or no fracture) is correct given the evidence present in the scan(s). This quantity is known as the posterior distribution over choices, which we denote $p(c|e)$ where c are the choices, and e is the evidence (i.e., $c =$ "fracture" or "no fracture", $e =$ the evidence, which in this case is the scan itself). The key computational question then concerns how to compute the most accurate posterior distribution, $p(c|e)$, so that the best decision can be made. The main goal of learning is indeed to improve the precision of such probabilistic inference. The very fact that trained radiologists are more competent at diagnosing subtle fractures and identifying them with more confidence exemplifies that proper training does lead to more accurate posterior distributions.

How then does this process occur in practice? During training, a young radiologist sees many scans and is explicitly told the state of the patient that generated the image (i.e., whether a fracture was present). Radiologists are trained on the statistics of the evidence $p(e|c)$ or the probability of the evidence (how the scan looks) given a known state of the world (fracture is present/absent). Through Bayes' rule, this value, $p(e|c)$, is proportional to the posterior probability $p(c|e)$ on which the decision will be based. This set of computations is termed probabilistic inference (Knill & Richards 1996). One of the major advances of systems and computational neuroscience over the past 10 years has been to show how such probabilistic computations may be implemented in networks of spiking neurons (Deneve 2008, Ma et al. 2006, Rao 2004).

Participants who first encounter a new task in the lab are not very different from young radiologists in training. Initially, there is no way for them to have perfect knowledge of the statistics of the evidence, which in turn means that the calculated posterior distribution over choices will be suboptimal. It is only through repeated exposure to the task that subjects can learn these statistics and, as a result, make decisions on the basis of a more accurate posterior distribution. By using a well-studied perceptual decision-making task (as described above in Benefits of Action Video Game Experience), we more directly tested whether action video game experience indeed results in improvements in probabilistic inference. In this task, subjects were asked to determine whether the main flow of motion within a random dot kinematogram was to the left or to the right. Using a neural model of the task (Beck et al. 2008), we showed that the pattern of behavior after game play is captured via a single change in the model: an increase in the connectivity between the model's sensory layer, where neurons code for direction of motion, and its integration layer, where neurons accumulate the information they receive over time from the sensory layer until a criterion is reached for decision and response. This increase in connectivity can be shown to be mathematically equivalent to improved statistical inference; game experience resulted in more accurate knowledge of the statistics of the evidence for the task or more accurate knowledge of $p(e|c)$ (**Figure 2*c***) (Green et al. 2010). This work ruled out simpler explanations for the variety of benefits observed after action game play, including simple speed–accuracy trade-offs or faster motor execution times in VGPs.

The modeling demonstrates that VGPs were making choices on the basis of a more accurate posterior distribution. Because these statistics could not have been directly taught by any action video games, but instead must have been learned through the course of the experiment, action video games thus taught these participants to learn the appropriate statistics for new tasks more quickly and more accurately than is possible for NVGPs.

The idea that VGPs have learned to learn the statistics of the task (or what we call later the generative model of the data) can be generalized beyond the specific example of perceptual decision making. Indeed, most of the behavioral tasks on which VGPs have been shown to excel can be formalized as instances of probabilistic inference. This includes tasks that are more classically categorized as attentional, such as multiple object tracking or locating a target among distractors, both of which have recently been modeled within the probabilistic framework, as well as more cognitive and perceptual tasks (Ma et al. 2011, Ma & Huang 2009, Vul et al. 2009).

COMPUTATIONAL PRINCIPLES OF LEARNING TO LEARN

Learning is commonly defined as a long-term improvement in performance due to training, a process that is often modeled as the tuning of parameters within a fixed architecture. The literature has countless examples wherein learning algorithms, such as gradient descent, are implemented within neural networks to tune weights to enhance performance on the trained task (Rumelhart et al. 1987, but see Gallistel & King 2009 and Gallistel 2008 for a different view on learning). For example, in classic orientation-discrimination training tasks, subjects have to learn through experience whether the image of a Gabor patch is oriented clockwise or counterclockwise from some reference angle (e.g., 45°). Human behavior on such tasks can be easily captured with two-layer networks, in which an input layer containing oriented filters projects to an output layer that produces a clockwise or counterclockwise decision (Dosher & Lu 1998, Pouget & Thorpe 1991). Learning in such models is typically produced via changes in the synaptic weights between the two layers. The pattern of synaptic weights that optimizes performance on this task is a simple instantiation of a discriminant function and thus is completely specific to the trained reference angle. Learning at one orientation does not transfer to other orientations. These models naturally capture the high degree of specificity typically seen in perceptual learning. However, such an implementation cannot be responsible for the types of benefits associated with learning to learn.

Transfer of learning is enabled by implementations that recognize the importance of resources, knowledge, and learning rules. These core computational elements are well illustrated by real-world tasks such as playing soccer, where players must keep track of an

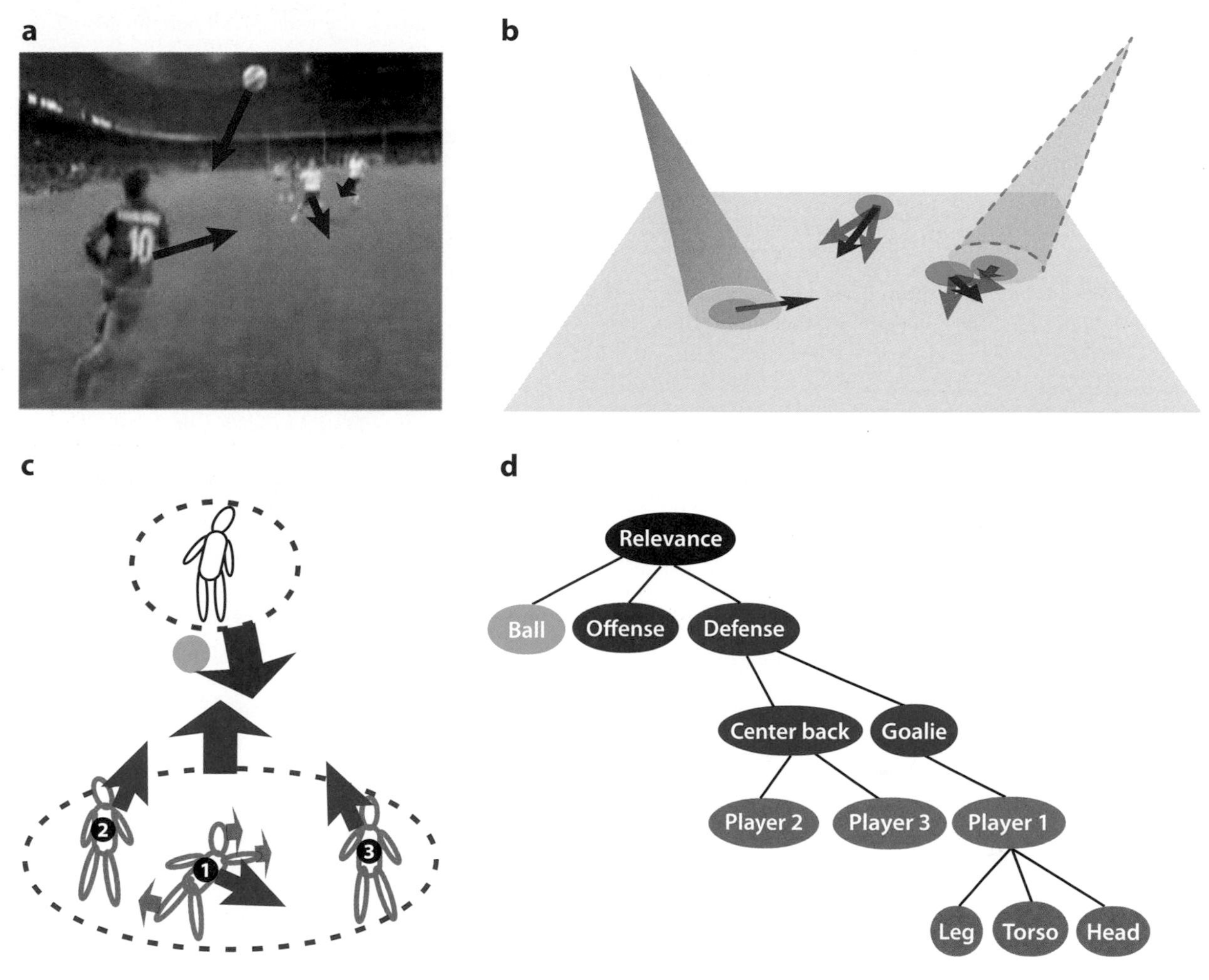

Figure 4

(*a*) In real-world tasks such as playing soccer, players must keep track of a complex array of moving objects, including the ball and all other players on the field. An expert player will readily recover the three-dimensional trajectories of various objects on the field, while maintaining their proper identities and roles at all times, allowing them to predict next game moves. This is possible as players gain experience with soccer play and build a rich internal model with the proper hierachical architecture to facilitate computation. (*b*) Schematic illustration of how resources can facilitate object tracking over time. Resources, shown as beams of light, are believed to be limited in number and require flexible allocation across objects as the number of objects increases. Recruiting additional resources, illustrated as the dashed beam, provides improved motion processing and tracking for more objects. (*c*, *d*) The dynamic of game play in a soccer game is best captured as the top node of a hierarchical architecture, whereby motion information is pooled across different levels of representations, from low level such as limb motion within a player to the motion of individual players to assemblies of players (offense versus defense). An appropriate abstraction allows hierarchical architectures to capture regularity across variations in task and stimuli, a key feature of learning transfer.

array of complex moving objects, including the ball and other players (**Figure 4*a***). As the player moves and scans the scene, the color, shape, and motion information they receive must be converted into the three-dimensional trajectories of objects while maintaining their identities and roles (opponent, teammate, ball, referee, spectator). This complex problem can be decomposed into a set of core skills, many of which have been psychophysically studied in isolation, including multiple object tracking, object identification, resource allocation, and eye-movement planning. Whereas the novice player will only partially achieve the coordination of these different demands, expert players have learned through experience how best to represent game play as a function of the different identities and roles of the players and how best to allocate resources among these different demands. Accordingly, top athletes can keep track of all the players on both teams, including those out of view (Cavanagh & Alvarez 2005).

Skill acquisition in such complex domains requires more than instantiating a simple discriminant function of the type discussed above. A neural architecture that incorporates knowledge critical for the given task is needed. In the case of soccer, this knowledge encompasses tracking objects, including features important for maintaining object identity and role, developing representations that are invariant to irrelevant internal limb motions, and representing the statistical motion behavior typical of different object types (motion of the ball versus that of players). Learning the proper knowledge for the given task is critically dependent not only on the chosen neural architecture but also on the learning rules that adapt the neural architecture with experience. Finally, the ability to adapt swiftly to the different demands of such complex tasks greatly benefits from improvement in the number or allocation of core resources such as attention. Changes in any or all of these three—knowledge, learning rules, and resources—could be produced by action video game experience and could account for widespread benefits observed in VGPs. As illustrated below (see Resources, Knowledge, and Learning Rules in Action Video Game Play), there is much evidence that action video game play improves resource allocation, and there are some pointers toward knowledge and/or learning rule changes as well. Before we discuss these, we lay out a general framework for the type of probabilistic inference at the core of learning to learn and discuss how resources, knowledge, and learning rules may come to account for the widespread benefits associated with learning to learn.

Resources

VGPs have increased attentional resources in multiple-object-tracking tasks, allowing for more accurate representation of motion and features, which in turn enables more accurate tracking and identification (Boot et al. 2008, Dye & Bavelier 2010, Green & Bavelier 2006b, Trick et al. 2005). As illustrated in **Figure 4*b***, this ability should be advantageous when playing soccer. However, such additional resources will do more than just improve performance on a single soccer action; they are also a key determinant of learning. Greater resources allow for finer-grain distinctions to be made, be they perceptual, conceptual, or in motor output. An increase in resources may therefore enable learners to achieve greater asymptotic performance (more learning capacity), or even faster learning, because critical distinctions will be more accessible to those learners with greater resources. The concept of resources we use here is closely related to that defined in the field of attention and executive control under the terms goal-directed attention or top-down control (Corbetta & Shulman 2002, Dosenbach et al. 2008, Koechlin & Summerfield 2007, Rougier et al. 2005). The cholinergic basal forebrain system may orchestrate such resource allocation. More specifically, it appears that the release of acetylcholine regulates the extent to which targets are processed and distractors are ignored (Baluch & Itti 2011, Herrero et al. 2008, Sarter et al. 2005) as well as enhances the precision of sensory representations (Goard & Dan 2009).

The role of acetylcholine in selective attention and resource allocation goes hand in hand with its recognized role in cortical map plasticity (Ji et al. 2001, Kilgard & Merzenich 1998). In a seminal study, Kilgard & Merzenich (1998) established that adult rats exposed to tones exhibited larger cortical map plasticity in the presence of episodic stimulation of the nucleus basalis. Early studies in humans also suggest a possible link between learning and the cholinergic system. For example, intake of a choline agonist leads to improved learning when participants are asked to learn lists of words. In contrast, a choline antagonist leads to reduced learning (Sitaram et al. 1978). In this view, resource allocation acts as a key gateway to learning by refining the distinction between signal and noise and enhancing the quality and precision of the to-be-learned information. The role that the distribution of attentional resources plays in guiding learning is well exemplified by a study by Polley et al. (2006), whereby rats were trained to discriminate tones on the basis of either their frequency or their intensity. When animals performed the frequency task, changes in frequency maps were observed that correlated with behavior. In contrast, no such changes were noted in neurons coding for intensity. The opposite pattern was observed when the task focused the animals' attention on the intensity task. A similar link is also supported in humans (see Saffell & Matthews 2003). In line with this view, Li et al. (2009) have shown that prefrontal areas, known to be involved in resource allocation, code for different decision criteria as a result of learning episodes that emphasized the relevant criterion. This neural signature was present only when participants were engaged in the trained task and not just exposed to the learned stimuli. More recently, Baldassare et al. (2012) demonstrated that resting state connectivity between prefrontal areas and the visual cortex was predictive of the amount of learning in an upcoming perceptual learning task. This study is a direct illustration of how efficiency in resource allocation comes to guide learning abilities at the individual level. As we highlight below (see Resources, Knowledge, and Learning Rules in Action Video Game Play), much evidence indicates that action video game play improves top-down resource allocation, thus leading to better selective attention and enhanced precision in representations. A direct link to the acetylcholine system, however, remains to be elucidated.

More computational and memory resources can also foster learning to learn by providing the capacity to explore alternative hypotheses and architectures to find those that best account for the knowledge being acquired. The burgeoning literature on the training of executive function testifies to this link (Klingberg 2010, Lustig et al. 2009). For example, as participants train on an N-back task and can handle more demanding updating in memory, they also show enhanced fluid intelligence as measured by the Raven's Advanced Progressive Matrices (RAPM) (Jaeggi et al. 2008). The RAPM presents visual patterns with one missing component that participants must complete. This test relies on the ability to search swiftly through many hypotheses to find the best match. Enhanced executive resources will allow such searches to be more efficient. Although executive functions appear trainable across the life span, some evidence indicates that transfer effects may be more limited in older age (Buschkuehl et al. 2008, Dahlin et al. 2008). At the neural level, working-memory training results in enhanced recruitment of the frontoparietal network associated with top-down attentional control, possibly increased myelination among related areas, as well as a decrease in cortical dopamine D1 binding potential in large parts of frontal and posterior areas (McNab et al. 2009, Olesen et al. 2004, Takeuchi et al. 2010). Whether action video game play effects are mediated by similar mechanisms remains to be assessed. Finally, having greater resources does not systematically guarantee more efficient learning; resource allocation needs to be guided by the presence of structured knowledge that helps select where useful information lies, as discussed below.

Knowledge

Knowledge here refers to the representational structure used to guide behavior. The realization that complex behaviors cannot be appropriately captured by a chain of stimulus-response associations is far from new (Lashley 1951) and has led to a keen interest in the role of hierarchical architecture in human action and cognition (see Botvinick 2008 for a review). Hierarchical behavior models capture the intuitive notion that tasks such as executing a play in soccer are organized into subtasks involving individual player moves, which are themselves decomposed into component actions of running, turning, kicking, and passing. In turn, knowing soccer makes learning a game such as hockey easier because the similarities between the games allow one to import skills and knowledge from one game to bootstrap learning of the other. Note that it is not just because both games share the category "goalies" that such generalization is possible. Rather, recent work in machine learning suggests that hierarchical architectures allow the decomposition of computations into multiple layers using greater and greater abstraction, a critical step for generalizable learning (Bengio 2009, Hinton 2007). In these architectures, categories such as "goalie" are distributed across multiple levels of abstraction, from sensory input to characteristic features to body plans to player types to the concept of a soccer player. In contrast, shallow architectures eschew abstraction and construct the category "goalie" by treating it as a classification problem that involves finding the right set of diagnostic features such as limb motion, jersey color or pattern, or position relative to the goal post. Although few examples of such deep coding in neuroscience exist as of yet, studies of categorization and object recognition point to the psychological grounding of deep architectures. For example, hierarchical architectures naturally account for the fact that categories appear organized around prototypes or most representative exemplars of a category, even when the prototypes do not reflect the frequency of diagnostic features (Rosch 1975). As work on the neural bases of object recognition progresses, characterizing how deep knowledge may constrain neural coding will become increasingly more relevant (DiCarlo & Cox 2007, Lee & Mumford 2003, Paupathy & Connor 2002, Yamane et al. 2008).

The shallow/deep distinction can be expressed in terms of Bayesian decision theory as the difference between learning a rich generative model that captures hidden structure in the data versus learning a discriminative model (a discriminant function) specific to a classification problem. Learning theory has shown that discriminative models require less data to learn the observed relationships but are task specific. For example, discriminative classifiers such as support vector machines learn boundaries in feature space. The fact that these boundaries can be quite complex does not guarantee generalizable knowledge. In fact, in the case of support vector machines, the learned boundaries typically depend intimately on the characteristics of the training set. Simple changes in team uniforms or players will require relearning a new boundary to classify them. Such implementations therefore appear ill adapted when considering learning to learn. In contrast, appropriate abstraction allows hierarchical architectures to capture regularity across variations in task and stimuli. **Figure 4*c*,*d*** illustrates a hierarchy for relevant motion in a soccer game. The tree structure shows how broader categories can be constructed by pooling across categories of motion at different levels, pooling limb motions to construct player motion, pooling player motions within a player type (such as goalie), pooling player-type motions within a team type (offense or defense), and pooling all game-relevant motions, including the ball, coaches, and referees, in opposition to nonrelevant motions such as spectators and billboards.

Abstract organization admits generalization to the extent that new tasks overlap and share structure and representations. For example, despite the fact that players use skates, hockey shares many of the same motion characteristics of the teams' offense and defense, roles of players (such as goalie), and requirements for passing and controlling the puck as does soccer. The generalization expected from a given architecture is not infinite, but will be determined instead by the level of abstraction at which knowledge is shared by the various tasks considered. Thus, for action video games to provide players with knowledge applicable to standard laboratory tasks, it must necessarily be the case that games and the laboratory tasks share structure at some level of abstraction. And perhaps, more importantly, training paradigms must necessarily share structure at some level of abstraction for them to affect performance in real-life tasks.

The need for structured representation to support more complex task performance has led to an interest into the neural bases of hierarchical tasks that embed goals and subgoals. Although this area of research is still fairly new, the available literature points to a crucial role of prefrontal cortex structures to enable hierarchical representations and coordinate resource allocation dynamically among these as task demands vary (Badre 2008, Botvinick 2008, Koechlin 2008, Sakai 2008). Indeed, the ability to enhance neural processing selectively by allocating more resources is central to behavior optimization, but it depends fundamentally on an internal understanding of which neural pool is important to augment at each point in time. The prefrontal cortex is a crucial candidate in the representation of these more complex behaviors. Prefrontal structures are key to maintaining context and goals over time with increasingly higher levels of behavioral abstraction being coded as one moves from caudal to rostral locations along the prefrontal cortex (Christoff & Keramatain 2007, Fuster 2004). There is certainly debate about the exact nature of the abstraction, with higher planning units, greater relational complexity, more branching control, or larger contextual integration being documented as one moves rostrally within the prefrontal cortex (see figure 2 in Badre 2008 for a summary). Yet, there is common agreement about the importance of hierarchical knowledge when considering how the brain learns from its successes and failures. Neural signatures for such an architecture are emerging, e.g., in the work of Ribas-Fernandes et al. (2011) who recently demonstrated that subgoals lead to reward-prediction signals even if they are not associated with primary reinforcers. Such an outcome is predicted under the assumption that the task is represented hierarchically but not otherwise. This research should be an important avenue for future research on learning, as the nature of the architecture fundamentally constrains the type of learning neural structures should implement.

Learning Algorithms

A common observation is that as an architecture becomes more complex, it becomes more critical that the learning algorithm can modify the representations in ways that are aligned with the nature of the problems to be solved. Simply put, knowledge and learning rules are intimately interdependent; different architectures typically call for different learning rules. A key question then concerns how improvement in learning algorithms and their related architecture may foster learning to learn. To facilitate generalization, learning algorithms require shared structure, a fact that constrains how learning rules should modify representations. Violating these constraints destroys transferrable knowledge because changes in representation that improve performance in one task can disrupt performance in another.

Over the past 20 years, this field has witnessed dramatic improvements in the basic set of machine-learning algorithms. For instance, algorithms such as those employed in Hinton's deep-belief networks have provided evidence that there is enough shared structure in the types of learning problems encountered by humans beings to allow for the use of flexible and generalizable learning algorithms (Hinton 2007, Hinton et al. 2006, Larochelle et al. 2007, Lee et al. 2009). In this case, speech recognition and handwritten character recognition can be achieved using initially similar architectures and the same learning rule. Of note, performance after learning rivals the best algorithms specifically developed for speech recognition or character recognition respectively (Mohamed et al. 2011). This approach is appealing because the lack of structural differentiation among cortical areas suggests a broadly shared learning algorithm, at least at the cortical level.

Fine-tuning of learning algorithms may occur through improvement in error detection, error computation, reward computation, and, more generally, synaptic learning rules. This fine tuning may involve neuromodulators, which are believed to be involved in the modulation of synaptic plasticity (Zhang et al. 2009) and in the computation of error signals (Schultz et al. 1997). Changes in the responses of the neurons synthesizing these neuromodulators could easily generalize across tasks and sensory modalities because these neurons tend to connect very wide cortical regions. Although few studies are available on this topic, action video game play likely alters patterns of neuromodulator release. For instance, an early positron emission tomography study indicated that a large release of striatal dopamine occurs when subjects play a basic shooting video game (Koepp et al. 1998, though see the call for further replication by Egerton et al. 2009). This should be a fruitful area of research for future studies.

RESOURCES, KNOWLEDGE, AND LEARNING RULES IN ACTION VIDEO GAME PLAY

All three factors that can contribute to learning to learn may, of course, be involved in the effects of video game training, but the experimental evidence so far has focused mainly on changes in resources and improvements in resource allocation. However, although no experiments have directly tested differences in knowledge or in learning rules, some data points do appear more consistent with these changes than does a resource-based account.

Attentional Resources

Enhanced resources. The proposal that VGPs benefit from greater attentional resources is supported by the observation that, in simple flanker compatibility tasks, VGPs' reaction times show a greater sensitivity to the identity of the flanker (Dye et al. 2009b; Green & Bavelier 2003, 2006a). Thus, reaction times on the main target task showed greater speeding (respectively, slowing down) when flankers were response compatible (respectively, response incompatible) in VGPs as compared with NVGPs. According to the work of Lavie and others (2005), in such flanker paradigms the resources utilized by the main target task determine the amount of processing the flanker receives and thus the size of the flanker compatibility effect. As the main target task exhausts more resources, the identity of the flanker affects reaction times to a lesser extent, leading to increasingly smaller compatibility effects. Green & Bavelier (2003) showed that although this pattern of results held in NVGPs, VGPs continued to show effects of flanker identity even for relatively high-load target tasks, thus suggesting the presence of greater attentional resources.

Selective attention. Action game play results in improved selective attention over space, over time, and to objects. VGPs exhibit more accurate spatial localization of a target whether presented in isolation as in Goldman perimetry (Buckley et al. 2010) or among distracting, irrelevant information as in the Useful Field of View task (Feng et al. 2007; Green & Bavelier 2003, 2006a; Spence et al. 2009; West et al. 2008). Action game play also enhances the ability to select relevant information over time. When viewing a stream of letters presented rapidly, one after the other, participants can typically identify the one white letter among black letters if asked. However, doing so creates a momentary blink in attention, leading the subject to be unaware of the next few items following the white letter. This effect, termed the attentional blink, is believed to measure a fundamental bottleneck in the dynamics of attention allocation (Shapiro et al. 1997). VGPs' blink is much less pronounced, with some VGPs failing to show any blink (at least at the rate of presentation tested; see Cohen et al. 2007, Green & Bavelier 2003). It is unlikely that action game play totally obliterates the attentional blink, but these data suggest a much faster rate of attentional recovery in VGPs. Such faster processing speed is in line with the reduced backward masking reviewed earlier and the report that VGPs perceive the timing of visual events more veridically than do NVGPs (Donohue et al. 2010, but see West et al. 2008 for a different view). A third aspect of selective attention documented to change for the better after action game play is attention to objects (Boot et al. 2008; Cohen et al. 2007; Green & Bavelier 2003, 2006b; Trick et al. 2005). Using the multiple-object-tracking task (Pylyshyn & Storm 1988), VGPs have been repeatedly shown to track more objects accurately than have NVGPs (Boot et al. 2008, Green & Bavelier 2006b, Trick et al. 2005). This skill requires efficient allocation of attentional resources as objects move and cross over each other in the display. The advantage noted in VGPs in this task is consistent with either large resources or a swifter allocation of resources as a result of game play.

The ability to focus attention on a target location and ignore irrelevant distracting information is a key determinant of selective attention. Converging evidence indicates that selective attention is a strong predictor of academic achievement in young children (Stevens & Bavelier 2012), illustrating the general benefits this skill may confer. Although seldom considered in conjunction with action game play, meditative activities such as mind-body training and eastern relaxation technique also act in part by enhancing selective attention (Lutz et al. 2008, Tang & Posner 2009). Whether these two rather different treatments achieve their effects through comparable neural modulation may be a fruitful avenue of investigation in the future.

Divided attention. The ability to divide attention between tasks or locations was one of the first attentional benefits noted in the early days of computer games (see Greenfield 2009 for a review). For example, when asked to report the appearance of a target in a high probability, low probability, or neutral location, VGPs did not show the expected decrement (slower reaction time compared with neutral) for the low probability location, suggesting a better strategy in spreading attention due to game expertise (Greenfield et al. 1994). More recently, VGPs have shown an enhanced ability at dividing attention between a peripheral localization task and a central identification task (Dye & Bavelier 2010, Green & Bavelier 2006a). VGPs also excelled when asked to detect a prespecified target in rapid sequences of Gabor stimuli presented in both visual fields. Concurrent event-related potential recordings indicated a larger occipital P1 component (latency 100–160 ms) in VGPs during this divided-attention task, consistent with the proposal of greater attentional modulation in early stages of processing such as the extrastriate visual cortex in VGPs (Khoe et al. 2010).

Sustained attention and impulsivity. Some evidence also supports the notion that sustained attention benefits from action video game play. Using the Test of Variables of Attention (T.O.V.A.), a computerized test often used to screen for attention deficit disorder, Dye et al. (2009a) found that VGPs responded faster and made no more mistakes than did NVGPs on this test. Briefly, this test requires participants to respond as fast as possible to shapes appearing at the target location, while ignoring the same shapes if they appear at another location. By manipulating the frequency of targets, the T.O.V.A. offers a measure of both impulsivity (is the observer able to withhold a response to a nontarget when most of the stimuli are targets?) and a measure of sustained attention (is the observer able to stay on task and respond quickly to a target when most of the stimuli are nontargets?). In all cases, VGPs were faster but no less accurate than were the NVGPs, indicating, if anything, enhanced performance on these aspects of attention as compared with NVGPs. It may be worth noting that VGPs were often so fast that the built-in data analysis software of the T.O.V.A. considered their reaction times to be anticipatory (200 ms or less). However, a close look at the reaction time distribution indicated that these anticipatory responses were nearly all correct responses (which would obviously be unlikely if they were driven by chance, rather than being visually evoked). In summary, VGPs are faster but not more impulsive than NVGPs and equally capable of sustaining their attention. Although some have argued for a link between technology use and attention-deficit disorder (Rosen 2007, but see Durkin 2010), in the case of playing action games this does not appear to be the case.

Resource allocation. The proposal that action game play enhances top-down aspects of attention by allowing VGPs to allocate their resources more flexibly where it matters for the given task is supported by several independent sources. First, some evidence indicates that VGPs may better ignore sources of distraction, a key skill for properly allocating resources. For example, VGPs have been shown to be more capable of overcoming attentional capture. Chisholm et al. (2010, Chisholm & Kingstone 2011) compared the performance of VGPs and NVGPs on a target-search task known to engage top-down attention, while concurrently manipulating whether a singleton distractor, known to capture attention automatically, was or was not present. When top-down and bottom-up attention interact in such a way, investigators noted that although the singleton distractor captured attention in both groups, it did so to a much lesser extent in VGPs than in NVGPs. Thus VGPs may better employ executive strategies to reduce the effects of distraction. More recently, Mishra et al. (2011) made use of the steady-state visual-evoked potentials technique to examine the neural bases of the attentional enhancements noted in VGPs. They found that VGPs more efficiently suppressed unattended, potentially distracting information when presented with highly taxing displays. Participants viewed four different streams of rapidly flashing alphanumeric characters. Each stream flashed at a distinct temporal frequency, allowing retrieval of the brain signals evoked by each stream independently at all times. Thus, brain activation evoked by the attended stream could be retrieved, as could brain activation evoked by each of the unattended and potentially distracting streams. VGPs suppressed irrelevant streams to a greater extent than did NVGPs, and the extent of the suppression predicted the speed of their responses. VGPs seem to focus better on the task at hand by ignoring other sources of potentially distracting information. This view is in line with the proposal in the literature of greater distractor suppression as one possible determinant of more efficient executive and attentional control (Clapp et al. 2011, Serences et al. 2004, Toepper et al. 2010).

VGPs may also benefit from greater flexibility in resource allocation. The finding that VGPs switch tasks more swiftly, whether or not switches are predictable, is consistent with this view (Green et al. 2012). A recent brain-imaging study involving VGPs and NVGPs also speaks to this issue. In this study, the recruitment of the frontoparietal network of areas hypothesized to control the flexible allocation of top-down attention was compared in VGPs and NVGPs. As attentional demands increased, NVGPs showed increased activation in this network as expected. In contrast, VGPs barely engaged this network despite a matched increase in attentional demands across groups. This reduced activity in the frontoparietal network is compatible with the proposal that resource allocation may be more automatic and swift in VGPs (Bavelier et al. 2011). Furthermore, many models of the multiple-object-tracking task, on which VGPs have been seen to excel, intimate a close relationship between the number of items that can be tracked and the efficiency of resource allocation (Ma & Huang 2009, Vul et al. 2010).

In sum, much evidence demonstrates that action video game play leads to not only enhanced resources, but also a more intelligent allocation of these resources given the goals at hand. This is one of the ways action game play may result in learning to learn. We now turn to other evidence for learning to learn after action game play that suggests changes in knowledge and/or learning algorithm.

Knowledge/Learning Rule

Direct evidence for changes in knowledge or learning rules is still lacking. However, the literature contains pointers suggestive of such changes. First, the observation that action game play leads to more accurate probabilistic inference, as discussed above (see A Common

Mechanism: Learning to Learn), suggests the development of new connectivity and knowledge to enable a more efficient architecture for the given task. Although some of these benefits could be driven initially by changes in resources, the fact that they can last for five months or longer suggests more profound and long-lasting changes in representation than what is typically afforded by attentional resources. Second, a number of experiments point to improvement despite little need for resources. This is true, for example, of very basic psychophysical skills such as visual acuity or contrast sensitivity (Green & Bavelier 2007, Li et al. 2009). These paradigms displayed only one target in isolation, and no uncertainty was present on either the place and/or time of the target's arrival (fixed target location, fixed SOA). In contrast, resources such as attention are typically called on when a target needs to be selected in time or space from distractors or when the time and/or place of arrival of the stimulus is uncertain (Carrasco et al. 2002). Third, an attentional explanation is not always in line with the noted changes. For example, contrast-sensitivity improvements were not larger as spatial frequency increased (and thus stimulus size decreased) as would be predicted by attentional accounts. Rather, the beneficial effect of action game play was maximal at intermediate frequencies. These considerations point to changes in representations likely to be mediated by enhancement either in learning rule or in knowledge.

Action game playing may act by enabling more generalizable knowledge through various abstractions, including the extent to which nontask-relevant information should be suppressed (Clapp et al. 2011, Serences et al. 2004, Toepper et al. 2010), how to modify performance to maximize reward rate (Simen et al. 2009), how to combine data across feature dimensions (which depends on the presence/type of dependencies between the dimensions; Perfors & Tenenbaum 2009, Stilp et al. 2010), and how to set a proper learning rate (which depends on the belief about whether generating distributions are stationary or drifting; Behrens et al. 2007). Although these avenues are only beginning to be explored, some support is emerging for distraction suppression and reward-rate changes after action game play. First, several studies document that action video game play alters the extent to which distracting information is suppressed. It is not always the case that VGPs show greater suppression of distractors. Rather, distractor processing appears to be greater in VGPs than in NVGPs under conditions of relatively low load. Under such conditions, task difficulty is sufficiently low for VGPs to remain efficient on the primary task and at the same time still be able to process distractors (Dye et al. 2009b, Green & Bavelier 2003). In contrast, VGPs show greater suppression of distractors under extremely high load conditions (Mishra et al. 2011). As task difficulty increases, suppression of distractors becomes necessary for VGPs to maintain their high performance on the primary task. Such differential effects may be understood if distractor processing has been encoded at a proper level of abstraction that is contingent on primary task success. Second, a key step in fostering learning is to engage the reward system (Bao et al. 2001, Koepp et al. 1998, Roelfsema et al. 2010). A distinguishing feature of action games is the layering of events/actions at many different timescales, resulting in a complex pattern of reward in time. This feature may explain, in part, why VGPs seem to maximize reward rate in a variety of tasks (A.F. Anderson, C.S. Green, D. Bavelier, manuscript in preparation; Green et al. 2010). Further investigation of the impact of action game play on knowledge and learning rules should provide an interesting test case, as the roles of resource control, hierarchical knowledge, and learning rules are refined.

Finally, the enhanced attentional control documented earlier is not, in this view, the proximal cause of the superior performance of VGPs, i.e., an end in and of itself. Instead, it is a means to an end, with that end being the development of more generalizable knowledge as one is faced with new tasks or new environments. Such learning to learn predicts that reasonably equivalent performance between groups should be seen early on when performing a new task, with the action gaming advantage appearing and then increasing through experience with the task. Such outcomes are being investigated in the domain of perceptual learning and decision making.

CONCLUDING REMARKS

In sum, we propose that action game play does not teach any one particular skill but instead increases the ability to extract patterns or regularities in the environment. As exemplified by their superior attentional control skills, VGPs develop the ability to exploit task-relevant information more efficiently while better suppressing irrelevant, potentially distracting sources of information. This may stem from the ability to discover the underlying structure of the task they face more readily, perform more accurate statistical inference over the data they experience, and thus exhibit superior performance on a variety of tasks previously thought to tap unrelated capacities. In this sense, action video game play may be thought of as fostering learning to learn.

The hypothesis that action video game play may improve learning to learn is appealing because it naturally accounts for the wide range of skills enhanced in VGPs. Yet, learning to learn does not guarantee better performance on all tasks. Instead, like all forms of learning, it comes with a number of constraints that determine the nature and extent of the generalization possible. For instance, changes in knowledge produce benefits only to the extent to which new tasks share structure with action video games. No benefits are expected in tasks that share no such structure. Furthermore, and again like all forms of learning, potential drawbacks exist. For instance, an increase in resources may predict a tendency to overexplore: to test models that are more elaborate than what is required by the task at hand, which could actually result in poorer performance and slower learning. Learning to learn does not predict that VGPs will excel at all tasks, but rather provides a theoretical framework to further our understanding of key concepts underlying generalization in learning as well as better characterizing the architecture, resources, and learning algorithms that generate the remarkable power of the human brain.

DISCLOSURE STATEMENT

D.B., C.S.G., and A.P. have patents pending concerning the use of video games for learning. D.B. is a consultant for PureTech Ventures, a company that develops approaches for various health areas, including cognition. D.B., C.S.G., and A.P. have a patent pending on action-video-game-based mathematics training; D.B. has a patent pending on action-video-game-based vision training.

ACKNOWLEDGMENTS

We thank Ted Jacques for his invaluable help with manuscript preparation. This work was made possible thanks to the support of the Office of Naval Research through a Multi University Research Initiative (MURI) that supported all four authors. Projects described here also received support in part from grants from the National Institutes of Health (EY- O16880 to D.B.; EY-0285 to P.S.; NIDA-0346785 to A.P.), from the National Science Foundation (BCS0446730 to A.P.), and from the James S. McDonnell Foundation (to D.B. and to A.P.).

LITERATURE CITED

Anderson AF, Bavelier D. 2011. Action game play as a tool to enhance perception, attention and cognition. In *Computer Games and Instruction*, ed. S Tobias, D Fleycher, pp. 307–29. Charlotte, NC: Information Age

Anderson AF, Kludt R, Bavelier D. 2011. *Verbal versus visual working memory skills in action video game players*. Poster presented at the Psychonomics Soc. Meet., Seattle

Andrews G, Murphy K. 2006. Does video-game playing improve executive function? In *Frontiers in Cognitive Psychology*, ed. MA Vanchevsky, pp. 145–61. New York: Nova Sci.

Badre D. 2008. Cognitive control, hierarchy, and the rostro-caudal organization of the frontal lobes. *Trends Cogn. Sci.* 12:193–200

Bailey K, West R, Anderson CA. 2010. A negative association between video game experience and proactive cognitive control. *Psychophysiology* 47:34–42

Baldassare A, Lewis CM, Committeri G, Snyder AZ, Romani GL, Corbetta M. 2012. Individual variability in functional connectivity predicts performance of a perceptual task. *Proc. Natl. Acad. Sci. USA* 109:3516–21

Baluch F, Itti L. 2011. Mechanisms of top-down attention. *Trends Neurosci.* 34:210–24

Bao S, Chan VT, Merzenich MM. 2001. Cortical remodelling induced by activity of ventral tegmental dopamine neurons. *Nature* 412:79–83

Bavelier D, Achtman RA, Mani M, Foecker J. 2011. Neural bases of selective attention in action video game players. *Vis. Res.* doi: 10.1016/j.visres.2011.08.007

Bavelier D, Levi DM, Li RW, Dan Y, Hensch TK. 2010. Removing brakes on adult brain plasticity: from molecular to behavioral interventions. *J. Neurosci.* 30:14964–71

Beck JM, Ma WJ, Kiani R, Hanks T, Churchland AK, et al. 2008. Bayesian decision making with probabilistic population codes. *Neuron* 60:1142–52

Behrens TE, Woolrich MW, Waldon ME, Rushworth MF. 2007. Learning the value of information in an uncertain world. *Nat. Neurosci.* 10:1214–21

Bengio Y. 2009. Learning deep architectures for AI. *Found. Trends Mach. Learn.* 2:1–127

Benson PG. 2001. Hawthorne effect. In *The Corsini Encyclopedia of Psychology and Behavioral Science*, ed. WE Craighead, CB Nemeroff, pp. 667–68. New York: Wiley

Binet A. 1984 (1909). *Les idées modernes sur les enfants*, transl. S Heisler. Menlo Park, CA: S. Heisler

Boot WR, Kramer AF, Simons DJ, Fabiani M, Gratton G. 2008. The effects of video game playing on attention, memory, and executive control. *Acta Psychol.* 129:387–98

Botvinick MM. 2008. Hierarchical models of behavior and prefrontal function. *Trends Cogn. Sci.* 12:201–8

Buckley D, Codina C, Bhardwaj P, Pascalis O. 2010. Action video game players and deaf observers have larger Goldmann visual fields. *Vis. Res.* 50:548–56

Buschkuehl M, Jaeggi SM, Hutchison S, Perrig-Chiello P, Däpp C, et al. 2008. Impact of working memory training on memory performance in old-old adults. *Psychol. Aging* 23:743–53

Cain MS, Landau AN, Shimamura AP. 2012. Action video game experience reduces the cost of switching tasks. *Atten. Percept. Psychophysics* 74:641–47

Carrasco M, Williams PE, Yeshurun Y. 2002. Covert attention increases spatial resolution with or without masks: support for signal enhancement. *J. Vis.* 2:467–79

Castel AD, Pratt J, Drummond E. 2005. The effects of action video game experience on the time course of inhibition of return and the efficiency of visual search. *Acta Psychol.* 119:217–30

Cavanagh P, Alvarez GA. 2005. Tracking multiple targets with multifocal attention. *Trends Cogn. Sci.* 9:349–54

Chase WG, Simon HA. 1973. The mind's eye in chess. In *Visual Information Processing*, ed. WG Chase, pp. 215–81. New York: Academic

Chisholm JD, Hickey C, Theeuwes J, Kingstone A. 2010. Reduced attentional capture in action video game players. *Atten. Percept. Psychophys.* 72:667–71

Chisholm JD, J, Kingstone A. 2011. Improved top-down control reduces oculomotor capture: the case of action video game players. *Atten. Percept. Psychophys.* 74:257–62

Christoff K, Keramatain K. 2007. Abstraction of mental representations: theoretical considerations and neuroscientific evidence. In *Perspectives on Rule-Guided Behavior*, ed. SA Bunge, JD Wallis, pp. 107–27. New York: Oxford Univ. Press

Clapp WC, Rubens MT, Sabharwal J, Gazzaley A. 2011. Deficit in switching between functional brain networks underlies the impact of multitasking on working memory in older adults. *Proc. Natl. Acad. Sci. USA* 108:7212–17

Cohen JE, Green CS, Bavelier D. 2007. Training visual attention with video games: Not all games are created equal. In *Computer Games and Adult Learning*, ed. H O'Neil, R Perez, pp. 205–27. Oxford, UK: Elsevier

Colzato LS, van Leeuwen PJA, van den Wildenberg WPM, Hommel B. 2010. DOOM'd to switch: superior cognitive flexibility in players of first person shooter games. *Front. Psychol.* 1:1–5

Corbetta M, Shulman G. 2002. Control of goal-directed and stimulus-driven attention in the brain. *Nat. Rev. Neurosci.* 3:201–15

Dahlin E, Nyberg L, Bäckman L, Neely AS. 2008. Plasticity of executive functioning in young and older adults: immediate training gains, transfer, and long-term maintenance. *Psychol. Aging* 23:720–30

Daily Mail. 2009. RAF jettisons its top guns: drones to fly sensitive missions over Afghanistan. *Daily Mail Online* Feb. 28: **http://www.dailymail.co.uk/news/worldnews/article-1158084/RAF-jettisons-Top-Guns-Drones-fly-sensitive-missions-Afghanistan.html**

Deneve S. 2008. Bayesian spiking neurons I: inference. *Neural Comput.* 20:91–117

DiCarlo JJ, Cox DD. 2007. Untangling invariant object recognition. *Trends Cogn. Sci.* 11:333–41

Donohue SE, Woldorff MG, Mitroff SR. 2010. Video game players show more precise multisensory temporal processing abilities. *Atten. Percept. Psychophys.* 72:1120–29

Dosenbach NUF, Fair DA, Cohen AL, Schlaggar BL, Petersen SE. 2008. A dual-networks architecture of top-down control. *Trends Cogn. Sci.* 12:99–105

Dosher BA, Lu Z-L. 1998. Perceptual learning reflects external noise filtering and internal noise reduction through channel reweighting. *Proc. Natl. Acad. Sci. USA* 95:13988–93

Durkin K. 2010. Videogames and young people with developmental disorders. *Rev. Gen. Psychol.* 14:122–40

Dye MW, Green CS, Bavelier D. 2009a. Increasing speed of processing with action video games. *Curr. Dir. Psychol. Sci.* 18:321–26

Dye MWG, Bavelier D. 2010. Differential development of visual attention skills in school-age children. *Vis. Res.* 50:452–59

Dye MWG, Green CS, Bavelier D. 2009b. The development of attention skills in action video game players. *Neuropsychologia* 47:1780–89

Egerton A, Mehta MA, Montgomery AJ, Lappin JM, Howes OD, et al. 2009. The dopaminergic basis of human behaviors: a review of molecular imaging studies. *Neurosci. Biobehav. Rev.* 33:1109–32

Erickson KI, Kramer AF. 2009. Aerobic exercise effects on cognitive and neural plasticity in older adults. *Br. J. Sports Med.* 43:22–24

Feng J, Spence I, Pratt J. 2007. Playing an action videogame reduces gender differences in spatial cognition. *Psychol. Sci.* 18:850–55

Frey SH, Fogassi L, Grafton S, Picard N, Rothwell JC, et al. 2011. Neurological principles and rehabilitation of action disorders: computation, anatomy, and physiology (CAP) model. *Neurorehabil. Neural Repair* 25(5):6S–20

Fuster JM. 2004. Upper processing stages of the perception-action cycle. *Trends Cogn. Sci.* 8:143–45

Gallistel CR. 2008. Learning and representation. In *Learning Theory and Behavior.* Vol. 1: *Learning and Memory: A Comprehensive Reference*, ed. R Menzel, pp. 529–48. Oxford: Elsevier

Gallistel CR, King AP. 2009. *Memory and the Computational Brain: Why Cognitive Science Will Transform Neuroscience*. New York: Wiley/Blackwell

Goard M, Dan Y. 2009. Basal forebrain activation enhances cortical coding of natural scenes. *Nat. Neurosci.* 12:1444–49

Godden D, Baddeley A. 1975. Context dependent memory in two natural environments. *Br. J. Psychol.* 66:325–31

Green CS, Bavelier D. 2003. Action video games modify visual selective attention. *Nature* 423:534–37

Green CS, Bavelier D. 2006a. Effects of action video game playing on the spatial distribution of visual selective attention. *J. Exp. Psychol.: Hum. Percept. Perform.* 32:1465–78

Green CS, Bavelier D. 2006b. Enumeration versus multiple object tracking: the case of action video game players. *Cognition* 101:217–45

Green CS, Bavelier D. 2007. Action video game experience alters the spatial resolution of vision. *Psychol. Sci.* 18:88–94

Green CS, Bavelier D. 2012. Learning, attentional control and action video games. *Curr. Biol.* 22:R167–206

Green CS, Pouget A, Bavelier D. 2010. Improved probabilistic inference as a general learning mechanism with action video games. *Curr. Biol.* 20:1573–79

Green CS, Sugarman MA, Medford K, Klobusicky E, Bavelier D. 2012. The effect of action video games on task switching. *Comput. Hum. Behav.* 12:984–94

Greenfield PM. 2009. Technology and informal education: What is taught, what is learned. *Science* 323:69–71

Greenfield PM, DeWinstanley P, Kilpatrick H, Kaye D. 1994. Action video games and informal education: effects on strategies for dividing visual attention. *J. Appl. Dev. Psychol.* 15:105–23

Halpern DF, Benbow CP, Geary DC, Gur RC, Hyde JS, Gernsbacher MA. 2007. The science of sex differences in science and mathematics. *Psychol. Sci. Public Interest* 8:1–51

Harlow HF. 1949. The formation of learning sets. *Psychol. Rev.* 56:51–65

Herrero JL, Roberts MJ, Delicato LS, Gieselmann MA, Dayan P, Thiele A. 2008. Acetylcholine contributes through muscarinic receptors to attentional modulation in V1. *Nature* 454:1110–14

Hillman CH, Erickson KI, Kramer AF. 2008. Be smart, exercise your heart: exercise effects on brain and cognition. *Nat. Rev. Neurosci.* 9:58–65

Hinton GE. 2007. Learning multiple layers of representation. *Trends Cogn. Sci.* 11:428–34

Hinton GE, Osindero S, Tej YH. 2006. A fast learning algorithm for deep belief nets. *Neural Comput.* 18:1527–54

Hubert-Wallander BP, Green CS, Sugarman M, Bavelier D. 2011. Altering the rate of visual search through experience: the case of action video game players. *Atten. Percept. Psychophys.* 73:2399–412

Jaeggi SM, Buschkuehl M, Jonides J, Perrig WJ. 2008. Improving fluid intelligence with training on working memory. *Proc. Natl. Acad. Sci. USA* 105:6829–33

Ji W, Gao E, Suga N. 2001. Effects of acetylcholine and atropine on plasticity of central auditory neurons caused by conditioning in bats. *J. Neurophysiol.* 86:211–25

Karle JW, Watter S, Shedden JM. 2010. Task switching in video game players: benefits of selective attention but not resistance to proactive interference. *Acta Psychol.* 134:70–78

Kemp C, Goodman ND, Tenenbaum JB. 2010. Learning to learn causal models. *Cogn. Sci.* 23:1–59

Khoe WW, Bavelier D, Hillyard SA. 2010. Enhanced attentional modulation of early visual processing in video game players. *Soc. Neurosci. Meet. Plann.* Abstr. 399.1

Kilgard MP, Merzenich MM. 1998. Cortical map reorganization enabled by nucleus basalis activity. *Science* 279:1714–18

Klingberg T. 2010. Training and plasticity of working memory. *Trends Cogn. Sci.* 14:317–24

Knill DC, Richards W. 1996. *Perception as Bayesian Inference*. Cambridge, UK: Cambridge Univ. Press

Koechlin E. 2008. The cognitive architecture of the human lateral prefrontal cortex. In *Attention and Performance*, ed. P Haggard, Y Rosetti, M Kawato, pp. 483–509. New York: Oxford Univ. Press

Koechlin E, Summerfield C. 2007. An information theoretical approach to prefrontal executive function. *Trends Cogn. Sci.* 11:229–35

Koepp MJ, Gunn RN, Lawrence AD, Cunningham VJ, Dagher A, et al. 1998. Evidence for striatal dopamine release during a video game. *Nature* 393:266–68

Larochelle H, Erhan D, Courville A, Bergstra J, Bengio Y. 2007. *An empirical evaluation of deep architectures on problems with many factors of variation*. Presented at Proc. Int. Conf. Mach. Learn., 24th, Corvallis, Or.

Lashley KS. 1951. The problem of serial order in behavior. In *Cerebral Mechanisms in Behavior*, ed. LA Jeffress, pp. 112–36. New York: Wiley

Lavie N. 2005. Distracted and confused? Selective attention under load. *Trends Cogn. Sci.* 9:75–82

Lee H, Grosse R, Ranganath R, Ng AY. 2009. *Convolutional deep belief networks for scalable unsupervised learning of hierarchical representations*. Presented at Proc. Int. Conf. Mach. Learn., 26th, Montreal

Lee TS, Mumford D. 2003. Hierarchical Bayesian inference in the visual cortex. *J. Opt. Soc. Am.* 20:1434–48

Legge GE, Rubin GS, Pelli DG, Schleske MM. 1985. Psychophysics of reading–II. Low vision. *Vis. Res.* 25:253–65

Li R, Ngo C, Nguyen J, Levi DM. 2011. Video game play induces plasticity in the visual system of adults with amblyiopia. *Public Libr. Sci.* 9(8):e1001135

Li R, Polat U, Makous W, Bavelier D. 2009. Enhancing the contrast sensitivity function through action video game training. *Nat. Neurosci.* 12:549–51

Li RW, Polat U, Scalzo F, Bavelier D. 2010. Reducing backward masking through action game training. *J. Vis.* 10:1–13

Li S, Mayhew SD, Kourtzi Z. 2009. Learning shapes the representation of behavioral choice in the human brain. *Neuron* 62:441–52

Lustig C, Shah P, Seidler R, Reuter PA. 2009. Aging, training, and the brain: a review and future directions. *Neuropsychol. Rev.* 19:504–22

Lutz A, Slagter HA, Dunne JD, Davidson RJ. 2008. Attention regulation and monitering in meditation. *Trends Cogn. Sci.* 12:163–69

Ma WJ, Beck JM, Latham PE, Pouget A. 2006. Bayesian inference with probabilistic population codes. *Nat. Neurosci.* 9:1432–38

Ma WJ, Huang W. 2009. No capacity limit in attentional tracking: evidence for probabilistic inference under a resource constraint. *J. Vis.* 9:1–30

McClurg PA, Chaille C. 1987. Computer games: environments for developing spatial cognition. *J. Educ. Comput. Res.* 3:95–111

McKinley RA, McIntire LK, Funke MA. 2011. Operator selection for unmanned aerial systems: comparing video game players and pilots. *Aviat. Space Environ. Med.* 82:635–42

McNab F, Varrone A, Farde L, Jucaite A, Bystritsky P, et al. 2009. Changes in cortical dopamine D1 receptor binding associated with cognitive training. *Science* 323:800–2

Mishra J, Zinni M, Bavelier D, Hillyard SA. 2011. Neural basis of superior performance of video-game players in an attention-demanding task. *J. Neurosci.* 31:992–98

Mohamed A, Dahl GE, Hinton G. 2011. Acoustic modeling using deep belief networks. *IEEE Trans. Audio, Speech, Lang. Proc.*

Okagaki L, Frensch PA. 1994. Effects of video game playing on measures of spatial performance: gender effects in late adolescence. *J. Appl. Dev. Psychol.* 15:33–58

Olesen P, Westerberg H, Klingberg T. 2004. Increased prefrontal and parietal activity after training of working memory. *Nat. Neurosci.* 7:75–79

Ophir E, Nass C, Wagner AD. 2009. Cognitive control in media multitaskers. *Proc. Natl. Acad. Sci. USA* 106:15583–87

Palmer J, Huk AC, Shadlen MN. 2005. The effect of stimulus strength on the speed and accuracy of a perceptual decision. *J. Vis.* 5:376–404

Paupathy A, Connor CE. 2002. Population coding of shape in area V4. *Nat. Neurosci.* 5:1332–38

Perfors AF, Tenenbaum JB. 2009. *Learning to learn categories*. Presented at Annu. Conf. Cogn. Sci. Soc., 31st, Amsterdam

Polley DB, Steinberg EE, Merzenich MM. 2006. Perceptual learning directs auditory cortical map reorganization through top-down influences. *J. Neurosci.* 26:4970–82

Pouget A, Thorpe SJ. 1991. Connectionist model of orientation identification. *Connect. Sci.* 3:127–42

Pylyshyn ZW, Storm RW. 1988. Tracking multiple independent targets: evidence for a parallel tracking mechanism. *Spat. Vis.* 3:179–97

Rao RP. 2004. Bayesian computation in recurrent neural circuits. *Neural Comput.* 16:1–38

Ratcliff R, McKoon G. 2008. The diffusion decision model: theory and data or two-choice decision tasks. *Neural Comput.* 20:873–922

Renner MJ, Rosenzweig MR. 1987. *Enriched and Impoverished Environments Effects on Brain and Behavior Recent Research in Psychology*. New York: Springer-Verlag

Ribas-Fernandes JJF, Solway A, Diuk C, McGuire JT, Barto AG, et al. 2011. A neural signature of hierarchical reinforcement learning. *Neuron* 71:370–79

Roelfsema PR, van Ooyen A, Watanabe T. 2010. Perceptual learning rules based on reinforcers and attention. *Trends Cogn. Sci.* 14:64–71

Rosch E. 1975. Cognitive representations of semantic categories. *J. Exp. Psychol. Gen.* 104:192–233

Rosen LD. 2007. *Me, MySpace, and I*. New York: Macmillan

Rosser JCJ, Lynch PJ, Cuddihy L, Gentile DA, Klonsky J, Merrell R. 2007. The impact of video games on training surgeons in the 21st century. *Arch. Surg.* 142:181–86

Rougier NP, Noelle DC, Braver TS, Cohen JD, O'Reilly RC. 2005. Prefrontal cortex and flexible cognitive control: rules without symbols. *Proc. Natl. Acad. Sci. USA* 102:7338–43

Rumelhart DE, McClelland JL, PDP Res. Group. 1987. *Parallel Distributed Processing–Vol. 1*. Boston, MA: MIT Press

Saffell T, Matthews N. 2003. Task-specific perceptual learning on speed and direction discrimination. *Vis. Res.* 43(12):1365–74

Sakai K. 2008. Task set and prefrontal cortex. *Annu. Rev. Neurosci.* 31:219–45

Sarter M, Hasselmo ME, Bruno JP, Givens B. 2005. Unraveling the attentional functions of cortical cholinergic inputs: interactions between signal-driven and cognitive modulation of signal detection. *Brain Res. Rev.* 48:98–111

Schellenberg EG. 2004. Music lessons enhance IQ. *Psychol. Sci.* 15:511–14

Schlickum MK, Hedman L, Enochsson L, Kjellin A, Fellander-Tsai L. 2009. Systematic video game training in surgical novices improves performance in virtual reality endoscopic surgical simulators: a prospective randomized study. *World J. Surg.* 33:2360–67

Schultz W, Dayan P, Montague PR. 1997. A neural substrate of prediction and reward. *Science* 275:1593–99

Serences JT, Yantis S, Culberson A, Awh E. 2004. Preparatory activity in visual cortex indexes distractor suppression during covert spatial orienting. *J. Neurophysiol.* 92:3538–45

Shapiro DC, Schmidt RA. 1982. The schema theory: recent evidence and developmental implications. In *The Development of Movement Control and Co-ordination*, ed. JAS Kelso, JE Clark, pp. 113–50. New York: Wiley

Shapiro KL, Arnell KM, Raymond JE. 1997. The attentional blink. *Trends Cogn. Sci.* 1:291–96

Simen P, Contreras D, Buck C, Hu P, Holmes P, Cohen D. 2009. Reward rate optimization in two-alternative decision making: empirical tests of theoretical predictions. *J. Exp. Psychol.: Hum. Percept. Perform.* 35:1865–97

Sitaram N, Weingartner H, Gillin JC. 1978. Human serial learning: enhancement with arecholine and choline impairment with scopolamine. *Science* 201:274–76

Spelke ES. 2005. Sex differences in intrinsic aptitude for mathematics and science?: A critical review. *Am. Psychol.* 60:950–58

Spence I, Feng J. 2010. Video games and spatial cognition. *Rev. Gen. Psychol.* 14:92–104

Spence I, Yu JJ, Feng J, Marshman J. 2009. Women match men when learning a spatial skill. *J. Exp. Psychol.: Learn. Mem. Cogn.* 35:1097–103

Stevens C, Bavelier D. 2012. The role of selective attention on academic foundations: a cognitive neuroscience perspective. *Dev. Cogn. Neurosci.* 2:S30–48

Stilp CE, Rogers TT, Kluender KR. 2010. Rapid efficient coding of correlated complex acoustic properties. *Proc. Natl. Acad. Sci. USA* 107:21914–19

Strobach T, Frensch PA, Schubert T. 2012. Video game practice optimizes executive control skills in dual tasks and task switching situations. *Acta Psycholog.* 140:13–24

Subrahmanyam K, Greenfield PM. 1994. Effect of video game practice on spatial skills in girls and boys. *J. Appl. Dev. Psychol.* 15:13–32

Takeuchi H, Sekiguchi A, Taki Y, Yokoyama S, Yomogida Y, et al. 2010. Training of working memory impacts structural connectivity. *J. Neurosci.* 30:3297–303

Tang YY, Posner MI. 2009. Attention training and attention state training. *Trends Cogn. Sci.* 13:222–27

Thorndike EL, Woodworth RS. 1901. The influence of improvement in one mental function upon the efficiency of other functions. (I). *Psychol. Rev.* 8:247–61

Toepper M, Gebhardt H, Beblo T, Thomas C, Driessen M, et al. 2010. Functional correlates of distractor suppression during spatial working memory encoding. *Neuroscience* 165:1244–53

Trick LM, Jaspers-Fayer F, Sethi N. 2005. Multiple-object tracking in children: the "Catch the Spies" task. *Cogn. Dev.* 20:373–87

Vul E, Frank MC, Alvarez GA, Tenenbaum JB. 2010. Explaining human multiple object tracking as resource-constrained approximate inference in a dynamic probabilistic model. *Adv. Neural Inf. Proc. Syst.* 32:1955–63

Vul E, Hanus D, Kanwisher N. 2009. Attention as inference: Selection is probabilistic; responses are all-or-none samples. *J. Exp. Psychol. Gen.* 138(4):546–60

West GL, Stevens SA, Pun C, Pratt J. 2008. Visuospatial experience modulates attentional capture: evidence from action video game players. *J. Vis.* 8:1–9

Yamane Y, Carlson ET, Bowman KC, Wang Z, Connor CE. 2008. A neural code for three-dimensional object shape in macaque inferotemporal cortex. *Nat. Neurosci.* 11:1352–60

Zhang J-C, Lau P-M, Bi G-Q. 2009. Gain in sensitivity and loss in temporal contrast of STDP by dopaminergic modulation at hippocampal synapses. *Proc. Natl. Acad. Sci. USA* 106:13028–33

RELATED RESOURCES

Bavelier D, Levi DM, Li RW, Dan Y, Hensch TK. 2010. Removing brakes on adult plasticity: from molecular to behavioral interventions. *J. Neurosci.* 30(45):14964–71

Botvinick MM. 2008. Hierarchical models of behavior and prefrontal function. *Trends Cogn. Sci.* 12:201–8

Green CS, Pouget A, Bavelier D. 2010. Improved probabilistic inference, as a general learning mechanism with action video games. *Curr. Biol.* 20:1573–79

Hinton GE. 2007. Learning multiple layers of representation. *Trends Cogn. Sci.* 11:428–34

Spence I, Feng J. 2010. Video games and spatial cognition. *Rev. Gen. Psychol.* 14(2):92–104

The Pathophysiology of Fragile X (and What It Teaches Us about Synapses)

Asha L. Bhakar,[1] Gül Dölen,[2] and Mark F. Bear[1]

[1]Howard Hughes Medical Institute, Picower Institute for Learning and Memory, Massachusetts Institute of Technology, Cambridge, Massachusetts 02139; email: mbear@mit.edu, abhakar@mit.edu

[2]Department of Psychiatry and Developmental Sciences, Stanford University School of Medicine, Palo Alto, California 94305; email: gul@stanford.edu

Annu. Rev. Neurosci. 2012. 35:417–43

First published online as a Review in Advance on April 5, 2012

The *Annual Review of Neuroscience* is online at neuro.annualreviews.org

This article's doi: 10.1146/annurev-neuro-060909-153138

0147-006X/12/0721-0417$20.00

Keywords

FMRP, metabotropic glutamate receptor, autism, mRNA translation, long-term depression

Abstract

Fragile X is the most common known inherited cause of intellectual disability and autism, and it typically results from transcriptional silencing of *FMR1* and loss of the encoded protein, FMRP (fragile X mental retardation protein). FMRP is an mRNA-binding protein that functions at many synapses to inhibit local translation stimulated by metabotropic glutamate receptors (mGluRs) 1 and 5. Recent studies on the biology of FMRP and the signaling pathways downstream of mGluR1/5 have yielded deeper insight into how synaptic protein synthesis and plasticity are regulated by experience. This new knowledge has also suggested ways that altered signaling and synaptic function can be corrected in fragile X, and human clinical trials based on this information are under way.

Contents

INTRODUCTION

This year we expect to learn the outcome of clinical trials for potentially disease-modifying treatments of fragile X (FX). Three important developments outside the realm of basic neuroscience paved the way for this progress: First, careful clinical observation defined the syndrome and suggested a genetic etiology (Martin & Bell 1943); second, mutations that silenced a single gene (*FMR1*) on the X chromosome were discovered to be the major cause (Pieretti et al. 1991, Verkerk et al. 1991); and third, the generation and widespread dissemination of an *Fmr1*-knockout (KO) mouse enabled studies of pathophysiology (Dutch-Belgian Fragile X Consort. 1994) (**Figure 1**). *FMR1* encodes fragile X mental retardation protein (FMRP), an mRNA-binding protein that is highly expressed in neurons. As with most neurobehavioral disorders of genetic origin, it was assumed that development of the brain in the absence of this key protein irrevocably alters neuronal connectivity to produce the devastating behavioral symptoms, including intellectual disability and autism, that are characteristic of this disease.

FX: fragile X

Synaptic plasticity: the ability of synapses to change in strength in response to activity; an important cellular mechanism for learning and memory

However, this dim view of FX has changed dramatically in the past ten years. It is now believed that many symptoms of FX could arise from modest changes in synaptic signaling—changes that can be corrected with targeted therapies such as those that are now in clinical trials. The origins of this new view can be traced to fundamental research on synaptic plasticity (Bear et al. 2004, Huber et al. 2002). Since this initial insight into how synaptic signaling is altered in FX, the progress toward developing therapeutics for FX has been explosive. It has been shown that seemingly unrelated symptoms of the disease can be corrected by manipulating a molecular target, mGluR5, that is amenable to drug therapy (Dolen et al. 2007). Furthermore, studies in multiple animal models of FX have shown that this core pathophysiology is evolutionarily conserved. This extraordinary progress has been the subject of a number of recent reviews (see e.g., Dolen et al. 2010, Krueger & Bear 2011, Levenga et al. 2010, Santoro et al. 2011).

Certainly research on synaptic plasticity has informed the understanding of FX pathophysiology; but it is also true that the biology of FX has informed the understanding of synaptic function and plasticity. This is the point of view we take in the present review.

OVERVIEW OF FRAGILE X

In the majority of FX patients, a trinucleotide (CGG) repeat expansion leads to hypermethylation and transcriptional silencing of the *FMR1* gene and subsequent loss of FMRP (Fu et al. 1991, Pieretti et al. 1991). In one identified patient, disease is caused by a point mutation in *FMR1* that alters protein function (De Boulle et al. 1993). Disease severity varies with the expression level of FMRP, which can fluctuate as a result of germline mosaicism and, in females, X inactivation (De Boulle et al. 1993, Hatton et al. 2006, Kaufmann et al. 1999, Loesch et al. 1995, Lugenbeel et al. 1995, Reiss & Dant 2003). Accordingly, understanding the cellular function of FMRP has become an obvious priority.

Epidemiological studies conservatively estimate that FX occurs in 1:5000 males (and approximately half as many females), making it the leading cause of inherited intellectual disability (Coffee et al. 2009). FX was also the first recognized genetic disorder associated with autism, and despite expanding diagnostic criteria and newly discovered candidate genes, FX remains the most common known inherited cause of autism (Wang et al. 2010b). In addition to moderate to severe intellectual disability and autistic features (social/language deficits and stereotyped/restricted behaviors), the disease is characterized by seizures and/or epileptiform activity, hypersensitivity to sensory stimuli, attention deficit and hyperactivity, motor incoordination, growth abnormalities, sleep disturbances, craniofacial abnormalities, and macroorchidism. Because FX is a monogenic and relatively common cause of autism, it has been a useful model for dissecting pathophysiology that may apply to genetically heterogeneous autisms.

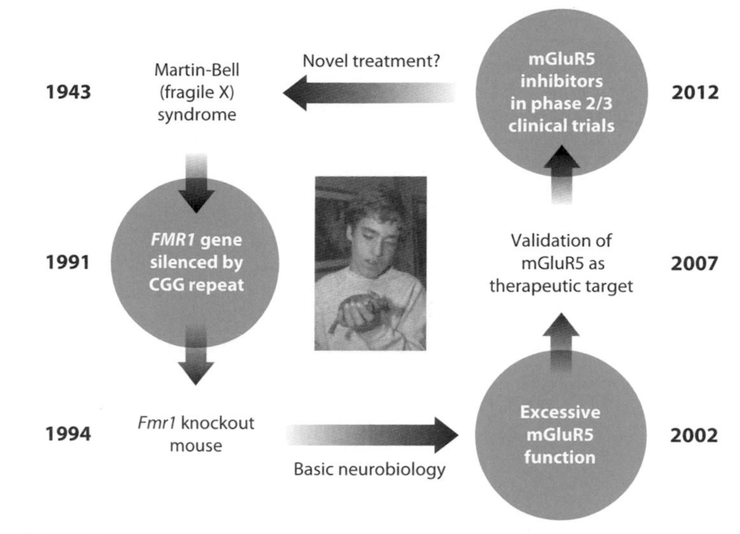

Figure 1

Fulfilling the promise of molecular medicine in FX. Martin & Bell (1943) described a group of patients characterized by a common set of features that included intellectual disability and social withdrawal. The causative gene mutation was discovered in 1991 (Pieretti et al. 1991, Verkerk et al. 1991). The *FMR1* gene on the X chromosome is silenced, and the protein FMRP is not produced. Shortly thereafter, the *Fmr1*-KO mouse model was generated (Dutch-Belgian Fragile X Consort. 1994) and has been intensively studied by neurobiologists interested both in the disease and FMRP. In 2002, it was discovered that a form of synaptic plasticity—mGluR LTD—was exaggerated in the *Fmr1* KO mouse (Huber et al. 2002). This led to the mGluR theory of fragile X (Bear et al. 2004), which posits that many symptoms of the disease are due to exaggerated responses to activation of mGluR5. The theory was definitively validated in 2007 with the demonstration that multiple FX phenotypes are corrected in the *Fmr1*-KO mouse by genetic reduction of mGluR5 protein production (Dolen et al. 2007). In addition, numerous animal studies showed that pharmacological inhibition of mGluR5 ameliorates FX mutant phenotypes. In 2009, inhibitors of mGluR5 entered into human phase 2 trials (**http://clinicaltrials.gov**). If successful, these trials will represent the first pharmacological treatment for a neurobehavioral disorder that was developed from the bottom up: from gene discovery to pathophysiology in animals to novel therapeutics in humans. Abbreviations: CGG, cytosine-guanine-guanine; FMRP, fragile X mental retardation protein; FX, fragile X; mGluR5, metabotropic glutamate receptor 5; KO, knockout; LTD, long-term synaptic depression. Image courtesy of FRAXA Research Foundation, with permission.

NEW INSIGHTS INTO THE BIOLOGY OF FMRP

Biochemical characterization of FMRP has provided key insights into the pathophysiology of FX, and after 20 years of research, we now know that FMRP is an RNA-binding protein that largely functions to negatively regulate protein synthesis in the brain. Recent work has led to the view that many symptoms of FX arise from a modest increase in synaptic protein synthesis, an aspect of cerebral metabolism that can continue to be corrected after birth to produce substantial benefit. Therefore, there is great interest in the question of how FMRP interacts with mRNA to regulate synaptic protein synthesis.

X inactivation: the process by which one of the two copies of the X chromosome present in female mammalian cells is transcriptionally silenced

FMRP Binds RNA

Sequence analysis first identified three common RNA-binding domains in the protein structure of FMRP, providing the first suggestion of a direct interaction between FMRP and

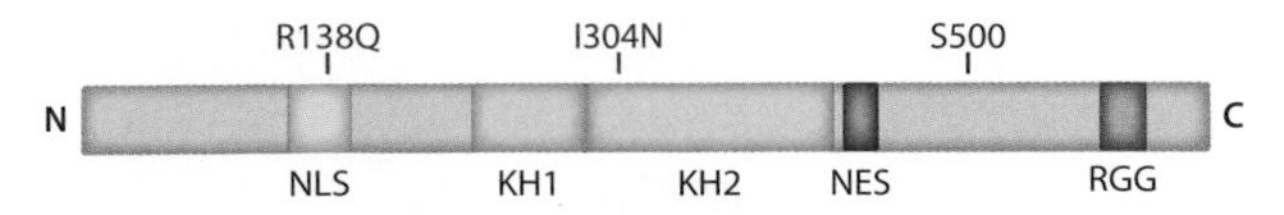

Figure 2

Functional domains of FMRP. Human FMRP, a 632 amino acid polypeptide (*gray bar*), has a nuclear localization signal (NLS; *light blue*), two K-homology domains (KH1 and KH2; *orange*), an RGG (arginine-glycine-glycine) box (*dark blue*), and a nuclear export sequence (NES; *red*). R138Q and I304N are naturally occurring mutations in patients with developmental delay and a severe form of FX, respectively. I304N abolishes polyribosome association. S500 is a major site of phosphorylation. Abbreviations: N, amino terminus; C, carboxy terminus; FMRP, fragile x mental retardation protein.

RNA (Ashley et al. 1993, Siomi et al. 1993). Two of the domains are hnRNP K-homology (KH) domains, and the third, located close to the C-terminal end, is an RGG box (**Figure 2**). KH domains are thought to recognize and bind "kissing-complex" tertiary motifs in RNA (Darnell et al. 2005), whereas the RGG box recognizes stem-G-quartet loops, possibly in a methylation-dependent manner (Blackwell et al. 2010). A stem loop SoSLIP motif, found in one target (Sod1 mRNA), has also been identified and can bind to the C-terminal RGG region (Bechara et al. 2009). In addition, U-rich sequences have been isolated as potential RNA-binding motifs, although no precise binding domains within FMRP have yet been described (Chen et al. 2003, Fahling et al. 2009).

How FMRP associates with specific mRNAs is still under active investigation. A point mutation (I304N) within the second KH domain leads to a severe clinical presentation of the disease and has provided the first evidence that binding to mRNA and this domain in particular are critical to the function of FMRP (De Boulle et al. 1993, Feng et al. 1997a). Recent work using ultraviolet light to crosslink FMRP with endogenous mRNA in situ revealed, surprisingly, that FMRP binds largely within the coding regions of many mRNAs instead of the 5′ or 3′ untranslated regions (Darnell et al. 2011). Although this study did not reveal a specific consensus motif, synthetic kissing-complex RNA was still effective in competing with these target mRNAs for binding to FMRP, confirming that KH domains and kissing-complex motifs are critically involved. It has been estimated that ~4% of total brain mRNA binds FMRP (Ashley et al. 1993, Brown et al. 2001, Darnell et al. 2011).

Polyribosome: a cluster of ribosomes all attached to a single mRNA molecule

FMRP May Regulate RNA Transport

FMRP also contains a nuclear localization sequence and a nuclear export sequence (Ashley et al. 1993), and although its expression is largely cytoplasmic (found in the cell body, dendrites, and synapses), some FMRP can be found shuttling in and out of the nucleus (Feng et al. 1997b). To date, few data exist to support a role for FMRP in regulating transcription or RNA processing, but FMRP can be found bound to nuclear mRNA, a nuclear exporter protein (Tap/NXF1), and to pre-mRNA while it is being transcribed (Kim et al. 2009). A novel missense mutation (R138Q) was detected in the nuclear localization sequence of *FMR1* in a patient with developmental delay (Collins et al. 2010), suggesting that nuclear FMRP is important for neuronal function.

Many in vitro studies have suggested a role for FMRP in transporting mRNA. The protein has been imaged in dynamic RNA granules that traffic from the soma to dendrites and axons (Antar et al. 2004, 2005, 2006; De Diego Otero et al. 2002). RNA granules are believed to be translationally repressed mRNP (messenger ribonucleoprotein) complexes. Granules are heterogeneous in their composition: P bodies and stress granules contain translational initiation machinery (e.g., monomeric ribosomal constituents, mRNA, and proteins) trapped before translational initiation, whereas high-density granules also contain elongation machinery (e.g., polyribosomes and ribosomal aggregates) whose translation has been stalled (Anderson & Kedersha 2006, Kiebler & Bassell 2006). Once localized to the synapse, mRNAs are released from the granules and subsequently translated in response to stimuli (Krichevsky & Kosik 2001). FMRP mRNPs have been found in all three types of RNA granules (Aschrafi et al. 2005, Barbee et al. 2006, Maghsoodi et al. 2008, Mazroui et al. 2002).

In some instances, FMRP trafficking into dendrites can be stimulated by neuronal

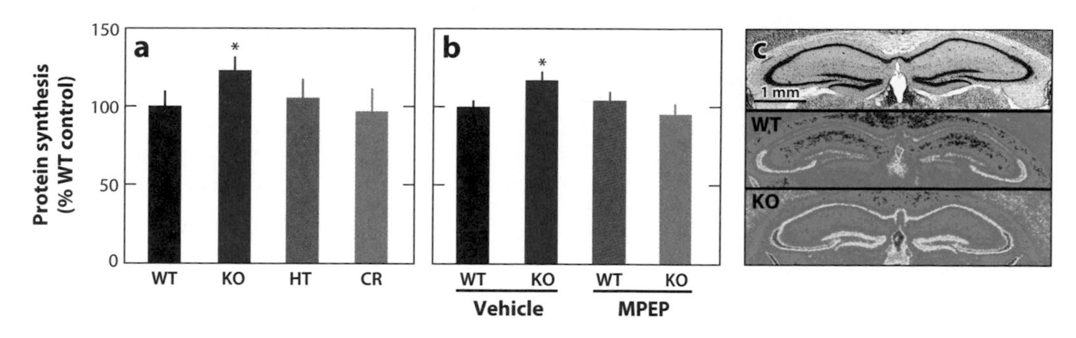

Figure 3

Excessive protein synthesis in the hippocampus of *Fmr1*-KO mice. Translation rates in the hippocampus measured by metabolic labeling in vitro (*a*,*b*) and in vivo (*c*) confirm that FMRP functions to negatively regulate protein synthesis in neurons. (*a*) Basal protein synthesis is significantly increased in *Fmr1*-KO hippocampal slices compared to control WT. Although there is no effect of reducing mGluR5 by 50% in *Grm5* heterozygous mice (HT), crossing these mice with *Fmr1*-KO mice (CR) is sufficient to correct the excessive protein synthesis (modified from Dolen et al. 2007). (*b*) Excessive protein synthesis in *Fmr1*-KO hippocampal slices is restored to normal levels by acute treatment with an mGluR5 inhibitor (MPEP), demonstrating it occurs downstream of constitutive mGluR5 activity (modified from Osterweil et al. 2010). (*c*) Nissl-stained coronal sections (*top panel*) and their corresponding pseudocolored autoradiograms (*middle and lower panels*) show quantitative increases in translation rates throughout the hippocampus of 6-month-old *Fmr1*-KO mice in vivo (*lower panel*) compared with WT controls (*middle panel*). Images courtesy of C.B. Smith (Qin et al. 2005). Hot colors represent higher rates of synthesis. Abbreviations: FMRP, fragile X mental retardation protein; KO, knockout; mGluR, metabotropic glutamate receptor; MPEP, 2-methyl-6-(phenylethynyl)-pyridine; WT, wild type.

activity (Antar et al. 2004, Gabel et al. 2004). However, it does not appear to be necessary for the steady-state maintenance or the constitutive localization of the majority of its target mRNAs in dendrites (Dictenberg et al. 2008, Steward et al. 1998). Indeed, most mRNAs that normally associate with FMRP are correctly targeted to the synapse in the absence of FMRP. Thus, another RNA-binding protein may be needed for the normal active transport of the majority of FMRP targets, and FMRP may be more of a passive passenger within the RNA transport granule.

FMRP Negatively Regulates Translation

Subcellular fractionation studies originally showed that the majority of FMRP-RNA complexes are in actively translating polyribosomal fractions, particularly in synaptic preparations (Aschrafi et al. 2005; Brown et al. 2001; Corbin et al. 1997; Eberhart et al. 1996; Feng et al. 1997a, 1997b; Khandjian et al. 1995; Stefani et al. 2004; Tamanini et al. 1996; Zalfa et al. 2007). These observations, together with the knowledge that both FMRP protein and mRNA are expressed in dendrites and dendritic spines, suggested that FMRP regulates local protein synthesis at the synapse.

Several independent lines of evidence support this hypothesis and show that FMRP functions to repress translation. First, purified recombinant FMRP added to rabbit reticulolysate or injected into Xenopus oocytes shows a dose-dependent suppression of mRNA translation that is abolished when FMRP-binding sequences are removed from target mRNA (Laggerbauer et al. 2001, Li et al. 2001). Second, an electrophysiological readout of synaptic protein synthesis in the hippocampus, metabotropic glutamate receptor (mGluR)-dependent long-term depression (LTD) (discussed below), is exaggerated in the absence of FMRP, consistent with increased protein synthesis (Huber et al. 2002). Third, direct measurement of the rate of protein synthesis in hippocampal slices or cortical synaptoneurosomes in vitro shows a significant elevation in the *Fmr1*-KO mouse (Dolen et al. 2007, Muddashetty et al. 2007, Osterweil et al. 2010) (**Figure 3**). Finally, similar measurements

mGluR: metabotropic glutamate receptor

LTD: long-term depression

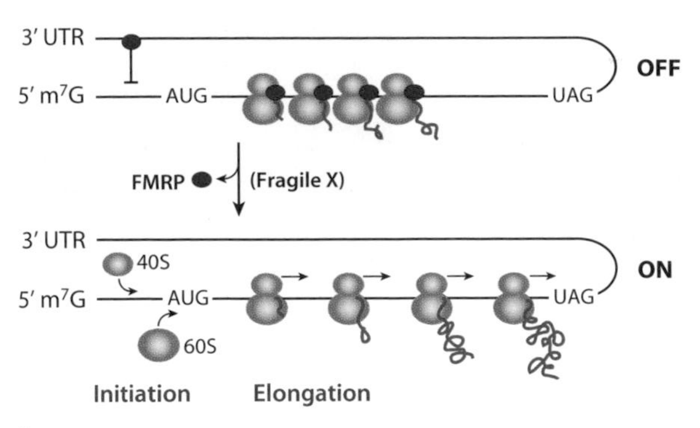

Figure 4

FMRP regulates mRNA translation. FMRP (*red ovals*) can be found bound to coding regions of mRNA in association with stalled ribosomes [complexes of 40S (*small gray ovals*) and 60S (*large gray ovals*) ribosomal subunits] and bound to 3′UTRs in association with inhibitory components of the initiation machinery (indicated by an inhibitory line). Data currently suggest that FMRP normally represses translation by stalling the elongation of actively translating ribosomes and by blocking the initiation of ribosome assembly. Loss of FMRP (as in fragile X) removes both of these inhibitory associations and leads to increased protein synthesis. Curly blue lines represent ribosomally synthesized polypeptide chains that lengthen as translation proceeds. Small arrows indicate active movement. Abbreviations: AUG, initiation codon; FMRP, fragile X mental retardation protein; m^7G, 7-methylguanylate cap; ON, translation on; OFF, translation off; UAG, termination codon; 3′UTR, 3 prime-end untranslated region.

performed in the KO mouse in vivo show a global increase in brain protein synthesis (Qin et al. 2005). The fact that increased protein synthesis can be observed in the intact animal in vivo has raised the possibility that measurements of protein synthesis could serve as a biomarker of disease (Bishu et al. 2008, Bishu et al. 2009). Indeed, studies are currently underway to test this hypothesis in human patients with FX (**http://www.clinicaltrials.gov**).

Mechanisms of Translational Regulation by FMRP

Run-off assay: a biochemical method to assess translational competence of polyribosomes initiated in vivo then completed in vitro

Although it is now appreciated that FMRP functions to negatively regulate protein synthesis, the mechanism by which repression is achieved remains controversial. Given that the majority of FMRP cosediments with polyribosomes, FMRP was originally suspected to repress translation by blocking elongation (Ceman et al. 2003, Feng et al. 1997a, Khandjian et al. 1996, Stefani et al. 2004, Tamanini et al. 1996). This hypothesis has received strong support in a recent study in which FMRP mRNA targets were identified following ultraviolet cross-linking (Darnell et al. 2011). The majority (66%) of mRNA binding was found within the coding sequence of the 842 transcripts cross-linked to FMRP in mouse brain polysomes. Ribosomal run-off assays on these transcripts demonstrated that FMRP is associated with transcripts on which ribosomes are stalled. These data support a model whereby FMRP dynamically represses translation in a complex consisting of target mRNAs and stalled ribosomes (**Figure 4**).

However, the presence of FMRP mRNPs in p bodies, stress granules, and high-density granules has suggested that FMRP represses translation throughout many phases of translational regulation. FMRP can cosediment with the monomeric 80S ribosomes and in light mRNP complexes with BC1 (brain cytoplasmic RNA 1), CYFIP1 (cytoplasmic FMRP-interacting protein), and translation initiation factors (Centonze et al. 2008, Gabus et al. 2004, Johnson et al. 2006, Lacoux et al. 2012, Laggerbauer et al. 2001, Napoli et al. 2008, Zalfa et al. 2007). These data suggest that FMRP also represses translation at the initiation stage. In this model, FMRP represses translation by inhibiting cap-dependent initiation through interactions with CYFIP1, a eukaryotic initiation factor 4E binding protein (4E-BP). Consistent with this proposal, genetic reduction of CYFIP1 levels increases the expression of several FMRP targets (Napoli et al. 2008). The in vivo relevance of these interactions, however, has been questioned (Iacoangeli et al. 2008a,2008b; Stefani et al. 2004; Wang et al. 2005).

Mechanisms to Stall Elongation

How FMRP cooperates with the translational machinery to stall elongation or block initiation is incompletely understood. Some data have suggested that association with the microRNA (miRNA) machinery may be involved. FMRP interacts with members of the RNA-induced silencing complex

(Bolduc et al. 2008, Caudy & Hannon 2004, Caudy et al. 2002, Cheever & Ceman 2009, Ishizuka et al. 2002, Jin et al. 2004, Muddashetty et al. 2011; but see Didiot et al. 2009) and several specific miRNAs (Edbauer et al. 2010, Plante et al. 2006, Xu et al. 2008, Yang et al. 2009) that function together to silence target mRNA, either by direct cleavage of transcripts or by translational repression (see, for a review, Schratt 2009). Because FMRP lacks a canonical miRNA-binding domain, it currently seems likely that this modulation occurs through protein-protein interactions between members of the RNA-induced silencing complex (e.g., Argonaute, Dicer) and FMRP, rather than direct binding to miRNAs. Still, the possibility remains that the kissing-complex structure, the putative ligand of the KH domain of FMRP, may be formed by miRNA and target mRNA together (Darnell et al. 2005, Plante et al. 2006).

Several post-translational modifications of FMRP have also been suggested to regulate translational repression. Methylation of FMRP on arginine residues can reduce FMRP binding to stem loop G-quartet structures (Stetler et al. 2006). Others have suggested that ubiquitin-proteasome degradation followed by resynthesis of FMRP may be a mechanism for transient derepression (Zhao et al. 2011), but some work has shown that FMRP synthesis increases upon stimulation prior to its degradation (Hou et al. 2006). FMRP can also be phosphorylated on a series of serine residues N terminal to the RGG box. Phosphorylation has been suggested to stall ribosomal translocation while preserving the association of FMRP with mRNA (Ceman et al. 2003, Coffee et al. 2011, Muddashetty et al. 2011). Thus, one way neural activity may gate translation is by regulating FMRP phosphorylation.

SYNAPTIC REGULATION OF PROTEIN SYNTHESIS

Although FMRP is expressed throughout the neuron, it has attracted particular attention as a regulator of protein synthesis at excitatory synapses. Because exaggerated protein synthesis is believed to be pathogenic in FX and possibly in other disorders associated with autism (Kelleher & Bear 2008, Darnell 2011), the question of how synaptic activity can trigger FMRP-regulated mRNA translation is of particular interest. Conversely, because neuronal protein synthesis has a fundamental role in synaptic plasticity and information storage (Kandel 2001), understanding how FMRP functions at the synapse has also become a high priority in basic neurobiology.

Interest in synaptically localized protein synthesis originated with the discovery that polyribosomes accumulate at the base of many dendritic spines that are postsynaptic to glutamatergic excitatory synapses (Steward & Levy 1982). These synaptic polyribosomes seemed to provide an ideal substrate for the structural changes that support long-term synaptic modifications, such as long-term potentiation (LTP) and LTD, that store memories. Consistent with this proposal, the transitions from early to late phases of LTP and LTD require new protein synthesis independent of transcription (Cracco et al. 2005, Huber et al. 2000, Kang & Schuman 1996). Furthermore, these modifications can be maintained by new translation in isolated dendrites, implicating pre-existing dendritically localized mRNA. Thus, glutamate release at individual synapses appears to stimulate local protein synthesis to maintain long-lasting synaptic change.

Translational Control at Glutamatergic Synapses

An understanding of the molecular mechanisms by which synaptic activity regulates local protein synthesis is beginning to emerge. Two types of postsynaptic glutamate receptors have been implicated: the calcium-permeable N-methyl-D-aspartate ionotropic receptors (NMDARs) and the Gq-coupled (group 1) mGluR1 and mGluR5. The mGluRs have complementary expression patterns: mGluR5 expression is highest in the forebrain and mGluR1 expression is highest in the cerebellum (Shigemoto

microRNA (miRNA): a short RNA molecule ~22 nucleotides long that binds to complementary sequences on target mRNAs, usually resulting in translational repression or target degradation

Glutamate: the major excitatory neurotransmitter in the nervous system

***N*-methyl-D-aspartate ionotropic glutamate receptor (NMDAR):** well known for triggering synaptic plasticity at the synapse

ERK: extracellular signal–regulated kinase

mTOR: mammalian target of rapamycin

et al. 1993). NMDARs are also widely expressed throughout the brain and stimulate the release of brain-derived neurotrophic factor, a ligand for TrkB receptors, which can contribute to synaptic protein synthesis (Kang & Schuman 1996, Schratt 2009).

Of particular interest in the context of FX is protein synthesis stimulated by activation of Gp1 mGluRs. Weiler & Greenough (1993) provided the first evidence that Gp1 mGluR agonists stimulate protein synthesis in biochemical preparations enriched for cortical synapses. It is now understood that Gp1 mGluRs couple to the synaptic translation machinery at synapses in many parts of the brain and that many functional consequences of Gp1 mGluR activation depend on new protein synthesis (see Krueger & Bear 2011 for a review).

Two intracellular signaling cascades have been proposed to couple mGluRs and other synaptic receptors to the translational machinery: (*a*) the mammalian target of rapamycin (mTOR) pathway and (*b*) the extracellular signal–regulated kinase (ERK) pathway. Both mTOR and ERK pathways can stimulate cap-dependent translation by regulating components of initiation. Initiation is the step during which the small ribosome subunit is recruited to the 5′ end of mRNA and scans toward the start codon to assemble into the complete ribosome (see Gebauer & Hentze 2004 for a review).

One key regulatory step in initiation is the recognition of the 5′ mRNA cap by eIF4E (**Supplemental Figure 1**. Follow the **Supplemental Material link** from the Annual Reviews home page at **http://www.annualreviews.org**), which leads to assembly of the eIF4F complex and recruitment of the small ribosomal subunit (Richter & Sonenberg 2005). A family of 4E-BPs inhibits this process by binding to eIF4E. This inhibition is relieved by phosphorylation of 4E-BPs by both mTOR and ERK or, in postnatal mammalian brain, by deamination (Bidinosti et al. 2010). The mTOR pathway can also facilitate initiation through phosphorylation of p70 ribosomal protein S6 kinases (S6Ks), leading to ribosomal protein S6 phosphorylation and phosphorylation of eIF4B. Similarly, the ERK pathway can facilitate initiation by phosphorylation of S6 and eIF4B through activation of p90 ribosomal protein S6 kinases (RSKs); however, it can also lead to phosphorylation of eIF4E through activation of MNK. Phosphorylation of eIF4B stimulates the eIF4F complex activity by potentiating the RNA-helicase activity of eIF4A. Phosphorylation of eIF4E generally decreases eIF4E affinity for the cap, however, and may function to reduce overall translation rates. Some researchers have hypothesized that this mechanism may allow for increases in the translation of a specific subset of mRNAs (Costa-Mattioli et al. 2009). This is likely to be one mechanism whereby specific pools of mRNAs are selected for translation (a topic we discuss below).

Another major regulatory step in initiation is the formation of the ternary complex (eIF2, Met-tRNA, and GTP) required to complete the 43S ribosomal complex. Phosphorylation of eIF2 inhibits the GDP/GTP exchange required to reconstitute a functional ternary complex, causing a decrease in general translation and an impairment in some forms of late-phase LTP and long-term memory (Costa-Mattioli et al. 2009). Curiously, however, eIF2 phosphorylation can also stimulate translation of a subset of mRNAs that contain short upstream open reading frames. Initiation can also be regulated at the mRNA 3′ end by CPEB (cytoplasmic polyadenylation element–binding protein), an RNA-binding protein that inhibits poly(A) tail addition and formation of the eIF4F complex. CPEB, similar to FMRP, is commonly found to repress the translation of dendritically transported mRNAs (Costa-Mattioli et al. 2009). How synaptic activity couples to eIF2 phosphorylation or CPEB regulation has yet to be fully explained.

Although initiation is usually the rate-limiting step in translation, in some instances excitatory synaptic stimulation can regulate the elongation phase of translation. Both mGluR5 and NMDAR, via activation of calcium/calmodulin-dependent eEF2 kinase, can increase phosphorylation of eEF2. Phospho-eEF2 stalls general elongation but allows translation of a subset of mRNAs (Scheetz et al. 2000), including those that encode the proteins Arc and MAP-1B (Park et al. 2008). Arc and MAP-1B are well-characterized targets of translation repression by FMRP. Below, we return to the question of how mGluRs couple specifically to FMRP-regulated protein synthesis.

THE MGLUR THEORY OF FRAGILE X

As mentioned above, it is now appreciated that Gp1 mGluRs couple to the translational machinery at many synapses in the brain. The mGluR theory of FX posits that many psychiatric and neurological aspects of FX are due to exaggerated downstream consequences of mGluR1/5 activation (Bear et al. 2004). The origins of this theory have been reviewed recently elsewhere (Krueger & Bear 2011). Briefly, Huber et al. (2000) showed that one protein synthesis-dependent consequence of Gp1 mGluR activation in the CA1 region of the hippocampus is a form of LTD, later shown to be expressed by internalization of AMPA-type glutamate receptors (Snyder et al. 2001). The early finding that FMRP can be synthesized in response to mGluR activation (Weiler et al. 1997) led to the study of LTD in the *Fmr1*-KO mouse (Huber et al. 2002). The prediction at that time was that absence of FMRP would result in impaired LTD, given the hypothesis that FMRP was one of the proteins synthesized to stabilize LTD. Instead, LTD was found to be exaggerated, suggesting that FMRP serves as a brake on mGluR-stimulated protein synthesis. As reviewed above, strong consensus now indicates that FMRP is a translational suppressor in vivo. The mGluR theory arose from the recognition that exaggerated consequences of mGluR activation at synapses throughout the nervous system could potentially provide a thread to connect seemingly unrelated FX phenotypes.

In the intervening decade, researchers have accumulated evidence that strongly supports the mGluR theory. The assumption that FMRP regulates varied responses triggered by mGluR-stimulated protein synthesis has been well validated (Auerbach & Bear 2010, Chuang et al. 2005, Dolen et al. 2007, Hou et al. 2006, Huber et al. 2002, Koekkoek et al. 2005, Lu et al. 2004, Muddashetty et al. 2007, Nosyreva & Huber 2006, Park et al. 2008, Ronesi & Huber 2008, Todd et al. 2003, Waung & Huber 2009, Westmark & Malter 2007, Zalfa et al. 2007, Zhang & Alger 2010, Zhao et al. 2005). Moreover, as summarized in **Table 1** and reviewed in greater detail elsewhere (Dolen et al. 2010, Krueger & Bear 2011), the important prediction that FX phenotypes can be corrected by reducing mGluR5 activity has been confirmed using both pharmacological and genetic approaches in evolutionarily distant animal models (Aschrafi et al. 2005; Bolduc et al. 2008; Chang et al. 2008; Choi et al. 2010, 2011; Chuang et al. 2005; de Vrij et al. 2008; Dolen et al. 2007; Hays et al. 2011; Koekkoek et al. 2005; Levenga et al. 2011; Liu et al. 2011; Malter et al. 2010; McBride et al. 2005; Meredith et al. 2011; Min et al. 2009; Nakamoto et al. 2007; Osterweil et al. 2010; Pan & Broadie 2007; Pan et al. 2008; Repicky & Broadie 2009; Su et al. 2011; Suvrathan et al. 2010; Tauber et al. 2011; Thomas et al. 2011, 2012; Tucker et al. 2006; Veloz et al. 2012; Yan et al. 2005). A way of conceptualizing the constellation of findings is that FX is a disorder of excess—an excess that develops as Gp1 mGluR-dependent signaling cascades operate unchecked and that can be corrected by intervening at the first step in the cascade, the mGluR. The evolutionarily conserved relationship of Gp1 mGluRs and FMRP has provided a strong rationale for studies in human FX (see review by Hagerman et al. 2012).

However, given when and where FMRP is normally expressed during development, it is clear that FX is a result of more than just altered mGluR signaling. Furthermore, because FMRP regulates signaling initiated by other neuronal receptors (Lee et al. 2011, Volk et al. 2007), reduction of Gp1 mGluR signaling seems unlikely to have a therapeutic benefit across all cognitive and somatic

Table 1 Phenotypes corrected by mGluR1/5 inhibition in animal models of FX*

Animal model	Fragile X phenotype (versus WT)	mGluR1/5 manipulation	Reference(s)
Mouse	Exaggerated mGluR-LTD	*Grm5*$^{+/-}$ cross Lithium	Dolen et al. 2007 Choi et al. 2011
Mouse	Increased AMPA receptor internalization	MPEP	Nakamoto et al. 2007
Mouse	Impaired spontaneous EPSCs in juvenile hippocampus	MPEP	Meredith et al. 2011
Mouse	Increased protein synthesis	*Grm5*$^{+/-}$ cross MPEP (mGluR5 NAM) Lithium	Dolen et al. 2007 Osterweil et al. 2010 Liu et al. 2011
Mouse	Decreased number of mRNA granules in whole brain	MPEP	Aschrafi et al. 2005
Mouse	Increased glycogen synthase kinase-3 activity	MPEP, Lithium	Min et al. 2009 Gross et al. 2010
Mouse	Increased beta amyloid	MPEP	Malter et al. 2010
Mouse	Increased dendritic spine/filopodia density	*Grm5*$^{+/-}$ cross Fenobam (mGluR5 NAM) MPEP AFQ056 (mGluR5 NAM)	Dolen et al. 2007 de Vrij et al. 2008 Su et al. 2011 Levenga et al. 2011
Mouse	Altered visual cortical plasticity	*Grm5*$^{+/-}$ cross	Dolen et al. 2007
Mouse	Exaggerated inhibitory avoidance extinction	*Grm5*$^{+/-}$ cross	Dolen et al. 2007
Mouse	Impaired eyelid conditioning	MPEP	Koekkoek et al. 2005
Mouse	Decreased initial performance on rotorod	MPEP	Thomas et al. 2012
Mouse	Associative motor-learning deficit	Fenobam	Veloz et al. 2012
Mouse	Increased audiogenic seizure	*Grm5*$^{+/-}$ cross MPEP Lithium JNJ16259685 (mGluR1 NAM)	Dolen et al. 2007 Thomas et al. 2012, Yan et al. 2005 Min et al. 2009 Thomas et al. 2012
Mouse	Prolonged epileptiform discharges in hippocampus	MPEP	Chuang et al. 2005
Mouse	Increased persistent activity states in neocortex	MPEP, *Grm5*$^{+/-}$ cross	Hays et al. 2011
Mouse	Increased open-field activity	MPEP Lithium Grm1 +/− cross	Min et al. 2009, Yan et al. 2005 Thomas et al. 2011
Mouse	Defective prepulse inhibition of acoustic startle	MPEP AFQ056	de Vrij et al. 2008 Levenga et al. 2011
Mouse	Abnormal social interaction with unfamiliar mouse	*Grm5*$^{+/-}$ cross	Thomas et al. 2011
Mouse	Increased marble burying (repetitive behavior)	JNJ16259685, MPEP	Thomas et al. 2012
Mouse	Impaired presynaptic function in amygdala	MPEP	Suvrathan et al. 2010

(Continued)

Table 1 (*Continued*)

Animal model	Fragile X phenotype (versus WT)	mGluR1/5 manipulation	Reference(s)
Mouse	Avoidance behavior deficit	Fenobam	Veloz et al. 2012
Mouse	Pubertal increase in body weight	*Grm5*$^{+/-}$ cross	Dolen et al. 2007
Zebrafish	Abnormal axon branching	MPEP	Tucker et al. 2006
Zebrafish	Craniofacial abnormalities	MPEP	Tucker et al. 2006
Zebrafish	Reduced number of trigeminal neurons	MPEP	Tucker et al. 2006
Fly	Increased synaptic transmission	*dmGluR-A* null cross	Repicky & Broadie 2009
Fly	Increased NMJ axon arborization	*dmGluR-A* null cross, MPEP	Pan et al. 2008
Fly	Increased NMJ presynaptic vesicle density	*dmGluR-A* null cross	Pan et al. 2008
Fly	Mushroom-body structural abnormalities	MPEP, lithium	McBride et al. 2005, Pan et al. 2008
Fly	Age-dependent cognitive decline	MPEP, lithium	Choi et al. 2010
Fly	Altered regulation of ionotropic glutamate receptor subtypes	*dmGluR-A* null cross	Pan & Broadie 2007
Fly	Decreased courtship/social learning	MPEP, lithium	McBride et al. 2005, Tauber et al. 2011
Fly	Decreased olfactory memory	MPEP	Bolduc et al. 2008
Fly	Increased embryonic lethality on glutamate enriched diet	MPEP	Chang et al. 2008
Fly	Increased roll-over (righting) time	*dmGluR-A* null cross	Pan et al. 2008

*Abbreviations: AMPA, α-amino-3-hydroxy-5-methyl-4-isoxazolepropionic acid; EPSC, excitatory postsynaptic currents; FX, fragile X; dmGluR, drosophila mGluR; LTD, long-term depression; mGluR, metabotropic glutamate receptor; MPEP, 2-methyl-6-(phenylethynyl)-pyridine; mRNA, messenger RNA; NAM, negative allosteric modulator; NMJ, neuromuscular junction; WT, wild type.

domains of what is a complex and pervasive neurodevelopmental disorder. Accordingly, efforts are under way to identify the aspects of FX pathophysiology that may not be related to mGluR function or that may arise before birth (Desai et al. 2006; Dolen et al. 2007; Suvrathan et al. 2010; Tauber et al. 2011; Thomas et al. 2011, 2012). Such knowledge is important in guiding therapy, both by defining the limits of what to expect from mGluR-based approaches and by suggesting additional therapeutic targets (see Fragile X Mental Retardation Protein and Neurogenesis, sidebar below).

HOW MGLUR5 COUPLES TO FMRP-REGULATED PROTEIN SYNTHESIS

Although the mGluR theory of FX has been well validated, it remains poorly understood how mGluR5 couples to protein synthesis and how this process is altered in the absence of FMRP to disrupt synaptic function. In addition to providing additional insight into FX pathophysiology and suggesting new therapeutic targets, investigating this question promises to shed light on long-standing but unresolved questions concerning how protein synthesis stabilizes LTD, LTP, and memory.

mGluR5 Signaling Pathways

Gp1 mGluRs were originally discovered on the basis of their ability to stimulate phospholipase C, the hydrolysis of phosphoinositides (PI), and the release of calcium from intracellular stores (Dudek et al. 1989, Nicoletti et al. 1988, Pin et al. 1992, Schoepp & Conn 1993). One phosphoinositide product, DAG (diacyl-glycerol), subsequently activates protein kinase C and

FRAGILE X MENTAL RETARDATION PROTEIN AND NEUROGENESIS

The metabotropic glutamate receptor (mGluR) theory has contributed to a paradigm shift in the way fragile X (FX) and other genetic disorders of brain development are viewed medically. The data now indicate that a constellation of seemingly unrelated and complex symptoms could be a consequence of altered cerebral metabolism—synaptic protein synthesis in the case of FX—that can be substantially improved by therapies begun after symptom onset, possibly even in adulthood. It is important to recognize, however, that *FMR1* is normally expressed early in embryogenesis (Devys et al. 1993, Hinds et al. 1993) and that full-mutation FX patients fail to express FMRP very early in gestation (Willemsen et al. 1996). FMRP is required for proper prenatal neurogenesis and neuronal differentiation (Callan et al. 2010, Castren et al. 2005, Eadie et al. 2009, Tervonen et al. 2009). Thus, the FX brain is different at birth.

However, neurogenesis occurs throughout life in the dentate gyrus of the hippocampus. Remarkably, hippocampus-dependent memory impairments have been rescued by re-expression of FMRP in adult neural stem cells in an *Fmr1* knockout (KO) mouse line (Guo et al. 2011). Moreover, these defects can be reversed in adults by treatment with an inhibitor of glycogen synthase kinase 3 (GSK3) (Guo et al. 2012). GSK3 activity is elevated in the *Fmr1* KO downstream of mGluR5 (Yuskaitis et al. 2009), suggesting that the mGluR theory may also be relevant to this aspect of FX pathophysiology.

protein kinase D (Krueger et al. 2010). This canonical signaling cascade does not appear to be critically involved in FMRP-regulated protein synthesis, however, as mGluR-LTD is insensitive to Ca^{2+} chelators and inhibitors of phospholipase C (Fitzjohn et al. 2001, Gallagher et al. 2004, Huber et al. 2001). Rather, signaling via the mTOR and ERK pathways is crucial for LTD and mGluR coupling to protein synthesis (**Figure 5**).

To activate the mTOR pathway, mGluR5 couples to Homer, a postsynaptic-density scaffolding protein that recruits the GTPase, PIKE-L, forming an mGluR-Homer-PIKE complex (Ahn & Ye 2005). PIKE directly enhances the lipid kinase activity of PI3K (phosphoinositide 3-kinase), leading to the phosphorylation of PIP2 (phosphatidylinositol-4,5-bisphosphate). PIP3 together with PDK (phosphoinositide-dependent kinase) activates the serine/threonine kinase Akt. Akt, in turn, can activate mTOR by direct phosphorylation and indirectly through inhibition of the tumor-suppressor complex, composed of TSC1 and TSC2 (for review see Han & Sahin 2011). The TSC1/2 complex has GAP activity against the small GTP-binding protein RHEB. When free of TSC1/2, RHEB activates mTOR within a rapamycin-sensitive protein complex called mTORC1. Activation of mTORC1 is best known for stimulating cap-dependent translation through its main effector proteins, namely the 4E-BPs and S6Ks (see above section).

The ERK cascade, as with all mitogen-activated protein kinase (MAPK) cascades, typically involves sequential activation of a small GTPase (Ras), a MAPK kinase kinase (Raf), and a MAPK kinase (MEK), to activate ERK. How mGluR5 couples to Ras or other downstream components of the ERK cascade is not fully understood. ERK activation is required for both mGluR LTD (Gallagher et al. 2004) and mGluR5 activation of FMRP-regulated mRNA translation (Osterweil et al. 2010).

Recent work on related G-protein-coupled receptors (GPCRs) suggest that mGluR5 may couple to the ERK cascade through β-arrestins. β-arrestins are scaffold proteins that are typically recruited to the receptor tails following serine/threonine phosphorylation by GPCR kinases (GRKs)—a response that is best understood for terminating the receptor's G-protein signaling (Ferguson 2001, Premont et al. 1995). However, more recent work has shown that β-arrestin binding to GPCRs may also serve to regulate mRNA translation by providing a scaffold for Raf, MEK, ERK, and MNK (DeWire et al. 2008).

Interestingly, FMRP appears to be a component of the signaling pathway that couples mGluR5 activation to protein synthesis. As mentioned above, dephosphorylation shifts FMRP from stalled to active polyribosomes (Ceman et al. 2003, Muddashetty et al. 2011), motivating a few groups to identify FMRP

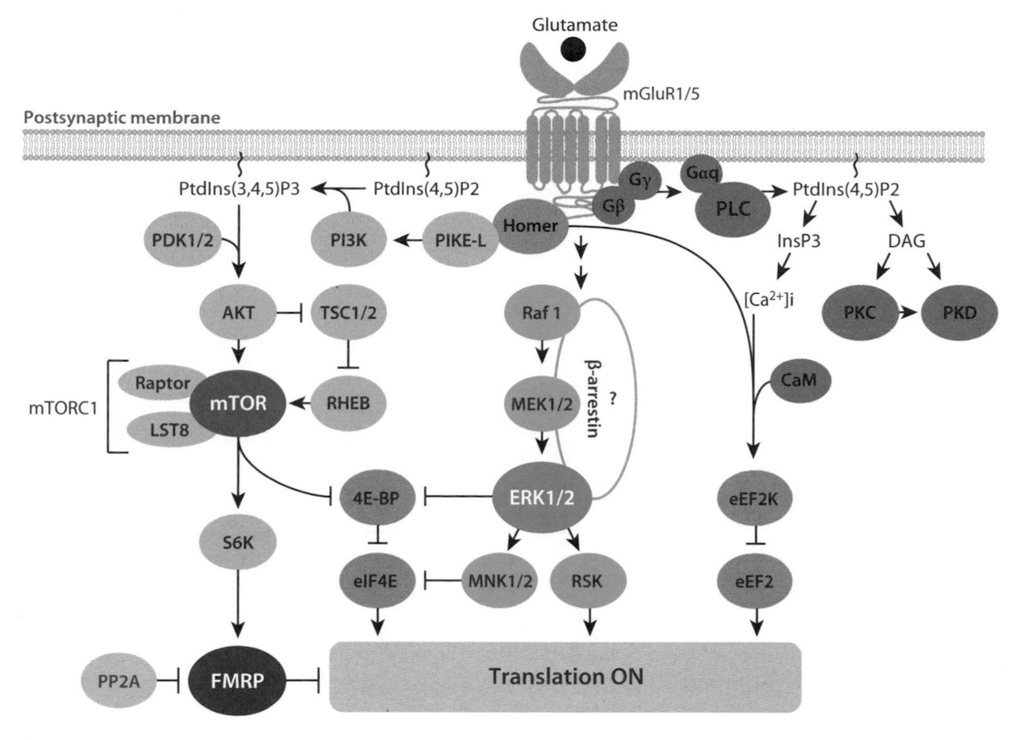

Figure 5

mGluR1/5 signaling pathways relevant to protein synthesis. Glutamate binding to Gp1 mGluRs activates three main pathways that couple the receptors to translational regulation: (*a*) the PLC/calcium-calmodulin pathway (*orange ovals*), (*b*) the mTOR pathway (*blue ovals*), and (*c*) the ERK pathway (*green ovals*). See main text for details. Key translational regulatory components implicated in these pathways are shown in brown. mGluR1/5 may also inhibit FMRP (*red oval*) function to regulate translation through a fourth pathway requiring stimulation of PP2A (*yellow oval*). Question marks indicate undetermined associations. Arrows indicate a positive consequence on downstream components; perpendicular lines indicate an inhibitory consequence. Abbreviations: $[Ca^{2+}]$i, calcium release from intracellular stores; CaM, calmodulin; ERK, extracellular signal–regulated kinase; FMRP, fragile X mental retardation protein; (Gαq, Gβ, Gγ), heterotrimeric G proteins; InsP3, inositol-1,4,5-triphosphate (InsP3); mGluR, metabotropic glutamate receptor; mTOR, mammalian target of rapamycin; PtdIns, phosphoinositides; PLC, phospholipase C; PP2A, protein phosphatase 2A; Raptor, regulatory-associated protein of mTOR.

phosphatases and kinases that lie downstream of mGluR5. S6K1 can phosphorylate FMRP on a conserved serine residue required for mRNA binding and PP2A can remove this phosphorylation (Mao et al. 2005, Narayanan et al. 2007, Narayanan et al. 2008, Wang et al. 2010a). Both enzymes are activated in response to mGluR5 stimulation, and one model proposes that activation of PP2A rapidly dephosphorylates FMRP to enable translation, followed by delayed translation suppression caused by S6K phosphorylation of FMRP downstream of mTOR (Santoro et al. 2011).

Regulation of mGluR5-dependent protein synthesis exclusively via FMRP is unlikely, however. In the *Fmr1*-KO mouse, which lacks FMRP, the excessive basal protein synthesis (and many other phenotypes) are rescued by inhibiting mGluR5 (**Figure 3**). If loss of FMRP completely uncoupled mGluR5 from protein synthesis regulation, there would be no effect of inhibiting mGluR5 on protein synthesis in the *Fmr1* KO. Therefore, mGluR5 stimulation of protein synthesis must occur via additional pathway(s) that are independent of FMRP (**Figure 6**).

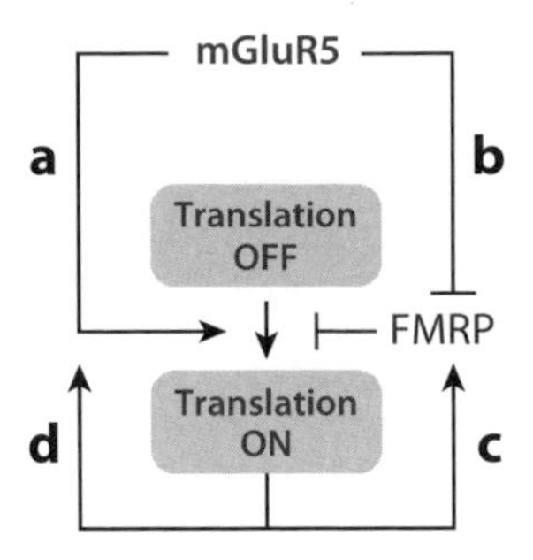

Figure 6

Schema for coupling mGluR5 to FMRP-regulated protein synthesis. Several lines of evidence suggest that mGluR5 couples to FMRP-regulated protein synthesis through multiple pathways. (*a*) Activation of mGluR5 directly stimulates mRNA translation through the ERK signaling pathway. (*b*) Additionally, activation of mGluR5 can trigger dephosphorylation of FMRP by PP2A, which derepresses translation. (*c*) FMRP is rapidly synthesized in response to mGluR5 activation, providing a negative-feedback loop to turn off protein synthesis. (*d*) Several FMRP target proteins are known components of mGluR5 signaling pathways, suggesting that positive feedback may occur, particularly in the context of FX. Abbreviations: ERK, extracellular signal–regulated kinase; FMRP, fragile X mental retardation protein; FX, fragile X; mGluR, metabotropic glutamate receptor; PP2A, protein phosphatase 2A.

Altered Signaling in the Absence of FMRP

Because both ERK and mTOR pathways can be activated by mGluR5 (Antion et al. 2008, Banko et al. 2006, Ferraguti et al. 1999, Gallagher et al. 2004, Hou et al. 2006, Ronesi & Huber 2008, Sharma et al. 2010) and both regulate protein synthesis, these two pathways have been most studied in the context of FX. One hypothesis has been that alterations in mGluR5 signaling through ERK or mTOR may be responsible for the excessive protein synthesis and exaggerated LTD in the *Fmr1*-KO mice. Consistent with the notion of altered signaling, mGluR5 receptors are less tightly associated with synaptic plasma membrane and Homer (Giuffrida et al. 2005), and they are unable to activate the mTOR pathway in *Fmr1*-KO mice (Ronesi & Huber 2008). Other reports suggest a basal increase in ERK activity (Hou et al. 2006), an aberrant mGluR-induced inactivation of ERK (Kim et al. 2008), and a basal increase in AKT/mTOR signaling that occludes further activation by mGluR stimulation (Gross et al. 2010, Sharma et al. 2010).

TSC: tuberous sclerosis complex

Although mGluR5 signaling is evidently altered in FX, it has not been shown that these alterations are responsible for the excessive protein synthesis that is believed to be the core pathogenic mechanism in FX. Indeed, one recent study examined ERK and mTOR pathways under the same experimental conditions that reveal excessive protein synthesis and exaggerated LTD and found no evidence for altered signaling (Osterweil et al. 2010). Protein synthesis rates could be restored to WT levels by acute partial inhibition of mGluR5 or ERK activity (but not mTOR), however, indicating that increased protein synthesis in FX occurs downstream of constitutive mGluR5/ERK activity (Osterweil et al. 2010). These data suggest that the excessive basal protein synthesis in *Fmr1*-KO mice is due to hypersensitivity of the translation machinery to normal mGluR signals (ERK, in particular), rather than to hyperactivity of the mGluR signaling pathways (**Figure 6**). If this model is correct, altered intracellular signaling in FX should be viewed as a consequence, rather than a cause, of the increased protein synthesis in this disease.

ERK and mTOR May Regulate Separate Pools of mRNA

Disentangling the contributions of ERK and mTOR signaling pathways to the protein synthesis required for mGluR-LTD has been difficult, but recent studies of a mouse model of tuberous sclerosis complex (TSC) have been illuminating. TSC is another single-gene disorder characterized by intellectual disability, seizures, and autism and is caused by heterozygous loss of function of either the *TSC1* or *TSC2* gene. The protein products of these genes form the TSC1/2 complex that normally represses mTOR signaling via inhibition of RHEB, as discussed above. Thus, TSC is caused by excessive mTOR signaling. If the excessive protein synthesis in FX were driven by the mTOR signaling pathway, one would expect *TSC* mutations to have similar effects on mGluR-dependent LTD. Very recently, three groups examined this hypothesis in the CA1 region of hippocampus using different but complementary animal models of TSC (Auerbach et al. 2011, Bateup et al. 2011, Chevere-Torres et al. 2012). The surprising result is that mouse *Tsc* mutants with excessive mTOR activity show impaired mGluR-LTD and basal protein synthesis, the exact opposite of what is observed in the *Fmr1*-KO. Moreover, synaptic, biochemical, and cognitive deficits in the *Tsc2*$^{+/-}$ mouse model were corrected by treatment with a positive allosteric modulator of mGluR5 as well as by introducing the FX mutation into the *Tsc2*$^{+/-}$ animals (Auerbach et al. 2011). These findings indicate that elevated mTOR signaling is not a proximal cause of FX pathophysiology.

The recent findings in *Tsc* mutants suggest that excessive mTOR signaling suppresses the synthesis of proteins required for LTD (Auerbach et al. 2011). One simple hypothesis is that elevated mTOR causes hyperphosphorylation of FMRP via activation of S6K1 (**Figure 5**), resulting in translational suppression of the FMRP-target mRNAs that gate LTD. However, this explanation is not easily reconciled with the observation that excess LTD in the *Fmr1*-KO mice (lacking FMRP) is rescued by crossing them with the *Tsc2*$^{+/-}$ mice. An alternative model is that mTOR stimulates translation of a pool of mRNA (call it Pool II) that competes with a second, ERK- and FMRP-regulated pool (Pool I) for access to the translation machinery (**Figure 7**) (see also Bear et al. 2004).

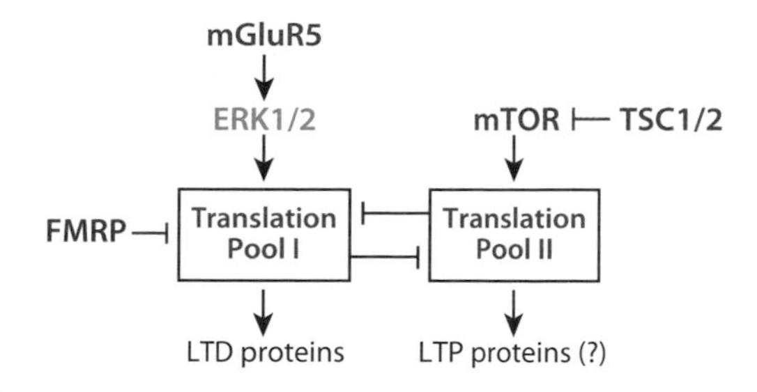

Figure 7

The two-pool hypothesis. A model to account for the opposing mGluR5 responses detected in the *Tsc2*$^{+/-}$ and *Fmr1*-KO mice proposes that activation of mGluR5 stimulates the translation of a pool of mRNAs (Pool I), through ERK- and FMRP-dependent pathways, that are in competition for the translational machinery with a second pool of mRNAs (Pool II) that are regulated by mTOR activation. Current data suggest that mRNAs translated in Pool I may comprise the proteins required to stabilize LTD (LTD proteins), whereas mRNAs within Pool II stabilize LTP (LTP proteins). Consistent with this proposal, derepression of Pool I in FX causes excessive LTD, whereas derepression of Pool II in TSC causes enhanced LTP. Arrows indicate a positive consequence on downstream components; perpendicular lines indicate an inhibitory consequence. Abbreviations: ERK, extracellular signal–regulated kinase; FMRP, fragile X mental retardation protein; FX, fragile X; KO, knockout; LTD, long-term synaptic depression; LTP, long-term synaptic potentiation; mGluR, metabotropic glutamate receptor; mTOR, mammalian target of rapamycin; TSC, tuberous sclerosis complex.

As mentioned above, there is considerable precedent for a "push-pull" regulation of translation by different pools of mRNA. Inhibition of what is often called general translation enables certain types of specific translation of mRNAs that can include FMRP targets. Although this can occur via multiple mechanisms, to illustrate consider regulation of translation via the elongation factor eEF2. Phosphorylation of eEF2 by eEF2 kinase occurs in response to mGluR5 activation and promotes translation of specific transcripts in Pool I (including those for the FMRP targets Arc and MAP1b) by inhibiting translation of Pool II transcripts (Park et al. 2008). Conversely, activation of the mTOR pathway causes inhibitory phosphorylation of the eEF2 kinase (via S6 kinase), which stimulates translation of Pool II and thereby inhibits translation of Pool I (Costa-Mattioli et al. 2009, Herbert & Proud 2007).

Two distinct effects on protein synthesis–dependent synaptic plasticity have been reported in *Tsc2* mutants with increased mTOR activity: (*a*) The persistence of late-phase LTP is increased, presumably by increasing translation of the proteins required to make synapses stronger (Ehninger et al. 2008), and (*b*) mGluR-LTD is inhibited by eliminating the protein synthesis required to make synapses weaker (Auerbach et al. 2011, Bateup et al. 2011). It is tempting to speculate that Pool II includes LTP proteins regulated by mTOR signaling and that Pool I comprises LTD proteins regulated by mGluR5, ERK, and FMRP. According to this

Allosteric modulator: a drug that modulates the function of a receptor by binding to a site that is different from the orthosteric ligand binding site

idea, derepression of Pool I in FX causes excessive LTD, whereas derepression of Pool II in TSC causes enhanced LTP.

Such simple models are useful if they generate hypotheses and stimulate experiments. If this conjecture is correct, for example, proteomic comparison of *Tsc2*$^{+/-}$ and *Fmr1*-KO hippocampus may be a fruitful path to discover the elusive plasticity gating proteins. Of course, the regulation of plasticity-related protein synthesis is unlikely to be this simple. For example, the model suggests that LTP may be impaired in FX owing to repression of Pool II translation. Although there are some reports of deficient LTP in the hippocampus of *Fmr1*-KO mice (Hu et al. 2008, Lauterborn et al. 2007, Lee et al. 2011, Meredith & Mansvelder 2010, Shang et al. 2009), many have found no difference in LTP threshold or long-term maintenance (Auerbach & Bear 2010, Godfraind et al. 1996, Zhang et al. 2009). Another element of the model that requires further clarification is how activity couples to the mTOR pathway. A recent study showed that inhibition of the mTOR pathway derepresses translation of the Pool I mRNA Kv4.2, but that this occurs via dephosphorylation of FMRP downstream of NMDA receptors instead of mGluRs (Lee et al. 2011). Other studies have shown mTOR is activated by Gp1 mGluR activation and is required for LTD (Hou & Klann 2004; but see Auerbach et al. 2011). One thing is certain: Intracellular signaling is complicated. Clarity will require that experiments be performed on the same synapses, prepared in the same way, and from animals that are at the same age.

These caveats notwithstanding, under identical experimental conditions, littermate mice carrying the *Fmr1* mutation, the *Tsc2* mutation, and both mutations show augmented, impaired, and WT levels, respectively, of mGluR-dependent LTD and protein synthesis (Auerbach et al. 2011). Of particular interest, both single mutants showed deficits in context-discrimination memory that were erased in the double mutants. These findings support the ideas that proper synaptic function requires an optimal level of mGluR-regulated protein synthesis and that deviations in either direction can yield similar behavioral disturbances that can include cognitive impairment (**Figure 8**).

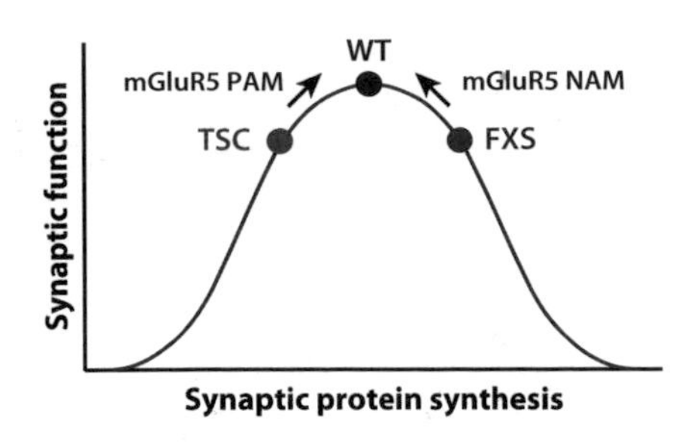

Figure 8

Mutations causing monogenic autism define an axis of synaptic pathophysiology. Recent data suggest that proper synaptic function requires an optimal level of mGluR-regulated protein synthesis and that deviations in either direction can produce similar impairments in cognitive function (Auerbach et al. 2011). Two types of monogenic autism, TSC and FXS, lie on opposite ends of this spectrum and, correspondingly, show reduced and increased protein synthesis rates, and respond to opposite alterations in mGluR5 activation (PAM and NAM, respectively). Abbreviations: FXS, fragile X syndrome; mGluR, metabotropic glutamate receptor; NAM, negative allosteric modulator; PAM, positive allosteric modulator; TSC, tuberous sclerosis complex; WT, wild type.

PATHOGENIC PROTEINS

Evidence suggests that synaptically controlled protein synthesis must be maintained in a normal range to ensure proper synaptic (and cognitive) function, and that important aspects of FX are a consequence of altered protein expression. Several-hundred mRNAs have been implicated as targets of FMRP (Darnell et al. 2011). Among these are the proteins that disrupt synaptic function in FX, and it is of great interest to identify those that are pathogenic.

Given the reversal of FX phenotypes by reducing mGluR1/5 stimulation, one way to prioritize the list of pathogenic proteins may be to determine which of the identified direct targets show (*a*) altered protein expression profiles in the *Fmr1*-KO mice, (*b*) translation under normal circumstances in response to mGluR1/5 activation, and (*c*) a contribution to the functional responses to activated mGluR5, e.g., mGluR LTD. For example, the plasticity protein Arc is an identified FMRP mRNA target, upregulated in the *Fmr1*-KO mouse and synthesized at the synapse in response to mGluR5 activation (Auerbach et al. 2011, Park et al. 2008, Waung et al. 2008). Similarly, the amyloid precursor protein (APP) and the brain-specific tyrosine phosphatase STEP are FMRP mRNA targets, synthesized in response to mGluR5 (Westmark & Malter 2007, Westmark et al. 2009, Zhang et al. 2008), and both APP cleavage products and STEP protein are overexpressed in the *Fmr1*-KO mouse (Goebel-Goody et al. 2011). Arc and STEP are both considered to be LTD proteins, involved in regulating AMPA-receptor membrane trafficking. The cleavage product of APP, β-amyloid, also triggers AMPA receptor internalization and LTD (Hsieh et al. 2006). Of particular interest, removing a single allele of APP in the *Fmr1* KO partially or completely corrects audiogenic seizure, anxiety, and mGluR LTD phenotypes (Westmark et al. 2011). Another FMRP target of interest is metalloproteinase 9 (MMP-9), also overexpressed in the *Fmr1*-KO downstream of mGluR5. MMP-9 is a secreted extracellular endopeptidase that, similar to Gp1 mGluR agonists (Vanderklish & Edelman 2002), elongates and thins dendritic spines (Michaluk et al. 2011). Treatment with the tetracycline analogue minocycline (among other actions) inhibits MMP-9 and corrects the spine phenotype in the *Fmr1*-KO mouse (Bilousova et al. 2009). Moreover, both minocycline and genetic reduction of MMP rescue circuit disruptions in the *dfmr1*-null fly model of FX (Siller & Broadie 2011).

Additional downstream consequences of altered synaptic protein expression may be dysregulation of the signaling components that normally control protein synthesis. For example, both the catalytic subunit of PI3K (p110b) and the PI3K enhancer PIKE-L are FMRP mRNA targets, translated in response to mGluR activation and elevated in the *Fmr1*-KO mice (Gross et al. 2010, Sharma et al. 2010). Indeed, 62% of the genes composing the mGluR5 postsynaptic proteome (Croning et al. 2009) are direct FMRP targets (Darnell et al. 2011). These findings fit with data showing abnormal mGluR5 signaling in FX.

The list of pathogenic proteins is sure to expand as additional research is conducted. Particularly interesting are those that can be targeted with small-molecule therapeutics. In addition to those mentioned above, interesting prospects include p21-activated kinase (Hayashi et al. 2007) and glycogen synthase kinase-3 (Mines & Jope 2011).

The overlap of FMRP targets and genes implicated in autism is intriguing. One-quarter of the SFARI database of autism risk genes (**http://gene.sfari.org**) are FMRP targets. Among these are *NLGN3*, *NRXN1*, *SHANK3*, *PTEN*, *TSC2*, and *NF1*, all of which encode proteins that control synaptic structure or protein synthesis. Rare mutations of these genes all cause autism (Zoghbi & Bear 2012). These findings reinforce the belief that the study of FX, the most common known genetic cause of autism, provides insight into the molecular pathophysiology of autism and associated intellectual disability of unknown etiology. The hope is that treatments developed for FX will be useful for treating autism of diverging etiologies, with the important caveat that it will be critical to understand where an individual is on the spectrum of altered synaptic protein synthesis to devise an appropriate therapy (Auerbach et al. 2011).

CONCLUDING REMARKS

Interest in FX has burgeoned in recent years. It is now appreciated to be a disease of the synapse, amenable to potentially disease-altering therapeutic interventions and relevant to understanding the pathophysiology of autism and intellectual disability more broadly. We appear to be close to fulfilling the promise of molecular medicine in FX (Krueger & Bear 2011). We have gone from identification of the gene to the discovery and validation of novel therapeutic targets, and there is good reason for optimism that new therapies will emerge that can greatly enhance the quality of life for affected individuals and their families (see **Figure 1**).

This field has grown so large that it is impossible to cover adequately all the developments given the space limitations of this review. We have chosen to focus on synaptic control of protein synthesis because it appears to be proximal to the biology of FMRP and the pathogenesis of the disease in multiple animal models. In addition to targeting synaptic protein synthesis, other approaches also show promise, for example, changing the balance of excitation to inhibition by enhancing GABA signaling (Hampson et al. 2011, Rooms & Kooy 2011). Whether different approaches will converge on the same pathophysiological processes or whether they will target distinct aspects of the disease remains to be determined. Regardless, understanding how synaptic transmission differs in FX holds the key to developing new therapies. Furthermore, the study of FX has greatly enriched our understanding of the neurobiology of synaptic transmission.

DISCLOSURE STATEMENT

M.B. has a financial interest in Seaside Therapeutics, Inc.

ACKNOWLEDGMENTS

The authors thank Dr. Emily Osterweil for help with the manuscript and the Howard Hughes Medical Institute, FRAXA Research Foundation, U.S. Department of Defense, National Institute of Child Health and Human Development, and National Institute of Mental Health for support.

LITERATURE CITED

Ahn JY, Ye K. 2005. PIKE GTPase signaling and function. *Int. J. Biol. Sci.* 1:44–50

Anderson P, Kedersha N. 2006. RNA granules. *J. Cell Biol.* 172:803–8

Antar LN, Afroz R, Dictenberg JB, Carroll RC, Bassell GJ. 2004. Metabotropic glutamate receptor activation regulates fragile X mental retardation protein and FMR1 mRNA localization differentially in dendrites and at synapses. *J. Neurosci.* 24:2648–55

Antar LN, Dictenberg JB, Plociniak M, Afroz R, Bassell GJ. 2005. Localization of FMRP-associated mRNA granules and requirement of microtubules for activity-dependent trafficking in hippocampal neurons. *Genes Brain Behav.* 4:350–59

Antar LN, Li C, Zhang H, Carroll RC, Bassell GJ. 2006. Local functions for FMRP in axon growth cone motility and activity-dependent regulation of filopodia and spine synapses. *Mol. Cell Neurosci.* 32:37–48

Antion MD, Hou L, Wong H, Hoeffer CA, Klann E. 2008. mGluR-dependent long-term depression is associated with increased phosphorylation of S6 and synthesis of elongation factor 1A but remains expressed in S6K-deficient mice. *Mol. Cell. Biol.* 28:2996–3007

Aschrafi A, Cunningham BA, Edelman GM, Vanderklish PW. 2005. The fragile X mental retardation protein and group I metabotropic glutamate receptors regulate levels of mRNA granules in brain. *Proc. Natl. Acad. Sci. USA* 102:2180–85

Ashley C, Wilkinson K, Reines D, Warren S. 1993. FMR1 protein: conserved RNP family domains and selective RNA binding. *Science* 262:563–66

Auerbach BD, Bear MF. 2010. Loss of the fragile X mental retardation protein decouples metabotropic glutamate receptor dependent priming of long-term potentiation from protein synthesis. *J. Neurophysiol.* 104:1047–51

Auerbach BD, Osterweil EK, Bear MF. 2011. Mutations causing syndromic autism define an axis of synaptic pathophysiology. *Nature* 480:63–68

Banko JL, Hou L, Poulin F, Sonenberg N, Klann E. 2006. Regulation of eukaryotic initiation factor 4E by converging signaling pathways during metabotropic glutamate receptor-dependent long-term depression. *J. Neurosci.* 26:2167–73

Barbee SA, Estes PS, Cziko AM, Hillebrand J, Luedeman RA, et al. 2006. Staufen- and FMRP-containing neuronal RNPs are structurally and functionally related to somatic P bodies. *Neuron* 52:997–1009

Bateup HS, Takasaki KT, Saulnier JL, Denefrio CL, Sabatini BL. 2011. Loss of Tsc1 in vivo impairs hippocampal mGluR-LTD and increases excitatory synaptic function. *J. Neurosci.* 31:8862–69

Bear MF, Huber KM, Warren ST. 2004. The mGluR theory of fragile X mental retardation. *Trends Neurosci.* 27:370–77

Bechara EG, Didiot MC, Melko M, Davidovic L, Bensaid M, et al. 2009. A novel function for fragile X mental retardation protein in translational activation. *PLoS Biol.* 7:e16

Bidinosti M, Ran I, Sanchez-Carbente MR, Martineau Y, Gingras AC, et al. 2010. Postnatal deamidation of 4E-BP2 in brain enhances its association with raptor and alters kinetics of excitatory synaptic transmission. *Mol. Cell* 37:797–808

Bilousova TV, Dansie L, Ngo M, Aye J, Charles JR, et al. 2009. Minocycline promotes dendritic spine maturation and improves behavioural performance in the fragile X mouse model. *J. Med. Genet.* 46:94–102

Bishu S, Schmidt KC, Burlin T, Channing M, Conant S, et al. 2008. Regional rates of cerebral protein synthesis measured with L-[1-11C]leucine and PET in conscious, young adult men: normal values, variability, and reproducibility. *J. Cereb. Blood Flow Metab.* 28:1502–13

Bishu S, Schmidt KC, Burlin TV, Channing MA, Horowitz L, et al. 2009. Propofol anesthesia does not alter regional rates of cerebral protein synthesis measured with L-[1-(11)C]leucine and PET in healthy male subjects. *J. Cereb. Blood Flow Metab.* 29:1035–47

Blackwell E, Zhang X, Ceman S. 2010. Arginines of the RGG box regulate FMRP association with polyribosomes and mRNA. *Hum. Mol. Genet.* 19:1314–23

Bolduc FV, Bell K, Cox H, Broadie KS, Tully T. 2008. Excess protein synthesis in Drosophila fragile X mutants impairs long-term memory. *Nat. Neurosci.* 11:1143–45

Brown V, Jin P, Ceman S, Darnell JC, O'Donnell WT, et al. 2001. Microarray identification of FMRP-associated brain mRNAs and altered mRNA translational profiles in fragile X syndrome. *Cell* 107:477–87

Callan MA, Cabernard C, Heck J, Luois S, Doe CQ, Zarnescu DC. 2010. Fragile X protein controls neural stem cell proliferation in the Drosophila brain. *Hum. Mol. Genet.* 19:3068–79

Castren M, Tervonen T, Karkkainen V, Heinonen S, Castren E, et al. 2005. Altered differentiation of neural stem cells in fragile X syndrome. *Proc. Natl. Acad. Sci. USA* 102:17834–39

Caudy AA, Hannon GJ. 2004. Induction and biochemical purification of RNA-induced silencing complex from Drosophila S2 cells. *Methods Mol. Biol.* 265:59–72

Caudy AA, Myers M, Hannon GJ, Hammond SM. 2002. Fragile X-related protein and VIG associate with the RNA interference machinery. *Genes Dev.* 16:2491–96

Ceman S, O'Donnell WT, Reed M, Patton S, Pohl J, Warren ST. 2003. Phosphorylation influences the translation state of FMRP-associated polyribosomes. *Hum. Mol. Genet.* 12:3295–305

Centonze D, Rossi S, Mercaldo V, Napoli I, Ciotti MT, et al. 2008. Abnormal striatal GABA transmission in the mouse model for the fragile X syndrome. *Biol. Psychiatry* 63:963–73

Chang S, Bray SM, Li Z, Zarnescu DC, He C, et al. 2008. Identification of small molecules rescuing fragile X syndrome phenotypes in Drosophila. *Nat. Chem. Biol.* 4:256–63

Cheever A, Ceman S. 2009. Phosphorylation of FMRP inhibits association with Dicer. *RNA* 15:362–66

Chen L, Yun SW, Seto J, Liu W, Toth M. 2003. The fragile X mental retardation protein binds and regulates a novel class of mRNAs containing U rich target sequences. *Neuroscience* 120:1005–17

Chevere-Torres I, Kaphzan H, Bhattacharya A, Kang A, Maki JM, et al. 2012. Metabotropic glutamate receptor-dependent long-term depression is impaired due to elevated ERK signaling in the DeltaRG mouse model of tuberous sclerosis complex. *Neurobiol. Dis.* 45(3):1101–10

Choi CH, McBride SM, Schoenfeld BP, Liebelt DA, Ferreiro D, et al. 2010. Age-dependent cognitive impairment in a Drosophila fragile X model and its pharmacological rescue. *Biogerontology* 11:347–62

Choi CH, Schoenfeld BP, Bell AJ, Hinchey P, Kollaros M, et al. 2011. Pharmacological reversal of synaptic plasticity deficits in the mouse model of fragile X syndrome by group II mGluR antagonist or lithium treatment. *Brain Res.* 1380:106–19

Chuang SC, Zhao W, Bauchwitz R, Yan Q, Bianchi R, Wong RK. 2005. Prolonged epileptiform discharges induced by altered group I metabotropic glutamate receptor-mediated synaptic responses in hippocampal slices of a fragile X mouse model. *J. Neurosci.* 25:8048–55

Coffee B, Keith K, Albizua I, Malone T, Mowrey J, et al. 2009. Incidence of fragile X syndrome by newborn screening for methylated FMR1 DNA. *Am. J. Hum. Genet.* 85:503–14

Coffee RL Jr, Williamson AJ, Adkins CM, Gray MC, Page TL, Broadie K. 2011. In vivo neuronal function of the fragile X mental retardation protein is regulated by phosphorylation. *Hum. Mol. Genet.* doi: 10.1093/hmg/ddr527

Collins SC, Bray SM, Suhl JA, Cutler DJ, Coffee B, et al. 2010. Identification of novel FMR1 variants by massively parallel sequencing in developmentally delayed males. *Am. J. Med. Genet. A* 152A:2512–20

Corbin F, Bouillon M, Fortin A, Morin S, Rousseau F, Khandjian EW. 1997. The fragile X mental retardation protein is associated with poly(A)$^{+}$ mRNA in actively translating polyribosomes. *Hum. Mol. Genet.* 6:1465–72

Costa-Mattioli M, Sossin WS, Klann E, Sonenberg N. 2009. Translational control of long-lasting synaptic plasticity and memory. *Neuron* 61:10–26

Cracco JB, Serrano P, Moskowitz SI, Bergold PJ, Sacktor TC. 2005. Protein synthesis-dependent LTP in isolated dendrites of CA1 pyramidal cells. *Hippocampus* 15:551–56

Croning MD, Marshall MC, McLaren P, Armstrong JD, Grant SG. 2009. G2Cdb: the Genes to Cognition database. *Nucleic Acids Res.* 37:D846–51

Darnell JC. 2011. Defects in translational regulation contributing to human cognitive and behavioral disease. *Curr. Opin. Genet. Dev.* 21:465–73

Darnell JC, Fraser CE, Mostovetsky O, Stefani G, Jones TA, et al. 2005. Kissing complex RNAs mediate interaction between the fragile-X mental retardation protein KH2 domain and brain polyribosomes. *Genes Dev.* 19:903–18

Darnell JC, Van Driesche SJ, Zhang C, Hung KY, Mele A, et al. 2011. FMRP stalls ribosomal translocation on mRNAs linked to synaptic function and autism. *Cell* 146:247–61

De Boulle K, Verkerk AJ, Reyniers E, Vits L, Hendrickx J, et al. 1993. A point mutation in the FMR-1 gene associated with fragile X mental retardation. *Nat. Genet.* 3:31–35

De Diego Otero Y, Severijnen LA, van Cappellen G, Schrier M, Oostra B, Willemsen R. 2002. Transport of fragile X mental retardation protein via granules in neurites of PC12 cells. *Mol. Cell. Biol.* 22:8332–41

de Vrij FM, Levenga J, van der Linde HC, Koekkoek SK, De Zeeuw CI, et al. 2008. Rescue of behavioral phenotype and neuronal protrusion morphology in Fmr1 KO mice. *Neurobiol. Dis.* 31:127–32

Desai NS, Casimiro TM, Gruber SM, Vanderklish PW. 2006. Early postnatal plasticity in neocortex of Fmr1 knockout mice. *J. Neurophysiol.* 96:1734–45

Devys D, Lutz Y, Rouyer N, Bellocq JP, Mandel JL. 1993. The FMR-1 protein is cytoplasmic, most abundant in neurons and appears normal in carriers of a fragile X premutation. *Nat. Genet.* 4:335–40

DeWire SM, Kim J, Whalen EJ, Ahn S, Chen M, Lefkowitz RJ. 2008. Beta-arrestin-mediated signaling regulates protein synthesis. *J. Biol. Chem.* 283:10611–20

Dictenberg JB, Swanger SA, Antar LN, Singer RH, Bassell GJ. 2008. A direct role for FMRP in activity-dependent dendritic mRNA transport links filopodial-spine morphogenesis to fragile X syndrome. *Dev. Cell* 14:926–39

Didiot MC, Subramanian M, Flatter E, Mandel JL, Moine H. 2009. Cells lacking the fragile X mental retardation protein (FMRP) have normal RISC activity but exhibit altered stress granule assembly. *Mol. Biol. Cell* 20:428–37

Dolen G, Carpenter RL, Ocain TD, Bear MF. 2010. Mechanism-based approaches to treating fragile X. *Pharmacol. Ther.* 127:78–93

Dolen G, Osterweil E, Rao BS, Smith GB, Auerbach BD, et al. 2007. Correction of fragile X syndrome in mice. *Neuron* 56:955–62

Dudek SM, Bowen WD, Bear MF. 1989. Postnatal changes in glutamate stimulated phosphoinositide turnover in rat neocortical synaptoneurosomes. *Brain Res. Dev. Brain Res.* 47:123–28

Dutch-Belgian Fragile X Consort. 1994. Fmr1 knockout mice: a model to study fragile X mental retardation. *Cell* 78:23–33

Eadie BD, Zhang WN, Boehme F, Gil-Mohapel J, Kainer L, et al. 2009. Fmr1 knockout mice show reduced anxiety and alterations in neurogenesis that are specific to the ventral dentate gyrus. *Neurobiol. Dis.* 36:361–73

Eberhart DE, Malter HE, Feng Y, Warren ST. 1996. The fragile X mental retardation protein is a ribonucleoprotein containing both nuclear localization and nuclear export signals. *Hum. Mol. Genet.* 5:1083–91

Edbauer D, Neilson JR, Foster KA, Wang CF, Seeburg DP, et al. 2010. Regulation of synaptic structure and function by FMRP-associated microRNAs miR-125b and miR-132. *Neuron* 65:373–84

Ehninger D, Han S, Shilyansky C, Zhou Y, Li W, et al. 2008. Reversal of learning deficits in a $Tsc2^{+/-}$ mouse model of tuberous sclerosis. *Nat. Med.* 14:843–48

Fahling M, Mrowka R, Steege A, Kirschner KM, Benko E, et al. 2009. Translational regulation of the human achaete-scute homologue-1 by fragile X mental retardation protein. *J. Biol. Chem.* 284:4255–66

Feng Y, Absher D, Eberhart DE, Brown V, Malter HE, Warren ST. 1997a. FMRP associates with polyribosomes as an mRNP, and the I304N mutation of severe fragile X syndrome abolishes this association. *Mol. Cell* 1:109–18

Feng Y, Gutekunst CA, Eberhart DE, Yi H, Warren ST, Hersch SM. 1997b. Fragile X mental retardation protein: nucleocytoplasmic shuttling and association with somatodendritic ribosomes. *J. Neurosci.* 17:1539–47

Ferguson SS. 2001. Evolving concepts in G protein–coupled receptor endocytosis: the role in receptor desensitization and signaling. *Pharmacol. Rev.* 53:1–24

Ferraguti F, Baldani-Guerra B, Corsi M, Nakanishi S, Corti C. 1999. Activation of the extracellular signal-regulated kinase 2 by metabotropic glutamate receptors. *Eur. J. Neurosci.* 11:2073–82

Fitzjohn SM, Palmer MJ, May JE, Neeson A, Morris SA, Collingridge GL. 2001. A characterisation of long-term depression induced by metabotropic glutamate receptor activation in the rat hippocampus in vitro. *J. Physiol.* 537:421–30

Fu YH, Kuhl DP, Pizzuti A, Pieretti M, Sutcliffe JS, et al. 1991. Variation of the CGG repeat at the fragile X site results in genetic instability: resolution of the Sherman paradox. *Cell* 67:1047–58

Gabel LA, Won S, Kawai H, McKinney M, Tartakoff AM, Fallon JR. 2004. Visual experience regulates transient expression and dendritic localization of fragile X mental retardation protein. *J. Neurosci.* 24:10579–83

Gabus C, Mazroui R, Tremblay S, Khandjian EW, Darlix JL. 2004. The fragile X mental retardation protein has nucleic acid chaperone properties. *Nucleic Acids Res.* 32:2129–37

Gallagher SM, Daly CA, Bear MF, Huber KM. 2004. Extracellular signal-regulated protein kinase activation is required for metabotropic glutamate receptor-dependent long-term depression in hippocampal area CA1. *J. Neurosci.* 24:4859–64

Gebauer F, Hentze MW. 2004. Molecular mechanisms of translational control. *Nat. Rev. Mol. Cell Biol.* 5:827–35

Giuffrida R, Musumeci S, D'Antoni S, Bonaccorso CM, Giuffrida-Stella AM, et al. 2005. A reduced number of metabotropic glutamate subtype 5 receptors are associated with constitutive homer proteins in a mouse model of fragile X syndrome. *J. Neurosci.* 25:8908–16

Godfraind JM, Reyniers E, De Boulle K, D'Hooge R, De Deyn PP, et al. 1996. Long-term potentiation in the hippocampus of fragile X knockout mice. *Am. J. Med. Genet.* 64:246–51

Goebel-Goody SM, Baum M, Paspalas CD, Fernandez SM, Carty NC, et al. 2011. Therapeutic implications for striatal-enriched protein tyrosine phosphatase (STEP) in neuropsychiatric disorders. *Pharmacol. Rev.* 64:65–87

Gross C, Nakamoto M, Yao X, Chan CB, Yim SY, et al. 2010. Excess phosphoinositide 3-kinase subunit synthesis and activity as a novel therapeutic target in fragile X syndrome. *J. Neurosci.* 30:10624–38

Guo W, Allan AM, Zong R, Zhang L, Johnson EB, et al. 2011. Ablation of FMRP in adult neural stem cells disrupts hippocampus-dependent learning. *Nat. Med.* 17:559–65

Guo W, Murthy AC, Zhang L, Johnson EB, Schaller EG, et al. 2012. Inhibition of GSK3β improves hippocampus-dependent learning and rescues neurogenesis in a mouse model of fragile X syndrome. *Hum. Mol. Genet.* 21:681–91

Hagerman R, Lauterborn J, Au J, Berry-Kravis E. 2012. Fragile x syndrome and targeted treatment trials. *Results Probl. Cell Differ.* 54:297–335

Hampson DR, Adusei DC, Pacey LK. 2011. The neurochemical basis for the treatment of autism spectrum disorders and Fragile X Syndrome. *Biochem. Pharmacol.* 81:1078–86

Han JM, Sahin M. 2011. TSC1/TSC2 signaling in the CNS. *FEBS Lett.* 585:973–80

Hatton DD, Sideris J, Skinner M, Mankowski J, Bailey DB Jr, et al. 2006. Autistic behavior in children with fragile X syndrome: prevalence, stability, and the impact of FMRP. *Am. J. Med. Genet. A* 140A:1804–13

Hayashi ML, Rao BS, Seo JS, Choi HS, Dolan BM, et al. 2007. Inhibition of p21-activated kinase rescues symptoms of fragile X syndrome in mice. *Proc. Natl. Acad. Sci. USA* 104:11489–94

Hays SA, Huber KM, Gibson JR. 2011. Altered neocortical rhythmic activity states in Fmr1 KO mice are due to enhanced mGluR5 signaling and involve changes in excitatory circuitry. *J. Neurosci.* 31:14223–34

Herbert TP, Proud CG. 2007. Regulation of translation elogation and the cotranslational protein target pathway. In *Translational Control in Biology and Medicine*, ed. MB Mathews, N Sonenberg, JWB Hershey, pp. 601–24. Cold Spring Harbor, NY: Cold Spring Harbor Press

Hinds HL, Ashley CT, Sutcliffe JS, Nelson DL, Warren ST, et al. 1993. Tissue-specific expression of FMR-1 provides evidence for a functional role in fragile X syndrome. *Nat. Genet.* 3:36–43

Hou L, Antion MD, Hu D, Spencer CM, Paylor R, Klann E. 2006. Dynamic translational and proteasomal regulation of fragile X mental retardation protein controls mGluR-dependent long-term depression. *Neuron* 51:441–54

Hou L, Klann E. 2004. Activation of the phosphoinositide 3-kinase-Akt-mammalian target of rapamycin signaling pathway is required for metabotropic glutamate receptor-dependent long-term depression. *J. Neurosci.* 24:6352–61

Hsieh H, Boehm J, Sato C, Iwatsubo T, Tomita T, et al. 2006. AMPAR removal underlies Aβ-induced synaptic depression and dendritic spine loss. *Neuron* 52:831–43

Hu H, Qin Y, Bochorishvili G, Zhu Y, van Aelst L, Zhu JJ. 2008. Ras signaling mechanisms underlying impaired GluR1-dependent plasticity associated with fragile X syndrome. *J. Neurosci.* 28:7847–62

Huber KM, Gallagher SM, Warren ST, Bear MF. 2002. Altered synaptic plasticity in a mouse model of fragile X mental retardation. *Proc. Natl. Acad. Sci. USA* 99:7746–50

Huber KM, Kayser MS, Bear MF. 2000. Role for rapid dendritic protein synthesis in hippocampal mGluR-dependent long-term depression. *Science* 288:1254–57

Huber KM, Roder JC, Bear MF. 2001. Chemical induction of mGluR5- and protein synthesis-dependent long-term depression in hippocampal area CA1. *J. Neurophysiol.* 86:321–5

Iacoangeli A, Rozhdestvensky TS, Dolzhanskaya N, Tournier B, Schutt J, et al. 2008a. On BC1 RNA and the fragile X mental retardation protein. *Proc. Natl. Acad. Sci. USA* 105:734–39

Iacoangeli A, Rozhdestvensky TS, Dolzhanskaya N, Tournier B, Schutt J, et al. 2008b. Reply to Bagni: on BC1 RNA and the fragile X mental retardation protein. *Proc. Natl. Acad. Sci. USA* 105:E29

Ishizuka A, Siomi MC, Siomi H. 2002. A Drosophila fragile X protein interacts with components of RNAi and ribosomal proteins. *Genes Dev.* 16:2497–508

Jin P, Zarnescu DC, Ceman S, Nakamoto M, Mowrey J, et al. 2004. Biochemical and genetic interaction between the fragile X mental retardation protein and the microRNA pathway. *Nat. Neurosci.* 7:113–17

Johnson EM, Kinoshita Y, Weinreb DB, Wortman MJ, Simon R, et al. 2006. Role of Pur alpha in targeting mRNA to sites of translation in hippocampal neuronal dendrites. *J. Neurosci. Res.* 83:929–43

Kandel ER. 2001. The molecular biology of memory storage: a dialogue between genes and synapses. *Science* 294:1030–38

Kang H, Schuman EM. 1996. A requirement for local protein synthesis in neurotrophin-induced hippocampal synaptic plasticity. *Science* 273:1402–6

Kaufmann WE, Abrams MT, Chen W, Reiss AL. 1999. Genotype, molecular phenotype, and cognitive phenotype: correlations in fragile X syndrome. *Am. J. Med. Genet.* 83:286–95

Kelleher RJ 3rd, Bear MF. 2008. The autistic neuron: troubled translation? *Cell* 135:401–6

Khandjian EW, Corbin F, Woerly S, Rousseau F. 1996. The fragile X mental retardation protein is associated with ribosomes. *Nat. Genet.* 12:91–93

Khandjian EW, Fortin A, Thibodeau A, Tremblay S, Cote F, et al. 1995. A heterogeneous set of FMR1 proteins is widely distributed in mouse tissues and is modulated in cell culture. *Hum. Mol. Genet.* 4:783–89

Kiebler MA, Bassell GJ. 2006. Neuronal RNA granules: movers and makers. *Neuron* 51:685–90

Kim M, Bellini M, Ceman S. 2009. Fragile X mental retardation protein FMRP binds mRNAs in the nucleus. *Mol. Cell. Biol.* 29:214–28

Kim SH, Markham JA, Weiler IJ, Greenough WT. 2008. Aberrant early-phase ERK inactivation impedes neuronal function in fragile X syndrome. *Proc. Natl. Acad. Sci. USA* 105:4429–34

Koekkoek SK, Yamaguchi K, Milojkovic BA, Dortland BR, Ruigrok TJ, et al. 2005. Deletion of FMR1 in Purkinje cells enhances parallel fiber LTD, enlarges spines, and attenuates cerebellar eyelid conditioning in Fragile X syndrome. *Neuron* 47:339–52

Krichevsky AM, Kosik KS. 2001. Neuronal RNA granules: a link between RNA localization and stimulation-dependent translation. *Neuron* 32:683–96

Krueger DD, Bear MF. 2011. Toward fulfilling the promise of molecular medicine in fragile X syndrome. *Annu. Rev. Med.* 62:411–29

Krueger DD, Osterweil EK, Bear MF. 2010. Activation of mGluR5 induces rapid and long-lasting protein kinase D phosphorylation in hippocampal neurons. *J. Mol. Neurosci.* 42:1–8

Lacoux C, Di Marino D, Pilo Boyl P, Zalfa F, Yan B, et al. 2012. BC1-FMRP interaction is modulated by 2′-O-methylation: RNA-binding activity of the tudor domain and translational regulation at synapses. *Nucleic Acids Res.* doi: 10.1093/nar/gkr1254

Laggerbauer B, Ostareck D, Keidel EM, Ostareck-Lederer A, Fischer U. 2001. Evidence that fragile X mental retardation protein is a negative regulator of translation. *Hum. Mol. Genet.* 10:329–38

Lauterborn JC, Rex CS, Kramar E, Chen LY, Pandyarajan V, et al. 2007. Brain-derived neurotrophic factor rescues synaptic plasticity in a mouse model of fragile X syndrome. *J. Neurosci.* 27:10685–94

Lee HY, Ge WP, Huang W, He Y, Wang GX, et al. 2011. Bidirectional regulation of dendritic voltage-gated potassium channels by the fragile X mental retardation protein. *Neuron* 72:630–42

Levenga J, de Vrij FM, Oostra BA, Willemsen R. 2010. Potential therapeutic interventions for fragile X syndrome. *Trends Mol. Med.* 16:516–27

Levenga J, Hayashi S, de Vrij FM, Koekkoek SK, van der Linde HC, et al. 2011. AFQ056, a new mGluR5 antagonist for treatment of fragile X syndrome. *Neurobiol. Dis.* 42:311–17

Li Z, Zhang Y, Ku L, Wilkinson KD, Warren ST, Feng Y. 2001. The fragile X mental retardation protein inhibits translation via interacting with mRNA. *Nucleic Acids Res.* 29:2276–83

Liu ZH, Huang T, Smith CB. 2011. Lithium reverses increased rates of cerebral protein synthesis in a mouse model of fragile X syndrome. *Neurobiol. Dis.* 45:1145–52

Loesch DZ, Huggins R, Petrovic V, Slater H. 1995. Expansion of the CGG repeat in fragile X in the FMR1 gene depends on the sex of the offspring. *Am. J. Hum. Genet.* 57:1408–13

Lu R, Wang H, Liang Z, Ku L, O'Donnell W T, et al. 2004. The fragile X protein controls microtubule-associated protein 1B translation and microtubule stability in brain neuron development. *Proc. Natl. Acad. Sci. USA* 101:15201–6

Lugenbeel KA, Peier AM, Carson NL, Chudley AE, Nelson DL. 1995. Intragenic loss of function mutations demonstrate the primary role of FMR1 in fragile X syndrome. *Nat. Genet.* 10:483–85

Maghsoodi B, Poon MM, Nam CI, Aoto J, Ting P, Chen L. 2008. Retinoic acid regulates RARα-mediated control of translation in dendritic RNA granules during homeostatic synaptic plasticity. *Proc. Natl. Acad. Sci. USA* 105:16015–20

Malter JS, Ray BC, Westmark PR, Westmark CJ. 2010. Fragile X syndrome and Alzheimer's disease: another story about APP and beta-amyloid. *Curr. Alzheimer Res.* 7:200–6

Mao L, Yang L, Arora A, Choe ES, Zhang G, et al. 2005. Role of protein phosphatase 2A in mGluR5-regulated MEK/ERK phosphorylation in neurons. *J. Biol. Chem.* 280:12602–10

Martin JP, Bell J. 1943. A pedigree of mental defect showing sex-linkage. *J. Neurol. Psychiatry* 6:154–57

Mazroui R, Huot ME, Tremblay S, Filion C, Labelle Y, Khandjian EW. 2002. Trapping of messenger RNA by Fragile X Mental Retardation protein into cytoplasmic granules induces translation repression. *Hum. Mol. Genet.* 11:3007–17

McBride SM, Choi CH, Wang Y, Liebelt D, Braunstein E, et al. 2005. Pharmacological rescue of synaptic plasticity, courtship behavior, and mushroom body defects in a Drosophila model of fragile X syndrome. *Neuron* 45:753–64

Meredith RM, de Jong R, Mansvelder HD. 2011. Functional rescue of excitatory synaptic transmission in the developing hippocampus in Fmr1-KO mouse. *Neurobiol. Dis.* 41:104–10

Meredith RM, Mansvelder HD. 2010. STDP and mental retardation: dysregulation of dendritic excitability in fragile X syndrome. *Front. Synaptic Neurosci.* 2:10

Michaluk P, Wawrzyniak M, Alot P, Szczot M, Wyrembek P, et al. 2011. Influence of matrix metalloproteinase MMP-9 on dendritic spine morphology. *J. Cell Sci.* 124:3369–80

Min WW, Yuskaitis CJ, Yan Q, Sikorski C, Chen S, et al. 2009. Elevated glycogen synthase kinase-3 activity in Fragile X mice: key metabolic regulator with evidence for treatment potential. *Neuropharmacology* 56:463–72

Mines MA, Jope RS. 2011. Glycogen synthase kinase-3: a promising therapeutic target for fragile x syndrome. *Front. Mol. Neurosci.* 4:35

Muddashetty RS, Kelic S, Gross C, Xu M, Bassell GJ. 2007. Dysregulated metabotropic glutamate receptor-dependent translation of AMPA receptor and postsynaptic density-95 mRNAs at synapses in a mouse model of fragile X syndrome. *J. Neurosci.* 27:5338–48

Muddashetty RS, Nalavadi VC, Gross C, Yao X, Xing L, et al. 2011. Reversible inhibition of PSD-95 mRNA translation by miR-125a, FMRP phosphorylation, and mGluR signaling. *Mol. Cell* 42:673–88

Nakamoto M, Nalavadi V, Epstein MP, Narayanan U, Bassell GJ, Warren ST. 2007. Fragile X mental retardation protein deficiency leads to excessive mGluR5-dependent internalization of AMPA receptors. *Proc. Natl. Acad. Sci. USA* 104:15537–42

Napoli I, Mercaldo V, Boyl PP, Eleuteri B, Zalfa F, et al. 2008. The fragile X syndrome protein represses activity-dependent translation through CYFIP1, a new 4E-BP. *Cell* 134:1042–54

Narayanan U, Nalavadi V, Nakamoto M, Pallas DC, Ceman S, et al. 2007. FMRP phosphorylation reveals an immediate-early signaling pathway triggered by group I mGluR and mediated by PP2A. *J. Neurosci.* 27:14349–57

Narayanan U, Nalavadi V, Nakamoto M, Thomas G, Ceman S, et al. 2008. S6K1 phosphorylates and regulates fragile X mental retardation protein (FMRP) with the neuronal protein synthesis-dependent mammalian target of rapamycin (mTOR) signaling cascade. *J. Biol. Chem.* 283:18478–82

Nicoletti F, Wroblewski JT, Fadda E, Costa E. 1988. Pertussis toxin inhibits signal transduction at a specific metabolotropic glutamate receptor in primary cultures of cerebellar granule cells. *Neuropharmacology* 27:551–56

Nosyreva ED, Huber KM. 2006. Metabotropic receptor-dependent long-term depression persists in the absence of protein synthesis in the mouse model of fragile X syndrome. *J. Neurophysiol.* 95:3291–95

Osterweil EK, Krueger DD, Reinhold K, Bear MF. 2010. Hypersensitivity to mGluR5 and ERK1/2 leads to excessive protein synthesis in the hippocampus of a mouse model of fragile X syndrome. *J. Neurosci.* 30:15616–27

Pan L, Broadie KS. 2007. Drosophila fragile X mental retardation protein and metabotropic glutamate receptor A convergently regulate the synaptic ratio of ionotropic glutamate receptor subclasses. *J. Neurosci.* 27:12378–89

Pan L, Woodruff E 3rd, Liang P, Broadie K. 2008. Mechanistic relationships between Drosophila fragile X mental retardation protein and metabotropic glutamate receptor A signaling. *Mol. Cell. Neurosci.* 37:747–60

Park S, Park JM, Kim S, Kim JA, Shepherd JD, et al. 2008. Elongation factor 2 and fragile X mental retardation protein control the dynamic translation of Arc/Arg3.1 essential for mGluR-LTD. *Neuron* 59:70–83

Pieretti M, Zhang FP, Fu YH, Warren ST, Oostra BA, et al. 1991. Absence of expression of the FMR-1 gene in fragile X syndrome. *Cell* 66:817–22

Pin JP, Waeber C, Prezeau L, Bockaert J, Heinemann SF. 1992. Alternative splicing generates metabotropic glutamate receptors inducing different patterns of calcium release in Xenopus oocytes. *Proc. Natl. Acad. Sci. USA* 89:10331–35

Plante I, Davidovic L, Ouellet DL, Gobeil LA, Tremblay S, et al. 2006. Dicer-derived microRNAs are utilized by the fragile X mental retardation protein for assembly on target RNAs. *J. Biomed. Biotechnol.* 2006:64347

Premont RT, Inglese J, Lefkowitz RJ. 1995. Protein kinases that phosphorylate activated G protein–coupled receptors. *FASEB J.* 9:175–82

Qin M, Kang J, Burlin TV, Jiang C, Smith CB. 2005. Postadolescent changes in regional cerebral protein synthesis: an in vivo study in the FMR1 null mouse. *J. Neurosci.* 25:5087–95

Reiss AL, Dant CC. 2003. The behavioral neurogenetics of fragile X syndrome: analyzing gene-brain-behavior relationships in child developmental psychopathologies. *Dev. Psychopathol.* 15:927–68

Repicky S, Broadie K. 2009. Metabotropic glutamate receptor-mediated use-dependent down-regulation of synaptic excitability involves the fragile X mental retardation protein. *J. Neurophysiol.* 101:672–87

Richter JD, Sonenberg N. 2005. Regulation of cap-dependent translation by eIF4E inhibitory proteins. *Nature* 433:477–80

Ronesi JA, Huber KM. 2008. Homer interactions are necessary for metabotropic glutamate receptor-induced long-term depression and translational activation. *J. Neurosci.* 28:543–47

Rooms L, Kooy RF. 2011. Advances in understanding fragile X syndrome and related disorders. *Curr. Opin. Pediatr.* 23:601–6

Santoro MR, Bray SM, Warren ST. 2011. Molecular mechanisms of fragile x syndrome: a twenty-year perspective. *Annu. Rev. Pathol.* 7:219–45

Scheetz AJ, Nairn AC, Constantine-Paton M. 2000. NMDA receptor-mediated control of protein synthesis at developing synapses. *Nat. Neurosci.* 3:211–16

Schoepp DD, Conn PJ. 1993. Metabotropic glutamate receptors in brain function and pathology. *Trends Pharmacol. Sci.* 14:13–20

Schratt G. 2009. microRNAs at the synapse. *Nat. Rev. Neurosci.* 10:842–49

Shang Y, Wang H, Mercaldo V, Li X, Chen T, Zhuo M. 2009. Fragile X mental retardation protein is required for chemically-induced long-term potentiation of the hippocampus in adult mice. *J. Neurochem.* 111:635–46

Sharma A, Hoeffer CA, Takayasu Y, Miyawaki T, McBride SM, et al. 2010. Dysregulation of mTOR signaling in fragile X syndrome. *J. Neurosci.* 30:694–702

Shigemoto R, Nomura S, Ohishi H, Sugihara H, Nakanishi S, Mizuno N. 1993. Immunohistochemical localization of a metabotropic glutamate receptor, mGluR5, in the rat brain. *Neurosci. Lett.* 163:53–57

Siller SS, Broadie K. 2011. Neural circuit architecture defects in a Drosophila model of Fragile X syndrome are alleviated by minocycline treatment and genetic removal of matrix metalloproteinase. *Dis. Model Mech.* 4:673–85

Siomi H, Siomi MC, Nussbaum RL, Dreyfuss G. 1993. The protein product of the fragile X gene, FMR1, has characteristics of an RNA-binding protein. *Cell* 74:291–98

Snyder EM, Philpot BD, Huber KM, Dong X, Fallon JR, Bear MF. 2001. Internalization of ionotropic glutamate receptors in response to mGluR activation. *Nat. Neurosci.* 4:1079–85

Stefani G, Fraser CE, Darnell JC, Darnell RB. 2004. Fragile X mental retardation protein is associated with translating polyribosomes in neuronal cells. *J. Neurosci.* 24:7272–76

Stetler A, Winograd C, Sayegh J, Cheever A, Patton E, et al. 2006. Identification and characterization of the methyl arginines in the fragile X mental retardation protein FMRP. *Hum. Mol. Genet.* 15:87–96

Steward O, Bakker CE, Willems PJ, Oostra BA. 1998. No evidence for disruption of normal patterns of mRNA localization in dendrites or dendritic transport of recently synthesized mRNA in FMR1 knockout mice, a model for human fragile-X mental retardation syndrome. *Neuroreport* 9:477–81

Steward O, Levy WB. 1982. Preferential localization of polyribosomes under the base of dendritic spines in granule cells of the dentate gyrus. *J. Neurosci.* 2:284–91

Su T, Fan HX, Jiang T, Sun WW, Den WY, et al. 2011. Early continuous inhibition of group 1 mGlu signaling partially rescues dendritic spine abnormalities in the Fmr1 knockout mouse model for fragile X syndrome. *Psychopharmacology* 215:291–300

Suvrathan A, Hoeffer CA, Wong H, Klann E, Chattarji S. 2010. Characterization and reversal of synaptic defects in the amygdala in a mouse model of fragile X syndrome. *Proc. Natl. Acad. Sci.* 107:11591–96

Tamanini F, Meijer N, Verheij C, Willems PJ, Galjaard H, et al. 1996. FMRP is associated to the ribosomes via RNA. *Hum. Mol. Genet.* 5:809–13

Tauber JM, Vanlandingham PA, Zhang B. 2011. Elevated levels of the vesicular monoamine transporter and a novel repetitive behavior in the Drosophila model of fragile X syndrome. *PLoS ONE* 6:e27100

Tervonen TA, Louhivuori V, Sun X, Hokkanen ME, Kratochwil CF, et al. 2009. Aberrant differentiation of glutamatergic cells in neocortex of mouse model for fragile X syndrome. *Neurobiol. Dis.* 33:250–59

Thomas AM, Bui N, Graham D, Perkins JR, Yuva-Paylor LA, Paylor R. 2011. Genetic reduction of group 1 metabotropic glutamate receptors alters select behaviors in a mouse model for fragile X syndrome. *Behav. Brain Res.* 223:310–21

Thomas AM, Bui N, Perkins JR, Yuva-Paylor LA, Paylor R. 2012. Group I metabotropic glutamate receptor antagonists alter select behaviors in a mouse model for fragile X syndrome. *Psychopharmacology* 219:47–58

Todd PK, Mack KJ, Malter JS. 2003. The fragile X mental retardation protein is required for type-I metabotropic glutamate receptor-dependent translation of PSD-95. *Proc. Natl. Acad. Sci. USA* 100:14374–78

Tucker B, Richards RI, Lardelli M. 2006. Contribution of mGluR and Fmr1 functional pathways to neurite morphogenesis, craniofacial development and fragile X syndrome. *Hum. Mol. Genet.* 15:3446–58

Vanderklish PW, Edelman GM. 2002. Dendritic spines elongate after stimulation of group 1 metabotropic glutamate receptors in cultured hippocampal neurons. *Proc. Natl. Acad. Sci. USA* 99:1639–44

Veloz MF, Buijsen R, Willemsen R, Cupido A, Bosman LW, et al. 2012. The effect of an mGluR5 inhibitor on procedural memory and avoidance discrimination impairments in Fmr1 KO mice. *Genes Brain Behav.* In press

Verkerk AJMH, Pieretti M, Sutcliffe JS, Fu Y-H, Kuhl DPA, et al. 1991. Identification of a gene (FMR-1) containing a CGG repeat coincident with a breakpoint cluster region exhibiting length variation in fragile X syndrome. *Cell* 65:905–14

Volk LJ, Pfeiffer BE, Gibson JR, Huber KM. 2007. Multiple Gq-coupled receptors converge on a common protein synthesis-dependent long-term depression that is affected in fragile X syndrome mental retardation. *J. Neurosci.* 27:11624–34

Wang H, Iacoangeli A, Lin D, Williams K, Denman RB, et al. 2005. Dendritic BC1 RNA in translational control mechanisms. *J. Cell Biol.* 171:811–21

Wang H, Kim SS, Zhuo M. 2010a. Roles of fragile X mental retardation protein in dopaminergic stimulation-induced synapse-associated protein synthesis and subsequent alpha-amino-3-hydroxyl-5-methyl-4-isoxazole-4-propionate (AMPA) receptor internalization. *J. Biol. Chem.* 285:21888–901

Wang LW, Berry-Kravis E, Hagerman RJ. 2010b. Fragile X: leading the way for targeted treatments in autism. *Neurotherapeutics* 7:264–74

Waung MW, Huber KM. 2009. Protein translation in synaptic plasticity: mGluR-LTD, fragile X. *Curr. Opin. Neurobiol.* 19:319–26

Waung MW, Pfeiffer BE, Nosyreva ED, Ronesi JA, Huber KM. 2008. Rapid translation of Arc/Arg3.1 selectively mediates mGluR-dependent LTD through persistent increases in AMPAR endocytosis rate. *Neuron* 59:84–97

Weiler IJ, Greenough WT. 1993. Metabotropic glutamate receptors trigger postsynaptic protein synthesis. *Proc. Natl. Acad. Sci. USA* 90:7168–71

Weiler IJ, Irwin SA, Klintsova AY, Spencer CM, Brazelton AD, et al. 1997. Fragile X mental retardation protein is translated near synapses in response to neurotransmitter activation. *Proc. Natl. Acad. Sci. USA* 94:5395–400

Westmark CJ, Malter JS. 2007. FMRP mediates mGluR5-dependent translation of amyloid precursor protein. *PLoS Biol.* 5:e52

Westmark CJ, Westmark PR, Malter JS. 2009. MPEP reduces seizure severity in Fmr-1 KO mice over expressing human Aβ. *Int. J. Clin. Exp. Pathol.* 3:56–68

Westmark CJ, Westmark PR, O'Riordan KJ, Ray BC, Hervey CM, et al. 2011. βPP/Aβ levels in *Fmr1*^{KO} mice. *PLoS ONE* 6:e26549

Willemsen R, Oosterwijk JC, Los FJ, Galjaard H, Oostra BA. 1996. Prenatal diagnosis of fragile X syndrome. *Lancet* 348:967–68

Xu XL, Li Y, Wang F, Gao FB. 2008. The steady-state level of the nervous-system-specific microRNA-124a is regulated by dFMR1 in Drosophila. *J. Neurosci.* 28:11883–89

Yan QJ, Rammal M, Tranfaglia M, Bauchwitz RP. 2005. Suppression of two major Fragile X syndrome mouse model phenotypes by the mGluR5 antagonist MPEP. *Neuropharmacology* 49:1053–66

Yang Y, Xu S, Xia L, Wang J, Wen S, et al. 2009. The bantam microRNA is associated with drosophila fragile X mental retardation protein and regulates the fate of germline stem cells. *PLoS Genet.* 5:e1000444

Yuskaitis CJ, Mines MA, King MK, Sweatt JD, Miller CA, Jope RS. 2009. Lithium ameliorates altered glycogen synthase kinase-3 and behavior in a mouse model of fragile X syndrome. *Biochem. Pharmacol.* 79:632–46

Zalfa F, Eleuteri B, Dickson KS, Mercaldo V, De Rubeis S, et al. 2007. A new function for the fragile X mental retardation protein in regulation of PSD-95 mRNA stability. *Nat. Neurosci.* 10:578–87

Zhang J, Hou L, Klann E, Nelson DL. 2009. Altered hippocampal synaptic plasticity in the FMR1 gene family knockout mouse models. *J. Neurophysiol.* 101:2572–80

Zhang L, Alger BE. 2010. Enhanced endocannabinoid signaling elevates neuronal excitability in fragile X syndrome. *J. Neurosci.* 30:5724–29

Zhang Y, Venkitaramani DV, Gladding CM, Kurup P, Molnar E, et al. 2008. The tyrosine phosphatase STEP mediates AMPA receptor endocytosis after metabotropic glutamate receptor stimulation. *J. Neurosci.* 28:10561–66

Zhao MG, Toyoda H, Ko SW, Ding HK, Wu LJ, Zhuo M. 2005. Deficits in Trace fear memory and long-term potentiation in a mouse model for fragile X syndrome. *J. Neurosci.* 25:7385–92

Zhao W, Chuang SC, Bianchi R, Wong RK. 2011. Dual regulation of fragile X mental retardation protein by group I metabotropic glutamate receptors controls translation-dependent epileptogenesis in the hippocampus. *J. Neurosci.* 31:725–34

Zoghbi HY, Bear MF. 2012. Synaptic dysfunction in neurodevelopmental disorders associated with autism and intellectual disabilities. *Cold Spring Harb. Perspect. Biol.* doi: 10.1101/cshperspect.a009886

RELATED RESOURCES

Darnell JC. 2011. Defects in translational regulation contributing to human cognitive and behavioral disease. Curr. Opin. Genet. Dev. 21:465–73

FRAXA Res. Found. **http://www.fraxa.org**

Krueger DD, Bear MF. 2011. Toward fulfilling the promise of molecular medicine in fragile X syndrome. Annu. Rev. Med. 62:411–29

Levenga J, de Vrij FM, Oostra BA, Willemsen R. 2010. Potential therapeutic interventions for fragile X syndrome. Trends Mol. Med. 16:516–27

Natl. Fragile X Found. **http://www.fragileX.org**

Central and Peripheral Circadian Clocks in Mammals

Jennifer A. Mohawk,[1] Carla B. Green,[1] and Joseph S. Takahashi[1,2]

[1]Department of Neuroscience, [2]Howard Hughes Medical Institute, University of Texas Southwestern Medical Center, Dallas, Texas 75390-9111;
email: Jennifer.Mohawk@UTSouthwestern.edu, Carla.Green@UTSouthwestern.edu, Joseph.Takahashi@UTSouthwestern.edu

Annu. Rev. Neurosci. 2012. 35:445–62

First published online as a Review in Advance on April 5, 2012

The *Annual Review of Neuroscience* is online at neuro.annualreviews.org

This article's doi:
10.1146/annurev-neuro-060909-153128

0147-006X/12/0721-0445$20.00

Keywords

clock genes, suprachiasmatic nucleus, oscillator coupling, metabolism, temperature

Abstract

The circadian system of mammals is composed of a hierarchy of oscillators that function at the cellular, tissue, and systems levels. A common molecular mechanism underlies the cell-autonomous circadian oscillator throughout the body, yet this clock system is adapted to different functional contexts. In the central suprachiasmatic nucleus (SCN) of the hypothalamus, a coupled population of neuronal circadian oscillators acts as a master pacemaker for the organism to drive rhythms in activity and rest, feeding, body temperature, and hormones. Coupling within the SCN network confers robustness to the SCN pacemaker, which in turn provides stability to the overall temporal architecture of the organism. Throughout the majority of the cells in the body, cell-autonomous circadian clocks are intimately enmeshed within metabolic pathways. Thus, an emerging view for the adaptive significance of circadian clocks is their fundamental role in orchestrating metabolism.

Contents

Circadian rhythm: an endogenously generated oscillation with a period of ~24 h; can entrain to external cues and is temperature compensated

Suprachiasmatic nucleus (SCN): the site of the master circadian pacemaker in mammals, comprising ~20,000 individual neurons that couple to form a robust oscillatory network

INTRODUCTION

Living systems possess an exquisitely accurate internal biological clock that times daily events ranging from sleep and wakefulness in humans to photosynthesis in plants (Takahashi et al. 2008). These circadian rhythms represent an evolutionarily conserved adaptation to the environment that can be traced back to the earliest life forms. In animals, circadian behavior can be analyzed as an integrated system—beginning with genes and leading ultimately to behavioral outputs. In the past 15 years, the molecular mechanism of circadian clocks has been uncovered by the use of phenotype-driven (forward) genetic analysis in a number of model systems (Lowrey & Takahashi 2011). Circadian oscillations are generated by a set of genes forming a transcriptional autoregulatory feedback loop. In mammals, these include *Clock*, *Bmal1*, *Per1*, *Per2*, *Cry1*, and *Cry2*. Researchers have identified another dozen candidate genes that play additional roles in the circadian gene network such as the feedback loop involving *Rev-erbα*.

Early work on mammalian rhythms used rhythmic behavior as a readout of the clock, and the hypothalamic suprachiasmatic nucleus (SCN) was identified as the dominant circadian pacemaker driving behavioral rhythms (Welsh et al. 2010). However, the discovery of "clock genes" led to the realization that the capacity for circadian gene expression is widespread throughout the body (Dibner et al. 2010). Using circadian gene reporter methods, one can demonstrate that most peripheral organs and tissues can express circadian oscillations in isolation yet still receive, and may require, input from the SCN in vivo. The cell-autonomous clock is ubiquitous, and almost every cell in the body contains a circadian clock (Balsalobre et al. 1998, Nagoshi et al. 2004, Welsh et al. 2004, Yoo et al. 2004). It is now evident that the circadian oscillators within individual cells respond differently to entraining signals, control different physiological outputs, and interact with each other and with the system as a whole. These discoveries have raised a number of questions concerning the synchronization and coherence of rhythms at the cellular level as well as the architecture of circadian clocks at the systems level (Hogenesch & Ueda 2011). Here we discuss recent work that addresses these organizational issues and examines a number of levels of complexity within the circadian system. We review mechanisms by which circadian clocks govern biological processes as well as mechanisms by which these processes feed back into the circadian system. Perhaps the most important example of this is the intimate and reciprocal interaction between the circadian clock system and fundamental metabolic pathways (Bass & Takahashi 2010, Green et al. 2008, Rutter et al. 2002). In addition, there exist additional oscillatory processes in the circadian time domain that are observable in the presence of scheduled meals or methamphetamine

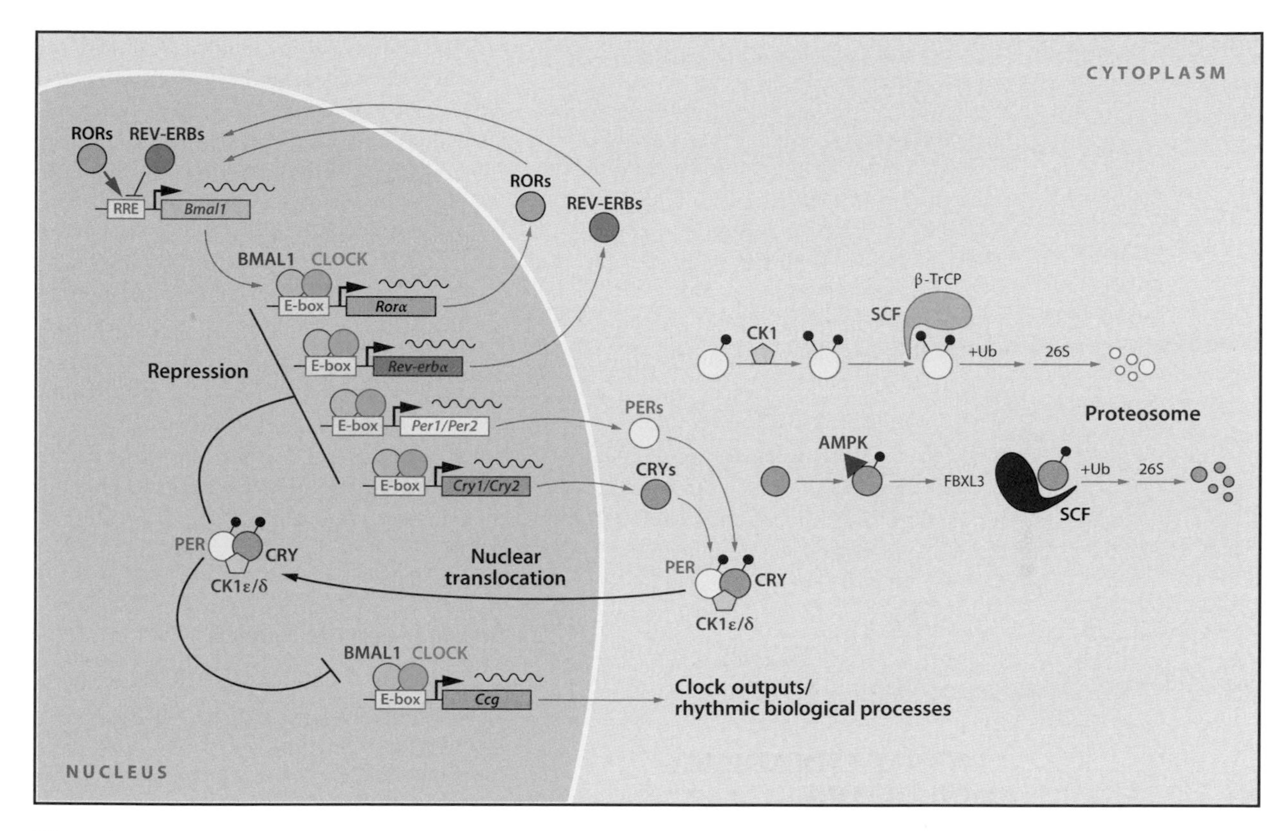

Figure 1

The molecular mechanism of the circadian clock in mammals. Constituting the core circadian clock is an autoregulatory transcriptional feedback loop involving the activators CLOCK and BMAL1 and their target genes *Per1*, *Per2*, *Cry1*, and *Cry2*, whose gene products form a negative-feedback repressor complex. In addition to this core transcriptional feedback loop, other feedback loops are also driven by CLOCK:BMAL1. One feedback loop involving *Rev-erbα* and *Rorα* that represses *Bmal1* transcription leads to an antiphase oscillation in *Bmal1* gene expression. CLOCK:BMAL1 also regulates many downstream target genes known as clock-controlled genes (Ccg). At a post-transcriptional level, the stability of the PER and CRY proteins is regulated by SCF (Skp1-Cullin-F-box protein) E3 ubiquitin ligase complexes involving β-TrCP and FBXL3, respectively. The kinases, casein kinase 1ε/δ (CK1ε/δ) and AMP kinase (AMPK), phosphorylate the PER and CRY proteins, respectively, to promote polyubiquitination by their respective E3 ubiquitin ligase complexes, which in turn tag the PER and CRY proteins for degradation by the 26S proteasome complex.

treatment. These oscillators can generate behavioral rhythms in vivo in the absence of the SCN (Honma & Honma 2009).

MOLECULAR MECHANISM OF THE CIRCADIAN CLOCK IN MAMMALS

In mammals, the mechanism of the circadian clock is cell autonomous and arises from an autoregulatory negative-feedback transcriptional network (Lowrey & Takahashi 2004, Takahashi et al. 2008) (**Figure 1**). At the core of this clock network are the transcriptional activators, CLOCK (and its paralog, NPAS2) and BMAL1, which positively regulate the expression of the *Period* (*Per1*, *Per2*) and *Cryptochrome* (*Cry1*, *Cry2*) genes at the beginning of the cycle. *Per* and *Cry* gene products accumulate, dimerize, and form a complex that translocates into the nucleus to interact with CLOCK and BMAL1, repressing their own transcription. This feedback cycle takes ~24 h, and the turnover of the PER and CRY proteins is tightly regulated by E3 ubiquitin ligase complexes. There are additional feedback loops interlocked with the core CLOCK-BMAL1/PER-CRY loop. Prominent among these is a loop involving *Rev-erbα* (*Nr1d1*) and *Rora*, which are also direct targets of CLOCK-BMAL1. The feedback effects of this loop impinge on the transcription of *Bmal1* (and to a lesser extent on *Clock*) to cause an antiphase oscillation of BMAL1. Other feedback loops involve the PAR-bZip family members, DBP, HLF, and TEF; the bZip protein, E4BP4 (*Nfil3*); and the bHLH proteins, DEC1 and DEC2 (*Bhlhb2*, *Bhlhb3*), all of which are transcriptional targets of CLOCK-BMAL1 (Gachon 2007, Lowrey & Takahashi 2004, Takahashi et al. 2008).

The discovery of a ubiquitous, cell-autonomous clock in mammals has led to a re-evaluation of central and peripheral oscillators: Are they fundamentally similar in mechanism, how do they function in different cellular contexts, and what role does coupling in the central SCN clock play in its functional properties?

CENTRAL CIRCADIAN OSCILLATORS

The hypothalamic SCN acts as a master pacemaker for the generation of circadian behavioral rhythms in mammals (for a review, see Welsh et al. 2010). Classic work not reviewed here has shown that the SCN is both necessary and sufficient for the generation of circadian activity rhythms in rodents. The SCN receives direct photic input from the retina from a recently discovered photoreceptor cell type termed the intrinsically photoreceptive retinal ganglion cell (ipRGC) (reviewed in Do & Yau 2010). These ipRGCs express a novel photopigment, melanopsin, that renders them intrinsically photosensitive to short-wavelength irradiation. Interestingly, ipRGCs are depolarizing photoreceptors that employ a phototransduction mechanism that is analogous to that seen in invertebrate photoreceptors. The photoresponse in ipRGCs has slow kinetics and a relatively high threshold to light, making them ideally suited to function as circadian photoreceptors, which must integrate light information over relatively long durations and must be insensitive to transient light signals that are not associated with the solar light cycle. Although ipRGCs appear to be optimal circadian photoreceptors, they do not act alone; rod and cone photoreceptors also have photic inputs to the SCN. Interestingly, these nonvisual inputs from rods and cones to the SCN are mediated by the ipRGCs (Chen et al. 2011, Guler et al. 2008). An emerging theme is that melanopsin-positive ipRGCs are involved in a surprisingly broad array of nonvisual photic responses in mammals. The complexity of the ipRGCs and their contribution to circadian rhythms and other behaviors are beyond the scope of this discussion, but recent reviews have covered this topic in depth (Do & Yau 2010, Schmidt et al. 2011).

Suprachiasmatic Nucleus

The SCN is composed of ~20,000 neurons, each of which is thought to contain a cell-autonomous circadian oscillator. The SCN functions as a network in which the population of SCN cells are coupled together and oscillate in a coherent manner (Herzog 2007). The dynamics of the spatial and temporal coordination of rhythms in the SCN have been studied recently, enabled by the advent of single-cell circadian reporter technology, which has revealed unexpected complexity in the temporal architecture of the nucleus (Evans et al. 2011, Foley et al. 2011, Yamaguchi et al. 2003). At the single-cell level, SCN neurons exhibit a wide range in cell-autonomous circadian periods that vary from 22 h to 30 h (Ko et al. 2010, Liu et al. 1997, Welsh et al. 1995). Intercellular coupling among SCN neurons acts to mutually couple the entire population to a much narrower range that corresponds to the circadian period of the locomotor activity rhythm, which is extremely precise (a standard deviation in period that is ~0.2 h or 12 min in mice) (Herzog et al. 2004). The heterogeneity in intrinsic period of the SCN cells confers at least two important functions: phase lability and phase plasticity. The phases of the rhythms of individual SCN neurons are highly stereotyped anatomically and appear as a wave that spreads across the nucleus over time (Evans et al. 2011, Foley et al. 2011, Yamaguchi et al. 2003). Intrinsically shorter-period cells have earlier phases and intrinsically longer-period cells have later phases within the SCN, reflecting phase lability (Yamaguchi et al. 2003). Under different photoperiods (e.g., long- versus short-photoperiod light cycles), the waveform of the SCN population rhythm is modulated such that in short photoperiods the SCN waveform is narrow and has a high amplitude, whereas in long photoperiods the SCN waveform is broad and has a low amplitude, reflecting phase plasticity (Inagaki et al. 2007, VanderLeest et al. 2007). In addition to the heterogeneity of SCN oscillator period and phase, it has been proposed that the cell-autonomous SCN oscillators are not intrinsically uniformly robust (Webb et al. 2009). Rather, the intercellular coupling of SCN neurons appears critical to the robustness of the SCN network oscillatory system.

With the discovery of peripheral oscillators (Balsalobre et al. 1998, Yamazaki et al. 2000, Yoo et al. 2004) and the apparent ubiquity of clock mechanisms (Yagita et al. 2001), a critical question arises concerning the similarity and differences in the SCN pacemaker as compared with peripheral oscillators. To address this question, Liu et al. (2007a) examined whether canonical clock mutations previously assessed in vivo affected the SCN and peripheral oscillators in a similar manner. Using *Per2::luciferase* reporter mice, they found that the effects of the *Period* and *Cryptochrome* loss-of-function mutations were the same in SCN explants as those seen previously at the behavioral level. By contrast, in peripheral tissues, single loss-of-function mutations that are subtle at the behavioral level, such as *Per1* or *Cry1* knockouts, produced very strong loss-of-rhythm phenotypes. Interestingly, the effects of these mutations are cell autonomous in both fibroblasts and in isolated SCN neurons, supporting the idea that the cell-autonomous clock is similar in these two cell types. However, when the SCN population is coupled, the effects of these mutations are non-cell autonomous. This occurs as a consequence of the intercellular coupling in the SCN network, which is capable of rescuing a cell-autonomous defect in the individual cells (**Figure 2**). This transformation of the oscillatory capability of SCN neurons from damped to self-sustained is an important illustration of the robustness of the SCN network. Indeed, Ko et al. (2010) have found that the SCN network is capable of generating oscillations in the circadian domain in the complete absence of cell-autonomous oscillatory potential. In *Bmal1*-knockout mice, which are arrhythmic at the behavioral level, SCN explants unexpectedly express stochastic oscillations in the circadian range that are highly variable. When the individual cells are no longer rhythmic, the coupling pathways within the SCN network can propagate stochastic rhythms that are a reflection of both feed-forward coupling mechanisms and intracellular noise. Thus, in a manner analogous to central pattern generators in neural circuits, rhythmicity can arise as an emergent property of the network in the absence of the component pacemaker or oscillator cells.

In addition to the generation of sustained oscillations by the SCN network, the SCN is also robust to perturbations from environmental inputs. In wild-type mice, the phase-resetting curve to light pulses is characteristic of Type 1 or weak resetting (low amplitude) (Vitaterna et al. 2006). This is a reflection of the robustness of the SCN pacemaker because inputs such as light can perturb the phase of the oscillation only to a limited extent. In contrast, genetic mutations that lower the amplitude of the molecular oscillation in the SCN lead to increases in the sensitivity to light-induced phase shifts (Type 0 resetting) without changing the strength of the light signals impinging on the SCN (Vitaterna et al. 2006). Similar effects are seen with temperature cycles. Peripheral oscillators are exquisitely sensitive to the phase-shifting effects of temperature and can be entrained strongly by low-amplitude temperature cycles that are equivalent to the circadian fluctuation in core body temperature

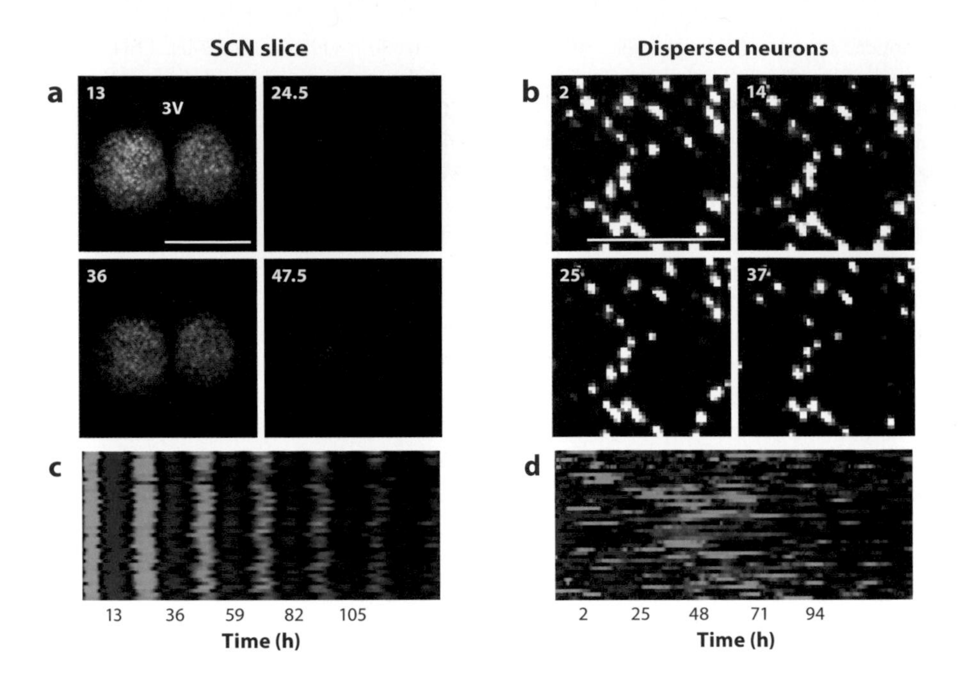

SCN slice
Dispersed neurons
a
13
3V
24.5
36
47.5
b
2
14
25
37
c
13
36
59
82
105
Time (h)
d
2
25
48
71
94
Time (h)

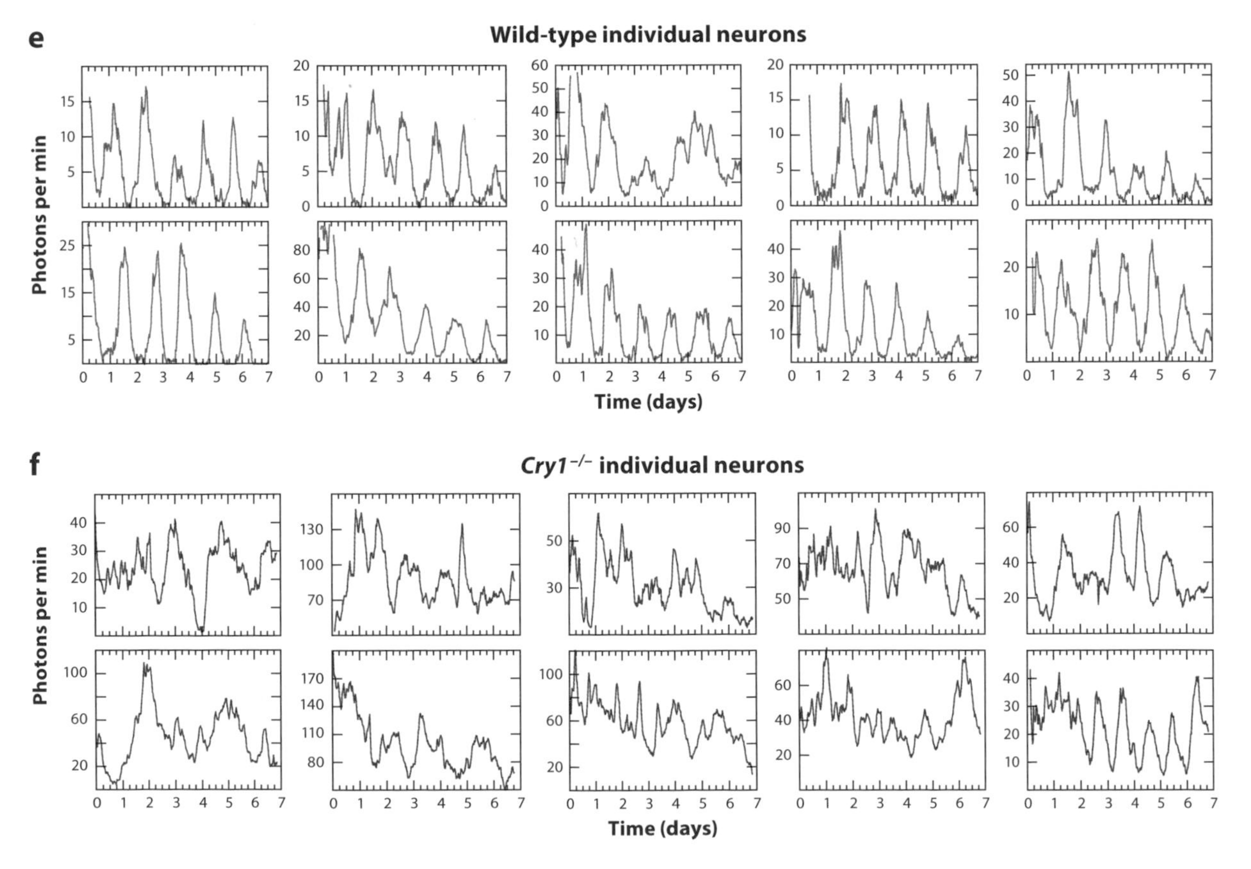

e
Wild-type individual neurons
Photons per min
Time (days)
f
Cry1−/− individual neurons
Photons per min
Time (days)

(~2.5°C oscillation in mice) (Brown et al. 2002, Buhr et al. 2010). Interestingly, the SCN is resistant to entrainment by low-amplitude temperature cycles, and this resistance depends on the intercellular coupling of SCN neurons (Abraham et al. 2010, Buhr et al. 2010). As was the case for genetic mutations on the generation of rhythms, the effects of temperature perturbations on the SCN are cell autonomous when intercellular coupling is eliminated. Thus, both SCN and peripheral oscillators are sensitive to temperature cycles at the cell-autonomous level; however, coupling within the SCN network confers robustness and makes the SCN network resistant to temperature perturbations. As discussed below, the temperature resistance of the SCN makes functional sense because the SCN drives the body-temperature rhythm, and this temperature signal, which can serve as an entraining signal for peripheral clocks, may otherwise feed back and interfere with the SCN (Buhr et al. 2010).

PERIPHERAL CLOCKS

Rhythms of clock gene and/or protein expression have been observed in cells and tissues throughout the body in mammals, and these rhythms persist in culture, demonstrating that non-SCN cells also contain endogenous circadian oscillators (Balsalobre et al. 1998, Yamazaki et al. 2000, Yoo et al. 2004). Although the core clock machinery is conserved in these different cellular clocks, there are significant differences in the relative contributions of the individual clock components, as well as in the manner in which these peripheral clocks are reset and in the output pathways that are under their control. These endogenous cellular clocks drive extensive rhythms of gene transcription, with 3–10% of all mRNAs in a given tissue showing circadian rhythms in steady-state levels (Akhtar et al. 2002, Duffield et al. 2002, Hughes et al. 2009, Miller et al. 2007, Panda et al. 2002, Storch et al. 2002). However, the genes that are under circadian control are largely nonoverlapping in each tissue, reflecting the need for temporal control of the cellular physiology relevant to each unique cell type. As a result, the circadian clock exerts broad-ranging control over many biological processes, including many aspects of metabolism such as xenobiotic detoxification (Gachon et al. 2006), glucose homeostasis (Lamia et al. 2008, Marcheva et al. 2010, So et al. 2009, Turek et al. 2005), and lipogenesis (Gachon et al. 2011, Le Martelot et al. 2009).

Entrainment: synchronization and phase control of a rhythm to a regularly occurring environmental cycle (usually ~24 h)

ORGANIZATION OF THE CIRCADIAN SYSTEM

For biological clocks to be effective, they must accurately keep time and adjust to environmental signals. In an organized circadian system, this requires SCN control of peripheral oscillators, and loss of the SCN results in peripheral circadian clocks that become desynchronized (Yoo et al. 2004). However, tissue-specific gene expression patterns are likely to be regulated by both "local" as well as central mechanisms. This concept was elegantly demonstrated through genetic disruption of the

Figure 2

Network and autonomous properties of suprachiasmatic nucleus (SCN) neurons. Network properties of the SCN can compensate for genetic defects affecting rhythmicity at the cell-autonomous level. (*a*) Bioluminescence images of a *Cry1*$^{-/-}$ SCN in organotypic slice culture. Note the stable, synchronized oscillations. Numbers indicate hours after start of imaging; 3V indicates the third ventricle. (*b*) Bioluminescence images of dissociated individual *Cry1*$^{-/-}$ SCN neurons showing cell-autonomous, largely arrhythmic patterns of high bioluminescence intensity. (*c,d*) Heat-map representations of bioluminescence intensity of individual *Cry1*$^{-/-}$ neurons in an (*a*) SCN slice and (*b*) dispersed culture. Values above and below the mean are shown in red and green, respectively, for 40 SCN neurons in each condition. (*e,f*) Ten single SCN neuron rhythms from (*e*) wild-type and (*f*) *Cry1*$^{-/-}$ mice. Imaging began immediately following a media change at day 0. Dissociated *Cry1*$^{-/-}$ SCN neurons are largely arrhythmic, whereas dissociated wild-type cells are rhythmic. By contrast, in organotypic slice cultures, both wild-type and *Cry1*$^{-/-}$ SCN cells are robustly rhythmic and tightly synchronized. Figure and legend adapted and reprinted from Liu et al. (2007a), with permission from Elsevier.

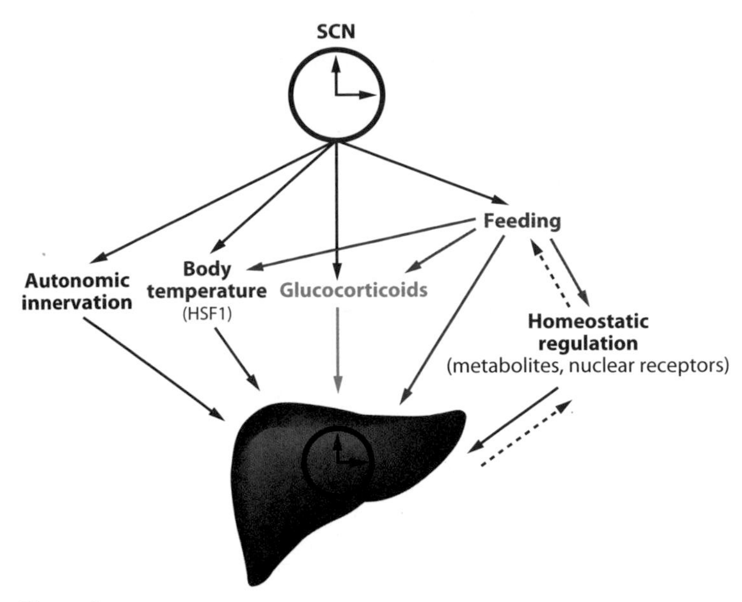

Figure 3

Pathways of peripheral clock entrainment. The master circadian pacemaker within the suprachiasmatic nucleus (SCN) relays temporal information to peripheral oscillators through autonomic innervation, body temperature, humoral signals (such as glucocorticoids), and feeding-related cues. Independent of the SCN, local signaling pathways can also affect peripheral oscillators.

circadian clock mechanism specifically in the hepatocytes of mice, while leaving the circadian clock intact in the SCN and other cell types throughout the body (Kornmann et al. 2007). Microarray analysis of mRNAs in the livers from these mice demonstrated that the disruption of the circadian molecular feedback loop specifically within the liver results in arrhythmicity of most hepatic transcripts. Thus, most circadian oscillations of hepatic function rely on an intact liver clock. However, a subset of transcripts, including the core clock component *Per2*, continued to cycle robustly even in the absence of a functional liver clock. In livers maintained in explant culture, rhythms in *Per2* transcription were observed in livers with intact clocks but were absent in livers with inactivated clocks. Thus, rhythmic gene expression can be driven by both local intracellular clocks and by extracellular systemic cues.

What are these systemic cues? The photically entrained SCN is thought to convey signals to light-insensitive peripheral clocks to synchronize these systems, and SCN transplant studies (Ralph et al. 1990, Silver et al. 1996) and parabiosis experiments in mice (Guo et al. 2005) have demonstrated that both humoral and nonhumoral pathways are important for SCN coordination of circadian output rhythms. In addition, complex feedback loops link the circadian clock with rhythmic metabolic networks, integrating these systems in a light-independent manner. Circadian control of metabolism occurs at the central (SCN) as well as local levels and involves clocks within a number of peripheral tissues including the liver, pancreas, skeletal muscle, intestine, and adipose tissue (for a review, see Bass & Takahashi 2010, Green et al. 2008). This intimate relationship between clocks and metabolism is an example of how the circadian "system" is integrated with, and influenced by, the physiology that is under its control. Therefore, organization of the circadian system requires a combination of (*a*) autonomic innervation of peripheral tissues, (*b*) endocrine signaling, (*c*) temperature, and (*d*) local signals (**Figure 3**).

Neural Control of Peripheral Oscillators: The Autonomic Nervous System

The SCN controls peripheral oscillators through both sympathetic and parasympathetic pathways (Kalsbeek et al. 2010, Ueyama et al. 1999). SCN projections through the paraventricular nucleus–superior cervical ganglia (PVN-SCG) pathway provide the dominant entraining signal for the submandibular salivary glands (Ueyama et al. 1999, Vujovic et al. 2008). Sympathetic innervation from the SCN to the PVN to the liver results in daily rhythms of plasma glucose, presumably by directly influencing the rhythm of hepatic gluconeogenesis (Cailotto et al. 2005, Kalsbeek et al. 2004).

Autonomic pathways from the SCN relay photic information to oscillators in the adrenal gland and liver (Buijs et al. 1999, Cailotto et al. 2009, Ishida et al. 2005). Sympathetic innervation also modulates the sensitivity of the adrenal to adrenocorticotropic hormone (ACTH) and directly influences glucocorticoid release (Buijs et al. 1999, Kalsbeek et al. 2010, Kaneko et al. 1981). Oscillators in both the adrenal cortex and medulla respond to neural inputs emanating from the SCN (Buijs et al. 1999, Mahoney et al. 2010). The adrenal clock is of particular interest given the strong case for glucocorticoids as a humoral entraining signal for peripheral clocks.

Hormonal Control of Peripheral Oscillators

Although a number of hormones may have roles in mammalian circadian organization, glucocorticoids have received the most attention. Rhythmic glucocorticoids result both from the sympathetic inputs discussed above and from an underlying rhythm of corticotropin releasing hormone (CRH) and ACTH function (Kaneko et al. 1980, Kaneko et al. 1981). The adrenal clock also provides temporal control of sensitivity to ACTH-induced glucocorticoid release (Oster et al. 2006).

The demonstration that dexamethasone (a glucocorticoid analog) could shift the phase of peripheral tissues in vivo (Balsalobre et al. 2000) provided the first definitive evidence that glucocorticoids were entraining signals for peripheral oscillators. Dexamethasone was initially shown to shift the phase of clock gene expression in the liver, kidney, and heart as well as cultured fibroblasts. What provides glucocorticoid input into the clock at the local level? Glucocorticoid-response elements exist in the regulatory regions of the core clock genes, *Bmal1*, *Cry1*, *Per1* (Reddy et al. 2007, Yamamoto et al. 2005), and *Per2* (So et al. 2009). These glucocorticoid-response elements may lead to the transcriptional activation of a number of clock genes and clock-controlled genes by glucocorticoids.

Glucocorticoids can synchronize circadian expression of much of the oscillatory component of the liver transcriptome in SCN-lesioned mice (Reddy et al. 2007). This is accomplished, in part, through activation of the nuclear receptor (and hepatic transcription factor) HNF4α. HNF4α is responsive to glucocorticoids (Reddy et al. 2007), contains E-boxes, which may allow for transcriptional control by CLOCK:BMAL1 (Reddy et al. 2007), and can interact with PER2 (Schmutz et al. 2010). Other metabolically relevant nuclear receptors, including peroxisome proliferator-activated receptor (PPAR)α, also respond to glucocorticoids (Lemberger et al. 1994). Nuclear receptor activation by clock components and glucocorticoids provides another point of circadian input to metabolic pathways as described below.

PPAR: peroxisome proliferator-activated receptor

HSF1: heat shock factor 1

Temperature

In most organisms, temperature is a powerful entraining agent for circadian rhythms. However, in mammals, external temperature cycles are very weak entraining agents (Refinetti 2010); this has been attributed to the fact that homeotherms regulate their body temperature and can defend their body temperature against environmental fluctuations. It has long been known that body temperature is circadian and the rhythm is driven by the SCN. Peripheral oscillators, including fibroblasts, liver, kidney, and lung, are exquisitely sensitive to temperature changes (Abraham et al. 2010, Brown et al. 2002, Buhr et al. 2010, Kornmann et al. 2007). These oscillators can be strongly reset by low-amplitude temperature pulses that mimic the range of circadian variation, and temperature profiles that match circadian body temperature rhythms strongly entrain peripheral clocks (Brown et al. 2002, Buhr et al. 2010). However, as described above, the SCN is resistant to temperature cycles in the circadian range. Because of this system design, the SCN is ideally situated to utilize circadian temperature cycles as a universal entraining signal for peripheral oscillators. The influence of temperature on peripheral oscillators likely occurs through the transcription factor heat shock factor 1 (HSF1). HSF1 transcriptional activity oscillates with a circadian rhythm in the liver and can be driven by temperature cycles (Reinke et al. 2008). HSF1 inhibitors block temperature-induced resetting in extra-SCN oscillators (Buhr et al. 2010). Because HSF1 is influenced by a wide range of signaling pathways in the cell (Akerfelt et al. 2010), temperature and HSF1 may form a final common pathway for the integration of resetting signals in peripheral clocks.

NAD: nicotinamide adenine dinucleotide

Behavioral and Homeostatic Regulation: Local Cues Feed Back into the Clock

In addition to controlling hormone secretion and body temperature directly, the SCN coordinates rhythms in behavioral processes, such as locomotor activity and feeding, which can influence endocrine function and body temperature. These behaviors, feeding in particular, can regulate peripheral clocks at the local level, modulating local signaling pathways and metabolic processes. Homeostatic signaling pathways also affect peripheral clocks and their function, allowing for extra SCN control of circadian processes. Studies in which mealtime has been experimentally manipulated to occur antiphase to the normal SCN-driven feeding rhythm have attempted to elucidate local mechanisms for controlling clocks in peripheral tissues.

The liver clock, unlike the SCN, is particularly sensitive to resetting by feeding. Hepatic rhythms of clock gene and protein expression rapidly shift their phase to follow the timing of a scheduled meal (Damiola et al. 2000, Stokkan et al. 2001). Similarly, livers of *Cry1/Cry2*-null mice display rhythms in many transcripts (including a number of transcripts involved in metabolic processes) when fed in regular 24-h intervals (Vollmers et al. 2009). Feeding appears to result in cues that bypass the core circadian feedback loop to drive these rhythms. These cues may include feeding-induced changes in temperature and HSF1 activity (Kornmann et al. 2007) or activation of other metabolically sensitive pathways.

A number of local mediators of both core clock components and clock-controlled rhythmic transcripts have been identified, and can respond to SCN-driven inputs as well as local signals related to homeostasis and metabolic state. These mediators include members of the nuclear receptor family of transcription factors, many of which exhibit circadian rhythms of transcription within the liver and other metabolically relevant tissues (Yang et al. 2006). These rhythmic nuclear receptors regulate transcription of downstream metabolic pathways. Among the rhythmic nuclear receptors are PPARs and members of the REV-ERB and ROR families. As described above, RORα and REV-ERBα participate directly in the clock mechanism by regulating *Bmal1* transcription (Preitner et al. 2002, Sato et al. 2004), but they are also important for many aspects of metabolic regulation. Similar to REV-ERBs and RORs, many of the other rhythmic nuclear receptors are regulators of clock function, providing a mechanism by which signals of metabolic status can influence rhythmicity. Glucocorticoid receptors, as discussed above, induce transcription of *Per* and potentially a number of other clock and clock-controlled genes (Reddy et al. 2007, So et al. 2009, Yamamoto et al. 2005). PPARα, which responds to lipid status and glucocorticoids, may also regulate *Bmal1* transcription (Canaple et al. 2006).

PPARγ coactivator-1α (PGC-1α, a transcriptional coactivator) provides a link between the clock and changes in metabolic status. PGC-1α is critical for adaptive responses to nutritional and metabolic state, particularly following fasting (reviewed in Lin et al. 2005). PGC-1α is rhythmic and activates expression of *Bmal1* and *Rev-erbα* through coactivation of RORs (Liu et al. 2007b). PGC-1α-null mice display disruptions in a number of circadian outputs including locomotor activity, oxygen consumption rate, and expression of both clock and metabolic genes (Liu et al. 2007b). PGC-1α also interacts with SIRTUIN 1 (SIRT1), a nicotinamide adenine dinucleotide (NAD)-dependent histone deacetylase (Rodgers et al. 2005).

Another mechanism by which metabolic signals can feed into the clock is through the adenosine monophosphate-activated protein kinase (AMPK) (Bass & Takahashi 2010). This kinase is a central mediator of metabolic signals. AMPK activity is robustly rhythmic in mouse liver and is regulated by nutrient status, as reflected in the ratio of AMP to ATP. Active AMPK directly regulates the central clock mechanism by phosphorylating and destabilizing the clock component CRY1 (Lamia et al. 2009).

Cellular redox state can also serve as a mechanism by which the metabolic status of the cell can impact the circadian system. NAD levels exhibit circadian oscillations in the liver, likely owing to transcriptional regulation of nicotinamide phosphoribosyltransferase (*Nampt*, encoding the rate-limiting enzyme in the NAD^+ salvage pathway) by CLOCK:BMAL1 (Nakahata et al. 2009, Ramsey et al. 2009). NAD levels also vary with cellular redox state as a consequence of metabolic changes, which, in turn, can directly impact clock function. The ratio of NAD^+ to NADH influences binding of NPAS2:BMAL1 and CLOCK:BMAL1 to DNA in vitro, suggesting one way in which NAD could interact with clock components (Rutter et al. 2001).

NAD^+-dependent SIRT1 also displays daily oscillations and feeds back onto the circadian clock. SIRT1 forms a complex with CLOCK:BMAL1, leading to the deacetylation of PER2 (Asher et al. 2008) and BMAL1 (Nakahata et al. 2008). SIRT1 also suppresses CLOCK:BMAL1-mediated transcription, resulting in decreased expression of *Per2* (Nakahata et al. 2008, Ramsey et al. 2009) and the clock-controlled gene *Dbp* (Nakahata et al. 2008).

Another molecule with NAD^+-sensitive activity, poly(ADP-ribose) polymerase 1 (PARP-1), was recently shown to interact with the circadian clock (Asher et al. 2010). PARP-1 activity is rhythmic in the liver, and this rhythm persists even in the absence of a functional hepatic circadian clock. The rhythm of PARP-1 activity can be entrained by scheduled meals, however, suggesting that the circadian activity of PARP-1 is driven by feeding-related cues. PARP-1 interacts with CLOCK:BMAL1 in a rhythmic fashion and inhibits DNA binding by the CLOCK:BMAL1 complex. PARP-1 also polyADP-ribosylates CLOCK and appears to temporally regulate the interaction of CLOCK:BMAL1 with PER2 and CRYs. The circadian regulation of PARP-1 by feeding, and subsequent consequences for the circadian clock, are likely not entirely mediated through NAD^+. Regardless of the mechanism, PARP-1 provides another way for metabolic signals to influence timekeeping by the core molecular clock.

OTHER OSCILLATORS: FOOD AND DRUGS

The circadian system consists of a web of interconnected oscillators and feedback loops. The core molecular clock within cells keeps time and responds to cues from the SCN (through neural, hormonal, and activity-driven pathways) as well as signals from the local cellular environment. These cell- and tissue-level clocks result in rhythms of physiologically relevant outputs, including glucose production, fat storage, and hormone production. These outputs, in turn, become circadian time-keeping cues relayed to other clocks throughout the body, likely ultimately feeding back to the central nervous system and the SCN. Under normal circumstances, the SCN maintains temporal organization of body temperature, activity, feeding, and neural output rhythms. This keeps local and systemic circadian signals aligned. In the absence of the SCN, however, the system becomes disorganized. Activity is nonrhythmic and peripheral tissues and cells drift out of phase with one another. There are two striking exceptions to this phenomenon. Scheduled, restricted feeding and chronic administration of methamphetamine, a psychostimulant drug of abuse, are both capable of organizing the circadian outputs in the absence of the SCN.

The Food-Entrainable Oscillator

It is not surprising that food and food-related cues are salient for many biological processes, including circadian rhythms. The ability of animals to anticipate food availability is well

FEO: food-entrainable oscillator

DMH: dorsomedial hypothalamus

Free-running period: the period of an oscillation in the absence of entraining signals; reflects the intrinsic period of the oscillator uninfluenced by environmental timing cues

MASCO: methamphetamine-sensitive circadian oscillator

established and persists even when food is provided at a time that is out of phase with the animal's normal feeding time (Richter 1922, Stephan et al. 1979). When food is temporally restricted to the daytime (the normal rest period), nocturnal rodents will anticipate the arrival of the meal with an increase in activity; if the timing of that meal is shifted, rats will display transients, gradually shifting their food-anticipatory activity bout each day until it again precedes the start of food availability.

In animals with lesions of the SCN, temporal food restriction will induce circadian rhythmicity of locomotor behavior (Stephan et al. 1979) and an accompanying temperature rhythm (Krieger et al. 1977). Food-anticipatory activity persists on days of total food deprivation, demonstrating that these cycles are not merely an hourglass phenomenon but are driven by an underlying oscillator (Stephan 2002). This food-entrainable oscillator (FEO) can take on pacemaking functions—organizing rhythms of activity, body temperature, and peripheral tissues in SCN-lesioned animals. Peripheral tissues from both SCN-intact and SCN-ablated mice are sensitive to temporally restricted feeding. In SCN-intact animals, phase desynchrony among peripheral oscillators can occur: Some tissues remain in phase with the (food-unaffected) SCN, and some follow the phase of food availability (Damiola et al. 2000, Pezuk et al. 2010). Cues related to the meal must be the dominant entraining signals in this latter group of tissues. In SCN-lesioned mice, food entrainment organizes rhythms throughout the periphery, and stable phase relationships are observed among tissues (Hara et al. 2001, Pezuk et al. 2010).

Interestingly, the FEO does not appear to require a functional molecular clock, as *Bmal1*$^{-/-}$ and *Per1/Per2*$^{-/-}$ mice can entrain to restricted feeding (Pendergast et al. 2009, Storch & Weitz 2009). The mechanism by which food drives oscillatory behavior throughout the organism is unknown. It is possible that the FEO exploits some of the same pathways used by the SCN, such as hormone- and temperature-dependent cues, to organize peripheral tissues. The locus (or loci) of the FEO is also unknown. A number of structures, including the olfactory bulbs (Davidson et al. 2001), the ventromedial hypothalamus (Mistlberger & Rechtschaffen 1984), the paraventricular thalamic nucleus (Landry et al. 2007), and a large portion of the digestive system (Davidson et al. 2003), have been ruled out.

The dorsomedial hypothalamus (DMH) has received considerable attention for its role in food entrainment. Data supporting a role for the DMH in the generation of food-anticipatory circadian activity are controversial (Gooley et al. 2006, Landry et al. 2006, Moriya et al. 2009), and the DMH may not be essential for the expression of the FEO (Landry et al. 2006, Moriya et al. 2009). The DMH does, however, interact with the SCN under conditions of food restriction and may influence the strength of the FEO output, particularly in SCN-intact animals (Acosta-Galvan et al. 2011).

The Methamphetamine-Sensitive Circadian Oscillator

Chronic or scheduled methamphetamine treatment affects circadian outputs in a manner similar to food restriction (Honma & Honma 2009). Methamphetamine, provided in the drinking water of rats and mice, is capable of driving circadian rhythms of locomotor behavior in the absence of the SCN (Honma et al. 1987, Tataroglu et al. 2006). In SCN-intact animals, this appears as a lengthening of the free-running period of locomotor activity, and in some cases, two activity components (relatively coordinated with each other) are observed. Much like the FEO, these rhythms persist when the stimulus (in this case, methamphetamine) is withdrawn (Tataroglu et al. 2006) as well as in the absence of a functional molecular clock (Honma et al. 2008, Mohawk et al. 2009). The methamphetamine-sensitive circadian oscillator (MASCO) is capable of functioning as a pacemaker driving rhythms in locomotor activity, body temperature, endocrine function, and the oscillators of peripheral tissues (Honma et al. 1988, Pezuk et al. 2010). In the presence of the SCN, methamphetamine results in desynchrony among internal oscillators, as some follow the SCN and some follow the presumed phase of the MASCO (Pezuk et al. 2010). When the SCN is ablated, however, the MASCO organizes oscillators in tissues throughout the organism, resulting in a coordinated system (Pezuk et al. 2010).

The site of the MASCO is also unknown. It is possible that the FEO and MASCO share an anatomical and mechanistic basis, or, indeed, that they represent a single oscillator. Research focused on understanding how the MASCO and FEO relay circadian information to peripheral tissues will likely uncover novel (or underappreciated) mechanisms that control circadian rhythms. The role of these oscillators in the absence of food restriction and methamphetamine must also be determined. It is unlikely that these oscillators are dormant under normal conditions; instead, the FEO, MASCO, and SCN probably cooperate in a hierarchically organized, perhaps necessarily redundant, timing network.

SUMMARY

It is now clear that there is feedback at nearly every level of the circadian system. "Outputs" such as body temperature and feeding become inputs to other oscillators and are capable of influencing the core molecular clockwork, generating complex interconnectivity between the circadian system and the biological outputs it controls. Reciprocity between the circadian and metabolic systems makes it likely that perturbations in one system affect the other. This idea is supported both genetically—circadian mutants have metabolic phenotypes—and environmentally—nutrient intake can modulate circadian rhythms (Bass & Takahashi 2010). In recent years, considerable progress has been made in unraveling the connections between the circadian clock and metabolism. The role of circadian clocks in governing many other physiological systems has been established, but is far less well characterized.

We still know very little about how oscillators and timing cues are integrated at the local and organismal levels to coordinate the circadian architecture of the animal. Peripheral clocks must balance (sometimes conflicting) inputs arising from the SCN with those signaling local cellular and metabolic state. Moreover, recent work has revealed that the cell-autonomous oscillator, which normally lies at the foundation of the circadian clockwork, is not absolutely crucial for the expression of rhythms by other components of the system. Within the SCN, coupling among individual neurons gives rise to a heterogeneous, yet elegantly organized, robust oscillatory network, which can overcome impaired rhythmicity at the cellular level (Ko et al. 2010, Liu et al. 2007a). Food- and drug-sensitive oscillators (FEO, MASCO) can influence circadian rhythms and drive rhythmic outputs in the absence of the core molecular clock mechanism (Honma et al. 2008, Mohawk et al. 2009, Pendergast et al. 2009, Storch & Weitz 2009). The ability of rhythmic circadian outputs to persist in the absence of the SCN necessitates that any model of the circadian network include alternative mechanisms for controlling circadian rhythms at the cell, tissue, and organism levels.

DISCLOSURE STATEMENT

The authors are not aware of any affiliations, memberships, funding, or financial holding that might be perceived as affecting the objectivity of this review.

LITERATURE CITED

Abraham U, Granada AE, Westermark PO, Heine M, Kramer A, Herzel H. 2010. Coupling governs entrainment range of circadian clocks. *Mol. Syst. Biol.* 6:438

Acosta-Galvan G, Yi CX, van der Vliet J, Jhamandas JH, Panula P, et al. 2011. Interaction between hypothalamic dorsomedial nucleus and the suprachiasmatic nucleus determines intensity of food anticipatory behavior. *Proc. Natl. Acad. Sci. USA* 108:5813–18

Akerfelt M, Morimoto RI, Sistonen L. 2010. Heat shock factors: integrators of cell stress, development and lifespan. *Nat. Rev. Mol. Cell Biol.* 11:545–55

Akhtar RA, Reddy AB, Maywood ES, Clayton JD, King VM, et al. 2002. Circadian cycling of the mouse liver transcriptome, as revealed by cDNA microarray, is driven by the suprachiasmatic nucleus. *Curr. Biol.* 12:540–50

Asher G, Gatfield D, Stratmann M, Reinke H, Dibner C, et al. 2008. SIRT1 regulates circadian clock gene expression through PER2 deacetylation. *Cell* 134:317–28

Asher G, Reinke H, Altmeyer M, Gutierrez-Arcelus M, Hottiger MO, Schibler U. 2010. Poly(ADP-ribose) polymerase 1 participates in the phase entrainment of circadian clocks to feeding. *Cell* 142:943–53

Balsalobre A, Brown SA, Marcacci L, Tronche F, Kellendonk C, et al. 2000. Resetting of circadian time in peripheral tissues by glucocorticoid signaling. *Science* 289:2344–47

Balsalobre A, Damiola F, Schibler U. 1998. A serum shock induces circadian gene expression in mammalian tissue culture cells. *Cell* 93:929–37

Bass J, Takahashi JS. 2010. Circadian integration of metabolism and energetics. *Science* 330:1349–54

Brown SA, Zumbrunn G, Fleury-Olela F, Preitner N, Schibler U. 2002. Rhythms of mammalian body temperature can sustain peripheral circadian clocks. *Curr. Biol.* 12:1574–83

Buhr ED, Yoo SH, Takahashi JS. 2010. Temperature as a universal resetting cue for mammalian circadian oscillators. *Science* 330:379–85

Buijs RM, Wortel J, Van Heerikhuize JJ, Feenstra MG, Ter Horst GJ, et al. 1999. Anatomical and functional demonstration of a multisynaptic suprachiasmatic nucleus adrenal (cortex) pathway. *Eur. J. Neurosci.* 11:1535–44

Cailotto C, La Fleur SE, Van Heijningen C, Wortel J, Kalsbeek A, et al. 2005. The suprachiasmatic nucleus controls the daily variation of plasma glucose via the autonomic output to the liver: Are the clock genes involved? *Eur. J. Neurosci.* 22:2531–40

Cailotto C, Lei J, van der Vliet J, van Heijningen C, van Eden CG, et al. 2009. Effects of nocturnal light on (clock) gene expression in peripheral organs: a role for the autonomic innervation of the liver. *PLoS One* 4:e5650

Canaple L, Rambaud J, Dkhissi-Benyahya O, Rayet B, Tan NS, et al. 2006. Reciprocal regulation of brain and muscle Arnt-like protein 1 and peroxisome proliferator-activated receptor alpha defines a novel positive feedback loop in the rodent liver circadian clock. *Mol. Endocrinol.* 20:1715–27

Chen SK, Badea TC, Hattar S. 2011. Photoentrainment and pupillary light reflex are mediated by distinct populations of ipRGCs. *Nature* 476:92–95

Damiola F, Le Minh N, Preitner N, Kornmann B, Fleury-Olela F, Schibler U. 2000. Restricted feeding uncouples circadian oscillators in peripheral tissues from the central pacemaker in the suprachiasmatic nucleus. *Genes Dev.* 14:2950–61

Davidson AJ, Aragona BJ, Werner RM, Schroeder E, Smith JC, Stephan FK. 2001. Food-anticipatory activity persists after olfactory bulb ablation in the rat. *Physiol. Behav.* 72:231–35

Davidson AJ, Poole AS, Yamazaki S, Menaker M. 2003. Is the food-entrainable circadian oscillator in the digestive system? *Genes Brain Behav.* 2:32–39

Dibner C, Schibler U, Albrecht U. 2010. The mammalian circadian timing system: organization and coordination of central and peripheral clocks. *Annu. Rev. Physiol.* 72:517–49

Do MT, Yau KW. 2010. Intrinsically photosensitive retinal ganglion cells. *Physiol. Rev.* 90:1547–81

Duffield GE, Best JD, Meurers BH, Bittner A, Loros JJ, Dunlap JC. 2002. Circadian programs of transcriptional activation, signaling, and protein turnover revealed by microarray analysis of mammalian cells. *Curr. Biol.* 12:551–57

Evans JA, Leise TL, Castanon-Cervantes O, Davidson AJ. 2011. Intrinsic regulation of spatiotemporal organization within the suprachiasmatic nucleus. *PLoS One* 6:e15869

Foley NC, Tong TY, Foley D, Lesauter J, Welsh DK, Silver R. 2011. Characterization of orderly spatiotemporal patterns of clock gene activation in mammalian suprachiasmatic nucleus. *Eur. J. Neurosci.* 33:1851–65

Gachon F. 2007. Physiological function of PARbZip circadian clock-controlled transcription factors. *Ann. Med.* 39:562–71

Gachon F, Leuenberger N, Claudel T, Gos P, Jouffe C, et al. 2011. Proline- and acidic amino acid-rich basic leucine zipper proteins modulate peroxisome proliferator-activated receptor alpha (PPARα) activity. *Proc. Natl. Acad. Sci. USA* 108:4794–99

Gachon F, Olela FF, Schaad O, Descombes P, Schibler U. 2006. The circadian PAR-domain basic leucine zipper transcription factors DBP, TEF, and HLF modulate basal and inducible xenobiotic detoxification. *Cell Metab.* 4:25–36

Gooley JJ, Schomer A, Saper CB. 2006. The dorsomedial hypothalamic nucleus is critical for the expression of food-entrainable circadian rhythms. *Nat. Neurosci.* 9:398–407

Green CB, Takahashi JS, Bass J. 2008. The meter of metabolism. *Cell* 134:728–42

Guler AD, Ecker JL, Lall GS, Haq S, Altimus CM, et al. 2008. Melanopsin cells are the principal conduits for rod-cone input to nonimage-forming vision. *Nature* 453:102–5

Guo H, Brewer JM, Champhekar A, Harris RB, Bittman EL. 2005. Differential control of peripheral circadian rhythms by suprachiasmatic-dependent neural signals. *Proc. Natl. Acad. Sci. USA* 102:3111–16

Hara R, Wan K, Wakamatsu H, Aida R, Moriya T, et al. 2001. Restricted feeding entrains liver clock without participation of the suprachiasmatic nucleus. *Genes Cells* 6:269–78

Herzog ED. 2007. Neurons and networks in daily rhythms. *Nat. Rev. Neurosci.* 8:790–802

Herzog ED, Aton SJ, Numano R, Sakaki Y, Tei H. 2004. Temporal precision in the mammalian circadian system: a reliable clock from less reliable neurons. *J. Biol. Rhythms* 19:35–46

Hogenesch JB, Ueda HR. 2011. Understanding systems-level properties: timely stories from the study of clocks. *Nat. Rev. Genet.* 12:407–16

Honma K, Honma S. 2009. The SCN-independent clocks, methamphetamine and food restriction. *Eur. J. Neurosci.* 30:1707–17

Honma K, Honma S, Hiroshige T. 1987. Activity rhythms in the circadian domain appear in suprachiasmatic nuclei lesioned rats given methamphetamine. *Physiol. Behav.* 40:767–74

Honma S, Honma K, Shirakawa T, Hiroshige T. 1988. Rhythms in behaviors, body temperature and plasma corticosterone in SCN lesioned rats given methamphetamine. *Physiol. Behav.* 44:247–55

Honma S, Yasuda T, Yasui A, van der Horst GT, Honma K. 2008. Circadian behavioral rhythms in Cry1/Cry2 double-deficient mice induced by methamphetamine. *J. Biol. Rhythms* 23:91–94

Hughes ME, DiTacchio L, Hayes KR, Vollmers C, Pulivarthy S, et al. 2009. Harmonics of circadian gene transcription in mammals. *PLoS Genet.* 5:e1000442

Inagaki N, Honma S, Ono D, Tanahashi Y, Honma K. 2007. Separate oscillating cell groups in mouse suprachiasmatic nucleus couple photoperiodically to the onset and end of daily activity. *Proc. Natl. Acad. Sci. USA* 104:7664–69

Ishida A, Mutoh T, Ueyama T, Bando H, Masubuchi S, et al. 2005. Light activates the adrenal gland: timing of gene expression and glucocorticoid release. *Cell Metab.* 2:297–307

Kalsbeek A, Bruinstroop E, Yi CX, Klieverik LP, La Fleur SE, Fliers E. 2010. Hypothalamic control of energy metabolism via the autonomic nervous system. *Ann. N.Y. Acad. Sci.* 1212:114–29

Kalsbeek A, La Fleur S, Van Heijningen C, Buijs RM. 2004. Suprachiasmatic GABAergic inputs to the paraventricular nucleus control plasma glucose concentrations in the rat via sympathetic innervation of the liver. *J. Neurosci.* 24:7604–13

Kaneko M, Hiroshige T, Shinsako J, Dallman MF. 1980. Diurnal changes in amplification of hormone rhythms in the adrenocortical system. *Am. J. Physiol.* 239:R309–16

Kaneko M, Kaneko K, Shinsako J, Dallman MF. 1981. Adrenal sensitivity to adrenocorticotropin varies diurnally. *Endocrinology* 109:70–75

Ko CH, Yamada YR, Welsh DK, Buhr ED, Liu AC, et al. 2010. Emergence of noise-induced oscillations in the central circadian pacemaker. *PLoS Biol.* 8:e1000513

Kornmann B, Schaad O, Bujard H, Takahashi JS, Schibler U. 2007. System-driven and oscillator-dependent circadian transcription in mice with a conditionally active liver clock. *PLoS Biol.* 5:e34

Krieger DT, Hauser H, Krey LC. 1977. Suprachiasmatic nuclear lesions do not abolish food-shifted circadian adrenal and temperature rhythmicity. *Science* 197:398–99

Lamia KA, Sachdeva UM, DiTacchio L, Williams EC, Alvarez JG, et al. 2009. AMPK regulates the circadian clock by cryptochrome phosphorylation and degradation. *Science* 326:437–40

Lamia KA, Storch KF, Weitz CJ. 2008. Physiological significance of a peripheral tissue circadian clock. *Proc. Natl. Acad. Sci. USA* 105:15172–77

Landry GJ, Simon MM, Webb IC, Mistlberger RE. 2006. Persistence of a behavioral food-anticipatory circadian rhythm following dorsomedial hypothalamic ablation in rats. *Am. J. Physiol. Regul. Integr. Comp. Physiol.* 290:R1527–34

Landry GJ, Yamakawa GR, Mistlberger RE. 2007. Robust food anticipatory circadian rhythms in rats with complete ablation of the thalamic paraventricular nucleus. *Brain Res.* 1141:108–18

Le Martelot G, Claudel T, Gatfield D, Schaad O, Kornmann B, et al. 2009. REV-ERBalpha participates in circadian SREBP signaling and bile acid homeostasis. *PLoS Biol.* 7:e1000181

Lemberger T, Staels B, Saladin R, Desvergne B, Auwerx J, Wahli W. 1994. Regulation of the peroxisome proliferator-activated receptor alpha gene by glucocorticoids. *J. Biol. Chem.* 269:24527–30

Lin J, Handschin C, Spiegelman BM. 2005. Metabolic control through the PGC-1 family of transcription coactivators. *Cell Metab.* 1:361–70

Liu AC, Welsh DK, Ko CH, Tran HG, Zhang EE, et al. 2007a. Intercellular coupling confers robustness against mutations in the SCN circadian clock network. *Cell* 129:605–16

Liu C, Li S, Liu T, Borjigin J, Lin JD. 2007b. Transcriptional coactivator PGC-1α integrates the mammalian clock and energy metabolism. *Nature* 447:477–81

Liu C, Weaver DR, Strogatz SH, Reppert SM. 1997. Cellular construction of a circadian clock: period determination in the suprachiasmatic nuclei. *Cell* 91:855–60

Lowrey PL, Takahashi JS. 2004. Mammalian circadian biology: elucidating genome-wide levels of temporal organization. *Annu. Rev. Genomics Hum. Genet.* 5:407–41

Lowrey PL, Takahashi JS. 2011. Genetics of circadian rhythms in Mammalian model organisms. *Adv. Genet.* 74:175–230

Mahoney CE, Brewer D, Costello MK, Brewer JM, Bittman EL. 2010. Lateralization of the central circadian pacemaker output: a test of neural control of peripheral oscillator phase. *Am. J. Physiol.* 299:R751–61

Marcheva B, Ramsey KM, Buhr ED, Kobayashi Y, Su H, et al. 2010. Disruption of the clock components CLOCK and BMAL1 leads to hypoinsulinaemia and diabetes. *Nature* 466:627–31

Miller BH, McDearmon EL, Panda S, Hayes KR, Zhang J, et al. 2007. Circadian and CLOCK-controlled regulation of the mouse transcriptome and cell proliferation. *Proc. Natl. Acad. Sci. USA* 104:3342–47

Mistlberger RE, Rechtschaffen A. 1984. Recovery of anticipatory activity to restricted feeding in rats with ventromedial hypothalamic lesions. *Physiol. Behav.* 33:227–35

Mohawk JA, Baer ML, Menaker M. 2009. The methamphetamine-sensitive circadian oscillator does not employ canonical clock genes. *Proc. Natl. Acad. Sci. USA* 106:3519–24

Moriya T, Aida R, Kudo T, Akiyama M, Doi M, et al. 2009. The dorsomedial hypothalamic nucleus is not necessary for food-anticipatory circadian rhythms of behavior, temperature or clock gene expression in mice. *Eur. J. Neurosci.* 29:1447–60

Nagoshi E, Saini C, Bauer C, Laroche T, Naef F, Schibler U. 2004. Circadian gene expression in individual fibroblasts: cell-autonomous and self-sustained oscillators pass time to daughter cells. *Cell* 119:693–705

Nakahata Y, Kaluzova M, Grimaldi B, Sahar S, Hirayama J, et al. 2008. The NAD^+-dependent deacetylase SIRT1 modulates CLOCK-mediated chromatin remodeling and circadian control. *Cell* 134:329–40

Nakahata Y, Sahar S, Astarita G, Kaluzova M, Sassone-Corsi P. 2009. Circadian control of the NAD^+ salvage pathway by CLOCK-SIRT1. *Science* 324:654–57

Oster H, Damerow S, Kiessling S, Jakubcakova V, Abraham D, et al. 2006. The circadian rhythm of glucocorticoids is regulated by a gating mechanism residing in the adrenal cortical clock. *Cell Metab.* 4:163–73

Panda S, Antoch MP, Miller BH, Su AI, Schook AB, et al. 2002. Coordinated transcription of key pathways in the mouse by the circadian clock. *Cell* 109:307–20

Pendergast JS, Nakamura W, Friday RC, Hatanaka F, Takumi T, Yamazaki S. 2009. Robust food anticipatory activity in BMAL1-deficient mice. *PLoS One* 4:e4860

Pezuk P, Mohawk JA, Yoshikawa T, Sellix MT, Menaker M. 2010. Circadian organization is governed by extra-SCN pacemakers. *J. Biol. Rhythms* 25:432–41

Preitner N, Damiola F, Lopez-Molina L, Zakany J, Duboule D, et al. 2002. The orphan nuclear receptor REV-ERBα controls circadian transcription within the positive limb of the mammalian circadian oscillator. *Cell* 110:251–60

Ralph MR, Foster RG, Davis FC, Menaker M. 1990. Transplanted suprachiasmatic nucleus determines circadian period. *Science* 247:975–78

Ramsey KM, Yoshino J, Brace CS, Abrassart D, Kobayashi Y, et al. 2009. Circadian clock feedback cycle through NAMPT-mediated NAD^+ biosynthesis. *Science* 324:651–54

Reddy AB, Maywood ES, Karp NA, King VM, Inoue Y, et al. 2007. Glucocorticoid signaling synchronizes the liver circadian transcriptome. *Hepatology* 45:1478–88

Refinetti R. 2010. Entrainment of circadian rhythm by ambient temperature cycles in mice. *J. Biol. Rhythms* 25:247–56

Reinke H, Saini C, Fleury-Olela F, Dibner C, Benjamin IJ, Schibler U. 2008. Differential display of DNA-binding proteins reveals heat-shock factor 1 as a circadian transcription factor. *Genes Dev.* 22:331–45

Richter C. 1922. A behavioristic study of the activity of the rat. *Comp. Psychol. Monogr.* 1:1–55

Rodgers JT, Lerin C, Haas W, Gygi SP, Spiegelman BM, Puigserver P. 2005. Nutrient control of glucose homeostasis through a complex of PGC-1α and SIRT1. *Nature* 434:113–18

Rutter J, Reick M, McKnight SL. 2002. Metabolism and the control of circadian rhythms. *Annu. Rev. Biochem.* 71:307–31

Rutter J, Reick M, Wu LC, McKnight SL. 2001. Regulation of clock and NPAS2 DNA binding by the redox state of NAD cofactors. *Science* 293:510–14

Sato TK, Panda S, Miraglia LJ, Reyes TM, Rudic RD, et al. 2004. A functional genomics strategy reveals Rora as a component of the mammalian circadian clock. *Neuron* 43:527–37

Schmidt TM, Chen SK, Hattar S. 2011. Intrinsically photosensitive retinal ganglion cells: many subtypes, diverse functions. *Trends Neurosci.* 34:572–80

Schmutz I, Ripperger JA, Baeriswyl-Aebischer S, Albrecht U. 2010. The mammalian clock component PERIOD2 coordinates circadian output by interaction with nuclear receptors. *Genes Dev.* 24:345–57

Silver R, LeSauter J, Tresco PA, Lehman MN. 1996. A diffusible coupling signal from the transplanted suprachiasmatic nucleus controlling circadian locomotor rhythms. *Nature* 382:810–13

So AY, Bernal TU, Pillsbury ML, Yamamoto KR, Feldman BJ. 2009. Glucocorticoid regulation of the circadian clock modulates glucose homeostasis. *Proc. Natl. Acad. Sci. USA* 106:17582–87

Stephan FK. 2002. The "other" circadian system: food as a Zeitgeber. *J. Biol. Rhythms* 17:284–92

Stephan FK, Swann JM, Sisk CL. 1979. Entrainment of circadian rhythms by feeding schedules in rats with suprachiasmatic lesions. *Behav. Neural. Biol.* 25:545–54

Stokkan KA, Yamazaki S, Tei H, Sakaki Y, Menaker M. 2001. Entrainment of the circadian clock in the liver by feeding. *Science* 291:490–93

Storch KF, Lipan O, Leykin I, Viswanathan N, Davis FC, et al. 2002. Extensive and divergent circadian gene expression in liver and heart. *Nature* 417:78–83

Storch KF, Weitz CJ. 2009. Daily rhythms of food-anticipatory behavioral activity do not require the known circadian clock. *Proc. Natl. Acad. Sci. USA* 106:6808–13

Takahashi JS, Hong HK, Ko CH, McDearmon EL. 2008. The genetics of mammalian circadian order and disorder: implications for physiology and disease. *Nat. Rev. Genet.* 9:764–75

Tataroglu O, Davidson AJ, Benvenuto LJ, Menaker M. 2006. The methamphetamine-sensitive circadian oscillator (MASCO) in mice. *J. Biol. Rhythms* 21:185–94

Turek FW, Joshu C, Kohsaka A, Lin E, Ivanova G, et al. 2005. Obesity and metabolic syndrome in circadian *Clock* mutant mice. *Science* 308:1043–45

Ueyama T, Krout KE, Nguyen XV, Karpitskiy V, Kollert A, et al. 1999. Suprachiasmatic nucleus: a central autonomic clock. *Nat. Neurosci.* 2:1051–53

VanderLeest HT, Houben T, Michel S, Deboer T, Albus H, et al. 2007. Seasonal encoding by the circadian pacemaker of the SCN. *Curr. Biol.* 17:468–73

Vitaterna MH, Ko CH, Chang AM, Buhr ED, Fruechte EM, et al. 2006. The mouse *Clock* mutation reduces circadian pacemaker amplitude and enhances efficacy of resetting stimuli and phase-response curve amplitude. *Proc. Natl. Acad. Sci. USA* 103:9327–32

Vollmers C, Gill S, DiTacchio L, Pulivarthy SR, Le HD, Panda S. 2009. Time of feeding and the intrinsic circadian clock drive rhythms in hepatic gene expression. *Proc. Natl. Acad. Sci. USA* 106:21453–58

Vujovic N, Davidson AJ, Menaker M. 2008. Sympathetic input modulates, but does not determine, phase of peripheral circadian oscillators. *Am. J. Physiol.* 295:R355–60

Webb AB, Angelo N, Huettner JE, Herzog ED. 2009. Intrinsic, nondeterministic circadian rhythm generation in identified mammalian neurons. *Proc. Natl. Acad. Sci. USA* 106:16493–98

Welsh DK, Logothetis DE, Meister M, Reppert SM. 1995. Individual neurons dissociated from rat suprachiasmatic nucleus express independently phased circadian firing rhythms. *Neuron* 14:697–706

Welsh DK, Takahashi JS, Kay SA. 2010. Suprachiasmatic nucleus: cell autonomy and network properties. *Annu. Rev. Physiol.* 72:551–77

Welsh DK, Yoo SH, Liu AC, Takahashi JS, Kay SA. 2004. Bioluminescence imaging of individual fibroblasts reveals persistent, independently phased circadian rhythms of clock gene expression. *Curr. Biol.* 14:2289–95

Yagita K, Tamanini F, van Der Horst GT, Okamura H. 2001. Molecular mechanisms of the biological clock in cultured fibroblasts. *Science* 292:278–81

Yamaguchi S, Isejima H, Matsuo T, Okura R, Yagita K, et al. 2003. Synchronization of cellular clocks in the suprachiasmatic nucleus. *Science* 302:1408–12

Yamamoto T, Nakahata Y, Tanaka M, Yoshida M, Soma H, et al. 2005. Acute physical stress elevates mouse period1 mRNA expression in mouse peripheral tissues via a glucocorticoid-responsive element. *J. Biol. Chem.* 280:42036–43

Yamazaki S, Numano R, Abe M, Hida A, Takahashi R, et al. 2000. Resetting central and peripheral circadian oscillators in transgenic rats. *Science* 288:682–85

Yang X, Downes M, Yu RT, Bookout AL, He W, et al. 2006. Nuclear receptor expression links the circadian clock to metabolism. *Cell* 126:801–10

Yoo SH, Yamazaki S, Lowrey PL, Shimomura K, Ko CH, et al. 2004. PERIOD2::LUCIFERASE real-time reporting of circadian dynamics reveals persistent circadian oscillations in mouse peripheral tissues. *Proc. Natl. Acad. Sci. USA* 101:5339–46

RELATED RESOURCES

Bass J, Takahashi JS. 2010. Circadian integration of metabolism and energetics. *Science* 330:1349–54

Dibner C, Schibler U, Albrecht U. 2010. The mammalian circadian timing system: organization and coordination of central and peripheral clocks. *Annu. Rev. Physiol.* 72:517–49

Do MT, Yau KW. 2010. Intrinsically photosensitive retinal ganglion cells. *Physiol. Rev.* 90:1547–81

Hogenesch JB, Ueda HR. 2011. Understanding systems-level properties: timely stories from the study of clocks. *Nat. Rev. Genet.* 12:407–16

Lowrey PL, Takahashi JS. 2011. Genetics of circadian rhythms in mammalian model organisms. *Adv. Genet.* 74:175–230

Welsh DK, Takahashi JS, Kay SA. 2010. Suprachiasmatic nucleus: cell autonomy and network properties. *Annu. Rev. Physiol.* 72:551–77

Decision-Related Activity in Sensory Neurons: Correlations Among Neurons and with Behavior

Hendrikje Nienborg,[1,*] Marlene R. Cohen,[2,*] and Bruce G. Cumming[3]

[1]Werner Reichardt Center for Integrative Neuroscience, 72076 Tuebingen, Germany; email: hnienb@gmail.com

[2]Department of Neuroscience and Center for the Neural Basis of Cognition, University of Pittsburgh, Pittsburgh, Pennsylvania 15213; email: cohenm@pitt.edu

[3]Laboratory of Sensorimotor Research, National Eye Institute, National Institutes of Health, Bethesda, Maryland 20892; email: bgc@lsr.nei.nih.gov

Annu. Rev. Neurosci. 2012. 35:463–83

First published online as a Review in Advance on April 5, 2012

The *Annual Review of Neuroscience* is online at neuro.annualreviews.org

This article's doi:
10.1146/annurev-neuro-062111-150403

0147-006X/12/0721-0463$20.00

*These authors contributed equally to this work.

Keywords

choice probability, decision making, noise correlation

Abstract

Neurons in early sensory cortex show weak but systematic correlations with perceptual decisions when trained animals perform at psychophysical threshold. These correlations are observed across repeated presentations of identical stimuli and cannot be explained by variation in external factors. The relationship between the activity of individual sensory neurons and the animal's behavioral choice means that even neurons in early sensory cortex carry information about an upcoming decision. This relationship, termed choice probability, may reflect the effect of fluctuations in neuronal firing rate on the animal's decision, but it can also reflect modulation of sensory responses by cognitive factors, or network properties such as variability that is shared among populations of neurons. Here, we review recent work clarifying the relationship among fluctuations in the responses of individual neurons, correlated variability, and behavior in a variety of tasks and cortical areas. We also discuss the possibility that choice probability may in part reflect the influence of cognitive factors on sensory neurons and explore the situations in which choice probability can be used to make inferences about the role of particular sensory neurons in the decision-making process.

Contents

INTRODUCTION

How the responses of individual neurons and their interactions with nearby neurons relate to perception is critical to understanding how the brain generates our mental world. Until the late 1980s, this question was addressed predominantly by studying a subject's perceptual performance and the response properties of individual sensory neurons in separate experiments (reviewed in Parker & Newsome 1998). Since then, numerous studies have simultaneously measured the activity of individual neurons and subjects' perceptual judgments.

Simultaneous neuronal recordings and behavioral measurements provide the opportunity to determine the relationship between the neuron's responses and the animal's behavioral choices. Cortical neurons respond variably to repeated presentations of the same stimulus (Tolhurst et al. 1983) (see **Figure 1**). For near-threshold perceptual judgments, observers make variable choices as well, making errors on some fraction of behavioral trials. Measuring the extent to which the fluctuations in the responses of an individual neuron predict perceptual judgments may reveal important information about the role played by single neurons in a behavioral task.

The relationship between the trial-to-trial fluctuations in the activity of individual sensory neurons and perceptual decisions is typically quantified using a measure called choice probability (CP). The idea of CP was first introduced by Britten and colleagues (1996), who recorded the activity of direction-selective neurons in the middle temporal area (MT) while monkeys performed a two-alternative forced-choice motion-direction-discrimination task (**Figure 1*a***). The authors noticed that they could predict, from the responses of single MT neurons, whether, for example, a monkey would report that a random dot stimulus containing no net motion was moving upward or downward (**Figure 1*b*,*c***).

The authors used CP to quantify the extent to which they could use the neuron's responses to a given stimulus to predict the animal's perceptual decisions. If, for example, the authors recorded from a neuron whose preferred direction was upward, they hypothesized that the animal would be more likely to report upward motion on trials in which that neuron fired more than its average. Conversely, they hypothesized that the animal would be more likely to report downward motion on trials in which the neuron fired less than its average. CP quantifies the accuracy of this prediction by representing the proportion of trials on which an ideal observer could predict the animal's choices given the firing rate of the neuron (**Figure 1*c***) (Shadlen et al. 1996). A CP of 1 would mean that the neuron always fired more on trials when the monkey reported upward motion than on trials

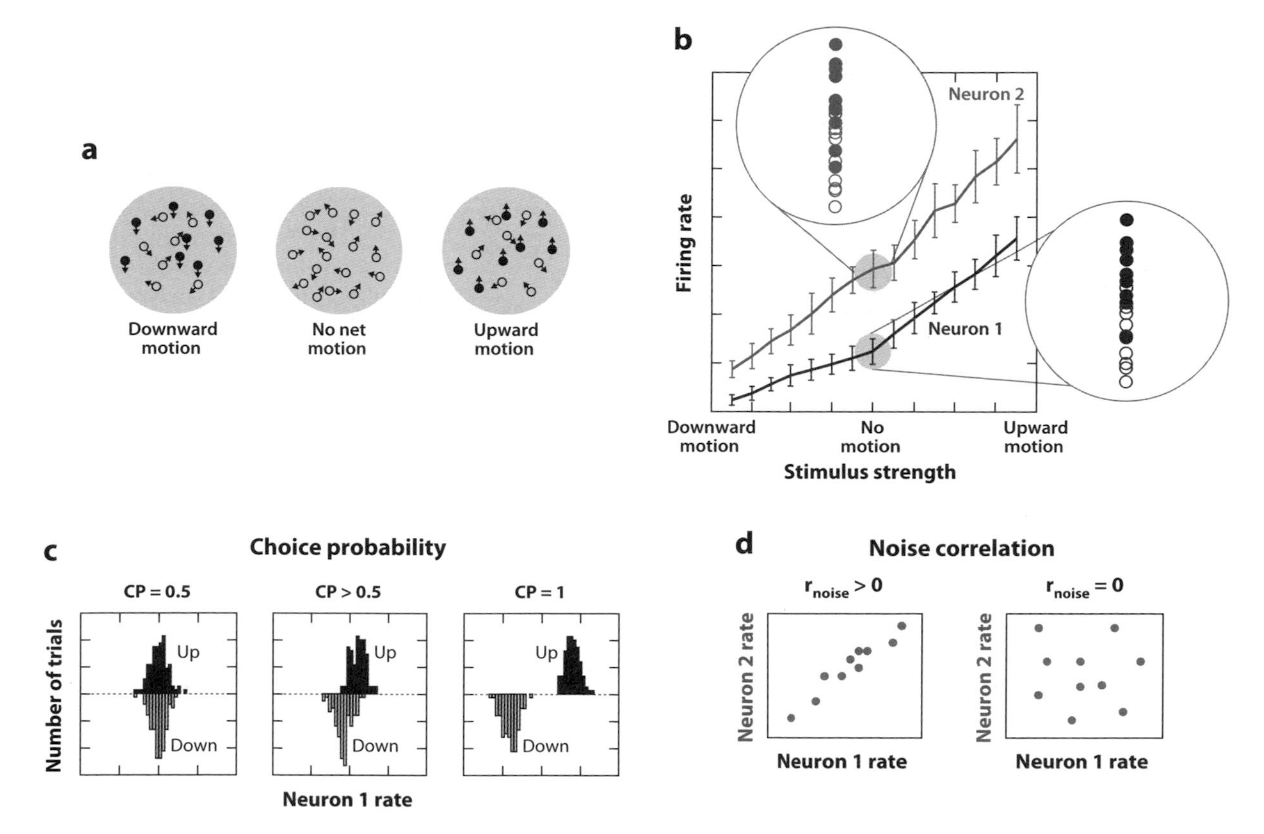

Figure 1

Measuring and calculating choice probability and noise correlation. (*a*) The original choice probability (CP) study (Britten et al. 1996) used a task in which monkeys reported whether the coherent motion in a random dot display moved in one of two opposite directions, illustrated here for an up-down discrimination. Signal strength was adjusted by varying the fraction of dots moving coherently in one direction (*filled circles*), whereas the remaining dots moved at random (*noise dots, open circles*). CP was usually calculated from responses to stimuli with no coherent motion in which all the dots moved randomly (*center panel*). (*b*) Average firing rate responses of two hypothetical middle temporal area (MT) neurons with upward preferred directions to random dot stimuli of various motion strengths. The circles zoom in on responses to stimuli with no net motion to illustrate that the responses of each neuron will be different on each stimulus presentation. These trial-to-trial fluctuations in neuronal response are the basis for CP (which measures the correlation between these fluctuations and behavior) and noise correlation (which measures the correlation between these fluctuations in a pair of neurons). In this illustration, filled circles represent responses on trials in which the animal reported upward motion, and open circles represent trials in which the animal reported downward motion. (*c*) Frequency histograms of responses of the hypothetical magenta neuron to stimuli with no net motion separated by whether the animal reported upward motion (*top, filled histograms*) or downward motion (*downward, open histograms*). CP measures the probability with which an ideal observer could predict the animal's choices from the responses of this neuron. CP is defined such that if the distributions of responses are identical for up and down choices (*left histograms*), CP = 0.5 because an ideal observer would be at chance, or 50% correct performance. If the histograms for the two choices did not overlap (*right histograms*), CP = 1 because the observer would be able to use this neuron to predict choices with 100% accuracy. Note that if the spike counts were lower when the animal reported the neuron's preferred direction (rates on up trials were less than on down trials in this case), CP would be less than 0.5. (*d*) Noise correlation is calculated using the same trial-to-trial fluctuations as used in CP but collapsing across choices. The correlation between two neurons is defined as the Pearson's correlation coefficient between the responses of a pair of neurons to repeated presentations of the same stimulus. This panel illustrates two potential noise correlations for the pair of hypothetical neurons in panel *b*, one in which fluctuations in the responses are correlated (*left panel*; each point represents responses on one presentation of a stimulus with no net motion) and one in which the responses are uncorrelated. In visual cortex, noise correlations tend to be weak but positive (for review, see Cohen & Kohn 2011).

when it reported downward motion, whereas a CP of 0 would indicate that the neuron always fired more when the monkey reported downward motion. A CP of 0.5 would indicate no relationship between the neuron's responses and the animal's decisions. Empirically, the authors found a mean CP of 0.56, meaning that the activity of MT neurons is weakly but consistently related to perceptual choices in this task (Britten et al. 1996).

Although CPs for sensory neurons are typically small (usually <0.6), it is remarkable that they are detectable at all, given that many thousands of cortical neurons are activated by a typical sensory stimulus. If the responses of all these thousands of neurons contribute independently to a decision, the relationship between the one that an experimenter happens to find with an electrode and the animal's behavior should be negligible. If, however, very few neurons are involved in the decision, the chances of finding one of them during a recording session should be very low.

Quantitative simulations have been invaluable for understanding how it is possible that the activity of any one neuron could correlate with behavior when so many neurons are potentially involved in the decision. The dominant theoretical framework has been a simple pooling model, developed to describe the data from the original study by Britten and colleagues (Shadlen et al. 1996). This model necessarily represents a considerable simplification of the cortical circuitry, but because of its simplicity, it has provided a useful and influential framework in which to think about CPs. In the motion direction–discrimination task, the model discriminates upward from downward motion by comparing the activity of two pools of direction-selective neurons, one preferring upward motion, the other preferring downward (**Figure 2*a***). The activity of the neurons in each pool is summed, and the model reports the direction of motion corresponding to the pool with greater activity. When the authors assumed that the responses of each neuron fluctuated independently of all other neurons, the model indicated that a mean CP as high as 0.56 (Britten et al. 1996) implies that the decision was based on a very small number of neurons (<10).

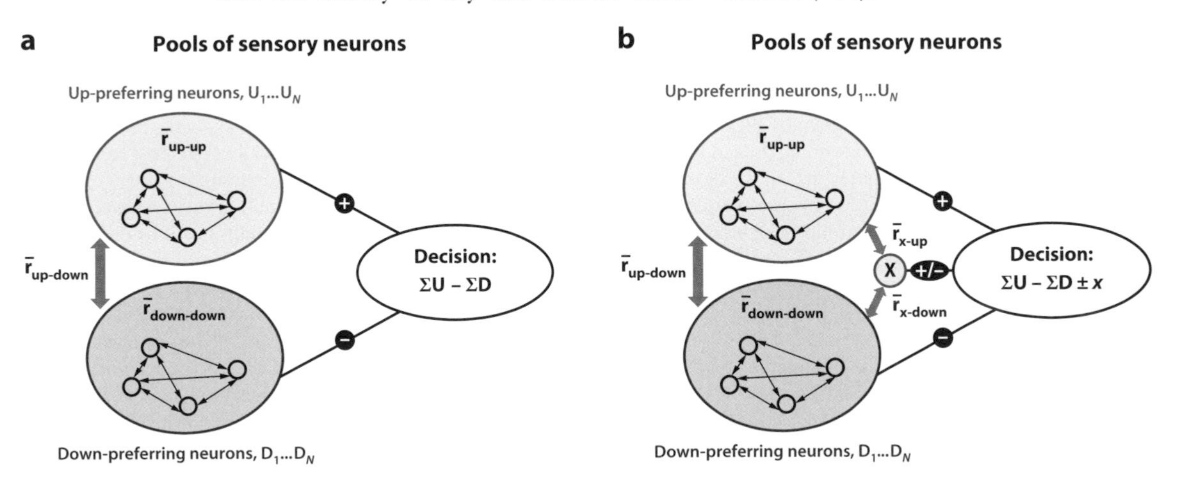

Figure 2

(*a*) Schematic of the pooling model. The spikes from one group of neurons, preferring upward motion (*blue*), are summed, as are the responses of a group preferring the opposite direction (*red*). The decision is based on the difference of these sums: $\sum U - \sum D$. Double-headed arrows indicate interneuronal noise correlations. (*b*) A single neuron (x) is added. Whether this neuron is part of the "up" pool (positive weight) the down pool (negative weight) or neither (0 weight) has negligible effect on the decision if the pools are large. If fluctuations in spike count of neuron X are correlated with those of neurons in one decision pool, then neuron X will be correlated with the decision. If neuron X is correlated with neurons in both pools to equal extents, then there will be no correlation with the decision.

The resolution to the apparent paradox between large numbers of neurons and measurable CPs lies in the observation that the responses of nearby cortical neurons tend to fluctuate in a correlated manner (**Figure 1*d***). CP measures the relationship between fluctuations in the response of one neuron to repeated presentations of the same stimulus and the animal's choices. The term noise correlation refers to the relationship between those same response fluctuations in pairs of nearby neurons (Bair et al. 2001). Noise correlation (also called spike count correlation or interneuronal correlation) is simply the Pearson's correlation between the trial-to-trial fluctuations in the responses of a pair of neurons. In sensory cortex, noise correlations among pairs of nearby neurons with similar tuning are typically weak but positive (for review, see Cohen & Kohn 2011).

The existence of positive noise correlations explains the existence of CPs in large pools of neurons. The pooling model hypothesizes that the animal's decision is related to the summed activity of many neurons. If the responses of all neurons fluctuated independently, the summed activity of all the neurons that contribute to a decision would be nearly constant because the noise would be averaged away. If, however, the responses of all neurons fluctuate up and down together (as is the case when noise correlations are positive), the summed activity of even thousands of neurons will fluctuate as well. In this scenario, the activity of any one neuron will be correlated with behavior simply because it is correlated with all the rest of the neurons and therefore with the monkey's percept. Measured CPs therefore depend critically on the presence of noise correlations within a population of neurons.

Since the original study by Britten and colleagues, our knowledge of the conditions under which CPs are observed, and our understanding of what they can tell us about the process of perceptual decision making, has greatly expanded. Our goals here are to review studies of CP in different systems and tasks, to explore the implications of CPs for population coding and decision making, and to discuss how recent and future experimental and theoretical work will shed light on the neuronal mechanisms underlying perception. We discuss experimental methods for determining the number of neurons involved in a perceptual decision, the role of noise correlations in determining observed CPs, evidence that CPs may at least partially reflect the impact of top-down signals on the activity of sensory neurons, and cases in which CPs can and cannot be used to infer the way that the responses of many neurons are combined to guide decisions.

DETERMINING THE NUMBER OF NEURONS INVOLVED IN A DECISION

Above, we describe two important factors that interact to determine CP: the number of neurons that contribute to the decision (pool size) and the extent to which fluctuations in activity are correlated between neurons. At one extreme is the possibility that fluctuations in activity are independent in different neurons. Under this assumption, the pooling model shows that typical CPs imply that very few neurons are involved in the decision (pool sizes of fewer than 10 neurons). Larger pool sizes require the existence of noise correlations to explain CP (Shadlen et al. 1996). Most studies reporting CP have not simultaneously measured interneuronal correlations from the same trials. In such studies, since noise correlations might have been very small, small pool sizes remain a viable theoretical explanation.

One difficulty with proposing small decision pools is explaining how it is possible to record significant CP on average in a large group of neurons sampled at random from the population. If all these neurons were recorded using the same physical stimulus, the observation of CP in many neurons would rule out the possibility of very small pools; if only very few neurons contribute to a decision, the probability of encountering such a neuron would be tiny. However, in most studies, a number of stimulus parameters are adjusted to make signals from a given cell especially relevant. If the stimulus is sufficiently tailored in this way, it becomes possible that for each of the specific stimuli used, the pool size underlying psychophysical responses is indeed small. Some studies that tried to ensure very specific matching of stimuli to neuronal preferences have argued in favor of such a scheme (Ghose & Harrison 2009), and such data are certainly compatible with small pool sizes. However, because noise correlations were not measured, the data are equally compatible with large pool sizes and correlated activity.

One experimental observation argues strongly against the idea of small pool sizes. In the widely used random dot direction task, CP changes quite gradually as the stimulus orientation moves away from a neuron's preferred direction (Bosking & Maunsell 2011, Cohen & Newsome 2009). In contrast, if the careful tailoring of stimulus parameters leads to CP because of small pool sizes, changing the stimulus should lead to substantial reductions in CP. To our knowledge, no study has demonstrated a rapid decrease in CP with changes in the stimulus, making it unlikely that small pools explain CP.

In the absence of clear evidence that small pools of cortical neurons underlie sensory decisions, the remainder of the review makes the assumption that large pools of neurons are responsible. As mentioned above, the existence of CPs when pool sizes are large implies that fluctuations in activity are correlated between neurons. We now discuss these correlations in more detail.

INTERNEURONAL NOISE CORRELATIONS AND CHOICE PROBABILITY

The Importance of Noise Correlation Structure

Based on the limited data available at the time, the original pooling model used a correlation structure (a relationship between noise correlation and tuning or pool identity) consisting of weak positive noise correlations between neurons within a pool (which we term $r_{up\text{-}up}$ to refer to correlations between pairs of neurons preferring upward motion) and no noise correlations between neurons in opposite pools (e.g., between an upward- and a downward-preferring neuron, $r_{up\text{-}down} = 0$). The CPs produced by the pooling model in this scheme depend on both the correlations between neurons and on how the neuronal signals are pooled (or read out by downstream neurons). To understand the relationship between readout, correlation, and CP, it is useful to consider what determines the CP found in any one model neuron (neuron "X" in **Figure 2*b***). A recent study (Haefner et al. 2012) derived an expression for this quantity and showed that

$$CP_i - 0.5 \propto \sum_{j=1}^{N} w_j r_{ij} / \beta N, \qquad 1.$$

where CPi is the CP observed in the ith neuron, w_j is the contribution of neuron j to the decision (pooling weight), and r_{ij} is the noise correlation between neuron i and neuron j. The term β depends on the weighted mean of all correlations but is the same for all neurons in a given model.

In fact, the relationship is not exactly proportional, but over the range of observed CP, Equation 1 is an excellent approximation. When the pool of neurons is large, the contribution of neuron i to the sum on the right-hand side of Equation 1 becomes negligible. Therefore, the pooling weight assigned to any one neuron has no impact on the CP recorded from that neuron. Rather, CP is determined by the mean correlation between that neuron and all other neurons, weighted according to pooling weight of each of the other neurons.

The idea that an individual neuron's pooling weight does not affect its CP can be illustrated in some informative scenarios. Consider a situation in which there are just two opposing sets of neurons ("up" and "down" pools). In this case, the two pools are given opposite weights. For example, all up-preferring neurons might have a weight of +1, and down-preferring neurons might have a weight of −1 (meaning that they contribute evidence against, rather than in support of, an up decision). In the simplest model used by Shadlen et al. (1996), correlations between all pairs of up-preferring neurons take the same value ($r_{up\text{-}up}$), whereas correlations between pairs involving one up-preferring neurons and one down-preferring neuron take a different value ($r_{up\text{-}down}$). If $r_{up\text{-}down}$ is zero, Equation 1 shows that CP depends only on $r_{up\text{-}up}$, the correlation between neurons belonging to a given pool, as found in Shadlen et al. (1996).

However, several studies have revealed that even neurons with different preferred directions tend to have weakly positive noise correlations (Cohen & Newsome 2008, Gutnisky & Dragoi 2008, Huang & Lisberger 2009, Jermakowicz et al. 2009, Kohn & Smith 2005, Smith & Kohn 2008, Zohary et al. 1994), so it is more realistic to consider the case where $r_{up\text{-}down}$ is also positive. In this case, Equation 1 shows that CP depends on $r_{up\text{-}up} - r_{up\text{-}down}$, as found in numerical simulations (Nienborg & Cumming 2010). CP does not simply depend on the overall level of correlations between all pairs of neurons in the decision pools. It depends on the structure of these correlations; there must be a specific relationship between pooling weight and noise correlation. The simplest case is that correlations between neurons within a single pool (neurons with weights of the same sign) are higher than correlations between neurons that belong to different pools (weights of opposite signs).

Because the pooling weight of any individual neuron does not affect that neuron's CP, Equation 1 accounts for simulations that included a population of model neurons making no contribution to a decision [i.e., have a weight of 0 (Cohen & Newsome 2009)]. These "irrelevant" neurons can have CP just as large as neurons that do contribute to the decision. As long as both types of neurons have the same correlations with the other neurons in the decision pool, they will have the same CP regardless of their pooling weight. In **Figure 2*b***, if the set of correlations between neuron "X" and all other neurons is fixed and N is large, then the CP observed in neuron "X" is the same whether the weight (w) is -1, 0, or 1. This lack of dependence on pooling weight occurs because the contribution of this neuron to the decision, $w_x x$ (where x is the response of neuron "X"), is negligible when N is large. Consequently any correlation between x and the decision arises only because of correlations between neuron "X" and the pools of upward- or downward-preferring neurons (U_i and D_i). For the same reasons, neurons that do contribute to a decision will not show CP if they lack the appropriate correlations. If the mean correlation between neuron "X" and all up neurons equals the mean correlation between "X" and all down neuron correlations, then CP will be 0.5, regardless of the weight w_x given to that neuron (Equation 1).

Neurons that carry the most reliable signals for discrimination (quantified using a neurometric threshold; for review, see Parker & Newsome 1998) tend to show larger CPs (Britten et al. 1996, Celebrini & Newsome 1994, Gu et al. 2008, Parker et al. 2002, Purushothaman & Bradley 2005, Romo et al. 2002, Uka & DeAngelis 2004). Because the weight assigned to a neuron does not influence its CP, this result cannot be explained by suggesting that the animals assign greater weight to these more reliable neurons. Rather, the relationship between neuronal sensitivity and CP has been explained by suggesting that neurons with weak signals also have weaker noise correlations with neurons carrying strong signals (Shadlen et al. 1996). Although this correlation between sensitivity and CP is commonly found, 20-fold changes in neuronal sensitivity have been reported across studies with little difference in CP (e.g., Britten et al. 1992, 1996; Cook & Maunsell 2002; Purushothaman & Bradley 2005). Conversely, studies finding higher CPs than typical (e.g., Dodd et al. 2001) do not show unusually high neuronal sensitivity (Parker et al. 2002). Thus, across studies, there is not a consistent relationship between CP and the ratio of neurometric threshold to psychometric threshold. This variation between studies can nonetheless be explained by differences in how neurons are weighted: Whereas psychometric thresholds depend on the relative weights of sensitive and less sensitive neurons in a pool, CPs depend on the correlation structure and only indirectly on the weights of the neurons within a pool (Equation 1). A study that looked at CP over several months of training found that as behavioral thresholds improved, CP for the more sensitive neurons gradually increased, whereas neurometric thresholds were unchanged (Law & Gold 2008). This finding was explained by suggesting that the animals learn to adjust pooling weights to favor more sensitive neurons. Equation 1 shows that this explanation requires that the less sensitive neurons that have lower weights also have weak noise correlations with more sensitive neurons. If this were not true, the terms $w_j r_{ij}$ contributed by the insensitive neurons would still produce high CP early in training.

At first sight, this dissociation between pooling weight and CP for individual neurons suggests that it is not possible to infer anything about the contributions of neurons to decisions based on CP. However, considering the entire population, CP does depend on both correlations and pooling weights. For this reason, it may still be possible to use measures of CP combined with measures of noise correlation to say something about the contribution of identified subpopulations of neurons to a decision (see Using Choice Probability to Infer Readout). Because this process still makes assumptions about what gives rise to a structured pattern of noise correlations, it is important first to consider how this pattern of correlations might arise.

Sources of Correlation Leading to Choice Probability

Equation 1 shows that in discrimination tasks, CP will arise when correlations within a pool are higher than those between pools. This finding suggests that neurons within a pool supporting one decision receive a common signal that is not shared between pools supporting opposite decisions. Because correlations are likely the largest determinant of CP, understanding the origin of this common signal is critical for understanding the origin of CPs.

Three potential (not mutually exclusive) sources of common signals could lead to CPs.

1. The common signal could reflect noise in shared, feed-forward sensory afferents. This includes sensory noise at the level of the sensory receptors and noise arising from action-potential propagation and synaptic transmission. As neurons with similar tuning preference share more inputs than do neurons with different tuning preferences the noise in such feed-forward afferents could cause higher correlations in similarly tuned neurons than in neurons with different tuning, as is required for CPs.

   The spatial structure of correlations in cortex suggests that sensory afferents are not the only inputs producing noise correlation. If they were, then noise correlations should be restricted to distances over which neurons receive common feed-forward input. In contrast, measurements in V1 have found that correlations are present over large distances of cortex (up to 10 mm; e.g., Smith & Kohn 2008), whereas correlations in retinal ganglion cells are spatially very localized and restricted largely to directly neighboring pairs (Greschner et al. 2011). This broad spatial structure suggests that, at least in area V1, shared noise in feed-forward inputs is not the only source of variability contributing to noise correlation. These spatially extended correlations in V1 mean that subsequent processing stages (e.g., MT) might show correlations over large distances that reflect correlations in the afferent input from V1.
2. The common signal could be generated within a sensory area (and perhaps transmitted in a feed-forward way to downstream areas). This signal could arise from mutual connectivity between neurons with similar tuning preference such that they share locally generated noise or from horizontal connections among functionally similar neurons (Ahmed et al. 2012, Malach et al. 1997). This source of noise likely contributes to CP, given that the majority of inputs in

cortex originate from intrinsic connections (Salin & Bullier 1995) and synaptic transmission is a substantial source of noise (Faisal et al. 2008).

These first two sources of correlation are compatible with the idea that noise correlations reflect hard-wired mechanisms that are a fixed property of the network. Such inputs likely make an important contribution to CPs, and many CP results can be well explained by these sources of correlations. More recent data, reviewed below, however, suggest that the structure of the correlation is more flexible and can depend, e.g., on the task an animal performs. Therefore, a third, more flexible, common signal likely contributes to CP.

3. The flexible common signal underlying noise correlations and CPs could reflect a top-down signal (i.e., a signal originating anywhere but the ascending pathway of the sensory-processing hierarchy preceding the sensory area from which CPs are recorded). This common signal could reflect a cognitive process such as feature attention and change with task instructions or any other downstream signal related to the decision. Such a top-down origin of the common signal implies that the correlation structure is not fixed but can change dynamically with the animal's cognitive state, e.g., with task instructions. Recent evidence suggests that top-down signals contribute to noise correlations, changes in the structure of the noise correlations, and CPs.

Several recent studies have found that noise correlations can be changed dynamically by the task instructions given to the animal. Two studies manipulating spatial attention found that noise correlation decreased when spatial attention was directed to the receptive field (Cohen & Maunsell 2009, Mitchell et al. 2009). However, although the changes in the mean noise correlation with spatial attention were dramatic, the structure of the correlation (the relationship between noise correlation and tuning curve similarity) was unchanged (Cohen & Maunsell 2009). Therefore, if spatial attention decreases overall noise correlation without changing the structure of the correlation in all tasks, it would not lead to CP changes. (Note that given that Cohen & Maunsell used a detection task, the observed changes in mean correlation have consequences for measurements of "detect probabilities" in their task, as addressed below in Detection Tasks: A Different Relationship Between Noise Correlations, Neuronal Activity, and Behavior.)

A different study looked explicitly at correlation in two discrimination tasks (Cohen & Newsome 2008) and found changes in the correlation structure that would lead to changes in CP. In interleaved trials, the animals were required to discriminate the direction of motion along orthogonal axes of motion direction, e.g., left versus right in one trial, then up versus down in a subsequent trial. The discrimination axes were carefully chosen for any given pair of neurons to ensure that for one axis the two MT neurons contributed to the same decision pool, whereas for the orthogonal axis they contributed to opposite decision pools. Cohen & Newsome found systematic changes in correlation as a function of the task. For neurons with preferred directions that differed by less than 135°, they found higher correlations when both neurons contributed to the same decision pool, compared with when they contributed to opposite pools. These correlation measures were all made using responses to a single visual stimulus, dots moving with zero coherence. Only the task context in which they were presented had changed, suggesting that these changes in correlation reflect some top-down input. The authors' simulations demonstrate that correlation change could result from fluctuations in feature-selective attention to the two choice directions. For example, while discriminating upward from downward motion, the animal may attend more strongly to upward motion on some trials and downward motion on others. These fluctuations would add a positively correlated signal to neurons belonging to the same decision pool and a negatively correlated signal to neurons in opposite pools. Using the basic pooling model, Nienborg & Cumming (2010) estimated that the signal coming from these fluctuations accounts for approximately half the observed CP in MT.

However, using the pooling model in this way makes an important assumption: that the changes in spike count produced by feature attention still contribute to the animal's decision. Alternately, the animal could know its attentional state on each trial and discount it when making a decision. If attentional changes are not discounted, and hence do contribute to decisions, there should be a systematic relationship between the attentional state and choice. Evidence in favor of this was found by Nienborg & Cumming (2009), who used a variant of reverse-correlation analysis to estimate changes in the neuronal response function with choice. They found that choices to a neuron's preferred stimulus feature were associated with an increase in neuronal response gain, similar to effects of attention.

If such a feature-selective signal is present before the decision is formed (possibly reflecting a bias or expectation), it may bias the decision via its effect on the sensory neurons. Nienborg & Cumming (2009) found indirect evidence to support this notion. On trials where small rewards were available, animals made less use of visual information in the stimulus, presumably relying more on biases. The trials with small reward were associated with slightly higher CP, suggesting that fluctuations in bias contributed to CP.

Alternatively, a feature-selective top-down signal could occur after the decision is formed (postdecision). The role of such a feature-selective feedback signal may be to serve perceptual stability, in particular when the sensory signals are weak or ambiguous, the situation in which CPs are typically measured. Nienborg & Cumming (2009) also found evidence that postdecision top-down signals may contribute to CPs. The study used a reverse-correlation approach ("psychophysical reverse correlation") to quantify how the monkeys weighted the relevant information in the visual stimulus and simultaneously measured CPs in V2. Although the weight that the animals gave to the visual stimulus decreased over the course of the trial, CPs did not decrease. As the animals give less weight to the visual inputs, stochastic variation in the neural representation of those inputs should also have less impact on choice, and hence CP should fall. Nienborg & Cumming proposed that a feature-selective feedback signal (reflecting the decision) supported CP at the end of the trial. If these changes in noise correlation during stimulus presentation were a fixed property of feedforward correlations, this might also explain the result. These changes in noise correlation during stimulus presentation being a property of fixed, feedforward noise correlations might also explain the result. Data on the structure of the correlation over time from untrained fixating or anesthetized animals are sparse and differ between studies (Samonds et al. 2009; M.A. Smith, M.A. Sommer, A. Kohn, unpublished observations), making it unclear whether the structure of feed-forward noise correlations changes in a way to account for the time course of CPs.

Further experiments will be required to understand the relationship between the effect of top-down signals on the structure of noise correlations, on the sensory representation, and on the readout of that representation. Without new data, it will be difficult to provide quantitative estimates of the contribution that top-down signals make to the structure of noise correlations, CP, and the extent to which these precede or follow the decision.

Choice Probability for More than One Stimulus Attribute Indicates Highly Structured or Flexible Correlations

Neurons in MT show CP for direction-discrimination tasks (Britten et al. 1996), disparity-discrimination tasks (Uka & DeAngelis 2004), speed-discrimination tasks (Liu & Newsome 2005), and a task that required subjects to identify the conjunctions of disparity and motion (differentiating "near and left" from "near and right"; Dodd et al. 2001). As laid out above (see The Importance of Noise Correlation Structure), the observation of CPs in each of these tasks implies that the correlation between neurons supporting the same decision is higher than the correlation between neurons supporting different decisions.

The CP observed in the disparity task implies that correlations between neurons with similar disparity preferences (e.g., both preferring "near" disparities) are higher than correlations between neurons with dissimilar disparity preferences (e.g., one preferring near disparity, one preferring far disparity) (**Figure 3*b***). For the direction-discrimination task, a mean CP in right-preferring neurons is observed because they have larger noise correlations with one another than with left-preferring neurons. But this population of right-preferring neurons includes both near- and far-preferring neurons (**Figure 3*a***), and correlations between these pairs need to be low to explain CP in the disparity task. In the

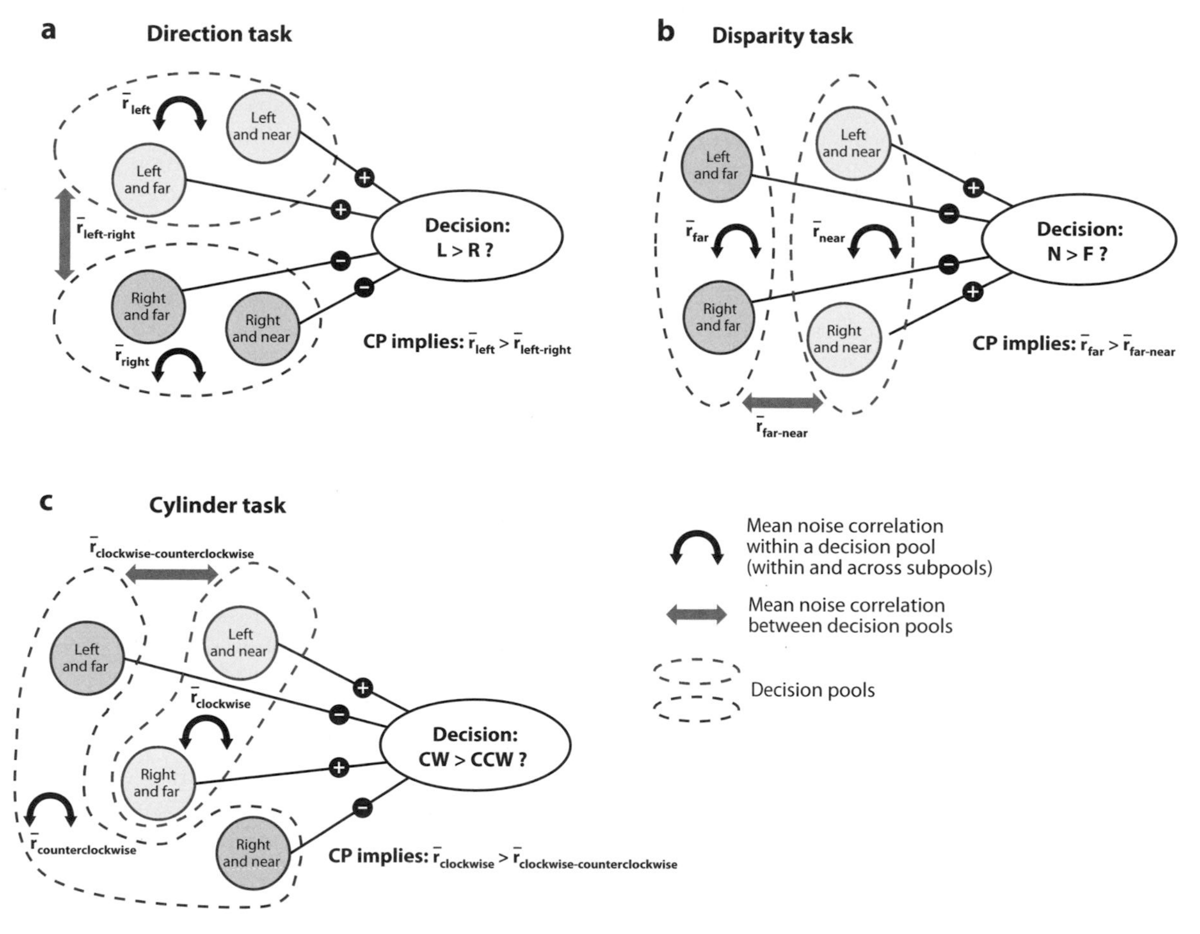

Figure 3

Using the same neurons for different tasks (motion and disparity). Four pools of neurons are shown with preferences for each conjunction: left and far, left and near, right and far, and right and near. Depending on the task being performed, these groups must be pooled differently. Dashed ellipses group together neurons in a decision pool in each case. (*a*) Direction discrimination. (*b*) Disparity discrimination. (*c*) A task involving two transparent surfaces forming the appearance of a rotating cylinder. Animals reported the conjunction of disparity and motion, corresponding to identifying whether the cylinder rotation was clockwise or counterclockwise. In each case, the requirements for interneuronal correlation to produce choice probability (CP) are different. A single pattern of interneuronal noise correlations that can explain CP in all three tasks needs to be highly structured.

conjunction task (**Figure 3*c***) near and right neurons are pooled with far and left neurons. These two groups belong to opposite pools in the direction task and in the disparity task. It is not clear that a single fixed set of interneuronal correlations can explain CP in all three tasks, and no quantitative studies have explored this.

This argument compares different studies, so it is important to confirm that CP for different tasks can be measured in the same neurons. Sasaki & Uka (2009) recorded from MT neurons in animals trained to perform both disparity- and depth-discrimination tasks and found significant CP for both tasks. They found that the direction-discrimination task produced substantially larger CPs when the animals' behavioral report involved the same eye movement used to report preferred disparity. This may be explained in principle if the animals were not reliably switching tasks according to their instructions, although the authors present evidence against this interpretation. If the effect is not explained by poor behavior, it implies that the strongest correlations exist among neurons with preferences for disparity and motion that follow the (arbitrary) convention chosen by the experimenters. It is difficult to imagine that this represents an innate property of MT. The alternative is to suggest that training has established a particular structure of interneuronal correlations in these animals. If this does happen, there would be no need to find a fixed set of noise correlations that can explain CPs for multiple tasks. Rather, the correlation structure changes with the task because of training. More work with animals trained on multiple tasks will be required to address this point. Of course, a significant top-down component that changes with task also removes the need for a single, fixed, correlation structure.

Detection Tasks: A Different Relationship Between Noise Correlations, Neuronal Activity, and Behavior

The impact of noise correlations on the relationship between the activity of any given neuron and behavior depends on the way that information from multiple neurons is combined to drive a perceptual decision. So far, our discussion has focused on discrimination tasks, in which the logical decision rule is to compute the difference in activity between two pools of neurons (each pool favoring one of the two possible choices). Other tasks, however, have different ideal algorithms for combining the activity of multiple cells.

One commonly studied task with a fundamentally different relationship among noise correlations, activity in single neurons, and behavior is a change-detection task. Rather than asking the subject to differentiate between, for example, upward from downward motion, these tasks require a subject to notice a change in a single stimulus (Bosking & Maunsell 2011; Cohen & Maunsell 2010, 2011; Cook & Maunsell 2002) or the onset of a weak stimulus (Palmer et al. 2007).

In certain detection tasks, ideal decoding often involves simply taking the (perhaps weighted) average response of a single pool of neurons. For example, the vast majority of cells respond more to a stimulus than to a uniform background, so the onset of a low-contrast stimulus could be detected simply by comparing the average response to some criterion level (Palmer et al. 2007). In other tasks, prolonged exposure to a base stimulus may lead to adaptation of neuronal responses, so most neurons will show increased responses to the changed stimulus when that adaptation is released (Cohen & Maunsell 2010, 2011).

Even when a decision is based on the average of a single neuronal pool, one may quantify the relationship between the activity of individual neurons and behavior in the same way as for a discrimination task. The detect probability (DP) is the probability that an ideal observer could predict whether an animal will detect a near-threshold change in a stimulus on the basis of the fluctuations in the responses of an individual neuron. The term detect probability was chosen to distinguish it from CP in discrimination tasks.

Unlike CPs, DPs in tasks in which the decision is based on the average activity in a single group of neurons are monotonically related to the average amount of noise correlation in the pool (rather than the difference in correlation between two pools). This result can be seen from Equation 1 if all the weights are positive. When noise correlations are low, the average response upon which the decision is based does not vary much, so the activity of one neuron is not very predictive of the animal's choices. When correlations are high, fluctuations in each neuron's responses will be correlated with the decision because they are correlated with the activity of the pool as a whole.

One important consequence of this different relationship between noise correlations and DPs is that cognitive factors and experimental artifacts can have a greater effect on DPs than CPs. Because CPs are affected only by the difference in the correlations within and between two pools, they can be affected only by a cognitive factor that affects the two pools differently. Suppose that an animal fails to direct spatial attention to the stimulus during one trial. This will reduce neuronal firing rate and reduce the probability that an animal detects any change, producing a correlation between firing rate and choice. In the direction-discrimination task, spatial attention should affect "up" and "down" pools equally and hence will not produce systematic CP. In this way, fluctuations in global factors such as arousal, alertness, or motivation or in many experimental artifacts such as fixational eye movements or blinks can introduce correlations in firing rate across the entire population and affect DP. Because CP does not depend on the overall correlation level, but on the differences in correlation between pools, these global factors should not affect CP. A study of the effects of fixational eye movements found that these did indeed affect DP, but not CP (Herrington et al. 2009). These cognitive and experimental factors therefore pose some difficulty for interpreting DPs in tasks in which the decision might be based on the average response of all neurons.

Detection tasks using stimulus changes that elicit increases in the responses of some neurons and decreases in others can be useful for determining the impact of global factors on DP. In this situation, the animal cannot detect the change simply by summing the response of all neurons, so must compare two pools. Because global factors will affect both pools, while also affecting psychophysical performance, they will produce different DP in the two groups. For example, changes in speed (Price & Born 2010) or the onset of coherent motion (Bosking & Maunsell 2011) elicit both increases and decreases in MT neurons, depending on the relationship between the particular stimulus and tuning of the neuron under study. Bosking & Maunsell (2011) found that neurons whose preferred direction matched the direction of the onset of a coherent motion stimulus had DP that was significantly greater than chance (0.58), whereas neurons with opposite tuning had a DP of 0.46 (significantly less than 0.5). Because the responses of neurons whose preferred direction matched or opposed the direction of the coherent motion are equally informative for solving this task, the responses of the two groups of neurons should ideally be given equal and opposite weight in the decision. If this were true and there were no influence of global factors on DPs, the DPs for the two groups should deviate from 0.5 in equal and opposite amounts. The observed difference in the magnitude of DP for two groups (0.58–0.5 = 0.08 for neurons whose preferred direction matched the stimulus direction and 0.5–0.46 = 0.04 for neurons whose preferred direction was opposite the stimulus direction), therefore, places an upper bound on the effects of global factors on DP.

USING CHOICE PROBABILITY TO INFER READOUT

The existence and pattern of CP are often used to make inferences about the nature of the mechanisms that read out the responses of sensory neurons to drive decisions, but CP's dependence on correlations makes these inferences difficult (Nienborg & Cumming 2007). As Equation 1 shows, for any given neuron, it is impossible to determine whether CP reflects a causal contribution to a decision or whether it reflects responses that are simply correlated with other neurons that do.

Nonetheless, CPs can place some constraints on possible readout mechanisms. For example, in MT, neurons with opposite preferred directions but similar spatial receptive fields tend to be weakly but positively correlated (Cohen & Newsome 2008, Huang & Lisberger 2009, Zohary et al. 1994). In Bosking & Maunsell's (2011) detection task, if the neurons whose preferred directions were opposite the stimulus direction were not involved in the decision, their positive correlation with neurons whose tuning matched the stimulus direction would have led to DPs greater than 0.5. That the authors observed DPs less than 0.5 suggests that the readout mechanism accounts for the fact that decreases, rather than increases, in the responses of these neurons signal motion onset. As described above (see The Importance of Noise Correlation Structure), this conclusion depends on both the observed DP and the interneuronal noise correlation. Because the latter was not measured, the authors' conclusion requires an assumption that the correlation structure during their task was the same as that reported by other investigators.

CP has also been used to assess the aspects of the neural code that are relevant for guiding behavior. Salinas et al. (2000) showed CP for spike counts (or rates) in primary sensory cortex in a tactile frequency-discrimination task. They did not observe significant CP for measures of the periodicity of the response (their estimate of a spike-timing signal). Comparing these results suggests that firing rates, rather than spike timing, affect decisions. However, this result may also reflect differences in how these different properties are correlated between neurons. We point out above (see The Importance of Noise Correlation Structure) that if a neuron's firing rate does not show appropriate correlations with other neurons, then that neuron will not show CP even if it contributes to the decision. The same argument applies to spiking periodicity: If trial-to-trial fluctuations in this periodicity are not correlated between neurons, or noise correlations in periodicity do not show the appropriate structure, then CP will be absent even if the information is being used to guide decisions. A study of choice-related responses in the rat olfactory bulb (Cury & Uchida 2011) found the opposite result: Fine temporal responses were correlated with the animal's reaction time in an olfactory discrimination task, but the mean rate was not correlated. In the future, we must identify the correlation structures for different measures of neuronal activity to determine whether rate and temporal codes play different roles in two tasks, in different brain areas, or across different species.

It Is Difficult to Infer the Time Window in Which Decisions Are Made

Ideally, CPs (and all analyses of neuronal responses contributing to perceptual decisions) would be calculated using spikes recorded over the same period during which the animal makes the perceptual decision. The timing and duration of this window likely depend on the stimulus: Tasks involving noisy stimuli or stimuli that evolve over time (Britten et al. 1996, Nienborg & Cumming 2006, 2009, Uka & DeAngelis 2006) may benefit from longer viewing durations than would studies that use high-contrast, noise-free stimuli (Ghose & Harrison 2009, Price & Born 2010). The decision window also likely depends on the task: Object recognition or discrimination may require longer viewing durations than simple onset detection tasks. In general, more difficult tasks require longer durations than easier ones do.

The decision window used by the animal can be short, variable, and difficult to determine. Many studies use stimulus presentations that last greater than one second, but evidence suggests that animals typically avail themselves of only a few hundred ms of the available stimulus duration, even in tasks in which integrating evidence over time is beneficial to performance (Cohen & Newsome 2009, Kiani et al. 2008, Roitman & Shadlen 2002). When longer stimuli are available, animals may ignore a stimulus during, for example, the period immediately after stimulus onset or after the decision has been made (Kiani et al. 2008, Nienborg & Cumming 2009). This issue is made more complicated for the experimenter because the dynamics of the decision may be different on every behavioral trial.

Reaction time tasks (in which subjects are free to respond as soon as they are ready) can be useful for identifying the decision period used by the subject because the response time provides an upper bound for the time used and because most animals are motivated to make quick decisions to receive a reward (or an easier trial) sooner. Viewing durations that more closely match the decision window provide more accurate measurements of the spikes that may be involved in the decision. In a reaction-time version of the motion-direction discrimination task, substantially less noise in the decision process is necessary to explain observed neuronal sensitivities and CP (Cohen & Newsome 2009). However, even reaction-time tasks inevitably include nondecision time, and the stimulus-presentation time may include motor-preparation delays (which often occur after stimulus offset in fixed-duration tasks). Behavioral and neuronal data can be useful for estimating these nondecision periods (Huk & Shadlen 2005, Janssen & Shadlen 2005, Mazurek et al. 2003, Palmer et al. 2005, Stanford et al. 2010). In practice, however, the amount of data required to estimate CP accurately and the fact that the animal's decision-making process likely varies from trial to trial make it difficult to estimate the decision window precisely.

The timescale of interneuronal noise correlations and autocorrelations in a single neuron's responses can affect how much misidentifying the decision period will affect CP measurements. If the timescale of interneuronal correlations is substantially shorter than the decision period and if fluctuations in neuronal responses are independent across time (i.e., the responses show little autocorrelation), then CP should be high during the decision period and at chance during other times. Therefore, counting spikes during nondecision periods should lead to lower-than-accurate measurements of CP. If, however, correlations have a long timescale or there are substantial autocorrelations, then firing rates outside the decision period should predict choices. In this case, longer measurement windows will not bias measurements of CP and may lead to more accurate measures because task-independent noise will be averaged out.

Several lines of evidence suggest that CPs can remain high during nondecision times. Most directly, several studies have examined the time course of CP by computing CP in sliding windows throughout the stimulus-presentation period. Although the time course varies somewhat across studies (and even across subjects within a study) and these calculations suffer from the weakness that the decision period may not be the same on every trial, most studies find that in fixed-duration tasks, CPs are high throughout the stimulus-presentation period. CP tends to rise after stimulus onset and remains above chance through the stimulus period (Britten et al. 1996, Celebrini & Newsome 1994, Nienborg & Cumming 2006, 2009, Uka & DeAngelis 2004). In reaction-time tasks, CP tends to decay before the saccade, which could be because the decision must be expedited to allow for motor preparation (Cohen & Newsome 2009, Cook & Maunsell 2002, Price & Born 2010).

Measurements of interneuronal correlations, autocorrelations, and fluctuations in top-down or cognitive factors are consistent with the idea that CPs could be high outside the decision period. Interneuronal correlations are typically dominated by fluctuations on the timescale of tens to hundreds of milliseconds (Bair et al. 2001, Ecker et al. 2010, Kohn & Smith 2005, Mitchell et al. 2009). MT neurons show a small amount of autocorrelation on a similar timescale (Bair et al. 2001), suggesting that CPs will remain high beyond the decision period. Two studies using briefly presented stimuli have demonstrated CPs that lasted less than 100 ms (Ghose & Harrison 2009, Price & Born 2010), challenging the above description. Such rapid decisions may be dominated by the activity of a very few neurons (Ghose &

Harrison 2009), in which case a component of CP no longer depends on correlations (see Determining the Number of Neurons Involved in a Decision). If the pools are small enough, timing of activity in single neurons could be reflected in CP. Alternatively, the temporal correlations in firing under these circumstances (not reported) may have been short enough to explain the results.

The timescale of fluctuations in top-down or cognitive factors is typically longer than that of a perceptual decision. Early attempts to measure fluctuations in spatial (Cohen & Maunsell 2010) and feature attention (Cohen & Maunsell 2011) based on the activity of a few dozen sensory neurons suggest that attention fluctuates on timescales ranging from a few hundred milliseconds to tens of seconds. Even tasks that push animals to shift attention as quickly as possible report relatively slow changes. In the absence of salient changes in a visual stimulus (exogenous attention), animals can only shift spatial attention approximately every 400 ms (Cheal & Lyon 1991, Muller et al. 1998, Muller & Rabbitt 1989). Even exogenously cued shifts of attention take at least 100–200 ms (Bisley & Goldberg 2003, Cheal & Lyon 1991, Herrington & Assad 2009, Krose & Julesz 1989, Muller & Rabbitt 1989, Nakayama & Mackeben 1989).

Together, these studies suggest that CP will remain high during nondecision times, which indicates that measuring CP is not an effective way to measure a decision period or estimate the algorithm by which an animal integrates evidence. On the flip side, the relative constancy of CP throughout a stimulus presentation indicates that it may be possible to measure CPs fairly accurately even though the precise decision window is difficult to determine.

Combining Measures of Noise Correlation and Choice Probability to Infer Readout

Observing CP in a single neuron does not imply that the neuron contributes to the decision. Therefore, it is difficult to use CP to infer the algorithms by which sensory information is read out to drive decisions. However, in cases when both CPs and noise correlations are measured within and between identified groups of neurons, one can make reliable inferences concerning how a population of neurons is read out.

A powerful example of this notion comes from a series of studies investigating how neurons in the dorsal subdivision of the medial superior temporal area (MSTd) are related to judgments of the direction of self-motion. Neurons in MSTd carry signals encoding the direction of self-motion, even when an animal is moved in total darkness (information believed to be derived from the vestibular system). These neurons have CPs during vestibular stimulation (Gu et al. 2007). MSTd also contains many cells that encode the direction of self-motion simulated by a visual stimulus when the animal is stationary. These cells also show CPs during the visual task (Gu et al. 2008).

Although many cells respond to both visual and vestibular signals, they do not necessarily signal the same motion direction for both cues. So-called incongruent cells are activated most strongly by visual stimuli indicating one direction but by vestibular stimulation indicating the opposite direction (Gu et al. 2008). A similar number of congruent cells prefer the same direction of motion regardless of the cue used. This heterogeneous population offers some unique insights into how the sensory information is used because congruent and incongruent neurons make distinctive contributions to CP in different scenarios.

Figure 4 illustrates the contribution of four groups of neurons (congruent and incongruent for each direction) to decisions in a self-motion direction-discrimination task. CP in these groups depends on both the correlations between the groups and on how the population is read out. Provided noise correlations are highest for pairs of neurons with similar tuning for both parameters (e.g., matched incongruent pairs have higher correlation than do congruent-incongruent pairs), then CP in incongruent neurons reflects the way they are read out. Gu and colleagues (2011) measured interneuronal correlations for these neurons

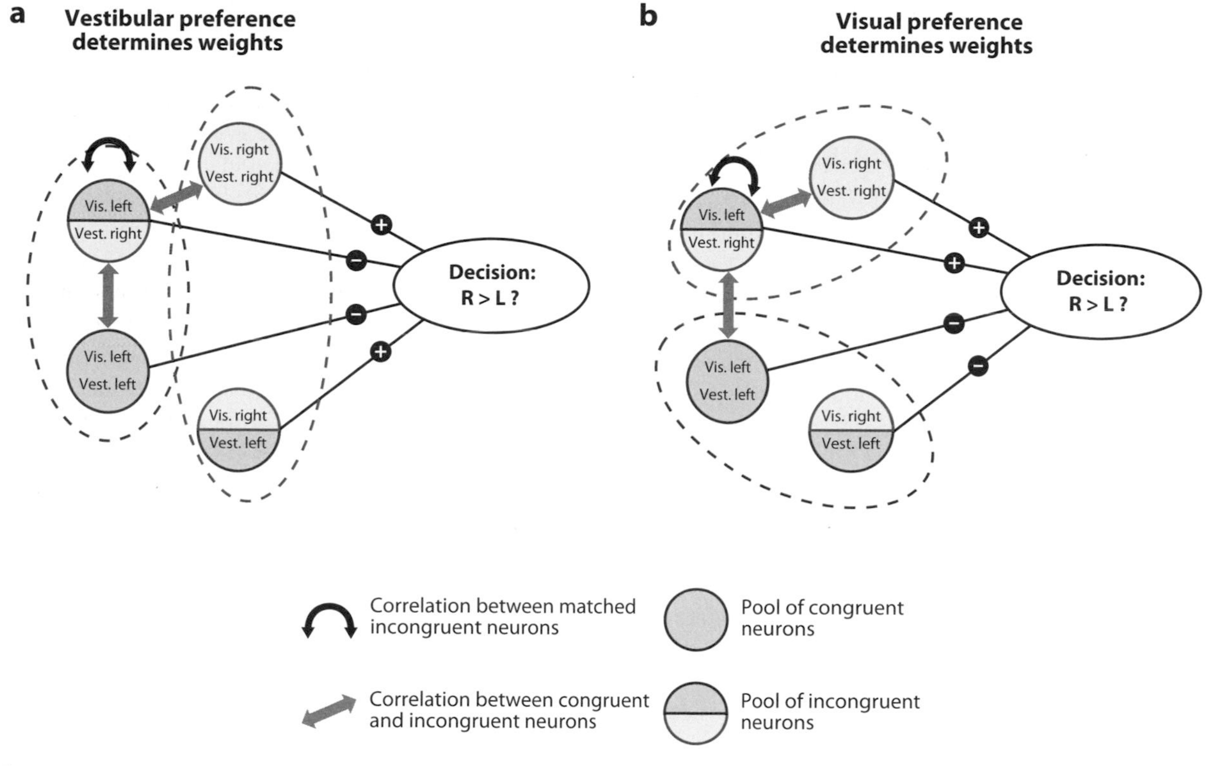

Figure 4

Combining measures of choice probability (CP) with measures of interneuronal correlation can test models of readout. Area MST (medial superior temporal) contains neurons that signal self-motion to visual and vestibular cues. Direction preference is shown by color (*Blue* = *right*, *Red* = *left*). Congruent neurons show the same preferred directions for both cues, whereas incongruent neurons show opposite preferences (*hence two colors*). Interneuronal noise correlations are shown with double-headed arrows. The darker color for correlations between neurons within a single pool indicates that these are stronger than correlations between congruent and incongruent neurons. For simplicity, only three arrows are shown, illustrating how the readout rule affects CP in visual left/vestibular right incongruent neurons. (*a*) Neurons are pooled according to vestibular preference (*dashed ellipses show pooling rule*). (*b*) Neurons are pooled according to visual preference. In both cases, there are weak correlations with congruent neurons in both pools. These correlations have opposite effects on the CP of neurons in the visual left/vestibular right group, which tend to cancel. For that reason, correlations with the other neurons in the same incongruent group largely determine the sign of CP in incongruent neurons. The result is that the sign of the CP in this group reflects how it is read out.

and found the typical relationship between tuning curve similarity and noise correlation in which correlations are highest for neurons with similar tuning. Gu and colleagues (2008) found that CP for incongruent cells was greater than 0.5 when calculated according to the vestibular preference and was therefore systematically less than 0.5 when calculated with reference to the visual preference. This observation suggests that these animals read out the activity of incongruent cells according to the direction of their vestibular preference. Of course, the results do not prove that the responses of these neurons causally contribute to the animal's decision. It is always possible that some other population of neurons forms the decision pool, and neurons in MSTd simply have appropriate noise correlations with that population. However, these experiments do demonstrate that the incongruent neurons are not read out according to their visual preference. Without measures of interneuronal correlation, even this statement would not be possible. This example illustrates that measures of CP combined with measures of interneuronal correlation do allow strong tests of hypotheses describing the way that neuronal populations are read out to guide decisions.

CONCLUSIONS

The most straightforward interpretation of the widely observed relationship between the activity of neurons in sensory cortex and animals' behavioral choices is that random fluctuations in the activity of sensory neurons influence perceptual decisions. If the decision is supported by large pools of neurons (more than $\sim$100 neurons), these random fluctuations must be correlated between members of a pool. Understanding which signals give rise to these correlations is therefore central to the interpretation of CPs. The results of multiple studies suggest that the correlation structure is not fixed but depends on the task an animal performs. Recent evidence suggests that, at least for some tasks, part of this signal reflects the influence of cognitive factors on sensory neurons, but there is currently no agreed upon method that allows the relative magnitude of flexible top-down and hard-wired bottom-up components to be quantified.

The difficulty of quantifying the relative influence of various sources of correlations on CP illustrates a wider problem: that any attempt to answer a question such as this depends on the model used. Cognitive factors are not part of the pooling model, which to date has dominated thinking about CP. More complex models of the relationship between sensory neurons and perceptual decisions will probably be required in the future. Nonlinear summation of neuronal responses (even simple nonlinearities such as thresholding) could lead to a situation in which correlated activity in some pairs of neurons has more impact on choices than it does in other pairs. Exploring correlations between identified neuronal subtypes may also help to identify populations with distinctive contributions to a decision. It may be important to consider recurrent interactions between neurons in a pool, neurons in different pools, and even neurons in different cortical areas, especially as these may have a profound effect on interneuronal correlations. Using different, more sophisticated models to study decision making may also change the questions that seem most relevant. For example, in models that include recurrent interactions, asking whether CPs and noise correlations arise from common feed-forward or top-down inputs may not be possible or even sensible.

Clearly, there is room for much more sophisticated models. The development of such models, however, must depend on empirical data that are not explained by simple pooling models. The paucity of such data illustrates the power of the pooling model. Simultaneous recordings from large groups of neurons in animals performing threshold psychophysics may be able to provide such data. New technologies for manipulating the activity of identified subgroups of neurons are also likely to be useful. Combining these new physiological methods with more refined ways of measuring animal behavior may provide a basis for models that go beyond simple pooling and clarify how activity in sensory neurons supports perceptual decisions.

DISCLOSURE STATEMENT

The authors are not aware of any affiliations, memberships, funding, or financial holdings that might be perceived as affecting the objectivity of this review.

LITERATURE CITED

Ahmed B, Cordery PM, McLelland D, Bair W, Krug K. 2012. Long-range clustered connections within extrastriate visual area V5/MT of the rhesus macaque. *Cereb. Cortex* 22:60–73

Bair W, Zohary E, Newsome WT. 2001. Correlated firing in macaque visual area MT: time scales and relationship to behavior. *J. Neurosci.* 21:1676–97

Bisley JW, Goldberg ME. 2003. Neuronal activity in the lateral intraparietal area and spatial attention. *Science* 299:81–86

Bosking WH, Maunsell JH. 2011. Effects of stimulus direction on the correlation between behavior and single units in area MT during a motion detection task. *J. Neurosci.* 31:8230–38

Britten KH, Shadlen MN, Celebrini S, Newsome WT, Movshon JA. 1996. A relationship between behavioural choice and the visual responses of neurons in macaque MT. *Vis. Neurosci.* 13:87–100

Britten KH, Shadlen MN, Newsome WT, Movshon JA. 1992. The analysis of visual motion: a comparison of neuronal and psychophysical performance. *J. Neurosci.* 12:4745–65

Celebrini S, Newsome WT. 1994. Neuronal and psychophysical sensitivity to motion signals in extrastriate area MST of the macaque monkey. *J. Neurosci.* 14:4109–24

Cheal M, Lyon DR. 1991. Central and peripheral precuing of forced-choice discrimination. *Q. J. Exp. Psychol. A* 43:859–80

Cohen MR, Kohn A. 2011. Measuring and interpreting neuronal correlations. *Nat. Neurosci.* 14:811–19

Cohen MR, Maunsell JH. 2009. Attention improves performance primarily by reducing interneuronal correlations. *Nat. Neurosci.* 12:1594–600

Cohen MR, Maunsell JH. 2010. A neuronal population measure of attention predicts behavioral performance on individual trials. *J. Neurosci.* 30:15241–53

Cohen MR, Maunsell JH. 2011. Using neuronal populations to study the mechanisms underlying spatial and feature attention. *Neuron* 70:1192–204

Cohen MR, Newsome WT. 2008. Context-dependent changes in functional circuitry in visual area MT. *Neuron* 60:162–73

Cohen MR, Newsome WT. 2009. Estimates of the contribution of single neurons to perception depend on timescale and noise correlation. *J. Neurosci.* 29:6635–48

Cook EP, Maunsell JH. 2002. Attentional modulation of behavioral performance and neuronal responses in middle temporal and ventral intraparietal areas of macaque monkey. *J. Neurosci.* 22:1994–2004

Cury KM, Uchida N. 2011. Robust odor coding via inhalation-coupled transient activity in the mammalian olfactory bulb. *Neuron* 68:570–85

Dodd JV, Krug K, Cumming BG, Parker AJ. 2001. Perceptually bistable three-dimensional figures evoke high choice probabilities in cortical area MT. *J. Neurosci.* 21:4809–21

Ecker AS, Berens P, Keliris GA, Bethge M, Logothetis NK, Tolias AS. 2010. Decorrelated neuronal firing in cortical microcircuits. *Science* 327:584–87

Faisal AA, Selen LP, Wolpert DM. 2008. Noise in the nervous system. *Nat. Rev. Neurosci.* 9:292–303

Ghose GM, Harrison IT. 2009. Temporal precision of neuronal information in a rapid perceptual judgment. *J. Neurophysiol.* 101:1480–93

Greschner M, Shlens J, Bakolitsa C, Field GD, Gauthier JL, et al. 2011. Correlated firing among major ganglion cell types in primate retina. *J. Physiol.* 589:75–86

Gu Y, Angelaki DE, Deangelis GC. 2008. Neural correlates of multisensory cue integration in macaque MSTd. *Nat. Neurosci.* 11:1201–10

Gu Y, DeAngelis GC, Angelaki DE. 2007. A functional link between area MSTd and heading perception based on vestibular signals. *Nat. Neurosci.* 10:1038–47

Gu Y, Liu S, Fetsch CR, Yang Y, Fok S, et al. 2011. Perceptual learning reduces interneuronal correlations in macaque visual cortex. *Neuron* 71:750–61

Gutnisky DA, Dragoi V. 2008. Adaptive coding of visual information in neural populations. *Nature* 452:220–24

Haefner RM, Gerwinn S, Macke JH, Bethge M. 2012. Inferring decoding strategy from choice probabilities in the presence of noise correlations. *Nat. Precedings.* **http://hdl.handle.net/10101/npre.2012.7014.1**

Herrington TM, Assad JA. 2009. Neural activity in the middle temporal area and lateral intraparietal area during endogenously cued shifts of attention. *J. Neurosci.* 29:14160–76

Herrington TM, Masse NY, Hachmeh KJ, Smith JE, Assad JA, Cook EP. 2009. The effect of microsaccades on the correlation between neural activity and behavior in middle temporal, ventral intraparietal, and lateral intraparietal areas. *J. Neurosci.* 29:5793–805

Huang X, Lisberger SG. 2009. Noise correlations in cortical area MT and their potential impact on trial-by-trial variation in the direction and speed of smooth-pursuit eye movements. *J. Neurophysiol.* 101:3012–30
Huk AC, Shadlen MN. 2005. Neural activity in macaque parietal cortex reflects temporal integration of visual motion signals during perceptual decision making. *J. Neurosci.* 25:10420–36
Janssen P, Shadlen MN. 2005. A representation of the hazard rate of elapsed time in macaque area LIP. *Nat. Neurosci.* 8:234–41
Jermakowicz WJ, Chen X, Khaytin I, Bonds AB, Casagrande VA. 2009. Relationship between spontaneous and evoked spike-time correlations in primate visual cortex. *J. Neurophysiol.* 101:2279–89
Kiani R, Hanks TD, Shadlen MN. 2008. Bounded integration in parietal cortex underlies decisions even when viewing duration is dictated by the environment. *J. Neurosci.* 28:3017–29
Kohn A, Smith MA. 2005. Stimulus dependence of neuronal correlation in primary visual cortex of the macaque. *J. Neurosci.* 25:3661–73
Krose BJ, Julesz B. 1989. The control and speed of shifts of attention. *Vis. Res.* 29:1607–19
Law CT, Gold JI. 2008. Neural correlates of perceptual learning in a sensory-motor, but not a sensory, cortical area. *Nat. Neurosci.* 11:505–13
Liu J, Newsome WT. 2005. Correlation between speed perception and neural activity in the middle temporal visual area. *J. Neurosci.* 25:711–22
Malach R, Schirman TD, Harel M, Tootell RB, Malonek D. 1997. Organization of intrinsic connections in owl monkey area MT. *Cereb. Cortex* 7:386–93
Mazurek ME, Roitman JD, Ditterich J, Shadlen MN. 2003. A role for neural integrators in perceptual decision making. *Cereb. Cortex* 13:1257–69
Mitchell JF, Sundberg KA, Reynolds JH. 2009. Spatial attention decorrelates intrinsic activity fluctuations in macaque area V4. *Neuron* 63:879–88
Muller HJ, Rabbitt PM. 1989. Reflexive and voluntary orienting of visual attention: time course of activation and resistance to interruption. *J. Exp. Psychol. Hum. Percept. Perform.* 15:315–30
Muller MM, Teder-Salejarvi W, Hillyard SA. 1998. The time course of cortical facilitation during cued shifts of spatial attention. *Nat. Neurosci.* 1:631–34
Nakayama K, Mackeben M. 1989. Sustained and transient components of focal visual attention. *Vis. Res.* 29:1631–47
Nienborg H, Cumming BG. 2006. Macaque V2 neurons, but not V1 neurons, show choice-related activity. *J. Neurosci.* 26:9567–78
Nienborg H, Cumming BG. 2007. Psychophysically measured task strategy for disparity discrimination is reflected in V2 neurons. *Nat. Neurosci.* 10:1608–14
Nienborg H, Cumming BG. 2009. Decision-related activity in sensory neurons reflects more than a neuron's causal effect. *Nature* 459:89–92
Nienborg H, Cumming B. 2010. Correlations between the activity of sensory neurons and behavior: How much do they tell us about a neuron's causality? *Curr. Opin. Neurobiol.* 20:376–81
Palmer C, Cheng SY, Seidemann E. 2007. Linking neuronal and behavioral performance in a reaction-time visual detection task. *J. Neurosci.* 27:8122–37
Palmer J, Huk AC, Shadlen MN. 2005. The effect of stimulus strength on the speed and accuracy of a perceptual decision. *J. Vis.* 5:376–404
Parker AJ, Krug K, Cumming BG. 2002. Neuronal activity and its links with the perception of multi-stable figures. *Philos. Trans. R. Soc. Lond. B Biol. Sci.* 357:1053–62
Parker AJ, Newsome WT. 1998. Sense and the single neuron: probing the physiology of perception. *Annu. Rev. Neurosci.* 21:227–77
Price NS, Born RT. 2010. Timescales of sensory- and decision-related activity in the middle temporal and medial superior temporal areas. *J. Neurosci.* 30:14036–45
Purushothaman G, Bradley DC. 2005. Neural population code for fine perceptual decisions in area MT. *Nat. Neurosci.* 8:99–106
Roitman JD, Shadlen MN. 2002. Response of neurons in the lateral intraparietal area during a combined visual discrimination reaction time task. *J. Neurosci.* 22:9475–89
Romo R, Hernandez A, Zainos A, Lemus L, Brody CD. 2002. Neuronal correlates of decision-making in secondary somatosensory cortex. *Nat. Neurosci.* 5:1217–25

Salin PA, Bullier J. 1995. Corticocortical connections in the visual system: structure and function. *Physiol. Rev.* 75:107–54

Salinas E, Hernandez A, Zainos A, Romo R. 2000. Periodicity and firing rate as candidate neural codes for the frequency of vibrotactile stimuli. *J. Neurosci.* 20:5503–15

Samonds JM, Potetz BR, Lee TS. 2009. Cooperative and competitive interactions facilitate stereo computations in macaque primary visual cortex. *J. Neurosci.* 29:15780–95

Sasaki R, Uka T. 2009. Dynamic readout of behaviorally relevant signals from area MT during task switching. *Neuron* 62:147–57

Shadlen MN, Britten KH, Newsome WT, Movshon JA. 1996. A computational analysis of the relationship between neuronal and behavioural responses to visual motion. *J. Neurosci.* 16:1486–510

Smith MA, Kohn A. 2008. Spatial and temporal scales of neuronal correlation in primary visual cortex. *J. Neurosci.* 28:12591–603

Stanford TR, Shankar S, Massoglia DP, Costello MG, Salinas E. 2010. Perceptual decision making in less than 30 milliseconds. *Nat. Neurosci.* 13:379–85

Tolhurst DJ, Movshon JA, Dean AF. 1983. The statistical reliability of signals in single neurons in cat and monkey visual cortex. *Vis. Res.* 23:775–85

Uka T, DeAngelis GC. 2004. Contribution of area MT to stereoscopic depth perception: choice-related response modulations reflect task strategy. *Neuron* 42:297–310

Uka T, DeAngelis GC. 2006. Linking neural representation to function in stereoscopic depth perception: roles of the middle temporal area in coarse versus fine disparity discrimination. *J. Neurosci.* 26:6791–802

Zohary E, Shadlen MN, Newsome WT. 1994. Correlated neuronal discharge rate and its implications for psychophysical performance. *Nature* 370:140–43

Compressed Sensing, Sparsity, and Dimensionality in Neuronal Information Processing and Data Analysis

Surya Ganguli[1] and Haim Sompolinsky[2,3]

[1]Department of Applied Physics, Stanford University, Stanford, California 94305; email: sganguli@stanford.edu

[2]Edmond and Lily Safra Center for Brain Sciences, Interdisciplinary Center for Neural Computation, Hebrew University, Jerusalem 91904, Israel; email: haim@fiz.huji.ac.il

[3]Center for Brain Science, Harvard University, Cambridge, Massachusetts 02138

Annu. Rev. Neurosci. 2012. 35:485–508

First published online as a Review in Advance on April 5, 2012

The *Annual Review of Neuroscience* is online at neuro.annualreviews.org

This article's doi:
10.1146/annurev-neuro-062111-150410

0147-006X/12/0721-0485$20.00

Keywords

random projections, connectomics, imaging, memory, communication, learning, generalization

Abstract

The curse of dimensionality poses severe challenges to both technical and conceptual progress in neuroscience. In particular, it plagues our ability to acquire, process, and model high-dimensional data sets. Moreover, neural systems must cope with the challenge of processing data in high dimensions to learn and operate successfully within a complex world. We review recent mathematical advances that provide ways to combat dimensionality in specific situations. These advances shed light on two dual questions in neuroscience. First, how can we as neuroscientists rapidly acquire high-dimensional data from the brain and subsequently extract meaningful models from limited amounts of these data? And second, how do brains themselves process information in their intrinsically high-dimensional patterns of neural activity as well as learn meaningful, generalizable models of the external world from limited experience?

Contents

INTRODUCTION

For most of its history, neuroscience has made wonderful progress by considering problems whose descriptions require only a small number of variables. For example, Hodgkin & Huxley (1952) discovered the mechanism of the nerve impulse by studying the relationship between two variables: the voltage and the current across the cell membrane. But as we have started to explore more complex problems, such as the brain's ability to process images and sounds, neuroscientists have had to analyze many variables at once. For example, any given gray-scale image requires N analog variables, or pixel intensities, for its description, where N could be on the order of 1 million. Similarly, such images could be represented in the firing-rate patterns of many neurons, with each neuron's firing rate being a single analog variable. The number of variables required to describe a space of objects is known as the dimensionality of that space; i.e., the dimensionality of the space of all possible images of a given size equals the number of pixels, whereas the dimensionality of the space of all possible neuronal firing-rate patterns in a given brain area equals the number of neurons in that area. Thus our quest to understand how networks of neurons store and process information depends crucially on our ability to measure and understand the relationships between high-dimensional spaces of stimuli and neuronal activity patterns.

However, the problem of measuring and finding statistical relationships between patterns becomes more difficult as their dimensionality increases. This phenomenon is known as the curse of dimensionality. One approach to addressing this problem is to somehow reduce the number of variables required to describe the patterns in question, a process known as dimensionality reduction. We can do this, for example, with natural images, which are a highly restricted subset of all possible images, so that they can be described by many fewer variables than the number of pixels. In particular, natural images are often sparse in the sense that if you view them in the wavelet domain (roughly as a superposition of edges), only a very small number of K wavelet coefficients will have significant power, where K can be on the order of 20,000 for a 1-million-pixel image. This observation underlies JPEG compression, which computes all possible wavelet coefficients and keeps only the K largest (Taubman et al. 2002). Similarly, neuronal activity patterns that actually occur are often a highly restricted subset of all possible patterns (Ganguli et al. 2008a, Yu et al. 2009, Machens et al. 2010) in the sense that they often lie along a low K-dimensional manifold embedded in N-dimensional firing-rate space; by this we mean that only K numbers are required to uniquely specify any observed activity pattern across N neurons, where K can be much smaller than N. As a concrete example, consider the set of visual activity patterns in N neurons in response to a bar presented at a variety of orientations. As the orientation varies, the elicited firing-rate responses trace out a circle, or a one-dimensional manifold in N-dimensional space.

More generally, given a class of apparently high-dimensional stimuli, or neuronal activity patterns, how can either we or neural systems extract a small number of variables to describe these patterns without losing too much important information? Machine learning provides a variety of algorithms to perform this dimensionality reduction, but they are often computationally expensive in terms of running time. Moreover, how neuronal circuits could implement many of these algorithms is not clear. However, recent advances in an emerging field of high-dimensional statistics (Donoho 2000, Baraniuk 2011) have revealed a surprisingly simple yet powerful method of performing dimensionality reduction: One can randomly project patterns into a lower-dimensional space. To understand the central concept of a random projection (RP), it is useful to think of the shadow of a wire-frame object in three-dimensional space projected onto a two-dimensional screen by shining a light beam on the object. For poorly chosen angles of light, the shadow may lose important information about the wire-frame object. For example, if the axis of light is aligned with any segment of wire, that entire length of wire will have a single point as its shadow. However, if the axis of light is chosen randomly, it is highly unlikely that the same degenerate situation will occur; instead, every length of wire will have a corresponding nonzero length of shadow. Thus the shadow, obtained by this RP, generically retains much information about the wire-frame object.

In the context of image acquisition, an RP of an image down to an M-dimensional space can be obtained by taking M measurements of the image, where each measurement consists of a weighted sum of all the pixel intensities, and allowing the weights themselves to be chosen randomly (for example, drawn independently from a Gaussian distribution). Thus the original image (i.e., the wire-frame structure) is described by M measurements (i.e., its shadow) by projecting against a random set of weights (i.e., a random light angle). Now, the field of compressed sensing (CS) (Candes et al. 2006, Candes & Tao 2006, Donoho 2006; see Baraniuk 2007, Candes & Wakin 2008, Bruckstein et al. 2009 for reviews) shows that the shadow can contain enough information to reconstruct the original image (i.e., all N pixel values) as long as the original image is sparse enough. In particular, if the space of the images in question can be described by K variables, then as long as M is slightly larger than K, CS provides an algorithm (called L_1 minimization, described below) to reconstruct the image. Thus for typical images, we can simultaneously sense and compress 1-million-pixel images with ~20,000 random measurements. As we review below, these CS results have significant implications for data acquisition in neuroscience.

Furthermore, in the context of neuronal information processing, an RP of neuronal activity in an upstream brain region consisting of N neurons can be achieved by synaptic mapping to a downstream region consisting of $M < N$ neurons, where the downstream neurons' firing rates are obtained by linearly summing the firing rates of the upstream neurons through a set of random synaptic weights. Thus the downstream activity constitutes a shadow of the upstream activity through an RP determined by the synaptic weights (i.e., angle of light). As we review below, the theory of CS and RPs can provide a theoretical framework for understanding one of the most salient aspects of neuronal information processing: radical changes in the dimensionality, and sometimes sparsity, of neuronal representations, often within a single stage of synaptic transformation.

Finally, another application of CS is the problem of modeling high-dimensional data. This is challenging because such models have high-dimensional parameter spaces, necessitating many example data points to learn the correct parameter values. Neural systems face a similar challenge in searching high-dimensional synaptic weight spaces to learn generalizable rules from limited experience. We review how regularization techniques (Tibshirani 1996, Efron et al. 2004) closely related to CS allow statisticians and neural systems alike to rapidly learn sparse models of high-dimensional data from limited examples.

ADVANCES IN THE THEORY OF HIGH-DIMENSIONAL STATISTICS

Before we describe the applicability of CS and RPs to the acquisition and analysis of data and to neuronal information processing and learning, we first give in this section a more precise overview of recent results in high-dimensional statistics. We begin by giving an overview of the CS framework and define the mathematical notation we use throughout this review. Subsequently, a reader who is interested mainly in applications can skip the rest of this section. Here, we discuss how to recover the sparse signals from small numbers of measurements, even in the presence of approximate sparsity and noise, and we discuss RPs and sparse regression in more detail. Finally, we discuss dictionary learning, an approach to find bases in which ensembles of signals are sparse.

The Compressed Sensing Framework: Incoherence and Randomness

We now formalize the intuitions given in the introduction and describe the mathematical notation that we use throughout this review (see also **Figure 1**). We let $\mathbf{u}^0$ be an N-dimensional signal that we wish to measure. Thus $\mathbf{u}^0$ is a vector with components $\mathbf{u}_i^0$ for $i = 1, \ldots, N$, where each $\mathbf{u}_i^0$ can take an analog value. In the example of an image, $\mathbf{u}_i^0$ would be the gray-scale intensity of the ith pixel. The M linear measurements of $\mathbf{u}^0$ are of the form $x_\mu = \mathbf{b}^\mu \cdot \mathbf{u}^0$ for $\mu = 1, \ldots, M$. Here we think of x_μ as an analog outcome of measurement μ obtained by computing the overlap or dot product between the unknown signal $\mathbf{u}^0$ and a measurement vector $\mathbf{b}^\mu$. We can summarize the relationship between the signal and the measurements via the matrix relationship $\mathbf{x} = \mathbf{B}\mathbf{u}^0$. Here $\mathbf{B}$ is an $M \times N$ measurement matrix, whose μth row is the vector $\mathbf{b}^\mu$, and $\mathbf{x}$ is a measurement vector whose μ'th component is x_μ. Now the true signal $\mathbf{u}^0$ is sparse in a basis given by the columns of an $N \times N$ matrix $\mathbf{C}$. By this we mean that $\mathbf{u}^0 = \mathbf{C}\mathbf{s}^0$, where $\mathbf{s}^0$ is a sparse N-dimensional vector, in the sense that it has a relatively small number K of nonzero elements, though we do not know ahead of time which K of the N components are nonzero. For example, when $\mathbf{u}^0$ is an image in the pixel basis, $\mathbf{s}^0$ could be the wavelet coefficients of that same image, and the columns of $\mathbf{C}$ would comprise a complete basis

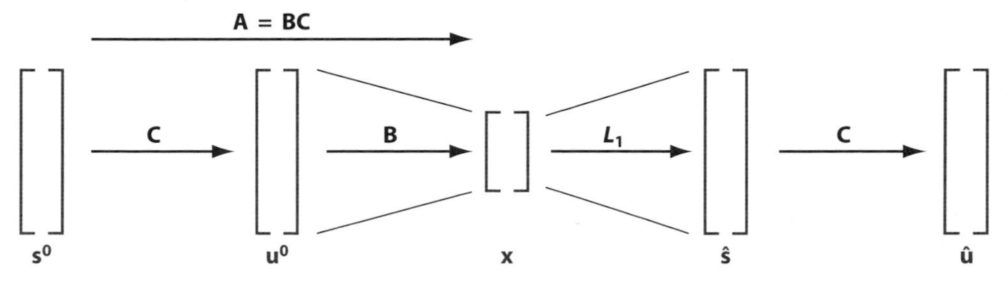

Figure 1

Framework of compressed sensing (CS). A high-dimensional signal $\mathbf{u}^0$ is sparse in a basis given by the columns of a matrix $\mathbf{C}$ so that $\mathbf{u}^0 = \mathbf{C}\mathbf{s}^0$, where $\mathbf{s}^0$ is a sparse coefficient vector. Through a set of measurements given by the rows of $\mathbf{B}$, $\mathbf{u}^0$ is compressed to a low-dimensional space of measurements $\mathbf{x}$. If the measurements are incoherent with respect to the sparsity basis, then L_1 minimization can recover a good estimate $\hat{\mathbf{s}}$ of the sparse coefficients $\mathbf{s}^0$ from $\mathbf{x}$, and then an estimate of $\mathbf{u}^0$ can be recovered by expanding in the basis $\mathbf{C}$.

of orthonormal wavelets. Finally, the overall relationship between the measurements and the sparse coefficients is given by $\mathbf{x} = \mathbf{A}\mathbf{s}^0$, where $\mathbf{A} = \mathbf{BC}$. We often refer to $\mathbf{A}$ also as the measurement matrix.

An important question is, given a sparsity basis $\mathbf{C}$, what should we choose as our measurement basis $\mathbf{B}$? Consider what might happen if we measured signals in the same basis in which they were sparse. For example, in the case of an image, one could directly measure M randomly chosen wavelet coefficients of the image in which M is just a little larger than K. The problem, of course, is that for any given image, it is highly unlikely that all the K coefficients with large power coincide with the M coefficients we chose to measure. So unless the number of measurements M equals the dimensionality of the image, N, we will inevitably miss important coefficients. In the wire-frame shadow example above, this is the analog of choosing a poor angle of light (i.e., measurement basis) that aligns with a segment of wire (i.e., sparsity basis), which causes information loss.

To circumvent this problem, one of the key ideas of CS is that we should make our measurements as different as possible from the domain in which the signal is sparse (i.e., shine light at an angle that does not align with any segment of wire frame). In particular, the measurements should have many nonzero elements in the domain in which the image is sparse. This notion of difference is captured by the mathematical definition of incoherence, or a small value of the maximal inner product between rows of $\mathbf{B}$ and columns of $\mathbf{C}$, so that no measurement vector should look like any sparsity vector. CS provides mathematical guarantees that one can achieve perfect recovery with a number of measurements M that is only slightly larger than K, as long as the M measurement vectors are sufficiently incoherent with respect to the sparsity domain (Candes & Romberg 2007).

An important observation is that any set of measurement vectors, which are themselves random, will be incoherent with respect to any fixed sparsity domain. For example, the elements of each such measurement vector can be drawn independently from a Gaussian distribution. Intuitively, it is highly unlikely for a random vector to look like a sparsity vector (i.e., just as it is unlikely for a random light angle to align with a wire segment). One of the key results of CS is that with such random measurement vectors, only

$$M > O(K \log(N/K)) \qquad 1.$$

measurements are needed to guarantee perfect signal reconstruction with high probability (Candes & Tao 2005, Baraniuk et al. 2008, Candes & Plan 2010). Thus random measurements constitute a universal measurement strategy in the sense that they will work for signals that are sparse in any basis. Indeed, the sparsity basis need not even be known yet when the measurements are chosen. Its knowledge is required only after measurements are taken, during the nonlinear reconstruction process. And remarkably, investigators have further shown that no measurement matrices and no reconstruction algorithm can yield sparse signal recovery with substantially fewer measurements (Candes & Tao 2006, Donoho 2006) than that shown in Equation 1.

L_1 Minimization: A Nonlinear Recovery Algorithm

Given only our measurements $\mathbf{x}$, how can we recover the unknown signal $\mathbf{u}^0$? One could potentially do this by inverting the relationship between measurements and signal by solving for an unknown candidate signal $\mathbf{u}$ in the equation $\mathbf{x} = \mathbf{Bu}$. This is a set of M equations, one for each measurement, with N unknowns, one for each component of the candidate signal $\mathbf{u}$. If the number of independent measurements M is greater than or equal to the dimensionality N of the signal, then the set of equations $\mathbf{x} = \mathbf{Bu}$ has a unique solution $\mathbf{u} = \mathbf{u}^0$; thus, solving these equations will recover the true signal $\mathbf{u}^0$. However, if $M < N$, the set of equations $\mathbf{x} = \mathbf{Bu}$ no longer has a unique solution. Indeed there is generically an $N - M$ dimensional space of candidate signals $\mathbf{u}$ that satisfy the measurement constraints. How might we find the true signal $\mathbf{u}^0$ in this large space of candidate signals?

If we know nothing further about the true signal $\mathbf{u}^0$, then the situation is indeed hopeless. However, if $\mathbf{u}^0 = \mathbf{Cs}^0$ where $\mathbf{s}^0$ is sparse, we can try to exploit this prior knowledge as follows (see **Figure 1**). First, the measurements are linearly related to the sparse coefficients $\mathbf{s}^0$ through the M equations $\mathbf{x} = \mathbf{As}^0$, where $\mathbf{A} = \mathbf{BC}$ is an $M \times N$ matrix. Again, when $M < N$, there is a large $N - M$ dimensional space of solutions $\mathbf{s}$ to the measurement constraint $\mathbf{x} = \mathbf{As}$. However, not all of them will be sparse, as we expect the true solution $\mathbf{s}^0$ to be. Thus one might try to construct an estimate $\hat{\mathbf{s}}$ of $\mathbf{s}^0$ by solving the optimization problem

$$\hat{\mathbf{s}} = \arg\min_{\mathbf{s}} \sum_{i=1}^{N} V(s_i) \quad \text{subject to } \mathbf{x} = \mathbf{As}, \quad 2.$$

where $V(s)$ is any cost function that penalizes nonzero values of s. A natural choice is $V(s) = 0$ if $s = 0$ and $V(s) = 1$ otherwise. With this choice, Equation 2 says that our estimate $\hat{\mathbf{s}}$ is obtained by searching, in the space of all candidate signals $\mathbf{s}$ that satisfy the measurement constraints $\mathbf{x} = \mathbf{As}$, for the one that has the smallest number of nonzero elements. This approach, while reasonable given the prior knowledge that the true signal $\mathbf{s}^0$ has a small number of nonzero coefficients, unfortunately yields a computationally intractable combinatorial optimization problem; to solve it, one must essentially search over all subsets of possible nonzero elements in $\mathbf{s}$.

An alternative approach, adopted by CS, is to solve a related and potentially easier problem, by choosing $V(s) = |s|$. The quantity $\sum_{i=1}^{N} |s_i|$ is known as L_1 norm of $\mathbf{s}$; hence, this method is called L_1 minimization. The advantage of this choice is that the L_1 norm is a convex function on the space of candidate signals, which implies that the optimization problem in Equation 2, with $V(s) = |s|$, has no (nonglobal) local minima, and there are efficient algorithms for finding the global minimum using methods of linear programming (Boyd & Vandenberghe 2004), message passing (Donoho et al. 2009), and neural circuit dynamics (see below). CS theory shows that with an appropriate choice of $\mathbf{A}$, L_1 minimization exactly recovers the true signal so that $\hat{\mathbf{s}} = \mathbf{s}^0$, with a number of measurements that is roughly proportional to the number of nonzero elements in the source, K, which can be much smaller than the dimensionality N of the signal.

A popular and even simpler reconstruction algorithm is L_2 minimization in which $V(s) = s^2$ in Equation 2. This result can arise as a consequence of oft-used Gaussian priors on the unknown signal and leads to an estimate that is simply linearly related to the measurements through the pseudoinverse relation

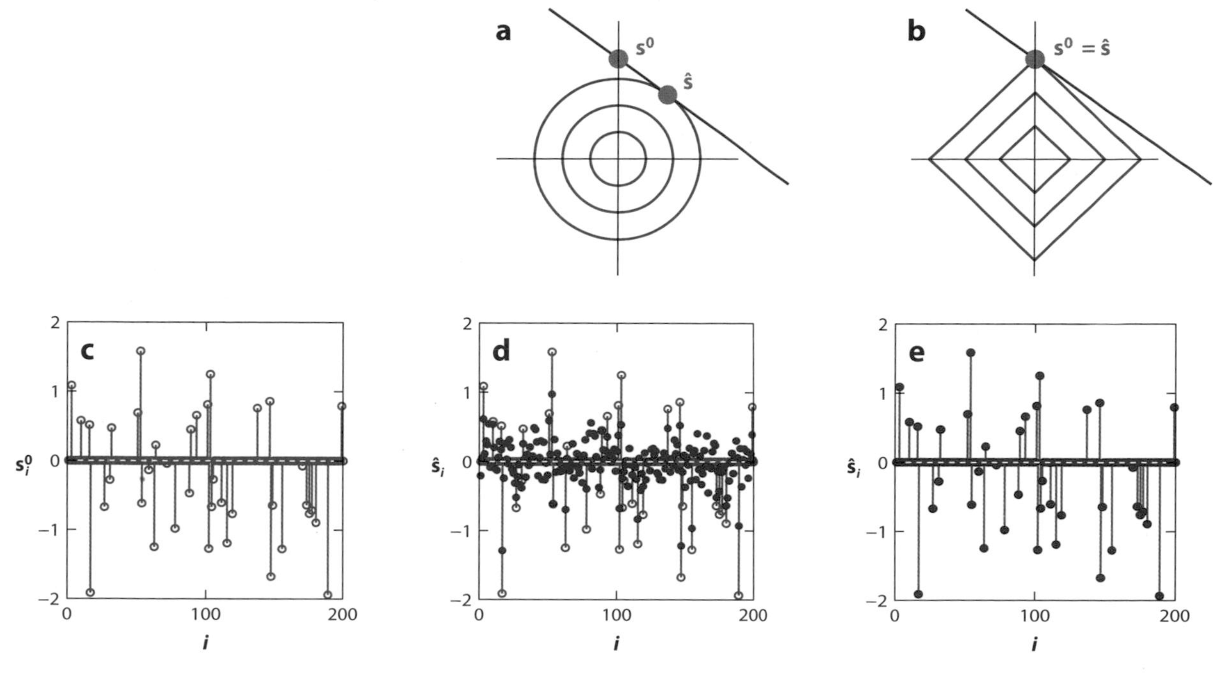

Figure 2

Geometry of compressed sensing (CS). (*a*) A geometric interpretation of L_2 minimization. An unknown $N = 2$ dimensional sparse signal $\mathbf{s}^0$ with $K = 1$ nonzero components is measured using $M = 1$ linear measurements, yielding a one-dimensional space of candidate signals consistent with the measurement constraints (*red line*). The estimate $\hat{\mathbf{s}}$ is the candidate signal with the smallest L_2 norm and can be found geometrically by expanding the locus of points with a fixed and increasing L_2 norm (*the olive circles*) until the locus first intersects the allowed space of candidate signals. This intersection point is the L_2 estimate $\hat{\mathbf{s}}$, which is different from the true signal $\mathbf{s}^0$. (*b*) In the identical scenario as in panel *a*, L_1 minimization recovers an estimate by expanding the locus of points with the same L_1 norm (*blue diamonds*), and in this case, the expanding locus first intersects the space of candidate signals at the true signal $\mathbf{s}^0$ so that perfect recovery $\hat{\mathbf{s}} = \mathbf{s}^0$ is achieved. Of course, a sparse signal could also have been located on the other coordinate axis, in which case L_1 minimization would have failed to recover $\mathbf{s}^0$ accurately. (*c*) An unknown sparse signal $\mathbf{s}^0$ of dimension $N = 200$, with $f = K/N = 0.2$, i.e., 20% of its elements are nonzero. (*d*) An estimate $\hat{\mathbf{s}}$ (*red dots*) recovered from $M = 120$ random linear measurements of $\mathbf{s}^0$ ($\alpha = N/T = 0.6$, or 60% subsampling) by L_2 minimization superimposed on the true signal $\mathbf{s}^0$. (*e*) From the same measurements in panel *d*, L_1 minimization yields an estimate $\hat{\mathbf{s}}$ (*red dots*) that coincides with the true signal. Note that the parameters of $f = 0.2$ and $\alpha = 0.6$ lie just above the phase boundary for perfect recovery in **Figure 3**.

$\hat{\mathbf{s}} = (\mathbf{A}^T\mathbf{A})^{-1}\mathbf{A}^T\mathbf{x}$. **Figure 2** provides heuristic intuition for the utility of L_1 minimization and its superior performance over L_2 minimization in the case of sparse signals.

An interesting observation is that the bound in Equation 1 represents a sufficient condition on the number of measurements M for perfect signal recovery. Alternately, recent work on the typical behavior of CS in the limit where M and N are large has revealed that the performance of CS is surprisingly insensitive to the details of the measurement matrix $\mathbf{A}$ and the unknown signal $\mathbf{s}^0$ and depends only on the degree of subsampling $\alpha = M/N$ and the signal sparsity $f = K/N$. In the $\alpha - f$ plane, there is a universal, critical phase boundary $\alpha_c(f)$ such that if $\alpha > \alpha_c(f)$, then L_1 minimization will typically yield perfect signal reconstruction, whereas if $\alpha < \alpha_c(f)$, it will yield a nonzero error (see **Figure 3**) (Donoho & Tanner 2005a,b, Donoho et al. 2009, Kabashima et al. 2009, Ganguli & Sompolinsky 2010b).

Dimensionality Reduction by Random Projections

The above CS results can be understood using the theory of RPs. Geometrically, the mapping

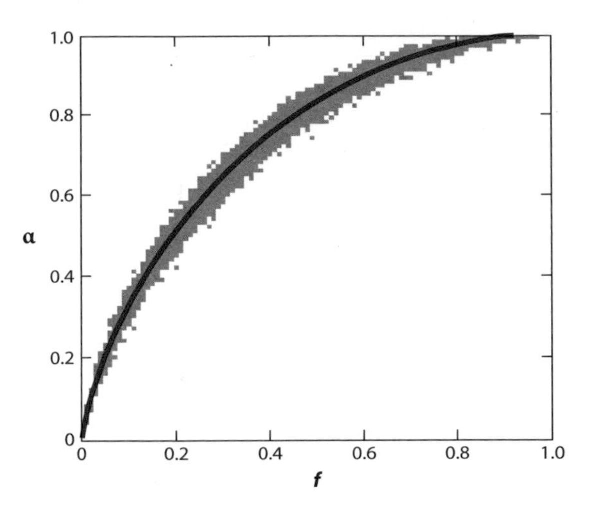

Figure 3

Phase transition in compressed sensing (CS) (reproduced from Ganguli & Sompolinsky 2010b). We use linear programming to solve Equation 2 50 times for each value of α and f in increments of 0.01, with $N = 500$. The grey transition region shows when the fraction of times perfect recovery occurs is neither 0 nor 1. The red curve is the theoretical phase boundary $\alpha_c(f)$. As $f \to 0$, this boundary is of the form $\alpha_c(f) = f \log 1/f$.

$\mathbf{x} = \mathbf{As}$ through a measurement matrix $\mathbf{A}$ can be thought of as a linear projection from a high N-dimensional space of signals down to a low M-dimensional space of measurements. In this geometric picture, the space of K-sparse signals consists of a low-dimensional (nonsmooth) manifold, which is the union of all K-dimensional linear spaces characterized by K nonzero values at specific locations, as in **Figure 4*a***. Candes & Tao (2005) show that any projection that preserves the geometry of all K-sparse vectors allows one to reconstruct these vectors from the low-dimensional projection efficiently and robustly using L_1 minimization. The power of compression by RPs lies in the fact that they preserve the geometrical structure of this manifold. In particular, Baraniuk et al. (2008) show that RPs down to an $M = O(K \log(N/K))$ dimensional space preserve the distance between any pair of K-sparse signals up to a small distortion.

However, we can move beyond sparsity and consider how well RPs preserve the geometric structure of other signal or data patterns that lie on more general low-dimensional manifolds embedded in a high-dimensional space. An extremely simple manifold is a point cloud consisting of a finite set of points, as in **Figure 4*b***. Suppose this cloud consists of P points $\mathbf{s}^\alpha$, for $\alpha = 1, \ldots, P$, embedded in an N-dimensional space, and we project them down to the points $\mathbf{x}^\alpha = \mathbf{As}^\alpha$ in a low M-dimensional space through an appropriately normalized RP. How small can we make M before the point cloud becomes distorted in the low-dimensional space so that pairwise distances in the low-dimensional space are no longer similar to the corresponding distances in the high-dimensional space?

The celebrated Johnson-Lindenstrauss (JL) lemma (Johnson & Lindenstrauss 1984, Indyk & Motwani 1998, Dasgupta & Gupta 2003) provides a striking answer. It states that RPs with $M > O(\log P)$ will yield, with high probability, only a small distortion in distance between all pairs of points in the cloud. Thus the number of projected dimensions M needs only be logarithmic in the number of points P independent of the embedding dimension of the source data, N.

Finally, we consider data distributed along a nonlinear K-dimensional manifold embedded in N-dimensional space, as in **Figure 4*c***. An example might be a set of images of a single object observed under different lighting conditions, perspectives, rotations, and scales. Another example would be the set of neural firing-rate vectors in a brain region in response to a continuous family of stimuli. Baraniuk & Wakin (2009) and Baraniuk et al. (2010) show that $M > O(K \log NC)$ RPs preserve the geometry of the manifold with small distortion. Here C is a number related to the curvature of the manifold so that highly curved manifolds require more projections. Overall, these results show that surprisingly small numbers of RPs, which can be chosen without any knowledge of the data distribution, can preserve geometric structure in data.

Compressed Computation

Although CS emphasizes the reconstruction of sparse high-dimensional signals from

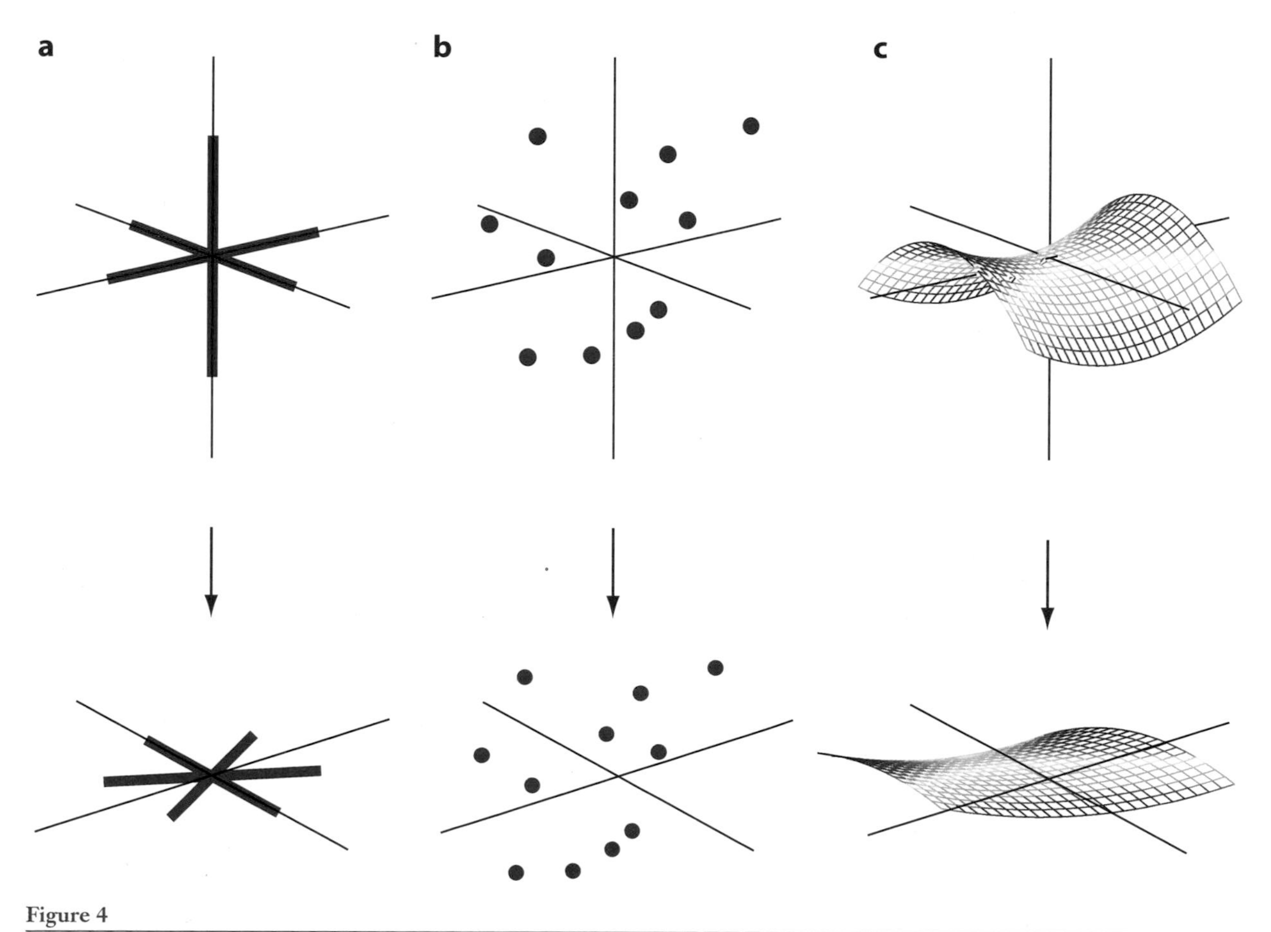

Figure 4

Random projections. (*a*) A manifold of K-sparse signals (red) in N-dimensional space is randomly projected down to an M-dimensional space (here $K = 1$, $N = 3$, $M = 2$). (*b*,*c*) Projection of a point cloud, and a nonlinear manifold respectively.

low-dimensional projections, many important problems in signal processing and learning can be accomplished by performing computations directly in the low-dimensional space without the need to first reconstruct the high-dimensional signal. For example, regression (Zhou et al. 2009), signal detection (Duarte et al. 2006), classification (Blum 2006, Haupt et al. 2006, Davenport et al. 2007, Duarte et al. 2007), manifold learning (Hegde et al. 2007), and nearest neighbor finding (Indyk & Motwani 1998) can all be accomplished in a low-dimensional space given a relatively small number of RPs. Moreover, task performance is often comparable to what can be obtained by performing the task directly in the original high-dimensional space. The reason for this remarkable performance is that these computations rely on the distances between data points, which are preserved by RPs. Thus RPs provide one way to cope with the curse of dimensionality, and as we discuss below, this can have significant implications for neuronal information processing and data analysis.

Approximate Sparsity and Noise

Above, we have assumed a definition of sparsity in which an N-dimensional signal $\mathbf{s}^0$ has $K < N$ nonzero elements, with the other elements being exactly 0. In reality, many of the coefficients of a signal may be small, but they are unlikely to be exactly zero. We thus expect signals not to be exactly sparse but to be well approximated by a K-sparse vector $\mathbf{s}_K^0$, which is obtained by keeping the K largest coefficients of $\mathbf{s}^0$ and setting the rest of them to 0. In addition, we have to allow for measurement noise so that $\mathbf{x} = \mathbf{A}\mathbf{s}^0 + \mathbf{z}$, where $\mathbf{z}$ is a noise vector whose μ'th component is zero mean Gaussian noise with a fixed variance.

In the presence of noise, it no longer makes sense to enforce perfectly the measurement constraints $\mathbf{x} = \mathbf{A}\mathbf{s}$. Instead, a common approach, known as the LASSO method, is to solve the alternate optimization problem

$$\hat{\mathbf{s}} = \arg\min_{\mathbf{s}} \left\{ \|\mathbf{x} - \mathbf{A}\mathbf{s}\|^2 + \lambda \sum_{i=1}^{T} V(s_i) \right\}, \quad 3.$$

where $V(s) = |s|$ (the absolute value function) and λ is a parameter to be optimized. The cost function minimized here allows deviations between $\mathbf{A}\mathbf{s}$, which are the noise-free measurement outcomes generated by a candidate signal $\mathbf{s}$, and the actual noisy measurements $\mathbf{x}$. However, such deviations are penalized by the quadratic term in Equation 3.

Several works (see e.g., Candes et al. 2006, Wainwright 2009, Bayati et al. 2010, Candes & Plan 2010) have addressed the performance of the LASSO in the combined situation of noise and departures from perfect sparsity. The main outcome is roughly that for an appropriate choice of λ, which depends on the signal-to-noise ratio (SNR), the same conditions that guaranteed exact recovery of K-sparse signals by L_1 minimization in the absence of noise also ensure good performance of the LASSO for approximately sparse signals in the presence of noise. In particular, whenever $\mathbf{s}_K^0$ is a good approximation to $\mathbf{s}^0$, the LASSO estimate $\hat{\mathbf{s}}$ in Equation 3 is a good approximation to $\mathbf{s}^0$, up to a level of precision that is allowed by the noise.

Sparse Models of High-Dimensional Data

L_1-based minimization can also be applied to the modeling of high-dimensional data. A simple example is sparse linear regression. Suppose that our data set consists of M N-dimensional vectors, $\mathbf{a}^\mu$, along with M scalar response variables x_μ. The regression model assumes that on each observation, μ, $x_\mu = \mathbf{a}^\mu \cdot \mathbf{s}^0 + z_\mu$, where $\mathbf{s}^0$ is an N-dimensional vector of unknown regression coefficients and z_μ is Gaussian measurement noise. This can be summarized in the matrix equation $\mathbf{x} = \mathbf{A}\mathbf{s}^0 + \mathbf{z}$, where the M rows of the $M \times N$ matrix $\mathbf{A}$ are the N-dimensional data points, $\mathbf{a}^\mu$. Now if the number of data points M is fewer than the dimensionality of the data N, it would seem hopeless to infer the regression coefficients. However, in many high-dimensional regression problems, we expect that the regression coefficients will be sparse. For example, $\mathbf{a}^\mu$ could be a vector of expression levels of $N = O(1000)$ genes measured in a microarray under experimental condition μ, and x_μ could be the response of a biological signal of interest. However, only a small fraction of genes are expected to regulate any given signal of interest, and hence we expect the regression coefficients $\mathbf{s}^0$ to be sparse.

This scenario is exactly equivalent to the case of CS with noise. Here the regression coefficients $\mathbf{s}^0$ play the role of an unknown sparse signal to be recovered, the input data points $\mathbf{a}^\mu$ play the role of the measurement vectors, and the scalar output or response x_μ plays the role of the measurement outcome in CS. The same LASSO algorithm described in Equation 3 can be used to infer the regression coefficients (Tibshirani 1996). Here, the parameter λ is not set by the SNR but rather is chosen to minimize some measure of the prediction error on a new input. This estimate can be obtained through cross validation, for example. Efron et al. (2004) have proposed efficient algorithms to compute $\hat{\mathbf{s}}$, optimizing over λ for a given data set $(\mathbf{A}, \mathbf{x})$.

The technique of L_1 regularization generalizes beyond linear regression to the problem of learning large statistical models with expected sparse parameter sets. Indeed it has been used successfully in learning logistic regression (Lee et al. 2006b) and in various graphical models (Lee et al. 2006a, Wainwright et al. 2007), as well as in point process models of neuronal spike trains (Kelly et al. 2010).

Dictionary Learning

As Equations 2 and 3 imply, to reconstruct a signal from a small number of random measurements using L_1 minimization, we need to know $\mathbf{A} = \mathbf{BC}$, which means that we need to know the basis $\mathbf{C}$ in which the signal is sparse. What if we have to work with a new ensemble of signals and we do not yet know of a basis in which these signals are sparse?

One approach is to perform dictionary learning (Olshausen et al. 1996; Olshausen & Field 1996a,b, 1997) on the ensemble of signals. Suppose $\{\mathbf{x}^\alpha\}$ for $\alpha = 1, \ldots, P$ is a collection of P M-dimensional signals. We imagine that each signal is well approximated by a sparse linear combination of the columns of an unknown $M \times N$ matrix $\mathbf{A}$, i.e., $\mathbf{x}^\alpha \approx \mathbf{A}\mathbf{s}^\alpha$ for all $\alpha = 1, \ldots, P$, where $\mathbf{s}^\alpha$ is an unknown sparse N-dimensional vector. We refer to the columns of $\mathbf{A}$ as the dictionary elements. Thus, the nonzero coefficients of $\mathbf{s}^\alpha$ indicate which dictionary elements linearly combine to form the signal $\mathbf{x}^\alpha$. Here N can be larger than M, in which case we are looking for an overcomplete basis, or dictionary, to represent the ensemble of signals. Given our training signals $\mathbf{x}^\alpha$, we wish to find the sparse codes $\mathbf{s}^\alpha$ and dictionary $\mathbf{A}$. These can potentially be found by minimizing the following energy function:

$$E(\mathbf{s}^1, \ldots, \mathbf{s}^P, \mathbf{A}) = \sum_{\alpha=1}^{P} (\|\mathbf{x}^\alpha - \mathbf{A}\mathbf{s}^\alpha\|^2 + \lambda\|\mathbf{s}^\alpha\|_1), \tag{4}$$

where $||\mathbf{s}^\alpha||_1$ denotes the L_1 norm of $\mathbf{s}^\alpha$. For each α, this second term enforces the sparsity of the code, whereas the first quadratic cost term enforces the fidelity of the code and the dictionary. Subsequent work (Kreutz-Delgado et al. 2003; Aharon et al. 2006a,b) has extended this basic formalism as well as derived efficient algorithms for solving Equation 4. Moreover, Aharon et al. (2006b), Isely et al. (2010), and Hillar & Sommer (2011) have recently shown that if the signals $\mathbf{x}^\alpha$ are indeed generated by sparse noiseless codes through a dictionary $\mathbf{A}$, under certain conditions related to CS, dictionary learning will recover $\mathbf{A}$, up to permutations and scalings of its columns.

COMPRESSED SENSING OF THE BRAIN

Rapid Functional Imaging

In many ways, magnetic resonance imaging (MRI) is a well-suited application for CS (Lustig et al. 2008). In MRI, a strong static magnetic field with a linear spatial gradient, $\Delta\mathbf{H}$, causes magnetic dipoles in a tissue sample to align with the magnetic field. A radio frequency excitation pulse then generates a transverse complex magnetic moment at location $\mathbf{r}$, with amplitude $m(\mathbf{r})$ and a phase $\phi(\mathbf{r})$ proportional to $\mathbf{r} \cdot \Delta\mathbf{H}$. Depending on the sample preparation, the amplitudes $m(\mathbf{r})$ correlate with various local properties of interest. For example, in functional MRI, it correlates with the concentration of oxygenated hemoglobin, which in turn increases in response to neural activity. Thus, the measurement goal is to extract the spatial profile of $m(\mathbf{r})$. A detector coil measures the spatial integral of the complex magnetization. Hence, it essentially measures a spatial Fourier transform of the profile with a Fourier wave vector $\mathbf{k} = (\mathbf{k}_x, \mathbf{k}_y, \mathbf{k}_z) \propto \Delta\mathbf{H}$.

The traditional approach to MR imaging has been to sample the image densely through a regular lattice in Fourier wave vector space, or $\mathbf{k}$-space, by generating a sequence of static linear gradient fields and radio frequency pulses. If the Fourier space is sampled at the Nyquist-Shannon rate, then one can perform a linear reconstruction of the image $m(\mathbf{r})$ simply by performing an inverse Fourier transform of the measurements. However, acquiring each Fourier sample can take time, so any method to reduce the number of such samples can dramatically reduce patient time in scanners, as well as increase the temporal resolution of dynamic imaging.

CS provides an interesting approach to reducing the number of measurements. In the CS framework, the measurement basis $\mathbf{B}$ in **Figure 1** consists of Fourier modes. CS will work well if the MRI image is sparse in a basis $\mathbf{C}$ that is incoherent with respect to $\mathbf{B}$. For example, many MRI images, such as angiograms, are sparse in the position, or pixel basis. For such images, one can subsample random trajectories in **k**-space and use nonlinear L_1 reconstruction to recover the image. For appropriately chosen random trajectories, one can obtain high-quality images using a tenth of the number of measurements required in the traditional approach (Lustig et al. 2008). Similarly, brain images are often sparse in a wavelet basis, and for such images, random trajectories in **k**-space can be found that speed up the rate at which images can be acquired by a factor of 2.4 compared with the traditional approach (Lustig et al. 2007). Moreover, dynamic movies of oscillatory phenomena that are sparse in the temporal frequency domain can be obtained at high temporal resolution by sampling randomly both in **k**-space and in time (Parrish & Hu 1995).

Fluorescence Microscopy

Simultaneously imaging the dynamics of multiple molecular species at both high spatial and temporal resolution is a central goal of cellular microscopy. CS-inspired technologies such as single-pixel cameras (Takhar et al. 2006, Duarte et al. 2008) combined with fluorescence microscopy techniques (Wilt et al. 2009, Taraska & Zagotta 2010) provide one promising route toward such a goal (Coskun et al. 2010; E. Candes, personal communication). In fluorescence imaging, multiple molecular species can be tagged with markers capable of emitting light at different frequencies. Imaging the molecules then requires two key steps: First, the sample must be illuminated with light, causing the tagged species to fluoresce, and second, the emitted photons from the fluorescent species must be detected. Traditionally, two main methods have been used to accomplish both steps. In widefield (WF) microscopy, the entire image is illuminated at once, and a large array of detectors records the emitted photons. In raster scan (RS) microscopy, each point of the image is illuminated in sequence, so only one detector is required to collect the emitted photons at any given time.

WF can achieve high temporal resolution but requires many photodetectors for high spatial resolution. This is problematic for imaging applications in which photons at many different frequencies, corresponding to different molecules, need to be simultaneously measured. This requires a prohibitively expensive high-density array of photodetectors that can perform hyperspectral imaging, i.e., measure many spectral channels at once. One could employ a single such detector in RS mode, but then achieving high spatial resolution comes at the cost of low temporal resolution because of the required number of raster scans.

The single-pixel-camera approach exploits the potential spatial sparsity of a fluorescence image to achieve both high spatial and temporal resolution. In this approach, the image is illuminated using a sequence of random light patterns. This can be achieved by a digital micromirror device (DMD), which consists of a spatial array of micrometer scale mirrors whose angles can be rapidly and individually adjusted. Light is reflected off this array into the sample, and on each trial, a different configuration of mirrors leads to a different pattern of illumination. A single hyperspectral photodetector (the single pixel) then measures the total emitted fluorescence. Owing to the randomness of the light patterns, the image can be reconstructed at the micrometer spatial resolution of the DMD using a number of measurements that is much smaller than the number of pixels (or resolvable spatial locations) in the image. Thus compressive imaging retains the relative speed and resolution of WF and the simplicity and achievable spectral range of RS. As such, this rapidly evolving method has the potential to open up new experimental windows into the dynamics of intracellular molecular cascades within neurons.

Gene-Expression Analysis

The use of microarrays to collect large-scale data sets of gene-expression levels across many brain regions is now a well-established enterprise in neuroscience. Suppose we want to measure a vector $\mathbf{s}^0$ of concentrations of N genetic sequences in a sample. A microarray consists of N spots, indexed by $i = 1, \ldots, N$, where each spot i contains a unique complementary sequence that will specifically bind with the sequence i in the sample. All N genetic sequences of interest in the sample are fluorescently tagged and exposed to all the spots. Each spot binds a specific sequence, and after the excess unbound DNA is washed off, the vector of concentrations $\mathbf{s}^0$ can be read off by imaging the fluorescence levels of the spots.

Often this procedure is highly inefficient because any particular sample will contain only a few genetic sequences of interest, i.e., the concentration vector $\mathbf{s}^0$ is sparse. Dai et al. (2009) proposed a CS-based approach in which one can use $M < N$ spots, where each spot contains a random subset of the N sequences of interest. Thus each spot, now indexed by $\mu = 1, \ldots, M$, is characterized by an N-dimensional measurement vector $\mathbf{a}^\mu$, where the component $\mathbf{a}_i^\mu$ reflects the binding affinity of sequence i in the sample to the contents of spot μ. After the CS microarray is exposed to the sample, the M-dimensional vector of fluorescence levels $\mathbf{x}$ is approximately related to the sample concentration $\mathbf{s}^0$ through the linear relation $\mathbf{x} = \mathbf{A}\mathbf{s}^0$, where the rows of $\mathbf{A}$ are the measurement vectors $\mathbf{a}^\mu$. Thus if each spot contains enough randomly chosen complementary sequences, such that the measurements are incoherent with regard to the basis of sequences, one can use the LASSO method in Equation 3 to recover the concentrations $\mathbf{s}^0$ from the fluorescence measurements $\mathbf{x}$. Dai et al. (2009) do a thorough analysis of this basic framework. Overall, reducing the number of spots required to collect gene expression data reduces both the cost and the size of the array, as well as the amount of biological sample material required to make accurate concentration measurements.

Compressed Connectomics

The problem of reconstructing functional circuit connectivity from recordings of neuronal postsynaptic responses presents a considerable challenge to neuroscience. Consider, for example, a simple scenario in which we have a population of N neurons that are potentially presynaptic to a given neuron whose membrane voltage x we can record intracellularly. The synaptic strengths from the N neurons to the recorded neuron is an unknown N-dimensional vector $\mathbf{s}^0$. The traditional approach to estimating this set of synaptic strengths is to excite each potential presynaptic neuron one by one and record the resultant postsynaptic membrane voltage x. Each such measurement reveals the strength of one synapse. This brute-force approach is highly inefficient because the synaptic connectivity $\mathbf{s}^0$ is often sparse, with only $K < N$ nonzero elements, where K/N is $\sim$10%. Thus most measurements would simply yield 0.

Hu & Chklovskii (2009) propose a CS-based approach to recovering $\mathbf{s}^0$ by randomly stimulating F neurons out of N on any given trial μ. This method corresponds to a random measurement matrix $\mathbf{A}$ characterized by F nonzero entries per row. Given that the true weight vector $\mathbf{s}^0$ is sparse, Hu & Chklovskii (2009) propose to use L_1 minimization in Equation 2 to recover $\mathbf{s}^0$ from knowledge of the inputs $\mathbf{A}$ and outputs $\mathbf{x}$. The authors find for a wide range of parameters that $F/N = 0.1$ minimizes the required number of measurements, M, and for this value of F, $M = O(K \log N)$ measurements are required to recover $\mathbf{s}^0$. Thus random stimulation of 10% of the population constitutes an effective measurement basis for CS of synaptic connectivity (Hu & Chklovskii 2009). Alternative ideas have been proposed for CS of connectivity using fluorescent synaptic markers (Mishchenko 2011).

COMPRESSED SENSING BY THE BRAIN

The problem of storing, communicating, and processing high-dimensional neural activity patterns, or external stimuli, presents a fundamental challenge to any neural system. This challenge is complicated by the widespread existence of convergent pathways, or bottlenecks, in which information stored in a large number of neurons is often compressed into a small number of axons, or neurons in a downstream system. For example, 1 million optic nerve fibers carry information about the activity of 100 times as many photoreceptors. Only 1 million pyramidal tract fibers carry information from motor cortex to the spinal cord. And corticobasal ganglia pathways undergo a 10–1,000-fold convergence. In this section we review how the theory of CS and RPs yields theoretical insight into how efficient storage, communication, and computation are possible despite drastic reductions in the dimensionality of neural representations through information bottlenecks.

Semantic Similarity and Random Projections

How much can a neural system reduce the dimensionality of its activity patterns without incurring a large loss in its ability to perform relevant computations? A plausible minimal requirement is that any reduction through a convergent pathway should preserve the similarity structure of the neuronal representations at the source area. This requirement is motivated by the observation that in higher perceptual or association areas in the brain semantically similar objects elicit similar neural activity patterns (Kiani et al. 2007). This similarity structure of the neural code is likely the basis of our ability to categorize objects and generalize appropriate responses to new objects (Rogers & McClelland 2004). Moreover, this similarity structure is remarkably preserved across monkeys and humans, for example, in image representations in the inferotemporal (IT) cortex (Kriegeskorte et al. 2008).

When a semantic task involves a finite number of activity patterns, or objects, the JL lemma discussed above implies that the required communication resources vary only logarithmically with the number of patterns, independent of how many neurons are involved in the source area. For example, suppose 20,000 images can be represented by the corresponding population activity patterns in the IT cortex. Then the similarity structure between all pairs of images can be preserved to 10% precision in a downstream area using only ~1000 neurons. Furthermore, this result can be achieved with a very simple dimensionality-reduction scheme, namely by a random synaptic connectivity matrix. Moreover, any computation that relies on similarity structure, and can be solved by the IT cortex, can also be solved by the downstream region.

A more stringent challenge occurs when convergent pathways must preserve the similarity structure of not just a finite set of neuronal activity patterns, but an arbitrarily large, possibly infinite, number of patterns, as is likely the case in any pathway that represents information about continuous families of stimuli. The theories of CS and RPs of manifolds discussed above reveal that again drastic compression is possible if the corresponding neural patterns are sparse or lie on a low-dimensional manifold (for example, as in **Figure 4*a–c***). In this case, the number of required neurons in a randomly connected downstream area is proportional to the intrinsic dimension of the ensemble of neural activity patterns and depends only weakly (logarithmically) on the number of neurons in the source area.

Hidden low-dimensional structure in neural activity patterns has been found in several systems (Ganguli et al. 2008a, Yu et al. 2009, Machens et al. 2010), and moreover, intrinsic spatiotemporal fluctuations exhibited in many models of recurrent neuronal circuits, including chaotic networks, are low dimensional (Rajan et al. 2010, Sussillo & Abbott 2009). The ubiquity of this low-dimensional structure in neuronal systems may be intimately related to the requirement of communication and computation through widespread anatomical bottlenecks.

Short-Term Memory in Neuronal Networks

Another bottleneck is posed by the task of working memory, where streams of sensory inputs must presumably be stored within the dynamic reverberations of neuronal circuits. This is a bottleneck from time into space: Long temporal streams of input must be stored in the instantaneous spatial activity patterns of a limited number of neurons. The influential idea of attractor dynamics (Hopfield 1982) suggests how single stimuli can be stored as stable patterns of activity, or fixed points, but such simple fixed points are incapable of storing temporal sequences of information, like an ongoing sentence, song, or motion trajectory. More recent proposals (Jaeger 2001, Maass et al. 2002, Jaeger & Haas 2004) suggest that recurrent networks could store temporal sequences of inputs in their ongoing, transient activity. This new paradigm raises several theoretical questions about how long memory traces can last in such networks, as functions of the network size, connectivity, and input statistics. Several studies have addressed these questions in the case of simple linear neuronal networks and Gaussian input statistics. These studies show that the duration of memory traces in any network cannot exceed the number of neurons (in units of the intrinsic time constant) (Jaeger 2001, White et al. 2004) and that no network can outperform an equivalent delay line or a nonnormal network, characterized by a hidden feedforward structure (Ganguli et al. 2008b).

However, a more ethologically relevant temporal input statistic is that of a sparse, non-Gaussian sequence. Indeed a wide variety of temporal signals of interest are sparse in some basis, for example, human speech in a wavelet basis. Recent work (Ganguli & Sompolinsky 2010a) has derived a connection between CS and short-term memory by showing that recurrent neuronal networks can essentially perform online, dynamical compressed sensing of an incoming sparse sequence, yielding sequence memory traces that are longer than the number of neurons, again in units of the intrinsic time constant. In particular, neuronal circuits with M neurons can remember sparse sequences, which have a probability f of being nonzero at any given time for an amount of time that is $O(\frac{M}{f\log(1/f)})$. This enhanced capacity cannot be attained by purely feedforward networks, or random Gaussian network connectivities, but requires antisymmetric connectivity matrices that generate complex transient activity patterns and diverse temporal filtering properties.

SPARSE EXPANDED NEURONAL REPRESENTATIONS

In the previous section, we have discussed how CS and RPs can explain how convergent pathways can compress neuronal representations. However, in many computations, neural systems may need to expand these low-dimensional compressed representations back into high-dimensional sparse ones. For example, such representations reduce the overlap between activity patterns, thereby simplifying the tasks of learning, discrimination, categorization, noise filtering, and multiscale stimulus representation. Indeed, like convergence, the expansion of neural representations through divergent pathways is a widespread anatomical motif. For example, information in 1 million optic nerve fibers is expanded into more than 100 million primary visual cortical neurons. Also in the cerebellum, a small number of mossy fibers target a large number of granule cells, creating a 100-fold expansion.

How do neural circuits transform compressed dense codes into expanded sparse ones? A simple mechanism would be to project the dense activity patterns into a larger pool of neurons via random divergent projections and use high spiking thresholds to ensure sparsity of the target activity patterns. Indeed, Marr (1969) suggested this mechanism in his influential hypothesis that the granule cell layer in the cerebellar cortex performs sparse coding of dense stimulus representations in incoming mossy fibers to facilitate learning of sensorimotor associations at the Purkinje cell layer. Although random expansion may work for some computations, sparse codes are generally most useful when they represent essential sparse features of the compressed signal. In the next sections, we review how CS methods for generating sparse expanded representations, which faithfully capture hidden structures in compressed data, can operate within neural systems.

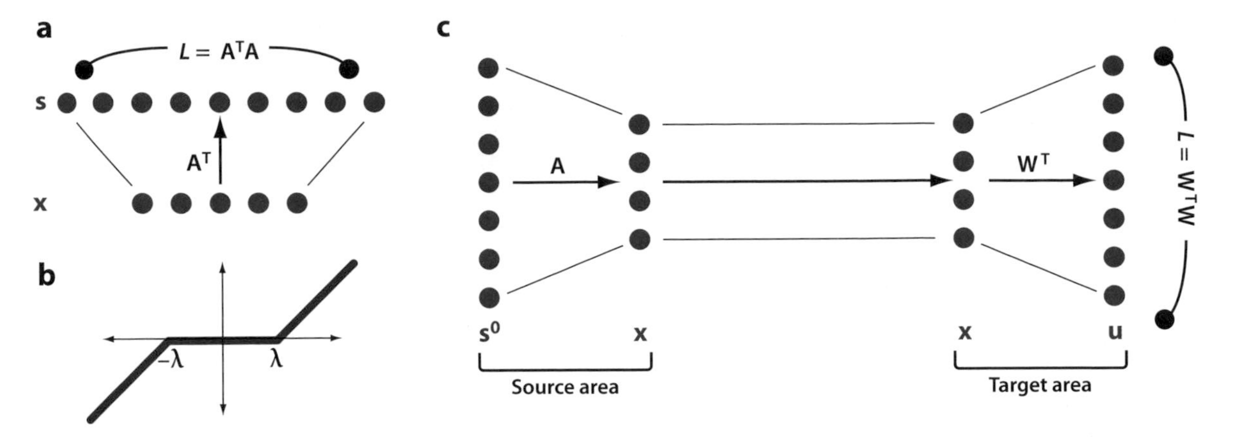

Figure 5

Neural L_1 minimization and long-range brain communication. (*a*) A two-layer circuit for performing L_1 minimization and dictionary learning. (*b*) Nonlinear transfer function from inputs to firing rates of neurons in the second layer in panel *a*. (*c*) A scheme for efficient long-range brain communication in which sparse activity $\mathbf{s}^0$ is compressed to a low-dimensional dense representation **x** in a source area and efficiently communicated downstream to a target area with a small number of axons, where it could be re-expanded into a new sparse representation **u** through a dictionary learning circuit as in panel *a*.

Neuronal Implementations of L1 Minimization

Given that solving the optimization problem in Equation 3 with $V(s) = |s|$ has proven to be an efficient method for sparse signal reconstruction, whether neuronal circuits can perform this computation is a natural question. Here we describe one plausible two-layer circuit solution (see **Figure 5*a***) proposed in Rozell et al. (2008), inspired by gradient descent in **s** on the cost function in Equation 3. Suppose that the low M-dimensional input **x** is represented in the first layer by the firing rates of a population of M neurons such that the μth input neuron has a firing rate $\mathbf{x}_\mu$. Now suppose that the reconstructed sparse signal is represented by a larger population of N neurons where s_i is the firing rate of neuron i. In this population, we denote the synaptic potential for each neuron by v_i, which determines the neuron's firing rate via a static nonlinearity F, $s_i = F(v_i)$.

The synaptic connectivity from the M input neurons to the N second-layer neurons computing the sparse representation **s** is given by the $N \times M$ matrix $\mathbf{A}^T$ such that the ith column of **A**, $\mathbf{a}^i$, denotes the set of M synaptic weights from the input neurons to neuron i in the second layer. Finally, assume there is lateral inhibition between any pair of neurons i and j in the second layer, governed by synaptic weights L_{ij}, which are related to the feedforward weight vectors to the pair of neurons, through $L_{ij} = \mathbf{a}^i \cdot \mathbf{a}^j$. Then the internal dynamics of the second-layer neurons obey the differential equations

$$\tau \frac{dv_i}{dt} = -v_i + \mathbf{a}^i \cdot \mathbf{x} - \sum_{j=1}^{T} L_{ij} s_j, \qquad 5.$$

where **x** is the activity of the input layer. Rozell et al. (2008) found that for an appropriate choice of the static nonlinearity, this dynamic is similar to a gradient descent on the cost function given by Equation 3. In particular, for L_1 minimization, the static nonlinearity F is simply a threshold linear function with threshold λ and gain 1 (see **Figure 5*b***).

To obtain a qualitative understanding of this circuit, consider what happens when the second-layer activity pattern is initially inactive so that $\mathbf{s} = 0$ and an input **x** occurs in the first layer. Then the internal variable $v_i(t)$ of each second-layer neuron i will charge up with a rate controlled by the overlap of the input **x** with the synaptic weight vector, $\mathbf{a}^i$, which is closely related to the receptive field (RF) of neuron

i. As neuron i's internal activation crosses the threshold λ, it starts to fire and inhibits neurons with RFs similar to $\mathbf{a}^i$. This sets up a competitive dynamic in which a small number of neurons with RFs similar to the input $\mathbf{x}$ come to represent it, yielding a sparse representation $\hat{\mathbf{s}}$ of the input $\mathbf{x}$, which is the solution to Equation 3. In the case of zero noise, the above circuit dynamic needs to be supplemented with an appropriate dynamic update of the threshold λ, which eventually approaches zero at the fixed point (Donoho et al. 2009). Finally, we note that several works (Olshausen et al. 1996, Perrinet 2010) have proposed synaptic Hebbian plasticity and homeostasis rules that supplement Equation 5 and allow the circuit to solve the full dictionary learning problem, Equation 4, without prior knowledge of $\mathbf{A}$.

An intriguing feature of the above dynamic is that the inhibitory recurrent connections are tightly related to the feedforward excitatory drive. Koulakov & Rinberg (2011) suggest that exactly this computation may be implemented in the rodent olfactory bulb. They propose that reciprocal dendrodendritic synaptic coupling between mitral cells and granule cells yields an effective lateral inhibition between granule cells that is related to the feedforward drive from mitral cells to granule cells, in accordance with the requirements of Equation 5. Thus the composite olfactory circuit builds up a sparse code for odors in the granule cell population. Likewise, Hu et al. (2011) proposed that sparse coding is implemented within the amacrine/horizontal cell layers in the retina.

Compression and Expansion in Long-Range Brain Communication

A series of papers (Coulter et al. 2010, Isely et al. 2010, Hillar & Sommer 2011) have integrated the dual aspects of CS theory: dimensionality reduction of sparse neural representations, and the recoding of stimuli in sparse overcomplete representations into a theory of efficient long-range brain communication (see also Tarifi et al. 2011). According to this theory (see **Figure 5*c***), each area in a long-range communication pathway has both dense and sparse representations. Local sparse representations are first compressed to communicate them using a small number of axons and potentially re-expanded in a downstream area.

Where in the brain might these transformations occur? Coulter et al. (2010) predict that this could occur within every cortical column, with compressive projections, possibly random, occurring between more superficial cortical layers and the output layer 5. A key testable physiological prediction would then be that activity in more superficial layers is sparser than activity in deeper output layers. Another possibility is the transformation from sparse high-dimensional representations of space in the CA3/CA1 fields of the hippocampus to denser, lower-dimensional representations of space in the subiculum, which constitutes the major output structure of the hippocampus. A functional explanation for this representational dichotomy could be that the hippocampus is performing an RP from CA3/CA1 to the subiculum, thereby minimizing the number of axons required to communicate the results of hippocampal computations to the rest of the brain.

Overall, these works suggest more generally that random compression and sparse coding can be combined to yield computational strategies for efficient use of the limited bandwidth available for long-range brain communication.

LEARNING IN HIGH-DIMENSIONAL SYNAPTIC WEIGHT SPACES

Learning new skills and knowledge is thought to be achieved by continuous synaptic modifications that explore the space of possible neuronal circuits, selecting through experience those that are well adapted to the given task. We review how regularization techniques used by statisticians to learn high-dimensional statistical models from limited amounts of data can also be employed by synaptic learning rules to search efficiently the high-dimensional space of synaptic patterns to learn appropriate rules from limited experience.

Neural Learning of Classification

A simple model of neural decision making and classification is a single-layer feedforward network in which the postsynaptic potential of the readout neuron is a sum of the activity of its afferents, weighted by a set of synaptic weights, and the decision is signaled by firing or not firing depending on whether the potential reaches threshold. Such a model is equivalent to the classical perceptron (Rosenblatt 1958). Computationally, this model classifies N-dimensional input patterns into two categories separated by a hyperplane determined by the synaptic weights. These weights are learned through experience-dependent modifications based on a set of M training input examples and their correct classifications. Of course, the goal of any organism is not to classify past experience correctly, but rather to generalize to novel experience. Thus an important measure of learning performance is the generalization error, or the probability of incorrectly classifying a novel input, and a central question of learning theory is how many examples M are required to achieve a good generalization error given a number of N unknown synaptic weights that need to be learned.

This question has been studied exhaustively (Gardner 1988, Seung et al. 1992) (see Engel & den Broeck 2001 for an overview), and the general consensus finds that for a wide variety of learning rules, a small generalization error can occur only when the number of examples M is larger than the number of synapses N. This result has striking implications because it suggests that learning may suffer from a curse of dimensionality: Given the large number of synapses involved in any task, this theory suggests we need an equally large number of training examples to learn any task.

Recent work (Lage-Castellanos et al. 2009) has considered the case when a categorization task can be realized by a sparse synaptic weight vector, meaning that only a subset of inputs are task relevant, though which subset is a priori unknown. The authors showed that a simple learning rule that involves minimization of the classification error on the training set, plus an L_1 regularization on the synaptic weights of the perceptron, yields a good generalization error even when the number of examples can be less than the number of synapses. Thus a sparsity prior is one route to combat the curse of dimensionality in learning tasks that are realizable by a sparse rule.

Optimality and Sparsity of Synaptic Weights

Consider again the perceptron learning to classify a finite set of M input patterns. In general, many synaptic weight vectors will classify these inputs correctly. We can, however, look for the optimal weight vector that maximizes the margin, or the minimal distance between input patterns and the category boundary. For such weights, the induced synaptic potentials are as far as possible from threshold, and the resultant classifications yield good generalization and noise tolerance (Vapnik 1998).

A remarkable theoretical result is that if synapses are constrained to be either excitatory or inhibitory, then near capacity, the optimal solution is sparse, with most of the synapses silent (Brunel et al. 2004), even if the input patterns themselves show no obvious sparse structure. This result has been proposed as a functional explanation for the abundance of silent synapses in the cerebellum and other brain areas.

When the sign of the weights are unconstrained, the optimal solutions are still sparse, but not in the basis of neurons. Instead, the optimal weight vector can be expressed as a linear combination of a small number of input patterns, known as support vectors, the number of support vectors being much smaller than their dimensionality. Indeed, several powerful learning algorithms, including support vector machines (SVMs) (see Burges 1998, Vapnik 1998, Smola 2000 for reviews), exploit this form of sparsity to achieve good generalization from relatively few high-dimensional examples.

Finally, because a sufficiently large number of RPs preserve Euclidean distances, they incur only a modest reduction in the margin of the optimal category boundary separating classes (Blum 2006). Hence, classification problems can also be learned directly in a low-dimensional space. In summary, there is an interesting interplay among sparsity, dimensionality, and the learnability of high-dimensional classification problems: Any such rapidly learnable problem (i.e., one with a large margin) is both (*a*) sparse, in the sense that its solution can be expressed in terms of a sparse linear combination of input patterns, and (*b*) low-dimensional in the sense that it can be learned in a compressed space after a RP.

DISCUSSION

Dimensionality Reduction: CS versus Efficient Coding

Efficient coding theories (Barlow 1961, Atick 1992, Atick & Redlich 1992, Barlow 2001) suggest that information bottlenecks in the brain perform optimal dimensionality reduction by maximizing mutual information between the low-dimensional output and the high-dimensional input (Linsker 1990). The predictions of such information maximization theories depend on assumptions about input statistics, neural noise, and metabolic constraints. In particular, infomax theories of early vision, based on Gaussian signal and noise assumptions, predict that high-dimensional spatiotemporal patterns of photoreceptor activation should be projected onto the linear subspace of their largest principal components. Furthermore, the individual projection vectors, i.e., retinal ganglion cell (RGC) RFs, depend on the stimulus SNR; in particular, at a high SNR, RFs should decorrelate or whiten the stimulus. This is consistent with the center-surround arrangement of RFs, which removes much of the low-frequency correlations in natural images (Atick 1992, Atick & Redlich 1992, Borghuis et al. 2008).

What is the relation between infomax theories and CS? According to CS theory, for sparse inputs, close to optimal dimensionality reduction is achieved when the projection vectors are maximally incoherent with respect to the basis in which the stimulus is sparse. Assuming visual stimuli are approximately sparse in a wavelet or Gabor-like basis, incoherent projections are likely to be spatially distributed. If sparseness is a prominent feature of natural visual spatiotemporal signals, how can we reconcile the observed RGC center-surround RFs with the demand for incoherence? Incoherent or random projections are optimal for signal ensembles composed of a combination of a few feature vectors in which the identity of these vectors varies across signals. This may be an adequate description of natural images after whitening. However, prewhitened natural images have strong second-order correlations, implying that they lie close to a low-dimensional linear space given by their principal components. Thus, the ensemble of natural images is characterized by both linear low-dimensional structure and sparse structure imposed by higher-order statistics. In such ensembles, whether sensory stimuli or neuronal activity patterns, when second-order correlations are strong enough, the optimal dimensionality reduction may indeed be close to that predicted by Gaussian-based infomax, as has been argued in recent work (Weiss et al. 2007).

Expansion and Sparsification: Compressed Sensing versus Independent Components Analysis

What does efficient coding theory predict regarding the recoding of signals through expansive transformations, for example, from the optic nerve to visual cortex? Several modern efficient coding theories, such as basis pursuit, independent components analysis (ICA), maximizing non-Gaussianity, and others, suggest that even after decorrelation, natural images include higher-order statistical dependencies that arise through linear mixing of statistically independent sources. The role of the cortical representation is to further reduce the redundancy of the signal by separating the mixed signal into its independent causes (i.e., an unmixing operation), essentially generating a factorial statistical representation of the signal.

The application of ICA to natural images and movies yields at the output layer, single-neuron response histograms, which are considerably sparser than those in the input layer. These responses have Gabor-like RFs similar to those of simple cells in V1 (Olshausen et al. 1996, Bell & Sejnowski 1997, van Hateren & Ruderman 1998, van Hateren & van der Schaaf 1998, Simoncelli & Olshausen 2001, Hyvarinen 2010). ICA algorithms have also been applied to natural sounds (Lewicki 2002), yielding a set of temporal filters, resembling auditory cortical RFs.

Although the algorithms and results of ICA and source extraction by CS are often similar, there are important differences. First, CS results in signals that are truly sparse, i.e., most of the coefficients are zero, whereas ICA algorithms generally yield signals with many small values, i.e., distributions with high kurtosis but no coefficients vanish (Olshausen et al. 1996, Bell & Sejnowski 1997, Hyvarinen 2010). Second, ICA emphasizes the statistical independence of the unmixed sources (Barlow 2001). Sparseness is a special case; ICA can be applied to reconstruct dense sources as well. In contrast, signal extraction by CS relies only on the assumed approximate sparseness of the signal, and not on any statistical priors, and is similar in spirit to the seminal work of Olshausen et al. (1996). Indeed, a recent study suggests that sparseness may be a more useful notion than independence and that the success of ICA in some applications is due to its ability to generate sparse representations rather than to discover statistically independent features (Daubechies et al. 2009).

Beyond Linear Projections: Neuronal Nonlinearities

The abundance of nonlinearities in neuronal signaling raises the question of the relevance of the CS linear projections to neuronal information processing. One fundamental nonlinearity is the input-output relation between synaptic potentials and action potential firing of individual neurons. This nonlinearity is often approximated by the linear-nonlinear (LN) model (Dayan & Abbott 2001, Ostojic & Brunel 2011) in which the firing rate of a neuron, x, is related to its input activity $\mathbf{a}$ through $x = \sigma(\mathbf{a} \cdot \mathbf{s}^0)$, where $\mathbf{s}^0$ is the neuron's spatiotemporal linear filter and $\sigma(\cdot)$ is a scalar sigmoidal function. As long as $\sigma(\cdot)$ is an invertible function of its input, the nonlinearity in the measurement can be undone to recover the fundamental linear relation between the synaptic input to the neuron and the source, given by $\mathbf{A}\mathbf{s}^0$; hence, the results of CS should hold. More generally, it will be an important challenge to evaluate the role of dimensionality reduction, expansion, and sparse coding in neuronal circuit models that incorporate additional nonlinearities, including nonlinear temporal coding of inputs, synaptic depression and facilitation, and nonlinear feedback dynamics through recurrent connections.

In summary, we have reviewed a relatively new set of surprising mathematical phenomena related to RPs of high-dimensional patterns. But far from being a set of intellectual curiosities, these phenomena have important practical implications for data acquisition and analysis and important conceptual implications for neuronal information processing. It is likely that more surprises await us, lurking in the properties of high-dimensional spaces and mappings, properties that could further change the way we measure, analyze, and understand the brain.

DISCLOSURE STATEMENT

The authors are not aware of any affiliations, memberships, funding, or financial holdings that might be perceived as affecting the objectivity of this review.

ACKNOWLEDGMENTS

S.G. and H.S. thank the Swartz Foundation, Burroughs Wellcome Foundation, Israeli Science Foundation, Israeli Defense Ministry (MAFAT), the McDonnell Foundation, and the Gatsby Charitable Foundation for support, and we thank Daniel Lee for useful discussions.

LITERATURE CITED

Aharon M, Elad M, Bruckstein A. 2006a. K-SVD: an algorithm for designing overcomplete dictionaries for sparse representation. *IEEE Trans. Signal Proc.* 54(11):4311

Aharon M, Elad M, Bruckstein A. 2006b. On the uniqueness of overcomplete dictionaries, and a practical way to retrieve them. *Linear Algebr. Appl.* 416(1):48–67

Atick J. 1992. Could information theory provide an ecological theory of sensory processing? *Netw. Comput. Neural Syst.* 3(2):213–51

Atick J, Redlich A. 1992. What does the retina know about natural scenes? *Neural Comput.* 4(2):196–210

Baraniuk R. 2007. Compressive sensing. *Signal Proc. Mag. IEEE* 24(4):118–21

Baraniuk R. 2011. More is less: signal processing and the data deluge. *Science* 331(6018):717–19

Baraniuk R, Cevher V, Wakin M. 2010. Low-dimensional models for dimensionality reduction and signal recovery: a geometric perspective. *Proc. IEEE* 98(6):959–71

Baraniuk R, Davenport M, DeVore R, Wakin M. 2008. A simple proof of the restricted isometry property for random matrices. *Constr. Approx.* 28(3):253–63

Baraniuk R, Wakin M. 2009. Random projections of smooth manifolds. *Found. Comput. Math.* 9(1):51–77

Barlow H. 1961. Possible principles underlying the transformation of sensory messages. In *Sensory Communication*, ed. WA Rosenblith, pp. 217–34. New York: Wiley

Barlow H. 2001. Redundancy reduction revisited. *Netw. Comput. Neural Syst.* 12(3):241–53

Bayati M, Bento J, Montanari A. 2010. The LASSO risk: asymptotic results and real world examples. *Neural Inf. Process. Syst. (NIPS)*

Bell A, Sejnowski T. 1997. The independent components of natural scenes are edge filters. *Vis. Res.* 37(23):3327–38

Blum A. 2006. Random projection, margins, kernels, and feature-selection. In *Subspace, Latent Structure and Feature Selection*, ed. C Saunders, M Grobelnik, S Gunn, J Shawe-Taylor, pp. 52–68. Heidelberg, Germ.: Springer

Borghuis B, Ratliff C, Smith R, Sterling P, Balasubramanian V. 2008. Design of a neuronal array. *J. Neurosci.* 28(12):3178–89

Boyd S, Vandenberghe L. 2004. *Convex Optimization*. New York: Cambridge Univ Press

Bruckstein A, Donoho D, Elad M. 2009. From sparse solutions of systems of equations to sparse modeling of signals and images. *Siam Rev.* 51(1):34–81

Brunel N, Hakim V, Isope P, Nadal J, Barbour B. 2004. Optimal information storage and the distribution of synaptic weights: perceptron versus Purkinje cell. *Neuron* 43(5):745–57

Burges C. 1998. A tutorial on support vector machines for pattern recognition. *Data Min. Knowl. Discov.* 2(2):121–67

Candes E, Plan Y. 2010. A probabilistic and RIPless theory of compressed sensing. *IEEE Trans. Inf. Theory* 57(11)7235–54

Candes E, Romberg J. 2007. Sparsity and incoherence in compressive sampling. *Invers. Probl.* 23(3):969–85

Candes E, Romberg J, Tao T. 2006. Stable signal recovery from incomplete and inaccurate measurements. *Commun. Pure Appl. Math.* 59(8):1207–23

Candes E, Tao T. 2005. Decoding by linear programming. *IEEE Trans. Inf. Theory* 51:4203–15

Candes E, Tao T. 2006. Near-optimal signal recovery from random projections: universal encoding strategies? *IEEE Trans. Inf. Theory* 52(12):5406–25

Candes E, Wakin M. 2008. An introduction to compressive sampling. *IEEE Sig. Proc. Mag.* 25(2):21–30

Coskun A, Sencan I, Su T, Ozcan A. 2010. Lensless wide-field fluorescent imaging on a chip using compressive decoding of sparse objects. *Opt. Express* 18(10):10510–23

Coulter W, Hillar C, Isley G, Sommer F. 2010. Adaptive compressed sensing—a new class of self-organizing coding models for neuroscience. Presented at *IEEE Int. Conf. Acoust. Speech Signal Process.* (*ICASSP*), pp. 5494–97

Dai W, Sheikh M, Milenkovic O, Baraniuk R. 2009. Compressive sensing DNA microarrays. *EURASIP J. Bioinf. Syst. Biol.* 2009:162824

Dasgupta S, Gupta A. 2003. An elementary proof of a theorem of Johnson and Lindenstrauss. *Random Struct. Algorithms* 22(1):60–65

Daubechies I, Roussos E, Takerkart S, Benharrosh M, Golden C, et al. 2009. Independent component analysis for brain fMRI does not select for independence. *Proc. Natl. Acad. Sci.* 106(26):10415–20

Davenport M, Duarte M, Wakin M, Laska J, Takhar D, et al. 2007. The smashed filter for compressive classification and target recognition. *Proc. Comput. Imaging V SPIE Electron Imaging*, San Jose, CA

Dayan P, Abbott L. 2001. *Theoretical Neuroscience. Computational and Mathematical Modelling of Neural Systems.* Cambridge, MA: MIT Press

Donoho D. 2000. High-dimensional data analysis: the curses and blessings of dimensionality. *AMS Math Challenges Lecture*, pp. 1–32

Donoho D. 2006. Compressed sensing. *IEEE Trans. Inf. Theory* 52(4):1289–306

Donoho D, Maleki A, Montanari A. 2009. Message-passing algorithms for compressed sensing. *Proc. Natl. Acad. Sci. USA* 106(45):18914–19

Donoho D, Tanner J. 2005a. Neighborliness of randomly projected simplices in high dimensions. *Proc. Natl. Acad. Sci. USA* 102:9452–57

Donoho D, Tanner J. 2005b. Sparse nonnegative solution of underdetermined linear equations by linear programming. *Proc. Natl. Acad. Sci. USA* 102:9446–51

Duarte M, Davenport M, Takhar D, Laska J, Sun T, et al. 2008. Single-pixel imaging via compressive sampling. *Signal Proc. Mag. IEEE* 25(2):83–91

Duarte M, Davenport M, Wakin M, Baraniuk R. 2006. Sparse signal detection from incoherent projections. *Proc. Acoust. Speech Signal Process.* (*ICASSP*) 3:III–III

Duarte M, Davenport M, Wakin M, Laska J, Takhar D, et al. 2007. Multiscale random projections for compressive classification. Presented at *IEEE Int. Conf. Image Process* (*ICIP*) *Int. Conf.* 6:VI161–64, San Antonio, TX

Efron B, Hastie T, Johnstone I, Tibshirani R. 2004. Least angle regression. *Ann. Stat.* 32(2):407–99

Engel A, den Broeck CV. 2001. *Statistical Mechanics of Learning*. London: Cambridge Univ. Press

Ganguli S, Bisley J, Roitman J, Shadlen M, Goldberg M, Miller K. 2008a. One-dimensional dynamics of attention and decision making in lip. *Neuron* 58(1):15–25

Ganguli S, Huh D, Sompolinsky H. 2008b. Memory traces in dynamical systems. *Proc. Natl. Acad. Sci. USA* 105(48):18970–74

Ganguli S, Sompolinsky H. 2010a. Short-term memory in neuronal networks through dynamical compressed sensing. *Neural Inf. Process. Syst.* (*NIPS*) 23:667–75

Ganguli S, Sompolinsky H. 2010b. Statistical mechanics of compressed sensing. *Phys. Rev. Lett.* 104(18):188701

Gardner E. 1988. The space of interactions in neural network models. *J. Phys. A* 21:257–70

Haupt J, Castro R, Nowak R, Fudge G, Yeh A. 2006. Compressive sampling for signal classification. Presented at *Conf. Signals, Syst. Comput.* (*ACSSC*), 40th, Asilomar, pp. 1430–34

Hegde C, Wakin M, Baraniuk R. 2007. Random projections for manifold learning. *Neural Inf. Process. Syst.* **http://books.nips.cc/papers/files/nips20/NIPS2007_1100.pdf**

Hillar CJ, Sommer FT. 2011. Ramsey theory reveals the conditions when sparse coding on subsampled data is unique. *ArXiv* abs/1106.3616

Hodgkin A, Huxley A. 1952. A quantitative description of membrane current and its application to conduction and excitation in nerve. *J. Physiol.* 117:500–44

Hopfield J. 1982. Neural networks and physical systems with emergent collective computational abilities. *Proc. Natl. Acad. Sci. USA* 79(8):2554–59

Hu T, Chklovskii D. 2009. Reconstruction of sparse circuits using multi-neuronal excitation (rescume). *Adv. Neural Inf. Proc. Syst.* 22:790–98

Hu T, Druckmann S, Chklovskii D. 2011. Early sensory processing as predictive coding: subtracting sparse approximations by circuit dynamics. *Front. Neurosci. Conf. Abs: COSYNE*

Hyvarinen A. 2010. Statistical models of natural images and cortical visual representation. *Top. Cogn. Sci.* 2:251–64

Indyk P, Motwani R. 1998. Approximate nearest neighbors: towards removing the curse of dimensionality. *Proc. Annu. ACM Symp. Theory Comput.*, 30th, pp. 604–13

Isely G, Hillar CJ, Sommer FT. 2010. Deciphering subsampled data: adaptive compressive sampling as a principle of brain communication. *Adv. Neural Inf. Proc. Syst. (NIPS)* 23:910–18

Jaeger H. 2001. Short term memory in echo state networks. *GMD Rep. 152*. Germ. Natl. Res. Cent. Inf. Technol., Bremen

Jaeger H, Haas H. 2004. Harnessing nonlinearity: predicting chaotic systems and saving energy in wireless communication. *Science* 304(5667):78–81

Johnson W, Lindenstrauss J. 1984. Extensions of Lipschitz mappings into a Hilbert space. *Contemp. Math.* 26:189–206

Kabashima Y, Wadayama T, Tanaka T. 2009. A typical reconstruction limit for compressed sensing based on l p-norm minimization. *J. Stat. Mech.* L09003

Kelly R, Smith M, Kass R, Lee T. 2010. Accounting for network effects in neuronal responses using l1 regularized point process models. *Neural Inf. Proc. Syst. (NIPS)* 23:1099–107

Kiani R, Esteky H, Mirpour K, Tanaka K. 2007. Object category structure in response patterns of neuronal population in monkey inferior temporal cortex. *J. Neurophysiol.* 97(6):4296–309

Koulakov A, Rinberg D. 2011. Sparse incomplete representations: a novel role for olfactory granule cells. *Neuron* 72(1):124–36

Kreutz-Delgado K, Murray J, Rao B, Engan K, Lee T, Sejnowski T. 2003. Dictionary learning algorithms for sparse representation. *Neural Comput.* 15(2):349–96

Kriegeskorte N, Mur M, Ruff D, Kiani R, Bodurka J, et al. 2008. Matching categorical object representations in inferior temporal cortex of man and monkey. *Neuron* 60(6):1126–41

Lage-Castellanos A, Pagnani A, Weigt M. 2009. Statistical mechanics of sparse generalization and graphical model selection. *J. Stat. Mech.: Theory Exp.* 2009:P10009

Lee S, Ganapathi V, Koller D. 2006a. Efficient structure learning of Markov networks using L1 regularization. *Neural Inf. Process. Syst. (NIPS)* 19:817–24

Lee S, Lee H, Abbeel P, Ng A. 2006b. Efficient l1 regularized logistic regression. *Proc. Natl. Conf. on Artif. Intell.* 21:401

Lewicki M. 2002. Efficient coding of natural sounds. *Nat. Neurosci.* 5(4):356–63

Linsker R. 1990. Perceptual neural organization: some approaches based on network models and information theory. *Annu. Rev. Neurosci.* 13(1):257–81

Lustig M, Donoho D, Pauly J. 2007. Sparse MRI: the application of compressed sensing for rapid MR imaging. *Magn. Reson. Med.* 58(6):1182–95

Lustig M, Donoho D, Santos J, Pauly J. 2008. Compressed sensing MRI. *Signal Proc. Mag. IEEE* 25(2):72–82

Maass W, Natschlager T, Markram H. 2002. Real-time computing without stable states: a new framework for neural computation based on perturbations. *Neural Comput.* 14(11):2531–60

Machens C, Romo R, Brody C. 2010. Functional, but not anatomical, separation of what and when in prefrontal cortex. *J. Neurosci.* 30(1):350–60

Marr D. 1969. A theory of cerebellar cortex. *J. Physiol.* 202(2):437–70

Mishchenko Y. 2011. Reconstruction of complete connectivity matrix for connectomics by sampling neural connectivity with fluorescent synaptic markers. *J. Neurosci. Methods* 196(2):289–302

Olshausen B, Field DJ. 1996a. Emergence of simple-cell receptive field properties by learning a sparse code for natural images. *Nature* 381(6583):607–9

Olshausen B, Field D. 1996b. Natural image statistics and efficient coding. *Netw. Comput. Neural Syst.* 7(2):333–39

Olshausen B, Field D. 1997. Sparse coding with an overcomplete basis set: a strategy employed by v1? *Vis. Res.* 37(23):3311–25

Ostojic S, Brunel N. 2011. From spiking neuron models to linear-nonlinear models. *PLoS Comput. Biol.* 7(1):e1001056

Parrish T, Hu X. 1995. Continuous update with random encoding (cure): a new strategy for dynamic imaging. *Magn. Reson. Med.* 33(3):326–36

Perrinet L. 2010. Role of homeostasis in learning sparse representations. *Neural Comput.* 22(7):1812–36

Rajan K, Abbott L, Sompolinsky H. 2010. Stimulus-dependent suppression of chaos in recurrent neural networks. *Phys. Rev. E* 82(1):011903

Rogers T, McClelland J. 2004. *Semantic Cognition: A Parallel Distributed Processing Approach*. Cambridge, MA: MIT Press

Rosenblatt F. 1958. The perceptron: a probabilistic model for information storage and organization in the brain. *Psychol. Rev.* 65(6):386–408

Rozell C, Johnson D, Baraniuk R, Olshausen B. 2008. Sparse coding via thresholding and local competition in neural circuits. *Neural Comput.* 20(10):2526–63

Seung H, Sompolinsky H, Tishby N. 1992. Statistical mechanics of learning from examples. *Phys. Rev. A* 45(8):6056–91

Simoncelli E, Olshausen B. 2001. Natural image statistics and neural representation. *Annu. Rev. Neurosci.* 24(1):1193–216

Smola A. 2000. *Advances in Large Margin Classifiers*. Cambridge, MA: MIT Press

Sussillo D, Abbott L. 2009. Generating coherent patterns of activity from chaotic neural networks. *Neuron* 63(4):544–57

Takhar D, Laska J, Wakin M, Duarte M, Baron D, et al. 2006. A new compressive imaging camera architecture using optical-domain compression. In *Proc. Comput. Imaging IV Imaging*, San Jose, CA

Taraska J, Zagotta W. 2010. Fluorescence applications in molecular neurobiology. *Neuron* 66(2):170–89

Tarifi M, Sitharam M, Ho J. 2011. Learning hierarchical sparse representations using iterative dictionary learning and dimension reduction. *ArXiv* 1106:0357

Taubman D, Marcellin M, Rabbani M. 2002. Jpeg2000: image compression fundamentals, standards and practice. *J. Electron. Imaging* 11:286

Tibshirani R. 1996. Regression shrinkage and selection via the LASSO. *J. R. Stat. Soc. Ser. B (Methodol.)* 58:267–88

van Hateren J, Ruderman D. 1998. Independent component analysis of natural image sequences yields spatio-temporal filters similar to simple cells in primary visual cortex. *Proc. R. Soc. Lond. Ser. B: Biol. Sci.* 265(1412):2315–20

van Hateren J, van der Schaaf A. 1998. Independent component filters of natural images compared with simple cells in primary visual cortex. *Proc. R. Soc. Lond. Ser. B: Biol. Sci.* 265(1394):359–66

Vapnik V. 1998. *Statistical Learning Theory*. New York: Wiley-Interscience

Wainwright M. 2009. Sharp thresholds for high-dimensional and noisy sparsity recovery using L1-constrained quadratic programming (LASSO). *Inf. Theory IEEE Trans.* 55(5):2183–202

Wainwright M, Ravikumar P, Lafferty J. 2007. High-dimensional graphical model selection using L1-regularized logistic regression. *Adv. Neural Inf. Proc. Syst.* 19:1465–72

Weiss Y, Chang H, Freeman W. 2007. Learning compressed sensing. Presented at *Allerton Conf.*, Urbana-Champaign, IL

White O, Lee D, Sompolinsky H. 2004. Short-term memory in orthogonal neural networks. *Phys. Rev. Lett.* 92(14):148102–5

Wilt B, Burns L, Ho E, Ghosh K, Mukamel E, Schnitzer M. 2009. Advances in light microscopy for neuroscience. *Annu. Rev. Neurosci.* 32:435–506

Yu B, Cunningham J, Santhanam G, Ryu S, Shenoy K, Sahani M. 2009. Gaussian-process factor analysis for low-dimensional single-trial analysis of neural population activity. *J. Neurophysiol.* 102(1):614–35

Zhou S, Lafferty J, Wasserman L. 2009. Compressed and privacy-sensitive sparse regression. *Inf. Theory IEEE Trans.* 55(2):846–66

The Auditory Hair Cell Ribbon Synapse: From Assembly to Function

Saaid Safieddine,[1,2,3,4] Aziz El-Amraoui,[1,2,3] and Christine Petit[1,2,3,5]

[1]Institut Pasteur, Unité de Génétique et Physiologie de l'Audition, F75015, Paris, France; email: saaid.safieddine@pasteur.fr, aziz.el-amraoui@pasteur.fr, christine.petit@pasteur.fr

[2]Inserm UMRS587, Paris, France

[3]UPMC-Paris 6, Paris, France

[4]CNRS, Paris, France

[5]Collège de France, Paris, France

Annu. Rev. Neurosci. 2012. 35:509–28

The *Annual Review of Neuroscience* is online at neuro.annualreviews.org

This article's doi:
10.1146/annurev-neuro-061010-113705

0147-006X/12/0721-0509$20.00

Keywords

synaptic vesicle, neurotransmitter release, temporal precision, Ca^{2+} sensors, genetics of hearing

Abstract

Cochlear inner hair cells (IHCs), the mammalian auditory sensory cells, encode acoustic signals with high fidelity by Graded variations of their membrane potential trigger rapid and sustained vesicle exocytosis at their ribbon synapses. The kinetics of glutamate release allows proper transfer of sound information to the primary afferent auditory neurons. Understanding the physiological properties and underlying molecular mechanisms of the IHC synaptic machinery, and especially its high temporal acuity, which is pivotal to speech perception, is a central issue of auditory science. During the past decade, substantial progress in high-resolution imaging and electrophysiological recordings, as well as the development of genetic approaches both in humans and in mice, has produced major insights regarding the morphological, physiological, and molecular characteristics of this synapse. Here we review this recent knowledge and discuss how it enlightens the way the IHC ribbon synapse develops and functions.

Contents

INTRODUCTION

Hearing contributes to animals' survival and helps them thrive in the environment by analyzing auditory scenes (Bregman 1994). However, hearing's main function is deeply rooted in social living, which requires an effective means of communication and that in many species is carried by sound. Human evolution is characterized by the development of a mode of acoustic communication: speech and language. Acoustic communication relies on the precise and rapid processing of the spectrotemporal properties of sounds all along the auditory pathway. This performance, in mammals, involves the unique properties of the inner hair cells (IHCs), the genuine sensory cells of the cochlea, which are functionally coupled to the outer hair cells (OHCs) to improve sound detection (**Figure 1*a–d***).

Sound-induced vibrations in the cochlear sensory epithelium reach their maximal amplitudes at specific positions along the cochlear longitudinal axis depending on the frequency of the stimulus. This frequency-place relationship underlies the cochlear tonotopic (frequency) map, whereby high- and low-frequency sounds are analyzed at the base and the apex of the cochlea, respectively. In response to sound-induced nanometer displacements of their hair bundles, IHCs undergo a mechanoelectrical transduction (MET) current (Fettiplace & Hackney 2006, Fuchs 2005, Richardson et al. 2011) (**Figure 1*c***) that drives graded variations of their membrane potential. Such a sensory-transduction mechanism, in which a synapse responds not to action potentials (AP) but to a graded receptor potential, is characteristic of all sensory cell types having synaptic active zones equipped with a presynaptic electron-dense structure known as ribbon, to which synaptic vesicles are tethered (**Figure 1*d***) (Smith & Sjostrand 1961). Also common to all ribbon synapses is the fact that they are glutamatergic, wherein glutamate is released at high and sustained rates. Glutamate released by the IHC synapses in response to the receptor potential drives the firing pattern of the primary auditory neurons upon binding to the AMPA (alpha-amino-3-hydroxy-5-methyl-4-isoxazole propionate) receptors (AMPAR) of their afferent boutons (Glowatzki & Fuchs 2002).

IHCs responding to low sound frequencies (up to a few kHz) display receptor potentials that vary in synchrony with the waveform of the sound stimulus. The resulting modulation of neurotransmitter release induces phase-locked spikes of their afferent auditory neurons. Each IHC is innervated by 10–30 auditory afferent bipolar neurons that, although individually limited in their firing frequency by their refractory periods of about 1 ms, together display a firing pattern in cadence with the sound up to a few kHz (Palmer & Russell 1986). At higher sound frequencies, likely as a result of the low-pass electrical filtering properties of the IHC membrane, the receptor potential varies over time with sound intensity but does not vary anymore in synchrony with sound waveform; the frequency coding for these high

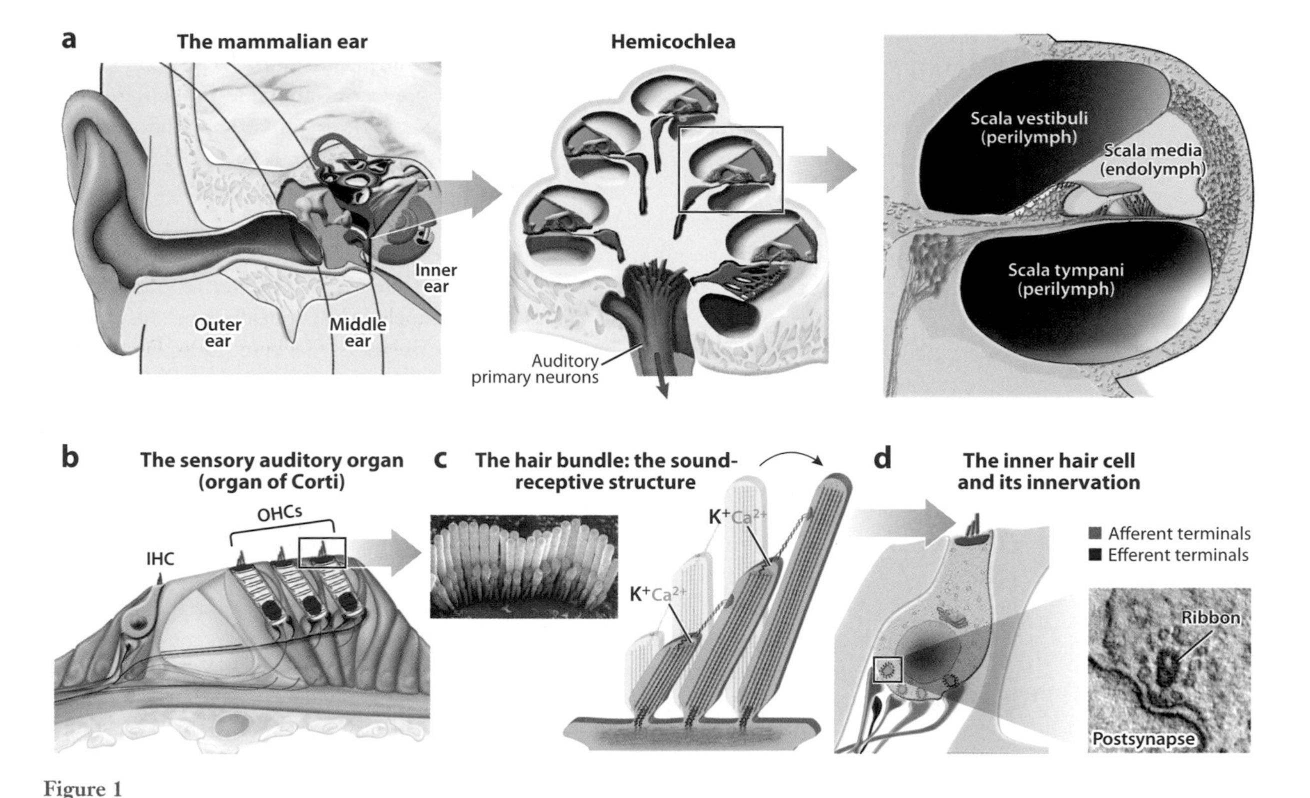

Figure 1

The mammalian ear and the different structures involved in sound processing: the inner ear auditory organ; and sections of the cochlea (hemicochlea) and of a cochlear turn (*a*); the sensory cells of the organ of Corti: inner (IHC) and outer (OHC) hair cells (*b*); the mechanosensitive organelle of IHCs, the hair bundle, with Ca^{2+} and K^+ influx through the mechanoelectrical transduction channels of the stereocilia upon hair bundle deflection (*c*); and an electron microscopy picture of an IHC ribbon surrounded by synaptic vesicles, together with a scheme of an IHC with its innervation (*d*).

frequencies relies on the cochlear tonotopic map. Because the primary afferent auditory neurons have different intensity thresholds and dynamic ranges (Liberman & Kiang 1978), they allow every IHC to encode a wide range of sound intensities regardless of sound frequency (**Figure 1*a–d***). Moreover, expansion of the IHC receptor potential that results from sound intensity augmentation increases the release probability of the neurotransmitter, which in turn increases the discharge rate of the afferent neurons. In response to low-frequency sounds, this occurs with the maintenance of the phase synchrony of their firing (Rose et al. 1967), whereas in response to high-frequency sounds, the frequency of their spiking increases.

Sound processing in the auditory system is characterized by an extreme temporal precision, in the microsecond range, which is critical to hearing. IHC synaptic activity, as the first relay, is entrusted with the same property. The phase-locking of the neuronal response is indeed used to compute the interaural time difference (also called binaural temporal delay) of the acoustic input signal (Rayleigh 1907), which allows animals with large enough heads to localize low-frequency sound sources in the azimuthal plane (Goldberg et al. 1990). In addition, the onset time of the neuronal response has a prominent role in acoustic directional sensitivity (Devore et al. 2009). The IHC temporal acuity is also essential to speech intelligibility because the rapid changes, onsets, offsets, and intensity variations of acoustic signals over time (sound envelope) are major cues of speech understanding. This temporal information, converted into rapid changes in IHC receptor potential, is encoded by rapid variations in the primary auditory neuron firing rate, whatever the sound frequency (see above).

During the past decade, the use of high-resolution imaging of the IHC synapse and the development of new electrophysiological approaches—specifically, the membrane capacitance measurements (C_m) (Moser & Beutner 2000), the simultaneous recordings of the Ca^{2+} currents in IHCs (the presynapse), and the response of its afferent boutons (the postsynapse)—have provided new physiological insights into this synapse (Glowatzki & Fuchs 2002, Goutman & Glowatzki 2007). In parallel, genetic approaches based on the study of mutant mice defective for known synaptic proteins and of mouse models of human deafness forms (Leibovici et al. 2008) have launched IHC synapse molecular physiology and have also led investigators to uncover new aspects of the developmental steps of this synapse. This review focuses on recent progress regarding the development and functioning of the IHC synapse, with a particular emphasis on its temporal performance.

MORPHOLOGICAL MATURATION OF THE IHC RIBBON SYNAPSE

In most altricial rodents, the onset of hearing occurs around postnatal day 12 (P12). However, before hearing onset, and already at P0, the primary auditory neurons spontaneously fire bursts of action potentials, which are thought to be driven by a spontaneous spiking activity of the immature IHCs (Johnson et al. 2011b, Tritsch & Bergles 2010). Both spiking activities develop as early as embryonic day 16 (E16) (Johnson et al. 2005). From E16 on, while the differentiation of the auditory sensory epithelium proceeds from the cochlear base to its apex, the IHC synaptic machinery undergoes a series of morphofunctional changes (Johnson et al. 2009, Kros 2007, Roux et al. 2009, Sobkowicz et al. 1986) (see **Figure 2*a***). In the newborn mouse up to P6, the immature ribbons are typically electron dense spheres surrounded by synaptic vesicles (Roux et al. 2009, Sobkowicz et al. 1986). The postsynaptic density (PSD) appears later, at around P8 (**Figure 2*a***), indicating that, as in CNS synapses, the presynaptic maturation precedes the postsynaptic one (Friedman et al. 2000). At P6, most of the ribbons are anchored by two tubular rootlets to a presynaptic thickening (**Figure 2*a***). Between P6 and P10, the IHC ribbons progressively acquire an oval shape and are attached to the presynaptic active zone by a single arcuate density (**Figure 2*a***). By P12, oval-shaped and larger ribbons with a single attachment are highly prevalent (**Figure 2*a***). Immature IHCs synapse with efferent neurons, the number of which decreases from P6 onward. The number of ribbons also

Figure 2

The morphofunctional postnatal maturation of the mouse inner hair cell (IHC) ribbon synapse. (*a*) Morphological maturation of the mouse IHC ribbon. (*b*) Different morphofunctional characteristics of the afferent innervation in the mature IHC synapse. Synapses at the modiolar side of the IHC involve afferent neurons that have smaller diameters, higher-threshold (HT) sensitivity, and lower spontaneous rate (LSR), and the presynaptic regions have larger ribbons (*violet*) and display higher Ca^{2+} inputs (*light green*). At the pillar side, the synapses display opposite characteristics. Other abbreviations: HSR, high spontaneous rate; LT, lower threshold. (*c*) Developmental changes in ΔC_m and I_{Ca} (adapted from Johnson et al. 2005). The upper and lower insets show the relation between I_{Ca} (presynaptic Ca^{2+} entry) and vesicle exocytosis (transfer function), and IHC innervation, respectively [note the disappearance by P12 of direct synapses between the IHC and the efferent endings (*blue*) and the decrease of the number of Ca_v1.3 channels (*red dots*) from immature to mature IHCs]. Of note, some maturation of exocytosis persists beyond P20 (Beutner & Moser 2001, Johnson et al. 2009). (*c*) Diagram showing the appearance of the $I_{k,f}$ current at about hearing onset (P12), concomitant with the disappearance of spiking activity. Neither synaptotagmin 1 nor synaptotagmin 2 proteins (*green*) persist in the IHC at hearing onset. Synaptotagmin 7 has been detected by single-cell Reverse Transcription Polymerase Chain Reaction (RT-PCR) at P2 and P8. Otoferlin is present as early as E16, but IHC synaptic exocytosis becomes otoferlin-dependent only from P5–P6 onward.

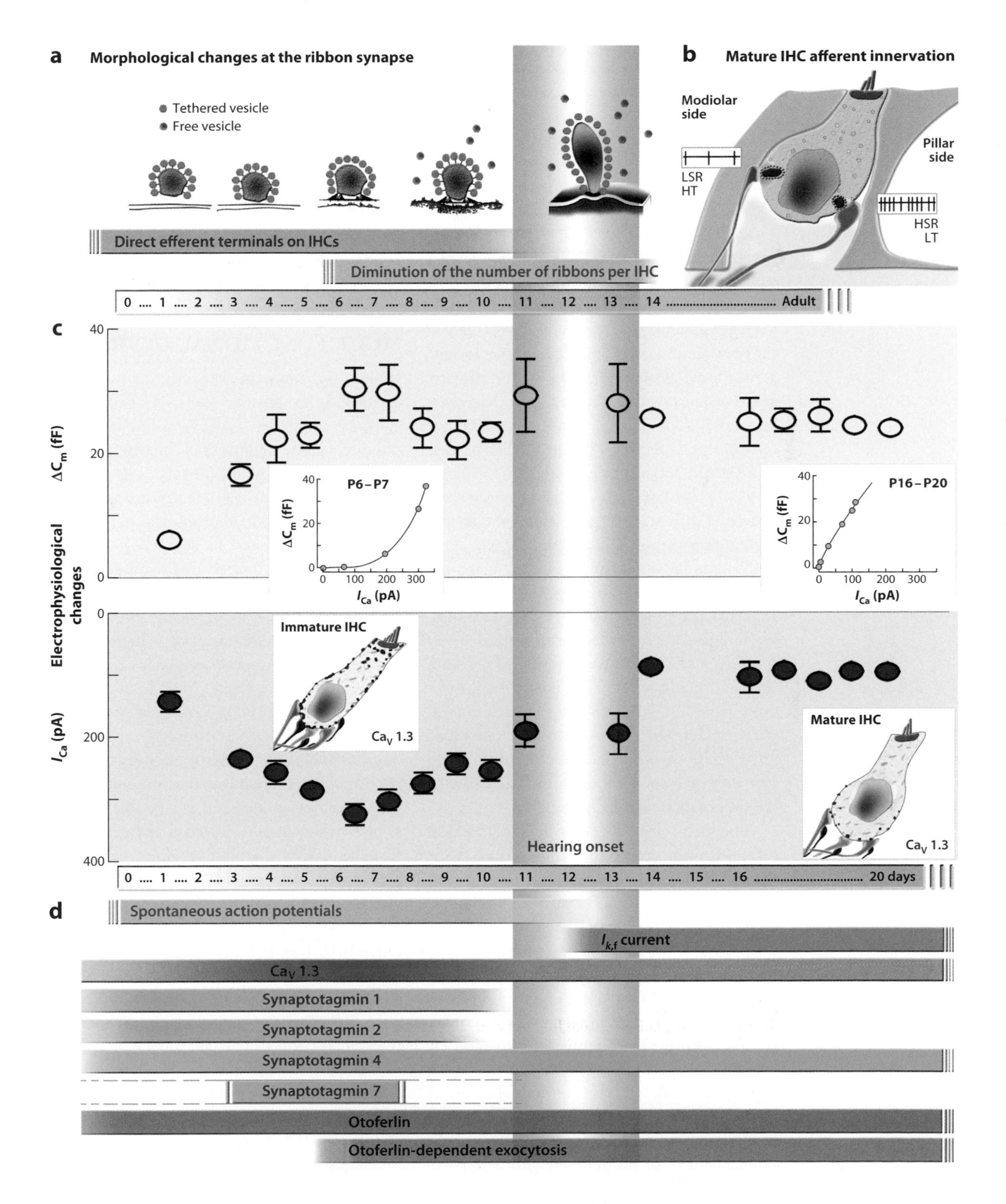
a
Morphological changes at the ribbon synapse
Tethered vesicle
Free vesicle
Direct efferent terminals on IHCs
Diminution of the number of ribbons per IHC
b
Mature IHC afferent innervation
Modiolar side
Pillar side
LSR
HT
HSR
LT
0 .... 1 .... 2 .... 3 .... 4 .... 5 .... 6 .... 7 .... 8 .... 9 .... 10 .... 11 .... 12 .... 13 .... 14 .... Adult
c
Electrophysiological changes
ΔCm (fF)
P6–P7
P16–P20
ICa (pA)
Immature IHC
CaV 1.3
Mature IHC
Hearing onset
0 .... 1 .... 2 .... 3 .... 4 .... 5 .... 6 .... 7 .... 8 .... 9 .... 10 .... 11 .... 12 .... 13 .... 14 .... 15 .... 16 .... 20 days
d
Spontaneous action potentials
Ik,f current
CaV 1.3
Synaptotagmin 1
Synaptotagmin 2
Synaptotagmin 4
Synaptotagmin 7
Otoferlin
Otoferlin-dependent exocytosis

declines (Roux et al. 2009, Sobkowicz et al. 1986) (**Figure 2*a***), likely as a result of the pruning of the immature branched afferent neurons (Huang et al. 2007, Pujol et al. 1998).

At mature stages, the afferent neurons of IHCs display not only different sensitivity thresholds but also different spontaneous firing rates (SR), the two parameters being inversely correlated (Liberman 1980). All these neurons exhibit temporal precision of their sound-evoked firing (Johnson 1980). The IHC synapses are spatially distributed according to a morphofunctional circumferential gradient at the basolateral area of the cell. Synapses located at the modiolar (neural) side of the IHC involve afferent neurons that have a higher sensitivity threshold, a lower SR, a wider dynamic range of response (Liberman 1980), smaller patches of AMPAR, and a smaller diameter (Liberman et al. 2011), and these synaptic regions are equipped with much larger ribbons and display higher presynaptic Ca^{2+} inputs (Meyer et al. 2009) than do the synapses located at the opposite, pillar (abneural) side (Liberman 1980, Liberman et al. 2011) (**Figure 2*b***). This finding indicates that the electrical response of the primary auditory neurons is shaped by both pre- and postsynaptic elements, which, however, have not yet been integrated in an explicit model. Of note, Goutman & Glowatzki (2007) have shown the adaptation of the auditory neuron response to be mainly achieved by the presynaptic release machinery.

The mature ribbon, regardless of its size or location, lines up a single and concave PSD, which largely exceeds the territory of the presynaptic active zone (Roux et al. 2009, Sobkowicz et al. 1982, Stamataki et al. 2006) (see **Figures 1*d*** and **2*a***). Such an unusually large PSD may contribute to the peculiar properties of this synapse. Thanks to a large number of glutamate receptors (Choquet & Triller 2003), it may secure the synaptic transmission by avoiding receptor saturation, thus contributing to synaptic temporal precision and reliability. A large pool of extrasynaptic glutamate receptors (away from those facing the ribbon), allowing rapid changes in the number and/or type of synaptic receptors, might also ensure a highly dynamic and finely tuned postsynaptic response (Borgdorff & Choquet 2002, Choquet 2010). Within the PSD of CNS synapses, the AMPARs, which dominate in IHC synaptic transmission (Glowatzki & Fuchs 2002), have indeed been shown to display movements fast enough to impact synaptic transmission in the millisecond time scale (Frischknecht et al. 2009).

THE RIBBON AS A MULTIFUNCTIONAL DEVICE

Ribbon synapses in vertebrates are found exclusively in sensory cells. These include auditory, vestibular, and lateral line hair cells, photoreceptors, and retinal bipolar neurons as well as pineal gland sensory cells and the sensory cells of the fish electrical organs. All have a high rate of neurotransmitter release that can be sustained in response to a long stimulation, even though this feature is more pronounced in IHCs. Insights into the synaptic function of IHCs have been initially obtained from the recordings of voltage-evoked changes in membrane electric capacitance (C_m) (Glowatzki & Fuchs 2002, Li et al. 2009, Moser & Beutner 2000). Excitatory postsynaptic currents (EPSCs) recording from afferent boutons of rat IHCs and frog auditory hair cells have subsequently shown that the increase in depolarization-induced C_m can be attributed to synaptic exocytosis (Glowatzki & Fuchs 2002, Li et al. 2009). Correlations were tentatively established between the number of synaptic vesicles present at a given location, specifically the vesicles docked below the ribbon and those tethered to it, and the two synaptic vesicle pools functionally defined by analogy with CNS synapse vesicular pools: a readily releasable pool (RRP), or fast component, and a slowly releasable pool (SRP), or slow component of exocytosis. Regardless of the IHC ribbon synapse explored, consensus is emerging that a morphologically defined population, specifically, the vesicles located at the plasma membrane below the ribbon, contributes to the RRP (Goutman & Glowatzki 2007, Graydon et al. 2011). In contrast, the correlation between the ribbon-associated vesicles and the SRP is far from being established (see Sterling & Matthews 2005 for review). Yet, Graydon and colleagues recently established a correlation in frog hair cells between the vesicles tethered to the ribbon and the SRP (Graydon et al. 2011). In the mouse, both RRP and SRP represent only a tiny fraction (1–2%) of the total number of vesicles, with depletion time constants of 10 ms and 200 ms, respectively (Nouvian et al. 2006). Remarkably, the IHC synaptic exocytosis can be maintained at an almost constant rate, with limited fatigue, in response to a sustained stimulation lasting seconds (Beurg et al. 2010, Dulon et al. 2009, Johnson et al. 2005, Moser & Beutner 2000, Schnee et al. 2011). This finding suggests that to avoid depression the RRP and SRP are rapidly replenished, faster than they are depleted, by recycling vesicles or reserve vesicles, which implies the existence of a fast and precisely regulated mode of vesicle retrieval, possibly coupled with the exocytotic pathway.

The ribbon has been proposed as underpinning the most prominent features of IHC synaptic exocytosis because of its high rate and sustained neurotransmitter release and also its temporal acuity and fidelity (Parsons & Sterling 2003). The ribbon has been portrayed as a conveyor belt that rapidly delivers the synaptic vesicles to the release sites. However, by combining electron microscopy quantification of synaptic vesicles and mathematical modeling, the ribbon has been proposed to slow down, rather than speed up, the delivery of vesicles to release sites and thus to limit the fusion rate (Jackman et al. 2009). Likewise, in hair cells of the amphibian auditory papilla, the majority of synaptic vesicles of the active zone are washed out after a strong depolarization, except those associated with the ribbon (Lenzi et al. 1999). The ribbon has been considered as a possible regulator of vesicle-vesicle or vesicle–plasma membrane fusion, namely an inhibitor and/or accelerator and possibly a synchronizer of the vesicle fusion process (LoGiudice et al. 2008, Nouvian et al. 2006, Paillart et al. 2003, Parsons & Sterling 2003, Sterling & Matthews 2005). Until recently, the sole informative animal model to address the role of ribbons was the mutant mouse deficient for bassoon, a multidomain protein involved in the assembly of functional active zones at the CNS synapses (Garner et al. 2000). The predominant reported feature in bassoon mutant mice was the loss of the ribbons at active zones (Dick et al. 2003, Khimich et al. 2005, Nemzou et al. 2006). Subsequent in-depth analysis revealed the heterogeneity of the synapses in this mutant, including normal, absent, and misplaced ribbons. To disentangle the contribution of the absence of bassoon and of the anomalies of the ribbons to the IHC physiological defects (Buran et al. 2010), the reduced number of exocytotic sites observed in IHCs lacking bassoon was tentatively attributed to the ribbon anomalies (Frank et al. 2010). This highlights the need for other genetic models, in which the ribbon itself would be the primary target of the deficit, for a better understanding of its roles. Snellman et al. (2011) recently found that an acute alteration of the ribbons by photodamage in mouse retinal bipolar cells and salamander cones results in a marked reduction of both RRP and SRP, even though neither their driving Ca^{2+} currents nor the ultrastructure of the ribbon or the distribution of synaptic vesicles are affected. These findings suggest that the ribbon is involved in the priming process, which makes the vesicles competent for fusion (Snellman et al. 2011). Whether priming proteins are associated with the ribbons is presently unknown (Gebhart et al. 2010, Uthaiah & Hudspeth 2010).

FUNCTIONAL MATURATION OF THE INNER HAIR CELL SYNAPSE

Major changes of biophysical properties of the IHC synapse parallel its morphological maturation. The spontaneous firing of neurons during early developmental stages is well documented in the retina, spinal cord, hippocampus, and cerebellum and has been known to ensure the survival and the differentiation of neurons (Meister et al. 1991; see Blankenship & Feller 2010 for a review). The immature IHC also generates spontaneous action potentials (APs) as early as E16 and up to P12 in the mouse (Marcotti et al. 2003) (see also **Figure 2*d***). These are Ca^{2+}-dependent APs that lead to neurotransmitter release, which in turn triggers bursts of APs in the primary auditory neurons. At first, the Ca^{2+} spikes are small and broad; they become faster around P6. Whereas guidance cues underlie the initial tonotopic organization of the primary auditory neuron projections in the cochlear nucleus (Huffman & Cramer 2007), recent evidence suggests that the prehearing bursts of APs in auditory neurons refine their tonotopic projections (Johnson et al. 2011b). This subsequent refinement fails to occur upon hair cell loss within the first postnatal week in both gerbils and cats (Leake et al. 2006, Mostafapour et al. 2000, Tierney et al. 1997).

The spontaneous APs of IHCs depend mainly on voltage-dependent Ca^{2+} channels to produce a regenerative depolarization and on voltage-gated, delayed rectifier K^+ channels to produce membrane repolarization. Both currents increase in amplitude during the first postnatal week (Marcotti et al. 2003). They have been proposed to stem from ATP released from a group of supporting cells located next to the IHCs (Tritsch et al. 2007, Tritsch & Bergles 2010). ATP release, through the activation of purinergic receptors on IHCs, could induce their Ca^{2+} spikes and may synchronize the activity of neighboring IHCs. A recent study, however, reached a different conclusion. The authors showed that ATP seems to be dispensable in the initiation of IHC spontaneous APs (Johnson et al. 2011b). Immature IHCs seem to be endowed with an intrinsic electrical activity to generate their spontaneous APs. The resting potential of IHCs, which is close to the threshold of voltage activation of Ca^{2+} channels, likely accounts for AP initiation. ATP and acetylcholine, released near the IHCs by supporting cells and by the IHC efferent nerve neurons, respectively, modulate these AP patterns (Johnson et al. 2011b). IHCs of the cochlear apex fire bursts of APs presumably modulated by acetylcholine, whereas IHCs of the cochlear base fire more frequent bursts of APs (Johnson et al. 2011b), possibly modulated by ATP acting on P2X3 receptors that are differentially expressed along the cochlea (Huang et al. 2006).

The final developmental step in the electrical activity of the IHCs is the switch from APs to graded variations of the membrane potential, starting at P10 at the cochlear base and reaching the apex by P12 in the mouse (Kros 2007). This final change is due mainly to the drop in number of voltage-gated Ca^{2+} channels (**Figure 2*d***) with the concomitant acquisition of big potassium (BK) channels underlying the large, fast, and noninactivating K^+ current $I_{k,f}$ (**Figure 2*d***) (Kros 2007, Zampini et al. 2010), which prevents regenerative membrane depolarization. This change also enables membrane potential variations in synchrony with sound stimulus waveforms, which drives a phase-locked response of the primary auditory neurons for sound frequencies up to a few kHz (Palmer & Russell 1986).

In both the pre- and posthearing periods, exocytosis at the IHCs is triggered by synaptic Ca^{2+} currents supported for more than 90% of their amplitude by Ca^{2+} influx through the L-type voltage-gated Ca^{2+} channel $Ca_v1.3$ (Brandt et al. 2003, Platzer et al. 2000) (see **Figure 3**). The Ca^{2+} currents recorded in immature IHCs, however, display much slower kinetics, slightly stronger inactivation, and much larger amplitude (Johnson & Marcotti 2008, Zampini et al. 2010) than do those of the mature IHCs (Johnson & Marcotti 2008) (**Figure 2*c***). Fulfilling the requirement of a precise temporal coding and a sustained synaptic exocytosis, $Ca_v1.3$ channels of the mature IHCs display very rapid gating (submillisecond activation kinetics), are activated near the resting membrane potential of the cell, which is about −60 mV (Johnson et al. 2011a), and show little inactivation (Grant & Fuchs 2008, Johnson & Marcotti 2008, Zampini et al. 2010). The gating properties of the Ca^{2+} current

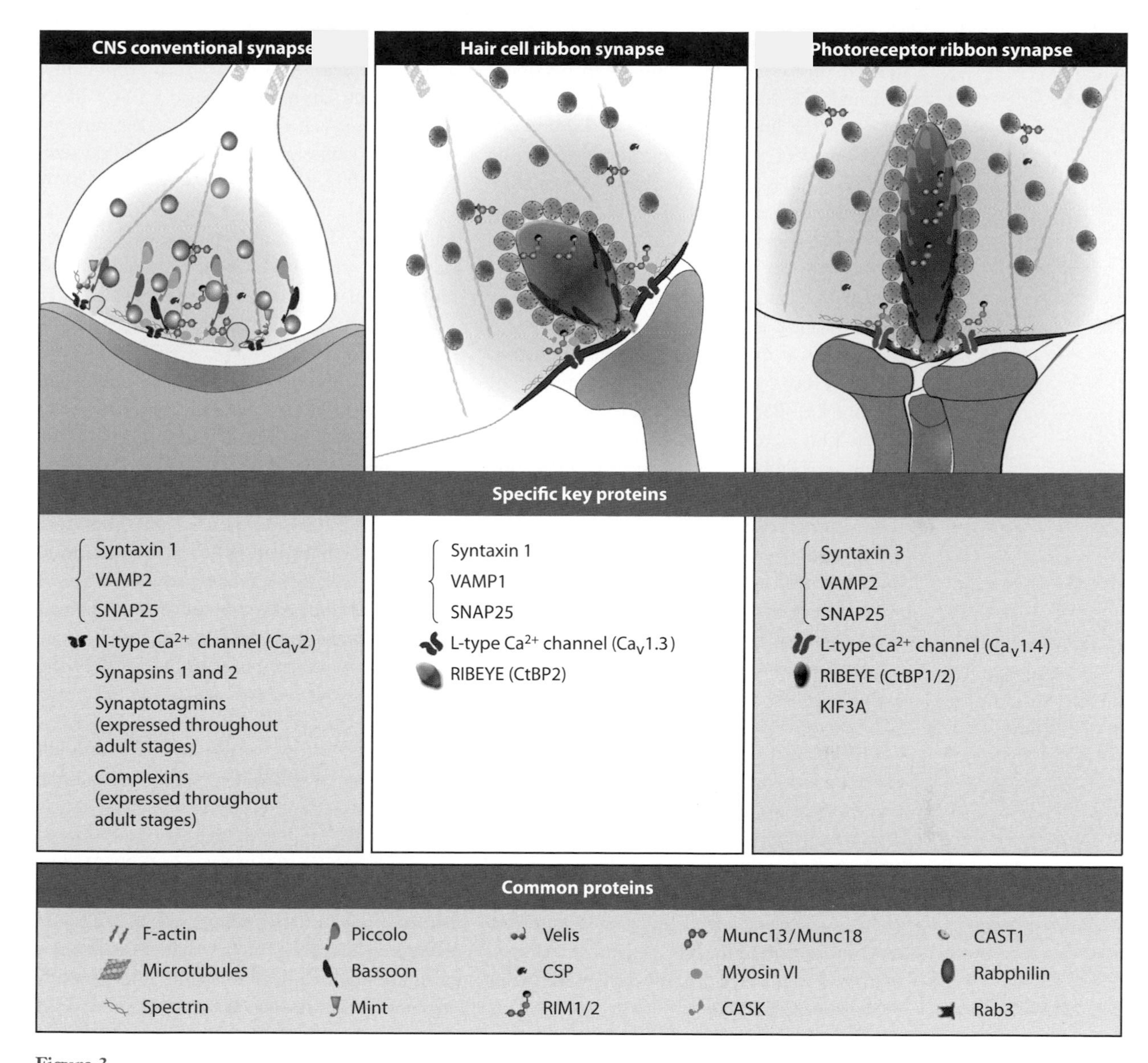

Figure 3

Schematic representation of the architecture of rodent mature synapses in the CNS, hair cells, and photoreceptor cells, and proteins common or specific to these synapses.

in mature IHCs vary along the tonotopic gradient, being faster in IHCs of the cochlear base, which are tuned to high-frequency sounds (Johnson & Marcotti 2008). Several interacting proteins have been reported as modulators of the $Ca_v1.3$ channels and may tune the channel activity along the cochlear tonotopic axis (Cui et al. 2007, Jenkins et al. 2010).

The transfer function, which defines the relationship between Ca^{2+} entry and the rate of synaptic exocytosis, also changes during IHC maturation. The transfer function, tested under physiological conditions of stimulation, switches from a nonlinear (exponential) relation in immature IHCs, indicating the involvement of a high Ca^{2+} cooperativity in the exocytotic process, to a linear one (insets in **Figure 2c**). This linearity preserved in the postsynaptic current implies that the auditory information is transmitted to the CNS with a very low threshold response and minimal distortion (Goutman & Glowatzki 2007, Keen & Hudspeth 2006). Moreover, the amplitude of IHC synaptic exocytosis at mature stages is larger than at immature stages, although triggered by Ca^{2+} currents of smaller amplitude (**Figure 2b**) (Beutner et al. 2001, Johnson et al. 2005). This increase in release efficiency and the switch to a linear Ca^{2+} dependency of mature IHC exocytosis [except for very-low-frequency IHCs (Johnson & Marcotti 2008)] may result from a spatial reorganization of the IHC synaptic machinery: a switch from a microdomain synaptic organization, involving the cooperation of several voltage-gated Ca^{2+} channels to trigger synaptic vesicle fusion, to a nanodomain organization, in which the opening of a single Ca^{2+} channel can drive a fusion event. In the nanodomain configuration, Ca^{2+} channels and Ca^{2+} sensors associated to the synaptic vesicles are thought to be tightly coupled within a distance, which has been estimated, using exogenous Ca^{2+} chelators, at 20 nm–40 nm (Goutman & Glowatzki 2007, Johnson et al. 2008). A developmental change of the molecular composition of the IHC synaptic machinery might also account for the observed switch (Johnson et al. 2008).

DEVELOPMENTAL TRANSITION IN MECHANISMS OF VESICLE FUSION AT THE IHC RIBBON SYNAPSE

The EPSCs recorded from afferent boutons of the auditory neurons on IHCs and frog auditory hair cells have shown monophasic and multiphasic waveforms even at a single bouton (Glowatzki & Fuchs 2002, Li et al. 2009). Both spontaneous and evoked monophasic EPSCs had variable amplitudes, some of them largely exceeding that expected from uniquantal release. The large monophasic EPSCs likely reflect the synchronized release of multiple vesicles at single ribbon synapses (Glowatzki & Fuchs 2002, Goutman & Glowatzki 2007, Grant et al. 2010, Li et al. 2009), whereas multiphasic EPSCs are considered to result from uncoordinated (or asynchronous) multivesicular fusion events (Glowatzki & Fuchs 2002, Li et al. 2009). Changes in the patterns of EPSCs during IHC maturation also occur. The ratio between the monophasic and multiphasic EPSCs and their kinetics varies greatly during cochlear development. The large monophasic EPSCs become predominant (75%), and the kinetics of the monophasic and multiphasic EPSCs are much faster in mature synapses (Grant et al. 2010). The large and fast monophasic EPSCs are likely to be critical for the precise temporal coding of sound. The role of the multiphasic EPSCs is less clear because they may create temporal jitter in the firing of the auditory neurons. In the bullfrog hair cells, only the large and monophasic EPSCs can trigger spikes in the primary afferent neurons that are phase-locked to the sinusoidal sound stimulation, whereas the multiphasic EPSCs seem unable to trigger the firing of these neurons (Li & von Gersdorff 2011). The simultaneous fusion of several vesicles could be accomplished in two different ways. Vesicles docked below the ribbon could fuse at once with the presynaptic membrane. Alternatively, vesicles could fuse with one another prior to their fusion with the plasma membrane, a process analogous to the compound fusion that occurs during secretion in mast cells (Alvarez de Toledo & Fernandez 1990). Electron microscopy imaging and physiological evidence argue in favor of compound fusion at the ribbon synapses of retinal goldfish bipolar cells subjected to strong stimulation (Matthews & Sterling 2008) and the calyces of Held (He et al. 2009), respectively (for a review, see also Matthews & Fuchs 2010). Simultaneous and compound fusion mechanisms may coexist in IHCs.

THE MOLECULAR COMPONENTS OF IHC SYNAPTIC EXOCYTOTIC MACHINERY: AN INTRIGUING SINGULARITY RESULTING FROM THE ABSENCE OF KEY PLAYERS

Deciphering of the molecular composition of the IHC ribbon synaptic machinery was initiated about a decade ago. Expression of N-ethylmaleimide sensitive factor attachment protein receptor (SNARE) complex proteins (see **Figure 3**) was investigated first. Investigators reported on IHC expression of the vesicular proteins synaptobrevin 1 and 2, also known as VAMP1 and VAMP2 (v- or R-SNAREs), and the presynaptic membrane proteins syntaxin 1 and SNAP-25 (t- or Q-SNAREs) (Chen & Scheller 2001, Sollner et al. 1993), which form the SNARE complex, the minimal molecular machinery that brings the membrane of synaptic vesicles and the plasma membrane into close apposition for fusion (Eybalin et al. 2002, Safieddine & Wenthold 1999, Uthaiah & Hudspeth 2010, Wenthold et al. 2002) (**Figure 3**). Dick et al. (2001) subsequently found the presence of bassoon and piccolo, proteins of the synaptic cytoskeleton matrix (see above). However, many molecular peculiarities were determined. The two major Ca^{2+} sensors involved in fast exocytosis in CNS synapses, specifically, synaptotagmins 1 and 2 (Safieddine & Wenthold 1999, Uthaiah & Hudspeth 2010, Wenthold et al. 2002) (see below), as well as synaptophysins and synapsins, were not detected in the mature IHCs (Eybalin et al. 2002, Mandell et al. 1990, Safieddine & Wenthold 1997) (**Figures 2*d*** and **3**). Beurg et al. (2010) recently confirmed the absence of synaptotagmins 1 and 2 at the adult stage using different approaches. Moreover, complexins (CPX I to IV), which regulate a late step in the synaptic release process, most likely by clamping the SNARE complex (Martin et al. 2011, Xu et al. 1999, Yang et al. 2010), were not found in IHCs either, and likewise, no IHC synaptic exocytic defects were observed in CPX-null mice (Strenzke et al. 2009). Recent results also challenged the implication of a neuronal SNARE complex in IHC synaptic exocytosis. Neither the botulinum neurotoxins that cleave SNAP25, syntaxins 1–3, and VAMP1–3, nor the absence of SNAP25, VAMP1, or VAMP2/3 affected IHC synaptic exocytosis (Nouvian et al. 2011). Moreover, no neuronal SNARE proteins could be immunodetected in IHCs (Nouvian et al. 2011). However, there are precedents of similar observations in neuronal SNARE-mediated exocytosis. A dramatic reduction of the efficacy of the toxin cleavage of VAMP2 has indeed been reported during synaptogenesis when a high rate of synaptic vesicle fusion is taking place (Daly & Ziff 1997, Matteoli et al. 1996), and IHC ribbon synapse has a particularly high synaptic vesicle turnover (Beutner et al. 2001, Griesinger et al. 2005). The functional redundancy between the SNARE components, which may account for these observations, is well documented (Bhattacharya et al. 2002, Borisovska et al. 2005, Liu & Barlowe 2002, Sadoul et al. 1997, Vilinsky et al. 2002), and such a redundancy may explain the observation of a neuronal SNARE-dependent exocytosis also resistant to the action of the botulinum toxin in hippocampal GABAergic synapses (Verderio et al. 2004). Finally, whereas neuronal SNAREs have not been observed in IHCs by immunolabeling, their transcripts have been detected (Nouvian et al. 2011, Sendin et al. 2007), and SNAREs have been found by proteomic analysis of chick hair cell ribbons (Uthaiah & Hudspeth 2010). Therefore, either a low abundance of the synaptic machinery proteins is a characteristic of the IHC ribbon synapse and their redundancy is particularly high (but as yet incompletely explored) or a set of proteins uncommon to CNS synapses comprises the IHC synaptic exocytotic machinery.

About 20 proteins have been identified as components of the mature IHC synapse (see **Figure 3**) of which six, namely myosin VI, bassoon, VGLUT3, $Ca_v1.3$, synaptotagmin 4, and otoferlin, were demonstrated as playing essential roles in synaptic exocytosis thanks to

① Hair bundle development and functioning

MYO7A (myosin VIIa, DFNB2/DFNA11, USH1B, OMIM276903)[a]
MYO15 (myosin XV, DFNB3, OMIM602666)
MYO6 (myosin VI, DFNB37/DFNA22, OMIM600970)
MYO3A (myosin IIIa, DFNB30, OMIM606808)
MYO1A (myosin Ia, DFNA48, OMIM601478)
MYO1C (myosin Ic, DFNAi, OMIM606538)
ACTG1 (γ-actin, DFNA20/26, OMIM102560)
RDX (radixin, DFNB24, OMIM179410)
TRIOBP (trio/F-actin-binding protein, DFNB28, OMIM609761)

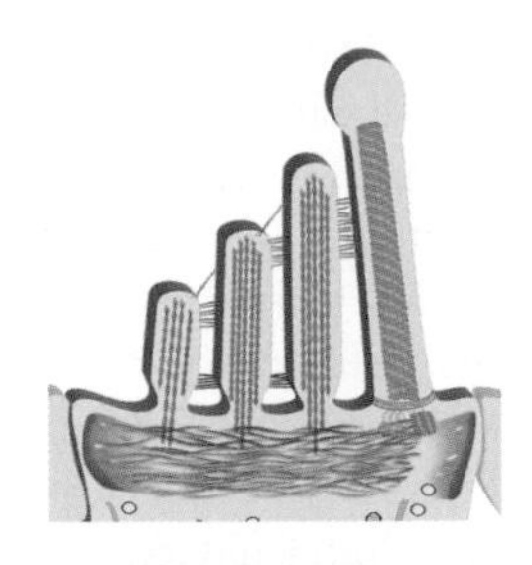

TMC1 (DFNB7/DFNB11/DFNA36, OMIM606706)
CDH23 (cadherin 23, DFNB12, USH1D, OMIM602092)[a]
STRC (stereocilin, DFNB16, OMIM606440)
USH1C (harmonin, DFNB18, OMIM605242)[a]
PCDH15 (protocadherin 15, DFNB23, USH1F, OMIM605514)[a]
GRXCR1 (glutaredoxin cys-rich 1, DFNB25, OMIM613283)
ESP (espin, DFNB36, OMIM606351)
WHRN (whirlin, DFNB31, USH2D, OMIM607928)[a]
TMHS (LHFPL5, tetraspanin protein, DFNB66/67, OMIM609427)
TPRN (taperin, DFNB79, OMIM613354)
PTPRQ (tyrosine phosphatase receptor, DFNB84, OMIM603317)

② Synaptic transmission

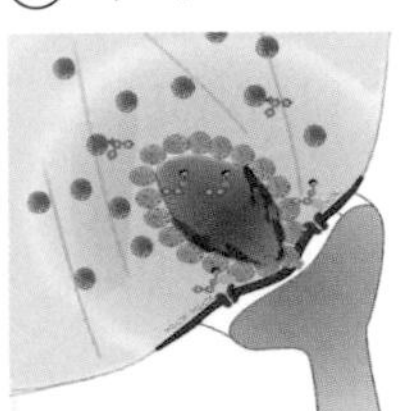

OTOF (otoferlin, DFNB9, OMIM603681)
PJVK (pejvakin, DFNB59, OMIM610219)
VGLUT3 (vesicular glutamate transporter, DFNA25, OMIM607557)

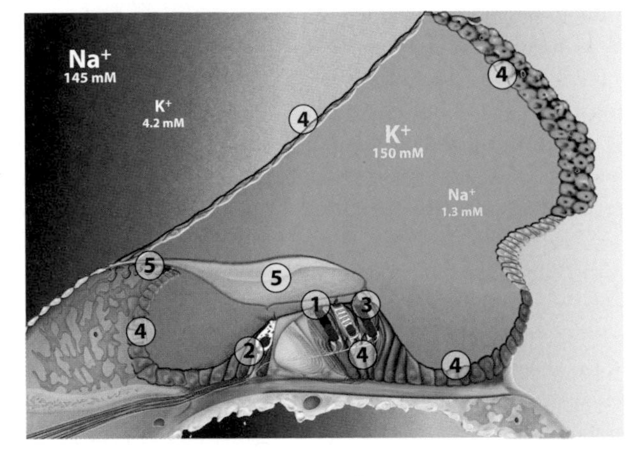

③ Cell-cell adhesion

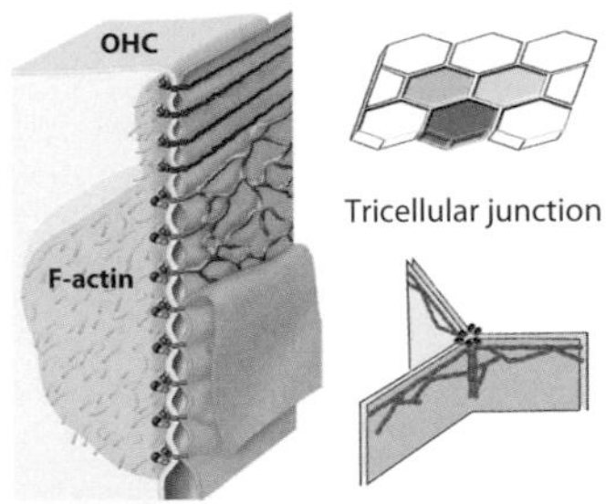

CLDN14 (claudin14, DFNB27, OMIM605608)
TRIC (tricellulin, DFNB49, OMIM610572)
VEZT (vezatin, adherens junction protein, DFNBi)

④ Ion homeostasis

Gap junction proteins ×6 Connexon

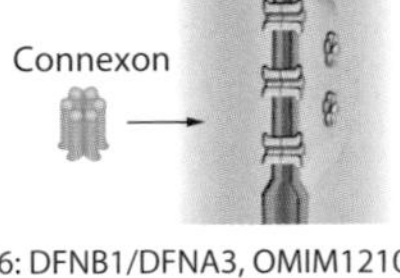

GJB2 (connexin 26: DFNB1/DFNA3, OMIM121011)
GJB3 (connexin 31: DFNB2b, OMIM604418)
GJB6 (connexin 30: DFNB1a/DFNA3b, OMIM604418)
KCNQ4 (DFNA2, OMIM603537)
PDS/SLC26A4 (pendrin, DFNB4, OMIM605646)
BSND (barttin, DFNB73, OMIM606412)

⑤ Extracellular matrix

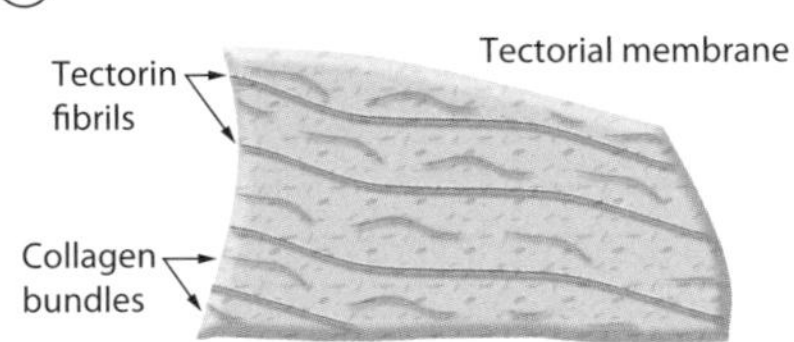

TECTA (α-tectorin, DFNB21/DFNA8/DFNA12, OMIM602574)
COL11A2 (collagen XIα2, DFNB53/DFNA13, OMIM120290)
COCH (cochlin, DFNA9, OMIM603196)
OTOA (otoancorin: DFNB22, OMIM607038)

Deafness genes involved in other, multiple, or unknown functions

Dominant deafness forms

DIAPH1 (DFNA1, OMIM602121)
DIAPH3 (AUNA1/DFNAi, OMIM609129)
MYH14 (DFNA4, OMIM608568)
DFNA5 (DFNA5, OMIM608798)
WFS1 (DFNA6/14/38, OMIM606201)
MYH9 (DFNA17, OMIM160775)
EYA4 (DFNA10, OMIM603550)
POU4F3 (DFNA15, OMIM602460)
TFCP2L3 (GRHL2, DFNA28, OMIM608576)
CCDC50 (DFNA44, OMIM611051)
MIRN6 (miR96, DFNA50, OMIM611606)
CRYM (DFNAi, OMIM123740)
TJP2 (DFNA51, OMIM607709)
SMAC/DIABLO (DFNA64, OMIM605219)

Recessive deafness forms

TMIE (DFNB6, OMIM607237)
TMPRSS3 (DFNB8/10, OMIM605511)
ESRRB (DFNB35, OMIM602167)
HGF (DFNB39, OMIM142409)
ILDR1 (DFNB42, OMIM609739)
LRTOMT (DFNB63, OMIM612414)
MSRB3 (DFNB74, OMIM613719)
GIPC3 (DFNB15/DFNB95/DFNB72, OMIM608792)
LOXHD1 (DFNB77, OMIM613072)
GPSM2 (DFNB82, OMIM609245)
SERPINB6 (DFNB91, OMIM173321)

X chromosome-linked deafness forms

PRPS1 (DFN2, OMIM311850)
POU3F4 (Brn4, DFN3, OMIM300039)
SMPX (DFN6, OMIM300226)

Mitochondrial deafness forms

MTRNR1 (12S, OMIM561000)
MTTS1 (tRNA Ser, OMIM590080)

genetic approaches in humans (see **Figure 4** for the list of the 66 presently identified human deafness genes involved in isolated forms of deafness) or in the mouse. Because myosin VI is essential to the hair bundle structure (Avraham et al. 1995), in itself accounting for deafness when defective (panel 1, **Figure 4**), it is only recently that the implication of this molecular motor in ribbon morphogenesis and synaptic exocytosis has been recognized (Roux et al. 2009). The lack of myosin VI in mutant mice results in a dramatic decrease in IHC synaptic exocytosis despite normal Ca^{2+} currents (Roux et al. 2009). The direct interaction of myosin VI and otoferlin, a synaptic vesicular protein (see below) (Heidrych et al. 2009, Roux et al. 2009), suggests that this motor protein, and consequently the actin cytoskeleton, may be involved in synaptic vesicle trafficking. The vesicular glutamate transporter Vglut3 (Slc17a8) was uncovered as the glutamate transporter of IHCs. IHCs in Vglut3-null mice completely lack glutamate release, despite unaffected synaptic vesicular fusion (Seal et al. 2008). A degeneration of afferent neurons occurs before hearing onset (Seal et al. 2008). The mechanisms underlying this neuronal degeneration are yet to be uncovered. Mutations in *VGLUT3* were subsequently detected in patients affected by a dominant form of auditory neuropathy (Ruel et al. 2008) (see panel 2, **Figure 4**).

The IHC neurotransmitter machinery involves proteins mediating synaptic vesicle exocytosis and recycling/resupplying and Ca^{2+} influx, as well as proteins controlling Ca^{2+} homeostasis. The $Ca_v1.3$ channels account for 90% of the maximal Ca^{2+} influx into the IHCs (see above). Pharmacological analysis of $Ca_v1.3^{-/-}$ IHCs indicated that $Ca_v1.4$ channels (Brandt et al. 2003) likely carry the residual Ca^{2+} current. Although the ribbons mature normally in $Ca_v1.3^{-/-}$ mice, despite the absence of a firing activity, IHCs fail to establish mature afferent synaptic contacts, and their efferent synapses persist more than four weeks after birth. Moreover, these IHCs lack functional BK and SK channels (Brandt et al. 2003). Baig et al. (2011) recently reported on patients with a syndromic deafness who carry a mutation in the *CACNA1D* gene, which encodes the pore-forming subunit (α1D) of $Ca_v1.3$ channels.

Which Ca^{2+} sensor(s) trigger(s) the synaptic exocytosis at the IHC active zone is a major focus of interest because of the intriguing absence of synaptotagmins 1 and 2, the most common Ca^{2+} sensors of fast synaptic exocytosis in CNS, in mature IHCs (**Figures 2*d*** and **3**). Of the 15 synaptotagmins reported in mammals, 8 are known to bind Ca^{2+} (synaptotagmins 1–3, 5–7, 9, and 10) (Pang & Südhof 2010). Single-cell reverse transcription polymerase chain reaction (RT-PCR) has shown high expression levels of four of them, synaptotagmins 1, 2, 6, and 7, in the IHCs during the prehearing period (Beurg et al. 2010). A developmental downregulation of synaptotagmin 1 and 2 expression, which are the closest homologs, occurs subsequently. Both proteins are detected at relatively high levels in all IHCs of prehearing mice but are absent from nearly all IHCs at around the onset of hearing (Beurg et al. 2010) (see **Figure 2*d***). In mice lacking synaptotagmin 2 or 7, IHC exocytosis is unaffected. In synaptotagmin 1-null mice, which can be analyzed only immediately after birth because they do not survive, the response to repetitive stimulations is dramatically reduced, indicating that the recruitment of synaptic vesicles is impeded.

Figure 4

Causative genes for human isolated deafness grouped into functional categories, hair bundle development, and functioning (panel 1), synaptic transmission (panel 2), cell-cell adhesion (panel 3), ion homeostasis (panel 4), and extracellular matrix (panel 5). The genes that have other, multiple, or unknown functions are listed in panel 6. [a]Mutations in the genes encoding myosin VIIa (USH1B), harmonin (USH1C), cadherin-23 (USH1D), protocadherin-15 (USH1F), Sans (USH1G), usherin (USH2A), Vlgr1 (USH2C), and whirlin (USH2D) cause Usher syndrome (USH), characterized by deaf-blindness. DFNAi and DFNBi denote autosomal-dominant and -recessive forms of deafness with undefined locus number. Abbreviation: OHC, outer hair cell.

This finding suggests that synaptotagmin 1 is involved in vesicle replenishment of the ribbon synapse, a function that synaptotagmin 2 fails to supply. Finally, synaptotagmin 4, which is expressed in both immature and mature IHCs (Johnson et al. 2010, Safieddine & Wenthold 1999) (**Figure 2*d***), may be required for the linearity of the relationship between Ca^{2+} loading and exocytosis (Johnson et al. 2009). However, synaptotagmin 4 does not bind Ca^{2+} ions and has no known Ca^{2+}-dependent activity (von Poser et al. 1997), and its proposed role through an interaction with synaptotagmin 2 is difficult to reconcile with the dispensable role of synaptotagmin 2 in IHC exocytosis as well as its absence in mature IHCs (Beurg et al. 2010, Safieddine & Wenthold 1999, von Poser et al. 1997) (see **Figure 2*d***). IHCs are thus endowed with a complex set of synaptotagmins. Their dynamic developmental pattern of expression (**Figure 2*d***) suggests their involvement in the progressive tuning of IHC synaptic activity, i.e., the increased rapidity, fidelity, efficiency, and linearity of its neurotransmitter release.

Finally, a member of the ferlin protein family, otoferlin, a six C2-domain containing transmembrane protein (the transmembrane domain being in the C-terminus while being in the N-terminus in synaptotagmins) has been identified as defective in a recessive form of human deafness (Roux et al. 2006, Yasunaga et al. 2000) (panel 2, **Figure 4**). In otoferlin-null ($Otof^{-/-}$) mice, IHC exocytosis of both RRP and SRP is defective, whereas the Ca^{2+} currents are normal (Roux et al. 2006). So far, the $Otof^{-/-}$ mutant is the sole knockout mouse in which RRP can be neither detected nor rescued by increasing the intracellular Ca^{2+} concentration. Otoferlin has been proposed as a Ca^{2+} sensor of synaptic exocytosis on the basis of the following arguments: (*a*) otoferlin is associated with synaptic vesicles; (*b*) in $Otof^{-/-}$ mice, ribbons develop a normal morphology, and synaptic vesicles are docked to the synaptic membrane as in wild-type mice, suggesting a defect in a downstream step; (*c*) otoferlin directly interacts with the proteins of the SNARE complex in a Ca^{2+}-dependent manner (Johnson & Chapman 2010, Ramakrishnan et al. 2009, Roux et al. 2006); (*d*) in the presence of the SNARE complex, otoferlin, of which five C2 domains bind to Ca^{2+}, promotes Ca^{2+}-dependent membrane-membrane fusion in in vitro assays (Johnson & Chapman 2010); (*e*) otoferlin binds to $Ca_v1.3$ channels (Ramakrishnan et al. 2009); and finally (*f*) the other ferlin family members (FER-1 in *C. elegans*, myoferlin and dysferlin in humans and mice), when defective, underlie vesicle fusion anomalies, and FER-1 is likely involved in Ca^{2+}-triggered vesicle-plasma membrane fusion (Washington & Ward 2006). The occurrence of spontaneous EPSCs, although at lower rates than in wild-type mice, indicates that neurotransmitter release can occur in IHCs in the absence of otoferlin (Pangrsic et al. 2010). Moreover, recent results demonstrate that otoferlin becomes indispensable for synaptic exocytosis only around P6, when it reaches its maximal expression (**Figure 2*d***). This switch to otoferlin-dependent synaptic exocytosis is concomitant with the maximal synaptic IHC exocytosis rate and Ca^{2+} currents (Beutner & Moser 2001, Johnson et al. 2005) (see **Figure 2**). The study of a mouse mutant, *pachanga* ($Otof^{Pga/Pga}$), carrying a missense mutation in the C2F domain of otoferlin showed a depression of IHC exocytosis and slowed RRP refilling, which suggests that otoferlin confers the high capacity for vesicle resupply to the IHC synapse (Pangrsic et al. 2010). Otoferlin has indeed been detected along the presynaptic plasma membrane and also throughout the cytosol of IHCs (Pangrsic et al. 2010, Seal et al. 2008), and interacts with the actin-based motor protein myosin VI (Heidrych et al. 2009, Roux et al. 2009). Otoferlin may thus be a multifunctional protein regulating both Ca^{2+}-dependent exocytosis and endocytosis vesicle recycling, possibly triggered independently according to the local Ca^{2+} concentration. Nonetheless, experimental evidence is needed to support its role as a Ca^{2+} sensor in synaptic vesicle–plasma membrane fusion, as has been provided for synaptotagmins (Fernandez-Chacon et al. 2001), and more direct observations are needed to support its role in vesicle-trafficking cells. Ca^{2+}-dependent vesicle fusion involving otoferlin may be a critical step of the IHC synaptic vesicle cycle. Whatever the precise roles of otoferlin in IHC synaptic exocytosis/endocytosis, the functional characterization of *Otof*$^{-/-}$ mice has identified a novel nosological entity: the auditory hair cell synaptopathies, which also include Vglut3 defects (Seal et al. 2008) (see panel 2, **Figure 4**).

CONCLUDING REMARKS

Over the past decade, great strides have been made toward understanding the machinery of the IHC ribbon synapse. Nonetheless, several of its elementary properties remain uncharacterized. Much of what we have learned so far is indeed based on C_m measurements and EPSC recordings in response to induced depolarization of the IHCs and not to the graded variations of its membrane potential elicited by mechanoelectrical transduction. Similar studies in response to hair bundle stimulation will explore the relationship between mechanoelectrical transduction and neurotransmitter release under physiological stimulation and identify IHC synaptic properties along the tonotopic cochlear axis. The temporal rapidity and fidelity of the IHC ribbon synapse, its major physiological feature, likely involve a multiscaled control, including the multivesicular mode of fusion, to allow for large monophasic EPSCs (highly synchronous or compound/homotypic fusion) and a high rate of synaptic refilling that stringently constrains synaptic retrieval of membranes and vesicle cycling (processes that are still poorly understood). Engineering recombinant mutant mice such that ribbons, Ca^{2+} influx fluctuations, and synaptic vesicle movements can be visualized by high-resolution optical imaging in live IHCs while monitoring their electrophysiological responses should help to address these issues. Finally, identifying the key molecular players of this synapse seems to be presently hampered by their functional redundancy, a difficulty that might be overcome by engineering a mouse mutant in which the corresponding genes have been coinactivated. New genetic tools must also be developed to properly address the roles of these molecules in the mature IHC, in particular to circumvent their potentially critical roles in IHC synapse development.

DISCLOSURE STATEMENT

The authors are not aware of any affiliations, memberships, funding, or financial holding that might be perceived as affecting the objectivity of this review.

ACKNOWLEDGMENTS

We thank Jean-Pierre Hardelin and Jacques Boutet de Monvel for their critical reading of the manuscript and Jacqueline Levilliers for her help. This work was supported by the French National Research Agency, ANR-07-Neuroscience (S.S), and Louis-Jeantet for Medicine Foundation (C.P).

LITERATURE CITED

Alvarez de Toledo G, Fernandez JM. 1990. Compound versus multigranular exocytosis in peritoneal mast cells. *J. Gen. Physiol.* 95:397–409

Avraham KB, Hasson T, Steel KP, Kingsley DM, Russell LB, et al. 1995. The mouse Snell's waltzer deafness gene encodes an unconventional myosin required for structural integrity of inner ear hair cells. *Nat. Genet.* 11:369–75

Baig SM, Koschak A, Lieb A, Gebhart M, Dafinger C, et al. 2011. Loss of Ca(v)1.3 (CACNA1D) function in a human channelopathy with bradycardia and congenital deafness. *Nat. Neurosci.* 14:77–84

Beurg M, Michalski N, Safieddine S, Bouleau Y, Schneggenburger R, et al. 2010. Control of exocytosis by synaptotagmins and otoferlin in auditory hair cells. *J. Neurosci.* 30:13281–90

Beutner D, Moser T. 2001. The presynaptic function of mouse cochlear inner hair cells during development of hearing. *J. Neurosci.* 21:4593–99

Beutner D, Voets T, Neher E, Moser T. 2001. Calcium dependence of exocytosis and endocytosis at the cochlear inner hair cell afferent synapse. *Neuron* 29:681–90

Bhattacharya S, Stewart BA, Niemeyer BA, Burgess RW, McCabe BD, et al. 2002. Members of the synaptobrevin/vesicle-associated membrane protein (VAMP) family in Drosophila are functionally interchangeable in vivo for neurotransmitter release and cell viability. *Proc. Natl. Acad. Sci. USA* 99:13867–72

Blankenship AG, Feller MB. 2010. Mechanisms underlying spontaneous patterned activity in developing neural circuits. *Nat. Rev. Neurosci.* 11:18–29

Borgdorff AJ, Choquet D. 2002. Regulation of AMPA receptor lateral movements. *Nature* 417:649–53

Borisovska M, Zhao Y, Tsytsyura Y, Glyvuk N, Takamori S, et al. 2005. v-SNAREs control exocytosis of vesicles from priming to fusion. *EMBO J.* 24:2114–26

Brandt A, Striessnig J, Moser T. 2003. CaV1.3 channels are essential for development and presynaptic activity of cochlear inner hair cells. *J. Neurosci.* 23:10832–40

Bregman AS. 1994. *Auditory Scene Analysis: The Perceptual Organization of Sound.* Cambridge, MA: MIT Press

Buran BN, Strenzke N, Neef A, Gundelfinger ED, Moser T, Liberman MC. 2010. Onset coding is degraded in auditory nerve fibers from mutant mice lacking synaptic ribbons. *J. Neurosci.* 30:7587–97

Chen YA, Scheller RH. 2001. SNARE-mediated membrane fusion. *Nat. Rev. Mol. Cell Biol.* 2:98–106

Choquet D. 2010. Fast AMPAR trafficking for a high-frequency synaptic transmission. *Eur. J. Neurosci.* 32:250–60

Choquet D, Triller A. 2003. The role of receptor diffusion in the organization of the postsynaptic membrane. *Nat. Rev. Neurosci.* 4:251–65

Cui G, Meyer AC, Calin-Jageman I, Neef J, Haeseleer F, et al. 2007. Ca^{2+}-binding proteins tune Ca^{2+}-feedback to Cav1.3 channels in mouse auditory hair cells. *J. Physiol.* 585:791–803

Daly C, Ziff EB. 1997. Post-transcriptional regulation of synaptic vesicle protein expression and the developmental control of synaptic vesicle formation. *J. Neurosci.* 17:2365–75

Devore S, Ihlefeld A, Hancock K, Shinn-Cunningham B, Delgutte B. 2009. Accurate sound localization in reverberant environments is mediated by robust encoding of spatial cues in the auditory midbrain. *Neuron* 62:123–34

Dick O, Hack I, Altrock WD, Garner CC, Gundelfinger ED, Brandstatter JH. 2001. Localization of the presynaptic cytomatrix protein Piccolo at ribbon and conventional synapses in the rat retina: comparison with Bassoon. *J. Comp. Neurol.* 439:224–34

Dick O, tom Dieck S, Altrock WD, Ammermuller J, Weiler R, et al. 2003. The presynaptic active zone protein bassoon is essential for photoreceptor ribbon synapse formation in the retina. *Neuron* 37:775–86

Dulon D, Safieddine S, Jones SM, Petit C. 2009. Otoferlin is critical for a highly sensitive and linear calcium-dependent exocytosis at vestibular hair cell ribbon synapses. *J. Neurosci.* 29:10474–87

Eybalin M, Renard N, Aure F, Safieddine S. 2002. Cysteine-string protein in inner hair cells of the organ of Corti: synaptic expression and upregulation at the onset of hearing. *Eur. J. Neurosci.* 15:1409–20

Fernandez-Chacon R, Konigstorfer A, Gerber SH, Garcia J, Matos MF, et al. 2001. Synaptotagmin I functions as a calcium regulator of release probability. *Nature* 410:41–49

Fettiplace R, Hackney CM. 2006. The sensory and motor roles of auditory hair cells. *Nat. Rev. Neurosci.* 7:19–29

Frank T, Rutherford MA, Strenzke N, Neef A, Pangrsic T, et al. 2010. Bassoon and the synaptic ribbon organize Ca^{2+} channels and vesicles to add release sites and promote refilling. *Neuron* 68:724–38

Friedman HV, Bresler T, Garner CC, Ziv NE. 2000. Assembly of new individual excitatory synapses: time course and temporal order of synaptic molecule recruitment. *Neuron* 27:57–69

Frischknecht R, Heine M, Perrais D, Seidenbecher CI, Choquet D, Gundelfinger ED. 2009. Brain extracellular matrix affects AMPA receptor lateral mobility and short-term synaptic plasticity. *Nat. Neurosci.* 12:897–904

Fuchs PA. 2005. Time and intensity coding at the hair cell's ribbon synapse. *J. Physiol.* 566:7–12

Garner CC, Kindler S, Gundelfinger ED. 2000. Molecular determinants of presynaptic active zones. *Curr. Opin. Neurobiol.* 10:321–27
Gebhart M, Juhasz-Vedres G, Zuccotti A, Brandt N, Engel J, et al. 2010. Modulation of Cav1.3 Ca^{2+} channel gating by Rab3 interacting molecule. *Mol. Cell Neurosci.* 44:246–59
Glowatzki E, Fuchs PA. 2002. Transmitter release at the hair cell ribbon synapse. *Nat. Neurosci.* 5:147–54
Goldberg JM, Lysakowski A, Fernandez C. 1990. Morphophysiological and ultrastructural studies in the mammalian cristae ampullares. *Hear. Res.* 49:89–102
Goutman JD, Glowatzki E. 2007. Time course and calcium dependence of transmitter release at a single ribbon synapse. *Proc. Natl. Acad. Sci. USA* 104:16341–46
Grant L, Fuchs P. 2008. Calcium- and calmodulin-dependent inactivation of calcium channels in inner hair cells of the rat cochlea. *J. Neurophysiol.* 99:2183–93
Grant L, Yi E, Glowatzki E. 2010. Two modes of release shape the postsynaptic response at the inner hair cell ribbon synapse. *J. Neurosci.* 30:4210–20
Graydon CW, Cho S, Li GL, Kachar B, von Gersdorff H. 2011. Sharp Ca^{2+} nanodomains beneath the ribbon promote highly synchronous multivesicular release at hair cell synapses. *J. Neurosci.* 31:16637–50
Griesinger CB, Richards CD, Ashmore JF. 2005. Fast vesicle replenishment allows indefatigable signalling at the first auditory synapse. *Nature* 435:212–15
He L, Xue L, Xu J, McNeil BD, Bai L, et al. 2009. Compound vesicle fusion increases quantal size and potentiates synaptic transmission. *Nature* 459:93–97
Heidrych P, Zimmermann U, Kuhn S, Franz C, Engel J, et al. 2009. Otoferlin interacts with myosin VI: implications for maintenance of the basolateral synaptic structure of the inner hair cell. *Hum. Mol. Genet.* 18:2779–90
Huang LC, Ryan AF, Cockayne DA, Housley GD. 2006. Developmentally regulated expression of the P2X3 receptor in the mouse cochlea. *Histochem. Cell Biol.* 125:681–92
Huang LC, Thorne PR, Housley GD, Montgomery JM. 2007. Spatiotemporal definition of neurite outgrowth, refinement and retraction in the developing mouse cochlea. *Development* 134:2925–33
Huffman KJ, Cramer KS. 2007. EphA4 misexpression alters tonotopic projections in the auditory brainstem. *Dev. Neurobiol.* 67:1655–68
Jackman SL, Choi SY, Thoreson WB, Rabl K, Bartoletti TM, Kramer RH. 2009. Role of the synaptic ribbon in transmitting the cone light response. *Nat. Neurosci.* 12:303–10
Jenkins MA, Christel CJ, Jiao Y, Abiria S, Kim KY, et al. 2010. Ca^{2+}-dependent facilitation of Cav1.3 Ca^{2+} channels by densin and Ca^{2+}/calmodulin-dependent protein kinase II. *J. Neurosci.* 30:5125–35
Johnson CP, Chapman ER. 2010. Otoferlin is a calcium sensor that directly regulates SNARE-mediated membrane fusion. *J. Cell Biol.* 191:187–97
Johnson DH. 1980. The relationship between spike rate and synchrony in responses of auditory-nerve fibers to single tones. *J. Acoust. Soc. Am.* 68:1115–22
Johnson SL, Beurg M, Marcotti W, Fettiplace R. 2011a. Prestin-driven cochlear amplification is not limited by the outer hair cell membrane time constant. *Neuron* 70:1143–54
Johnson SL, Eckrich T, Kuhn S, Zampini V, Franz C, et al. 2011b. Position-dependent patterning of spontaneous action potentials in immature cochlear inner hair cells. *Nat. Neurosci.* 14:711–17
Johnson SL, Forge A, Knipper M, Munkner S, Marcotti W. 2008. Tonotopic variation in the calcium dependence of neurotransmitter release and vesicle pool replenishment at mammalian auditory ribbon synapses. *J. Neurosci.* 28:7670–78
Johnson SL, Franz C, Knipper M, Marcotti W. 2009. Functional maturation of the exocytotic machinery at gerbil hair cell ribbon synapses. *J. Physiol.* 587:1715–26
Johnson SL, Franz C, Kuhn S, Furness DN, Ruttiger L, et al. 2010. Synaptotagmin IV determines the linear Ca^{2+} dependence of vesicle fusion at auditory ribbon synapses. *Nat. Neurosci.* 13:45–52
Johnson SL, Marcotti W. 2008. Biophysical properties of CaV1.3 calcium channels in gerbil inner hair cells. *J. Physiol.* 586:1029–42
Johnson SL, Marcotti W, Kros CJ. 2005. Increase in efficiency and reduction in Ca^{2+} dependence of exocytosis during development of mouse inner hair cells. *J. Physiol.* 563:177–91
Keen EC, Hudspeth AJ. 2006. Transfer characteristics of the hair cell's afferent synapse. *Proc. Natl. Acad. Sci. USA* 103:5537–42

Khimich D, Nouvian R, Pujol R, Tom Dieck S, Egner A, et al. 2005. Hair cell synaptic ribbons are essential for synchronous auditory signalling. *Nature* 434:889–94

Kros CJ. 2007. How to build an inner hair cell: challenges for regeneration. *Hear. Res.* 227:3–10

Leake PA, Hradek GT, Chair L, Snyder RL. 2006. Neonatal deafness results in degraded topographic specificity of auditory nerve projections to the cochlear nucleus in cats. *J. Comp. Neurol.* 497:13–31

Leibovici M, Safieddine S, Petit C. 2008. Mouse models of human hereditary deafness. *Curr. Top. Dev. Biol.* 84:385–429

Lenzi D, Runyeon JW, Crum J, Ellisman MH, Roberts WM. 1999. Synaptic vesicle populations in saccular hair cells reconstructed by electron tomography. *J. Neurosci.* 19:119–32

Li G, von Gersdorff H. 2011. *Synaptic mechanisms of phase-locking at a mature hair cell ribbon synapse*. Presented at ARO MidWinter Meet., 34th, Baltimore, 34:580

Li GL, Keen E, Andor-Ardo D, Hudspeth AJ, von Gersdorff H. 2009. The unitary event underlying multiquantal EPSCs at a hair cell's ribbon synapse. *J. Neurosci.* 29:7558–68

Liberman LD, Wang H, Liberman MC. 2011. Opposing gradients of ribbon size and AMPA receptor expression underlie sensitivity differences among cochlear-nerve/hair-cell synapses. *J. Neurosci.* 31:801–8

Liberman MC. 1980. Morphological differences among radial afferent fibers in the cat cochlea: an electron-microscopic study of serial sections. *Hear. Res.* 3:45–63

Liberman MC, Kiang NY. 1978. Acoustic trauma in cats. Cochlear pathology and auditory-nerve activity. *Acta Otolaryngol. Suppl.* 358:1–63

Liu Y, Barlowe C. 2002. Analysis of Sec22p in endoplasmic reticulum/Golgi transport reveals cellular redundancy in SNARE protein function. *Mol. Biol. Cell* 13:3314–24

LoGiudice L, Sterling P, Matthews G. 2008. Mobility and turnover of vesicles at the synaptic ribbon. *J. Neurosci.* 28:3150–58

Mandell JW, Townes-Anderson E, Czernik AJ, Cameron R, Greengard P, De Camilli P. 1990. Synapsins in the vertebrate retina: absence from ribbon synapses and heterogeneous distribution among conventional synapses. *Neuron* 5:19–33

Marcotti W, Johnson SL, Rusch A, Kros CJ. 2003. Sodium and calcium currents shape action potentials in immature mouse inner hair cells. *J. Physiol.* 552:743–61

Martin JA, Hu Z, Fenz KM, Fernandez J, Dittman JS. 2011. Complexin has opposite effects on two modes of synaptic vesicle fusion. *Curr. Biol.* 21:97–105

Matteoli M, Verderio C, Rossetto O, Iezzi N, Coco S, et al. 1996. Synaptic vesicle endocytosis mediates the entry of tetanus neurotoxin into hippocampal neurons. *Proc. Natl. Acad. Sci. USA* 93:13310–15

Matthews G, Fuchs P. 2010. The diverse roles of ribbon synapses in sensory neurotransmission. *Nat. Rev. Neurosci.* 11:812–22

Matthews G, Sterling P. 2008. Evidence that vesicles undergo compound fusion on the synaptic ribbon. *J. Neurosci.* 28:5403–11

Meister M, Wong RO, Baylor DA, Shatz CJ. 1991. Synchronous bursts of action potentials in ganglion cells of the developing mammalian retina. *Science* 252:939–43

Meyer AC, Frank T, Khimich D, Hoch G, Riedel D, et al. 2009. Tuning of synapse number, structure and function in the cochlea. *Nat. Neurosci.* 12:444–53

Moser T, Beutner D. 2000. Kinetics of exocytosis and endocytosis at the cochlear inner hair cell afferent synapse of the mouse. *Proc. Natl. Acad. Sci. USA* 97:883–88

Mostafapour SP, Cochran SL, Del Puerto NM, Rubel EW. 2000. Patterns of cell death in mouse anteroventral cochlear nucleus neurons after unilateral cochlea removal. *J. Comp. Neurol.* 426:561–71

Nemzou NR, Bulankina AV, Khimich D, Giese A, Moser T. 2006. Synaptic organization in cochlear inner hair cells deficient for the CaV1.3 (alpha1D) subunit of L-type Ca^{2+} channels. *Neuroscience* 141:1849–60

Nouvian R, Beutner D, Parsons TD, Moser T. 2006. Structure and function of the hair cell ribbon synapse. *J. Membr. Biol.* 209:153–65

Nouvian R, Neef J, Bulankina AV, Reisinger E, Pangrsic T, et al. 2011. Exocytosis at the hair cell ribbon synapse apparently operates without neuronal SNARE proteins. *Nat. Neurosci.* 14:411–13

Paillart C, Li J, Matthews G, Sterling P. 2003. Endocytosis and vesicle recycling at a ribbon synapse. *J. Neurosci.* 23:4092–99

Palmer AR, Russell IJ. 1986. Phase-locking in the cochlear nerve of the guinea-pig and its relation to the receptor potential of inner hair-cells. *Hear. Res.* 24:1–15

Pang ZP, Südhof TC. 2010. Cell biology of Ca^{2+}-triggered exocytosis. *Curr. Opin. Cell Biol.* 22:496–505

Pangrsic T, Lasarow L, Reuter K, Takago H, Schwander M, et al. 2010. Hearing requires otoferlin-dependent efficient replenishment of synaptic vesicles in hair cells. *Nat. Neurosci.* 13:869–76

Parsons TD, Sterling P. 2003. Synaptic ribbon. Conveyor belt or safety belt? *Neuron* 37:379–82

Platzer J, Engel J, Schrott-Fischer A, Stephan K, Bova S, et al. 2000. Congenital deafness and sinoatrial node dysfunction in mice lacking class D L-type Ca^{2+} channels. *Cell* 102:89–97

Pujol R, Lavigne-Rebillard M, Lenoir M. 1998. *Development of Sensory and Neural Structures in the Mammalian Cochlea*. New York: Springer

Ramakrishnan NA, Drescher MJ, Drescher DG. 2009. Direct interaction of otoferlin with syntaxin 1A, SNAP-25, and the L-type voltage-gated calcium channel Cav1.3. *J. Biol. Chem.* 284:1364–72

Rayleigh SJW. 1907. On our perception of sound direction. *Philos. Mag.* 13:214–32

Richardson GP, de Monvel JB, Petit C. 2011. How the genetics of deafness illuminates auditory physiology. *Annu. Rev. Physiol.* 73:311–34

Rose JE, Brugge JF, Anderson DJ, Hind JE. 1967. Phase-locked response to low-frequency tones in single auditory nerve fibers of the squirrel monkey. *J. Neurophysiol.* 30:769–93

Roux I, Hosie S, Johnson SL, Bahloul A, Cayet N, et al. 2009. Myosin VI is required for the proper maturation and function of inner hair cell ribbon synapses. *Hum. Mol. Genet.* 18:4615–28

Roux I, Safieddine S, Nouvian R, Grati M, Simmler M-C, et al. 2006. Otoferlin, defective in a human deafness form, is essential for exocytosis at the auditory ribbon synapse. *Cell* 127:277–89

Ruel J, Emery S, Nouvian R, Bersot T, Amilhon B, et al. 2008. Impairment of SLC17A8 encoding vesicular glutamate transporter-3, VGLUT3, underlies nonsyndromic deafness DFNA25 and inner hair cell dysfunction in null mice. *Am. J. Hum. Genet.* 83:278–92

Sadoul K, Berger A, Niemann H, Weller U, Roche PA, et al. 1997. SNAP-23 is not cleaved by botulinum neurotoxin E and can replace SNAP-25 in the process of insulin secretion. *J. Biol. Chem.* 272:33023–27

Safieddine S, Wenthold RJ. 1997. The glutamate receptor subunit delta1 is highly expressed in hair cells of the auditory and vestibular systems. *J. Neurosci.* 17:7523–31

Safieddine S, Wenthold RJ. 1999. SNARE complex at the ribbon synapses of cochlear hair cells: analysis of synaptic vesicle- and synaptic membrane-associated proteins. *Eur. J. Neurosci.* 11:803–12

Schnee ME, Santos-Sacchi J, Castellano-Munoz M, Kong JH, Ricci AJ. 2011. Calcium-dependent synaptic vesicle trafficking underlies indefatigable release at the hair cell afferent fiber synapse. *Neuron* 70:326–38

Seal RP, Akil O, Yi E, Weber CM, Grant L, et al. 2008. Sensorineural deafness and seizures in mice lacking vesicular glutamate transporter 3. *Neuron* 57:263–75

Sendin G, Bulankina AV, Riedel D, Moser T. 2007. Maturation of ribbon synapses in hair cells is driven by thyroid hormone. *J. Neurosci.* 27:3163–73

Smith CA, Sjostrand FS. 1961. Structure of the nerve endings on the external hair cells of the guinea pig cochlea as studied by serial sections. *J. Ultrastruct. Res.* 5:523–56

Snellman J, Mehta B, Babai N, Bartoletti TM, Akmentin W, et al. 2011. Acute destruction of the synaptic ribbon reveals a role for the ribbon in vesicle priming. *Nat. Neurosci.* 14:1135–41

Sobkowicz HM, Rose JE, Scott GE, Slapnick SM. 1982. Ribbon synapses in the developing intact and cultured organ of Corti in the mouse. *J. Neurosci.* 2:942–57

Sobkowicz HM, Rose JE, Scott GL, Levenick CV. 1986. Distribution of synaptic ribbons in the developing organ of Corti. *J. Neurocytol.* 15:693–714

Sollner T, Whiteheart SW, Brunner M, Erdjument-Bromage H, Geromanos S, et al. 1993. SNAP receptors implicated in vesicle targeting and fusion. *Nature* 362:318–24

Stamataki S, Francis HW, Lehar M, May BJ, Ryugo DK. 2006. Synaptic alterations at inner hair cells precede spiral ganglion cell loss in aging C57BL/6J mice. *Hear. Res.* 221:104–18

Sterling P, Matthews G. 2005. Structure and function of ribbon synapses. *Trends Neurosci.* 28:20–29

Strenzke N, Chanda S, Kopp-Scheinpflug C, Khimich D, Reim K, et al. 2009. Complexin-I is required for high-fidelity transmission at the endbulb of Held auditory synapse. *J. Neurosci.* 29:7991–8004

Tierney TS, Russell FA, Moore DR. 1997. Susceptibility of developing cochlear nucleus neurons to deafferentation-induced death abruptly ends just before the onset of hearing. *J. Comp. Neurol.* 378:295–306

Tritsch NX, Bergles DE. 2010. Developmental regulation of spontaneous activity in the Mammalian cochlea. *J. Neurosci.* 30:1539–50

Tritsch NX, Yi E, Gale JE, Glowatzki E, Bergles DE. 2007. The origin of spontaneous activity in the developing auditory system. *Nature* 450:50–55

Uthaiah RC, Hudspeth AJ. 2010. Molecular anatomy of the hair cell's ribbon synapse. *J. Neurosci.* 30:12387–99

Verderio C, Pozzi D, Pravettoni E, Inverardi F, Schenk U, et al. 2004. SNAP-25 modulation of calcium dynamics underlies differences in GABAergic and glutamatergic responsiveness to depolarization. *Neuron* 41:599–610

Vilinsky I, Stewart BA, Drummond J, Robinson I, Deitcher DL. 2002. A Drosophila SNAP-25 null mutant reveals context-dependent redundancy with SNAP-24 in neurotransmission. *Genetics* 162:259–71

von Poser C, Ichtchenko K, Shao X, Rizo J, Südhof TC. 1997. The evolutionary pressure to inactivate. A subclass of synaptotagmins with an amino acid substitution that abolishes Ca^{2+} binding. *J. Biol. Chem.* 272:14314–19

Washington NL, Ward S. 2006. FER-1 regulates Ca^{2+}-mediated membrane fusion during *C. elegans* spermatogenesis. *J. Cell Sci.* 119:2552–62

Wenthold RJ, Safieddine S, Ly CD, Wang YX, Lee HK, et al. 2002. Vesicle targeting in hair cells. *Audiol. Neurootol.* 7:45–48

Xu T, Rammner B, Margittai M, Artalejo AR, Neher E, Jahn R. 1999. Inhibition of SNARE complex assembly differentially affects kinetic components of exocytosis. *Cell* 99:713–22

Yang X, Kaeser-Woo YJ, Pang ZP, Xu W, Südhof TC. 2010. Complexin clamps asynchronous release by blocking a secondary Ca^{2+} sensor via its accessory alpha helix. *Neuron* 68:907–20

Yasunaga S, Grati M, Chardenoux S, Smith TN, Friedman TB, et al. 2000. OTOF encodes multiple long and short isoforms: genetic evidence that the long ones underlie recessive deafness DFNB9. *Am. J. Hum. Genet.* 67:591–600

Zampini V, Johnson SL, Franz C, Lawrence ND, Munkner S, et al. 2010. Elementary properties of CaV1.3 Ca^{2+} channels expressed in mouse cochlear inner hair cells. *J. Physiol.* 588:187–99

Multiple Functions of Endocannabinoid Signaling in the Brain

István Katona and Tamás F. Freund

Institute of Experimental Medicine, Hungarian Academy of Sciences, 1051 Budapest, Hungary; email: katona@koki.hu, freund@koki.hu

Annu. Rev. Neurosci. 2012. 35:529–58

First published online as a Review in Advance on April 17, 2012

The *Annual Review of Neuroscience* is online at neuro.annualreviews.org

This article's doi:
10.1146/annurev-neuro-062111-150420

0147-006X/12/0721-0529$20.00

Keywords

retrograde signaling, feedback inhibition, synaptic plasticity, G protein–coupled receptors, diacylglycerol lipase

Abstract

Despite being regarded as a hippie science for decades, cannabinoid research has finally found its well-deserved position in mainstream neuroscience. A series of groundbreaking discoveries revealed that endocannabinoid molecules are as widespread and important as conventional neurotransmitters such as glutamate or GABA, yet they act in profoundly unconventional ways. We aim to illustrate how uncovering the molecular, anatomical, and physiological characteristics of endocannabinoid signaling has revealed new mechanistic insights into several fundamental phenomena in synaptic physiology. First, we summarize unexpected advances in the molecular complexity of biogenesis and inactivation of the two endocannabinoids, anandamide and 2-arachidonoylglycerol. Then, we show how these new metabolic routes are integrated into well-known intracellular signaling pathways. These endocannabinoid-producing signalosomes operate in phasic and tonic modes, thereby differentially governing homeostatic, short-term, and long-term synaptic plasticity throughout the brain. Finally, we discuss how cell type– and synapse-specific refinement of endocannabinoid signaling may explain the characteristic behavioral effects of cannabinoids.

Contents

THE "GRASS ROUTE" TO THE DISCOVERY OF THE ENDOCANNABINOID SYSTEM

Δ⁹-THC: (–)-Δ⁹-tetrahydrocannabinol

Predator-prey competition is a major driving force behind evolution. For example, most plants have developed a dedicated repertoire of chemical molecules to distract consumption. These allelochemicals often mimic or perturb endogenous signaling pathways in the nervous system, thereby becoming behaviorally effective. The underlying evolutionary processes appear robust enough to invent receptor agonists or antagonists as well as allosteric activators or inhibitors of enzymes with excellent potency and affinity. This natural treasure trove served traditional medicine for several thousand years and remains a major frontier for drug discovery (Li & Vederas 2009). Moreover, neuroscience research has also greatly profited from deciphering the mechanisms by which these plant products are behaviorally active, paving the way for the discovery of several endogenous signaling systems primarily active in the brain (Prisinzano 2009).

A prime example is the cannabis plant (*Cannabis sativa* L.) and the discovery of the endocannabinoid system in animals. Cannabis plants produce a unique mixture of chemical constituents; the most famous products are the C_{21} terpenophenolic compounds, which are collectively called phytocannabinoids. Detailed chemical analysis has identified approximately 70 molecular species of phytocannabinoids (ElSohly & Slade 2005), among which one molecule and its discovery stands out. Until recently, this aromatic terpenoid had been more famous (or infamous, depending on the individual being asked) among the public than it had been among neuroscientists. Termed (–)-Δ^9-tetrahydrocannabinol (Δ^9-THC), this compound was isolated from confiscated hashish by Yechiel Gaoni & Raphael Mechoulam (1964) (**Figure 1*a***) and was shown to account for the psychotropic effects of cannabis preparations in rhesus monkeys (Mechoulam et al. 1970). This seminal discovery transformed cannabinoid research from an anecdote-based practice into an evidence-based modern research field. Using the chemically defined Δ^9-THC molecule, researchers were able to obtain qualitatively and quantitatively reproducible pharmacological, physiological, or behavioral data, which then helped them to uncover the neurobiological substrates of the psychoactive effects of cannabis.

The second major breakthrough in cannabinoid research provided an answer to the conceptual question of why our brain reacts to cannabis. Using [^{3}H]-CP55,940, a potent radioactively labeled synthetic cannabinoid, Bill Devane, Allyn Howlett, and their colleagues obtained the first unequivocal evidence for the presence of a specific cannabinoid receptor, which inhibits adenylate cyclase via G_i protein signaling in the brain (**Figure 1*b***) (Devane et al. 1988, Bidaut-Russell et al. 1990). This discovery is also considered as the first direct evidence for the existence of the endocannabinoid system. Subsequent qualitative and quantitative radioligand-binding studies revealed the distribution of cannabinoid receptors in the brain (**Figure 1*c***) (Herkenham et al. 1990). First, lesion experiments showed that the vast majority of cannabinoid-binding sites in the brain are located on neurons, most likely on their axonal bundles (Herkenham et al. 1991). Second, the quantitative distribution pattern fit well with the brain regions underlying the behavioral effects of cannabis. Third, this pattern was remarkably similar across species, indicating a conserved physiological function for cannabinoid receptors. Finally, and most importantly, the density of cannabinoid receptors in the brain was comparable to the levels of glutamate, gamma-aminobutyric acid (GABA), or striatal dopamine receptors (Herkenham et al. 1990). Thus, these observations collectively predicted that cannabinoid receptors are components of chemical synapses as ubiquitous as conventional neurotransmitter receptors.

Coinciding with the golden age for the cloning of G protein–coupled receptors, these discoveries were soon followed by the molecular identification of the first cannabinoid receptor (Matsuda et al. 1990). The CB_1 cannabinoid receptor turned out to be a class A, G protein–coupled receptor and has a notably similar sequence (97–99% amino acid sequence identity) across mammalian species, further supporting a phylogenetically conserved function for CB_1. In situ hybridization confirmed neuronal expression and revealed a heterogeneous distribution pattern largely corresponding to the ligand-binding sites (Matsuda et al. 1990). Owing to significant homology (44% at the amino acid level), a second cannabinoid receptor was also discovered (Munro et al. 1993). These two receptors originate from a common ancestor, and research indicates that a phylogenetically closely related third cannabinoid receptor is unlikely to be found (Pertwee et al. 2010).

GABA: gamma-aminobutyric acid

Compelling evidence shows that CB_1 receptors are the major neurobiological substrates for Δ^9-THC effects on the human brain. The acute psychological consequences of marijuana smoking such as the subjective "high" experience was efficiently blocked by pretreatment with the CB_1 antagonist rimonabant in healthy human subjects (Huestis et al. 2001). Moreover, the development of novel inverse agonist radioligands for positron emission tomography now allows the monitoring of CB_1 receptor availability in living human brain (Burns et al. 2007) (**Figure 1*d***). The tremendous potential of this new approach is reflected by the emerging data showing robust changes in CB_1 receptor availability in patients with Huntington's disease or temporal-lobe epilepsy (Van Laere et al. 2010, Goffin et al. 2011) or by the demonstration of cortex-specific downregulation of CB_1 availability in chronic cannabis smokers, which is a long-suspected mechanism of cannabis tolerance (Hirvonen et al. 2011).

Whereas CB_1 receptors are considered as primarily neuronal receptors, CB_2 receptors are highly expressed in the spleen and regarded as the predominant cannabinoid receptor of the immune system. This somewhat simplified concept has been useful in providing a framework for the potential therapeutic exploitation of the endocannabinoid system; one particularly exciting approach results from the finding showing that the beneficial effects of Δ^9-THC in multiple sclerosis are mediated in tandem by neuronal CB_1 receptors and by CB_2 receptors on autoreactive T cells (Maresz et al. 2007). Accumulating data also support important physiological and pathophysiological functions for peripheral CB_1 receptors (Kunos et al. 2009), and the central effects of CB_2 receptors are well documented in studies of emesis regulation or the rewarding effects of cocaine (Van Sickle et al. 2005, Xi et al. 2011).

The discovery of cannabinoid receptors initiated a quest for their endogenous ligands,

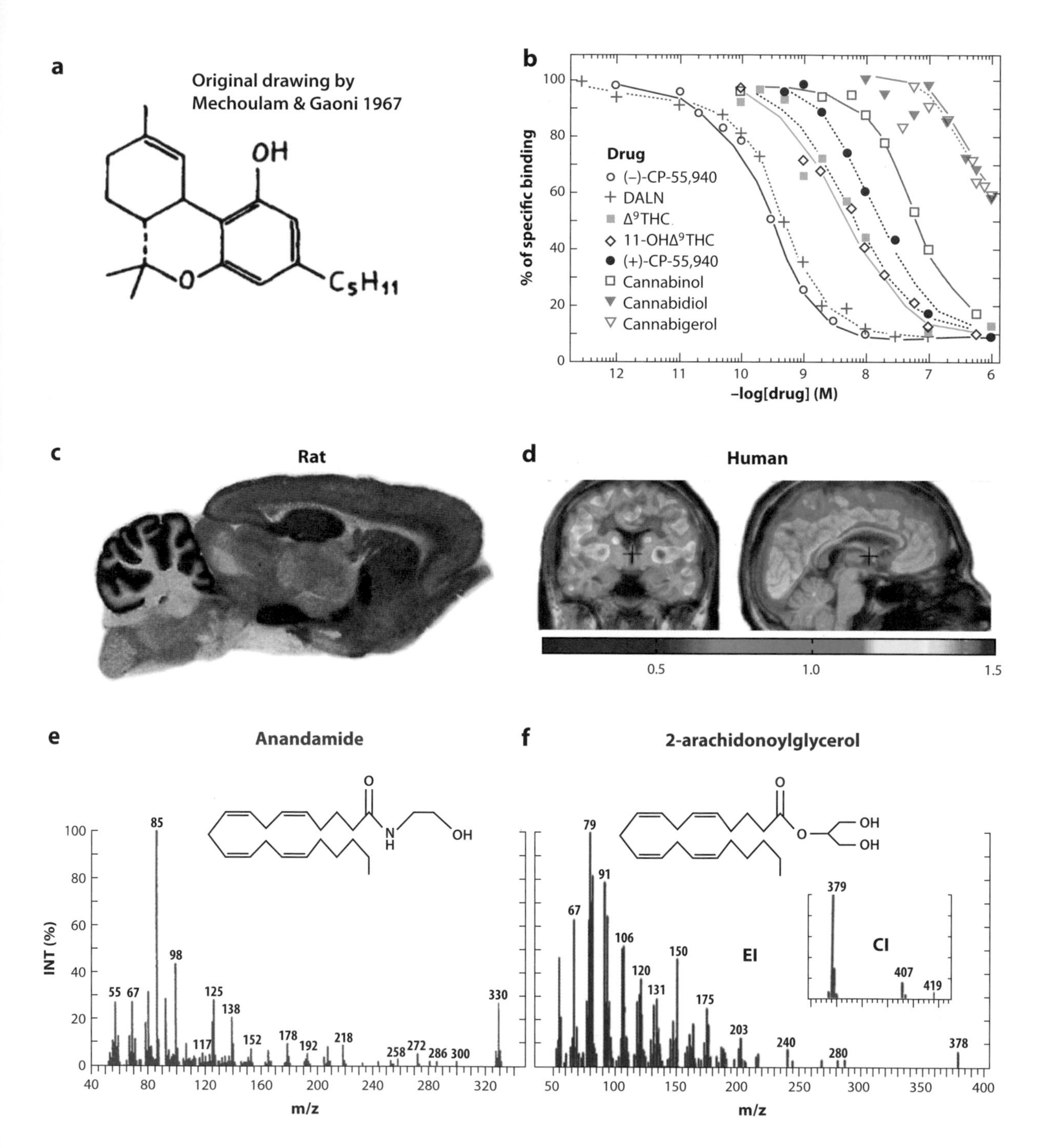

the so-called endocannabinoids. Because of the lipophilic nature of phytocannabinoids, Devane, Mechoulam, and colleagues argued that lipid-soluble fractions of the brain should contain the putative endocannabinoid molecule, which finally led them to its chemical identification (**Figure 1*e***) (Devane et al. 1992). The compound N-arachidonoylethanolamide was termed anandamide on the basis of the Sanskrit word *ananda* meaning "inner bliss." This name reflects good foresight, as anandamide plasma levels are significantly reduced in patients with major depression (Hill et al. 2009) and acute blockade of anandamide hydrolysis exerts robust antidepressant-like effects (Gobbi et al. 2005). Research has shown that anandamide is a partial agonist of the two cannabinoid receptors, which is unusual for an endogenous natural ligand, and is found at low concentrations (pmol per gram of tissue) in the brain (Pertwee et al. 2010). Because anandamide pharmacological activity did not fully recapitulate the behavioral effects of Δ^9-THC (Smith et al. 1994), the existence of a second endocannabinoid was postulated and soon identified as 2-arachidonoylglycerol (2-AG) (**Figure 1*f***) (Mechoulam et al. 1995, Sugiura et al. 1995). 2-AG is a full agonist at both CB_1 and CB_2 receptors and is present at much higher levels (nmol per gram tissue) in the brain (Stella et al. 1997). With the discovery of anandamide, 2-AG, and the cannabinoid receptors, the "grass route" has gloriously ended after three decades of research, giving neuroscientists the green light to continue pursuing hypothesis-driven questions regarding the function of endocannabinoid signaling, which has culminated in a paradigm shift in neuroscience.

THE "RETROGRADE ROUTE" OF ENDOCANNABINOIDS IN THE BRAIN

Whereas anterograde synaptic transmission has been extensively studied for decades, much less information about retrograde signaling pathways at chemical synapses is known. Several candidates such as gases (e.g., nitric oxide), peptides (e.g., dynorphin), growth factors (e.g., brain-derived neurotrophic factor), or conventional amino acid transmitters such as glutamate or GABA are released by the somatodendritic domain of postsynaptic neurons and then act on the axon terminals of presynaptic neurons (Regehr et al. 2009). However, most of these retrograde messengers operate at specific synapses or are restricted to a few cell types. Thus, none could be used to explain fully the most common forms of retrograde synaptic communication. Given high anandamide synthase activity in the hippocampus and the then-prevailing view that the chemically related lipid molecule arachidonic acid is a retrograde messenger in hippocampal long-term potentiation (Williams et al. 1989), Devane & Axelrod (1994) proposed first that anandamide may also play a retrograde messenger role on axonal CB_1 receptors. In accordance, the first immunohistochemical studies visualizing CB_1 protein localization reported a dense meshwork of fibers throughout the brain at the light microscopic level (Egertova et al. 1998, Tsou et al. 1998). These fibers proved to be axons

2-AG: 2-arachidonoylglycerol

Retrograde signaling: retrograde messengers are released from the somatodendritic domain of neurons and then modify release properties of afferent axon terminals or regulate activity in nearby glial processes

Figure 1

A tribute to the discoveries unraveling the endocannabinoid system. (*a*) First precise description of the chemical structure and its absolute configuration of the psychoactive compound in marijuana, (−)-Δ^9-tetrahydrocannabinol (Δ^9-THC) (Gaoni & Mechoulam 1967). (*b*) First demonstration by competitive inhibition of the existence of a high-affinity, stereoselective, pharmacologically distinct cannabinoid receptor in brain tissue (Devane et al. 1988). (*c*) Localization of CB_1 by high-affinity receptor binding and autoradiography in the rat (Herkenham et al. 1990) and (*d*) by positron emission tomography in human brain (Burns et al. 2007). (*e*) Original mass spectra demonstrating the existence of anandamide (Devane et al. 1992) and (*f*) 2-arachidonoylglycerol (2-AG), together with the chemical structures of the two major endocannabinoids (Mechoulam et al. 1995). The individual figures have been modified from the originals with permission from the authors.

Depolarization-induced suppression of inhibition or excitation (DSI or DSE): depolarized neurons release endocannabinoids that transiently inhibit GABA or glutamate release, respectively, from their afferent synaptic terminals

FAAH: fatty acid amide hydrolase

MGL: monoacylglycerol lipase

DGL: diacyglycerol lipase

in both the rat and human brain, as revealed by high-resolution immunogold staining and electron microscopy (Katona et al. 1999, Katona et al. 2000). In fact, the majority of CB_1 receptors accumulate presynaptically on axon terminals of GABAergic interneurons in the hippocampus (**Figure 2*a***). Furthermore, CB_1 receptor agonists reduced electrically evoked GABA release (**Figure 2*b***), in accordance with a proposed retrograde-messenger function of endocannabinoid signaling (Katona et al. 1999). The presynaptic localization of CB_1 receptors and its inhibitory effect on neurotransmitter release have proved to be a general feature of most axon terminals in the central (Kano et al. 2009) and peripheral nervous system (Vizi et al. 2001).

The most important support for the retrograde scenario came subsequently from the study of an electrophysiological paradigm in which selective depolarization of a postsynaptic neuron induces short-term depression of GABA release from axon terminals innervating the same postsynaptic neuron [depolarization-induced suppression of inhibition (DSI)] (Llano et al. 1991, Pitler and Alger, 1992). This robust phenomenon premised the existence of a bona fide retrograde messenger and thus, the demonstration that three independent antagonists of CB_1 receptors block DSI at hippocampal GABAergic synapses (**Figure 2*c,d***) (Ohno-Shosaku et al. 2001, Wilson & Nicoll 2001), together with the presynaptic localization of CB_1 receptors (Katona et al. 1999), were in agreement with the plausible scenario that endocannabinoids may be the long-awaited retrograde messengers. Importantly, depolarization-induced suppression of excitation (DSE) was also inhibited by a CB_1 antagonist at cerebellar excitatory synapses (Kreitzer & Regehr 2001). Several other forms of synaptic plasticity were subsequently reported to be CB_1-dependent, including, e.g., long-term depression (**Figure 2*e,f***) (Gerdeman et al. 2002, Marsicano et al. 2002, Robbe et al. 2002). Collectively, these findings indicate that endocannabinoid signaling plays a conceptually similar role at distinct types of synapses throughout the brain and represent the turning point when endocannabinoid research finally became a major focus for mainstream neuroscience.

Although the most parsimonious scenario is the retrograde route for endocannabinoids, the manner of how anandamide and/or 2-AG get through the extracellular space is still an unresolved issue considering that these lipid messengers are fairly hydrophobic with logP values of 5.1 and 5.39 for anandamide and 2-AG, respectively (Maccarrone 2008). Although converging evidence from the combination of biochemical, anatomical, pharmacological, and physiological experiments fully supports this scenario, the manner by which anandamide and/or 2-AG get through the extracellular space remains unresolved. As a first step toward deciphering this puzzle, molecular identification of enzymes involved in endocannabinoid metabolism helped researchers to define which endocannabinoid molecule is responsible for given forms of endocannabinoid-mediated plasticity. For example, the inactivation of anandamide is carried out by a serine hydrolase called fatty acid amide hydrolase (FAAH) (Cravatt et al. 1996). Thus, the lack of effect of FAAH inhibitors in a given paradigm indicates that anandamide is not involved in that particular phenomenon as was stated, for example, for hippocampal DSI (Kim & Alger 2004). The second candidate, 2-AG, is synthesized by two isoforms of diacylglycerol lipase, α and β (Bisogno et al. 2003), and a monoacylglycerol lipase (MGL) degrades the majority (85%) of 2-AG in the brain (Dinh et al. 2002, Blankman et al. 2007). Genetic inhibition of MGL consistently enhances short-term synaptic depression (Pan et al. 2011, Straiker et al. 2011), and genetic inactivation of diacylglycerol lipase (DGL)-α fully eliminates all forms of endocannabinoid-mediated synaptic plasticity in the prefrontal cortex (Yoshino et al. 2011), hippocampus (Gao et al. 2010, Tanimura et al. 2010), striatum, and cerebellum (Tanimura et al. 2010). Thus, the indispensable involvement of DGL-α and the regulatory role of MGL in short-term synaptic

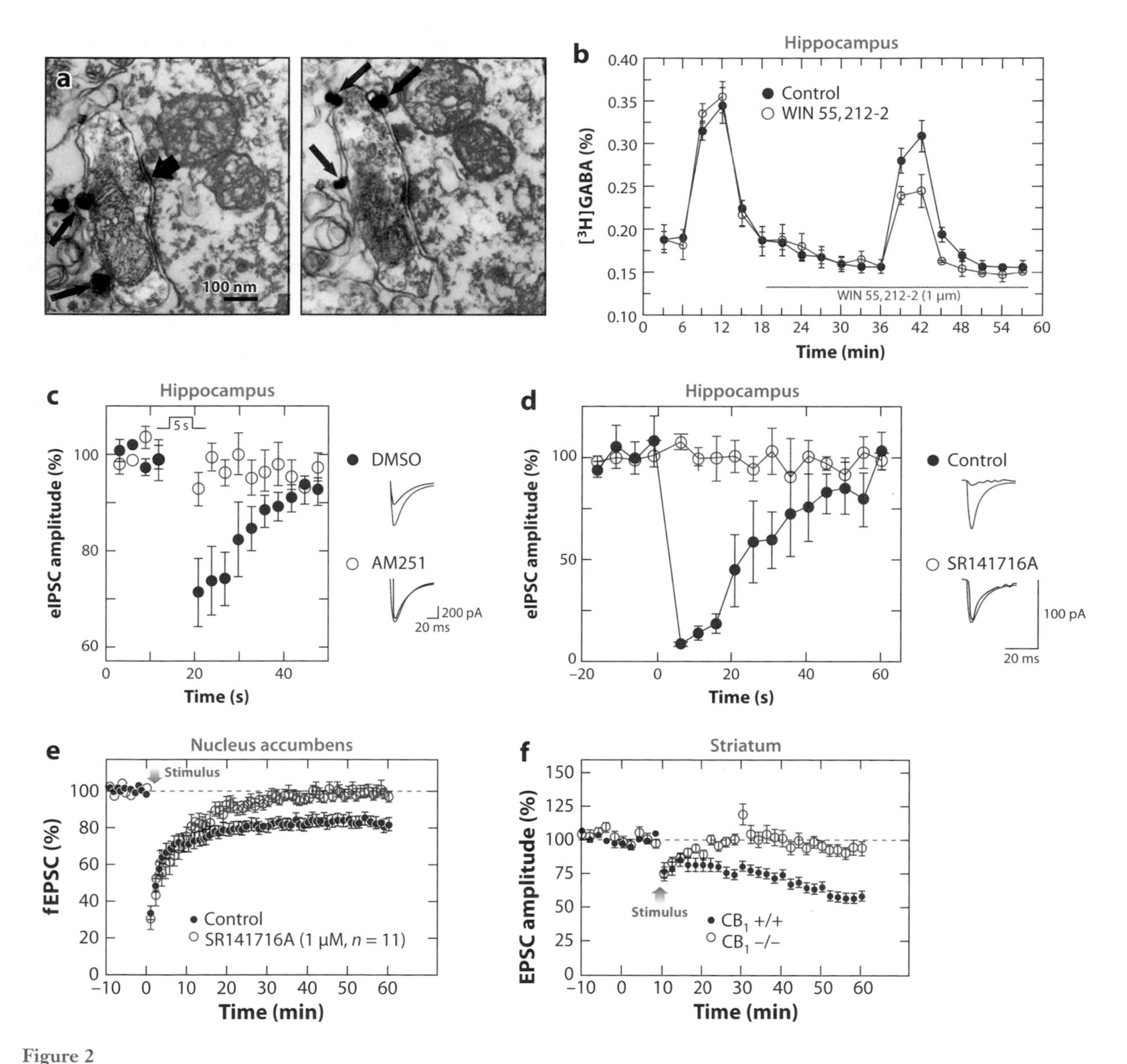

Figure 2

The retrograde route of endocannabinoid signaling in the brain. (*a*) Selective localization of CB_1 receptors on axon terminals predicting a retrograde direction of endocannabinoid action (Katona et al. 1999). Thin arrows point to immunogold particles representing the localization of CB_1 receptors; thick arrow labels the synaptic junction. (*b*) Administration of the synthetic cannabinoid WIN55,212-2 reduces ^{3}H-GABA release from hippocampal slices, demonstrating that presynaptic CB_1 receptors play an inhibitory role on neurotransmitter release (Katona et al. 1999). (*c,d*) First direct evidence that endocannabinoids mediate depolarization-induced suppression of inhibition (DSI) in the hippocampus (Ohno-Shosaku et al. 2001, Wilson & Nicoll 2001). (*e*) Long-term depression of glutamatergic transmission is mediated by endocannabinoids via CB_1 receptors in the nucleus accumbens (Robbe et al. 2002) and (*f*) the striatum (Gerdeman et al. 2002). The individual figures have been modified from the originals with permission from the authors.

depression clearly support the view that 2-AG is the enigmatic synaptic endocannabinoid.

However, these pharmacological and genetic perturbations do not exclude a scenario in which 2-AG plays an autocrine role on axon terminals where CB_1 receptors are located. Additional support is derived from anatomical observations, which reveal that the subcellular segregation of endocannabinoid-metabolizing enzymes paves a retrograde path for 2-AG throughout the brain. DGL-α was found postsynaptically at several synapse types in the spinal cord (Nyilas et al. 2009), cerebellum (Yoshida et al. 2006), ventral tegmental area (Mátyás et al. 2008), striatum (Uchigashima et al. 2007), basolateral amygdala (Yoshida et al. 2011), hippocampus (Katona et al. 2006, Yoshida et al. 2006), prefrontal cortex (Lafourcade et al. 2007) in rodents, but also in the human hippocampus (Ludányi et al. 2011). Conversely, MGL was observed presynaptically in axon terminals throughout the brain (Gulyás et al. 2004, Ludányi et al. 2011, Uchigashima et al. 2011, Yoshida et al. 2011). Taken together, these anatomical and physiological experiments outline the most parsimonious scenario for retrograde synaptic signaling as involving 2-AG, which is synthesized and released postsynaptically and then acts on CB_1 receptors located on nearby presynaptic axon terminals.

COMMON PRINCIPLES OF ANTEROGRADE AMINO ACID TRANSMISSION AND RETROGRADE ENDOCANNABINOID TRANSMISSION

In their influential review, Sudhof & Malenka (2008) argue that one of the most important advances in our understanding of synaptic transmission in the past 20 years was the discovery that endocannabinoids are the principal mediators of retrograde synaptic communication. Since the breakthrough discoveries 10 years ago, several-hundred studies have dealt with the role of endocannabinoids in synaptic transmission, and the retrograde scenario has become widely accepted. In most biological signaling systems, information flow must be precisely controlled by feedback mechanisms. Thus, it is not surprising that synaptic transmission in chemical synapses requires a similar feedback mechanism, although it was clearly unforeseen that the consensus molecule for this function would be an endocannabinoid. Also unexpected was the finding that the basic operational principles of retrograde endocannabinoid signaling have so many features in common with conventional anterograde synaptic transmission. In the following sections, we aim to highlight the striking conceptual similarities between classical amino acid transmitter-mediated neurotransmission and retrograde endocannabinoid signaling, arguing that the endocannabinoid system is a component of chemical synapses as basic as the conventional neurotransmitter systems (**Table 1**). It is impossible in a single review to detail these biological phenomena; hence, we highlight a few notable examples to summarize key features of endocannabinoid signaling together with some physiological and pathophysiological implications focusing on some burning questions in endocannabinoid research.

MOLECULAR COMPLEXITY OF ENDOCANNABINOID MOBILIZATION AND DEGRADATION

Why does the brain need (at least) two endocannabinoid molecules? Neurons exploit several messengers to operate anterograde synaptic transmission in the brain, predominantly the classical amino acids glutamate and GABA, although glycine and the aromatic amino acid derivative monoamines such as dopamine, serotonin, or noradrenaline also have important functions. As retrograde messenger, 2-AG is the key player in most forms of homo- and heterosynaptic short-term depression and in some forms of long-term depression

Table 1 Comparison between anterograde and retrograde transmission*

	Anterograde transmission	Retrograde transmission
	Classical amino acid transmitters	Endocannabinoids
Multiple transmitters	Glutamate and GABA	Anandamide and 2-AG
Basic biomolecules	Amino acids are also used as protein building blocks	Glycerol, arachidonic acid, and ethanolamine are widely used for energy production and as metabolic intermediaries
Can be metabolized to each other	Decarboxylation of glutamate leads to GABA	2-AG is degraded to arachidonic acid, which is conjugated with ethanolamine to form anandamide. Anandamide level is reduced upon perturbation of 2-AG biosynthesis
Multiple synthesizing enzymes	Glutamate can be synthesized by cytosolic aspartate aminotransferase or glutaminase in neurons. GABA can be synthesized by either GAD65 or GAD67	Five biosynthetic routes were postulated for anandamide DAG as a precursor for 2-AG is hydrolyzed by DGL-α or DGL-β. Lyso-PI may also be a precursor for 2-AG
More than one receptor family	Glutamate: three ionotropic receptor families, eight mGluRs. GABA: $GABA_A$, $GABA_B$, $GABA_C$	Anandamide: full agonist on $TRPV_1$, partial on CB_1 and CB_2 receptors. Acts on potassium channels, NMDA, or glycine receptors. 2-AG: full agonist on CB_1 and CB_2; desensitizes glycine receptors
Several receptor compositions	Subunit composition of ionotropic glutamate receptors define, e.g., Ca^{2+}-permeability. $GABA_A$ subunits underlie tonic or phasic GABA signaling	Heterodimerization of CB_1, e.g., with angiotensin, orexin, $GABA_B$ dopamine, or opiate receptors
Additional receptor regulation	Glycine and D-serine are endogenous allosteric regulators of NMDA receptors; neurosteroids act on $GABA_A$ receptors	Endogenous antagonist of CB_1: virodhamine (but an agonist of CB_2). Hemopressin: inverse agonist of CB_1
Activity-dependent receptor regulation	Lateral trafficking or internalization of AMPA subunits in LTP and LTD are well-characterized examples	Lateral movement and internalization regulate CB_1 availability on axons. Decoupling of $G_{q/11}$-coupled GPCRs from DGL-α by short Homer isoform
Tonic regulation of neuronal activity	Ambient GABA targets specific $GABA_A$ receptors and evokes tonic inhibition of target cells	Tonic endocannabinoid levels regulate transmitter release probability in several brain areas and cell types
Homosynaptic effects and its plasticity	Glutamatergic and GABAergic transmission are involved and undergo homosynaptic plasticity	2-AG is a consensus mediator of homosynaptic short-term depression
Heterosynaptic effects and its plasticity	Both glutamateric and GABAergic transmission are known to mediate several forms of heterosynaptic plasticity	2-AG mediates heterosynaptic depression between glutamatergic and GABAergic synapses in several brain areas
Synaptic homeostasis	Tonic $GABA_A$ replaces I(h) current in synaptic homeostasis. Increased mEPSC frequency is a hallmark of synaptic gain	Anandamide and 2-AG both mediate synaptic homeostasis
Autocrine effects	Glutamate affects its own release via group II and III mGluRs. GABA via $GABA_B$	Slow self-inhibition characterizes the autocrine effects of 2-AG
Neuron-glia communication	Glutamate and GABA are accepted gliotransmitters; glial cell types express several glutamate and GABA receptors	Glial cells release both anandamide and 2-AG. Neuronal 2-AG controls Ca^{2+} signaling via CB_1 receptors on astrocytes

*Analogies can be observed in many respects between anterograde transmission mediated by classical amino acid transmitters and endocannabinoid-mediated retrograde signaling. Abbreviations: 2-AG, 2-arachidonoylglycerol; DAG, diacylglycerol; DGL, diacyglycerol lipase; GABA, gamma-aminobutyric acid; GAD, glutamic acid decarboxylase; mGluR, metabotropic glutamate receptor; LTD, long-term depression; lyso-PI, lysophosphatidylinositol; mEPSC, miniature excitatory postsynaptic current; NMDA, *N*-methyl-D-aspartate; TRPV, transient receptor potential vanilloid.

COX-2: cyclooxygenase-2

(Heifets & Castillo 2009, Kano et al. 2009). Anandamide also acts through CB_1 receptors in several neurobiological paradigms (Kinsey et al. 2009, Clapper et al. 2010, Straiker et al. 2011) where it may function as a retrograde synaptic messenger and mediate certain forms of synaptic homeostasis and plasticity in a presynaptic CB_1 receptor–dependent manner (Gerdeman et al. 2002, Kim & Alger 2010). Alternatively, anandamide's synaptic function can also be the activation of postsynaptic, transient receptor potential (TRP)V_1 receptors (Chavez et al. 2010, Grueter et al. 2010). The division of labor between anandamide and 2-AG may be reflected in spatial segregation, if anandamide and 2-AG serve as messengers at different synapses, in distinct microcircuits, or in separate brain regions. This is supported by observations that specific FAAH and MGL inhibitors recapitulate only subsets of behavioral components of cannabinoid effects in vivo (Long et al. 2009a). These behaviors are regulated by CB_1 receptors located in distinct cell types and brain circuits (Monory et al. 2007). Spatial isolation also occurs at the subcellular level, because MGL and FAAH are segregated into the presynaptic and postsynaptic domains of neurons, respectively (Gulyás et al. 2004). The division of labor may also happen at different timescales. Phasic endocannabinoid signaling, such as DSI, is mediated by 2-AG (Kim & Alger 2004, Tanimura et al. 2010), whereas tonic endocannabinoid signaling involves the mobilization of both 2-AG (Hashimotodani et al. 2007) and anandamide (Kim & Alger 2010) at hippocampal GABAergic synapses, though likely under different physiological conditions. Finally, the two signaling systems may also interplay in certain behavioral processes. Neither FAAH nor MAGL inhibitors alone recapitulate catalepsy or drug discrimination, both of which are typical CB_1 agonist-evoked behavioral effects. However, a dual FAAH/MGL inhibitor produces catalepsy as well as a Δ^9-THC-like drug-discrimination response (Long et al. 2009b). Thus, the combined action of 2-AG and anandamide may be required to engage CB_1 receptors fully at specific synapses, and if the mobilization of two messenger molecules requires different physiological signals, then presynaptic CB_1 receptors may operate as coincidence detectors to underlie synaptic plasticity analogously as postsynaptic NMDA (*N*-methyl-D-aspartate) receptors.

The level of complexity in the operational principles and the functional significance of retrograde endocannabinoid signaling are further increased by the emerging concept that, in contrast to the primarily amino acid–based anterograde transmission, a multifaceted lipid signaling system evolved to fulfill the complex physiological tasks reliant on retrograde signaling. For example, both anandamide and 2-AG are oxygenated by cyclooxygenase-2 (COX-2) at postsynaptic sites (Yu et al. 1997, Kozak et al. 2000, Kim & Alger 2004, Straiker et al. 2011). The resulting prostanoids, e.g., prostaglandin E2 glycerol ester, increase neurotransmitter release from axon terminals (Yang et al. 2008), which is opposite to the effect of 2-AG. Thus, COX-2, which is transported to synapses in an activity-dependent manner, may be an important molecular switch to change the direction of synaptic plasticity. Therapeutically important in vivo manifestation of this phenomenon occurs in supraspinal and spinal pain circuitries, where 2-AG has an overall antinociceptive effect (Hohmann et al. 2005, Nyilas et al. 2009). By contrast, 2-AG-derived prostaglandin E2 glycerol ester causes hyperalgesia (Hu et al. 2008). Remarkably, the highly potent analgesic effects of nonsteroidal anti-inflammatory drugs such as paracetamol (acetaminophen), a COX-2 inhibitor showing significant selectivity over COX-1 (Hinz et al. 2008), or ibuprofen, a potent inhibitor of 2-AG oxidation by COX-2 (Prusakiewicz et al. 2009), require the activation of CB_1 receptors (Ahn et al. 2007, Telleria-Diaz et al. 2010). Thus, these drugs may partially exhibit their analgesic effects by vetoing the metabolism from antinociceptive 2-AG to hyperalgesic prostaglandin caused by hyperalgesia-induced elevations in COX-2 activity.

The discovery of 2-epoxyeicosatrienoylglycerols (2-EGs) also revealed a link to lipoxygenase and cytochrome P450 pathways. These novel lipids are present in the brain (especially 2-11,12-EG), and, surprisingly, they are potent endogenous ligands of both cannabinoid receptors in vivo (Chen et al. 2008). Anandamide can be transformed to 5′,6′-epoxyeicosatrienoic acid by CP450 epoxygenases, which then activates $TRPV_4$ channels at submicromolar concentrations (Watanabe et al. 2003a), whereas the related eicosanoid 12-(S)-hydroperoxyeicosatetraenoic acid is the retrograde mediator of $TRPV_1$-dependent long-term depression at hippocampal Schaffer collateral-interneuron synapses (Gibson et al. 2008). Although the puzzling question of why lipids are utilized so extensively as retrograde messengers in contrast to the predominantly amino acid–based anterograde transmitters may not be easy to answer, this perplexing diversity of lipid signaling pathways should definitely be a focus for neuroscience.

A striking example of parsimony in biology is the way in which neurons exploit amino acids such as glutamate or glycine, which are otherwise basic building blocks of proteins, as neurotransmitters. These amino acids can be metabolized to each other to produce additional messengers; for example, GABA is synthesized from glutamate by decarboxylation. This parsimony is also shared by the endocannabinoid system. Glycerol, ethanolamine, and arachidonic acid are at the crossroads of several metabolic pathways. They are also included in several major components of biological membranes and provide energy for various cellular functions. Arachidonic acid, the common constituent of 2-AG and anandamide, is a conditionally essential n-6 polyunsaturated fatty acid and makes up a significant fraction of brain lipids (Rapoport 2008). It is present in the diet together with its precursor linoleic acid; consequently, when their dietary concentrations are increased, anandamide and 2-AG levels also increase (Watanabe et al. 2003b). Remarkably, this has an impact on synaptic endocannabinoid signaling, because changing the ratio of dietary n-6/n-3 polyunsaturated fatty acid abolishes endocannabinoid-mediated long-term depression in the prefrontal cortex and nucleus accumbens (Lafourcade et al. 2011). Although the underlying mechanistic process is not fully understood, chronic elevation of 2-AG levels also disrupts CB_1-mediated signaling, highlighting the importance of precise metabolic regulation of 2-AG levels in the brain (Chanda et al. 2010, Schlosburg et al. 2010). Anandamide and 2-AG can also be metabolized to each other similar to amino acid transmitters, and this metabolic pathway can be region and subcellular-domain specific, as demonstrated by several findings. First, anandamide levels are decreased in the hippocampus but not in the prefrontal cortex of DGL-α-knockout animals (Gao et al. 2010, Tanimura et al. 2010, Yoshino et al. 2011). Second, free arachidonic acid can be released from 2-AG by at least three serine hydrolases (Blankman et al. 2007), one of which, ABHD6 (α/β-hydrolase-6) regulates synaptic endocannabinoid signaling in long-term depression (Marrs et al. 2010). Moreover, ABHD6 colocalizes postsynaptically in dendrites with FAAH, which conjugates ethanolamine with arachidonic acid to form anandamide (Mukhopadhyay et al. 2011).

PL: phospholipase

The biogenesis of endocannabinoids is also highly complex. Anandamide and related N-acylethanolamines have been postulated to be synthesized by at least five metabolic pathways. Using N-acyl-phosphatidylethanolamines as precursors, the phospholipase (PL)A route involves PLA_2 and lyso-PLD (Sun et al. 2004), the PLB route includes α/β-hydrolase 4 and glycerophosphodiesterase 1 (Simon & Cravatt 2006, 2008), the PLC route is mediated by a PLC and protein tyrosine phosphatase type-22 (Liu et al. 2006), and the PLD route exploits N-acyl-phosphatidylethanolamines–hydrolyzing PLD (Okamoto et al. 2004). In addition, anandamide can be formed by a conjugation of arachidonic acid and ethanolamine in brain synaptosomes (Devane & Axelrod 1994). Of great importance for endocannabinoid research is to understand the functional significance of this metabolic diversity. Spatial segregation of distinct anandamide- and N-acylethanolamine-synthesizing pathways may underlie a division

DAG: diacylglycerol

PSM: perisynaptic signaling machinery

of labor. Indeed, whereas FAAH is distributed in the somatodendritic domain of neurons (Egertova et al. 1998, Gulyas et al. 2004), N-acyl-phosphatidylethanolamines–hydrolyzing PLD is found inside glutamatergic axon terminals (Nyilas et al. 2008). Another indication of functional diversity is the different kinetics of anandamide synthesis, which suggest that the PLC route may operate on a timescale different from that of the PLB route (Liu et al. 2008).

The biosynthesis of 2-AG may seem to be more simple, but this approach can be misleading. Although the role of DGL-α in synaptic plasticity is unequivocal (Tanimura et al. 2010), the precursor diacylglycerol (DAG) can be synthesized in several ways. The canonical pathway includes PLC-βs, which hydrolyze phosphatidylinositol 4,5-bisphosphate to inositol 1,4,5-triphosphate and DAG (Stella et al. 1997). Most receptor-driven 2-AG synthesis may go through this route. However, PLCβ_1 is not required for depolarization-induced forms of endocannabinoid-mediated synaptic plasticity such as hippocampal DSI (Hashimotodani et al. 2005). DAG-independent 2-AG biosynthesis may occur via PLA_1 and lysophosphatidylinositol-specific PLC activity, the latter of which is found in brain synaptosome preparations (Tsutsumi et al. 1994, Sugiura et al. 1995). Finally, brain homogenates can also synthesize 2-AG from 2-arachidonoyl-lysophosphatidic acid by an unidentified phosphatase (Nakane et al. 2002). Thus, additional important tasks for endocannabinoid research are to dissect which biochemical pathways are responsible for 2-AG mobilization at given locations in neuronal networks and to identify which physiological or pathophysiological stimuli activate these pathways.

A spatial segregation indicating functional division of labor has already been reported for the two isoforms of DGL (Bisogno et al. 2003). Whereas DGL-α is found in the plasma membrane (Katona et al. 2006, Yoshida et al. 2006, Jung et al. 2011), DGL-β is restricted to intracellular membrane segments, including peri-nuclear lipid droplets (Jung et al. 2011). This spatial segregation suggests that DGL-α may play a role in intercellular 2-AG signaling, whereas DGL-β has an intracellular function. In parallel, basal 2-AG levels in the brain of DGL-α-knockout mice were dropped by 80% but they remained unaffected or only partially reduced in DGL-β-knockout mice (Gao et al. 2010, Tanimura et al. 2010). In contrast to findings in DGL-α-knockout mice, synaptic endocannabinoid signaling was not impaired by genetic inactivation of DGL-β (Gao et al. 2010, Tanimura et al. 2010). It is interesting to note that neurite outgrowth triggered by overexpression of DGL-α could be blocked by a CB_1 receptor antagonist, whereas neuritogenesis induced by overexpression of DGL-β was CB_1 receptor independent (Jung et al. 2011). Similarly, adult neurogenesis is impaired in DGL-α- but not in DGL-β-knockout mice (Gao et al. 2010).

ENDOCANNABINOID SIGNALOSOMES

The convincing demonstration that DGL-α is indispensable for various forms of retrograde synaptic signaling (Tanimura et al. 2010) calls for investigations of how this important enzyme is integrated into neuronal operations. At hippocampal glutamatergic synapses, DGL-α accumulates around the postsynaptic density at the edge of the synapses (Katona et al. 2006, Yoshida et al. 2006). Group I metabotropic glutamate receptors (mGluR) are located within the same perisynaptic annulus (Lujan et al. 1996), and their agonists evoke 2-AG release through the canonical $G_{q/11}$ and PLC-β pathway (Jung et al. 2005). Because both mGluR and DGL-α contain binding motifs for the synaptic scaffold protein Homer (Jung et al. 2007), we proposed that they form a macromolecular complex around the postsynaptic density, the so-called perisynaptic signaling machinery (PSM), which evolved to translate excess presynaptic activity—glutamate spillover, in this case—into a negative-feedback signal (Katona & Freund 2008). Since then, new findings have confirmed and extended the PSM concept: PLC-β_1, another molecular constituent of this pathway, was also found in the perisynaptic annulus (Fukaya et al. 2008). Moreover, the long-isoform Homer$_{2b}$ turned out to be necessary for mGlu-triggered 2-AG release (Won et al. 2009, Roloff et al. 2010), whereas the activity-dependent short isoform (Homer$_{1a}$) dismantled the PSM and ablated this process (Roloff et al. 2010).

Importantly, DGL-α not only may function as a 2-AG-synthesizing enzyme, but could also play another role in regulating DAG levels. This function may be phylogenetically more ancient, because insects lack cannabinoid receptors but express a DGL-α ortholog encoded by the *inaE* gene, which is necessary for the opening of TRP channels by DAG (Leung et al. 2008). In mammals, TRP channels such as TRPC1 or TRPC3 are anchored by Homer, concentrated perisynaptically, and stimulated by DAG (Kim et al. 2003, Yuan et al. 2003). In addition, activation of TRPC channels accounts for group I mGluR-triggered feed-forward enhancement of excitability of postsynaptic neurons (Kim et al. 2003, Hartmann et al. 2008). Taken together, these findings indicate that some molecular players within the PSM are primarily responsible for feed-forward excitation. However, this signal is controlled by DGL-α, which may act as a molecular switch by transforming a feed-forward excitatory DAG signal into a negative-feedback signal via 2-AG (**Figure 3**). Perisynaptic mGluRs cannot be activated by single synaptic volleys in the way that intrasynaptic ionotropic glutamate receptors can be. Instead, mGluRs require elevated, usually bursting, presynaptic population activity (Tempia et al. 1998, Fan et al. 2010). Therefore, we propose that the essential physiological function of the PSM domain of excitatory synapses is to monitor the magnitude of presynaptic activity. Increased presynaptic activity may need to be transformed to a feed-forward excitatory response to increase the excitability of the postsynaptic neuron and to support synaptic potentiation. However, excess presynaptic activity can also become pathological, at which point efficient negative-feedback mechanisms should kick in. We suggest that DGL-α may be involved in this mechanism in a strikingly parsimonious manner by terminating feed-forward excitation and initating feedback inhibition, in other words, by simultaneously eliminating the cause and the consequence of the excess presynaptic activity (**Figure 3*a*,*b***).

We recently termed the second 2-AG leg of the pathway as a "synaptic circuit breaker" to indicate that this process not only may happen at single synapses in a homosynaptic manner, but also may be a general network mechanism that has a pivotal role in regulating the overall level of network excitability under pathophysiological conditions with an excess glutamatergic tone (Katona & Freund 2008). Neuronal insults, e.g., convulsions or closed-head injury, evoke 2-AG release (Panikashvili et al. 2001, Wettschureck et al. 2006), whereas perturbations of the synaptic circuit breaker lead to reduced seizure thresholds and increased incidence of epileptic seizures. Double $G_{q/11}$ and PLC-β_1-knockout animals die at a young age as a result of spontaneous seizures (Kim et al. 1997, Wettschureck et al. 2006), whereas glutamatergic cell–specific overexpression or deletion of CB_1 receptors are protective or convulsive, respectively (Marsicano et al. 2003, Monory et al. 2006, Guggenhuber et al. 2010). Finally, breakdown of the circuit breaker also occurs in human patients with chronic, intractable temporal-lobe epilepsy whose hippocampus has reduced levels of DGL-α and CB_1 receptors (Ludányi et al. 2008). Taken together, these findings indicate that the significance of the PSM and retrograde 2-AG signaling is also reflected at the pathophysiological level and may be exploited for therapy in the future.

A salient emerging concept is that the PSM at glutamatergic synapses may be a specialized case of a much more fundamental cell-physiological mechanism involving the same macromolecular complex and retrograde

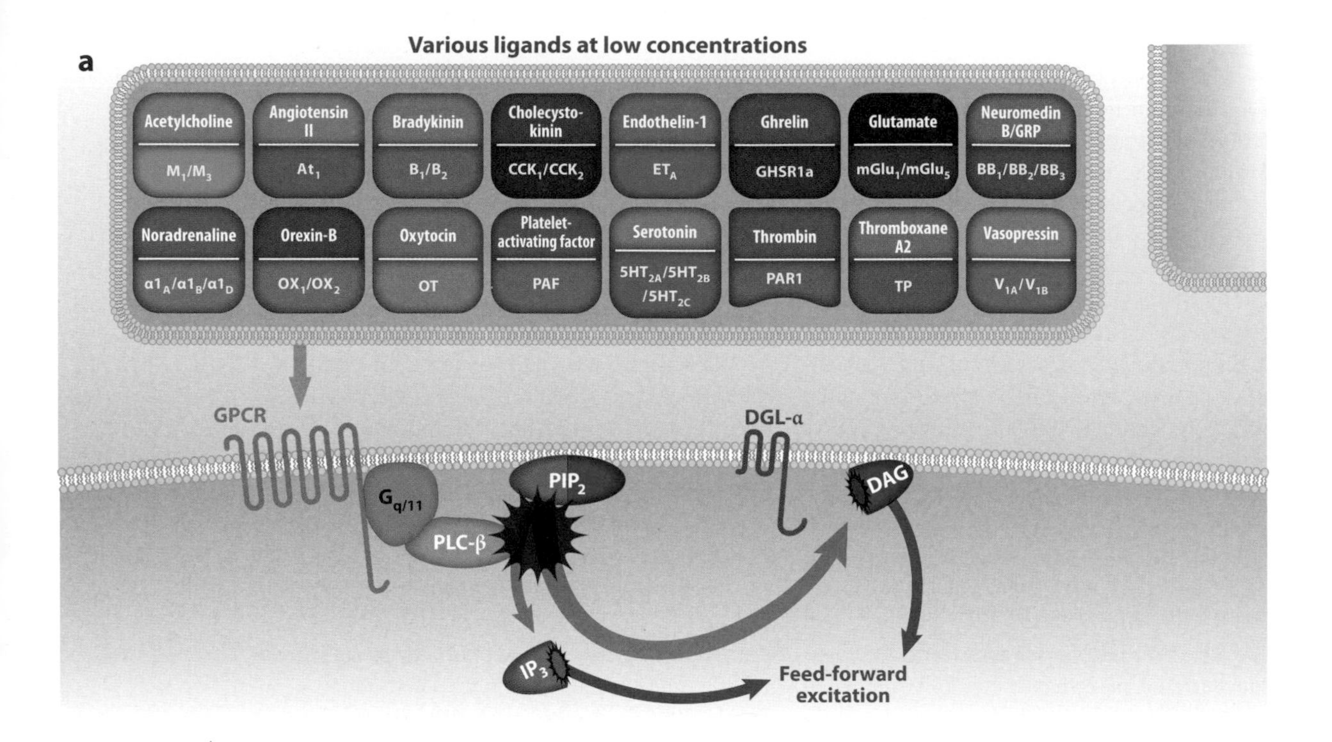
a
Various ligands at low concentrations
Acetylcholine
M1/M3
Angiotensin II
At1
Bradykinin
B1/B2
Cholecysto-kinin
CCK1/CCK2
Endothelin-1
ETA
Ghrelin
GHSR1a
Glutamate
mGlu1/mGlu5
Neuromedin B/GRP
BB1/BB2/BB3
Noradrenaline
α1A/α1B/α1D
Orexin-B
OX1/OX2
Oxytocin
OT
Platelet-activating factor
PAF
Serotonin
5HT2A/5HT2B/5HT2C
Thrombin
PAR1
Thromboxane A2
TP
Vasopressin
V1A/V1B
GPCR
Gq/11
PLC-β
PIP2
DGL-α
DAG
IP3
Feed-forward excitation

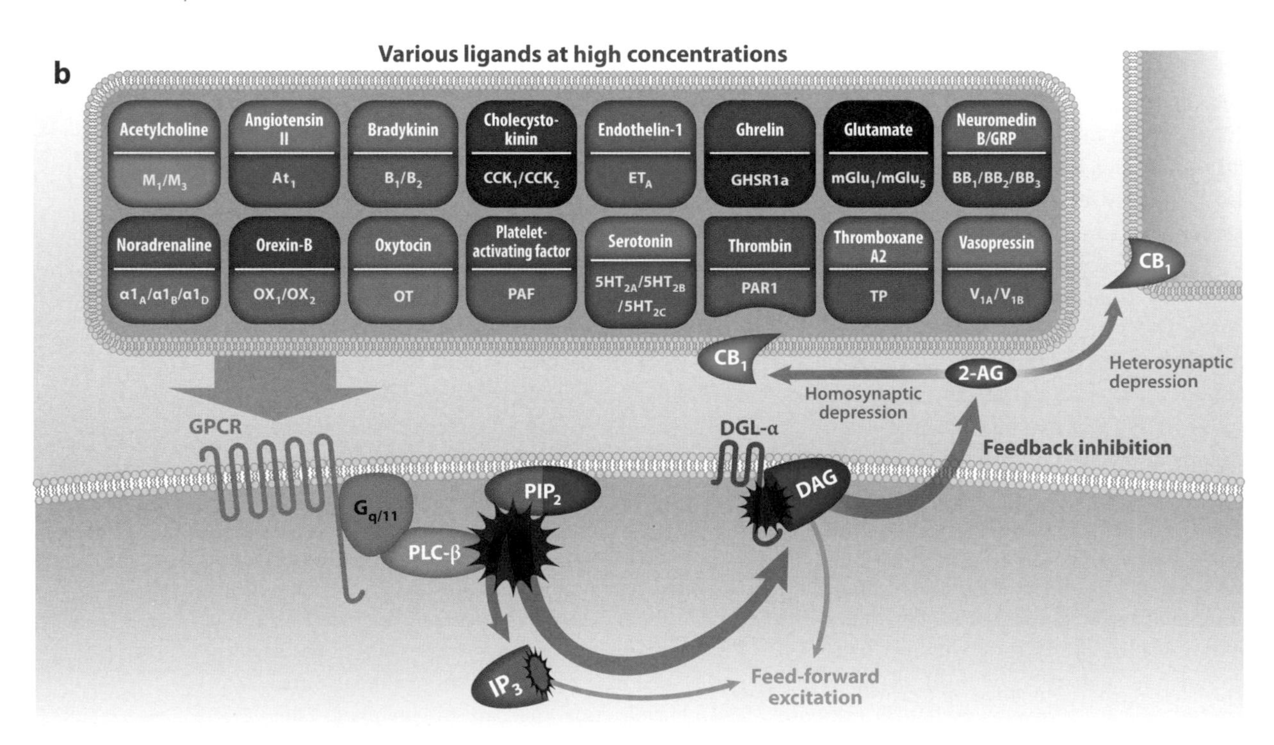
b
Various ligands at high concentrations
Acetylcholine
M1/M3
Angiotensin II
At1
Bradykinin
B1/B2
Cholecysto-kinin
CCK1/CCK2
Endothelin-1
ETA
Ghrelin
GHSR1a
Glutamate
mGlu1/mGlu5
Neuromedin B/GRP
BB1/BB2/BB3
Noradrenaline
α1A/α1B/α1D
Orexin-B
OX1/OX2
Oxytocin
OT
Platelet-activating factor
PAF
Serotonin
5HT2A/5HT2B/5HT2C
Thrombin
PAR1
Thromboxane A2
TP
Vasopressin
V1A/V1B
CB1
CB1
2-AG
Homosynaptic depression
Heterosynaptic depression
Feedback inhibition
GPCR
Gq/11
PLC-β
PIP2
DGL-α
DAG
IP3
Feed-forward excitation

endocannabinoid signaling. Following the initial observations that $G_{q/11}$-coupled, postsynaptic $mGlu_1$ and $mGlu_5$ receptors can elicit synaptic endocannabinoid signaling in the cerebellum and hippocampus (Maejima et al. 2001, Varma et al. 2001), researchers showed that at least 16 other neurotransmitter molecules trigger 2-AG release and CB_1 receptor activation via $G_{q/11}$-coupled receptors, PLC-β and DGL-α (**Figure 3**). Notably, besides the homosynaptic feedback processes, upstream activation of this macromolecular complex can often lead to heterosynaptic depression of the release of another neurotransmitter, as is the case for serotonin and $5HT_{2A}/5HT_{2B}/5HT_{2C}$ receptors in the inferior olive and elsewhere (Parrish & Nicolls 2006, Best & Regehr 2008). These signalosomes may not always be restricted to the perisynaptic zone of synapses, but their subcellular location reflects the source and chemical nature of the given transmitter. For example, acetylcholine, which reaches its receptors primarily by volume transmission, can evoke endocannabinoid signaling in several brain regions through M_1 receptors, which are distributed throughout the somatodendritic surface of postsynaptic neurons (Kim et al. 2002, Uchigashima et al. 2007, Yamasaki et al. 2010). Some of these signaling mechanisms can be surprisingly cell-type specific as has been shown for the neuropeptide cholecystokinin and CCK_2 receptors in the hippocampus (Lee & Soltesz 2011, Lee et al. 2011). Others such as the wake-promoting peptide orexin-B and its $G_{q/11}$-coupled receptors OX_1 and OX_2 (Haj-Dahmane and Shen 2005), endothelin-1 and its ET_A receptor (Zampronio et al. 2010), oxytocin and its OT receptor (Oliet et al. 2007), or ghrelin and its receptor (Kola et al. 2008) may convey information about the general physiological or metabolic state of the animal. They may then regulate feedback hormonal responses via the modification of synaptic weights in subcortical and cortical circuits. Particularly interesting is how these signalosomes may be involved in pathophysiological processes upon injury to the central nervous system. Thrombin-induced arachidonic acid release led to the original discovery of a PLC-DGL pathway (Bell et al. 1979), and thrombin regulates GABAergic synaptic currents through PAR-1 receptors and retrograde 2-AG signaling in the hippocampus (Hashimotodani et al. 2011). Other functionally related pathways also stimulate endocannabinoid signaling as has been demonstrated for thromboxane A2 via the prostanoid receptor TP (Rademacher et al. 2005) and for the platelet-activating factor through its receptor PAF (Berdyshev et al. 2001). Some other G protein–coupled receptor endocannabinoid signalosomes are also expected to be found in the central nervous system, as they are widely distributed throughout the body: Examples include angiotensin II and its AT_1 receptor (Turu et al. 2009), the bombesin's neuromedin B, gastrin-releasing peptide and its $BB_1/BB_2/BB_3$ receptors (Shimizu et al. 2011), bradykinin and its B_1 and B_2 receptors (Turu et al. 2009), noradrenaline and the adrenoceptors $\alpha 1_A$ $\alpha 1_B$ $\alpha 1_D$ (Turu et al. 2009), and vasopressin and its V_{1A} and V_{1B} receptors (Turu et al. 2009). It may be too early

Figure 3

Diacylglycerol lipase (DGL)-α acts as a molecular switch between G protein–coupled receptor (GPCR)-mediated feed-forward excitation and feedback inhibition. (*a*) The cascade of events triggered by a multitude of ligands via the activation of GPCRs coupled to $G_{q/11}$ and phospholipase (PL)C-β at low agonist concentrations: PLC-β will split phosphatidylinositol 4,5-bisphosphate (PIP_2) into inositol 1,4,5-triphosphate (IP_3) and diacylglycerol (DAG), both of which primarily enhance excitability of the target cell, e.g., by stimulating calcium release from intracellular stores or via transient receptor potential (TRP) channels, respectively. (*b*) The cascade of events triggered by a multitude of ligands via the activation of GPCRs coupled to $G_{q/11}$ and PLC-β at high agonist concentrations: A larger amount of DAG is produced, in which case DGL-α will step in and convert DAG to 2-arachidonoylglycerol (2-AG), which will mediate feedback inhibition in the form of homo- or heterosynaptic depression of transmitter release.

to claim that the DGL-α–2-AG–CB_1 (and maybe also CB_2) endocannabinoid signaling pathway is a built-in feature of the downstream signaling pathway upon $G_{q/11}$-coupled receptor activation; nevertheless, when present, it can efficiently translate a feed-forward signal into a feedback signal to regulate cellular functions.

MOLECULAR COMPLEXITY OF ENDOCANNABINOID-TARGETED RECEPTORS

Providing another conceptual similarity with classical amino acid transmitters, which act on several ionotropic (AMPA, NMDA, kainate or $GABA_A$, $GABA_C$, and Gly) and metabotropic (mGlu or $GABA_B$) receptors to accomplish their diverse responsibilities, these lipid messengers can also interact with several other molecular targets besides CB_1 and CB_2 cannabinoid receptors. Among these potential targets are ligand-gated ion channels such as 5-HT_3, glycine, and nicotinic acetylcholine receptors; nonselective cation channels such as $TRPV_1$, TRPA1, or TRPM8; voltage-gated ion channels such as T-type calcium channels or the TASK potassium channels; and metabotropic receptors such as GPR55 (for a review, see Pertwee et al. 2010). An especially exciting research direction is to delineate the neurobiological significance of these interactions, which, despite some promising progress (Chavez et al. 2010, Grueter et al. 2010), is largely unknown (Pertwee et al. 2010).

An additional level of signaling complexity for anterograde transmission comes from the variable subunit compositions of amino acid receptors, some well-known examples of which are the synapse-specific segregation of calcium-permeable AMPA receptors determined by the absence of the GluR2 subunit (Tóth & McBain 1998) or the observation that tonic and phasic modes of GABA signaling are mediated by $GABA_A$ receptors with different subunit compositions (Glykys & Mody 2007). A potential mechanism to increase the complexity of endocannabinoid signaling may be the phenomenon of receptor heteromerization. Heterodimers of CB_1 receptors have been observed, for example, with D_2 dopamine receptors (Kearn et al. 2005), μ-opioid receptors (Rios et al. 2006), or OX_1 orexin receptors (Ellis et al. 2006). Heteromerization may impact ligand sensitivity, downstream signaling, and compartmentalization of a given receptor, which all contribute to the ultimate physiological role of the receptor complex. Thus, it will be a very important task to exploit the latest available microscopy techniques offering appropriate spatial resolution such as super-resolution microscopy to characterize the cell- and synapse-type-specific distributions of given cannabinoid receptor heterodimers in brain circuits.

It is widely accepted that, besides their primary ligands, the activity of most receptors is controlled by endogenous allosteric modulators. Some famous examples in case of the ionotropic glutamate receptors include the function of glycine or D-serine in the regulation of NMDA receptors (Johnson & Ascher 1987, Kleckner & Dingledine 1988) as well as the profound impact of neurosteroids on δ-subunit-containing $GABA_A$ receptors (Stell et al. 2003). Similar endogenous modulators for cannabinoid receptors are starting to appear. The lipid virodhamine was the first reported endogenous antagonist of CB_1 receptors, which surprisingly acts as an agonist on CB_2 receptors (Porter et al. 2002). An unexpected new family of CB_1 receptor modulators comprises the hemoglobin-derived nonapeptide hemopressin and its longer congeners, which act as an inverse agonist or agonist, respectively (Heimann et al. 2007, Gomes et al. 2009). The potential presence of allosteric binding sites on CB_1 receptors and the design of selective agents targeting these sites would be especially advantageous, because full agonists evoke robust internalization of cannabinoid receptors (Jin et al. 1999), which lead to in vivo tolerance (Tappe-Theodor et al. 2007), and renders pharmacological exploitation difficult. By contrast, internalization of presynaptic CB_1 receptors on axon terminals offers a new level of physiological control (Coutts et al. 2001), whereby the efficacy of endocannabinoid-mediated synaptic plasticity can be dynamically adjusted in an activity-dependent manner. In addition to internalization, lateral movement of CB_1 receptors on the surface of axon terminals is also well suited to regulate CB_1 receptor availability and desensitization (Mikasova et al. 2008). Notably, lateral mobility and internalization of postsynaptic AMPA receptors are also key underlying mechanisms for experience-dependent plasticity of anterograde excitatory synaptic transmission (Newpher & Ehlers 2009).

FUNCTIONAL COMPLEXITY OF ENDOCANNABINOID-MEDIATED SYNAPTIC PLASTICITY

The most spectacular evidence for the profound neurobiological significance of endocannabinoid signaling may be the wide repertoire of synaptic physiological processes that are mediated by endocannabinoids. It is conceivable to suppose that nature followed the same rule of parsimony as that employed for glutamatergic and GABAergic neurotransmission by adapting the activity of the same conserved molecular players throughout the central and peripheral nervous systems to accomplish so many different synaptic physiological tasks to enable the proper operation at various brain-circuit levels. It is out of the scope of this review to summarize the hundreds of studies describing the specific functions of retrograde endocannabinoid signaling from the spinal cord to the neocortex, but we recommend two excellent reviews for further reading (Heifets & Castillo 2009, Kano et al. 2009). Instead, we aim to illustrate with a few select examples the conceptual similarities by which the brain exploits endocannabinoids for retrograde signaling processes and classical amino acid transmitters for anterograde communication.

In terms of the two major modes of endocannabinoid signaling, one has to differentiate between tonic and phasic actions. Just as ambient GABA has a crucial role in establishing the excitability of postsynaptic neurons through extrasynaptic $GABA_A$ receptors (Glykys & Mody 2007), the pivotal role of ambient extracellular endocannabinoid concentrations in determinating neurotransmitter release has begun to unfold. Paired recordings from a special subtype of GABAergic interneuron and postsynaptic CA3 pyramidal neurons in the hippocampus revealed that tonic endocannabinoid signaling can mute GABA release from axon terminals through CB_1 receptor activation (Losonczy et al. 2004). Subsequent research has shown that this phenomenon is not due to the constitutive activity of CB_1 receptors per se, but instead depends on the constitutive release of endocannabinoids from the postsynaptic neuron, as postsynaptic BAPTA chelation of intracellular Ca^{2+} signals abolished the tonic endocannabinoid signaling (Hentges et al. 2005, Neu et al. 2007). Remarkably, tonic endocannabinoid signaling is also cell-type specific. It can depend on the type of the postsynaptic neuron, e.g., proopiomelanocortin neurons, but not the neighboring non-proopiomelanocortin neurons, can also release endocannabinoids constitutively and regulate their incoming GABAergic inputs in the arcuate nucleus of the hypothalamus, although both cell populations could produce endocannabinoids in a stimulation-dependent manner (Hentges et al. 2005). Alternatively, the same postsynaptic neuron can also regulate GABAergic inputs in a different manner, as has been elegantly demonstrated in the hippocampus where CA1 pyramidal neurons regulate perisomatic inhibition via endocannabinoids in both a tonic and phasic manner, whereas only phasic endocannabinoid signaling was found at the dendritic inhibitory inputs (Lee et al. 2010). This latter observation at unitary connections also confirms that constitutive endocannabinoid release can act in a homosynaptic and subcellularly restricted manner (Neu et al. 2007, Lee et al. 2010). Tonic endocannabinoid signaling may involve both anandamide and 2-AG, although probably at different timescales (Hashimotodani et al. 2007, Kim & Alger 2010). The presence of tonic endocannabinoid control of neurotransmitter release indicates that physiological signals must exist to override it whenever necessary. By contrast, this phenomenon can also serve to integrate the efficacy of a dedicated unitary connection into network activity, because high-frequency firing of the presynaptic neuron can eliminate the tonic endocannabinoid blockade (Losonczy et al. 2004, Foldy et al. 2006).

Salient features of synaptic endocannabinoid signaling ideally support the induction of changes in synaptic strength in a phasic, activity-dependent manner. The governing rules for these retrograde forms of synaptic plasticity also follow a similar logic, and these mechanisms are nicely integrated into several well-known forms of anterograde synaptic plasticity mediated by glutamate or GABA. Homosynaptic short-term synaptic depression of excitation and inhibition was the first described form of endocannabinoid-mediated synaptic plasticity (Kreitzer & Regehr 2001, Ohno-Shosaku et al. 2001, Wilson & Nicoll 2001) and was supposed to be mediated predominantly by 2-AG (Hashimotodani et al. 2007, Gao et al. 2010, Tanimura et al. 2010, Yoshino et al. 2011). Although the extracellular spread of endocannabinoids is limited (Wilson & Nicoll 2001), endocannabinoid-mediated short-term depression is also involved indirectly in heterosynaptic forms of plasticity (Lourenco et al. 2010). Endocannabinoid-mediated forms of long-term depression were first described in the dorsal and ventral striatum and in the amygdala, but they are also present at most synapses in the brain (Gerdeman et al. 2002, Marsicano et al. 2002, Robbe et al. 2002, Heifets & Castillo 2009, Kano et al. 2009). A special, potentially homosynaptic form is the so-called spike-timing-dependent long-term depression, which was described first in the neocortex and requires presynaptic NMDA or postsynaptic mGluRs (Sjostrom et al. 2003, Nevian & Sakmann 2006). Heterosynaptic forms of long-term depression also exploit endocannabinoids (Chevaleyre & Castillo 2003).

Another special form of synaptic plasticity serves to readjust synaptic gain in response to persistent changes of neuronal activity. This phenomenon of synaptic homeostasis involves not only an increase in the action-potential-independent release probability of glutamate and in the density of postsynaptic glutamate receptors, but also a decrease in the number of postsynaptic GABA receptors to restore circuit activity (Turrigiano 2007). There are likely to be multiple forms of synaptic scaling; correspondingly, both anandamide and 2-AG contribute to homeostatic regulation of GABAergic synapses in the hippocampus (Zhang et al. 2009, Kim & Alger 2010). In addition, just as glutamate and GABA act in an autocrine manner to regulate their own release, compelling evidence supports the idea that both endocannabinoids may also have autocrine functions. Researchers postulate that anandamide plays a postsynaptic role in long-term depression in the hippocampus and striatum (Chavez et al. 2010, Grueter et al. 2010) and that 2-AG is the mediator of the autocrine phenomenon of slow self-inhibition in neocortical interneurons (Marinelli et al. 2008). Although it took some time for neuroscientists to accept that glutamate can also be a gliotransmitter (multiple types of glutamate and GABA receptors are found on glial cells), the idea that molecules can be utilized for multiple physiological functions gained wider recognition. In parallel, exciting new discoveries revealed that endocannabinoid signaling not only depresses, but also potentiates glutamatergic transmission via a novel form of neuron-glia cross talk (Navarrete & Araque 2010).

CELL-TYPE- AND SYNAPSE-SPECIFIC DIFFERENCES IN ENDOCANNABINOID SIGNALING UNDERLYING CIRCUIT-DEPENDENT BEHAVIORS

In the above section, we aim to illustrate the perplexing chemical and functional diversity of the endocannabinoid system in the brain and to highlight that the same neurobiological principles may govern anterograde synaptic transmission and retrograde endocannabinoid signaling. In closing, we consider the idea that subtle, but important, refinements in the logic of endocannabinoid signaling have evolved to provide the most optimal contribution of retrograde communication to the functional operation of microcircuits. All brain circuits probably require detailed studies for researchers to delineate fully how given endocannabinoid signalosomes, in particular subcellular compartments, mediate certain forms of synaptic plasticity and thereby regulate network activity and behavioral processes. Here, we describe a few striking examples from the hippocampus (**Figure 4**).

In contrast to the qualitative uniformity of synaptic 2-AG signaling, each glutamatergic and GABAergic synapse is quantitatively different in terms of the density of the molecular components of the 2-AG pathway. DGL-α has the highest density in the inner molecular layer at mossy cell–granule cell synapses, is abundant at the Schaffer collateral–CA1 pyramidal neuron synapses, and is localized at other glutamatergic synapses throughout the hippocampal formation (Katona et al. 2006) (**Figure 4*a***). The functional consequence of this input-specific pattern is reflected in the distinct thresholds necessary to evoke 2-AG-mediated DSE at different glutamatergic synapses (Uchigashima et al. 2011). This complexity is further increased at the ultrastructural level and may underlie the contribution of 2-AG to different homo- or heterosynaptic forms of synaptic plasticity. DGL-α is concentrated in a perisynaptic annulus in the head of dendritic spines in the CA1 subfield (Katona et al. 2006, Yoshida et al. 2006); conversely, it is accumulated around the necks of spines of dentate gyrus granule cells (Uchigashima et al. 2011). In contrast to its presence in glutamatergic synapses, only a very small amount of DGL-α is present at hippocampal or neocortical GABAergic synapses (I. Katona, personal communication), as is clearly reflected by the lack of DGL-α immunostaining in the cell-body layers (**Figure 4*a*,*b***). By contrast, GABAergic synapses in the basolateral amygdala show an extremely high density of DGL-α (**Figure 4*c***) (Yoshida et al. 2011).

CB_1 receptors also exhibit a characteristic expression pattern and layer-specific distribution (**Figure 4*e*,*f***). In contrast to DGL-α, CB_1 receptors have the highest density on GABAergic synapses derived from the CCK-positive class of interneurons in the rodent and human hippocampus (Katona et al. 1999, Katona et al. 2000). However, distinct types of CCK-positive interneurons, e.g., the perisomatic basket cells and the Schaffer collateral-associated dendritic inhibitory cells, differ in their CB_1 content, and this matches the distinct efficacy of endocannabinoid-mediated synaptic plasticity at these synapses (Lee et al. 2010). CB_1 receptors are also present at glutamatergic synapses, albeit at much lower levels (Katona et al. 2006, Kawamura et al. 2006). The distribution of the degrading enzyme MGL displays a more surprising pattern. Whereas axon terminals of GABAergic interneurons and recurrent collaterals of CA3 pyramidal neurons bear a high density of MGL (**Figure 4*i*,*j***), Schaffer collaterals derived from the same CA3 pyramidal cells may contain less MGL (**Figure 4*k*,*l***) (Gulyás et al. 2004, Uchigashima et al. 2011). Moreover, MGL density is strikingly low in the inner molecular layer on the axon terminals of mossy cells (**Figure 4*g*,*h***) (Ludányi et al. 2011, Uchigashima et al. 2011). Although the physiological consequences of these quantitative differences are not yet fully understood, they suggest that the contribution of 2-AG signaling to distinct behavioral phenomena and pathophysiological processes may be qualitatively different at specific synapses and microcircuits.

The development of cell-type-specific CB_1 receptor–knockout models represents the key innovation to elucidating how endocannabinoid signaling at specific microcircuit locations contributes to network activity and behavior (Marsicano et al. 2003; Monory et al. 2006, 2007). Although GABAergic axon terminals carry many more CB_1 receptors than do their glutamatergic counterparts, they are not involved in seizure susceptibility (Monory et al. 2006). Instead, these receptors play a pivotal role in Δ^9-THC-induced long-term memory deficits (Puighermanal et al. 2009) and protect against age-related cognitive decline (Albayram et al. 2011). A similar cell-type-specific functional dichotomy was observed in the regulation of feeding and energy balance, during which CB_1 receptors on striatal GABAergic neurons reduce food intake, whereas those on forebrain glutamatergic axons convey an orexigenic signal (Bellocchio

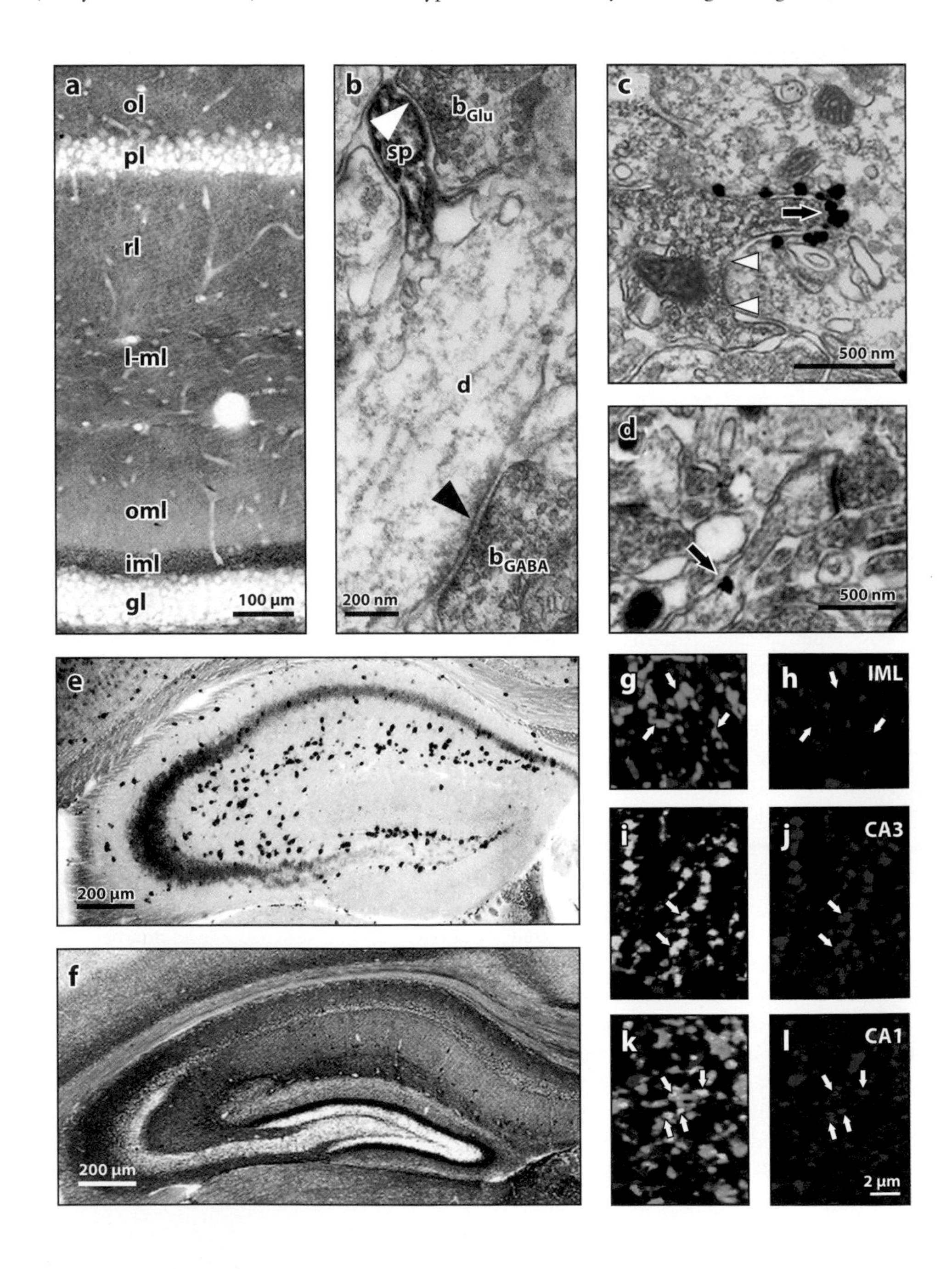

et al. 2010). An opposite function of CB_1 receptors on different cell types is also exemplified in the pain-transmission circuitry, where deletion of CB_1 from primary nociceptive neurons in the dorsal root ganglia proved to be pronociceptive, whereas removal from the GABA/glycinergic terminals provides protection from central hyperalgesia (Agarwal et al. 2007, Pernia-Andrade et al. 2009). Collectively, these findings demonstrate that endocannabinoid signaling at different synapses contributes to distinct behavioral components controlled by certain neuronal circuits.

CLOSING REMARKS

The ultimate mission of the life sciences in the postgenomic era of biology is to provide a full understanding of the function of all molecular players encoded in our genome together with all small-molecule metabolites composing our metabolome. In neuroscience, we will need to integrate these emerging data with a cell-type catalog of the brain and functional connectomics. One admitted expectation is that the enormous data sets generated by large-scale community efforts will support systems biology approaches to uncover conceptually new principles of biology; another major force fueling these efforts is the tremendous potential for evidence-based novel therapeutics. The unfolding of the molecular, anatomical, physiological, and behavioral features of endocannabinoid signaling in the past decade fully justifies this expectation, because it has led to our appreciation of the fundamental role of retrograde communication in the brain. In addition to the multiple functions of the endocannabinoid system, its molecular complexity and selective impairment in distinct brain disorders further fuel the hope that this system may be exploited therapeutically.

Figure 4

Quantitatively differential distribution of the molecular players of retrograde 2-arachidonoylglycerol (2-AG) signaling in some cortical areas. (*a*) Laminar distribution of diacylglycerol lipase (DGL)-α predicts association with glutamatergic synapses in the hippocampus (Katona et al. 2006). (*b*) Neurons target DGL-α preferentially into dendritic spine heads (sp) receiving asymmetric synapses (*white arrowhead*) from glutamatergic axon terminals (b_{GLU}). In contrast, dendritic shafts (d) or symmetric synapses (*black arrowhead*) formed by GABAergic boutons (b_{GABA}) are largely devoid of DGL-α-immunoreactivity (courtesy of B. Dudok, original is submitted for publication). (*c,d*) In the amygdala, a high density of DGL-α (*depicted by arrows*) occurs next to (*c*) invaginating gamma-aminobutyric acid (GABA)ergic synapses (*white arrowheads*) as well as (*d*) the neck of dendritic spines receiving glutamatergic axospinous contacts (Yoshida et al. 2011). (*e*) In situ hybridization in the hippocampus reveals high levels of CB_1 mRNA in interneurons, lower levels in CA3, and even lower levels in CA1 pyramidal cells. Dentate granule cells are devoid of labeling. (*f*) Immunostaining for the CB_1 receptor protein reveals a striking laminar pattern associated with both GABAergic and glutamatergic terminals (Katona et al. 2006). (*g–l*) Great variability in monoacylglycerol lipase (MGL) content between glutamatergic pathways is revealed by double labeling for $vGluT_1$ (*green*) and MGL (*red*). No colocalization occurs in the dentate inner molecular layer (*g,h*), but a high degree of overlap is found in recurrent axon terminals in the CA3 (*i,j*) and moderate colocalization is observed in Schaffer collateral terminals in the CA1 (*k,l*). (Uchigashima et al. 2011). White arrows in (*g–l*) point to characteristic axon terminal examples. Abbreviations: ol (stratum oriens); pl (stratum pyramidale); rl (stratum radiatum); l-ml (stratum lacunosum-moleculare); oml (outer part of stratum moleculare); iml (inner part of stratum moleculare); gl (stratum granulosum). The individual figures have been modified from the originals with permission from the authors.

DISCLOSURE STATEMENT

The authors are not aware of any affiliations, memberships, funding, or financial holding that might be perceived as affecting the objectivity of this review.

ACKNOWLEDGMENTS

The authors are grateful to M. Herkenham, R. Hargraeves, A. Howlett, M. Kano, D. Lovinger, O. Manzoni, R. Mechoulam, R. Nicoll, Y. Shim, K. Van Laere, and M. Watanabe for permission to modify figures from their original work. We thank Balázs Baksa for the artwork; Drs. Chris Henstridge, Ewen Legg, Barna Dudok, Eszter Horváth, and Balázs Pintér for help with the preparation of the manuscript; and members of the Katona and Freund labs for discussions. The authors were supported by grants from the Swiss Contribution (SH7/2/18), the Hungarian Scientific Research Fund (NK77793), the Norwegian Financial Mechanism Joint Program (NNF 78918), the European Research Council (243153), and the National Institutes of Health (MH 54671 and NS30549). I.K. is the recipient of a Wellcome Trust International Senior Research Fellowship.

LITERATURE CITED

Agarwal N, Pacher P, Tegeder I, Amaya F, Constantin CE, et al. 2007. Cannabinoids mediate analgesia largely via peripheral type 1 cannabinoid receptors in nociceptors. *Nat. Neurosci.* 10:870–79

Ahn DK, Choi HS, Yeo SP, Woo YW, Lee MK, et al. 2007. Blockade of central cyclooxygenase (COX) pathways enhances the cannabinoid-induced antinociceptive effects on inflammatory temporomandibular joint (TMJ) nociception. *Pain* 132:23–32

Albayram O, Alferink J, Pitsch J, Piyanova A, Neitzert K, et al. 2011. Role of CB1 cannabinoid receptors on GABAergic neurons in brain aging. *Proc. Natl. Acad. Sci. USA* 108:11256–61

Bell RL, Kennerly DA, Stanford N, Majerus PW. 1979. Diglyceride lipase: a pathway for arachidonate release from human platelets. *Proc. Natl. Acad. Sci. USA* 76:3238–41

Bellocchio L, Lafenetre P, Cannich A, Cota D, Puente N, et al. 2010. Bimodal control of stimulated food intake by the endocannabinoid system. *Nat. Neurosci.* 13:281–83

Berdyshev EV, Schmid PC, Krebsbach RJ, Schmid HH. 2001. Activation of PAF receptors results in enhanced synthesis of 2-arachidonoylglycerol (2-AG) in immune cells. *FASEB J.* 15:2171–78

Best AR, Regehr WG. 2008. Serotonin evokes endocannabinoid release and retrogradely suppresses excitatory synapses. *J. Neurosci.* 28:6508–15

Bidaut-Russell M, Devane WA, Howlett AC. 1990. Cannabinoid receptors and modulation of cyclic AMP accumulation in the rat brain. *J. Neurochem.* 55:21–26

Bisogno T, Howell F, Williams G, Minassi A, Cascio MG, et al. 2003. Cloning of the first sn1-DAG lipases points to the spatial and temporal regulation of endocannabinoid signaling in the brain. *J. Cell Biol.* 163:463–68

Blankman JL, Simon GM, Cravatt BF. 2007. A comprehensive profile of brain enzymes that hydrolyze the endocannabinoid 2-arachidonoylglycerol. *Chem. Biol.* 14:1347–56

Burns HD, Van Laere K, Sanabria-Bohorquez S, Hamill TG, Bormans G, et al. 2007. [18F]MK-9470, a positron emission tomography (PET) tracer for in vivo human PET brain imaging of the cannabinoid-1 receptor. *Proc. Natl. Acad. Sci. USA* 104:9800–5

Chanda PK, Gao Y, Mark L, Btesh J, Strassle BW, et al. 2010. Monoacylglycerol lipase activity is a critical modulator of the tone and integrity of the endocannabinoid system. *Mol. Pharmacol.* 78:996–1003

Chavez AE, Chiu CQ, Castillo PE. 2010. TRPV1 activation by endogenous anandamide triggers postsynaptic long-term depression in dentate gyrus. *Nat. Neurosci.* 13:1511–18

Chen JK, Chen J, Imig JD, Wei S, Hachey DL, et al. 2008. Identification of novel endogenous cytochrome p450 arachidonate metabolites with high affinity for cannabinoid receptors. *J. Biol. Chem.* 283:24514–24

Chevaleyre V, Castillo PE. 2003. Heterosynaptic LTD of hippocampal GABAergic synapses: a novel role of endocannabinoids in regulating excitability. *Neuron* 38:461–72

Clapper JR, Moreno-Sanz G, Russo R, Guijarro A, Vacondio F, et al. 2010. Anandamide suppresses pain initiation through a peripheral endocannabinoid mechanism. *Nat. Neurosci.* 13:1265–70

Coutts AA, Anavi-Goffer S, Ross RA, MacEwan DJ, Mackie K, et al. 2001. Agonist-induced internalization and trafficking of cannabinoid CB1 receptors in hippocampal neurons. *J. Neurosci.* 21:2425–33

Cravatt BF, Giang DK, Mayfield SP, Boger DL, Lerner RA, Gilula NB. 1996. Molecular characterization of an enzyme that degrades neuromodulatory fatty-acid amides. *Nature* 384:83–87

Devane WA, Axelrod J. 1994. Enzymatic synthesis of anandamide, an endogenous ligand for the cannabinoid receptor, by brain membranes. *Proc. Natl. Acad. Sci. USA* 91:6698–701

Devane WA, Dysarz FA 3rd, Johnson MR, Melvin LS, Howlett AC. 1988. Determination and characterization of a cannabinoid receptor in rat brain. *Mol. Pharmacol.* 34:605–13

Devane WA, Hanus L, Breuer A, Pertwee RG, Stevenson LA, et al. 1992. Isolation and structure of a brain constituent that binds to the cannabinoid receptor. *Science* 258:1946–49

Dinh TP, Carpenter D, Leslie FM, Freund TF, Katona I, et al. 2002. Brain monoglyceride lipase participating in endocannabinoid inactivation. *Proc. Natl. Acad. Sci. USA* 99:10819–24

Egertova M, Giang DK, Cravatt BF, Elphick MR. 1998. A new perspective on cannabinoid signalling: complementary localization of fatty acid amide hydrolase and the CB1 receptor in rat brain. *Proc. Biol. Sci.* 265:2081–85

Ellis J, Pediani JD, Canals M, Milasta S, Milligan G. 2006. Orexin-1 receptor-cannabinoid CB1 receptor heterodimerization results in both ligand-dependent and -independent coordinated alterations of receptor localization and function. *J. Biol. Chem.* 281:38812–24

Elsohly MA, Slade D. 2005. Chemical constituents of marijuana: the complex mixture of natural cannabinoids. *Life Sci.* 78:539–48

Fan W, Ster J, Gerber U. 2010. Activation conditions for the induction of metabotropic glutamate receptor-dependent long-term depression in hippocampal CA1 pyramidal cells. *J. Neurosci.* 30:1471–75

Foldy C, Neu A, Jones MV, Soltesz I. 2006. Presynaptic, activity-dependent modulation of cannabinoid type 1 receptor-mediated inhibition of GABA release. *J. Neurosci.* 26:1465–69

Fukaya M, Uchigashima M, Nomura S, Hasegawa Y, Kikuchi H, Watanabe M. 2008. Predominant expression of phospholipase Cβ1 in telencephalic principal neurons and cerebellar interneurons, and its close association with related signaling molecules in somatodendritic neuronal elements. *Eur. J. Neurosci.* 28:1744–59

Gao Y, Vasilyev DV, Goncalves MB, Howell FV, Hobbs C, et al. 2010. Loss of retrograde endocannabinoid signaling and reduced adult neurogenesis in diacylglycerol lipase knock-out mice. *J. Neurosci.* 30:2017–24

Gaoni Y, Mechoulam R. 1964. Isolation, structure and partial synthesis of an active constituent of hashish. *J. Am. Chem. Soc.* 86:1646–47

Gaoni Y, Mechoulam R. 1967. The absolute configuration of delta-1-tetrahydrocannabinol, the major active constituent of hashish. *Tetrahedron Lett.* 12:1109–11

Gerdeman GL, Ronesi J, Lovinger DM. 2002. Postsynaptic endocannabinoid release is critical to long-term depression in the striatum. *Nat. Neurosci.* 5:446–51

Gibson HE, Edwards JG, Page RS, Van Hook MJ, Kauer JA. 2008. TRPV1 channels mediate long-term depression at synapses on hippocampal interneurons. *Neuron* 57:746–59

Glykys J, Mody I. 2007. Activation of GABAA receptors: views from outside the synaptic cleft. *Neuron* 56:763–70

Gobbi G, Bambico FR, Mangieri R, Bortolato M, Campolongo P, et al. 2005. Antidepressant-like activity and modulation of brain monoaminergic transmission by blockade of anandamide hydrolysis. *Proc. Natl. Acad. Sci. USA* 102:18620–25

Goffin K, Van Paesschen W, Van Laere K. 2011. In vivo activation of endocannabinoid system in temporal lobe epilepsy with hippocampal sclerosis. *Brain* 134:1033–40

Gomes I, Grushko JS, Golebiewska U, Hoogendoorn S, Gupta A, et al. 2009. Novel endogenous peptide agonists of cannabinoid receptors. *FASEB J.* 23:3020–29

Grueter BA, Brasnjo G, Malenka RC. 2010. Postsynaptic TRPV1 triggers cell type-specific long-term depression in the nucleus accumbens. *Nat. Neurosci.* 13:1519–25

Guggenhuber S, Monory K, Lutz B, Klugmann M. 2010. AAV vector-mediated overexpression of CB1 cannabinoid receptor in pyramidal neurons of the hippocampus protects against seizure-induced excitoxicity. *PLoS One* 5:e15707

Gulyas AI, Cravatt BF, Bracey MH, Dinh TP, Piomelli D, et al. 2004. Segregation of two endocannabinoid-hydrolyzing enzymes into pre- and postsynaptic compartments in the rat hippocampus, cerebellum and amygdala. *Eur. J. Neurosci.* 20:441–58

Haj-Dahmane S, Shen RY. 2005. The wake-promoting peptide orexin-B inhibits glutamatergic transmission to dorsal raphe nucleus serotonin neurons through retrograde endocannabinoid signaling. *J. Neurosci.* 25:896–905

Hartmann J, Dragicevic E, Adelsberger H, Henning HA, Sumser M, et al. 2008. TRPC3 channels are required for synaptic transmission and motor coordination. *Neuron* 59:392–98

Hashimotodani Y, Ohno-Shosaku T, Kano M. 2007. Presynaptic monoacylglycerol lipase activity determines basal endocannabinoid tone and terminates retrograde endocannabinoid signaling in the hippocampus. *J. Neurosci.* 27:1211–19

Hashimotodani Y, Ohno-Shosaku T, Tsubokawa H, Ogata H, Emoto K, et al. 2005. Phospholipase Cβ serves as a coincidence detector through its Ca^{2+} dependency for triggering retrograde endocannabinoid signal. *Neuron* 45:257–68

Hashimotodani Y, Ohno-Shosaku T, Yamazaki M, Sakimura K, Kano M. 2011. Neuronal protease-activated receptor 1 drives synaptic retrograde signaling mediated by the endocannabinoid 2-arachidonoylglycerol. *J. Neurosci.* 31:3104–9

Heifets BD, Castillo PE. 2009. Endocannabinoid signaling and long-term synaptic plasticity. *Annu. Rev. Physiol.* 71:283–306

Heimann AS, Gomes I, Dale CS, Pagano RL, Gupta A, et al. 2007. Hemopressin is an inverse agonist of CB1 cannabinoid receptors. *Proc. Natl. Acad. Sci. USA* 104:20588–93

Hentges ST, Low MJ, Williams JT. 2005. Differential regulation of synaptic inputs by constitutively released endocannabinoids and exogenous cannabinoids. *J. Neurosci.* 25:9746–51

Herkenham M, Lynn AB, de Costa BR, Richfield EK. 1991. Neuronal localization of cannabinoid receptors in the basal ganglia of rat. *Brain Res.* 547:267–74

Herkenham M, Lynn AB, Little MD, Johnson MR, Melvin LS, et al. 1990. Cannabinoid receptor localization in brain. *Proc. Natl. Acad. Sci. USA* 87:1932–36

Hill MN, Miller GE, Carrier EJ, Gorzalka BB, Hillard CJ. 2009. Circulating endocannabinoids and N-acyl ethanolamines are differentially regulated in major depression and following exposure to social stress. *Psychoneuroendocrinology* 34:1257–62

Hinz B, Cheremina O, Brune K. 2008. Acetaminophen (paracetamol) is a selective cyclooxygenase-2 inhibitor in man. *FASEB J.* 22:383–90

Hirvonen J, Goodwin RS, Li CT, Terry GE, Zoghbi SS, et al. 2011. Reversible and regionally selective downregulation of brain cannabinoid CB(1) receptors in chronic daily cannabis smokers. *Mol. Psychiatry*. Epub ahead of print; doi:10.1038/mp.2011.82

Hohmann AG, Suplita RL, Bolton NM, Neely MH, Fegley D, et al. 2005. An endocannabinoid mechanism for stress-induced analgesia. *Nature* 435:1108–12

Hu SS, Bradshaw HB, Chen JS, Tan B, Walker JM. 2008. Prostaglandin E2 glycerol ester, an endogenous COX-2 metabolite of 2-arachidonoylglycerol, induces hyperalgesia and modulates NFkappaB activity. *Br. J. Pharmacol.* 153:1538–49

Huestis MA, Gorelick DA, Heishman SJ, Preston KL, Nelson RA, et al. 2001. Blockade of effects of smoked marijuana by the CB1-selective cannabinoid receptor antagonist SR141716. *Arch. Gen. Psychiatry* 58:322–28

Jin W, Brown S, Roche JP, Hsieh C, Celver JP, et al. 1999. Distinct domains of the CB1 cannabinoid receptor mediate desensitization and internalization. *J. Neurosci.* 19:3773–80

Johnson JW, Ascher P. 1987. Glycine potentiates the NMDA response in cultured mouse brain neurons. *Nature* 325:529–31

Jung KM, Astarita G, Thongkham D, Piomelli D. 2011. Diacylglycerol lipase-alpha and -beta control neurite outgrowth in neuro-2a cells through distinct molecular mechanisms. *Mol. Pharmacol.* 80:60–67

Jung KM, Astarita G, Zhu C, Wallace M, Mackie K, Piomelli D. 2007. A key role for diacylglycerol lipase-alpha in metabotropic glutamate receptor-dependent endocannabinoid mobilization. *Mol. Pharmacol.* 72:612–21

Jung KM, Mangieri R, Stapleton C, Kim J, Fegley D, et al. 2005. Stimulation of endocannabinoid formation in brain slice cultures through activation of group I metabotropic glutamate receptors. *Mol. Pharmacol.* 68:1196–202

Kano M, Ohno-Shosaku T, Hashimotodani Y, Uchigashima M, Watanabe M. 2009. Endocannabinoid-mediated control of synaptic transmission. *Physiol. Rev.* 89:309–80

Katona I, Freund TF. 2008. Endocannabinoid signaling as a synaptic circuit breaker in neurological disease. *Nat. Med.* 14:923–30

Katona I, Sperlagh B, Magloczky Z, Santha E, Kofalvi A, et al. 2000. GABAergic interneurons are the targets of cannabinoid actions in the human hippocampus. *Neuroscience* 100:797–804

Katona I, Sperlagh B, Sik A, Kafalvi A, Vizi ES, et al. 1999. Presynaptically located CB1 cannabinoid receptors regulate GABA release from axon terminals of specific hippocampal interneurons. *J. Neurosci.* 19:4544–58

Katona I, Urban GM, Wallace M, Ledent C, Jung KM, et al. 2006. Molecular composition of the endocannabinoid system at glutamatergic synapses. *J. Neurosci.* 26:5628–37

Kawamura Y, Fukaya M, Maejima T, Yoshida T, Miura E, et al. 2006. The CB1 cannabinoid receptor is the major cannabinoid receptor at excitatory presynaptic sites in the hippocampus and cerebellum. *J. Neurosci.* 26:2991–3001

Kearn CS, Blake-Palmer K, Daniel E, Mackie K, Glass M. 2005. Concurrent stimulation of cannabinoid CB1 and dopamine D2 receptors enhances heterodimer formation: a mechanism for receptor cross-talk? *Mol. Pharmacol.* 67:1697–704

Kim D, Jun KS, Lee SB, Kang NG, Min DS, et al. 1997. Phospholipase C isozymes selectively couple to specific neurotransmitter receptors. *Nature* 389:290–93

Kim J, Alger BE. 2004. Inhibition of cyclooxygenase-2 potentiates retrograde endocannabinoid effects in hippocampus. *Nat. Neurosci.* 7:697–98

Kim J, Alger BE. 2010. Reduction in endocannabinoid tone is a homeostatic mechanism for specific inhibitory synapses. *Nat. Neurosci.* 13:592–600

Kim J, Isokawa M, Ledent C, Alger BE. 2002. Activation of muscarinic acetylcholine receptors enhances the release of endogenous cannabinoids in the hippocampus. *J. Neurosci.* 22:10182–91

Kim SJ, Kim YS, Yuan JP, Petralia RS, Worley PF, Linden DJ. 2003. Activation of the TRPC1 cation channel by metabotropic glutamate receptor mGluR1. *Nature* 426:285–91

Kinsey SG, Long JZ, O'Neal ST, Abdullah RA, Poklis JL, et al. 2009. Blockade of endocannabinoid-degrading enzymes attenuates neuropathic pain. *J. Pharmacol. Exp. Ther.* 330:902–10

Kleckner NW, Dingledine R. 1988. Requirement for glycine in activation of NMDA-receptors expressed in Xenopus oocytes. *Science* 241:835–37

Kola B, Farkas I, Christ-Crain M, Wittmann G, Lolli F, et al. 2008. The orexigenic effect of ghrelin is mediated through central activation of the endogenous cannabinoid system. *PLoS One* 3:e1797

Kozak KR, Rowlinson SW, Marnett LJ. 2000. Oxygenation of the endocannabinoid, 2-arachidonylglycerol, to glyceryl prostaglandins by cyclooxygenase-2. *J. Biol. Chem.* 275:33744–49

Kreitzer AC, Regehr WG. 2001. Retrograde inhibition of presynaptic calcium influx by endogenous cannabinoids at excitatory synapses onto Purkinje cells. *Neuron* 29:717–27

Kunos G, Osei-Hyiaman D, Batkai S, Sharkey KA, Makriyannis A. 2009. Should peripheral CB(1) cannabinoid receptors be selectively targeted for therapeutic gain? *Trends Pharmacol. Sci.* 30:1–7

Lafourcade M, Elezgarai I, Mato S, Bakiri Y, Grandes P, Manzoni OJ. 2007. Molecular components and functions of the endocannabinoid system in mouse prefrontal cortex. *PLoS One* 2:e709

Lafourcade M, Larrieu T, Mato S, Duffaud A, Sepers M, et al. 2011. Nutritional omega-3 deficiency abolishes endocannabinoid-mediated neuronal functions. *Nat. Neurosci.* 14:345–50

Lee SH, Foldy C, Soltesz I. 2010. Distinct endocannabinoid control of GABA release at perisomatic and dendritic synapses in the hippocampus. *J. Neurosci.* 30:7993–8000

Lee SH, Soltesz I. 2011. Requirement for CB1 but not GABAB receptors in the cholecystokinin mediated inhibition of GABA release from cholecystokinin expressing basket cells. *J. Physiol.* 589:891–902

Lee SY, Foldy C, Szabadics J, Soltesz I. 2011. Cell-type-specific CCK2 receptor signaling underlies the cholecystokinin-mediated selective excitation of hippocampal parvalbumin-positive fast-spiking basket cells. *J. Neurosci.* 31:10993–1002

Leung HT, Tseng-Crank J, Kim E, Mahapatra C, Shino S, et al. 2008. DAG lipase activity is necessary for TRP channel regulation in Drosophila photoreceptors. *Neuron* 58:884–96

Li JW, Vederas JC. 2009. Drug discovery and natural products: end of an era or an endless frontier? *Science* 325:161–65

Liu J, Wang L, Harvey-White J, Huang BX, Kim HY, et al. 2008. Multiple pathways involved in the biosynthesis of anandamide. *Neuropharmacology* 54:1–7

Liu J, Wang L, Harvey-White J, Osei-Hyiaman D, Razdan R, et al. 2006. A biosynthetic pathway for anandamide. *Proc. Natl. Acad. Sci. USA* 103:13345–50

Llano I, Leresche N, Marty A. 1991. Calcium entry increases the sensitivity of cerebellar Purkinje cells to applied GABA and decreases inhibitory synaptic currents. *Neuron* 6:565–74

Long JZ, Li W, Booker L, Burston JJ, Kinsey SG, et al. 2009a. Selective blockade of 2-arachidonoylglycerol hydrolysis produces cannabinoid behavioral effects. *Nat. Chem. Biol.* 5:37–44

Long JZ, Nomura DK, Vann RE, Walentiny DM, Booker L, et al. 2009b. Dual blockade of FAAH and MAGL identifies behavioral processes regulated by endocannabinoid crosstalk in vivo. *Proc. Natl. Acad. Sci. USA* 106:20270–75

Losonczy A, Biro AA, Nusser Z. 2004. Persistently active cannabinoid receptors mute a subpopulation of hippocampal interneurons. *Proc. Natl. Acad. Sci. USA* 101:1362–67

Lourenco J, Cannich A, Carta M, Coussen F, Mulle C, Marsicano G. 2010. Synaptic activation of kainate receptors gates presynaptic CB(1) signaling at GABAergic synapses. *Nat. Neurosci.* 13:197–204

Ludanyi A, Eross L, Czirjak S, Vajda J, Halasz P, et al. 2008. Downregulation of the CB1 cannabinoid receptor and related molecular elements of the endocannabinoid system in epileptic human hippocampus. *J. Neurosci.* 28:2976–90

Ludanyi A, Hu SS, Yamazaki M, Tanimura A, Piomelli D, et al. 2011. Complementary synaptic distribution of enzymes responsible for synthesis and inactivation of the endocannabinoid 2-arachidonoylglycerol in the human hippocampus. *Neuroscience* 174:50–63

Lujan R, Nusser Z, Roberts JD, Shigemoto R, Somogyi P. 1996. Perisynaptic location of metabotropic glutamate receptors mGluR1 and mGluR5 on dendrites and dendritic spines in the rat hippocampus. *Eur. J. Neurosci.* 8:1488–500

Maccarrone M. 2008. Good news for CB1 receptors: endogenous agonists are in the right place. *Br. J. Pharmacol.* 153:179–81

Maejima T, Hashimoto K, Yoshida T, Aiba A, Kano M. 2001. Presynaptic inhibition caused by retrograde signal from metabotropic glutamate to cannabinoid receptors. *Neuron* 31:463–75

Maresz K, Pryce G, Ponomarev ED, Marsicano G, Croxford JL, et al. 2007. Direct suppression of CNS autoimmune inflammation via the cannabinoid receptor CB1 on neurons and CB2 on autoreactive T cells. *Nat. Med.* 13:492–97

Marinelli S, Pacioni S, Bisogno T, Di Marzo V, Prince DA, et al. 2008. The endocannabinoid 2-arachidonoylglycerol is responsible for the slow self-inhibition in neocortical interneurons. *J. Neurosci.* 28:13532–41

Marrs WR, Blankman JL, Horne EA, Thomazeau A, Lin YH, et al. 2010. The serine hydrolase ABHD6 controls the accumulation and efficacy of 2-AG at cannabinoid receptors. *Nat. Neurosci.* 13:951–57

Marsicano G, Goodenough S, Monory K, Hermann H, Eder M, et al. 2003. CB1 cannabinoid receptors and on-demand defense against excitotoxicity. *Science* 302:84–88

Marsicano G, Wotjak CT, Azad SC, Bisogno T, Rammes G, et al. 2002. The endogenous cannabinoid system controls extinction of aversive memories. *Nature* 418:530–34

Matsuda LA, Lolait SJ, Brownstein MJ, Young AC, Bonner TI. 1990. Structure of a cannabinoid receptor and functional expression of the cloned cDNA. *Nature* 346:561–64

Matyas F, Urban GM, Watanabe M, Mackie K, Zimmer A, et al. 2008. Identification of the sites of 2-arachidonoylglycerol synthesis and action imply retrograde endocannabinoid signaling at both GABAergic and glutamatergic synapses in the ventral tegmental area. *Neuropharmacology* 54:95–107

Mechoulam R, Ben-Shabat S, Hanus L, Ligumsky M, Kaminski NE, et al. 1995. Identification of an endogenous 2-monoglyceride, present in canine gut, that binds to cannabinoid receptors. *Biochem. Pharmacol.* 50:83–90

Mechoulam R, Shani A, Edery H, Grunfeld Y. 1970. Chemical basis of hashish activity. *Science* 169:611–12

Mikasova L, Groc L, Choquet D, Manzoni OJ. 2008. Altered surface trafficking of presynaptic cannabinoid type 1 receptor in and out synaptic terminals parallels receptor desensitization. *Proc. Natl. Acad. Sci. USA* 105:18596–601

Monory K, Blaudzun H, Massa F, Kaiser N, Lemberger T, et al. 2007. Genetic dissection of behavioural and autonomic effects of Delta(9)-tetrahydrocannabinol in mice. *PLoS Biol.* 5:e269

Monory K, Massa F, Egertova M, Eder M, Blaudzun H, et al. 2006. The endocannabinoid system controls key epileptogenic circuits in the hippocampus. *Neuron* 51:455–66

Mukhopadhyay B, Cinar R, Yin S, Liu J, Tam J, et al. 2011. Hyperactivation of anandamide synthesis and regulation of cell-cycle progression via cannabinoid type 1 (CB1) receptors in the regenerating liver. *Proc. Natl. Acad. Sci. USA* 108:6323–28

Munro S, Thomas KL, Abu-Shaar M. 1993. Molecular characterization of a peripheral receptor for cannabinoids. *Nature* 365:61–65

Nakane S, Oka S, Arai S, Waku K, Ishima Y, et al. 2002. 2-arachidonoyl-sn-glycero-3-phosphate, an arachidonic acid-containing lysophosphatidic acid: occurrence and rapid enzymatic conversion to 2-arachidonoyl-sn-glycerol, a cannabinoid receptor ligand, in rat brain. *Arch Biochem. Biophys.* 402:51–58

Navarrete M, Araque A. 2010. Endocannabinoids potentiate synaptic transmission through stimulation of astrocytes. *Neuron* 68:113–26

Neu A, Foldy C, Soltesz I. 2007. Postsynaptic origin of CB1-dependent tonic inhibition of GABA release at cholecystokinin-positive basket cell to pyramidal cell synapses in the CA1 region of the rat hippocampus. *J. Physiol.* 578:233–47

Nevian T, Sakmann B. 2006. Spine Ca^{2+} signaling in spike-timing-dependent plasticity. *J. Neurosci.* 26:11001–13

Newpher TM, Ehlers MD. 2009. Spine microdomains for postsynaptic signaling and plasticity. *Trends Cell Biol.* 19:218–27

Nyilas R, Dudok B, Urban GM, Mackie K, Watanabe M, et al. 2008. Enzymatic machinery for endocannabinoid biosynthesis associated with calcium stores in glutamatergic axon terminals. *J. Neurosci.* 28:1058–63

Nyilas R, Gregg LC, Mackie K, Watanabe M, Zimmer A, et al. 2009. Molecular architecture of endocannabinoid signaling at nociceptive synapses mediating analgesia. *Eur. J. Neurosci.* 29:1964–78

Ohno-Shosaku T, Maejima T, Kano M. 2001. Endogenous cannabinoids mediate retrograde signals from depolarized postsynaptic neurons to presynaptic terminals. *Neuron* 29:729–38

Okamoto Y, Morishita J, Tsuboi K, Tonai T, Ueda N. 2004. Molecular characterization of a phospholipase D generating anandamide and its congeners. *J. Biol. Chem.* 279:5298–305

Oliet SH, Baimoukhametova DV, Piet R, Bains JS. 2007. Retrograde regulation of GABA transmission by the tonic release of oxytocin and endocannabinoids governs postsynaptic firing. *J. Neurosci.* 27:1325–33

Pan B, Wang W, Zhong P, Blankman JL, Cravatt BF, et al. 2011. Alterations of endocannabinoid signaling, synaptic plasticity, learning, and memory in monoacylglycerol lipase knock-out mice. *J. Neurosci.* 31:13420–30

Panikashvili D, Simeonidou C, Ben-Shabat S, Hanus L, Breuer A, et al. 2001. An endogenous cannabinoid (2-AG) is neuroprotective after brain injury. *Nature* 413:527–31

Parrish JC, Nichols DE. 2006. Serotonin 5-HT(2A) receptor activation induces 2-arachidonoylglycerol release through a phospholipase c-dependent mechanism. *J. Neurochem.* 99:1164–75

Pernia-Andrade AJ, Kato A, Witschi R, Nyilas R, Katona I, et al. 2009. Spinal endocannabinoids and CB1 receptors mediate C-fiber-induced heterosynaptic pain sensitization. *Science* 325:760–64

Pertwee RG, Howlett AC, Abood ME, Alexander SP, Di Marzo V, et al. 2010. International Union of Basic and Clinical Pharmacology. LXXIX. Cannabinoid receptors and their ligands: beyond CB and CB. *Pharmacol. Rev.* 62:588–631

Pitler TA, Alger BE. 1992. Postsynaptic spike firing reduces synaptic GABAA responses in hippocampal pyramidal cells. *J. Neurosci.* 12:4122–32

Porter AC, Sauer JM, Knierman MD, Becker GW, Berna MJ, et al. 2002. Characterization of a novel endocannabinoid, virodhamine, with antagonist activity at the CB1 receptor. *J. Pharmacol. Exp. Ther.* 301:1020–24

Prisinzano TE. 2009. Natural products as tools for neuroscience: discovery and development of novel agents to treat drug abuse. *J. Nat. Prod.* 72:581–87

Prusakiewicz JJ, Duggan KC, Rouzer CA, Marnett LJ. 2009. Differential sensitivity and mechanism of inhibition of COX-2 oxygenation of arachidonic acid and 2-arachidonoylglycerol by ibuprofen and mefenamic acid. *Biochemistry* 48:7353–55

Puighermanal E, Marsicano G, Busquets-Garcia A, Lutz B, Maldonado R, Ozaita A. 2009. Cannabinoid modulation of hippocampal long-term memory is mediated by mTOR signaling. *Nat. Neurosci.* 12:1152–58

Rademacher DJ, Patel S, Ho WS, Savoie AM, Rusch NJ, et al. 2005. U-46619 but not serotonin increases endocannabinoid content in middle cerebral artery: evidence for functional relevance. *Am. J. Physiol. Heart Circ. Physiol.* 288:H2694–701

Rapoport SI. 2008. Arachidonic acid and the brain. *J. Nutr.* 138:2515–20

Regehr WG, Carey MR, Best AR. 2009. Activity-dependent regulation of synapses by retrograde messengers. *Neuron* 63:154–70

Rios C, Gomes I, Devi LA. 2006. mu opioid and CB1 cannabinoid receptor interactions: reciprocal inhibition of receptor signaling and neuritogenesis. *Br. J. Pharmacol.* 148:387–95

Robbe D, Kopf M, Remaury A, Bockaert J, Manzoni OJ. 2002. Endogenous cannabinoids mediate long-term synaptic depression in the nucleus accumbens. *Proc. Natl. Acad. Sci. USA* 99:8384–88

Roloff AM, Anderson GR, Martemyanov KA, Thayer SA. 2010. Homer 1a gates the induction mechanism for endocannabinoid-mediated synaptic plasticity. *J. Neurosci.* 30:3072–81

Schlosburg JE, Blankman JL, Long JZ, Nomura DK, Pan B, et al. 2010. Chronic monoacylglycerol lipase blockade causes functional antagonism of the endocannabinoid system. *Nat. Neurosci.* 13:1113–19

Shim JY. 2009. Transmembrane helical domain of the cannabinoid CB1 receptor. *Biophys. J.* 96:3251–62

Shimizu T, Lu L, Yokotani K. 2011. Endogenously generated 2-arachidonoylglycerol plays an inhibitory role in bombesin-induced activation of central adrenomedullary outflow in rats. *Eur. J. Pharmacol.* 658:123–31

Simon GM, Cravatt BF. 2006. Endocannabinoid biosynthesis proceeding through glycerophospho-N-acyl ethanolamine and a role for alpha/beta-hydrolase 4 in this pathway. *J. Biol. Chem.* 281:26465–72

Simon GM, Cravatt BF. 2008. Anandamide biosynthesis catalyzed by the phosphodiesterase GDE1 and detection of glycerophospho-N-acyl ethanolamine precursors in mouse brain. *J. Biol. Chem.* 283:9341–49

Sjostrom PJ, Turrigiano GG, Nelson SB. 2003. Neocortical LTD via coincident activation of presynaptic NMDA and cannabinoid receptors. *Neuron* 39:641–54

Smith PB, Compton DR, Welch SP, Razdan RK, Mechoulam R, Martin BR. 1994. The pharmacological activity of anandamide, a putative endogenous cannabinoid, in mice. *J. Pharmacol. Exp. Ther.* 270:219–27

Stell BM, Brickley SG, Tang CY, Farrant M, Mody I. 2003. Neuroactive steroids reduce neuronal excitability by selectively enhancing tonic inhibition mediated by delta subunit-containing GABAA receptors. *Proc. Natl. Acad. Sci. USA* 100:14439–44

Stella N, Schweitzer P, Piomelli D. 1997. A second endogenous cannabinoid that modulates long-term potentiation. *Nature* 388:773–78

Straiker A, Wager-Miller J, Hu SS, Blankman JL, Cravatt BF, Mackie K. 2011. COX-2 and FAAH can regulate the time course of depolarization induced suppression of excitation. *Br. J. Pharmacol.* 164:1672–83

Sudhof TC, Malenka RC. 2008. Understanding synapses: past, present, and future. *Neuron* 60:469–76

Sugiura T, Kondo S, Sukagawa A, Nakane S, Shinoda A, et al. 1995. 2-arachidonoylglycerol: a possible endogenous cannabinoid receptor ligand in brain. *Biochem. Biophys. Res. Commun.* 215:89–97

Sun YX, Tsuboi K, Okamoto Y, Tonai T, Murakami M, et al. 2004. Biosynthesis of anandamide and N-palmitoylethanolamine by sequential actions of phospholipase A2 and lysophospholipase D. *Biochem. J.* 380:749–56

Tanimura A, Yamazaki M, Hashimotodani Y, Uchigashima M, Kawata S, et al. 2010. The endocannabinoid 2-arachidonoylglycerol produced by diacylglycerol lipase alpha mediates retrograde suppression of synaptic transmission. *Neuron* 65:320–27

Tappe-Theodor A, Agarwal N, Katona I, Rubino T, Martini L, et al. 2007. A molecular basis of analgesic tolerance to cannabinoids. *J. Neurosci.* 27:4165–77

Telleria-Diaz A, Schmidt M, Kreusch S, Neubert AK, Schache F, et al. 2010. Spinal antinociceptive effects of cyclooxygenase inhibition during inflammation: involvement of prostaglandins and endocannabinoids. *Pain* 148:26–35

Tempia F, Miniaci MC, Anchisi D, Strata P. 1998. Postsynaptic current mediated by metabotropic glutamate receptors in cerebellar Purkinje cells. *J. Neurophysiol.* 80:520–28

Toth K, McBain CJ. 1998. Afferent-specific innervation of two distinct AMPA receptor subtypes on single hippocampal interneurons. *Nat. Neurosci.* 1:572–78

Tsou K, Brown S, Sanudo-Pena MC, Mackie K, Walker JM. 1998. Immunohistochemical distribution of cannabinoid CB1 receptors in the rat central nervous system. *Neuroscience* 83:393–411

Tsutsumi T, Kobayashi T, Ueda H, Yamauchi E, Watanabe S, Okuyama H. 1994. Lysophosphoinositide-specific phospholipase C in rat brain synaptic plasma membranes. *Neurochem. Res.* 19:399–406

Turrigiano G. 2007. Homeostatic signaling: the positive side of negative feedback. *Curr. Opin. Neurobiol.* 17:318–24

Turu G, Varnai P, Gyombolai P, Szidonya L, Offertaler L, et al. 2009. Paracrine transactivation of the CB1 cannabinoid receptor by AT1 angiotensin and other Gq/11 protein–coupled receptors. *J. Biol. Chem.* 284:16914–21

Uchigashima M, Narushima M, Fukaya M, Katona I, Kano M, Watanabe M. 2007. Subcellular arrangement of molecules for 2-arachidonoyl-glycerol-mediated retrograde signaling and its physiological contribution to synaptic modulation in the striatum. *J. Neurosci.* 27:3663–76

Uchigashima M, Yamazaki M, Yamasaki M, Tanimura A, Sakimura K, et al. 2011. Molecular and morphological configuration for 2-arachidonoylglycerol-mediated retrograde signaling at mossy cell–granule cell synapses in the dentate gyrus. *J. Neurosci.* 31:7700–14

Van Laere K, Casteels C, Dhollander I, Goffin K, Grachev I, et al. 2010. Widespread decrease of type 1 cannabinoid receptor availability in Huntington disease in vivo. *J. Nucl. Med.* 51:1413–17

Van Sickle MD, Duncan M, Kingsley PJ, Mouihate A, Urbani P, et al. 2005. Identification and functional characterization of brainstem cannabinoid CB2 receptors. *Science* 310:329–32

Varma N, Carlson GC, Ledent C, Alger BE. 2001. Metabotropic glutamate receptors drive the endocannabinoid system in hippocampus. *J. Neurosci.* 21:RC188

Vizi ES, Katona I, Freund TF. 2001. Evidence for presynaptic cannabinoid CB(1) receptor-mediated inhibition of noradrenaline release in the guinea pig lung. *Eur. J. Pharmacol.* 431:237–44

Watanabe H, Vriens J, Prenen J, Droogmans G, Voets T, Nilius B. 2003a. Anandamide and arachidonic acid use epoxyeicosatrienoic acids to activate TRPV4 channels. *Nature* 424:434–38

Watanabe S, Doshi M, Hamazaki T. 2003b. n-3 polyunsaturated fatty acid (PUFA) deficiency elevates and n-3 PUFA enrichment reduces brain 2-arachidonoylglycerol level in mice. *Prostaglandins Leukot. Essent. Fatty Acids* 69:51–59

Wettschureck N, van der Stelt M, Tsubokawa H, Krestel H, Moers A, et al. 2006. Forebrain-specific inactivation of Gq/G11 family G proteins results in age-dependent epilepsy and impaired endocannabinoid formation. *Mol. Cell. Biol.* 26:5888–94

Williams JH, Errington ML, Lynch MA, Bliss TV. 1989. Arachidonic acid induces a long-term activity-dependent enhancement of synaptic transmission in the hippocampus. *Nature* 341:739–42

Wilson RI, Nicoll RA. 2001. Endogenous cannabinoids mediate retrograde signalling at hippocampal synapses. *Nature* 410:588–92

Won YJ, Puhl HL 3rd, Ikeda SR. 2009. Molecular reconstruction of mGluR5a-mediated endocannabinoid signaling cascade in single rat sympathetic neurons. *J. Neurosci.* 29:13603–12

Xi ZX, Peng XQ, Li X, Song R, Zhang HY, et al. 2011. Brain cannabinoid CB_2 receptors modulate cocaine's actions in mice. *Nat. Neurosci.* 14:1160–66

Yamasaki M, Matsui M, Watanabe M. 2010. Preferential localization of muscarinic M1 receptor on dendritic shaft and spine of cortical pyramidal cells and its anatomical evidence for volume transmission. *J. Neurosci.* 30:4408–18

Yang H, Zhang J, Andreasson K, Chen C. 2008. COX-2 oxidative metabolism of endocannabinoids augments hippocampal synaptic plasticity. *Mol. Cell Neurosci.* 37:682–95

Yoshida T, Fukaya M, Uchigashima M, Miura E, Kamiya H, et al. 2006. Localization of diacylglycerol lipase-alpha around postsynaptic spine suggests close proximity between production site of an endocannabinoid, 2-arachidonoyl-glycerol, and presynaptic cannabinoid CB1 receptor. *J. Neurosci.* 26:4740–51

Yoshida T, Uchigashima M, Yamasaki M, Katona I, Yamazaki M, et al. 2011. Unique inhibitory synapse with particularly rich endocannabinoid signaling machinery on pyramidal neurons in basal amygdaloid nucleus. *Proc. Natl. Acad. Sci. USA* 108:3059–64

Yoshino H, Miyamae T, Hansen G, Zambrowicz B, Flynn M, et al. 2011. Postsynaptic diacylglycerol lipase alpha mediates retrograde endocannabinoid suppression of inhibition in mouse prefrontal cortex. *J. Physiol.* 589:4857–84

Yu M, Ives D, Ramesha CS. 1997. Synthesis of prostaglandin E2 ethanolamide from anandamide by cyclooxygenase-2. *J. Biol. Chem.* 272:21181–86

Yuan JP, Kiselyov K, Shin DM, Chen J, Shcheynikov N, et al. 2003. Homer binds TRPC family channels and is required for gating of TRPC1 by IP3 receptors. *Cell* 114:777–89

Zampronio AR, Kuzmiski JB, Florence CM, Mulligan SJ, Pittman QJ. 2010. Opposing actions of endothelin-1 on glutamatergic transmission onto vasopressin and oxytocin neurons in the supraoptic nucleus. *J. Neurosci.* 30:16855–63

Zhang SY, Xu M, Miao QL, Poo MM, Zhang XH. 2009. Endocannabinoid-dependent homeostatic regulation of inhibitory synapses by miniature excitatory synaptic activities. *J. Neurosci.* 29:13222–31

Circuits for Skilled Reaching and Grasping

Bror Alstermark[1] and Tadashi Isa[2]

[1]Department of Integrative Medical Biology, Section of Physiology, Umeå University, S-901 87 Umeå, Sweden; email: Bror.Alstermark@physiol.umu.se

[2]Department of Developmental Physiology, National Institute for Physiological Sciences, Myodaiji, Okazaki 444-8585, Japan; email: tisa@nips.ac.jp

Annu. Rev. Neurosci. 2012. 35:559–78

First published online as a Review in Advance on April 9, 2012

The *Annual Review of Neuroscience* is online at neuro.annualreviews.org

This article's doi: 10.1146/annurev-neuro-062111-150527

0147-006X/12/0721-0559$20.00

Keywords

corticomotoneuronal pathways, motoneuron, propriospinal neuron, segmental interneuron

Abstract

From an evolutionary perspective, it is clear that basic motor functions such as locomotion and posture are largely controlled by neural circuitries residing in the spinal cord and brain-stem. The control of voluntary movements such as skillful reaching and grasping is generally considered to be governed by neural circuitries in the motor cortex that connect directly to motoneurons via the corticomotoneuronal (CM) pathway. The CM pathway may act together with several brainstem systems that also act directly with motoneurons. This simple view was challenged by work in the cat, which lacks the direct CM system, showing that the motor commands for reaching and grasping could be mediated via spinal interneurons with input from the motor-cortex and brain-stem systems. It was further demonstrated that the spinal interneurons mediating the descending commands for reaching and grasping constitute separate and distinct populations from those involved in locomotion and posture. The aim of this review is to describe populations of spinal interneurons that are involved in the control of skilled reaching and grasping in the cat, monkey, and human.

Contents

INTRODUCTION

Neural circuits for skilled reaching and grasping are considered to be mainly located in the motor cortex and exert their effects in motoneurons (MNs) via the "direct" corticomotoneuronal (CM) pathway (**Figure 1*a*,*b***). This 60-year-old idea has been firmly established because the direct CM pathway is lacking in lower species and becomes more prominent in primates together with a parallel increase in manual dexterity (Bernhard & Bohm 1954). However, most of the descending control from the motor cortex to MNs is mediated via "indirect" CM pathways with intercalated neurons in the brain stem and spinal cord. The general understanding of interneuronal (IN) circuits in the brain stem and spinal cord is that they mainly control automatic movements such as locomotion and scratching in addition to posture, reflexes, and tonus (**Figure 1*a***) and are of less importance for skilled movements. In classical behavioral studies in the macaque monkey using pyramidal lesions at the brain-stem level (Lawrence & Kuypers 1968a, 1968b), it was concluded that highly fractionated finger movements in primates primarily depend on the direct CM pathway. From these studies, researchers proposed that a major advantage of the direct CM system was its ability to bypass the INs, which were considered to have too widespread connections with MNs innervating distal hand muscles to be able to mediate the command for independent digit movements (Kuypers 1982). As a result, researchers recently further suggested that the direct CM pathway has replaced one of the indirect CM pathways (Lemon 2008). However, in this review, we describe specialized neural circuits of spinal pre-MN centers controlling reaching and grasping. The focus is on the functional organization of these circuits, with less attention to the details regarding synaptic effects. For the latter, see reviews by Alstermark & Lundberg (1992), Alstermark et al. (2007), and Isa et al. (2007).

Sherrington (1906) together with his students laid the foundation for circuit analysis in the spinal cord with classical work on the organization of the stretch reflex, reciprocal inhibition, scratch reflex, flexor and crossed extensor reflexes (Creed et al. 1932), and locomotion (Brown 1911) as shown in **Figure 1*a***. Although such circuits included the MNs as the final common path (**Figure 1*b***), Sherrington also included INs in the common path (**Figure 1*c***) in the sense that it represented the "highest degree of communism" (Sherrington 1906). The great advantage of INs is that a major integration of sensory and descending signals can be made at a stage just prior to the MNs (Lundberg 1999). Descending supraspinal systems may facilitate or inhibit the signals from the primary afferents to the MNs. By using either excitatory or inhibitory INs, the sign of an effect can be reversed from excitation to inhibition or the reverse (disinhibition). We refer to INs with direct monosynaptic projection to MNs as last-order INs.

Modern investigations of the neuronal organization of descending pathways began

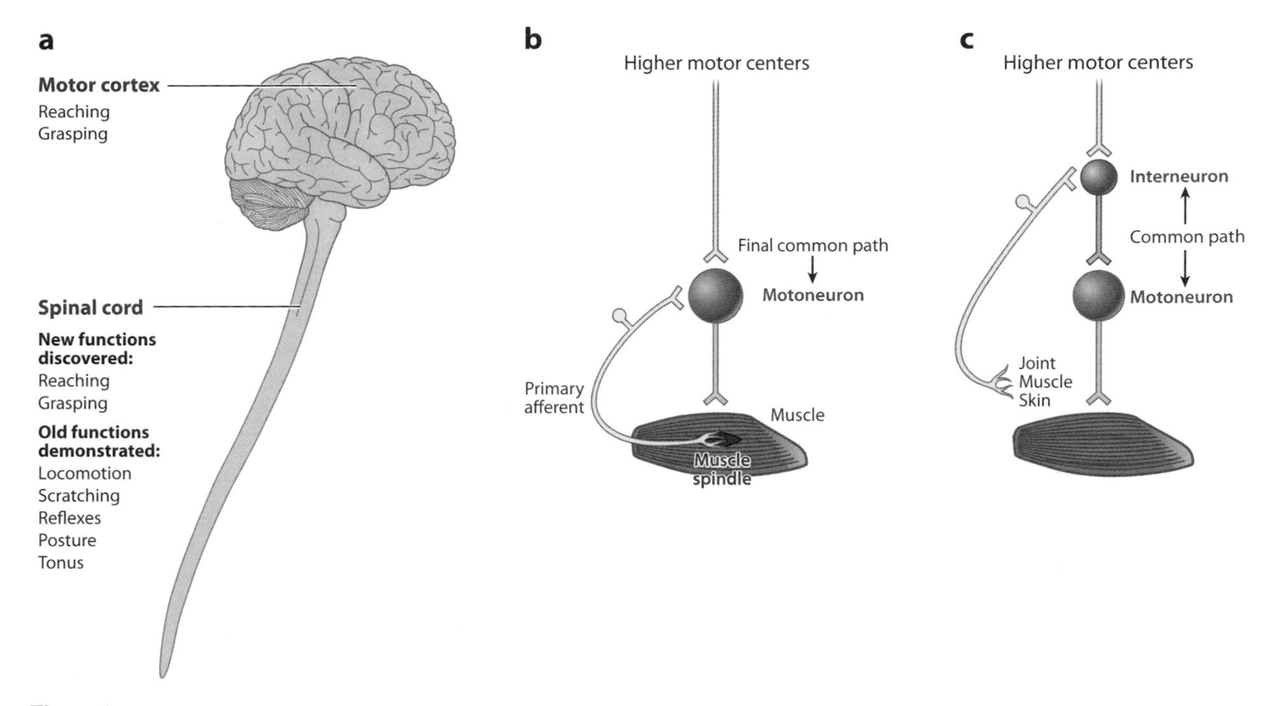

Figure 1

Motor principles. (*a*) Skilled voluntary movements such as reaching and grasping are commonly considered to be dependent on neural circuits mainly located in the motor cortex, whereas stereotyped motor behavior such as locomotion, scratching, reflexes, posture, and tonus are controlled by spinal circuits. However, as described in this review, the descending command for skilled reaching and grasping can also be controlled by spinal circuits. (*b*) Schematic neural diagram showing monosynaptic input from muscle spindle Ia afferents and descending pathways from higher motor centers in the cortex and brain stem. Sherrington (1906) considered motoneurons to be the "final common path." (*c*) Schematic neural diagram showing disynaptic input, via an interneuron, from afferents and higher motor centers to motoneurons. Sherrington (1906) also included the interneurons together with the motoneurons in the "common path."

with Lloyd's (1941) study of pyramidal effects in the cat spinal cord. He used the technique of conditioning monosynaptic test reflexes (Renshaw 1940) and showed that temporal facilitation was required to evoke pyramidal excitation in MNs, suggesting that intercalated neurons also mediated the effect for descending systems (illustrated in **Figure 2*a***). A similar approach was used by Lundberg & Voorhoeve (1962), who recorded intracellularly from MNs and found facilitation from the sensorimotor cortex of excitatory and inhibitory segmental reflexes. The final proof of a disynaptic excitatory CM pathway was provided in experiments by Lundberg and colleagues recording from forelimb MNs in the cat (Illert et al. 1977). They showed that disynaptic excitation could be mediated by a group of INs located in the upper cervical spinal cord segments C3-C4 and termed them C3-C4 propriospinal neurons (PNs) as shown in **Figure 2*b***. The C3-C4 PNs are characterized by an extensive convergence from several descending pathways: the cortico-, rubro-, reticulo-, and tectospinal tracts. These descending systems may also give feed-forward inhibition of subpopulations of PNs via feedforward inhibitory INs (**Figure 2*b***). Whereas the dominating descending input is excitatory, the input from forelimb afferents is predominantly inhibitory via feedback inhibitory INs. In addition to the MN projection, the C3-C4 PNs have an ascending projection to the lateral reticular nucleus (LRN). The LRN mediates the major mossy-fiber input to the cerebellum from the spinal cord. The ascending LRN projection may thus provide the cerebellum with an efferent copy of the command to the MNs (see below; **Figure 6**). On the basis of this information, the cerebellum may then interact quickly with the cortical and brain-stem systems to regulate activity in the C3-C4 PNs (see below). Some inhibitory C3-C4 PNs can also mediate disynaptic inhibition in MNs (**Figure 2*b***). These inhibitory PNs have the same extensive descending convergence as the excitatory ones, and they also have an ascending projection to the LRN (see references in Alstermark et al. 2007).

Disynaptic CM excitation can also be mediated via INs located in the same segments as the MNs (Illert & Wiedeman 1984, Alstermark & Sasaki 1985, Kitazawa et al. 1993). Such INs are termed segmental interneurons (sINs) and are shown in **Figure 3**. To visualize the relative contributions of PNs and sINs to reaching and grasping, two groups of cats were compared using activity-dependent transneuronal uptake of wheat-germ agglutinated horseradish peroxidase (WGA-HRP) into last-order INs following injection into the motor axons of the deltoid muscle (Alstermark & Kümmel

a Electrophysiological technique for analyzing circuitry

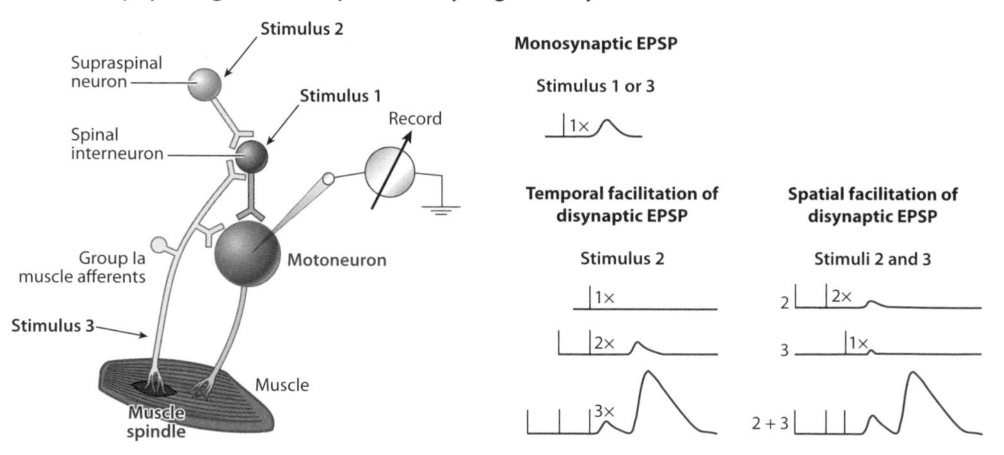

b The C3–C4 propriospinal system

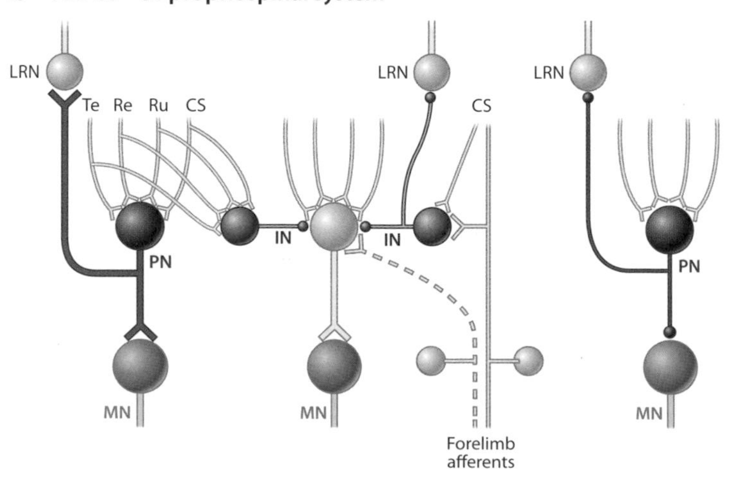

1990). Investigators pursued this study because the deltoid MNs are active in the reaching, grasping, as well as locomoting groups, but they may also be differentially controlled. One group performed unrestrained locomotion, and the other combined target reaching and grasping to remove a morsel of food from a small tube with the forelimb. As shown in **Figure 3**, reaching consists of rapid elbow flexion followed by combined shoulder and elbow extension. During this lifting and protraction phase of reaching, the forepaw is dorsi-flexed and the toes adducted just prior to the insertion into the tube. Inside the tube, the digits are abducted and flexed to grasp the morsel of food. After grasping, the forepaw is supinated during retraction out from the tube and then brought to the mouth (Alstermark et al. 1981b).

Figure 3 also shows the transverse and longitudinal distributions of last-order labeled spinal INs. In the locomoting group, only a few labeled cells were found in the C2-C4 segments, whereas many labeled cells were located in C5-Th1 with two peaks in the C6 and C7 segments. In contrast, labeled cells were found from C2 to Th1 in the cats performing reaching and grasping with a peak in the C6 segment. These results suggest that spinal INs devoted to the control of reaching and grasping exist all the way from C2 to Th1, whereas INs involved in the generation of locomotion are mainly confined to the C5-Th1 segments. Why the spinal INs involved in the control of reaching and grasping are also distributed rostral to the forelimb segments is still not known. The sINs are not as well characterized with respect to the descending systems; also unclear is whether they project to the LRN.

Although transneuronal labeling of last-order INs with WGA-HRP revealed a greater number of sINs than C3-C4 PNs (Alstermark & Kümmel 1990), the synaptic strength is similar for the CM pathway via the C3-C4 PNs (Alstermark & Sasaki 1985). Apparently, the sINs constitute heterogeneous groups, with only a part receiving strong corticospinal input. Disynaptic CM pathways via C3-C4 PNs have also been shown in the macaque monkey (Alstermark et al. 1999) and in humans (Malmgren & Pierrot-Deseilligny 1988), but they are absent in the rat (Alstermark et al. 2004) and mouse (Alstermark & Ogawa 2004). As illustrated in **Figure 3**, corticoreticulospinal (Peterson et al. 1979, Alstermark & Sasaki 1986, Riddle et al. 2009) and corticorubrospinal (Fujito & Aoki 1995) pathways can also mediate disynaptic CM excitation; these pathways constitute phylogenetically old

Figure 2

Electrophysiological techniques for analyzing circuitry. The excitability of motoneurons (MNs) can be tested by using the monosynaptic Ia reflex pathway as described by Renshaw (1940) and Lloyd (1941). Although this is usually technically easier than stimulating last-order interneurons (INs) directly, researchers were able to stimulate last-order INs in case of the C3-C4 propriospinal neurons (PNs) because of their ascending projection to the lateral reticular nucleus (LRN) (Alstermark et al. 1981a) (not shown in figure). However, last-order INs can also be activated synaptically, as in the case of the C3-C4 PN, by stimulating one of their descending inputs, e.g., the corticospinal tract (indicated here as a supraspinal neuron). Such a disynaptic pathway usually requires temporal summation of excitatory postsynaptic potentials (EPSPs) in the intercalated neurons (in this case, the C3-C4 PNs) to make them fire. When firing, a disynaptic EPSP can be recorded in the MNs. To investigate the convergence onto common intercalated spinal INs, two different inputs to the last-order INs can be stimulated together, producing spatial facilitation indicated by the large EPSP in the MNs. Within the C3-C4 propriospinal system, PNs receive monosynaptic excitation from the cortico-, rubro-, reticulo-, and tectospinal tracts. Via activation from these descending systems, the PNs may evoke disynaptic excitation in forelimb MNs. In addition, the PNs have ascending collaterals to the LRN. All the descending systems may evoke disynaptic inhibition in the PNs, denoted as feed-forward inhibition. There is only a weak and sparse monosynaptic excitation from the forelimb afferents. The major effect from the forelimb afferents is disynaptic inhibition, denoted as feedback inhibition. In addition to the excitatory PNs, a population of inhibitory PNs has a descending convergence similar to that of excitatory PNs. Inhibitory PNs, as well as the inhibitory INs mediating feedback inhibition, also project to the LRN.

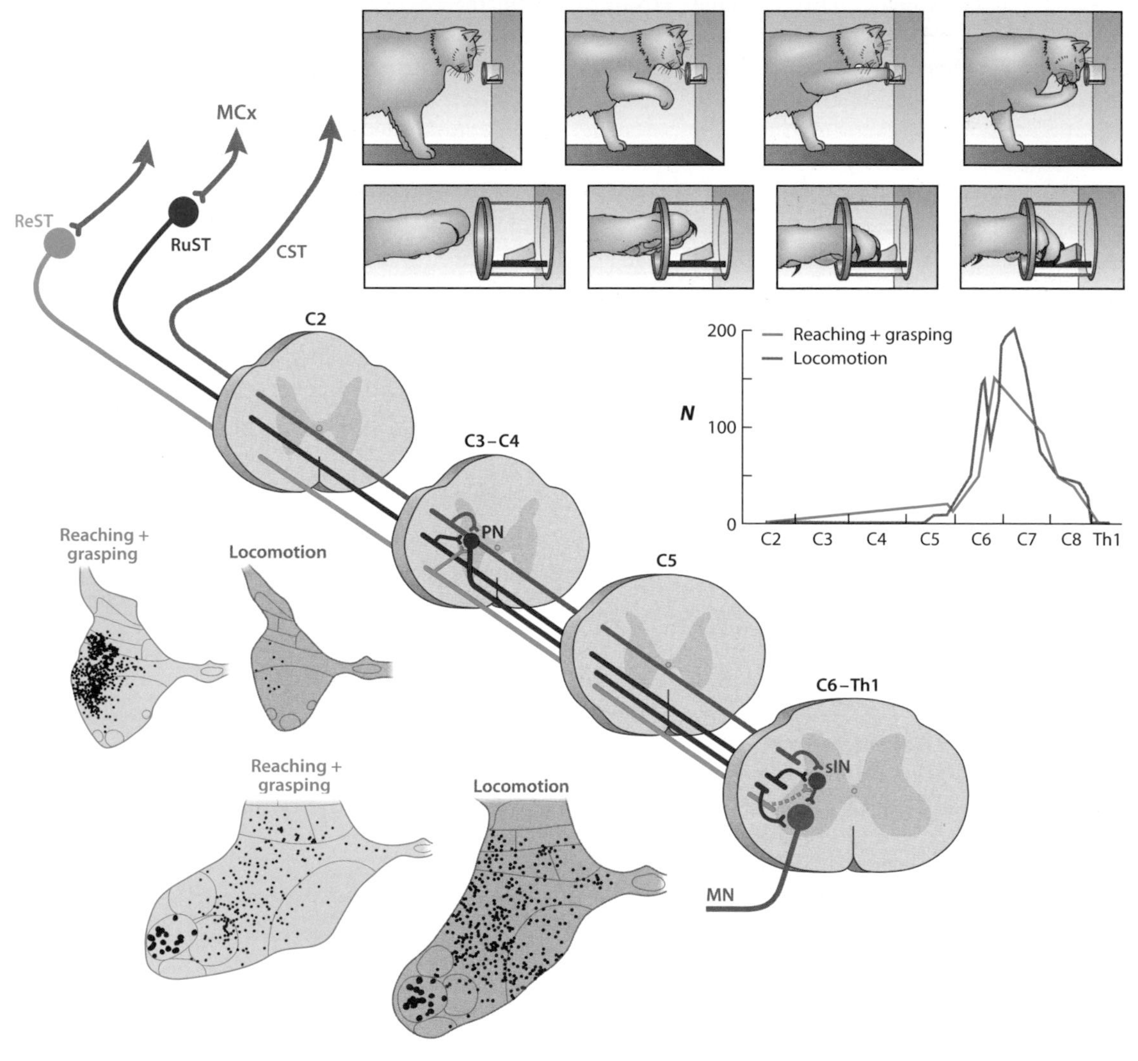

Figure 3

Longitudinal and transverse location of last-order spinal interneurons (INs) active during combined reaching and grasping or locomotion. (*Center diagonal image*) The organization of descending pathways controlling propriospinal and segmental interneurons (PNs and sINs, respectively); (*upper right corner*) the reaching and grasping movement sequences (from Alstermark et al. 1981b) and the longitudinal distribution of labeled INs in two groups of cats performing either combined reaching and grasping (*light blue line*) or unrestrained overground locomotion (*dark yellow line*); (*lower left corner*) the transverse distribution of labeled INs for the two groups of cats. The detailed images of the paw are courtesy of L.G. Pettersson. Wheat-germ agglutinated horseradish peroxidase was injected into the motor nerve of the deltoid muscle and was then taken up transneuronally from the motoneurons (MN) into last-order INs in an activity-dependent manner. The direct corticomotoneuronal (CM) pathway and monosynaptic MN connections from the rubrospinal (RuST) and reticulospinal (ReST) tracts are not shown. Other abbreviation: CST, corticospinal tract. Adapted with permission from Alstermark & Kümmel (1990).

systems (Shapovalov 1975). In primates, there is also a direct CM pathway (Bernhard & Bohm 1954, Landgren et al. 1962). It appears that an evolution of the CM pathways has taken place via indirect corticobulbospinal and corticopropriospinal pathways as well as a direct CM pathway. We describe evidence showing how these direct and indirect CM pathways are used to control reaching and grasping.

REACHING AND GRASPING IN THE CAT AND MACAQUE MONKEY

The above experiments using WGA-HRP suggest that spinal INs may be active during both reaching and grasping. To investigate the relative contribution of C3-C4 PNs and sINs to reaching and grasping, the corticospinal and rubrospinal tracts were lesioned in C5 in the cat (Alstermark et al. 1981a, Alstermark & Sasaki 1985) and the macaque monkey (Alstermark et al. 1999, Sasaki et al. 2004, Alstermark et al. 2011) (**Figure 4**). This lesion eliminates the cortico- and rubrospinal inputs to the sINs but spares them to the C3-C4 PNs. The lesions were confirmed electrophysiologically. In the cat, intracellular recording from MNs showed that the disynaptic pyramidal excitatory postsynaptic potential (EPSP) could still be evoked after the C5 lesion, although the amplitude was decreased owing to the interruption of the input to the sINs. In the cat, fast and accurate reaching of the forelimb could be observed during the first postoperative week (**Figure 4**). However, after the forepaw was inserted into the tube, the cat could not use its digits to grasp the morsel of food and instead made a raking movement. In the monkey, the lesion interrupted the inputs to the sINs as well as the direct CM pathways. In this species, after a complete lesion to the corticospinal tract that spared the majority of the propriospinal axons, both reaching to the tube and the precision grip formed by flexion and opposition of the index finger and thumb, which were observed preoperatively, could be performed during the first postoperative week (see **Supplemental Movie 1**. Follow the **Supplemental Material link** from the Annual Reviews home page at **http://www.annualreviews.org**). Transient deficits during the first two weeks were seen in the preshaping of the digits before insertion and coflexion of the other digits. These findings suggest that, in the cat, C3-C4 PNs can mediate the command for reaching, whereas the command for grasping requires sINs. In contrast, in the macaque monkey, the C3-C4 PNs seem capable of mediating the commands for reaching and grasping. The early deficits in digit control indicate that the sINs and the direct CM pathway play a critical role in preshaping and also in fractionated digit movements.

To test whether the reaching that remained in the cat and the reaching and grasping in the macaque monkey could be ascribed to transmission via the C3-C4 PNs or via the ventral corticoreticulospinal system, a dorsal lesion was made in C2 as shown in **Figure 5**. This lesion interrupts the cortico- and rubrospinal tracts to the C3-C4 PNs as well as to the sINs, but it spares the corticoreticular input. This lesion abolished the disynaptic pyramidal EPSP in forelimb MNs in both the cat and monkey; it also abolished the monosynaptic pyramidal EPSP in the monkey. Both reaching and grasping were severely impaired in the cat and monkey. In both species, reaching toward the tube could still be initiated, but precision was impaired in aiming toward the tube opening and in the coordination between joints. Although reaching recovered after approximately one month, fine-digit grasping was still abolished (see **Supplemental Movie 2**).

Taken together, the results following the dorsal C5 and C2 lesions show that C3-C4 PNs can mediate the command for reaching in the cat and for reaching and grasping in the macaque monkey. A lesion of the propriospinal axons made by transecting the ventral part of the lateral funiculus in C5 in the cat (Alstermark et al. 1981b, Pettersson et al. 1997) resulted in ataxia during reaching near the target (see section below) (**Figure 6*d***), even though grasping was still intact. On the basis

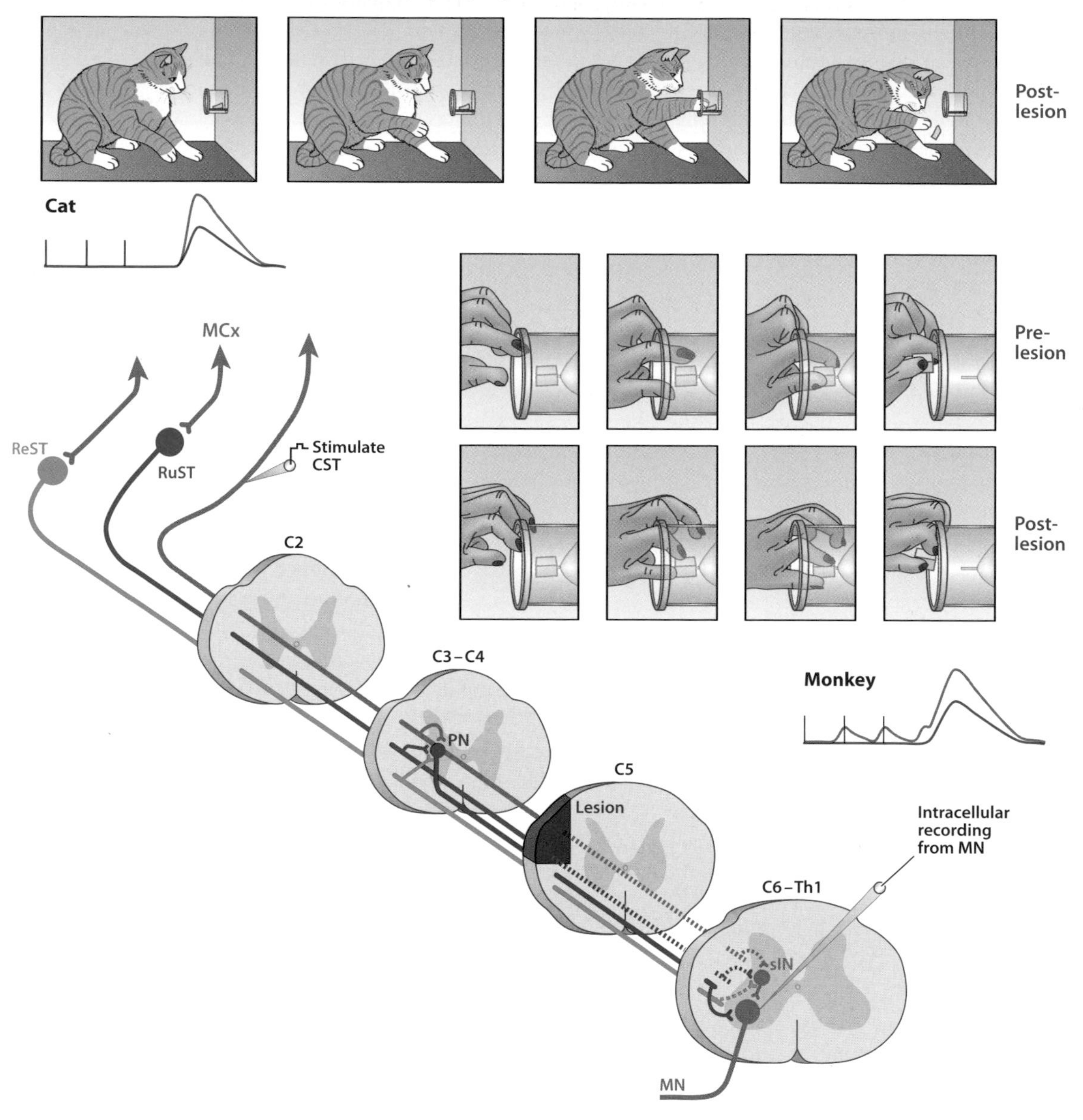

Figure 4

Behavioral and electrophysiological testing of propriospinal versus segmental interneurons (PNs and sINs, respectively) in the control of reaching and grasping after lesion of the cortico- and rubrospinal tracts in the C5 segment. The transection of these pathways (*center diagonal image*; *red*) eliminated the input to the sINs but spared the input to the C3-C4 PNs. In the cat (*top panels*), reaching was normal, whereas grasping with the digits was abolished. In the monkey (*middle panels*), reaching was normal and precision grip could be performed with some deficits in preshaping and conjoint extension in the other digits: (*middle upper row*) preoperative condition and (*middle lower row*) postoperative day 1. Abbreviations: CST, corticospinal tract; MCx, motor cortex; MN, motoneuron; ReST, reticulospinal tract; RuST, rubrospinal tract. Adapted from Alstermark et al. (1981b) and Sasaki et al. (2004) with permission.

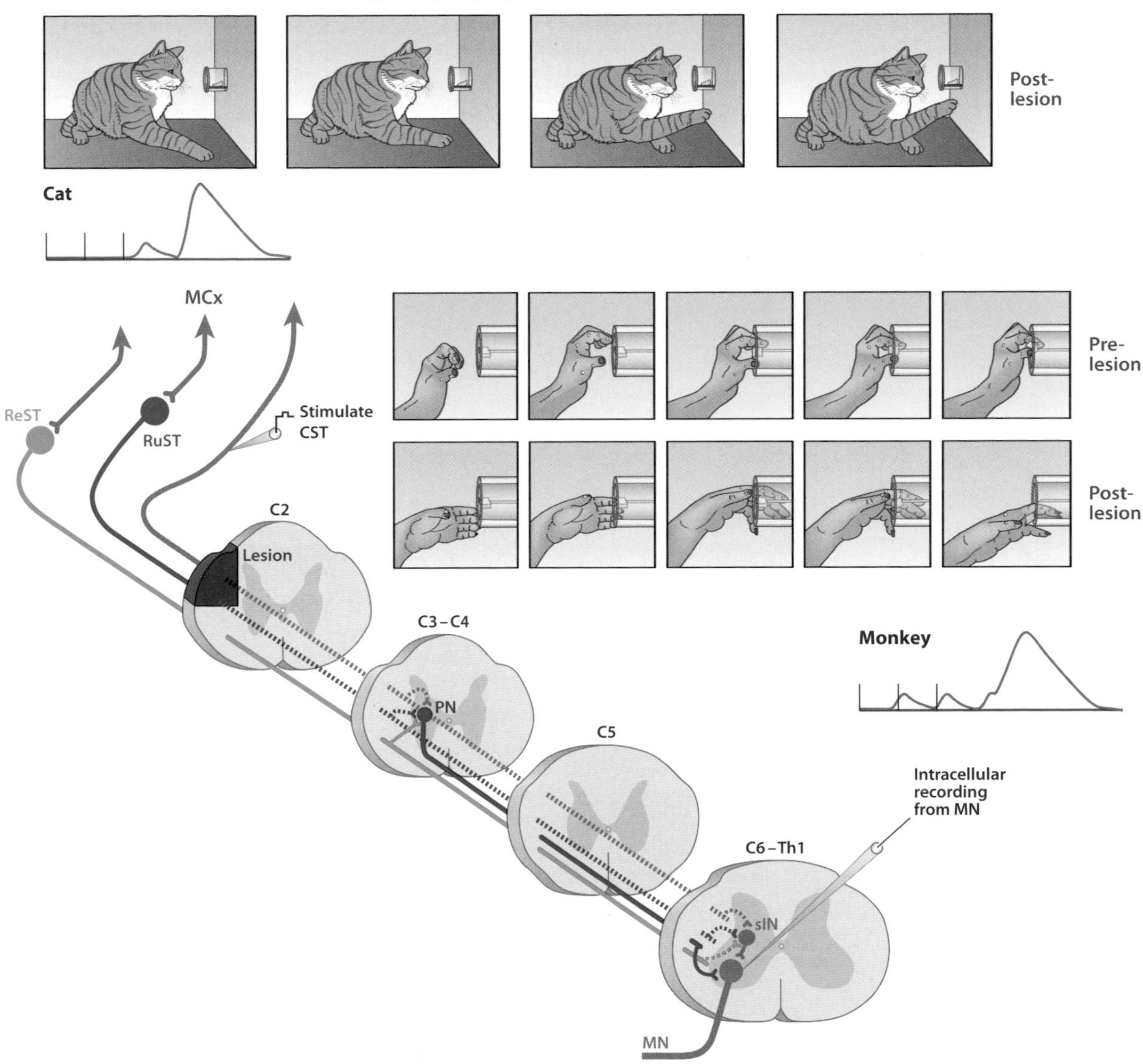

Figure 5

Behavioral and electrophysiological testing of propriospinal and segmental interneurons (PNs and sINs, respectively) versus corticoreticulospinal pathway in the control of reaching and grasping after lesion of the cortico- and rubrospinal tracts in C2. The transection of these pathways (*bottom diagonal image*; *red*) eliminated the input to both the C3-C4 PNs and sINs but spared the input to the corticoreticulospinal pathway. In the cat (*top panels*), reaching and grasping were defective. In the monkey (*center panels*), reaching was impaired and precision grip could not be performed: (*center upper row*) preoperative condition and (*center lower row*) postoperative day 1. Abbreviations: CST, corticospinal tract; MCx, motor cortex; MN, motoneuron; ReST, reticulospinal tract; RuST, rubrospinal tract. Adapted from Alstermark et al. (1981b, 2011) with permission.

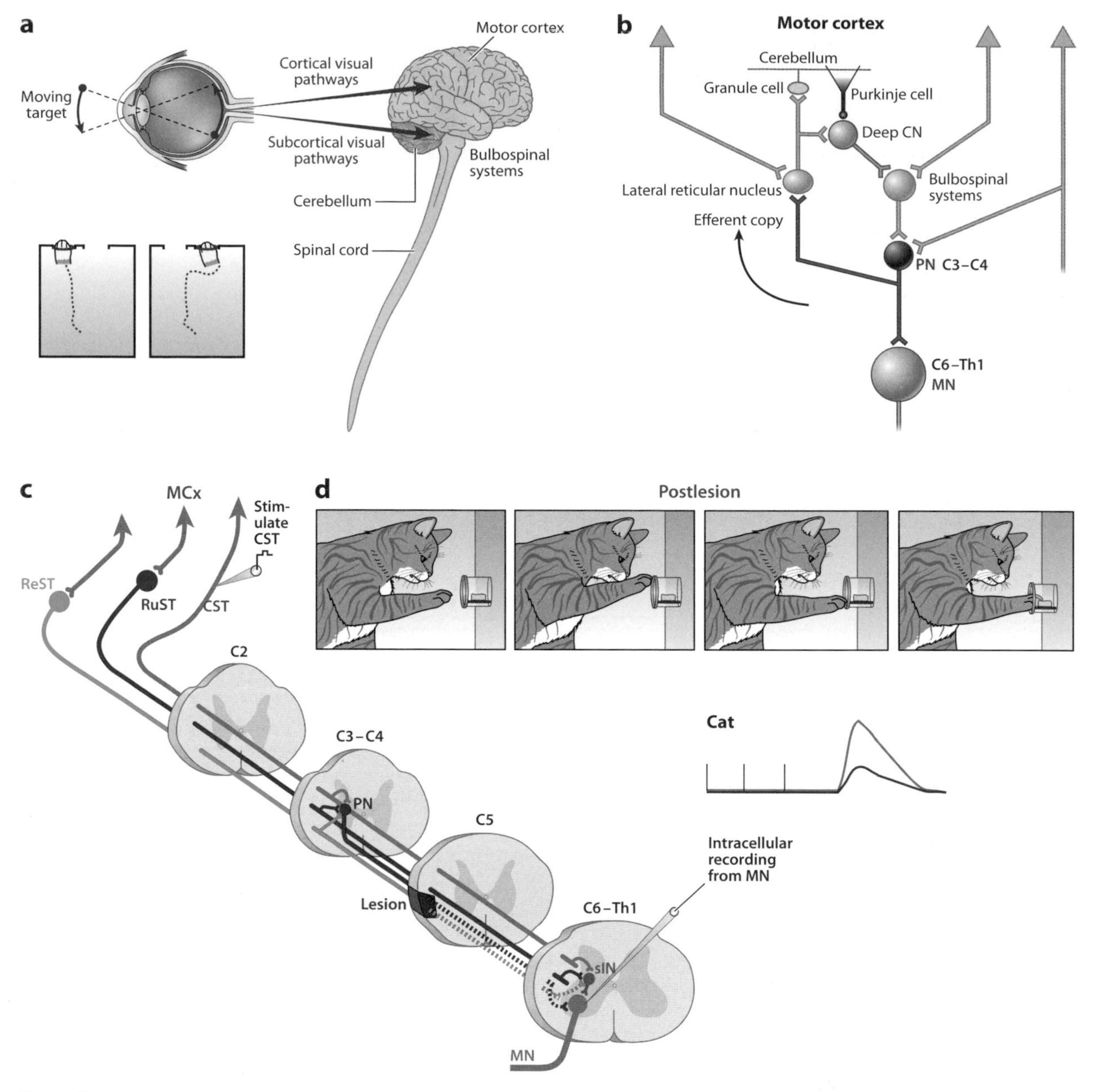

Figure 6

Subcortical visual updating via C3-C4 propriospinal neurons. (*a*) New visual information about changing target position can be mediated via both cortical and subcortical pathways: (*blue boxes*) reaching to the left target and switching from the left to the right target. (*b*) Schematic circuit consisting of the C3-C4 propriospinal system, efferent loop via the cerebellar cortex, and bulbospinal systems and the motor cortex. This circuit may be used to update the cortical command for reaching. (*c*) Ventral lesion in C5 of the reticulospinal and propriospinal pathways sparing the cortico- and rubrospinal pathways to segmental interneurons. Intracellular recordings from motorneurons (MNs) revealed reduced disynaptic corticospinal tract (CST) excitation. (*d*) Reaching behavior after the ventral C5 lesion, showing ataxia near the tube opening. Other abbreviations: CST, corticospinal tract; MCx, motor cortex; ReST, reticulospinal tract; RuST, rubrospinal tract.

of this observation, researchers proposed that sINs can mediate the command for grasping in the cat (Alstermark et al. 1981b). However, these results do not exclude the possibility that sINs can also help control reaching, because reaching, although defective, could be executed even when the axons of the PNs had been transected. Thus, the dual control of reaching and grasping by the PNs in the monkey may have developed because precision grip requires fine control of the upper arm (Alstermark et al. 2011).

In addition to its important role in controlling skilled reaching and grasping via the direct CM pathway in primates, the spinal cord contains separate neural circuits that are capable of conveying the descending motor commands from higher motor centers to the MNs. These spinal centers are presumably complementary to the higher centers in the motor cortex and brain stem. If so, what advantage could such integration at a stage just prior to the "final common path" provide? Illert et al. (1977) proposed that the descending cortical command could be updated on its way to the MNs via subcortical systems that have convergent input to the C3-C4 PNs and sINs.

SUBCORTICAL UPDATING OF CORTICAL COMMAND TO THE C3-C4 PROPRIOSPINAL NEURONS

During reaching, if the target suddenly changes its position in space, the movement trajectory has to be corrected using visual information. There are visual pathways to both the cortex and subcortical systems including the cerebellum and bulbospinal systems (**Figure 6*a***). Cortical visual pathways play an important role in the control of reaching (Goodale & Westwood 2004), usually with latencies longer than 200 msec for correction of an ongoing movement. However, shorter latencies in the range of 120-160 msec have been proposed to be subcortically mediated in humans (Day & Lyon 2000) as well as in subjects with complete corpus callosum agenesis (Day & Brown 2001) or whose primary visual area was temporarily blocked (Christensen et al. 2008). Recently, using electromyography recordings, researchers observed latencies around 100 msec in humans (Pruszynki et al. 2010).

To test the hypothesis of subcortical updating via the C3-C4 propriospinal system, cats were trained to perform reaching to one of two or three tubes. The correct tube was illuminated, and during reaching, the illumination could be switched to another tube with variable delay. The cats (**Figure 6*a***) showed remarkably short latencies for the correction of the movement trajectory, down to 50–60 msec after the light shift (Alstermark et al. 1990, Pettersson et al. 1997, Pettersson & Perfiliev 2002). Latencies longer than 100 msec were also observed. If transmission in the bulbospinal systems was interrupted by lesions in C2 of the rubro- and reticulospinal tracts, the cats could still correct the trajectory but only with latencies longer than 100 msec (Pettersson et al. 1997). In those cases, the new visual information was most likely processed at a cortical level.

These results are compatible with the neural circuitry known to control the C3-C4 propriospinal system (**Figure 6*b***), because bulbospinal systems receive very fast visual information (Werner 1993). However, a major problem is how to correct the ongoing cortical descending command via the C3-C4 PNs. As described above, the PNs, in addition to the MN projection, have ascending collaterals to the LRN in the lower brain stem (Illert & Lundberg 1978, Alstermark et al. 1981a, Alstermark & Ekerot 1992, Isa et al. 2006). The LRN is a major mossy-fiber input from the spinal cord to the cerebellum (Clendenin et al. 1974, Ekerot 1990). The cortical input to the LRN is separate from the corticospinal tract (Alstermark & Lundberg 1982, Alstermark & Ekerot 1992, Matsuyama & Drew 1997), which may control transmission via specific ascending spinocerebellar pathways such as the C3-C4 propriospinal system. The rubro- and reticulospinal neurons with projection to the C3-C4 PNs receive input from the motor cortex and the deep cerebellar nuclei (Illert

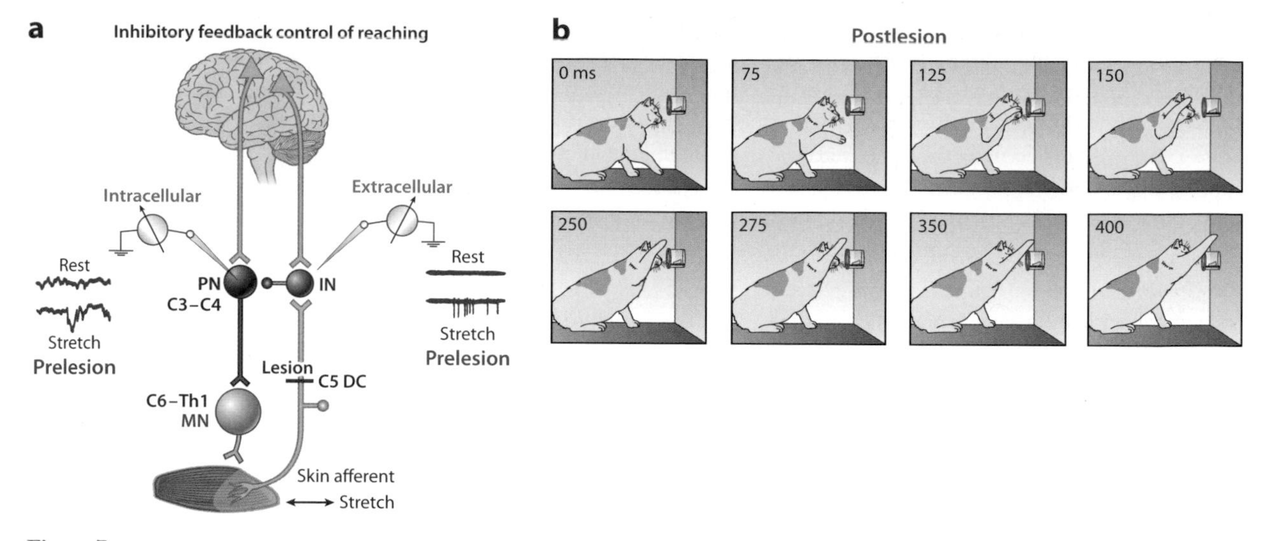

Figure 7

Feedback inhibition of the C3-C4 propriospinal neurons (PNs) during reaching. (*a*) Recording from PNs and feedback inhibitory interneurons (INs) in the C3-C4 segments during skin stretching revealed strong inhibition in the PNs. (*b*) Lesion of the dorsal column in C5 eliminates the feedback inhibition and results in the hypermetria in reaching. Abbreviation: MN, motorneuron.

et al. 1977). The ascending projection from the C3-C4 PNs may provide the cerebellum with an efferent copy of the descending command for reaching to the forelimb MNs (Alstermark et al. 1981b, Alstermark 1983). Using the new visual information, the cerebellum may then calculate a correction of the ongoing movement trajectory and issue new additional inputs to the C3-C4 PNs via the bulbospinal systems. The corrective signal will be matched by the new efferent copy until the activity level of the PNs has changed to correct for the new trajectory.

This hypothesis has received indirect support from behavioral testing of reaching following transection of the propriospinal axons in C5 (**Figure 6*c***). After this lesion, the cats had no problem initiating reaching toward the tube opening, but near the tube opening, they displayed severe ataxia during deceleration (**Figure 6*d***) (Alstermark et al. 1981b). In this case, transmission via the C3-C4 PNs to the MNs was interrupted, but because of the remaining ascending projection to the LRN, the cerebellum may be misinformed, resulting in the ataxia (Alstermark 1983).

FEEDBACK INHIBITORY CONTROL OF REACHING VIA C3-C4 PROPRIOSPINAL NEURONS

In addition to reaching, aiming and final positioning of the paw are critical for a successful grasp. Such control may depend heavily on afferent information. The C3-C4 PNs receive excitation from forelimb muscle and skin afferents, but the dominating effect is inhibition (Alstermark et al. 1984a, 1984b, 1986b). As shown in **Figure 7*a***, stretching the skin of the forelimb evokes firing of feedback inhibitory INs and concomitant inhibition in the PNs in the C3-C4 segments. To investigate the function of inhibitory feedback control of the C3-C4 PNs, researchers compared cats performing target reaching after dorsal column transection in C5 or C2. The C5 lesion interrupted the feedback inhibition to the PNs, although it was spared after the C2 lesion. Both lesions abolished the ascending information to higher centers. Interestingly, severe hypermetria in reaching was observed after the C5, but not after the C2, dorsal column lesion; the C2 lesion resulted only in dysmetria near the tube opening (Alstermark et al. 1986a) (**Figure 7*b***). The hypermetria was not compensated until after approximately one week of testing, presumably with the help of vision. These results suggest that feedback inhibition of PNs may be used conjointly with the descending command controlling reaching. Deceleration and correct termination of the forepaw may be difficult to control with only vision and may require multimodal control (Alstermark et al. 1986a). The INs mediating the inhibition of the PNs receive cortical input. Thus, the motor and sensory cortices could activate them as well as the PNs during reaching. The INs may then become activated by spatial facilitation from forelimb afferents when reaching close to the target. Similar to the PNs, the medial INs providing feedback inhibition of the PNs have ascending collaterals to the LRN and to the cuneate nucleus (Alstermark et al. 1984b). In analogy with the discussion above on the role of the efferent copy from the PNs to the LRN, the C5 dorsal column lesion may cause erroneous signaling from the feedback INs that could also contribute to the hypermetria. The dorsal column transections in both C5 and C2 eliminate afferent information to higher centers, which may be the cause of the dysmetria mainly observed after the C2 lesion. The additional hypermetria after the C5 dorsal column lesion suggest that the inhibitory INs are essential for inhibition of the C3-C4 PNs during reaching in its terminal phase close to the target.

PROPRIOSPINAL PATHWAYS IN HUMANS

Using indirect electrophysiological techniques, researchers found propriospinal pathways in the cervical and lumbar spinal cord segments of humans (for a review, see Pierrot-Deseilligny & Burke 2005). The cervical propriospinal pathway shares several properties with those in the cat and monkey, e.g., descending inhibition and strong feedback inhibition evoked by skin afferents. A difficulty in the study of humans when using indirect techniques is to determine the segments in which the PNs are located. However, in one patient with a spinal cord injury at the C6-C7 segmental junction restricted to the ventral and lateral funicles on one side, the propriospinally mediated excitation was abolished on the side of the lesion but remained intact on the nonlesioned side (Marchand-Pauvert et al. 2001). This finding, together with the systematically longer latency of propriospinally mediated excitation to more caudally located motor nuclei, suggests that the cervical PNs in humans are located rostral to C6. In the patient with a restricted lesion of the propriospinal pathway on one side, reaching and precision grip were normal on the nonlesioned side. On the lesioned side, researchers observed deficits in reaching close to the target with transient ataxia as well as difficulties in holding small objects with the thumb and index finger (B. Alstermark & L.G. Pettersson, unpublished observations). Surprisingly, transcranial magnetic stimulation revealed a similar short-latency-evoked cortical electromyography response on both the lesioned and the nonlesioned sides, indicating that transmission in the direct CM pathway was not significantly impaired (Marchand-Pauvert et al. 2001). These findings agree well with the behavioral observations described above for the monkey. Transmission via sINs in humans was not accessible within a similar investigation, but it likely plays an essential role as does the direct CM pathway.

NEW TECHNIQUES FOR INVESTIGATING THE FUNCTION OF NEURAL CIRCUITS

As described in Reaching and Grasping in the Cat and Macaque Monkey (section above), most of the behavioral work was performed using selective spinal cord lesions (but see above for a description of the technique using activity-dependent transneuronal labeling) (Alstermark & Kümmel 1990). The lesioning technique has several limitations such as causing changes in excitability in both the acute and chronic states and with compensatory mechanisms in the nonlesioned systems (see section below). With this technique, it is also difficult for researchers to make exactly similar lesions from one experiment to another. The lesion resulting from this technique will also unavoidably affect other axons with a location similar to that of the targets. To demonstrate directly the function of the C3-C4 PNs in the intact state, a novel technique to manipulate the activity of the C3-C4 PNs without a lesion would be desirable. For this purpose, researchers established a novel genetic tool of pathway-specific and reversible transmission block and applied it to macaque monkeys (M. Kinoshita, R. Matsui, S. Kato, T. Hasegawa, H. Kasahara, K. Isa, A. Watakabe, T. Yamamori, Y. Nishimura, B. Alstermark, D. Watanabe, D. Kobayashi, and T. Isa, unpublished observations) as shown in **Figure 8**. First, researchers injected a highly efficient retrogradely transported (HiRet) lentiviral vector carrying the enhanced GFP (EGFP) and enhanced tetanus neurotoxin (eTeNT) in the downstream of the tetracycline responsive element into the motor nuclei of the C6-Th1 segments. This caused infection through the axonal terminals of the fibers terminating in the motor nuclei.

Seven to ten days later, the second vector, the adeno associated vector carrying a highly efficient Tet-ON sequence, rtTAV16, was injected in a section ranging from the caudal C2 to the rostral C5 segments. These injections caused infection via the cell somata, but it was not transported retrogradely to the motor cortex. Together, the lentivirus (via axons) and adenovirus (somata) would specifically infect INs with somata located in C2-C5 with projection to forelimb motor nuclei in the C6-Th1 segments. In this condition, the enhanced GFP and eTeNT were transcribed under the presence of doxycycline (Dox), a tetracycline derivative, in the C3-C4 PNs with double infection. One to two months after the virus injections, Dox was orally administered to the monkeys. Two to five days after Dox application, allowing time for the expression of eTeNT, the monkeys showed deficits in reaching and/or grasping movements. The monkeys exhibited weakness in the index finger and thumb precision grip, grasping movements became slower, and, in some monkeys, reaching showed dysmetria. Terminal electrophysiological experiments under anesthesia showed that the synaptic transmission through the C3-C4 PNs was markedly reduced, presumably owing to the eTeNT that reduced synaptic release (Yamamoto et al. 2003). As shown in **Figure 8*c***, stimulation of the corticospinal tract evoked mainly the monosynaptic field but not the disynaptic field mediated via the C3–C4 PNs. Also, stimulation of the ascending axons of the PNs in the LRN revealed a large block in transmission of the monosynaptic field. These electrophysiological findings also showed that the direct CM pathway was not affected by the virus injections. In addition, with anti-GFP immunohistochemistry, researchers were able to visualize the location of the somata of the C3-C4 PNs whose transmission had been blocked. These somata were located mainly in the lateral part of the intermediate zone of the C3-C4 segments, which transneuronal labeling with WGA-HRP revealed was also where the C3-C4 PNs were located (Alstermark & Kümmel 1990). Another advantage of using this technique is its ability to visualize the axons all the way through the somata to the terminals. We found that the axons terminated on presumed MNs in the C6-Th1 segments and on neurons in the LRN. These results confirmed that the neurons affected by eTeNT during Dox administration belong to the group of C3-C4 PNs. Taken together with the previous behavioral findings, these new results demonstrate that C3-C4 PNs play a critical role in the control of reaching and grasping movements in macaque monkeys both under normal conditions and after spinal cord lesions.

FUNCTIONAL RECOVERY FOLLOWING SPINAL CORD INJURY

Although the macaque monkeys could perform the precision-grip exercise following lesion

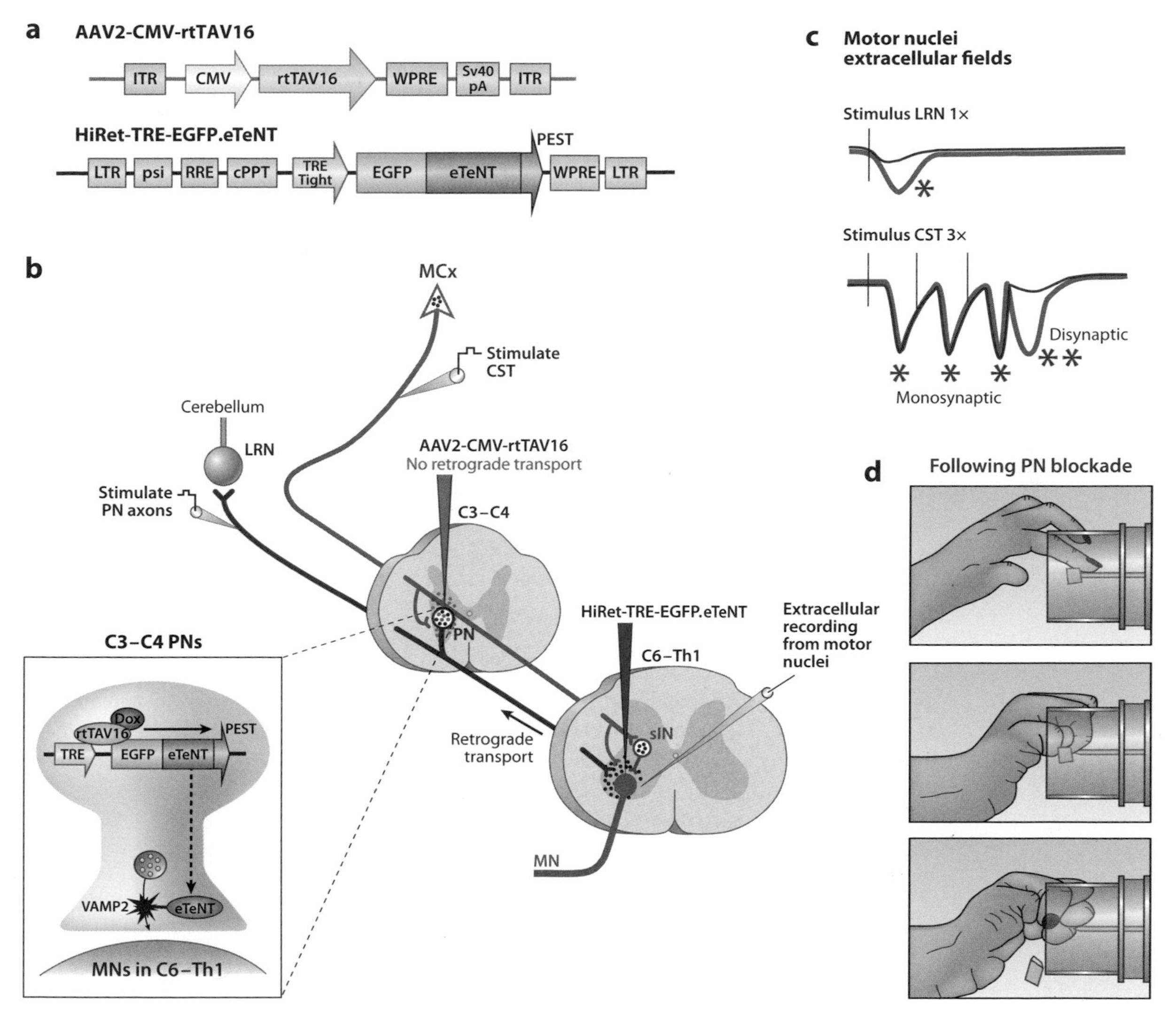

Figure 8

Selective and reversible blockade of synaptic transmission through the C3-C4 PNs via a double viral vector system. (*a*) Schematic diagram of the vector injections, showing how HiRet-TRE-eGFP.eTeNT (*red dots*) and AAV2-CMV-rtTAV16 (*green dots*) interact in the double-infected cells. The former vector could infect PNs, sINs, and corticospinal neurons in the MCx; the double infection occurred only in the C3-C4 PNs. The AAV2-CMV-rtTAV16 vector is not transported retrogradely. (*b*) Injection of vectors in the motor nuclei in C6-Th1 (*red*) and close to the PNs in C3-C4 (*green*). Schematic diagram showing that the eGFP and eTeNT that were transcribed in the presence of Dox and eTeNT cleave VAMP2, leading to the blockade of the synaptic transmission. (*c*) Extracellular recordings from the motor nuclei in C6-Th1, obtained while stimulating the axons of the PNs directly in the LRN or by stimulating the corticospinal tract in the contralateral pyramid. Single and double asterisks indicate the monosynaptic and disynaptic field potentials, respectively. (*d*) Impairment of the precision grip on the second day after initiation of Dox administration in a monkey with double vector infection. Abbreviations: Dox, doxycycline; CST, corticospinal tract; eGFP, enhanced GFP; eTeNT, enhanced tetanus neurotoxin (eTeNT); HiRet, highly efficient retrogradely transported; LRN, lateral reticular nucleus; MCx, motor cortex; MN, motorneuron; PN, propriospinal neuron; sIN, segmental interneuron.

of the cortico- and rubrospinal tracts in C5 (as described above), they showed deficits in their grasping movements: Independent finger control and their ability to preshape their digits accurately were both impaired. With daily training, lesioned monkeys showed a marked recovery, and all showed near-full recovery of precision grip within two weeks to two months post lesion (Sasaki et al. 2004; Nishimura et al. 2007, 2009). However, after a similar lesion in C2, monkeys displayed only partial, at best, recovery of precision grip. These results suggest that the C3-C4 PNs are, in addition to their normal function, involved in the postlesion recovery of dexterous hand movements without the direct CM pathway. In the cat, which lacks direct CM connections, the postlesion recovery of digit grasping with lesion in the cortico- and rubrospinal tracts in C5-C6 depends on takeover via C3-C4 PNs and the corticoreticulospinal pathway (Alstermark et al. 1987, Blagovechtchenski et al. 2000, Pettersson et al. 2007).

Researchers have used brain imaging with positron emission tomography to study cerebral activation during functional recovery after lesion of the cortico- and rubrospinal tracts in C5 in the macaque monkey (Nishimura et al. 2007). During the early phase of recovery (1 month), increased activation was present bilaterally in the primary motor cortex (M1), and during the late stage of recovery (3-4 months after lesion), activation was present in an extended area of the contralateral M1 and bilaterally in the ventral premotor cortex. These results suggest that a number of different pathways may be involved during the recovery process. Studies of humans who have had a stroke suggest that the corticoreticulospinal with projection to the propriospinal system may contribute to recovery (Pierrot-Deseilligny & Burke 2005).

To summarize, the experimental models reviewed here offer good opportunities to extend the research on motor recovery of voluntary movements after spinal cord injury. Recently, a vast literature (see Flynn et al. 2011, Nishimura & Isa 2011) has argued for the involvement of PNs in functional recovery after spinal cord injury. Researchers may be able to take advantage of the ability of PNs to control object-oriented reaching and dexterous hand movements to develop future therapeutic strategies.

SUMMARY POINTS AND FUTURE ISSUES

1. The control of skilled reaching and grasping involves not only cortical circuits with direct CM connection, but also indirect CM pathways via PNs and sINs in the spinal cord. In the cat, the C3-C4 PNs are specialized to control reaching, whereas grasping is more dependent on the sINs. In the macaque monkey, the C3-C4 PNs can also convey the command for precision grip involving the thumb and index fingers. These spinal circuits are different from other spinal networks involved in locomotion.
2. The motor cortex yields parallel control of the PNs and sINs via the corticospinal and bulbospinal pathways. Sending visual information to the bulbospinal pathway, the spinal circuits can quickly update the descending cortical command. An efferent copy from the PNs to the cerebellum via the LRN may play a critical role during reaching in updating the cortical command to a moving target. In addition to visual control, feedback inhibition of the C3-C4 PNs sent from the cutaneous receptors in the forelimb is critical for correct deceleration and accurate termination of reaching at the target location.

3. Future work remains to be performed on the sINs to investigate the descending convergence pattern. Research needs to determine whether sINs also have an ascending projection to the LRN. Molecular work must also contend with the challenges of characterizing and possibly finding differential markers for the C3-C4 PNs and sINs. From an evolutionary point of view, it is also interesting to understand why the PNs are located rostral to the forelimb segments.
4. In humans and rodents, a cervical propriospinal system also exists, but its normal function remains to be clarified.
5. So far, most behavioral work has been conducted using lesions. A new technique based on a specific infection of PNs using double viral vectors has enabled pathway-selective and reversible inactivation of the C3-C4 propriospinal system. In monkeys, this inactivation happened within days, resulting in deficits in reaching and grasping. It is also necessary to develop new techniques in which reversible inactivation of discrete circuits can be performed at a millisecond time resolution. The newly established optogenetic technique may be most important in this respect. Genetic techniques based on unique cell identity may also offer a new possibility for manipulating specific neural circuits.
6. The behavioral work using lesions has given new interesting results with respect to plastic changes during the early phase of recovery. These findings may be used to improve functional recovery after spinal cord injury in humans.

DISCLOSURE STATEMENT

The authors are not aware of any affiliations, memberships, funding, financial holdings, or any other conflicts of interest that might be perceived as affecting the objectivity of this review.

ACKNOWLEDGMENTS

B.A. is supported by the Swedish Research Council and Human Frontier Science Program. T.I. is supported by the Strategic Research Program of Brain Sciences by the Ministry of Education, Culture, Sports, Science, and Technology of Japan. The authors thank Dr. Pettersson for constructive criticism on an earlier version of this article.

LITERATURE CITED

Alstermark B. 1983. *Functional role of propriospinal neurons in the control of forelimb movements; a behavioural and electrophysiological study*. PhD thesis. Dep. Physiol., Univ. Göteborg. 32 pp.

Alstermark B, Ekerot CF. 1992. Organization of the ascending projection from C3-C4 propriospinal neurones to cerebellum via the lateral reticular nucleus. *Acta Physiol. Scand.* 146(Suppl. 608):P2.35, 151

Alstermark B, Górska T, Johannisson T, Lundberg A. 1986a. Hypermetria in forelimb target-reaching after interruption of the inhibitory pathway from forelimb afferents to C3-C4 propriospinal neurones. *Neurosci. Res.* 5:457–61

Alstermark B, Górska T, Lundberg A, Pettersson LG. 1990. Integration in descending motor pathways controlling the forelimb in the cat. 16. Visually guided switching of target-reaching. *Exp. Brain Res.* 80:1–11

Alstermark B, Isa T, Ohki Y, Saito Y. 1999. Disynaptic pyramidal excitation in forelimb motoneurons mediated via C3-C4 propriospinal neurons in the Macaca fuscata. *J. Neurophysiol.* 82:3580–95

Alstermark B, Isa T, Pettersson LG, Sasaki S. 2007. The C3-C4 propriospinal system in the cat and monkey: a spinal pre-motoneuronal centre for voluntary motor control. *Acta Physiol. Oxf.* 189:123–40

Alstermark B, Johannisson T, Lundberg A. 1986b. The inhibitory feedback pathway from the forelimb to C3-C4 propriospinal neurones investigated with natural stimulation. *Neurosci. Res.* 5:451–56

Alstermark B, Kümmel H. 1990. Transneuronal transport of wheat germ agglutinin conjugated horseradish peroxidase into last order spinal interneurones projecting to acromio- and spinodeltoideus motoneurones in the cat. 2. Differential labelling of interneurones depending on movement type. *Exp. Brain Res.* 80:96–103

Alstermark B, Lindström S, Lundberg A, Sybirska E. 1981a. Integration in descending motor pathways controlling the forelimb in the cat. 8. Ascending projection to the lateral reticular nucleus from C3-C4 propriospinal also projecting to forelimb motoneurones. *Exp. Brain Res.* 42:282–98

Alstermark B, Lundberg A. 1982. Electrophysiological evidence against the hypothesis that corticospinal fibres send collaterals to the lateral reticular nucleus. *Exp. Brain Res.* 47:148–50

Alstermark B, Lundberg A. 1992. The C3-C4 propriospinal system: target-reaching and food-taking. In *Muscle Afferents and Spinal Control of Movement*, ed. L Jami, E Pierrot-Deseilligny, D Zytnicki, pp. 327–54. Oxford/New York/Seoul/Tokyo: Pergamon

Alstermark B, Lundberg A, Norrsell U, Sybirska E. 1981b. Integration in descending motor pathways controlling the forelimb in the cat. 9. Differential behavioural defects after spinal cord lesions interrupting defined pathways from higher centres to motoneurones. *Exp. Brain Res.* 42:299–318

Alstermark B, Lundberg A, Pettersson LG, Tantisira B, Walkowska M. 1987. Motor recovery after serial spinal cord lesions of defined descending pathways in cats. *Neurosci. Res.* 5:68–73

Alstermark B, Lundberg A, Sasaki S. 1984a. Integration in descending motor pathways controlling the forelimb in the cat. 10. Inhibitory pathways to forelimb motoneurones via C3-C4 propriospinal neurones. *Exp. Brain Res.* 56:279–92

Alstermark B, Lundberg A, Sasaki S. 1984b. Integration in descending motor pathways controlling the forelimb in the cat. 11. Inhibitory pathways from higher motor centres and forelimb afferents to C3-C4 propriospinal neurones. *Exp. Brain Res.* 56:293–307

Alstermark B, Ogawa J. 2004. In vivo recordings of bulbospinal excitation in adult mouse forelimb motoneurones. *J. Neurophysiol.* 92:1958–62

Alstermark B, Ogawa J, Isa T. 2004. Lack of monosynaptic corticomotoneuronal EPSPs in rats: disynaptic EPSPs mediated via reticulospinal neurones and polysynaptic EPSPs via segmental interneurones. *J. Neurophysiol.* 91:1832–39

Alstermark B, Pettersson LG, Nishimura Y, Yoshino-Saito K, Tsuboi F, et al. 2011. Motor command for precision grip in the macaque monkey can be mediated by spinal interneurons. *J. Neurophysiol.* 106:122–26

Alstermark B, Sasaki S. 1985. Integration in descending motor pathways controlling the forelimb in the cat. 13. Corticospinal effects in shoulder, elbow, wrist, and digit motoneurones. *Exp. Brain Res.* 59:353–64

Alstermark B, Sasaki S. 1986. Integration in descending motor pathways controlling the forelimb in the cat. 15. Comparison of the projection from excitatory C3-C4 propriospinal neurones to different species of forelimb motoneurones. *Exp. Brain Res.* 63:543–56

Bernhard CG, Bohm E. 1954. Cortical representation and functional significance of the corticomotoneuronal system. *Arch. Neurol. Psychiat.* 72:473–502

Blagovechtchenski E, Pettersson LG, Perfiliev S, Krasnochokova E, Lundberg A. 2000. Control of digits via C3-C4 propriospinal neurones in cats; recovery after lesions. *Neurosci. Res.* 38:103–7

Brown TG. 1911. The intrinsic factors in the act of progression in the mammal. *Proc. R. Soc. Lond. Ser. B* 84:309–19

Christensen MS, Kristiansen L, Rowe JB, Nielsen JB. 2008. Action-blindsight in healthy subjects after transcranial magnetic stimulation. *Proc. Natl. Acad. Sci. USA* 105:1353–57

Clendenin M, Ekerot CF, Oscarsson O. 1974. The lateral reticular nucleus in the cat. III. Organization of component activated from the ipsilateral forelimb tract. *Exp. Brain Res.* 21:501–13

Creed RS, Denny-Brown DE, Eccles JC, Liddell EGT, Sherrington CS. 1932. *Reflex Activity of the Spinal Cord.* London: Oxford Univ. Press

Day BL, Brown P. 2001. Evidence for subcortical involvement in the visual control of human reaching. *Brain* 124:1832–40

Day BL, Lyon IN. 2000. Voluntary modification of automatic arm movements evoked by motion of a visual target. *Exp. Brain Res.* 130:159–68

Ekerot CF. 1990. The lateral reticular nucleus in the cat. VI. Excitatory and inhibitory paths. *Exp. Brain Res.* 79:109–19

Flynn JR, Graham BA, Galea MP, Callister RJ. 2011. The role of propriospinal interneurons in recovery from spinal cord injury. *Neuropharmacology* 60:809–22

Fujito Y, Aoki M. 1995. Monosynaptic rubrospinal projections to distal forelimb motoneurons in the cat. *Exp. Brain Res.* 105:181–90

Goodale MA, Westwood DA. 2004. An evolving view of duplex vision: separate but interacting cortical pathways for perception and action. *Curr. Opin. Neurobiol.* 2:203–11

Illert M, Lundberg A. 1978. Collateral connections to the lateral reticular nucleus from cervical propriospinal neurones projecting to forelimb motoneurones in the cat. *Neurosci. Lett.* 7:167–72

Illert M, Lundberg A, Tanaka R. 1977. Integration in descending motor pathways controlling the forelimb in the cat. 3. Convergence on propriospinal neurones transmitting disynaptic excitation from the corticospinal tract and other descending tracts. *Exp. Brain Res.* 29:323–46

Illert M, Wiedemann E. 1984. Pyramidal actions in identified radial motor nuclei of the cat. *Pflügers Arch.* 41:132–42

Isa T, Ohki Y, Alstermark B, Pettersson LG, Sasaki S. 2007. Direct and indirect cortico-motoneuronal pathways and control of hand/arm movements. *Physiology* 22:145–52

Isa T, Ohki Y, Seki K, Alstermark B. 2006. Properties of propriospinal neurons in the C3-C4 segments mediating disynaptic pyramidal excitation to forelimb motoneurones in the Macaque monkey. *J. Neurophysiol.* 95:3674–85

Kinoshita M, Matsui R, Kato S, Hasegawa T, Kasahara H, et al. 2012. Genetic dissection of the circuit for hand dexterity in primates. *Nature.* In press

Kitazawa S, Ohki Y, Sasaki M, Xi M, Hongo T. 1993. Candidate premotor neurons of skin reflex pathways to T1 forelimb motoneurons of the cat. *Exp. Brain Res.* 95:291–307

Kuypers HG. 1982. A new look at the organization of the motor system. *Prog. Brain Res.* 57:381–403

Landgren S, Phillips CG, Porter R. 1962. Cortical fields of origin of the monosynaptic pyramidal pathways to some alpha motoneurones of the baboon's hand and forearm. *J. Physiol.* 161:112–25

Lawrence DG, Kuypers HGJM. 1968a. The functional organization of the motor system in the monkey. I. The effects of bilateral pyramidal lesions. *Brain* 91:1–14

Lawrence DG, Kuypers HGJM. 1968b. The functional organization of the motor system in the monkey. II. The effects of lesions of the descending brain-stem pathways. *Brain* 91:15–36

Lemon RN. 2008. Descending pathways in motor control. *Annu. Rev. Neurosci.* 31:195–218

Lloyd DPC. 1941. The spinal mechanism of the pyramidal system in cats. *J. Neurophysiol.* 4:525–46

Lundberg A. 1999. Descending control of forelimb movements in the cat. *Brain Res. Bull.* 50:323–24

Lundberg A, Voorhoeve P. 1962. Effects from the pyramidal tract on spinal reflex arcs. *Acta Physiol. Scand.* 56:201–19

Malmgren K, Pierrot-Deseilligny E. 1988. Evidence for non-monosynaptic Ia excitation of human wrist flexor motoneurones, possibly via propriospinal neurones. *J. Physiol.* 405:747–64

Marchand-Pauvert V, Mazevet D, Pradat-Diehl P, Alstermark B, Pierrot-Deseilligny E. 2001. Interruption of a relay of corticospinal excitation by a spinal lesion at C6-C7. *Muscle Nerve* 24:1554–61

Matsuyama K, Drew T. 1997. Organization of the projections from the pericruciate cortex to the pontomedullary brainstem of the cat: a study using the anterograde tracer *Phaseolus vulgaris*-leucoagglutinin. *J. Comp. Neurol.* 389:617–41

Nishimura Y, Isa T. 2012. Cortical and subcortical compensatory mechanisms after spinal cord injury in monkeys. *Exp. Neurol.* 235:152–61

Nishimura Y, Morichika Y, Isa T. 2009. A subcortical oscillatory network contributes to recovery of hand dexterity after spinal cord injury. *Brain* 132:709–21

Nishimura Y, Onoe H, Morichika Y, Perfiliev S, Tsukada H, Isa T. 2007. Time-dependent central compensatory mechanisms of finger dexterity after spinal cord injury. *Science* 318:1150–55

Peterson B, Pitts NG, Fukushima K. 1979. Reticulospinal connections with limb and axial motoneurons. *Exp. Bran Res.* 36:1–20

Pettersson LG, Alstermark B, Blagovechtchenski E, Isa T, Sasaki S. 2007. Skilled digit movements in feline and primate—recovery after selective spinal cord lesions. *Acta Physiol.* 189:141–54

Pettersson LG, Lundberg A, Alstermark B, Isa T, Tantisira B. 1997. Effect of spinal cord lesions on forelimb target-reaching and on visually guided switching of target-reaching in the cat. *Neurosci. Res.* 29:241–56

Pettersson LG, Perfiliev S. 2002. Descending pathways controlling visually guided updating of reaching in cats. *Eur. J. Neurosci.* 16:1349–60

Pierrot-Deseilligny E, Burke D. 2005. *The Circuitry of the Human Spinal Cord.* Cambridge, UK: Cambridge Univ. Press

Pruszynki JA, King GL, Boisse L, Scott S, Flanagan JR, Munoz DP. 2010. Stimulus-locked responses on human arm muscles reveal rapid neural pathway linking visual input to arm motor output. *Eur. J. Neurosci.* 32:1049–57

Renshaw B. 1940. Activity in the simplest spinal reflex pathways. *J. Neurophysiol.* 3:373–87

Riddle CN, Edgley SA, Baker SN. 2009. Direct and indirect connections with upper limb motoneurons from the primate reticulospinal tract. *J. Neurosci.* 29:4993–99

Sasaki S, Isa T, Pettersson LG, Alstermark B, Naito K, et al. 2004. Dexterous finger movements in primate without monosynaptic corticomotoneuronal excitation. *J. Neurophysiol.* 92:3142–47

Shapovalov AI. 1975. Neuronal organization and synaptic mechanisms of supraspinal motor control in vertebrates. *Rev. Physiol. Biochem. Pharmacol.* 72:1–54

Sherrington CS. 1906. *The Integrative Action of the Nervous System.* New Haven, CT/London: Yale Univ. Press

Werner W. 1993. Neurons in the primate superior colliculus are active before and during arm movements to visual targets. *Eur. J. Neurosci.* 5:335–40

Yamamoto M, Wada N, Kitabatake Y, Watanabe D, Anzai M, et al. 2003. Reversible suppression of glutamatergic neurotransmission of cerebellar granule cells in vivo by genetically manipulated expession of tetanus neurotoxin light chain. *J. Neurosci.* 23:6759–67

Cumulative Indexes

Contributing Authors, Volumes 26–35

D

E

F

G

H

I

J

S

T

V

W

Y

Z

Chapter Titles, Volumes 26–35

Cognitive Neuroscience

Computational Approaches to Neuroscience

Emotion

Learning and Memory

Miscellaneous

Molecular Neuroscience

Neural Networks

Neuroethology

Neurogenetics

Neuroscience Techniques

Neurotransmitters

Olfaction/Taste

Pain